An Introduction to
Geometrical
Physics

Second Edition

An Introduction to
Geometrical
Physics

Second Edition

Ruben Aldrovandi • José Geraldo Pereira

Instituto de Física Teórica - UNESP, Brazil

W World Scientific

NEW JERSEY · LONDON · SINGAPORE · BEIJING · SHANGHAI · HONG KONG · TAIPEI · CHENNAI · TOKYO

Published by

World Scientific Publishing Co. Pte. Ltd.

5 Toh Tuck Link, Singapore 596224

USA office: 27 Warren Street, Suite 401-402, Hackensack, NJ 07601

UK office: 57 Shelton Street, Covent Garden, London WC2H 9HE

Library of Congress Cataloging-in-Publication Data

Names: Aldrovandi, R. (Ruben) author. | Pereira, J. G., author.

Title: An introduction to geometrical physics / by Ruben Aldrovandi
    (Instituto de Física Teórica - UNESP, Brazil), José Geraldo Pereira
    (Instituto de Física Teórica - UNESP, Brazil) World Scientific.

Description: 2nd edition. | New Jersey : World Scientific, 2016. |
    Includes bibliographical references and index.

Identifiers: LCCN 2016030547| ISBN 9789813146808 (hardcover : alk. paper) |
    ISBN 9789813146815 (pbk : alk. paper)

Subjects: LCSH: Geometry, Differential. | Mathematical physics.

Classification: LCC QC20.7.D52 A53 2016 | DDC 530.15/636--dc23

LC record available at https://lccn.loc.gov/2016030547

British Library Cataloguing-in-Publication Data

A catalogue record for this book is available from the British Library.

To our parents

Nice   and   José

Dina   and   Tito

Preface to the First Edition

This book grew out of courses given at the Instituto de Física Teórica for many years. As the title announces, it is intended as a first, elementary approach to "Geometrical Physics" — to be understood as a chapter of Mathematical Physics. Mathematical Physics is a moving subject, and has moved faster in recent times. From the study of differential equations and related special functions, it has migrated to the more qualitative realms of topology and algebra. The bridge has been the framework of geometry. The passage supposes an acquaintance with concepts and terms of a new kind, to which this text is a tentative introduction.

In its technical uses, the word "geometry" has since long lost its metric etymological meaning. It is the science of space, or better, of spaces. Thus, the name should be understood as a study of those spaces which are of interest in Physics. This emphasis on the notion of space has dominated the choice of topics — they will have in common the use of "spaces". Some may seem less geometric than others, but a space is always endowed with a few basic, irreducible properties enabling some kind of analysis, allowing a discussion of relations between its different parts.

Up until the sixties, General Relativity was seen as the prototype of geometrical intrusion in Physics. Gravitation stood apart as the sole example of geometrization amongst the fundamental interactions. Things changed a lot with the discovery of the renormalizability of gauge theories, and with their subsequent coming to age as the theories for all the other fundamental interactions. At least as a general principle, the gauge principle appears nowadays as a vast unifying idea. It came as an astonishing surprise that gauge theories are still more geometrical in character and structure than

General Relativity.

Landau's criterium, so well proven for Physics as for good wines, has been loosely adopted: results less than ten years old in the field are given only limited attention. It is true that some acceleration in new developments is easily felt, but it is always graceful to present some seemingly objective alibi for the authors' limitations. Let it be so, and let us, like those real Bourgogne connoisseurs, leave some recent vintages to rest in the shadows for some more time.

The text falls into five parts. Parts I, II and III constitute the main text. Part IV is made up of mathematical topics and Part V of physical topics. The main text is intended as an introductory, largely descriptive presentation of the basic notions of geometry from a physicist's point of view. Pedagogically, it would provide for a first one-semester course. But it is also a kind of guide to the topical chapters. It refers to Part IV for further details, and uses Part V as a source of examples. The main text is divided into chapters, these in sections and these in paragraphs. For instance, "§4.2.5" means chapter 4 of the main text, section 2, paragraph 5. Calls like "§n" refer to a numbered paragraph of the same section. Paragraphs with additional examples and complements, which can be skipped in a first lecture, are printed in smaller characters. The topical chapters are simply divided in sections. For instance, "Phys.7.5" means section 5 of the topical chapter Phys.7.

The topical sections oscillate between the "fascicule des résultats" and application-almost-exercise styles. The mathematical topics should not be mistaken as anything more than a guide for further study. They are a physicists' presentation of mathematical subjects in broad brushstrokes, intending to illustrate and complement the main text. The physical topics are similarly introductory, always in a physicists' view, trying never to say twain for two, and avoiding as long as possible the gobbledygook theoreticians became prone to in recent times. It tries to present well known subjects from a point of view which is more geometrical than the usual treatments. Some general references are given at the end of each topic, the footnotes being left for those of more immediate character.

Mathematics is, amongst many other things, the language of Physics. Repetitions are unavoidable and, as with all language apprenticeship, they may be even desirable. The pedagogical bias justifies some duplications, as that of indices at the beginning of each topical section, emphatically advocated by our students.

Like any text coming up from lecture notes, these pages owe a lot to many. To the financing agencies, like CNPq, FAPESP and CAPES, which supported the authors during the period. To our colleagues of São Paulo and Paris, in special C.E. Harle, B.M. Pimentel, D. Galetti, G. Francisco, R.A. Kraenkel, G.A. Matsas, L.A. Saeger, A.L. Barbosa, M. Dubois-Violette, R. Kerner and J. Madore. Of course, to the many consulted authors, hopefully all quoted in the reference list. And, last but not least, to our attentive and long-suffering editor, Tina Jeavons.

Preface to the Second Edition

A second edition is concomitantly a honour and a challenge. The two decades since the first edition of this book have shown unparalleled change in everything technological — and in everything on which technology may have a strong impact, like experimental science. And this revolution keeps its momentum at present, reserving to many things an uncertain future. Books themselves, in their traditional paper form, have been promised a honourable extinction ...

The marriage between geometry and physics is best perceived when we notice that the geometry of Nature is probed by the physical dynamics. Even our simplest everyday experience shows the pervasive presence of geometry. It is enough to recall that at least three metrics are at work inside a simple water-bucket, and that each one requires a different dynamical probe to be recognised: a sound, a ray of light, a stone.

As will be repeatedly said in the text, we are more and more conscious that our knowledge even of our everyday-life geometry is doubtful. We are by now sure that the standard euclidian one which has been taken for granted for millennia is not valid at distances of interest at the quantum level. The last few decades have brought to light doubts about its validity also at very large, cosmological distances. Adapting Montaigne's well-known utterance on the mountains around him to our present-day horizons:[1]

---

[1] *Quelle vérité que ces montagnes bornent, qui est mensonge au monde qui se tient au-delà?*, Michel de Montaigne, *Essays*, Book II, Chap.12, p. 241, Éditions du Seuil, Paris, 1967 (first published in 1580). In a free English translation, it reads: *What truth is it that these mountains bound, and is a lie in the world beyond?*

> *What truth is it that these nearby galaxies bound,*
> *and is a lie in the universe beyond?*

It seems clear by now that Special Relativity is unable to cope with the very high energies of the Planck scale. Since General Relativity is umbilically connected to Special Relativity through the equivalence principle, it seems equally unable to deal with gravitation at that scale. In the opposite scale of small energies, the behavior of the universe as we see it today cannot be explained by our current theories without additional ingredients, either in the form of new theories or in the form of new universe constituents, usually labelled *dark* due to our ignorance about their nature.

We have endeavoured to bring forward some of these recent questions, adding some new sections and chapters. Besides an introduction to the present-day Cosmology Standard Model, we have included a chapter on Teleparallel Gravity, an alternative theory for gravitation that is fully equivalent to General Relativity in what concerns observed results, but exhibits many conceptual differences with respect to it. One of the most important is that, in contrast to General Relativity, it has the geometric structure of a gauge theory. In another chapter, we have introduced the fundamentals of a de Sitter-invariant special relativity. As widely known, ordinary Special Relativity can be seen as a generalization of Galilei Relativity for velocities comparable to the speed of light. Similarly, the de Sitter-invariant special relativity can be seen as a generalization of Special Relativity for energies comparable to those turning up at the Planck scale. It is, consequently, universal in the sense that it is able to describe the spacetime kinematics at any energy scale. Some implications of this theory for gravitation and cosmology are also briefly sketched. Of course, these theories are not intended as a panacea for the current problems of Physics, but rather as alternatives towards their solutions. It is important to have available for handle and trial as many geometrical variant possibilities as possible. For that reason also a brief account is given of the Einstein–Cartan theory. It is an article of scientific faith that experiment and observation will someday decide about their importance.

We are very grateful to all those many readers who have taken seriously our first edition — and have communicated their doubts and criticism, suggested improvements in content, style and presentation. We thank our colleagues already named in the Preface of the first edition, who have all kept their comprehensive support. We owe a lot to the financing Brazilian agencies CNPq, FAPESP and CAPES, which have significantly aided the authors. Of course, also to the many consulted authors, hopefully all quoted in the reference list. And, last but not least, to the editors of World Scientific, in special Lim Swee Cheng, Tan Rok Ting and Pan Suqi, for their meticulous assistance and patience during the completion of the book.

Contents

Preamble: Space and Geometry

What stuff 'tis made of,
whereof it is born,
I am to learn.

Merchant of Venice

The simplest geometrical setting used — consciously or not — by physicists in their everyday work is the 3-dimensional euclidean space $\mathbb{E}^3$. It consists of the set $\mathbb{R}^3$ of ordered triples of real numbers such as $\boldsymbol{p} = (p^1, p^2, p^3)$, $\boldsymbol{q} = (q^1, q^2, q^3)$, etc, and is endowed with a very special characteristic, a metric defined by the distance function

$$d(\boldsymbol{p}, \boldsymbol{q}) = \left[\sum_{i=1}^{3} (p^i - q^i)^2 \right]^{1/2}.$$

It is the space of ordinary human experience and the starting point of our geometric intuition. Studied for two-and-a-half millenia, it has been the object of celebrated controversies, the most famous concerning the minimum number of properties necessary to define it completely.

From Aristotle and Euclid to Newton, through Galileo and Descartes, the very word *space* has been reserved to $\mathbb{E}^3$. Only in the 19-th century has it become clear that other, different spaces could be thought of, and mathematicians have since greatly amused themselves by inventing all kinds of them. For physicists, the age-long debate shifted to another question: how can we recognize, amongst such innumerable possible spaces, that *real* space chosen by Nature as the stage-set of its processes? For example, suppose the space of our everyday experience consists of the same set $\mathbb{R}^3$ of triples above, but with a different distance function, such as

$$d(\boldsymbol{p}, \boldsymbol{q}) = \sum_{i=1}^{3} |p^i - q^i|.$$

This, written in terms of moduli, would define a different metric space, in principle as good as that given above. Were it only a matter of principle, it would be as good as any other space given by any distance function with $\mathbb{R}^3$ as set point. It so happens, however, that Nature has chosen the former and not the latter space for us to live in. To know which one is *the* real space is not a simple question of principle — something else is needed. What else? The answer may seem rather trivial in the case of our home space, though less so in other spaces singled out by Nature in the many different situations which are objects of physical study. It was given by Riemann in his famous Inaugural Address:[2]

> "...*those properties which distinguish space from other*
> *conceivable triply extended quantities can*
> *only be deduced from experience*".

Thus, *from experience*! It is experiment that tells us in which space we actually live in. When we measure distances, we find them to be independent of the direction of the straight lines joining the points. And this isotropy property rules out the second proposed distance function, while admitting the metric of the euclidean space.

In reality, Riemann's statement implies an epistemological limitation: it will never be possible to ascertain *exactly* which space is the real one. Other isotropic distance functions are, in principle, admissible and more experiments are necessary to decide between them. In Riemann's time already other geometries were known (those found by Lobachevsky and Boliyai) that could be as similar to the euclidean geometry as we might wish in the restricted regions experience is confined to. In honesty, all we can say is that $\mathbb{E}^3$, as a model for our ambient space, is strongly favored by present day experimental evidence in scales ranging from (say) human dimensions down to about 10^{-15} cm. Our knowledge on smaller scales is limited by our capacity to probe them. For larger scales, according to General Relativity, the validity of this model depends on the presence and

---

[2] A translation of Riemann's Address can be found in Spivak 1970, vol. II. Clifford's translation (Nature, **8** (1873), 14 and 36), as well as the original transcribed by David R. Wilkins, can be found in the site http://www.emis.de/classics/Riemann/.

strength of gravitational fields: $\mathbb{E}^3$ is good only as long as gravitational fields are very weak.

> *"These data are — like all data — not logically necessary,*
> *but only of empirical certainty ... one can therefore*
> *investigate their likelihood, which is certainly very great*
> *within the bounds of observation, and afterwards decide*
> *upon the legitimacy of extending them beyond the*
> *bounds of observation, both in the direction of the*
> *immeasurably large and in the direction*
> *of the immeasurably small."*

The only remark we could add to these words, pronounced in 1854, is that the "bounds of observation" have greatly receded with respect to the values of Riemann times.

> *... geometry presupposes the concept of space, as well as*
> *assuming the basic principles for constructions in space.*

In our ambient space, we use in reality a lot more of structure than the simple metric model: we take for granted a vector space structure, or an affine structure; we transport vectors in such a way that they remain parallel to themselves, thereby assuming a connection. Which one is the minimum structure, the irreducible set of assumptions really necessary to the introduction of each concept? Physics should endeavour to establish on empirical data not only the basic space to be chosen but also the structures to be added to it. At present, we know for example that an electron moving in $\mathbb{E}^3$ under the influence of a magnetic field *feels* an extra connection (the electromagnetic potential), to which neutral particles may be insensitive.

Experimental science keeps a very special relationship with Mathematics. Experience counts and measures. But Science requires that the results be inserted in some logically ordered picture. Mathematics is expected to provide the notion of number, so as to make countings and measurements meaningful. But Mathematics is also expected to provide notions of a more qualitative character, to allow for the modeling of Nature. Thus, concerning numbers, there seems to be no result comforting the widespread prejudice by which we measure *real* numbers. We work with integers, or with rational numbers, which is fundamentally the same. No direct measurement will sort out a Dedekind cut. We must suppose, however, that real numbers exist: even from the strict experimental point of view, it does not matter whether objects like "π" or "e" are simple names or are endowed with some kind of

an sich reality: we cannot afford to do science without them. This is to say
that even pure experience needs more than its direct results, presupposes
a wider background for the insertion of such results. Real numbers are a
minimum background. Experience, and *logical necessity*, will say whether
they are sufficient or not.

From the most ancient extant treatise going under the name of Physics:[3]

> *"When the objects of investigation, in any subject,*
> *have first principles, foundational conditions, or basic con-*
> *stituents, it is through acquaintance with these that knowl-*
> *edge, scientific knowledge, is attained. For we cannot say*
> *that we know an object before we are acquainted with its*
> *conditions or principles, and have carried our analysis as*
> *far as its most elementary constituents.*
>
> *The natural way of attaining such a knowledge is to*
> *start from the things which are more knowable and obvious*
> *to us and proceed towards those which are clearer and more*
> *knowable by themselves ... "*

Euclidean spaces have been the starting spaces from which the basic geometrical and analytical concepts have been isolated by successive, tentative,
progressive abstractions. It has been a long and hard process to remove the
unessential from each notion. Most of all, as will be repeatedly emphasized,
it was a hard thing to put the idea of metric in its due position.

Structure is thus to be added step by step, under the control of experiment. Only once experiment has established the basic ground will internal
coherence, or logical necessity, impose its own conditions.

---

[3] Aristotle, *Physics* I.1.

Part 1: MANIFOLDS

Chapter 1

General Topology

In simple words, the purely qualitative properties of spaces.

1.1 Introductory remarks

§ **1.1.1** Let us consider again our ambient 3-dimensional euclidean space $\mathbb{E}^3$. In order to introduce ideas like proximity between points, boundedness of subsets, convergence of point sequences and — the dominating notion — continuity of mappings between $\mathbb{E}^3$ and other point sets, elementary real analysis starts by defining[1] open r-balls around a point p:

$$B_r(p) = \left\{ q \in \mathbb{E}^3 \text{ such that } d(q,p) < r \right\}.$$

The same is done for n-dimensional euclidean spaces $\mathbb{E}^n$, with open r-balls of dimension n. The question worth raising here is whether or not the real analysis so obtained depends on the chosen distance function. Or, putting it in more precise words: of all the usual results of analysis, how much is dependent on the metric and how much is not? As said in the Preamble, Physics should use experience to decide which one (if any) is the convenient metric in each concrete situation, and this would involve the whole body of properties consequent to this choice. On the other hand, some spaces of physical relevance, such as the space of thermodynamical variables, are not

---

[1] Defining balls requires the notion of distance function, which is a function d taking any pair (p,q) of points of a set X into the real line $\mathbb{R}$ and satisfying four conditions: (i) $d(p,q) \geq 0$ for all pairs (p,q); (ii) $d(p,q) = 0$ if and only if $p = q$; (iii) $d(p,q) = d(q,p)$ for all pairs (p,q); (iv) $d(p,r) + d(r,q) \geq d(p,q)$ for any three points p, q, r. It is actually a mapping $d : X \times X \to \mathbb{R}_+$, from a cartesian set product into the positive real line. Though it is better to separate the two notions, d is sometimes called a metric, because a positive-definite metric tensor does determine a distance function. A space on which a distance function is defined is a *metric space*. On vector spaces this expression is usually reserved to translation-invariant distance functions.

explicitly endowed with any metric. Are we always using properties coming from some implicit underlying notion of distance?

§ **1.1.2** There is more: physicists are used to "metrics" which in reality do not lead to good distance functions. Think of Minkowski space, which is $\mathbb{R}^4$ with the Lorentz metric η:

$$\eta(p, q) = \left[(p^0 - q^0)^2 - (p^1 - q^1)^2 - (p^2 - q^2)^2 - (p^3 - q^3)^2\right]^{1/2} .$$

It is not possible to define open balls with this pseudo-metric, which allows vanishing "distances" between distinct points on the light cone, and even purely imaginary "distances". If continuity, for example, depends upon the previous introduction of balls, then when would a function be continuous on Minkowski space?

§ **1.1.3** Actually, most of the properties of space are quite independent of any notion of distance. In particular, the above mentioned ideas of proximity, convergence, boundedness and continuity can be given precise meanings in spaces on which the definition of a metric is difficult, or even forbidden. Metric spaces are in reality very particular cases of more abstract objects, the *topological spaces*, on which only the minimal structure necessary to introduce those ideas is present. That minimal structure is a *topology*, and answers for the general qualitative properties of space.

§ **1.1.4** Consider the usual 2-dimensional surfaces immersed in $\mathbb{E}^3$. To begin with, there is something shared by all spheres, of whatever size. And also something which is common to all toruses, large or small; and so on. Something makes a sphere deeply different from a torus and both different from a plane, and that independently of any measure, scale or proportion. A hyperboloid sheet is quite distinct from the sphere and the torus, and also from the plane $\mathbb{E}^2$, but less so for the latter: we feel that it can be somehow unfolded without violence into a plane. A sphere can be stretched so as to become an ellipsoid but cannot be made into a plane without losing something of its "spherical character". Topology is that primitive structure which will be the same for spheres and ellipsoids; which will be another one for planes and hyperboloid sheets; and still another, quite different, for toruses. It will be that set of qualities of a space which is preserved under gentle stretching, bending, twisting. The study of this primitive structure makes use of very simple concepts: points, sets of points, mappings between sets of points. But the structure itself may be very involved and may leave an important (eventually dominant) imprint on the physical objects present in the space under consideration.

§ **1.1.5** The word "topology" is — like the word "algebra" — used in two different senses. One more general, naming the mathematical *discipline* concerned with spacial qualitative relationships, and another, more particular, naming that *structure* allowing for such relationships to be well–defined. We shall be using it almost exclusively with the latter, more technical, meaning.

Let us proceed to make the basic ideas a little more definite. In order to avoid leaving too many unstated assumptions behind, we shall feel justified in adopting a rather formal approach,[2] starting modestly with point sets.

1.2 Topological spaces

§ **1.2.1** Experimental measurements being inevitably of limited accuracy, the constants of Nature (such as Planck's constant $\hbar$, the light velocity c, the electron charge e, etc.) appearing in the fundamental equations are not known with exactitude. The process of building up Physics presupposes this kind of "stability": it assumes that, if some value for a physical quantity is admissible, there must be always a range of values around it which is also acceptable. A wavefunction, for example, will depend on Planck's constant. Small variations of this constant, within experimental errors, would give other wavefunctions, by necessity equally acceptable as possible.

It follows that, in the modeling of nature, each value of a mathematical quantity must be surrounded by other admissible values. Such neighbouring values must also, by the same reason, be contained in a set of acceptable values.

We come thus to the conclusion that values of quantities of physical interest belong to sets enjoying the following property: every acceptable point has a neighbourhood of points equally acceptable, each one belonging to another neighbourhood of acceptable points, etc, etc. Sets endowed with this property, that around each one of its points there exists another set of the same kind, are called "open sets". This is actually the old notion of open set, abstracted from euclidean balls: a subset U of an "ambient" set S is open if around each one of its points there is another set of points of

---

[2] A commendable text for beginners, proceeding constructively from unstructured sets up to metric spaces, is Christie 1976. Another readable account is the classic Sierpiński 1956.

S entirely contained in U.

All physically admissible values are, therefore, necessarily members of open sets:

Physics needs open sets.

Furthermore, we talk frequently about "good behaviour" of functions, or that they "tend to" some value, thereby loosely conveying ideas of continuity and limit. Through a succession of abstractions, the mathematicians have formalized the idea of open set while inserting it in a larger, more comprehensive context.

Open sets appear then as members of certain families of sets, the topologies, and the focus is concentrated on the properties of these families — not on those of its members. This enlarged context provides a general and abstract concept of open set and gives a clear meaning to the above rather elusive word "neighbourhood", while providing the general background against which the fundamental notions of continuity and convergence acquire well-defined contours.

§ 1.2.2 A space will be, to begin with, a set endowed with some decomposition allowing us to talk about its parts. Although the elements belonging to a space may be vectors, matrices, functions, other sets, etc, they will be called, to simplify the language, "points". Thus, a space will be a set S of points plus a structure leading to some kind of organization, such that we may speak of its relative parts and introduce "spatial relationships". This structure is introduced as a well-performed division of S, as a convenient family of subsets. There are various ways of dividing a set, each one designed to accomplish a definite objective.

We shall be interested in getting appropriate notions of neighbourhood, distinguishability of points, continuity and, later, differentiability. How is a fitting decomposition obtained? A first possibility might be to consider S with *all* its subsets. This conception, though acceptable in principle, is too particular: it leads to a quite disconnected space, every two points belonging to too many unshared neighbourhoods. It turns out (see Section 1.3) that any function would be continuous on such a "pulverized" space and in consequence the notion of continuity would be void. The family of subsets is too large, the decomposition would be too "fine-grained". In the extreme opposite, if we consider only the improper subsets, that is, the whole point set S and the empty set $\emptyset$, there would be no real decomposition and again no useful definition of continuity (subsets distinct from $\emptyset$ and S are called *proper* subsets). Between the two extreme choices of taking a family with all the subsets or a family with no subsets at all, a compromise has been found: good families are defined as those respecting a few well chosen, suitable conditions. Each one of such well-bred families of subsets is called a *topology*.

Given a point set S, a *topology* is a family of subsets of S (which are called, *by definition*, its *open sets*) respecting the 3 following conditions:

(i) The whole set S and the empty set $\emptyset$ belong to the family.

(ii) Given a *finite* number of members of the family, say $U_1, U_2, U_3, \ldots, U_n$, their intersection $\bigcap_{i=1}^{n} U_i$ is also a member.

(iii) Given *any* number (finite or infinite) of open sets, their union belongs to the family.

Thus, a topology on S is a collection of subsets of S to which belong the union of any subcollection and the intersection of any finite subcollection, as well as $\emptyset$ and the set S proper. The paradigmatic open balls of $\mathbb{E}^n$ satisfy, of course, the above conditions. Both the families suggested above, the family including all subsets and the family including no proper subsets, respect the above conditions and are consequently accepted in the club: they are topologies indeed (called respectively the *discrete topology* and the *indiscrete topology* of S), but very peculiar ones. We shall have more to say about them later (see below, § 1.2.18 and § 1.4.6). Now:

> a *topological space is a point set* S *on*
> *which a topology is defined.*

Given a point set S, there are in general many different families of subsets with the above properties, i.e., many different possible topologies. Each such family will make of S a different topological space. Rigour would then require that a name or symbol be attributed to that particular family of subsets (say, T) and the topological space be given name and surname, being denoted by the pair (S, T).

Some well known topological spaces have historical names. When we say "euclidean space", the set $\mathbb{R}^n$ of real n-uples with the usual topology of open balls is meant. The members of a topology are called "open sets" precisely by analogy with the euclidean case, but notice that they are determined by the specification of the family: an open set of (S, T) is not necessarily an open set of (S, T') when $T \neq T'$. Think of the point set of $\mathbb{E}^n$, which is $\mathbb{R}^n$, but with the discrete topology including all subsets: the set $\{p\}$ containing only the point p of $\mathbb{R}^n$ is an open set of the topological space $(\mathbb{R}^n$, discrete topology), but not of the euclidean space $\mathbb{E}^n = (\mathbb{R}^n$, topology of n-dimensional balls).

§ 1.2.3 Finite space. A very simple topological space is given by the set of four letters $S = \{a, b, c, d\}$ with the family of subsets

$$T = \{\{a\}, \{a, b\}, \{a, b, d\}, S, \emptyset\}.$$

The choice is not arbitrary: the family of subsets

$$\{\{a\}, \{a, b\}, \{b, c, d\}, S, \emptyset\},$$

for example, does not define a topology, because the intersection

$$\{a, b\} \cap \{b, c, d\} = \{b\}$$

is not an open set, as it is not included in T.

§ 1.2.4 Given a point $p \in S$, any set U containing an open set belonging to T which includes p is a *neighbourhood* of p. Notice that U itself is not necessarily an open set of T: it simply includes[3] some open set(s) of T. Of course any point will have at least one neighbourhood, S itself.

§ 1.2.5 Metric spaces are the archetypal topological spaces. The notion of topological space has evolved conceptually from metric spaces by abstraction: properties unnecessary to the definition of continuity were progressively forsaken. Topologies generated from a notion of distance (*metric topologies*) are the most usual in Physics. As an experimental science, Physics plays with countings and measurements, the latter in general involving some (at least implicit) notion of distance. Amongst metric spaces, a fundamental role will be played by the first example we have met, the euclidean space.

§ 1.2.6 The euclidean space $\mathbb{E}^n$. The point set is the set $\mathbb{R}^n$ of n-uples

$$\boldsymbol{p} = (p^1, p^2, \ldots, p^n), \quad \boldsymbol{q} = (q^1, q^2, \ldots, q^n), \ldots$$

of real numbers. The distance function is given by

$$d(\boldsymbol{p}, \boldsymbol{q}) = \left[\sum_{i=1}^{n} (p^i - q^i)^2\right]^{1/2}.$$

The topology is formed by the set of the open balls. It is a standard practice to designate a topological space by its point set when there is no doubt as to which topology is meant. That is why the euclidean space is frequently denoted simply by $\mathbb{R}^n$. We shall, however, insist on the notational

---

[3] Some authors (Kolmogorov & Fomin 1977, for example) do define a neighbourhood of p as an open set of T to which p belongs. In our language, a neighbourhood which is also an open set of T will be an "open neighbourhood".

difference: the euclidean space $\mathbb{E}^n$ will be $\mathbb{R}^n$, the set of ordered n-uples $x = (x^1, x^2, ..., x^n)$ of real numbers, *plus* the ball topology. $\mathbb{E}^n$ is the basic, starting space, as even differential manifolds will be presently defined so as to generalize it. We shall see later that the introduction of coordinates on a general space S requires that S resemble some $\mathbb{E}^n$ around each one of its points. It is important to notice, however, that many of the most remarkable properties of the euclidean space come from its being, besides a topological space, something else. Indeed, one must be careful to distinguish properties of purely topological nature from those coming from additional structures usually attributed to $\mathbb{E}^n$, the main one being that of a vector space.

§ **1.2.7** In metric spaces, any point p has a countable set of open neighbourhoods $\{N_i\}$ such that for any set U containing p there exists at least one N_j included in U. Thus, any set U containing p is a neighbourhood. This is not a general property of topological spaces. Those for which this happens are said to be *first-countable* spaces (Figure 1.1).

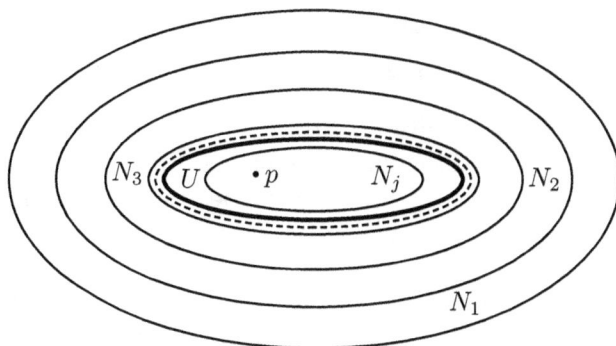

Fig. 1.1   In first-countable spaces, every point p has a countable set of open neighbourhoods $\{N_k\}$, of which at least one is included in a given $U \ni p$. We say that "all points have a local countable basis". All metric spaces are of this kind.

§ **1.2.8 Topology basis.** In order to specify a topological space, one has to fix the point set and tell which amongst all its subsets are to be taken as open sets. Instead of giving each member of the family T (which is frequently infinite to a very high degree), it is in general much simpler to give a subfamily from which the whole family can be retraced. A *basis* for a topology T is a collection B of its open sets such that any member of

T can be obtained as the union of elements of B. A general criterium for $B = \{U_\alpha\}$ to be a basis is stated in the following theorem:

$B = \{U_\alpha\}$ is a basis for T iff, for any open set $V \in T$ and all $p \in V$, there exists some $U_\alpha \in B$ such that $p \in U_\alpha \subset V$.

The open balls of $\mathbb{E}^n$ constitute a prototype basis, but one might think of open cubes, open tetrahedra, etc. It is useful, to get some insight, to think about open disks, open triangles and open rectangles on the euclidean plane $\mathbb{E}^2$. No two distinct topologies may have a common basis, but a fixed topology may have many different basis. On $\mathbb{E}^2$, for instance, we could take the open disks, or the open squares or yet rectangles, or still the open ellipses. We would say intuitively that all these different basis lead to the same topology and we would be strictly correct.

As a topology is most frequently introduced via a basis, it is useful to have a criterium to check whether or not two basis correspond to the same topology. This is provided by another theorem:

B and B' are basis defining the same topology iff, for every $U_\alpha \in B$ and every $p \in U_\alpha$, there exists some $U'_\beta \in B'$ such that $p \in B'_\beta \subset U_\alpha$, and vice-versa.

Again, it is instructive to give some thought to disks and rectangles in $\mathbb{E}^2$. A basis for the real euclidean line $\mathbb{E}^1$ is provided by all the open intervals of the type $(r - 1/n, r + 1/n)$, where r runs over the set of rational numbers and n over the set of the integer numbers. This is an example of *countable* basis. When a topology has at least one countable basis, it is said to be *second-countable*. Second countable topologies are always first-countable (§ 1.2.7 above) but the inverse is not true. We have said above that all metric spaces are first-countable. There are, however, metric spaces which are not second countable (Figure 1.2). We see here a first trial to classify topological spaces. Topology frequently resorts to this kind of practice, trying to place the space in some hierarchy.

In the study of the anatomy of a topological space, some variations are sometimes helpful. An example is a small change in the concept of a basis, leading to the idea of a 'network'. A network is a collection N of subsets such that any member of T can be obtained as the union of elements of N. Similar to a basis, but accepting as members also sets which are not open sets of T.

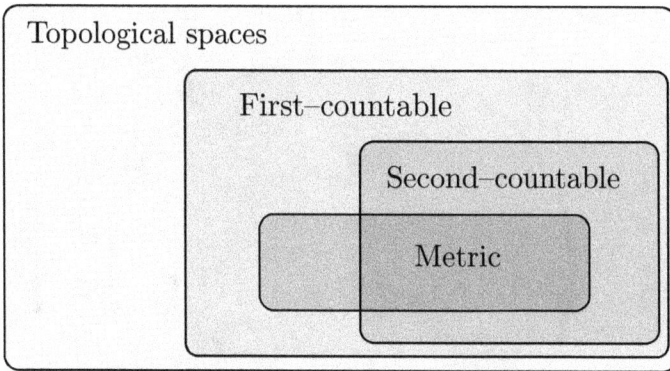

Fig. 1.2   A partial hierarchy: not all metric spaces are second-countable, but all of them are first-countable.

§ 1.2.9 Induced topology. The topologies of the usual surfaces immersed in $\mathbb{E}^3$ are obtained by intersecting them with the open 3-dimensional balls. This procedure can be transferred to the general case: let (S, T) be a topological space and X a subset of S. A topology can be defined on X by taking as open sets the intersections of X with the open sets belonging to T. This is called an *induced* (or *relative*) topology, denoted $X \cap T$. A new topological space $(X, X \cap T)$ is born in this way. An n-sphere S^n is the set of points of $\mathbb{E}^{n+1}$ satisfying

$$\sum_{i=1}^{n+1}(p^i)^2 = 1,$$

with the topology induced by the open balls of $\mathbb{E}^{n+1}$ (Figure 1.3). The set of real numbers can be made into the euclidean topological space $\mathbb{E}^1$ (popular names: "the line" and — rather oldish — "the continuum"), with the open intervals as 1-dimensional open balls. Both the set $\mathbb{Q}$ of rational numbers and its complement, the set $\mathbb{J} = \mathbb{E}^1 \backslash \mathbb{Q}$ of irrational numbers, constitute topological spaces with the topologies induced by the euclidean topology of the line.

§ 1.2.10 The upper-half space $\mathbb{E}^n_+$. Its point set is

$$\mathbb{R}^n_+ = \left\{ p = (p^1, p^2, \dots, p^n) \in \mathbb{R}^n \text{ such that } p^n \geq 0 \right\}. \qquad (1.1)$$

The topology is that induced by the ball-topology of $\mathbb{E}^n$. This space, which will be essential to the definition of *manifolds-with-boundary* in § 4.1.1, is

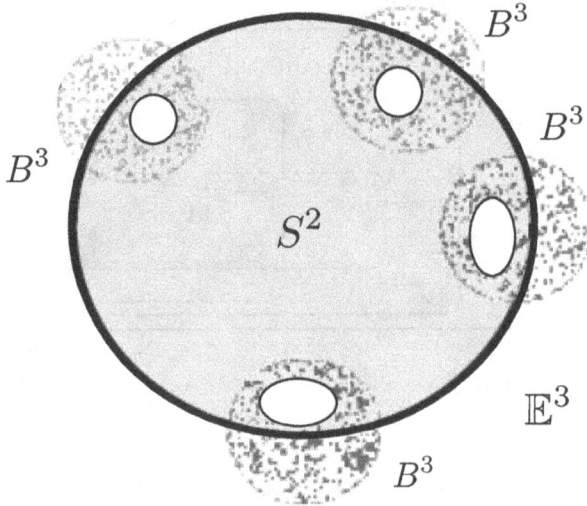

Fig. 1.3   The sphere S^2 with some of its open sets, which are defined as the intersections of S^2 with the open balls of the euclidean 3-dimensional space.

not second-countable. A particular basis is given by sets of two kinds: (i) all the open balls entirely included in $\mathbb{R}^n_+$; (ii) for each ball tangent to the hyperplane $p^n = 0$, the union of that ball with (the set containing only) the tangency point.

§ **1.2.11** Notice that, for the 2-dimensional case (the "upper-half plane", Figure 1.4) for example, sets of type ◻ , including intersections with the horizontal line, are not open in $\mathbb{E}^2$ but are open in $\mathbb{E}^2_+$. One speaks of the above topology as the "swimmer's topology": suppose a fluid flows upwardly from the horizontal borderline into the space with a monotonously decreasing velocity which is unit at the bottom. A swimmer with a constant unit velocity may start swimming in any direction at any point of the fluid. In a unit interval of time the set of all possible swimmers will span a basis.

§ **1.2.12** A cautionary remark: the definitions given above (and below) may sometimes appear rather queer and irksome, as if invented by some skew-minded daemon decided to hide simple things under tangled clothes. They have evolved, however, by a series of careful abstractions, start-

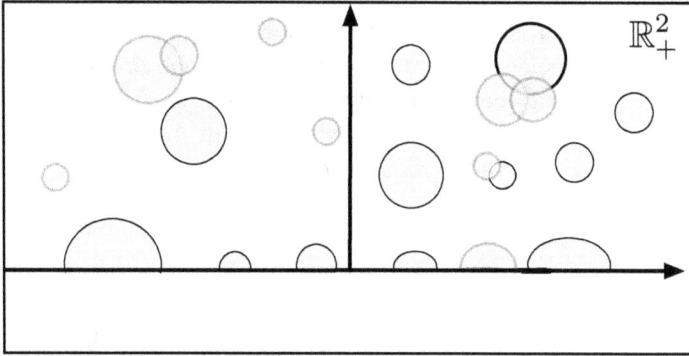

Fig. 1.4   The upper-half plane $\mathbb{E}^2_+$, whose open sets are the intersections of the point set $\mathbb{R}^2_+$ with the open disks of $\mathbb{E}^2$.

ing from the properties of metric spaces and painstakingly checked to see whether they lead to useful, meaningful concepts. Fundamental definitions are, in this sense, the products of "Experimental Mathematics". If a simpler, more direct definition seems possible, the reader may be sure that it has been invalidated by some counter-example (see as an example the definition of a continuous function in §1.4.5).

§ **1.2.13** Consider two topologies T_1 and T_2 defined on the same point set S. We say that T_1 is *weaker* than T_2 if every member of T_1 belongs also to T_2. The topology T_1 is also said to be *coarser* than T_2, and T_2 is *finer* than T_1 (or T_2 is a *refinement* of T_1, or still T_2 is *stronger* than T_1). The topology T for the finite space of § 1.2.3 is clearly weaker than the discrete topology for the same point set.

§ **1.2.14 A problem: the topology of spacetime.** We have said that a topology for Minkowski space time cannot be obtained from the Lorentz metric, which is unable to define balls. The specification of a topology is of fundamental importance because (as will be seen later) it is presupposed every time we talk of a continuous (say, wave) function. We could think of using an $\mathbb{E}^4$ topology, but this would be wrong because (besides other reasons) no separation would then exist between spacelike and timelike vectors. The fact is that *we do not know the real topology of spacetime*. We would like to retain euclidean properties both in the space sector and on the time axis.

Zeeman[4] has proposed an appealing topology: it is defined as the finest topology defined on $\mathbb{R}^4$ which induces an $\mathbb{E}^3$ topology on the space sector and an $\mathbb{E}^1$ topology on the time axis. It is not first-countable and, consequently, cannot come from any metric. In their everyday practice, physicists usually adopt an ambiguous behaviour and use the balls of $\mathbb{E}^4$ whenever they talk of continuity and/or convergence.

§ **1.2.15** Given the subset C of S, its *complement* is the set

$$C' = \{p \in S \text{ such that } p \notin C\}.$$

The subset C is a *closed* set in the topological space (S, T) if C' is an open set of T. Thus, the complement of an open set is (by definition) closed. It follows that $\emptyset$ and S are closed (and open!) sets in all topological spaces.

§ **1.2.16 Closedness** is a relative concept: a subset C of a topological subspace Y of S can be closed in the induced topology even if open in S. For instance, Y itself will be closed (and open) in the induced topology, even if Y is an open set of S.

Retain that "closed", just as "open", depends on the chosen topology.
A set which is open in a topology may be closed in another.

§ **1.2.17 A connected space** is a topological space in which no proper subset is simultaneously open and closed. The name comes from the fact that, in this case, S cannot be decomposed into the union of two disjoint open sets.

One should not confuse this concept with *path-connectedness*, to be defined later (§ 1.4.16) and which, intuitively, means that one can walk continuously between any two points of the space on a path entirely contained in it. Path-connectedness implies connectedness, but not vice-versa. Clearly the line $\mathbb{E}^1$ is connected, but the "line-minus-zero" space $\mathbb{E}^1 - \{0\}$ (another notation: $\mathbb{E}^1 \backslash \{0\}$) is not. The finite space of § 1.2.3 is connected.

§ **1.2.18 Discrete topology.** It turns up when the whole set S and all its subsets are taken as open sets. The set of all subsets of a set S is called its *power set*, denoted $P(S)$,[5] so that we are taking the topological space $(S, P(S))$. This does yield a quite adequate topological space. For each

---

[4] Zeeman 1964, 1967; later references may be traced from Fullwood 1992.

[5] Also indicated by the notation e^S. If S is finite with n elements, $P(S)$ has 2^n elements. Also for S infinite, $P(S)$ is "larger" than S (Cantor theorem).

point p, $\{p\}$ is open. All open sets are also closed and so we have extreme unconnectedness.

Lest the reader think this example to be physically irrelevant, we remark that the topology induced on the light cone by the above mentioned Zeeman's topology for spacetime (§ 1.2.14) is precisely of this type. Time is usually supposed to be a parameter running in $\mathbb{E}^1$ and a trajectory on some space S is a mapping attributing a point of S to each "instant" in $\mathbb{E}^1$. It will be seen later (Section 1.4) that no function from $\mathbb{E}^1$ to a discrete space may be continuous. A denizen of the light cone, like the photon, would not travel continuously through spacetime but "bound" from point to point. The discrete topology is, of course, the finest possible topology on any space.

A space is *metrizable* if its topology may be generated by the balls defined by some distance function. It is always first-countable. Curiously enough, the discrete topology can be obtained from a metric, the so-called *discrete metric*:

$$d(p,q) = 1 \text{ if } p \neq q \quad \text{and} \quad d(p,q) = 0 \text{ if } p = q.$$

For topological vector spaces, the *indiscrete* (or *trivial*) topology is

$$T = \{\emptyset, S\}.$$

It is the weakest possible topology on any space and — being not first-countable — the simplest example of topology which cannot be given by a metric. By the way, this is an illustration of the complete independence of topology from metrics: a non-metric topology may have finer topologies which are metric, and a metric topology can have finer non-metric topologies. And a non-metric topology may have weaker topologies which are metric, and a metric topology can have weaker non-metric topologies.

§ 1.2.19 Topological product. Given two topological spaces A and B, their topological product (or *cartesian product*) $A \times B$ is the set of pairs (p,q) with $p \in A$ and $q \in B$, and a topology for which a basis is given by all the pairs of type $U \times V$, U being a member of a basis in A and V a member of a basis in B. Thus, the cartesian product of two topological spaces is their cartesian set product endowed with a "product" topology. The usual torus imbedded in $\mathbb{E}^3$, denoted $\mathbb{T}^2$, is the cartesian product of two 1-dimensional spheres (or circles) S^1. The n-torus $\mathbb{T}^n$ is the product of S^1 by itself n times.

§ 1.2.20 We have clearly separated topology from metric and found examples of non-metric topologies, but it remains true that a metric does define a topology. A reverse point of view comes from asking the following question: are all the conditions imposed in the definition of a distance function necessary to lead to a topology? The answer is no. Much less is needed. A *prametric* suffices. On a set S, a prametric is a mapping

$$\rho : S \times S \to \mathbb{R}_+ \text{ such that } \rho(p,p) = 0 \text{ for all } p \in S.$$

Once endowed with a prametric, the space S is a *prametric space*, and $\rho(p,q)$ is the prametric distance between p and q. It is acceptable to have $\rho(p,q) = 0$ even if $p \neq q$. If $\rho(p,q) \neq 0$ implies $p \neq q$, then the prametric is *separating*. If for all pairs of arguments $\rho(p,q) = \rho(q,p)$, the prametric is symmetric. A *symmetric* is a separating symmetric prametric. Metrics are very special cases of symmetrics and are always separating. Notice that the Lorentz "metric" on Minkowski spacetime is not a mapping into $\mathbb{R}_+$.

§ **1.2.21** The consideration of spatial relationships requires a particular way of dividing a space into parts. We have chosen, amongst all the subsets of S, particular families satisfying well chosen conditions to define topologies. A family of subsets of S is a topology if it includes S itself, the empty set $\emptyset$, all unions of subsets and all intersections of a finite number of them. A topology is that simplest, minimal structure allowing for precise non-trivial notions of convergence and continuity.

Other kinds of families of subsets are necessary for other purposes. For instance, the detailed study of convergence in a non-metric topological space S requires cuttings of S not including the empty set, called filters. And, in order to introduce measures and define integration on S, still another kind of decomposition is essential: a σ-algebra. In order to make topology and integration compatible, a particular σ-algebra must be defined on S, the Borel σ-algebra. A sketchy presentation of these questions is given in Chapter 15.

1.3   Kinds of texture

We have seen that, once a topology is provided, the point set acquires a kind of elementary texture, which can be very tight (as in the indiscrete topology), very loose (as in the discrete topology), or intermediate. We shall see now that there are actually optimum decompositions of spaces. The best-behaved spaces have not too many open sets: they are "compact". Nor too few: they are of "Hausdorff type".

There are many ways of probing the topological makeup of a space. We shall later examine two "external" approaches: one of them (homology) tries to decompose the space into its "building bricks" by relating it to the decomposition of euclidean space into triangles, tetrahedra and the like. The other (homotopy) examines loops (of 1 or more dimensions) in the space and their continuous deformations. Both methods use relationships with other spaces and have the advantage of providing numbers ("topological numbers") to characterize the space topology.

For the time being, we shall study two "internal" ways of probing a space (S, T). One considers subsets of S, the other subsets of T. The first considers samples of isolated points, or sequences, and gives a complete characterization of the topology. The second consists of testing the texture by rarefying the family of subsets and trying to cover the space with a smaller number of them. It reveals important qualitative traits. We shall start by introducing some concepts which will presently come in handy.

§ 1.3.1 Closure, interior, boundary.

Consider a topological space (S, T). Given an arbitrary set $U \subset S$, not necessarily belonging to T, in general there will be some closed sets C_α which contain U. The intersection $\cap_\alpha C_\alpha$ of all closed sets containing U is the *closure* of U, denoted $\bar{U}$. An equivalent, more intuitive definition is $\bar{U} = \{p$ such that every neighbourhood of p has nonvanishing intersection with $U\}$. The best-known example is that of an open interval (a, b) in $\mathbb{E}^1$, whose closure is the closed interval $[a, b]$. The closure of a closed set V is V itself, and V being the closure of itself implies that V is closed.

Given an arbitrary set $W \subset S$, not necessarily belonging to T, its *interior*, denoted "int W" or W^0, is the largest open subset of W. Given all the open sets O_α contained in W, then

$$W^0 = \cup_\alpha O_\alpha.$$

W^0 is the set of points of S for which W is an open neighbourhood. The *boundary* $b(U)$ of a set U is the complement of its interior in its closure,

$$b(U) = \bar{U} - U^0 = \bar{U} \backslash U^0.$$

It is also true that $U^0 = \bar{U} \backslash b(U)$. If U is an open set of T, then $U^0 = U$ and $b(U) = \bar{U} \backslash U$. If U is a closed set, then

$$\bar{U} = U \quad \text{and} \quad b(U) = U \backslash U^0.$$

These definitions correspond to the intuitive view of the boundary as the "skin" of the set. From this point of view, a closed set includes its own skin. The sphere S^2, imbedded in $\mathbb{E}^3$, is its own interior and closure and consequently has no boundary. A set has empty boundary when it is both open and closed. This allows a rephrasing of the definition of connectedness: a space S is connected if, except for $\emptyset$ and S itself, it has no subset whose boundary is empty.

Let again S be a topological space and U a subset. A point $p \in U$ is an *isolated* point of U if it has a neighbourhood that contains no other point of U. A point p of S is a *limit point* of U if each neighbourhood of p contains at least one point of U distinct of p. The set of all the limit points of U is called the *derived set* of U, written $D(U)$. A theorem says that

$$\bar{U} = U \cup D(U).$$

It means that we may obtain the closure of a set by adding to it all its limiting points. U is closed iff it already contains them all, $U \supseteq D(U)$. When every neighbourhood of p contains infinite points of U, p is an *accumulation point* of U (when such infinite points are not countable, p is a *condensation point*). Though we shall not be using all these notions in what follows, they appear frequently in the literature and give a taste of the wealth and complexity of the theory coming from the three simple axioms of § 1.2.2.

§ 1.3.2 Let U and V be two subsets of a topological space S. The subset U is said to be *dense in V* if $\bar{U} \supset V$. The same U will be *everywhere dense* if $\bar{U} = S$. In this case, each neighbourhood of any $p \in S$ has nonvanishing intersection with U. A famous example is the set $\mathbb{Q}$ of rational numbers, which is dense in the real line $\mathbb{E}^1$ of real numbers. This can be generalized: the set of n-uples $(p^1, p^2, \ldots, p^n)$ of *rational* numbers is dense in $\mathbb{E}^n$. This is a fortunate property indeed. We (and digital computers alike) work ultimately only with rational (actually, integer) numbers (a terminated decimal is, of course, always a rational number). The property says that we can do it even to work with real numbers, as rational numbers lie arbitrarily close to them. A set U is a *nowhere dense* subset when the interior of its closure is empty: $\bar{U}^0 = \emptyset$. An equivalent definition is that the complement to its closure is everywhere dense in S. The boundary of any open set in S is nowhere dense. The space $\mathbb{E}^1$, seen as subset, is nowhere dense in $\mathbb{E}^2$.

§ 1.3.3 The double oscillator.   As a physical example of a lower-dimensional set which is everywhere dense in a higher-dimensional one, consider a double oscillator, say, a particle whose motion is the composition of a horizontal oscillator with amplitude A_1 and frequency ω_1, and a vertical oscillator of amplitude A_2 and frequency ω_2. As long as the ratio

ω_1/ω_2 is a rational number, its trajectory will be a closed Lissajous curve on the rectangle with sides A_1 and A_2. If that ratio is not a rational number, however, the trajectory will be an open unending curve which, in the long term, will pass as near as we may wish of any point on the rectangle. That is, it will traverse any neighbourhood of each point. It will be everywhere dense in the rectangle.[6] Many examples of this kind, concerning dense orbits in a confined domain, follow[7] from the celebrated *éternel retour* theorem by Poincaré (see § 6.4.20).

§ 1.3.4 The above denseness of a countable subset in the line extends to a whole class of spaces. S is said to be a *separable* space if it has a countable everywhere dense subset. This "separability" (a name kept for historical reasons) by denseness is *not* to be confused with the other concepts going under the same name (first-separability, second-separability, etc — see § 1.3.15 on), which constitute another hierarchy of topological spaces. The present concept is specially important for dimension theory (Section 4.1.2) and for the study of infinite-dimensional spaces. Intuitively, it means that S has a countable set P of points such that each open set contains at least one point of P. In metric spaces, this separability is equivalent to second-countability.

§ 1.3.5 The Cantor set. A remarkable example of closed set is the Cantor ternary set.[8] Take the closed interval $I = [0,1]$ in $\mathbb{E}^1$ with the induced topology and delete its middle third, the open interval $(1/3, 2/3)$, obtaining the closed interval

$$E_1 = [0, 1/3] \cup [2/3, 1].$$

Next delete from E_1 the two middle thirds $(1/9, 2/9)$ and $(7/9, 8/9)$. The remaining closed space E_2 is composed of four closed intervals. Then delete the next four middle thirds to get another closed set E_3. And so on to get sets E_n for any n. Call $I = E_0$: the Cantor set is the intersection

$$E = \bigcap_{n=0}^{\infty} E_n.$$

E is closed because it is the complement of a union of open sets. Its interior is empty, so that it is nowhere dense. This "emptiness" is coherent with the following: at the j-th stage of the building process, we delete 2^{j-1} intervals,

[6] See Connes 1994.
[7] Arnold 1976.
[8] See Kolmogorov & Fomin 1970 and/or Christie 1976.

each of length $(1/3^j)$, so that the sum of the deleted intervals is 1. On the other hand, it is possible to show that a one-to-one correspondence exists between E and I, so that this "almost" empty set has the power of the continuum. The dimension of E is discussed in § 4.1.5.

§ 1.3.6 Sequences are countable subsets $\{p_n\}$ of a topological space S. A sequence $\{p_n\}$ is said to *converge* to a point $p \in S$ (we write "$p_n \to p$ when $n \to \infty$") if any open set U containing p contains also all the points p_n for n large enough. Clearly, if W and T are topologies on S, and W is weaker than T, every sequence which is convergent in T is convergent in W; but a sequence may converge in W without converging in T. Convergence in the stronger topology forces convergence in the weaker. Whence, by the way, come these designations.

We may define the q-th *tail* t_q of the sequence $\{p_n\}$ as the set of all its points p_n for $n \geq q$, and say that the sequence converge to p if any open set U containing p traps some of its tails. It can be shown that, on first-countable spaces, each point of the derivative set $D(U)$ is the limit of some sequence in U, for arbitrary U.

Recall that we can *define* real numbers as the limit points of sequences of rational numbers. This is possible because the subset of rational numbers $\mathbb{Q}$ is everywhere dense in the set $\mathbb{R}$ of the real numbers with the euclidean topology (which turns $\mathbb{R}$ into $\mathbb{E}^1$). The set $\mathbb{Q}$ has derivative $D(\mathbb{Q}) = \mathbb{R}$ and interior $\mathbb{Q}^0 = \emptyset$. Its closure is $\mathbb{R}$ itself, and is the same as the closure of its complement, which is the set

$$\mathbb{J} = \mathbb{R} \backslash \mathbb{Q}$$

of irrational numbers. As said in § 1.2.9, both $\mathbb{Q}$ and $\mathbb{J}$ are topological subspaces of $\mathbb{R}$.

On a general topological space, it may happen that a sequence converges to more than one point. Convergence is of special importance in metric spaces, which are always first-countable. For this reason, metric topologies are frequently defined in terms of sequences. On metric spaces, it is usual to introduce *Cauchy sequences* (or *fundamental sequences*) as those $\{p_n\}$ for which, given any tolerance $\varepsilon > 0$, an integer k exists such that, for $n, m > k$,

$$d(p_n, p_m) < \varepsilon.$$

Every convergent sequence is a Cauchy sequence, but not vice-versa. If every Cauchy sequence is convergent, the metric space is said to be a *complete space*. If we add to a space the limits of all its Cauchy sequences, we obtain its *completion*. Euclidean spaces are complete. The space $\mathbb{J}$ of irrational numbers with the euclidean metric induced from $\mathbb{E}^1$ is incomplete.

On general topological spaces the notion of proximity of two points, clearly defined on metric spaces, becomes rather loose. All we can say is that the points of a convergent sequence get progressively closer to its limit, when this point is unique.

§ **1.3.7** Roughly speaking, *linear spaces*, or *vector spaces*, are spaces allowing for addition and rescaling of their members. We leave the definitions and the more algebraic aspects to Chapter 13, the details to Chapter 16, and concentrate here in some of their topological possibilities. What imports here is that a linear space over the set of complex numbers $\mathbb{C}$ may have a *norm*, which is a distance function and defines consequently a certain topology called the *norm topology* (for reasons to be given later, it is also known as uniform, or strong topology). Once endowed with a norm, a vector space V is a metric topological space. For instance, a norm may come from an *inner product*, a mapping from the cartesian set product $V \times V$ into $\mathbb{C}$,

$$V \times V \to \mathbb{C}, \quad (v, u) \to <v, u>$$

with suitable properties. In this case the number

$$||v|| = (<v, v>)^{1/2}$$

will be the norm of v induced by the inner product. This is a special norm, as norms may be defined independently of inner products. Actually, one must impose certain compatibility conditions between the topological and the linear structures (see Chapter 16).

§ **1.3.8 Hilbert space.**[9] Every physicist knows Hilbert spaces from (at least) Quantum Mechanics courses. They are introduced there as spaces of wavefunctions, on which it is defined a scalar product and a consequent norm. There are basic wavefunctions, in terms of which any other may be expanded. This means that the set of functions belonging to the basis is dense in the whole space. The scalar product is an inner product and defines a topology. In Physics textbooks two kinds of such spaces appear, according to whether the wavefunctions represent bound states, with a discrete spectrum, or scattering states. In the first case the basis is formed by a discrete set of functions, normalized to the Kronecker delta. In the

---

[9] Halmos 1957.

second, the basis is formed by a continuum set of functions, normalized to the Dirac delta. The latter are sometimes called Dirac spaces.

Formally, a Hilbert space is an inner product space which is complete under the inner product norm topology. It was originally introduced as an infinite space $\mathbb{H}$ endowed with a infinite but discrete basis $\{v_i\}_{i \in N}$, formed by a countably infinite orthogonal family of vectors. This family is dense in $\mathbb{H}$ and makes of $\mathbb{H}$ a separable space. Each member of the space can be written in terms of the basis:

$$X = \sum_{i=1}^{\infty} X^i v_i.$$

The space L^2 of all absolutely square integrable functions on the interval $(a, b) \subset \mathbb{R}$,

$$L^2 = \left\{ f \text{ on } [a, b] \text{ with } \int_a^b |f(x)|^2 dx < \infty \right\},$$

is a separable Hilbert space.

Historical evolution imposed the consideration of non-separable Hilbert spaces. These would come out if, in the definition given above, instead of $\{v_i\}_{i \in N}$ we had $\{v_\alpha\}_{\alpha \in \mathbb{R}}$: the family is not indexed by a natural number, but by a number belonging to the continuum. This definition would accommodate Dirac spaces. The energy eigenvalues, for the discrete or the continuum spectra, are precisely the indexes labeling the family elements, the wavefunctions or kets. Thus, bound states belong to separable Hilbert spaces while scattering states require non-separable Hilbert spaces. There are nevertheless new problems in this continuum-label case: the summations $\sum_{i=1}^{\infty}$ used in the expansions become integrals. As said in § 1.2.21, additional structures are necessary in order to define integration (a σ-algebra and a measure, discussed in Chapter 15).

It is possible to show that $\mathbb{E}^n$ is the cartesian topological product of $\mathbb{E}^1$ taken n times, and in a way such that

$$\mathbb{E}^{n+m} = \mathbb{E}^n \times \mathbb{E}^m.$$

The separable Hilbert space is isomorphic to $\mathbb{E}^\infty$, that is, the product of $\mathbb{E}^1$ an infinite (but countable) number of times. The separable Hilbert space is consequently the natural generalization of euclidean spaces to infinite dimension. This intuitive result is actually fairly non-trivial and has been demonstrated not long ago.

§ 1.3.9 Infinite dimensional spaces, specially those endowed with a linear structure, are a privileged arena for topological subtlety. Hilbert spaces are particular cases of normed vector spaces, particularly of Banach spaces, on which a little more is said in Chapter 16. An internal product like that above does define a norm, but there are norms which are not induced by an internal product. A Banach space is a normed vector space which is complete under the norm topology.

§ 1.3.10 Compact spaces. The idea of finite extension is given a precise formulation by the concept of *compactness*. The simplest example of a space confined within limits is the closed interval $\boldsymbol{I} = [0, 1]$ included in $\mathbb{E}^1$, but its finiteness may seem at first sight a relative notion: it is limited *within* $\mathbb{E}^1$, by which it is contained. The same happens with some closed surfaces in our ambient space $\mathbb{E}^3$, such as the sphere, the ellipsoid and the torus: they are contained in finite portions of $\mathbb{E}^3$, while the plane, the hyperboloid and the paraboloid are not. It is possible, however, to give an intrinsic characterization of finite extension, dependent only on the internal properties of the space itself and not on any knowledge of larger spaces containing it. We may guess from the above examples that spaces whose extensions are limited have a "lesser" number of open sets than those which are not. In fact, in order to get an intrinsic definition of finite extension, it is necessary to restrict the number of open sets in a certain way, imposing a limit to the divisibility of space. And, to arrive at that restriction, the preliminary notion of covering is necessary.

§ 1.3.11 Suppose a topological space S and a collection $C = \{U_\alpha\}$ of open sets such that S is their union:

$$S = \bigcup_\alpha U_\alpha.$$

The collection C is called an open *covering* of S. The interval $\boldsymbol{I}$ has a well known property, which is the Heine–Borel lemma: with the topology induced by $\mathbb{E}^1$, every covering of $\boldsymbol{I}$ has a finite subcovering. An analogous property holds in any euclidean space: a subset is bounded and closed iff any covering has a finite subcovering. The general definition of compactness is thereby inspired.

§ 1.3.12 Compactness. A topological space S is a *compact space* if each covering of S contains a *finite* subcollection of open sets which is also a covering. Cases in point are the historical forerunners, the closed balls in euclidean spaces, the spheres S^n and, as expected, all the bounded surfaces in $\mathbb{E}^3$. Spaces with a finite number of points (as that in § 1.2.3) are automatically compact.

In Physics, compactness is usually introduced through coordinates with ranges in suitably closed or half-closed intervals. It is, nevertheless, a purely topological concept, quite independent of the very existence of coordinates. As we shall see presently, not every kind of space accepts coordinates. And most of those which do accept require, in order to be completely described, the use of many distinct coordinate systems. It would not be possible to characterize the finiteness of a general space by this method.

On a compact space, every sequence contains a convergent subsequence, a property which is equivalent to the given definition and is sometimes used instead: in terms of sequences,

a space is compact if, from any sequence of its points,
one may extract a convergent subsequence.

§ **1.3.13** Compact spaces are mathematically simpler to handle than non-compact spaces. Many of the topological characteristics physicists became recently interested in (such as the existence of integer "topological numbers") only hold for them. In Physics, we frequently start working with a compact space with a boundary (think of quantization in a box), solve the problem and then push the bounds to infinity. This is quite inequivalent to starting with a non-compact space (recall that going from Fourier series to Fourier integrals requires some extra "smoothing" assumptions). Or, alternatively, by choosing periodic boundary conditions we somehow manage to make the boundary to vanish. We shall come to this later.

More recently, it has become fashionable to "compactify" non-compact spaces. For example: field theory supposes that all information is contained in the fields, which represent the degrees of freedom. When we suppose that all observable fields (and their derivatives) go to zero at infinity of (say) an euclidean space, we identify all points at infinity into one only point. In this way, by imposing a suitable behaviour at infinity, a field defined on the euclidean space $\mathbb{E}^4$ becomes a field on the sphere S^4. This procedure of "compactification" is important in the study of instantons[10] and is a generalization of the well known method by which one passes from the complex plane to the Riemann sphere. However, it is not always possible.

---

[10] Coleman 1977; Atiyah et al. 1978; Atiyah 1979.

§ 1.3.14 Local compactness. A topological space is *locally compact* if each one of its points has a neighbourhood with compact closure. Every compact space is locally compact, but not the other way round: $\mathbb{E}^n$ is not compact but is locally compact, as any open ball has a compact closure. The *compactification* above alluded to is possible only for a locally compact space and corresponds to adjoining a single point to it (see § 1.4.21).[11]

A subset U of the topological space S is *relatively compact* if its closure is compact. Thus, a space is locally compact if every point has a relatively compact neighbourhood.

Locally compact spaces are of particular interest in the theory of integration, when nothing changes by adding a set of zero measure. On topological groups (§ 1.5.9), local compactness plus separability are sufficient conditions for the existence of a left- and a right-invariant Haar measure (see § 18.3.2), which makes integration on the group possible. Such measures, which are unique up to real positive factors, are essential to the theory of group representations and general Fourier analysis.

Unlike finite-dimensional euclidean spaces, Hilbert spaces are not locally compact. They are infinite-dimensional, and there are fundamental differences between finite-dimensional and infinite-dimensional spaces. One of the main distinctive properties comes out precisely here:

*Riesz theorem: a normed vector space is locally compact
if and only if its dimension is finite.*

§ 1.3.15 Separability. Compactness imposes, as announced, a limitation on the number of open sets: a space which is too fine-grained will find a way to violate its requirements. As we consider finer and finer topologies, it becomes more and more possible to have a covering without a finite subcovering. In this way, compactness somehow limits the number of open sets. On the other hand, we must have a minimum number of open sets, as we are always supposed to be able to distinguish between points in spaces of physical interest: between neighbouring states in a phase space, between close events in spacetime, etc. Such values belong to open sets (§ 1.2.1).

Can we distinguish points by using only the notions above introduced? It seems that the more we add open sets to a given space, the easier it will be to separate (or distinguish) its points. We may say things like:

*p is distinct from q because p belongs to the
neighbourhood U while q does not.*

---

[11] For details, see Simmons 1963.

Points without even this property are practically indistinguishable:

$$p = \text{Tweedledee} \quad \text{and} \quad q = \text{Tweedledum}.$$

But we might be able to say still better:

> *p is quite distinct from q because p belongs to the*
> *neighbourhood U, q belongs to the neighbourhood V,*
> *and U and V are disjoint.*

To make these ideas precise and operational is an intricate mathematical problem coming under the general name of *separability*. We shall not discuss this question in any detail, confining ourselves to a strict minimum. The important fact is that separability is not an automatic property of all spaces and the possibility of distinguishing between close points depends on the chosen topology. There are in reality several different kinds of possible separability and which one (if any) is present in a space of physical significance is once again a matter to be decided by experiment.

Technically, the two phrases quoted above correspond respectively to first-separability and second-separability. A space is said to be first-separable when, given any two points, each one will have some neighbourhood not containing the other and vice-versa. The finite space of § 1.2.3 is not first-separable. Notice that in first-separable spaces the involved neighbourhoods are not necessarily disjoint. If we require the existence of *disjoint* neighbourhoods for every two points, we have *second-separability*, a property more commonly named after Hausdorff.

§ 1.3.16 Hausdorff character. A topological space S is said to be a *Hausdorff space* if every two distinct points $p, q \in S$ have disjoint neighbourhoods. There are consequently $U \ni p$ and $V \ni q$ such that $U \cap V = \emptyset$. This property is so important that spaces of this kind are simply called "separated" by many people (the term "separable" being then reserved to the separability by denseness of § 1.3.4).

We have already met a counter-example in the trivial topology (§ 1.2.18). Another non-Hausdorff space is given by two copies of $\mathbb{E}^1$, X and Z (Figure 1.5), of which we identify all (and only!) the points which are strictly negative:

$$p_X \equiv p_Z \text{ iff } p < 0.$$

The points $p_X = 0$ and $p_Z = 0$ are distinct, p_X lying in the region of X not identified with Z and p_Z lying in Z. But they have no disjoint neighbourhoods. The space has a "Y" aspect in this case, but not all non-Hausdorff

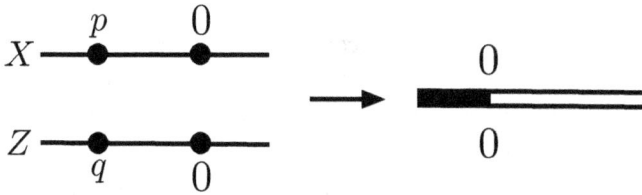

Fig. 1.5   An example of non-Hausdorff space.

spaces exhibit such a bifurcation. All Hausdorff spaces are of course necessarily first-separable, but they go much further, allowing to discern points in an way ideal for physicists, after the discussion of § 1.2.1. Actually, each point is a closed set. The Hausdorff property is also highly praised by analysts: it ensures the uniqueness of a converging sequence limit point. And it is fortunate that most spaces appearing in Physics are Hausdorff, as only on such spaces are the solutions of differential equations (with fixed initial conditions) assured to be unique — the Hausdorff character is included in the hypotheses necessary to prove the unicity-of-solution theorem. On non-Hausdorff spaces, solutions are only *locally* unique.[12] It would seem that physicists should not worry about possible violations of so desirable a condition, but non-Hausdorff spaces turn up in some regions of spacetime for certain solutions of Einstein's equations,[13] giving rise to causal anomalies.[14] Although the Hausdorff character is also necessary to the distinction of events in spacetime,[15] Penrose has speculated on its possible violation.[16]

An open set can be the union of disjoint point sets. Take the interval $I = [0, 1]$. Choose a basis containing I, $\emptyset$ and all the sets obtained by omitting from I at most a countable number of points. A perfect — though rather pathological — topological space results. It is clearly second-countable. Given two points p and q, there is always a neighbourhood of p not containing q and vice-versa. It is, consequently, also first-separable. The trouble is that two such neigbourhoods are not always disjoint: the space is not a Hausdorff space.

---

[12] Arnold 1973.

[13] Hajicek 1971.

[14] Hawking & Ellis 1973.

[15] Geroch & Horowitz in Hawking & Israel 1979. An interesting article on the topology of the Universe.

[16] Penrose in Hawking & Israel 1979, mainly in those pages dedicated to psychological time (591-596).

Topological spaces may have very distinct properties concerning count-ability and separability and are accordingly classified. We shall avoid such a analysis of the "systematic zoology" of topological spaces and only talk loosely about some of these properties, sending the more interested reader to the specialized bibliography.[17]

A Hausdorff space which is a compact (adjective) space is called *a compact* (noun). A closed subspace of a compact space is compact. But a compact subspace is necessarily closed only if the space is a Hausdorff space.

§ **1.3.17** A stronger condition is the following (Figure 1.6): S is *normal* if it is first-countable and every two of its closed disjoint sets have disjoint open neighbourhoods including them. Every normal space is Hausdorff but not vice-versa.[18] Every metric space is normal and so Hausdorff, but not vice–versa: there are normal spaces whose topology is not metrizable.

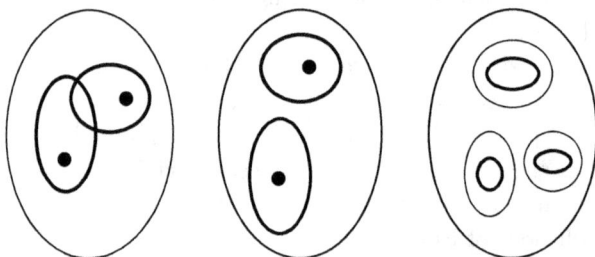

Fig. 1.6   First-separable, second-separable and normal spaces. Left: first separable — every two points have exclusive neighbourhoods. Center: Hausdorff — every two points have disjoint neighbourhoods. Right: normal — disjoint closed sets are included in disjoint open sets.

The upper-half plane $\mathbb{E}^2_+$ of Figure (1.4) is not normal and consequently non-metric.

Putting together, countability and separability may lead to many inter-esting results. Let us here only state *Urysohn's theorem*:

a topological space endowed with a countable basis (that is, second-countable) is metric iff it is normal.

[17] For instance, Kolmogorov & Fomin, 1977, Chap. II. A general résumé with many (counter) examples is Steen & Seebach 1970.
[18] For an example of Hausdorff but not normal space, see Kolmogorov & Fomin 1970, page 86.

We are not going to use much of these last considerations in the following. Our aim has been only to give a slight idea of the strict conditions a topology must satisfy in order to be generated by a metric. In order to prove that a topology T is non-metric, it suffices to show, for instance, that it is not normal.

§ 1.3.18 "Bad" $\mathbb{E}^1$, or Sorgenfrey line. The real line $\mathbb{R}^1$ with its proper (that is, non-vanishing) *closed* intervals does not constitute a topological space because the second defining property of a topology goes wrong. However, the *half-open* intervals of type $[p, q)$ on the real line do constitute a basis for a topology. The resulting space is unconnected (the complement of an interval of type $[$—$)$ is of type —$)[$— , which can be obtained as a union of an infinite number of half-open intervals) and not second-countable (because in order to cover —$)$, for example, one needs a number of $[$—$)$'s which is an infinity with the power of the continuum). It is, however, first-countable: given a point p, amongst the intervals of type $[p, r)$ with r rational, there will be some one included in any U containing p. The Sorgenfrey topology is finer than the usual euclidean line topology, though it remains separable (by denseness). This favorite pet of topologists is non-metrizable.

§ 1.3.19 Paracompactness. Consider a covering $\{U_\alpha\}$ of S. It is a *locally finite covering* if each point p has a neighbourhood $U \ni p$ such that U intersects only a finite number of the U_α. A covering $\{V_i\}$ is a *refinement* of $\{U_\alpha\}$ if, for every V_i, there is a U_α such that $U_\alpha \supset V_i$. A space is *paracompact* if it is Hausdorff and all its coverings have local finite refinements. Notice: finite subcoverings lead to compactness and finite refinements (plus Hausdorff) to paracompactness. A connected Hausdorff space is paracompact if it is second-countable. Figure 1.7 is a scheme of the separability hierarchy. Every metric space is paracompact. Every paracompact space is normal. Paracompactness is a condition of importance for integration, as it is sufficient for attributing a partition of unity (see Section 15.1.5) to any locally finite covering.

Paracompactness is also necessary to the existence of linear connections on the space.[19] Paracompact spaces are consequently essential to General Relativity, in which these connections are represented by the Christoffel symbols. The Lorentz metric on a Hausdorff space implies its paracom-

---

[19] Hawking & Ellis 1973.

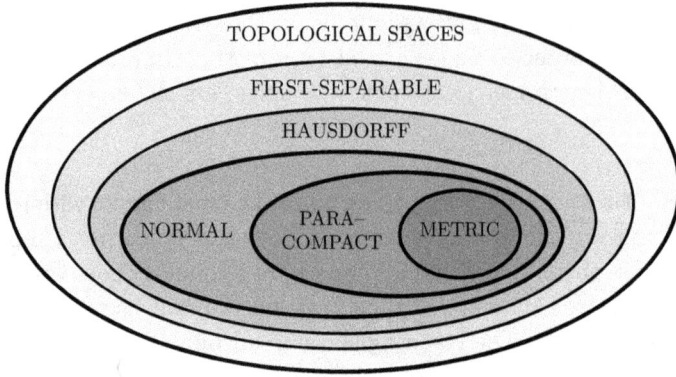

Fig. 1.7   A second hierarchy: all metric spaces are normal.

pactness.[20]

1.4   Functions

Continuity of functions is the central notion of general topology. Functions, furthermore, allow us to compare the properties of different spaces.

§ **1.4.1** The word "function" is used here — unless otherwise explicitly stated — in the modern sense of *monodromous* function. Perhaps the simplest examples of functions are the permutations of the elements of a finite set (Section 14.1.4).

§ **1.4.2** Just for the record. A function is a mapping with a single value in the target space for each point of its domain. As a point set function, $f : A \to B$ will be

(i) *Surjective* (or *onto*) if $f(A) = B$; that is, the values of f for all points of A cover the whole of B.

(ii) *Injective* (or *one-to-one*) if, for all a and $a' \in A$, the statement $f(a) = f(a')$ implies $a = a'$; that is, it takes distinct points of A into distinct points of B.

(iii) *Bijective* (also called a *condensation*) if it is both onto and one-to-one.

---

[20] Geroch 1968.

(iv) If $B \subset A$, an *inclusion* (indicated by the "hooked arrow" symbol $\hookrightarrow$ by some authors) is an injective map $i : B \to A$ with $i(p) = p$ if $p \in B$. The set $f(A)$ of points of B which are the image by f of some point of A is the *graph* of f.

§ 1.4.3 Mathematics deals basically with sets and functions. The sequences previously introduced as countable subsets $\{p_n\}$ of a topological space S are better defined as functions

$$p : \mathbb{N} \to S, \quad n \to p_n,$$

from the set of natural numbers $\mathbb{N}$ into S. Only to quote a famous fundamental case, two sets have the same *power* if there exists some bijective function between them. In this sense, set theory uses functions in "counting". We have said in § 1.2.18 that the power set $P(S)$ of a set S is the set of all its subsets. For S finite with n points, $P(S)$ will have $2^n (> n)$ elements and for this reason $P(S)$ is sometimes indicated by 2^S. $P(S)$ is larger than S, as there are surjective functions, but no injective functions, from $P(S)$ to S. This notion of a relative "size" of a set was shown by Cantor to keep holding for infinite sets ($P(S)$ is *always* larger than S) and led to his infinite hierarchy of infinite numbers.

§ 1.4.4 Let $f : A \to B$ be a function between two topological spaces. The *inverse image* of a subset X of B by f is

$$f^{<-1>}(X) = \{a \in A \text{ such that } f(a) \in X \}.$$

§ 1.4.5 The function f is *continuous* if the inverse images of all the open sets of the target space are open sets of the domain space. This is the notion of continuity on general topological spaces. And here we have a very good opportunity to illustrate the cautionary remark made in § 1.2.12. At first sight, the above definition is of that skew-minded type alluded to. We could try to define a continuous function "directly", as a function mapping open sets into open sets, but the following example shows that using the inverse as above is essential. Consider the function $f : \mathbb{E}^1 \longrightarrow \mathbb{E}^1$ given by

$$f(x) = \begin{cases} x & \text{for } x \le 0 \\ x + 1 & \text{for } x > 0. \end{cases}$$

which is shown in Figure 1.8. In reality, the target space is not $\mathbb{E}^1$. The function is actually

$$f : \mathbb{E}^1 \to (-\infty, \, 0] \cup (1, \, \infty),$$

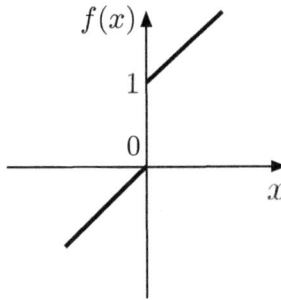

Fig. 1.8   A standard discontinuous function.

the latter with the induced topology. It has an obvious discontinuity at $x = 0$. It is indeed discontinuous by the definition given above, but it is easy to check that it would be continuous if the "direct" definition was used. With the induced topology, $(-\infty, 0]$ is open in the image space. The function f will take open sets of $\mathbb{E}^1$ into open sets of the union, but its inverse $f^{<-1>}$ will take $(-\infty, 0]$ into the same interval in $\mathbb{E}^1$, where it is not open. By the way, this also illustrates the wisdom of carefully and clearly stating the definition and value domains of functions, as mathematicians "pedantly" do!

§ **1.4.6** *One should specify the topology whenever one speaks of a continuous function.* Any function defined on a space with the discrete topology is automatically continuous. On a space with the indiscrete topology, no function of interest can be continuous. That is what we meant when we said that no useful definition of the continuity concept was possible in these two cases.

§ **1.4.7** In Zeeman's topology for spacetime, every function on the light cone is continuous, as any function will be continuous if the domain space is discrete. With these repeated references to Zeeman's work we want to stress the following point: people talk frequently of continuous wavefunctions and fields on spacetime but nobody really knows the meaning of that, as the real topology of spacetime is unknown. Curiously enough, much more study has been dedicated to the topological properties of functional spaces (such as the spaces of quantum fields) than to those of the most important space of all Physics. By the way, some other topologies have been proposed for spacetime,[21] whose properties would influence those of the fields. Actually, these tentatives proceed in the inverse way: they look for a topology in which better known quantities (such as classical paths on which integrations are to be performed in Feynman's formalism, or Green's functions) are continuous.[22]

§ 1.4.8 If a function is continuous on a space with topology T, it will be continuous in any refinement of T. Zeeman's topology is a refinement of that of $\mathbb{E}^4$, but a function which is continuous in his spacetime could appear as discontinuous upon "euclideanization". In this way "euclideanizations", procedures by which quantities of physical interest (such as Green's functions and S matrix elements), defined on spacetime, are transformed by some kind of analytical continuation into analogous quantities on $\mathbb{E}^4$, may provide constraints on spacetime possible topologies.

§ 1.4.9 Isometry is a distance-preserving mapping between metric spaces:

$$f : X \to Y \text{ such that } d_Y[f(p), f(q)] = d_X(p, q).$$

If a function g is such that target and domain spaces coincide, then g is an *automorphism*. Particular cases of continuous automorphisms are the *inner isometries* of metric spaces, when $X \equiv Y$. In this case, the old name "motions" is physically far more telling. On spacetimes, motions are governed by a special differential equation, the Killing equation (see § 6.6.13).

§ 1.4.10 Consider $\mathbb{E}^n$, taking into account its usual structure of vector space (Chapter 13). This means, of course, that we know how to add two points of $\mathbb{E}^n$ to get a third one, as well as to multiply them by external real scalars. We denote points in $\mathbb{E}^n$ by

$$p = (p^1, p^2, \ldots, p^n), \quad a = (a^1, a^2, \ldots, a^n), \quad \ldots .$$

Then, translations $t : p \to p + a$, homothecies $h : p \to \alpha p$ with $\alpha \neq 0$, and the invertible linear automorphisms are continuous. Consider in particular real functions $f : \mathbb{E}^1 \to \mathbb{E}^1$. Then

$$f(x) = x + 1$$

is surjective and injective (consequently, bijective). On the other hand,

$$f(x) = x^2$$

---

[21] Hawking, King & McCarthy 1976; Göbel 1976 a, b; Malament 1977.
[22] For a proposal of topology for Minkowski spacetime, which would allow an asymmetry in time evolution, see Wickramasekara 2001.

is neither, but becomes bijective if we restrict both sets to the set $\mathbb{R}_+$ of positive real numbers. The function

$$\exp : [0, 2\pi) \to S^1, \quad f(\alpha) = \exp(i\alpha),$$

is bijective. With the topologies induced by those of $\mathbb{E}^1$ and $\mathbb{E}^2$, respectively, it is also continuous. A beautiful result:

> *every continuous mapping of a topological space*
> *into itself has a fixed point.*

§ **1.4.11** The image of a connected space by a continuous function is connected. The image of a compact space by a continuous function is compact. Therefore, continuous functions preserve topological properties. But only in one way: the inverse images of connected and/or compact domains are not necessarily connected and/or compact. In order to establish a complete equivalence between topological spaces we need functions preserving topological properties both ways.

> *A bijective function $f : A \to B$ will be a homeomorphism between*
> *the topological spaces A and B if it is continuous and has a*
> *continuous inverse. Thus, it takes open sets into*
> *open sets and its inverse does the same.*

Two spaces are *homeomorphic* when there exists a homeomorphism between them. Notice that if

$$f : A \to B \quad \text{and} \quad g : B \to C$$

are continuous, then the composition

$$(f \circ g) : A \to C$$

is also continuous. If f and g are homeomorphisms, so is their composition. By the very definition, the inverse of a homeomorphism is a homeomorphism.

A homeomorphism is an equivalence relation: it establishes a complete topological equivalence between two topological spaces, as it preserves all the purely topological properties. We could in reality define a topology as an equivalence class under homeomorphisms. And the ultimate (perhaps too pretentious?) goal of "Topology" as a discipline is the classification of all topological spaces, of course up to such equivalences.

Intuitively, "A is homeomorphic to B" means that A can be deformed, without being torn or cut, to look just like B. Under a homeomorphism, images and pre-images of open sets are open, and images and pre-images of closed sets are closed. A sphere S^2 can be stretched to become an ellipsoid or an egg-shaped surface or even a tetrahedron. Such surfaces are indistinguishable from a purely topological point of view. They are the same topological space.

The concept of homeomorphism gives in this way a precise meaning to those rather loose notions of smooth deformations we have been talking about in § 1.1.4. Of course, there is no homeomorphism between either the sphere, the ellipsoid, or the tetrahedron and (say) a torus T^2, which has quite different topological properties.

From the point of view of sequences, a homeomorphism is a mapping

$$h : A \to B$$

such that, if $\{p_n\}$ is a sequence in A converging to a point p, then the sequence $\{h(p_n)\}$ in B converges to $h(p)$. And vice-versa: if the sequence $\{q_n\}$ in B converges to a point q, then the sequence given by $\{h^{-1}(q_n)\}$ in A converges to $\{h^{-1}(q)\}$. Two more things worth saying:

- A *condensation* is a mapping $f : A \to B$ which is one-to-one ($x \neq y \to f(x) \neq f(y)$) and onto ($f(A) = B$).
- A *homeomorphism* is a condensation whose inverse is also a condensation.

The study of homeomorphisms between spaces can lead to some surprises. It was found that a complete metric space (§ 1.3.6) can be homeomorphic to an incomplete space. Consequently, the completeness of metric spaces is not really a topological characteristic. Another result: the space of rational numbers with the usual topology is homeomorphic to any 1-dimensional metric countable space without isolated points.

Commentary 1.1   The main objective of Zeeman's cited papers on spacetime was to obtain a topology whose automorphic homeomorphisms preserve the causal structure and constitute a group including the Poincaré and/or the conformal group. ◄

§ 1.4.12 Take again the euclidean vector space $\mathbb{E}^n$. Any isometry will be a homeomorphism, in particular any translation. Also homothecies with reason $\alpha \neq 0$ are homeomorphisms. From these two properties it follows that any two open balls are homeomorphic to each other, and any open ball is homeomorphic to the whole space. As a hint of the fundamental role which euclidean spaces will come to play, suppose that a space S has some open set U which is by itself homeomorphic to an open set (a ball) in some

$\mathbb{E}^n$: there is a homeomorphic mapping

$$f : U \to \text{ball}, \quad (p \in U) = x \quad \text{with} \quad x = (x^1, x^2, \dots, x^n).$$

Such a homeomorphism is a *local homeomorphism*. Because the image space is $\mathbb{E}^n$, f is called a *coordinate function* and the values x^k are *coordinates* of p. Coordinates will be formally introduced in Section 4.2.

We are now in condition to explain better a point raised in § 1.3.8. The separable Hilbert space is actually homeomorphic to $\mathbb{E}^\infty$. Once this is granted, the same topology is given by another metric,

$$d(\boldsymbol{v}, \boldsymbol{u}) = \sum_k \frac{|v_k - u_k|}{2^k \left(1 + |v_k - u_k|\right)}.$$

The metric space so obtained is a Fréchet space, name given to any metrizable complete topological vector space.[23] Such spaces are important because a theorem says that any separable metric space is homeomorphic to some subspace of a Fréchet space.

§ **1.4.13** A counter-example: take S^1 as the unit circle on the plane,

$$S^1 = \{(x, y) \in \mathbb{R}^2 \quad \text{such that} \quad x^2 + y^2 = 1\},$$

with the topology induced by the usual topology of open balls (here, open disks) of $\mathbb{E}^2$. The open sets will be the open arcs on the circle (Figure 1.9). Take then the set consisting of the points in the semi-open interval $[0, 2\pi)$ of the real line, with the topology induced by the $\mathbb{E}^1$ topology. The open sets will be of two kinds: all the open intervals of $\mathbb{E}^1$ strictly included in $[0, 2\pi)$, and those intervals of type $[0, \beta)$, with $\beta \leq 2\pi$, which would be semi-open in $\mathbb{E}^1$. The function given by

$$f(\alpha) = \exp[i\alpha],$$

or

$$x = \cos\alpha \quad \text{and} \quad y = \sin\alpha,$$

is bijective and continuous (§ 1.4.10). But f takes open sets of the type [ ) into semi-open arcs, which are not open in the topology defined on S^1. Consequently, its inverse is not continuous. The function f is not a homeomorphism. In reality, none of such exist, as S^1 is not homeomorphic to the interval $[0, 2\pi)$ with the induced topology. It is possible, however, to define on $[0, 2\pi)$ another topology which makes it homeomorphic to S^1 (see § 1.5.3).

---

[23] Attention: the name "Fréchet space" is also used with another, more fundamental and quite distinct meaning: a set to each element p of which is associated a family of subsets including p. See, for instance, Sierpiński 1956.

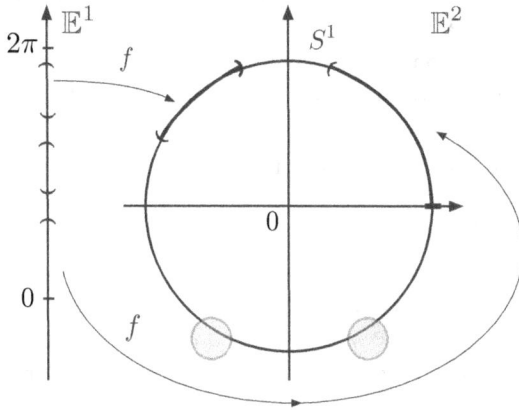

Fig. 1.9 The function $f : [0, 2\pi) \to S^1$ included in $\mathbb{E}^2$, given by $f(\alpha) = (\cos\alpha, \sin\alpha)$, is bijective and continuous; but it takes some open sets into sets which are not open; consequently its inverse $f^{<-1>}$ is not continuous and f is not a homeomorphism.

§ **1.4.14** Of fundamental importance is the following kind of function:

> A *curve* on a topological space S is a function α taking the compact interval $\boldsymbol{I} = [0, 1]$ into S. If $\alpha : \boldsymbol{I} \to S$ is a continuous function, then α is a *continuous curve*, or *path*.

Notice the semantic shift from the usual meaning of "curve", which would rather refer to the set of values of the mapping and not the mapping itself. A point on the curve will in general be represented by $\alpha(t)$, where the parameter $t \in \boldsymbol{I}$, so that $\alpha : t \to \alpha(t)$. Notice also that we reserve the word *path* to a continuous curve. When $\alpha(0) = \alpha(1)$, α is a *closed curve*, or a *loop*, which can be alternatively defined as a function from the circle S^1 into S.

§ **1.4.15 Function spaces.** Take two spaces X and Y and consider the set of functions between them, $\{f_\alpha : X \to Y\}$. This set is usually indicated by Y^X. Suppose X compact and Y metric with a distance function d. We may then define the distance between two functions f_1, f_2 as

$$\text{dist}\,(f_1, f_2) = \text{least upper bound }\{d[f_1(p), f_2(p)]\}$$

for all $p \in X$. It can be shown that this distance turns the set of functions into a metric (hence topological) space, whose topology depends on the topologies of X and Y but not in reality on the distance function d (that is: any other distance function leading to the same topology for Y will give the same topology for Y^X).

The rather strict conditions above can be softened in the following elaborate way: take any two topologies on X and Y. Call C the set of compact subsets of X and O the set of open subsets of Y. For each $c \in C$ and $o \in O$, call (c, o) that subset of Y^X whose members are the functions f such that $o \supset f(c)$. Call (C, O) the collection of such (c, o). A topology on Y^X is then generated from a basis consisting of all the finite intersections of the sets $(c, o) \in (C, O)$. This is the *compact-open topology*, which coincides with the previous one when X is compact and Y is metric. Other topologies may be defined on function spaces and their choice is a matter of convenience.

A point worth emphasizing is that, besides other requirements, a topology is presupposed in functional differentiation and integration. Function spaces are more complicated because their dimension is (highly) infinite and many of the usual properties of finite spaces do not carry over to them.[24] As stated in § 1.3.14, there is one remarkable difference: when the topology is given by a norm, an infinite-dimensional space is never locally compact.

§ 1.4.16 Connection by paths. A topological space S is *path-connected* (or *arcwise-connected*) if,

> *for every two points p, q in S there exists a*
> *path α with $\alpha(0) = p$ and $\alpha(1) = q$.*

We have said that path-connectedness implies connectedness but there are connected spaces which are not path-connected. They are, however, very peculiar spaces and in most applications the word "connected" is used for path-connectedness. Some topological properties of a space can be grasped through the study of its possible paths. This is the subject matter of homotopy theory, of which some introductory notions will be given in Chapter 3.

Commentary 1.2   Hilbert spaces are path-connected. ◀

§ 1.4.17 Suppose the space S is not path-connected. The *path-component* of a point p is that subset of S whose points can be linked to p by continuous curves. Of course "path" is not a gratuitous name. It comes from its most conspicuous example, the path of a particle in its configuration space. The evolution of a physical system is, most of times, represented by a curve in the space of its possible states (see, for example, Chapter 25). Suppose such

---

[24] Differences between finite- and infinite-dimensional spaces are summarized in DeWitt-Morette, Masheshwari & Nelson 1979, Appendix A.

space of states is not path-connected: time evolution being continuous, the system will never leave the path-component of the initial state. Topological conservation laws are then at work, the conserved quantities being real functions on the state space which are constant on each path-component. This leads to a relationship[25] between topological conservation laws and superselection rules[26] and is the underlying idea of the notion of kinks.[27] The relation

$$pRq = \text{"there exists a path on } S \text{ joining } p \text{ to } q\text{"}$$

is an equivalence relation. The path-component to which a point p belongs can be seen as the class of p.

§ 1.4.18 There are two kinds of conserved quantities in Physics: those coming from Noether's theorem — which come from the symmetry of the physical system — and the topological invariants — which come from the global, topological properties of the space of all possible states. The *true* Noether invariants (see Chapter 31) are functions which are constant on the solutions, while the topological invariants are constant on each path-component of the space of states. By a convenient choice of topology, also the Noether invariants can be made into topological invariants.[28] Let us stress: a topological invariant is any property (quality or number) of a topological space which is shared by all spaces homeomorphic to it. This is to say, invariant under homeomorphisms.

§ 1.4.19 Spaces of paths. They are particular cases of function spaces. Spaces of paths between two fixed end-points,

$$q : [t_1, t_2] \to \mathbb{E}^3$$

are used in Feynman's formulation of non-relativistic quantum mechanics. Integrations on these spaces presuppose additional structures. Notice that such spaces can be made into vector spaces if

$$q(t_1) = q(t_2) = 0.$$

This is a simple example of infinite-dimensional topological vector space.[29]

---

[25] Ezawa 1978; Ezawa 1979.

[26] Streater & Wightman 1964.

[27] Skyrme 1962; Finkelstein 1966.

[28] Dittrich 1979.

[29] DeWitt-Morette, Masheshwari & Nelson 1979.

§ 1.4.20 Measure and probability. A measure is a special kind of set function, that is, a function attributing values to sets (see Chapter 15). We actually consider a precise family of subsets, a σ-algebra $\mathcal{A}$, including the empty set and the finite unions of its own members. A (positive) measure is a function attributing to each subset a probability, that is, a positive real value. A good example is the Lebesgue measure on $\mathbb{E}^1$: the σ-algebra is that generated by the open intervals (a, b) with $b \geq a$ and the measure function is

$$m[(a, b)] = b - a.$$

A set with a sole point has zero measure. The Cantor set E of § 1.3.5 has $m(E) = 0$. Measure spaces are easily extended to Cartesian product spaces, so that the Lebesgue measure goes easily over higher dimensional euclidean spaces.

§ 1.4.21 Compactification. Let us give an example of a more technical use of homeomorphisms and the equivalences they engender. We have talked about locally compact spaces and their possible compactification. Given a locally compact space X, we define formally the *Alexandrov's compactified* of X as the pair (X', f') with the conditions:

(i) X' is a compact space.
(ii) f' is a homeomorphism between X and the complement in X' of a point p.
(iii) If (X'', f'') is another pair satisfying (i) and (ii), there exists a unique homeomorphism $h : X' \to X''$ such that $f'' = h \circ f'$.

We say then that p is the *point at infinity* of X', and that the compact X' is obtained from the locally compact space X by "adjunction of an infinity point". The Alexandrov's compactified of the plane $\mathbb{E}^2$ is the pair (sphere S^2, stereographic projection). For an example of a projection from the north pole, see Section 23.3. With enlarged stereographic projections (Section 37.2), the sphere S^n comes out from the compactification of $\mathbb{E}^n$.

Physically, such a process of compactification is realized when the following two steps are taken:

(i) Suppose all the physics of the system is contained in some functions; for example, in field theory the fields are the degrees of freedom, the coordinates of spacetime being reduced to mere parameters.
(ii) The functions or fields are supposed to vanish at all points of infinity, which makes them all equivalent.

Any bound-state problem of nonrelativistic Quantum Mechanics in $\mathbb{E}^3$, in which the wavefunction is zero outside a limited region, has actually S^3 as configuration space. In the relativistic case, it is frequent that we first "euclideanize" the Minkowski space, turning it into $\mathbb{E}^4$, and then suppose all involved fields to have the same value at infinity, thereby compactifying $\mathbb{E}^4$ to S^4.

1.5   Quotients and groups

§ **1.5.1 Quotient spaces.** Consider a spherically symmetric physical system in $\mathbb{E}^3$ (say, a central potential problem). We usually (for example, when solving the potential problem by separation of variables) perform some manipulations to reduce the space to sub-spaces and arrive finally to an equation in the sole variable "r". All the points on a sphere of fixed radius r will be equivalent, so that each sphere will correspond to an equivalence class. We actually merge all the equivalent points into one of them, labelled by the value of "r", which is taken as the representative of the class. On the other hand, points on spheres of different radii are nonequivalent, will correspond to distinct equivalence classes. The radial equation is thus an equation on the space of equivalence classes.

A physical problem in which a large number of points are equivalent reduces to a problem on the space of equivalence classes (or spaces whose "points" are sub-sets of equivalent points). Such spaces may have complicated topologies. These are called *quotient topologies*. Mostly, they come up when a symmetry is present, a set of transformations which do not change the system in any perceptible way.

The merging of equivalent points into one representative is realized by a mapping, called a *projection*. In the above spherical example, the whole set of equivalence classes will correspond to the real positive line $\mathbb{E}^1_+$, on which r takes its values. The projection will be

$$\pi : \mathbb{E}^3 \to \mathbb{E}^1_+, \quad \pi : p = (r, \theta, \varphi) \to r.$$

An open interval in $\mathbb{E}^1_+$, say,

$$J = (r - \varepsilon, r + \varepsilon),$$

will be taken back by $\pi^{<-1>}$ into the region between the two spheres of radiis $(r - \varepsilon)$ and $(r + \varepsilon)$. This region is an open set in $\mathbb{E}^3$, so that J is an open set in the quotient space if π is supposed continuous. As distinctions between equivalent points are irrelevant, the physical configuration space reduces to $\mathbb{E}^1_+$. The symmetry transformations constitute the rotation group in $\mathbb{E}^3$, the special orthogonal group $SO(3)$. In such cases, when the equivalence is given by symmetry under a transformation group G, the quotient space is denoted S/G. Here,

$$\mathbb{E}^1_+ = \mathbb{E}^3/SO(3).$$

§ **1.5.2** Inspired by the example above, we now formalize the general case of spaces of equivalence classes. Suppose a topological space S is given on which an equivalence relation R is defined: two points p and q are equivalent, $p \approx q$, if linked by the given relation. We can think of S as the configuration space of some physical system, of which R is a symmetry: two points are equivalent when attainable from each other by a symmetry transformation. All the points obtainable from a given point p by such a transformation constitute the equivalence class of p, indicated in general by $[p]$ when no simpler label is at hand. This class may be labelled by any of its points instead of p, of course, and points of distinct classes cannot be related by transformations. The set of equivalence classes, which we call $\{[p]\} = S/R$, can be made into a topological space, whose open sets are defined as follows: let $\pi : S \to S/R$ be the projection

$$\pi : p \to [p]$$

with $[p]$ the class to which p belongs. Then a set U contained in S/R is *defined* to be open iff $\pi^{<-1>}(U)$ is an open set in S. Notice that π is automatically continuous. The space S/R is called the *quotient space* of S by R and the topology is the *quotient topology*. The simplest example is the plane with points supposed equivalent when placed in the same vertical line. Each vertical line is an equivalence class and the quotient space is the horizontal axis. In another case, if the plane is the configuration space of some physical system which is symmetric under rotations around the origin, all the points lying on the same circle constitute an equivalence class. The quotient will be the space whose members are the circles.

§ **1.5.3** Take the real line $\mathbb{E}^1$ and the equivalence

$$p \approx q \ \text{ iff } \ p - q = n \in \mathbb{Z}$$

where $\mathbb{Z}$ is the set of all integers, an additive group. The space $\mathbb{E}^1/R$, or $\mathbb{E}^1/\mathbb{Z}$, has as set point the interval $[0,1)$, of which each point represents a class. The open sets of the quotient topology are now of two types:

(i) those of $\mathbb{E}^1$ included in the interval.

(ii) The neighbourhoods of the point 0, now the unions $u \cup v$ of intervals
as in the Figure 1.10.

This topology is able to "close" the interval $[0,1)$ on itself. The same
function f which in § 1.4.13 failed to take the neighbourhoods of 0 into
open sets of S^1 will do it now. Consequently, the same interval $[0, 2\pi)$ (we
could have used $[0, 1)$ instead), becomes, once endowed with the quotient
topology, homeomorphic to the circle. It acquires then all the topological
properties of S^1, for instance that of being compact.

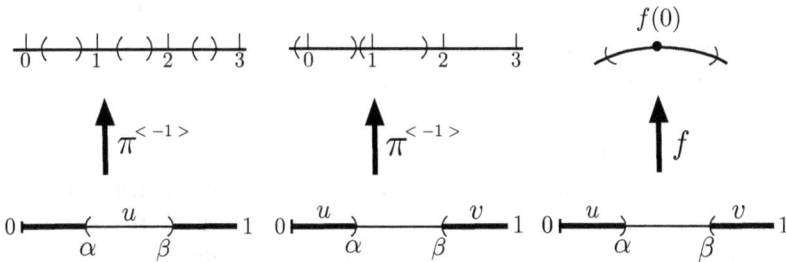

Fig. 1.10 Kinds of open sets in the quotient topology: (left) first-kind: $\pi^{<-1>}(u) = \cup_{n \in \mathbb{Z}}(\alpha + n, \beta + n)$; (center) second-kind: $\pi^{<-1>}(u \cup v) = \cup_{n \in \mathbb{Z}}(\alpha + n, \beta + n - 1)$; (right) with the quotient topology, $f(x) = \exp[i2\pi x]$ becomes a homeomorphism.

We shall see later that coordinates are necessarily related to homeomorphisms. The use, for the circle, of an angular real coordinate $\varphi \in \mathbb{E}^1$, which
repeats itself every time it arrives at equivalent points of $\mathbb{E}^1$, is just a manifestation of the "quotient" relation between $\mathbb{E}^1$ and S^1. Actually, the angle
φ does belong to the interval $[0, 2\pi)$, but with the quotient topology. We
shall also see that φ is not really a good coordinate, because coordinates
must belong to euclidean spaces.

§ 1.5.4 Frequent use is made of *congruencies* (or structures similar to,
or yet inspired by them) in getting quotient spaces. Only to have the
definitions at hand, recall their simplest examples: for a fixed positive
integer n, the number r $(0 \le r < n)$ such that

$$h = nq + r$$

for some q is *congruent to h modulo n*. Notation: $h = r(\mathrm{mod}\, n)$. The
number r such that

$$h + k = nq + r = r(\mathrm{mod}\, n)$$

is the *sum modulo* n of h and k. Adapted to multiplication: the *multiplication modulo* n of two integers p and q is the remainder of their usual product when divided by n, the number m such that

$$pq = m(\bmod\, n).$$

§ 1.5.5 The Möbius band. Even rap-singers are by now acquainted with the usual definition of this amazing one-sided (or one-edged) object: take a sheet as in Figure 1.11, and identify $a = b'$ and $b = a'$. This corresponds to taking the product of the intervals $(a, a') \times (a, b)$ and twisting it once to attain the said identification. It is possible to use $\mathbb{E}^1$ instead of the limited interval (a, a'). To simplify, use $\mathbb{E}^1 \times (-1, 1)$ and the equivalence given by (see Figure 1.12): $(p^1, p^2) \approx (q^1, q^2)$ iff

(i) $p^1 - q^1 = n \in \mathbb{Z}$ and   (ii) $p^2 = (-)^n q^2$.

Experiment with a paper sheet is commendable. Begin with the usual definition and then take sheets which are twice, thrice, etc, as long. A simple illustration of the "quotient" definition comes up. A simple cylinder comes out if we use the condition $p^2 = q^2$ instead of condition (ii) above. The cylinder topology so introduced coincides with that of the topological (or cartesian) product $\mathbb{E}^1 \times S^1$. The Möbius band, on the other hand, is not a topological product! Also experiments with a waist belt are instructive to check the one-edgedness and the fact that, turning twice instead of once before identifying the extremities, a certain object is obtained which can be deformed into a cylinder.

We can examine free quantum fields in the original sheet (quantization in a plane box). The use of periodic boundary conditions for the vertical coordinates corresponds to quantization on the cylinder. Quantization on the Möbius band is equivalent to antiperiodic boundary conditions and leads to quite distinct (and interesting) results.[30] For example, the vacuum (the lowest energy state, see Section 28.2.2) in the Möbius band is quite different from the vacuum in the cylinder.

§ 1.5.6 The torus. Take $\mathbb{E}^2$ and the equivalence $(p^1, p^2) \approx (q^1, q^2)$ iff

$$p^1 - q^1 = n \in \mathbb{Z} \quad \text{and} \quad p^2 - q^2 = m \in \mathbb{Z}.$$

A product of two intervals, homeomorphic to the torus

$$T^2 = S^1 \times S^1,$$

---

[30] Avis & Isham, in Lévy & Deser 1979.

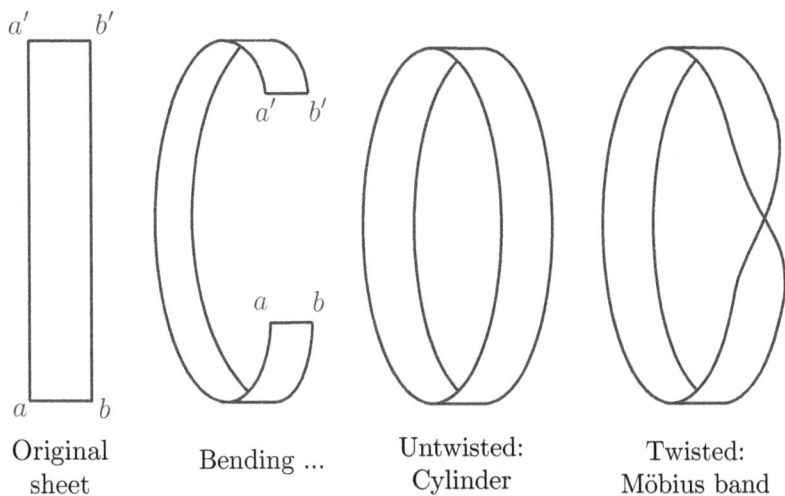

| Original sheet | Bending ... | Untwisted: Cylinder | Twisted: Möbius band |

Fig. 1.11   The cylinder and the Möbius strip.

is obtained. As T^2 is obtained by "dividing" the plane by twice the group of integers $\mathbb{Z}$, we write $T^2 = \mathbb{E}^2/\mathbb{Z}^2$. The "twisted" version is obtained by modifying the second condition to

$$p^2 - (-)^{n-1}q^2 = m \in \mathbb{Z}.$$

The resulting quotient space is the *Klein bottle*. Experiments with a paper leaf will still be instructive, but frustating — the real construction of the bottle will show itself impossible because we live in $\mathbb{E}^3$, and that would be possible only in $\mathbb{E}^4$! Higher dimensional toruses

$$T^n = S^1 \times S^1 \ldots \times S^1 \quad \text{(n times)}$$

are obtained equally as quotients, $T^n = \mathbb{E}^n/\mathbb{Z}^n$.

§ 1.5.7 The above examples and considerations on compactification illustrate the basic fact: for spaces appearing in physical problems, the topological characteristics are as a rule fixed by boundary conditions and/or symmetry properties.

§ 1.5.8 As said before, it is customary to write G instead of R in the quotient, S/G, when the relation is given by a group G — as with the toruses, the spherical case $\mathbb{E}^1_+ = \mathbb{E}^3/SO(3)$, and $S^1 = \mathbb{E}^1/\mathbb{Z}$ in § 1.5.3. Consider now

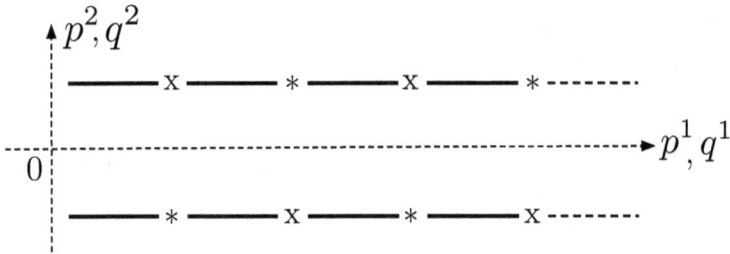

Fig. 1.12   Scheme of $\mathbb{E}^1 \times (-1, +1)$ showing the equivalent points to form a Möbius band.

the configuration space of a system of n particles in $\mathbb{E}^3$. From the classical point of view, it is simply $\mathbb{E}^{3n}$, the product of the configuration spaces of the n particles. If we now take into account particle indistinguishability, the particle positions become equivalent. The configuration space will be the same, but with all the particle positions identified, or "glued together". In an ideal gas, for example, the points occupied by the n particles are to be identified. The space is insensitive to the exchange of particles. These exchanges constitute the symmetric group S_n, the group of all permutations of n objects (see Section 14.1.4) and the final configuration space is $\mathbb{E}^{3n}/S_n$, a rather messy space.[31]

Many topological spaces of great relevance possess much more structure than a simple topology. The additional structure is usually of algebraic nature, as for example, that of a group.

§ **1.5.9 Topological groups.**   The euclidean space $\mathbb{E}^n$ is a topological space with the standard euclidean ball topology. But, being also a vector space, it is furthermore an abelian group, with the addition operation. A natural question arises: is there any compatibility relation between the algebraic group structure and the topological structure? The answer is positive: the mappings (representing the group operation and the obtention of the inverse)

$$(X, Y) \longrightarrow X + Y$$

and

$$X \longrightarrow -X$$

---

[31] See for instance Laidlaw & DeWitt-Morette 1971.

are continuous functions $\mathbb{E}^n \times \mathbb{E}^n \to \mathbb{E}^n$ and $\mathbb{E}^n \to \mathbb{E}^n$ respectively. In rough words, the group operations are continuous maps in the underlying topology. The notion of topological group comes from this compatibility between algebraic and topological structures.

Let us rephrase the definition of group (see Chapter 13). A group is a set G (whose members are its "elements") with an operation

$$m : G \times G \to G,$$

given by

$$m : (a, b) \to m(a, b),$$

and defined in such a way that:

(i) $m(a, b) \in G$ for all pairs a, b.
(ii) There exists a neutral element e such that $m(a, e) = m(e, a) = a$.
(iii) Every $a \in G$ has an inverse element a^{-1} which is such that
$m(a, a^{-1}) = m(a^{-1}, a) = e$.
(iv) $m(a, m(b, c)) = m(m(a, b), c)$ for all $a, b, c \in G$.

We can think of a mapping 'inv' which takes each element into its inverse:

$$\text{inv} : G \to G, \quad \text{inv} : a \to a^{-1}.$$

Suppose now a topological space G endowed with such a group structure.

§ **1.5.10** A *topological group* is a group G whose elements are members of a topological space in whose topology both the mappings

$$m : G \times G \to G, \quad (a, b) \to m(a, b)$$

and

$$\text{inv} : G \to G, \quad a \to a^{-1}$$

are continuous (with the product topology for $G \times G$).[32]

§ **1.5.11** The theory of topological groups has three main chapters, concerning their algebraic structure, their topological structure and their representations. But even the algebraic aspects become frequently clearer when something else, such as a differential structure, is added. Representation theory (see Chapter 18) involves functions and measures defined on topological groups, which are the subject of modern harmonic analysis.

---
[32] A classical text on topological groups is Pontryagin 1939.

§ 1.5.12 A topological group is compact if it is compact as a manifold. This means that from any sequence $\{g_n\}$ of its elements one can extract a finite convergent sub-sequence. Any abstract group is a topological group with respect to the discrete topology.

§ 1.5.13 The additive group of real numbers (or 1-dimensional translational group) is an abelian non-compact group whose underlying group-space is the infinite real line. It is commonly denoted by $\mathbb{R}$, but $(\mathbb{R},+)$, indicating also which operation is supposed, would be a better notation.

§ 1.5.14 The sets $\mathbb{R}\backslash\{0\}$ and $\mathbb{C}\backslash\{0\}$ of nonvanishing real and complex numbers are topological groups with the product operation. The set

$$S^1 \subset \mathbb{C}, \quad S^1 = \{z \in \mathbb{C} \text{ such that } |z| = 1\}$$

is a topological group with the operation of multiplication and the induced topology. S^1 is actually a subgroup of $\mathbb{C}\backslash\{0\}$.

§ 1.5.15 This example illustrates a more general result:

If G is any topological group and H a subgroup of G, then H is a topological group with the induced topology of the subspace.

The circle S^1 is thus a topological group and the set of positive real numbers $\mathbb{R}_+$ is a topological subgroup of $\mathbb{R}\backslash\{0\}$. As suggested by the example of $\mathbb{E}^n$, normed vector spaces in general (reviewed in Chapter 16), and consequently their vector subspaces, are topological groups with the addition operation

$$m(a,b) = a + b.$$

§ 1.5.16 The set Q of quaternions is a vector space on $\mathbb{R}$ with basis $\{1,i,j,k\}$. It constitutes topological groups with respect to the addition and multiplication operations, and is isomorphic to $\mathbb{R}^4$ as a vector space. An element of Q can be written as

$$q = a \times 1 + bi + cj + dk,$$

with $a,b,c,d \in \mathbb{R}$. The basis member 1 acts as the identity element. The multiplications of the other members are given in the table of Figure 1.13.

§ 1.5.17 Let $S^3 \subset Q$, where

$$S^3 = \{q \in Q \text{ such that } |q| = \sqrt{a^2 + b^2 + c^2 + d^2} = 1\}.$$

It turns out that S^3 is a topological group.

	i	j	k
i	-1	$+k$	$-j$
j	$-k$	-1	$+i$
k	$+j$	$-i$	-1

Fig. 1.13   Multiplication table for quaternions.

§ 1.5.18 Linear groups. Let $GL(m,K)$ be the set of all non-singular matrices $m \times m$, with entries belonging to a field $K = \mathbb{R}, \mathbb{C}$ or Q.[33] In short,

$$GL(m,K) = \{(m \times m) \text{ matrices } g \text{ on } K \text{ such that det } g \neq 0\}.$$

Given a vector space over the field K, the group of all its invertible linear transformations is isomorphic to $GL(m,K)$. The subsets

$$GL(m,\mathbb{R}) \subset \mathbb{R}^{m^2},$$
$$GL(m,\mathbb{C}) \subset \mathbb{R}^{(2m)^2},$$
$$GL(m,Q) \subset \mathbb{R}^{(4m)^2},$$

are open sets and also topological groups with the operation of matrix multiplication and the topologies induced by the inclusions in the respective euclidean spaces. These linear groups are neither abelian nor compact.

Generalizing a bit: let V be a vector space. The sets of automorphisms and endomorphisms of V are respectively,

Aut $V = \{$f: $V \to V$, such that f is linear and invertible$\}$,

End $V = \{$f: $V \to V$, such that f is linear$\}$.

Then, Aut $V \subset$ End V is a topological group with the composition of linear mappings as operation. If we represent the linear mappings by their matrices, this composition is nothing more than the matrix product, and we have precisely the case $GL(m,K)$.

The groups $O(n) \subset GL(m,\mathbb{R})$ of orthogonal $n \times n$ matrices are other examples. The special orthogonal groups $SO(n)$ of orthogonal matrices with det $= +1$ are ubiquitous in Physics and we shall come to them later.

§ 1.5.19 The set of all matrices of the form

$$\begin{pmatrix} L & t \\ 0 & 1 \end{pmatrix}$$

[33] The word "field" is a rather abused name. We are here using it in its purely algebraic sense, to be described in Section 13.3.

where $L \in GL(n, \mathbb{R})$ and $t \in \mathbb{R}^n$, with the matrix product operation, is a topological group called the *affine group* of dimension n. It is denoted $A(n, \mathbb{R})$.

§ **1.5.20 Linear projective groups.** Take the set M_n of all $n \times n$ matrices with entries in field $K = \mathbb{R}, \mathbb{C}$ or Q. It constitutes more than a vector space — it is an algebra.[34] Each matrix $A \in GL(n, K)$ defines an automorphism of the form AMA^{-1}, for all $M \in M_n$. This automorphism will be the same if A is replaced by $cA = cI_n A$, for any $c \in K$, and where I_n is the unit $n \times n$ matrix. The subgroup formed by the matrices of the type cI_n is indeed ineffective on M_n. The group of actual automorphisms is the quotient $GL(n, K)/\{cI_n\}$.

Consider the subalgebra of M_n formed by the n projectors P_k = matrix whose only nonvanishing element is the diagonal k-th entry, which is 1:

$$(P_k)_{rs} = \delta_{rs}\delta_{ks}.$$

The P_k's are projectors into 1-dimensional subspaces. They satisfy the relations $P_i P_j = \delta_{ij} P_i$. Each one of the above automorphisms transforms this set of projectors into another set of projectors with the same characteristics. For this reason the automorphisms are called projective transformations, and the group of automorphisms $GL(n, K)/\{cI_n\}$ is called the projective group, denoted $PL(n, K)$. The n-dimensional space of projectors, which is taken into an isomorphic space by the transformations, is the projective space, indicated by KP^n. There are, however, other approaches to this type of space.

§ **1.5.21 Projective spaces.** Every time we have a problem involving only the directions (not the senses) of vectors and in which their lengths are irrelevant, we are involved with a projective space. Given an n-dimensional vector space V over the field K, its corresponding projective space KP^n is the space formed by all the 1-dimensional subspaces of V. Each point of KP^n is the set formed by a vector v and all the vectors proportional to v. We may be in a finite-dimensional vector space, or in an infinite-dimensional space like a Hilbert space.

Quantum Mechanics describes pure states as rays in a Hilbert space, and rays are precisely phase-irrelevant. Thus, pure states are represented

---

[34] The word "Algebra" (we shall be repeating this) is routinely used in two senses: (i) a huge chapter of Mathematics and (ii) a very particular algebraic structure; the latter meaning, which is the one of interest here, is examined in detail in § 13.5.1.

by members of a projective Hilbert space.

Take on the sphere S^n the equivalence R given by:

$p \approx q$ if either p is identical to q, or p and q are antipodes.

For S^1 drawn on the plane,

$$(x, y) \approx (-x, -y).$$

On S^2 imbedded in $\mathbb{E}^3$,

$$(x, y, z) \approx (-x, -y, -z),$$

and so on. The quotient space S^n/R is the n-dimensional real projective space RP^n. Because pairs of antipodes are in one-to-one correspondence with straight lines through the origin, these lines can be thought of as the points of RP^n. The space RP^1 is called the "real projective line", and RP^2, the "real projective plane". It is the space of the values of "orientation fields" (see Section 28.3.4). It can be shown that RP^n is a connected Hausdorff compact space of empty boundary. The "antipode relation" can be related to a group, the cyclic group (Section 14.1.3) of second order $\mathbb{Z}_2$, so that

$$RP^n = S^n/\mathbb{Z}_2.$$

There are many beautiful (and not always intuitive) results concerning these spaces. For instance:

- RP^0 is homeomorphic to a point.
- RP^1 is homeomorphic to S^1.
- The complement of RP^{n-1} in RP^n is $\mathbb{E}^n$.
- RP^3 is homeomorphic to the group $SO(3)$ of rotations in $\mathbb{E}^3$.

Complex projective spaces CP^n are defined in an analogous way and also for them curious properties have been found:

- CP^1 is homeomorphic to S^2.
- The space CP^n is homeomorphic to S^{2n+1}/S^1 (recall that S^1 is indeed a group).
- The complement of CP^{n-1} in CP^n is $\mathbb{E}^{2n}$.

These spaces are ubiquitous in Mathematics and have also emerged in many chapters of Physics: model building in field theory, classification of instantons,[35] twistor formalism,[36] and many more.

---

[35] Atiyah, Drinfeld, Hitchin & Manin 1978; Atiyah 1979.
[36] Penrose in Isham, Penrose & Sciama 1975; Penrose & MacCallum 1972; Penrose 1977.

Another, equivalent definition is possible which makes no use of the vector structure of the host spaces $\mathbb{E}^n$ or $\mathbb{C}^n$. It is also a "quotient" definition. Consider, to fix the ideas, the topological space $\mathbb{C}^{n+1}$ of ordered $(n+1)$-tuples of complex numbers,

$$\mathbb{C}^{n+1} = \{z = (z_1, z_2, \dots, z_n, z_{n+1})\},$$

with the ball topology given by the distance function

$$d(z, z') = \sqrt{|z_1 - z_1'|^2 + |z_2 - z_2'|^2 + \dots + |z_n - z_n'|^2 + |z_{n+1} - z_{n+1}'|^2}\ .$$

The product of z by a complex number c is

$$cz = (cz_1, cz_2, \dots, cz_n, cz_{n+1}).$$

Define an equivalence relation R as follows: z and z' are equivalent if some non-zero c exists such that $z' = cz$. Then,

$$CP^n = \mathbb{C}^{n+1}/R,$$

that is, the quotient space formed by the equivalence classes of the relation. Any function f on $\mathbb{C}^{n+1}$ such that $f(z) = f(cz)$ (that is, a function which is homogeneous of degree zero) is actually a function defined on CP^n.

§ 1.5.22 Real Grassmann spaces. Real projective spaces are generalized in the following way: given the euclidean space $\mathbb{E}^N$ with its vector space structure, its d-th Grassmann space $G_{Nd}(\mathbb{E})$ [another usual notation is $G_d(\mathbb{E}^N)$] is the set of all d-dimensional vector subspaces of $\mathbb{E}^N$. Notice that, as vectors subspaces, they all include the origin (the zero vector) of $\mathbb{E}^N$. All these "Grassmannians" are compact spaces (see § 8.1.11). Projective spaces are particular cases:

$$RP^1 = G_1(\mathbb{E}^2),\ RP^2 = G_1(\mathbb{E}^3),\ \dots,\ RP^n = G_1(\mathbb{E}^{n+1}).$$

Projectors are in a one-to-one correspondence with the subspaces of a vector space. Those previously used were projectors of rank one. In the present case the euclidean structure is to be preserved, so that the projectors must be skew-symmetric endomorphisms, or matrices p satisfying $p = -p^T$ and $p^2 = p$. If they project into a d-dimensional subspace, they must furthermore be matrices of rank d. Consequently, G_{Nd} may be seen also as the space of such projectors:

$$G_{Nd} = \{p \in \text{End}(\mathbb{E}^N) \text{ such that } p^2 = p = -p^T,\ \text{rank } p = d\}.$$

Notice that $G_{Nd} = G_{N,N-d}$ and the dimension is dim $G_{Nd} = d(N - d)$. The set of orthogonal frames in the d-dimensional subspaces form another space, the Stiefel space (see § 8.1.12), which is instrumental in the general classification of fiber bundles (Section 9.7).

§ 1.5.23 Complex Grassmann spaces. Projective spaces are generalized to the complex case in an immediate way. One starts from a euclideanized (as above) complex vector space $\mathbb{C}^{n+1} = (z_1, z_2, \ldots, z_n, z_{n+1})$. Now, projectors must be hermitian endomorphisms and the space is

$$G_d(\mathbb{C}^N) = G_{Nd}(\mathbb{C}) = \{p \in \text{End}(\mathbb{C}^N) \text{ such that } p^2 = p = p^\dagger, \text{ rank } p = d\}.$$

Space $G_{Nd}(\mathbb{C})$ is a compact space whose "points" are complex d–dimensional planes in $\mathbb{C}^N$.

§ 1.5.24 We have tried in this chapter to introduce some notions of what is usually called *general topology*. The reader will have noticed the purely qualitative character of the discussed properties. The two forthcoming chapters are devoted to some notions coming under the name of *algebraic topology*, which lead to the computation of some numbers of topological significance. Roughly speaking, one looks for "defects", such as holes and forbidden subspaces in topological spaces, while remaining inside them. One way to find such faults comes from the observation that defects are frequently related to closed subspaces which do not bound other subspaces, as it happens with some of the closed lines on the torus. Such subspaces give origin to discrete groups and are studied in homology theory (Chapter 2). Another way is by trying to lasso them, drawing closed curves and their higher dimensional analogues to see whether or not they can be continuously reduced to points (or to other closed curves) in the space. Such loops are also conveniently classified by discrete groups. This is the subject of homotopy theory (Chapter 3). Both kinds of groups lead to some integer numbers, invariant under homeomorphisms, which are examples of *topological numbers*. We shall later introduce more structure on topological spaces and sometimes additional structure can be used to signal topological aspects. Within differentiable structure, vector and tensor fields arise which are able to reveal topological defects (singular points, see Chapter 21) through their behaviour around them, in a way analogous to the velocity field of a fluid inside an irregular container. Later, in Section 7.5, a little will be said about *differential topology*.

Chapter 2

Homology

Dissection of a space into cellular building bricks provides information about its topology.

2.1 Introductory remarks

The study of the detailed topological characteristics of a given space can be a very difficult task. Suppose, however, that we are able to decompose the space into subdomains, each one homeomorphic to a simpler space, whose properties are easier to be worked out. Suppose further that this analysis is done in such a way that rules emerge allowing some properties of the whole space to be obtained from those of these "components". Homology theory is concerned precisely with such a program: the dissection of topological spaces into certain basic cells called "chains", which in a way behave as their building bricks. The circle is homeomorphic to the triangle and the sphere is equivalent to the tetrahedron. The triangle and the tetrahedron can be build up from their vertices, edges and faces. It will be much easier to study the circle and the sphere through such "rectified" versions, which furthermore can be decomposed into simpler parts. Once in possession of the basic cells we can, by combining them according to specific rules, get back the whole space, and that with a vengeance: information is gained in the process. Chains come out to be elements of certain vector spaces (and so, of some abelian groups). Underlying algebraic structures come forth in this way, which turn out to be topological invariants: homeomorphic spaces have such structures in common.

As will be seen later on, chains are closely related to integration domains in the case of differentiable manifolds, and their algebraic properties will find themselves reflected in analogous properties of differential forms. Of course, only a scant introduction will find its place here.[1]

2.2   Graphs

Spaces usually appearing in Physics have much more structure than a mere topology. Exceptions are the graphs (or diagrams) used in perturbation techniques of Field Theory, cluster expansions in Statistical Mechanics, circuit analysis, etc, whose interest comes in part from their mere topological properties. Graph Theory is a branch of Mathematics by itself, with important applications in traffic problems, management planning, electronics, epidemic propagation, computer design and programming, and everywhere else. Only a very sketchy outline of the subject will be given here, although hopefully pedantic enough to prepare for ensuing developments.

Graphs provide a gate into homological language. The first part below introduces them through a mere formalization of intuitive notions. The second rephrases (and extends) the first: its approach allows the introduction of the ideas of chain and boundary, and sets the stage for the basic notions of simplex and complex.

2.2.1   *Graphs, first way*

§ **2.2.1** We shall call (closed) *edge* any subspace in $\mathbb{E}^3$ which is homeomorphic to the interval $\boldsymbol{I} = [0, 1]$ with the induced topology. Indicating the edge itself by $\bar{e}$, and by $h : \boldsymbol{I} \to \bar{e}$ the homeomorphism, the points $h(0)$ and $h(1)$ will be the *vertices* of the edge. A *graph* $G \subset \mathbb{E}^3$ is a topological space defined in the following way:

(i) the *point set* consists of the points of a union of edges satisfying the condition that the intersection of any two edges is either $\emptyset$ or one common vertex;

(ii) the *open sets* are subsets $X \subset G$ whose intersection with each edge is open in that edge.

§ **2.2.2** Because edges are homeomorphic to $\boldsymbol{I}$ with the induced topology (and not, for example, with the quotient topology of § 1.5.3), no isolated one-vertex bubbles like ⟡ are admitted. Notice that knots (Section 14.3)

---

[1] The subject is a whole exceedingly beautiful chapter of high Mathematics, and an excellent primer may be found in Chapters 19 to 22 of Fraleigh 1974. An extensive mathematical introduction, easily understandable if read from the beginning, is Hilton & Wylie l967.

are defined as subspaces in $\mathbb{E}^3$ homeomorphic to the circle S^1. So, absence of bubbles means that no knots are parts of graphs.

Graphs can be defined in a less restrictive way with no harm to the results involving only themselves, but the above restrictions are essential to the generalization to be done later on (to simplexes and the like). From (ii), a subset is closed if its intersection with each edge is closed in that edge.

The *open edges*, obtained by stripping the closed edges of their vertices, are open sets in G. The set of vertices is discrete and closed in G. A graph is compact iff it is finite, that is, if it has a finite number of edges. It is connected iff any two of its vertices can be linked by a sequence of non-disjoint edges belonging to the graph.

Take an open edge e and call e' its complement,

$$e' = G - e$$

(G with extraction of e). If, for every edge e, e' is unconnected, then G is a *tree graph*. Otherwise, it contains at least one *loop*, which is a finite sequence of edges

$$\bar{e}_1, \bar{e}_2, \ldots, \bar{e}_n$$

each one like $\bar{e}_i = u_i \bullet\!\!-\!\!\!-\!\!\bullet v_i$, with no repeated edges and only one repeated vertex, $v_n = u_1$, all the remaining ones satisfying $v_i = u_{i+1}$. These are, of course, complicated but precise definitions of usual objects. Notice: graphs are drawn in $\mathbb{E}^3$ with no intersections. It can be rigorously proved that any graph can be realized in $\mathbb{E}^3$ without crossing.

In the Feynman diagram quantization technique of field theory, the number of loops in a diagram is just the order of its contribution[2] in Planck's constant $\hbar$. A diagram with L loops will contribute to the order $\hbar^L$ in perturbation theory. The semiclassical approximation turns up when only "tree diagrams" (zero order in $\hbar$, zero loops) are considered. A tree edge is called a branch and an edge taking part in a loop is called a *chord*.

Euler number

§ 2.2.3 Let us make a few comments on *planar graphs*, those that *can* be drawn in $\mathbb{E}^2$ (although always considered as above, built in $\mathbb{E}^3$). The simplest one is $\bullet\!\!-\!\!\!-\!\!\bullet$. Call V the number of vertices, E the number of

---

[2] See for example Itzykson & Zuber 1980.

edges, L the number of loops, and define

$$\chi(G) = V - E + L.$$

For $\bullet\!\!-\!\!\bullet$, $\chi(G) = 1$. In order to build more complex graphs one has to
add edges one by one, conforming to the defining conditions. To obtain a
connected graph, each added edge will have at least one vertex in common
with the previous array. A trivial checking is enough to see that $\chi(G)$
remains invariant in this building process. For non-connected graphs, $\chi(G)$
will have a contribution as above for each connected component. Writing
N for the number of connected components,

$$V - E + L - N$$

is an invariant.

For connected compact graphs, $\chi(G)$ is called the *Euler number*. Being an integer number, it will not change under continuous deformations.
In other words, $\chi(G)$ is invariant under homeomorphisms, it is a topological invariant. It will be seen later that $\chi(G)$ can be defined on general
topological spaces. The number of loops, with the relation

$$L = 1 - V + E,$$

was first used in physics by Kirchhoff in his well known analysis of DC
circuits, and called "cyclomatic number" by Maxwell.

The result is also valid for Feynman diagrams and, with some goodwill,
for an island in the sea: take a diagram as that pictured in Figure 2.1, keep
the external edges fixed on the plane and pull the inner vertices up — the

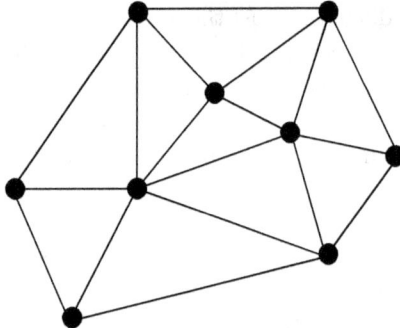

Fig. 2.1   Scheme of an island, with peaks (vertices), passes (edges) and valleys (loops).

number of peaks (V) minus the number of passes (E) plus the number of valleys (L) equals one! To compare with the non-planar case, consider the tetrahedron of Figure 2.2. A simple counting shows that

$$\chi(G) = V - E + L = 2.$$

Its plane, "flattened" version beside has $\chi(G) = 1$ because the lower face is no more counted. Notice that only "elementary" loops, those not containing sub-loops, are counted (see § 2.3.2).

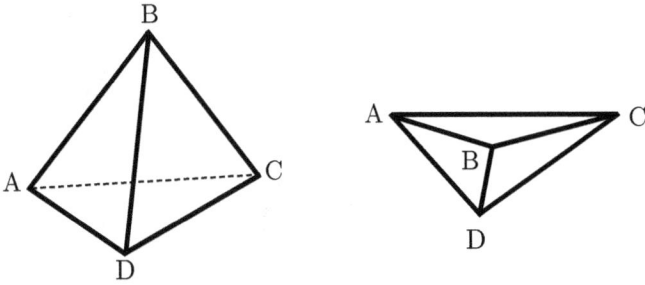

Fig. 2.2   The tetrahedron and one of its planar projections.

2.2.2 *Graphs, second way*

Consider now $\mathbb{E}^3$ as a vector space, and choose any three of its vectors v_0, v_1, v_2, imposing however that $(v_1 - v_0)$ and $(v_2 - v_0)$ be linearly independent:

$$k^1(v_1 - v_0) + k^2(v_2 - v_0) = 0 \quad \text{implies} \quad k^1 = k^2 = 0.$$

Defining

$$a^1 = k^1, \quad a^2 = k^2, \quad a^0 = -(k^1 + k^2),$$

this is equivalent to saying that the two conditions

$$a^0 v_0 + a^1 v_1 + a^2 v_2 = 0$$

and

$$a^0 + a^1 + a^2 = 0$$

imply

$$a^0 = a^1 = a^2 = 0.$$

Such conditions ensure that no two vectors are colinear, that (v_0, v_1, v_2) constitute a triad. Let us define a *vector dependent on* the triad (v_0, v_1, v_2) by the two conditions

$$b = \sum_{i=0}^{2} b^i v_i \quad \text{and} \quad \sum_{i=0}^{2} b^i = 1. \tag{2.1}$$

The points determined by the *barycentric coordinates* b^i describe a plane in $\mathbb{E}^3$ (Figure 2.3).

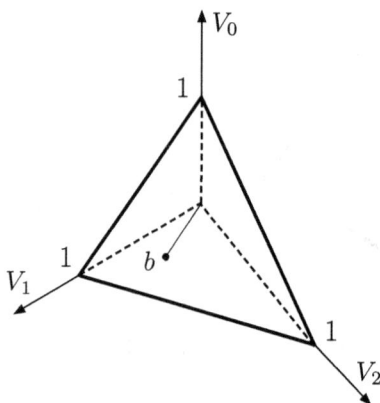

Fig. 2.3   The barycentric coordinates.

§ **2.2.4** Suppose we consider only two of the vectors, say v_0 and v_1. They are of course linearly independent. We can in this case define, just as above, their *dependent vectors* by

$$e = b^0 v_0 + b^1 v_1 \quad \text{with} \quad b^0 + b^1 = 1.$$

Now, the coordinates b^0 and b^1 determine points on a straight line. In the same way, we can take only one of the vectors, say v_0, and its dependent vector as $v = v_0$ itself: this will determine a point.

Add now an extra condition on the dependent vectors: that each coordinate be strictly positive, $b^i > 0$. Then, the vector b above will span the interior of a triangle; the vector e will span an open edge; and v_0 again will "span" an isolated point, or a vertex. Notice that the coordinates related to the vector e give actually a homeomorphism between the interval $(0,1)$ and a line segment in $\mathbb{E}^3$, justifying the name *open edge*. If instead we allow $b^i \geq 0$, a segment homeomorphic to the closed interval $[0, 1]$ results, a *closed edge*.

With these edges and vertices, graphs may now be defined as previously. An edge can be indicated by the pair (v_i, v_k) of its vertices, and a graph G by a set of vertices plus a set of pairs. An *oriented graph* is obtained when all the pairs are taken to be ordered pairs — which is a formal way of putting arrows on the edges. A *path* from vertex v_1 to vertex v_{n+1} is a sequence of edges $\bar{e}_1 \bar{e}_2 \bar{e}_3 \dots \bar{e}_n$ with

$$\bar{e}_i = (v_i, v_{i+1}) \quad \text{or} \quad \bar{e}_i = (v_{i+1}, v_i).$$

We have said that a graph is *connected* when, given two vertices, there exists at least one path connecting them. It is *multiply-connected* when there are at least two independent, nonintersecting paths connecting any two vertices. It is *simply-connected* when it is connected but not multiply-connected. In Physics, multiply-connected graphs are frequently called "irreducible" graphs.

§ 2.2.5 A path is not supposed to accord itself to the senses of the vector arrows. A path is called *simple* if all its edges are distinct (one does not go twice through the same edge) and all its vertices are distinct except possibly $v_1 = v_{n+1}$. In this last case, it is a *loop*. An *Euler path* on G is a path with all edges distinct and going through all the vertices in G. The number n_k of edges starting or ending at a vertex v_k is its *degree* ("coordination number" would be more to the physicist's taste; chemists would probably prefer "valence"). Clearly the sum of all degrees on a graph is even, as

$$\sum_1^V n_i = 2E.$$

The number of odd vertices (that is, those with odd degrees) is consequently even.

The Bridges of Königsberg

§ 2.2.6 Graph theory started up when Euler faced this problem. There were two islands in the river traversing Kant's town, connected between each other and to the banks by bridges as in the scheme of Figure 2.4. People wanted to know whether it was possible to do a tour traversing all the bridges *only once* and finishing back at the starting point. Euler found that it was impossible: he reasoned that people should have a departure

for each arrival at every point, so that all degrees should be even — which
was not the case.

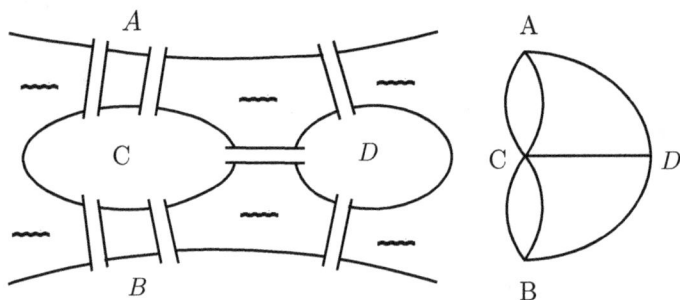

Fig. 2.4   Scheme of downtown Königsberg, and the corresponding graph.

§ 2.2.7 Graph Theory has scored a beautiful victory for Mathematics with
the seventies' developments on the celebrated (see Section 28.2.5) four-color
problem.[3] The old conjecture, to which a computer–assisted demonstration
has been given, was that four colors were sufficient for the coloring of a
(planar) map. This is a problem of graph theory.

As said above, graphs are also of large use in many branches of Physics.
Through Feynman's diagram technique, they have a fundamental role as
guidelines for perturbation calculations in field theory and in the many-
body problem. In Statistical Mechanics, besides playing an analogous role
in cluster expansions, graphs are basic personages in lattice models. In the
Potts model, for example, where the underlying lattice can be any graph,
they become entangled (sorry!) with knots (Section 28.2.3). They also
appear in the generalized use of the Cayley tree and the related Bethe
lattice in approximations to more realistic lattice models (Section 28.2.4).

§ 2.2.8 To the path $\bar{e}_1\bar{e}_2\bar{e}_3\ldots\bar{e}_n$ we can associate a formal sum

$$\varepsilon_1 e_1 + \varepsilon_2 e_2 + \ldots + \varepsilon_n e_n,$$

with $\varepsilon_i = +1$ if $\bar{e}_i = (v_i, v_{i+1})$, and $\varepsilon_i = -1$ if $\bar{e}_i = (v_{i+1}, v_i)$. The sum
is thus obtained by following along the path and taking the (+) sign when

---

[3] A readable account is Appel & Haken 1977. For the history, see Eric W. Weisstein:
"Four-Color Theorem", http://mathworld.wolfram.com/Four-ColorTheorem.html.

going in the sense of the arrows and the $(-)$ sign when in the opposite sense. The sum is called the *chain* of the path and ε_i is the *incidence number* of $\bar{e}_i$ in the chain.

§ 2.2.9 A further formal step, rather gratuitous at first sight, is to generalize the ε_i's to coefficients which are any integer numbers: a *1-chain* on the graph G is a formal sum

$$\sum_i m_i e_i = m_1 e_1 + m_2 e_2 + \cdots + m_n e_n,$$

with $m_j \in \mathbb{Z}$.

§ 2.2.10 We can define the sum of two 1-chains by

$$\sum_i m_i e_i + \sum_i m'_j e_j = \sum_k (m_k + m'_k) e_k.$$

Calling "0" the 1-chain with zero coefficients (the *zero 1-chain*), the set of 1-chains of G constitutes an abelian group, the *first order chain group* on G, usually denoted $C_1(G)$. In a similar way, a *0-chain* on G is a formal sum

$$r_1 v_1 + r_2 v_2 + \cdots + r_p v_p,$$

with $r_j \in \mathbb{Z}$. Like the 1-chains, the 0-chains on G form an abelian group, the *zeroth chain group* on G, denoted $C_0(G)$. Of course, $C_0(G)$ and $C_1(G)$ are groups because $\mathbb{Z}$ is itself a group: it was just to obtain groups that we have taken the formal step $\varepsilon_i \rightarrow m_j$ above. Groups of chains will be seen to be of fundamental importance later on, because some of them will show up as topological invariants.

§ 2.2.11 Take the oriented edge $\bar{e}_j = (v_j, u_j)$. It is a 1-chain by itself. We define the (oriented) *boundary* of $\bar{e}_j$ as the 0-chain

$$\partial \bar{e}_j = u_j - v_j.$$

In the same way, the boundary of a general 1-chain is defined as

$$\partial \sum_i m_i e_i = \sum_i m_i \partial e_i,$$

which is a 0-chain.

§ 2.2.12 The mapping

$$\partial : C_1(G) \longrightarrow C_0(G)$$

preserves the group operation and is called the *boundary homomorphism*. A 1-*cycle* on G is a loop, a closed 1-chain. It has no boundary and is formally defined as an element $c \in C_1(G)$ for which $\partial c = 0$ (that is, the zero 0-chain). The set of 1-cycles on G form a subgroup, denoted $Z_1(G)$.

§ **2.2.13** Consider the examples of Figure 2.5. Take first the graph at the

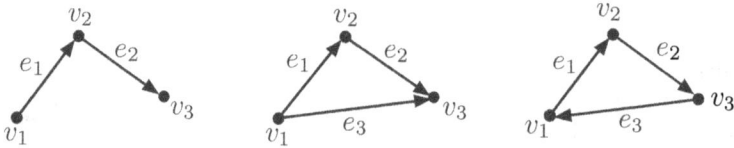

Fig. 2.5   Examples of low-dimensional chains.

left in the figure: clearly,

$$\partial(e_1 + e_2) = v_3 - v_1$$

and

$$\partial(me_1 + ne_2) = nv_3 - mv_1 + (m - n)v_2.$$

In the middle graph,

$$\partial(e_1 + e_2 + e_3) = 2v_3 - 2v_1.$$

Now, in the graph at the right, $(e_1 + e_2 + e_3)$ is clearly a cycle (which illustrates the extreme importance of orientation). On this graph,

$$C_1(G) = \{me_1 + ne_2 + re_3 \text{ with } m, n, r \in \mathbb{Z}\},$$
$$C_0(G) = \{pv_1 + qv_2 + sv_3 \text{ with } p, q, s \in \mathbb{Z}\},$$
$$Z_1(G) = \{m(e_1 + e_2 + e_3) \in C_1(G)\}.$$

Commentary 2.1   A very interesting survey of graph theory, including old classical papers (by Euler, Kirchhoff, ...) is Biggs, Lloyd & Wilson 1977. An introduction to Homology, most commendable for its detailed treatment starting with graphs, is Gibling 1977. An introduction to graphs with applications ranging from puzzles to the four color problem is Ore 1963. Finally, more advanced texts are Graver & Watkins 1977, Trudeau 1993 and Berge 2001. ◀

2.3 The first topological invariants

2.3.1 *Simplexes, complexes and all that*

Let us now proceed to generalize the previous considerations to objects which are more than graphs. Consider $\mathbb{E}^n$ with its structure of vector space. A set of vertices is said to be linearly independent if, for the set of vectors $(v_0,\, v_1,\, v_2,\, \ldots,\, v_p)$ fixing them, the two conditions

$$a^0 v_0 + a^1 v_1 + \cdots + a^p v_p = 0,$$

$$a^0 + a^1 + \cdots + a^p = 0,$$

imply $a^0 = a^1 = \cdots = a^p = 0$. This means that the vectors $(v_i - v_0)$ are linearly independent. We define a vector b "dependent on the vectors $v_0, v_1, \ldots, v_p$" by

$$b = \sum_{i=0}^{p} b^i v_i, \qquad \sum_{i=0}^{p} b^i = 1.$$

The points determined by the barycentric coordinates b^i describe a p-dimensional subspace of $\mathbb{E}^n$, more precisely, an euclidean subspace of $\mathbb{E}^n$.

§ **2.3.1** A (closed) *simplex* of dimension p (or a *p-simplex*) with vertices v_0, $v_1, v_2, \ldots, v_p$ is the set of points determined by the barycentric coordinates satisfying the conditions $b^i \geq 0$ for $i = 0, 1, 2, \ldots, p$. Special cases are points (0-simplexes), closed intervals (1-simplexes), triangles (2-simplexes) and tetrahedra (3-simplexes). A p-simplex is indicated by s_p and is said to be "generated" by its vertices. The points with all nonvanishing barycentric coordinates are *interior* to the simplex, their set constituting the *open simplex*. The boundary ∂s_p of s_p is the set of points with at least one vanishing coordinate $b^i = 0$. Given s_p generated by $v_0, v_1, v_2, \ldots, v_p$, any subset of vertices will generate another simplex: if s_q is such a subsimplex, we use the notation $s_q \langle s_p$. It is convenient to take the empty set $\emptyset$ as a subsimplex of any simplex. Dimension theory gives the empty set the dimension (-1), so that the empty simplex is designated by s_{-1}. The edge in the left branch of Figure 2.6 is a 1-simplex in $\mathbb{E}^3$. Its boundary is formed by the vertices v_0 and v_1, which are also taken as subsimplexes. In a standard notation,

$$v_0 \langle s_1 \quad \text{and} \quad v_1 \langle s_1 \, .$$

Also s_1 is taken to be a subsimplex of itself: $s_1 \langle s_1$. The empty set is by

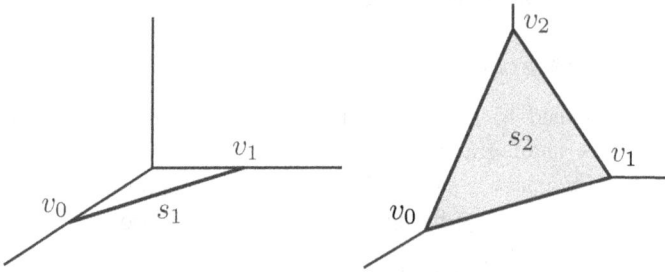

Fig. 2.6   The most trivial simplexes in $\mathbb{E}^3$.

convention a subsimplex of any simplex,

$$s_{-1}\langle s_1, \quad s_{-1}\langle v_o, \quad s_{-1}\langle v_1 \,.$$

The (full) triangle is a 2-simplex in $\mathbb{E}^3$. Its boundary is the set of points belonging to the three edges. And so on.

§ **2.3.2** The highest possible dimension of a simplex in $\mathbb{E}^n$ is $(n-1)$. If we are to "see" loops, even planar graphs are to be considered in $\mathbb{E}^3$. That is why we have not counted loops encircling other loops in our discussion of graphs. A (full) tetrahedron is a 3-simplex in $\mathbb{E}^n$ for $n \geq 4$.

Notice that a simplex is always a *convex* set in $\mathbb{E}^n$. It contains, together with any two of its points, all the points of the straight segment of line between them. Simplexes will be used as building bricks to construct more complex objects (fittingly enough called "complexes") and this convexity property is very convenient for the purpose. Notice that, due to condition (ii) of § 2.2.1, the vertices loose their individuality when considered as points of the boundary of a full triangle: the requirement that the intersection with every edge must be open excludes half-open intervals.

§ **2.3.3** Up to now, we have been generalizing to higher dimensions the notions of edge and vertex we have seen in the case of graphs. Let us proceed to the extension of the very idea of graph. A *simplicial complex* (or *cellular complex*) is a set K of simplexes in $\mathbb{E}^n$ satisfying the two conditions:

(i) If $s_p \in K$ and $s_q \langle s_p$, then $s_q \in K$.
(ii) If $s_r \in K$ and $s_t \in K$, then their intersection $s_r \cap s_t$ is either empty or a subsimplex common to both.

§ 2.3.4 The *dimension* of K is the maximal dimension of its simplexes. A graph is a 1-dimensional simplicial complex. Notice that K is the set of the building blocks, not the set of their points. The set of points of $\mathbb{E}^n$ belonging to K with the induced topology is the *polyhedron* of K, indicated by $P(K)$. Conversely, the complex K is a *triangulation* or *dissection* of its polyhedron. Due to the way in which it was constructed, a polyhedron inherits much of the topology of $\mathbb{E}^n$. In particular, its topology is metrizable.

§ 2.3.5 We now come to the main idea: suppose that a given topological space S is homeomorphic to some $P(K)$. Then, S will have the same topological properties of K. We shall see below that many topological properties of $P(K)$ are relatively simple to establish. These results for polyhedra can then be transferred to more general spaces via homeomorphism. Suppose h is the homeomorphism, $h(K) = S$. The points of S constitute then a *curvilinear simplex*. For each simplex s_p, the image $h(s_p) \subset S$ will keep the properties of s_p. The image $h(s_p)$ will be called a *p-simplex on S*. Again, the set of curvilinear simplexes on S will be a triangulation (or curvilinear complex) of S. A simple example is the triangulation determined on Earth's surface by the meridians and the equator.

§ 2.3.6 The boundary of a triangle is homeomorphic to the circle S^1. This can be seen by first deforming the triangle into an equilateral one, inscribing a circle and then projecting radially. This projection is a homeomorphism. An analogous procedure shows that the sphere S^n is homeomorphic to the boundary of an $(n+1)$-tetrahedron, or $(n+1)$-simplex.

§ 2.3.7 An important point is the following: above we have considered a *homeomorphism* h. However, it is possible in many cases to use a less stringent function to relate the topological properties of some topological space to a simplicial complex. Suppose, for instance, that S is a differentiable manifold (to be defined later: it is endowed with enough additional structure to make differential calculus possible on it). Then, it is enough that h be a differentiable function, which in many aspects is less stringent than a homeomorphism: its inverse has not necessarily a good behaviour and h itself may be badly behaved (singular) in some regions.

By taking simplexes from K to S via a differentiable function, a certain homology can be introduced on S, called the *singular homology*. In reality, many homologies can be introduced in this way, by choosing different conditions on h. It can be shown that, at least for compact differentiable manifolds, all these homologies are equivalent, that is, give the same topological invariants. A differentiable manifold turns out to be homeomorphic to a polyhedron, even if h is originally introduced only as a differentiable function from some polyhedron of $\mathbb{E}^n$ into it (Cairns' theorem).

§ **2.3.8** We have said that the region inscribed in a loop is not a simplex in $\mathbb{E}^2$. That is why graphs are to be considered as 1-simplexes in $\mathbb{E}^3$. The dimension of the surrounding space is here of fundamental importance. Let us repeat what has been said in § 2.2.3 on the tetrahedron and its "flattening". Take the graph of Figure 2.7 and pull the inner vertex up. In order to obtain the boundary of a tetrahedron (a simplex in $\mathbb{E}^4$, that is, a new 2-complex), an extra bottom face has to be added and the Euler number becomes 2. The boundary of a tetrahedron being homeomorphic to the sphere S^2, it follows that also

$$\chi(S^2) = 2.$$

The same will be true for any surface homeomorphic to the sphere. If the surface of the Earth (or Mars, or Venus, or the Moon) could be obtained by continuous deformations from the sphere, the number of peaks minus passes plus valleys would be two. Of course, this would neglect steep cliffs and any other singularities.

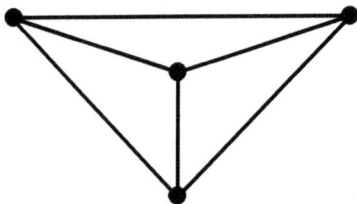

Fig. 2.7   To get a tetrahedron, a face must be added.

§ **2.3.9** We are now in condition to introduce a topological invariant which generalizes the Euler number. Call N_i the number of i-simplexes in a given n-dimensional complex K. The *Euler–Poincaré characteristic* of K is defined by

$$\chi(K) = \sum_{i=0}^{n} (-)^i N_i.$$

Being an integer number, it cannot be changed by continuous deformations of the space: it is an example of *topological number*, an integer number which is characteristic of a space and necessarily the same for spaces homeomorphic to it.

§ 2.3.10 Notice that non-homeomorphic spaces can have some topological numbers in common, their equality being only a *necessary* condition for topological equivalence. The circle S^1 and all the toruses

$$T^n = S^1 \times S^1 \times \cdots \times S^1$$

(topological product of n circles) have $\chi = 0$ but are not homeomorphic. They have other topological characteristics which are different, as for example the fundamental group, a homotopic character to be examined in Chapter 3. We shall see that also homology, like homotopy, provides only a partial characterization of the involved topology.

§ 2.3.11 In order to obtain more invariants and a deeper understanding of their emergence, it is necessary beforehand to generalize the oriented graphs of § 2.2.4. A simplex s_p is generated by its vertices and can be denoted by their specification:

$$s_p = (v_0, v_1, v_2, \ldots, v_p).$$

Suppose now that $(v_0, v_1, v_2, \ldots, v_p)$ is an ordered $(p+1)$-tuple of vertices. Each chosen order is an *orientation* of the simplex. For instance, the edge s_1 can be given the orientations $v_0 v_1$ and $v_1 v_0$ (it is usual to forget the commas and represent simplexes in this way, as a "product" of letters). It is only natural to consider these orientations as "opposite", and write

$$v_0 \bullet\!\!-\!\!\!-\!\!\bullet v_1 = v_0 v_1 = -v_1 v_0 = -(v_1 \bullet\!\!-\!\!\!-\!\!\bullet v_0).$$

§ 2.3.12 It is also convenient to think of a 2-simplex as a (full) triangle oriented via a fixed sense of rotation, say counter-clockwise (Figure 2.8). Here a problem arises: the edges are also oriented simplexes and we would like to have for the boundary an orientation coherent with that of the 2-simplex. The figure suggests that the faces coherently oriented with respect to the triangle are $v_0 v_1$, $v_1 v_2$ and $v_2 v_0$, so that the oriented boundary is

$$\partial(v_0 v_1 v_2) = v_0 v_1 + v_1 v_2 + v_2 v_0. \tag{2.2}$$

§ 2.3.13 Notice that the opposite orientation would be coherent with the opposite orientation of the edges. All the possible orientations are easily found to be equivalent to one of these two. As a general rule, for a p-simplex

$$(v_0 v_1 v_2 \ldots v_p) = \pm(v_{i_0} v_{i_1} v_{i_2} \ldots v_{i_p}),$$

$$v_2$$

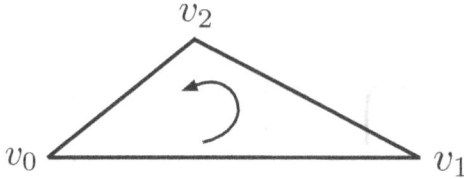

$$v_0 \qquad\qquad\qquad\qquad\qquad v_1$$

Fig. 2.8   Triangle orientation: counter-clockwise convention.

the sign being $+$ or $-$ according to whether the permutation

$$\begin{pmatrix} 0 & 1 & 2 & \dots & p \\ i_0 & i_1 & i_2 & \dots & i_p \end{pmatrix}$$

is even or odd. An equivalent device is to think of $(v_0 v_1 v_2 \dots v_p)$ as an anti-symmetric product of the vertices, or to consider that vertices anticommute. This is consistent with the boundary of an edge,

$$\partial(v_0 v_1) = v_1 - v_0.$$

As a mnemonic rule,

$$\partial(v_0 v_1 v_2 \dots v_p) = v_1 v_2 \dots v_p - v_0 v_2 \dots v_p + v_0 v_1 v_3 \dots v_p - \dots$$

The successive terms are obtained by skipping each time one vertex in the established order and alternating the sign. For a 0-simplex, the boundary is defined to be 0 (which is the zero chain). Each term in the sum above, with the corresponding sign, is a (oriented) *face* of s_p.

§ 2.3.14 Let us go back to Eq.(2.2), giving the boundary of the simplex $(v_0 v_1 v_2)$. What is the boundary of this boundary? An immediate calculation shows that it is 0. The same calculation, performed on the above general definition of boundary, gives the same result because, in $\partial\partial(v_0 v_1 v_2 \dots v_p)$, each $(p-2)$-simplex appears twice and with opposite signs. This is a result of fundamental significance:

$$\partial\partial s_p \equiv 0. \qquad (2.3)$$

The boundary of a boundary of a complex is always the zero complex.

§ 2.3.15 A complex K is *oriented* if every one of its simplexes is oriented. Of course, there are many possible sets of orientations and that one which is chosen should be specified. Let us consider the set of all oriented simplexes belonging to an oriented complex K,

$$s_0^1, s_0^2, \dots, s_0^{N_0}, s_1^1, s_1^2, \dots, s_1^{N_1}, , \dots, s_p^1, s_p^2, \dots, s_p^{N_p}$$

where N_i = number of i-simplexes in K.

§ **2.3.16** A *p-chain* c_p of K is a formal sum

$$c_p = m_1 s_p^1 + m_2 s_p^2 + \ldots + m_{N_p} s_p^{N_p} = \sum_{j=1}^{N_p} m_j s_p^j,$$

with $m_j \in \mathbb{Z}$. The number p is the *dimension* of c_p. Just as seen for graphs in § 2.2.10, the p-chains of K form a group, denoted $C_p(K)$. The *boundary* of the chain c_p is

$$\partial c_p = \sum_{j=1}^{N_p} m_j \partial s_p^j.$$

Each term in the right hand side is a *face* of c_p. Equation (2.3) implies that, also for chains,

$$\partial \partial c_p \equiv 0.$$

This is one of the many avatars of one of the most important results of all Mathematics, called for historical reasons *Poincaré lemma*. We shall meet it again, under other guises. In the present context, it is not far from intuitive. Think of the sphere S^2, the boundary of a 3-dimensional ball, or of the torus T^2 which bounds a full-torus, or of any other usual boundary in $\mathbb{E}^3$: they have themselves no boundary.

2.3.2 *Topological numbers*

There are topological invariants of various types: some are general qualities of the space, like connectedness and compactness. Other are algebraic structures related to it, as the homotopic groups to be seen in Chapter 3; still other are numbers, like the Euler number and the dimension. Whatever they may be, their common point is that they are preserved by (or are invariant under) homeomorphisms. Homology provides invariants of two kinds: groups and numbers.

§ **2.3.17** The p-chains satisfying

$$\partial c_p = 0$$

are called *closed p-chains*, or *p-cycles*. The zero p-chain is a trivial p-cycle. The p-cycles constitute a subgroup of $C_p(K)$, denoted $Z_p(K)$.

§ **2.3.18** We have been using integer numbers for the chain coefficients: $m_j \in \mathbb{Z}$. In fact, any abelian group can be used instead of $\mathbb{Z}$. Besides the *integer homology* we have been considering, other cases are of great importance, in particular the *real homology* with coefficients in $\mathbb{R}$. As we can also multiply chains by numbers in $\mathbb{Z}$ (or $\mathbb{R}$), chains constitute actually a vector space. To every vector space V corresponds its dual space V^* formed by all the linear mappings taking V into $\mathbb{Z}$ (or $\mathbb{R}$). V^* is isomorphic to V (as long as V has finite dimension) and we can introduce on V^* constructs analogous to chains, called *cochains*. A whole structure dual to homology, *cohomology*, can then be defined in a purely algebraic way. This would take us a bit too far. We shall see later that chains, in the case of differentiable manifolds, are fundamentally integration domains. They are dual to differential forms and to every property of chains corresponds an analogous property of differential forms. Taking the boundary of a chain will correspond to taking the differential of a form, and the Poincaré lemma will correspond to the vanishing of the differential of a differential. Differential forms will have the role of cochains and we shall leave the study of cohomology to that stage, restricting the treatment here to a minimum.

§ **2.3.19** Given a chain c_p, its coboundary $\tilde{\partial} c_p$ is the sum of all $(p+1)$-chains of which c_p is an oriented face. Although this is more difficult to see, the Poincaré lemma holds also for $\tilde{\partial}$:

$$\tilde{\partial}\tilde{\partial} c_p \equiv 0.$$

The coboundary operator

$$\tilde{\partial} : C_p(K) \to C_{p+1}(K)$$

is a linear operator. Chains satisfying

$$\tilde{\partial} c_p = 0$$

are *p-cocycles* and also constitute a subgroup of $C_p(K)$.

§ **2.3.20** An operator of enormous importance is the *laplacian* Δ, defined by

$$\Delta c_p := (\partial\tilde{\partial} + \tilde{\partial}\partial) c_p = (\partial + \tilde{\partial})^2 c_p.$$

As ∂ takes a p-chain into a $(p-1)$-chain and $\tilde{\partial}$ takes a p-chain into a $(p+1)$-chain, Δ takes a p-chain into another p-chain. A p-chain satisfying

$$\Delta c_p = 0$$

is a *harmonic p*-chain. Just as ∂ and $\tilde{\partial}$, Δ preserves the group structure.

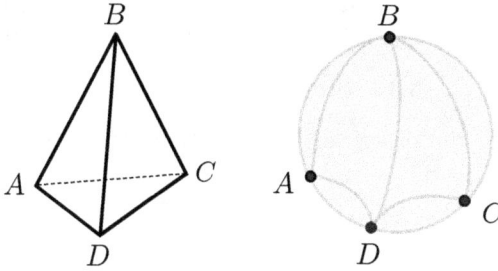

Fig. 2.9   A tetrahedron (left) and a triangulation of the sphere S^2 (right).

§ **2.3.21** Harmonic p-chains are, for finite K, simultaneously p-cycles and p-cocycles. They constitute still another subgroup of $C_p(K)$, denoted $B_p(K)$, the most important of all such groups because it is a topological invariant. The rank (number of generators) of this group is a topological number, called the p-th *Betti number*, denoted b_p.

§ **2.3.22** Consider the complex formed by the surface of a tetrahedron (Figure 2.9). The vertices A, B, C and D can be used to define the edges AB, AC, BD, etc. Let us begin by listing and commenting on some results:

(i) $\partial A = 0$ and $\partial(AB) = B - A$.

   Notice: B is a face of AB, while A is not. Only the oriented vertex $(-A)$ is a face.

(ii) $\tilde{\partial} A = BA + CA + DA$ (notice the correct signs!).

(iii) $\partial^2(AB) = 0$ and $\tilde{\partial}^2(AB) = 0$.

(iv) $\Delta(AB) = 4\, AB$.

The triangle ABD is bounded by

$$T = \partial(ABD) = AB + BD + DA.$$

But T is also the boundary of another chain:

$$T = \partial(ABC + ACD + BDC).$$

Thus, a chain can be simultaneously the boundary of two different chains. All 1-chains are generated by the edges. With coefficients a, b, c, etc in $\mathbb{Z}$ (or $\mathbb{R}$), a general 1-chain is written

$$U = aAB + bAC + cAD + dBC + eBD + fCD.$$

From the result above, $\Delta U = 4U$. As a consequence, $\Delta U = 0$ only if $U = 0$ and there exists no non-trivial harmonic 1-chain on the tetrahedron. The dimension of the space of harmonic 1-chains vanishes: b_1(tetrahedron)$= 0$. Are there harmonic 0-chains? In order to know it, we start by calculating

$$\Delta A = \partial \tilde{\partial} A = 3A - B - C - D,$$

and similarly for the other vertices. We look then for a general 0-chain which is harmonic:

$$\Delta(aA + bB + cC + dD) = 0.$$

This means that

$$(3a - b - c - d)A + (3b - a - c - d)B$$
$$+ (3c - a - b - d)C + (3d - a - b - c)D = 0.$$

The four coefficients must vanish simultaneously. Cramer's rule applied to these four equations will tell that a simple infinity of solutions exists, which can be taken to be $a = b = c = d$. Thus, the chain $a(A + B + C + D)$ is harmonic for any a in $\mathbb{Z}$ (or $\mathbb{R}$). The dimension of the space (or group) of harmonic 0-chains is one, b_0(tetrahedron) $= 1$. Proceeding to examine 2-chains, we start by finding that

$$\Delta(ABD) = 3ABD + BDC + ABC + ACD,$$

and similarly for the other triangles. Looking for a general harmonic 2-chain in the form

$$aBCD + bACD + cABD + dABC,$$

we find in the same way as for 0-chains that there is a simple infinity of solutions, so that b_2(tetrahedron) $= 1$. The tetrahedron is a triangulation of the sphere S^2 (and of the ellipsoid and other homeomorphic surfaces), as sketched in Figure 2.9 (right). With some abuse of language, we say that S^2 is one of the tetrahedron's polyhedra. As a consequence, the same Betti numbers are valid for the sphere:

$$b_0(S^2) = 1, \quad b_1(S^2) = 0, \quad b_2(S^2) = 1.$$

§ **2.3.23** We could think of using finer triangulations, complexes with a larger number of vertices, edges and triangles. It is a theorem that the Betti numbers are independent of the particular triangulation used. Notice, however, that not everything is a triangulation: the conditions defining a cellular complex must be respected. Take for instance the circle S^1: a triangulation is a juxtaposition of 1-simplexes joined in such a way that the resulting complex is homeomorphic to it. Now, an edge must have two

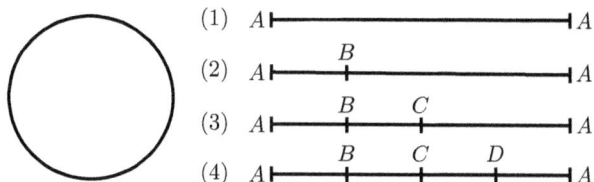

Fig. 2.10 Only the two lower simplexes are real triangulations of the circle.

distinct vertices, so that simplex (1) in Figure 2.10 is not suitable. The two edges in (2) are supposed to be distinct but they have two identical vertices. The complexes (3) and (4) are good triangulations of S^1. Take the case (3). It is easily seen that:

$$\tilde{\partial}A = CA + BA$$
$$\tilde{\partial}B = AB + CB$$
$$\tilde{\partial}C = BC + AC$$
$$\partial(AB) = B - A = \partial(CB + AC)$$
$$\Delta(AB) = 2AB - CA - BC, \quad \text{etc.}$$

Looking for solutions to

$$\Delta(aBC + bCA + cAB) = 0,$$

we arrive at

$$a = b = c, \quad \text{for any } a.$$

There is so a single infinity of solutions and $b_1(S^1) = 1$. In the same way we find $b_0(S^1) = 1$.

§ **2.3.24** Triangulations, as seen, reduce the problem of obtaining the Betti numbers to algebraic calculations. The examples above are of the simplest kind, chosen only to illustrate what is done in Algebraic Topology.

§ **2.3.25** Let us again be a bit formal. Taking the boundary of chains induces a group homomorphism from $C_p(K)$ into $C_{p-1}(K)$, the *boundary homomorphism*

$$\partial_p : C_p(K) \to C_{p-1}(K).$$

The kernel of ∂_p,

$$\ker \partial_p = \{c_p \in C_p(K) \text{ such that } \partial_p c_p = 0\}$$

is, of course, the set of p-cycles. We have already said that it constitutes a group, which we shall denote by $Z_p(K)$. Consider p-cycles

$$\alpha_p, \ \beta_p, \ \gamma_p \in Z_p(K),$$

and $(p+1)$-chains

$$\varepsilon_{p+1}, \ \eta_{p+1} \in C_{p+1}(K).$$

If

$$\alpha_p = \beta_p + \partial \varepsilon_{p+1} \quad \text{and} \quad \beta_p = \gamma_p + \partial \eta_{p+1}$$

then

$$\alpha_p = \gamma_p + \partial(\varepsilon_{p+1} + \eta_{p+1}) \ .$$

Consequently, the relation between p-cycles which differ by a boundary is an equivalence, and divides $Z_p(K)$ into equivalence classes. Each class consists of a p-cycle and all other p-cycles which differ from it by a boundary. The equivalence classes can be characterized by those α_p, β_p such that no η_{p+1} exists for which

$$\alpha_p - \beta_p = \partial \eta_{p+1}.$$

When such a η_{p+1} does exist, α_p and β_p are said to be *homologous* to each other. The relation between them is a *homology* and the corresponding classes, *homology classes*. Let us be formal once more: consider the image of ∂_{p+1}, the operator ∂ acting on $(p+1)$-chains:

$$\text{Im} \ \partial_{p+1} = \{\text{those } c_p \text{ which are boundaries of some } c_{p+1}\} \ .$$

The set of these p-boundaries form still another group, denoted $B_p(K)$. From the Poincaré lemma, $B_p(K) \subset Z_p(K)$. Every boundary is a cycle although not vice-versa. $B_p(K)$ is a subgroup of $Z_p(K)$ and there is a quotient group

$$H_p(K) = Z_p(K)/B_p(K) \ .$$

This quotient group is precisely the set of homology classes referred to above. Roughly speaking, it "counts" how many independent p-cycles exist which are not boundaries.

We have been talking of general complexes, which can in principle be homeomorphic to some topological spaces. If we restrict ourselves to finite complexes, which can only be homeomorphic to compact spaces, then it can be proved that the ranks of all the groups above are finite numbers. More important still, for finite complexes,

> *the groups $H_p(K)$ are isomorphic to the*
> *groups of harmonic p-cycles.*

Consequently,

$$b_p(K) = \text{rank } H_p(K) .$$

§ **2.3.26** These $H_p(K)$ are the *homology groups* of K and, of course, of any space homeomorphic to K. Let us further state a few important results:

(i) $b_0(S)$ is the number of connected components of S.

(ii) The Poincaré duality theorem: in an n-dimensional space S,

$$b_{n-p}(S) = b_p(S) .$$

(iii) the Euler–Poincaré characteristic is related to the Betti numbers by:

$$\chi(S) = \sum_{j=0}^{n} (-)^j b_j(S).$$

This expression is sometimes used as a definition of $\chi(S)$. Notice that for the circle, $\chi(S^1) = 0$.

For the sake of completeness: also ker $\tilde{\partial}_p$ constitutes a group, the group Z^p of p-cocycles. It contains the subgroup Im $\tilde{\partial}_{p-1}$ of those p-cocycles which are coboundaries of $(p-1)$-chains. The quotient group

$$H^p = \text{ker } \tilde{\partial}_p / \text{Im } \tilde{\partial}_{p-1}$$

is the p-th *cohomology group* of S. For finite complexes or compact spaces, it is isomorphic to the homology group H_p.

§ **2.3.27** Figure 2.11 shows a torus T^2 and a possible triangulation. The torus is obtained from the complex simply by identifying the vertices and edges with the same names. After such identification is done, it is easy to see that:

(i) The simplex $(e_1 + e_2 + e_3)$ is a cycle, but not a boundary.

(ii) The simplex $(e_4 + e_5 + e_6)$ is a cycle, quite independent of the previous one, but which is also not a boundary.

(iii) There are many cycles which are boundaries, such as $(e_4 + e_1 - e_7 - e_8)$.

Each cycle-not-boundary can give chains which are n-times itself: we can go along them n times. It can be shown, using algebraic manipulations analogous (though lengthier) to those used above for the circle and the tetrahedron, that there are two independent families of such 1-cycles which

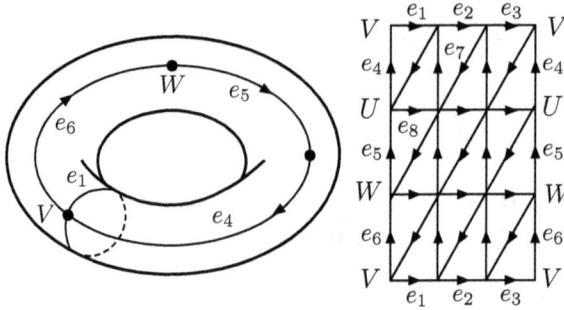

Fig. 2.11   A possible triangulation of T^2.

are not boundaries, so that $H_1(T^2)$ is isomorphic to $\mathbb{Z} \times \mathbb{Z}$. The Betti number $b_1(T^2) = 2$. Going on with an intuitive reasoning, we could ask about the 2-cycles. The torus itself is one such, as are the chains obtained by covering it n times. Such cycles are not boundaries and there are no other cases at sight. We would guess (correctly) that $b_2(T^2) = 1$, which would also come from the meaning of b_0 plus Poincaré duality. Calculations of course are necessary to confirm such results, but they are simple (though tedious) adaptations of what was done for the tetrahedron. The Poincaré duality, of course, reduce the calculations to (about) one half. The Euler characteristic may be found either by counting in the triangulation

$$\sum_i (-)^i N_i = 9 - 27 + 18 = 0,$$

or from

$$\sum_i (-)^i b_i = 1 - 2 + 1 = 0.$$

It is a general result for toruses that $\chi(T^n) = 0$ for any n.

§ 2.3.28 Genus. One of the oldest topological invariants. Denoted g, it is the largest number of closed *non-intersecting* continuous curves which can be drawn on a surface without separating it into distinct domains. It is zero for S^2, one for the torus, two for the double torus, etc. One may always think of any 2-dimensional connected, compact and orientable manifold as consisting of a sphere with n "handles", for some value of n, including $n = 0$. In this case, $g = n$. In general, the genus counts the number of toruses of the surface. It is also half the first Betti number, $2g = b_1$.

§ 2.3.29 We have considered a homeomorphism $h : K \to S$ between a complex and a space (or between polyhedra) to establish a complete identity of all homology groups. We might ask what happens if h were instead only a continuous function. The answer is the following. A continuous function $f : P \to P'$ between two polyhedra induces homomorphisms

$$f_{*k} : H_k(P) \longrightarrow H_k(P')$$

between the corresponding homology groups. Such homomorphisms become isomorphisms when f is a homeomorphism.

The homology group H_1 is the abelianized subgroup of the more informative fundamental homotopy group π_1, to be examined in Section 3.2.2.

§ 2.3.30 Once the Betti numbers for a space S are found, they may be put together as coefficients in the *Poincaré polynomial*

$$P_S(t) = \sum_{j=0}^{n} b_j(S)\, t^j,$$

in which t is a dummy variable. In the example of the sphere S^2,

$$P_{S^2}(t) = 1 + t^2.$$

For the torus T^2,

$$P_{T^2}(t) = 1 + 2t + t^2.$$

And so on. Of course, $\chi(S) = P_S(-1)$.

There is more than mere bookkeeping here. Polynomials which do not change under transformations are called "invariant polynomials" and are largely used, for example, in the classification of knots. Of course, distinct polynomials are related to spaces not equivalent under the transformations of interest. Poincaré polynomials are invariant under homeomorphic transformations.

§ 2.3.31 "Latticing" a space provides a psychological frame, helping to grasp its general profile. We have seen how it gives a clue to its homological properties. We shall see in Chapter 3 that the same happens for homotopical characteristics. In Physics, lattices provide the framework convenient to treat solid media but, more than that, they give working models for real systems. In the limit of very small spacing, they lead to a description of continuum media. But lattices are regular patterns, only convenient to modeling well-ordered systems such as crystals and metallic solids. Introducing defects into lattices leads to the description of amorphous media. We are thus led to examine (static) continuous media, elastic or not. Starting from crystals, the addition of defects allows a first glimpse into the qualitative structure of glasses. The very word "tensor" rings of Elasticity Theory, where in effect it had its origin. Deformed crystals provide the most "material" examples of tensor fields such as curvature and torsion (another resounding name). Physical situations appear usually in 3 dimensions, but 2 dimensional cases provide convenient modeling and there is nowadays a growing interest even in 2 dimensional physical cases, both for physical surfaces and biological membranes (see Chapter 28).

2.3.3 *Final remarks*

Homology is an incredibly vast subject and we have to stop somewhere. It has been intermittently applied in Physics, as in the analysis of Feynman diagrams[4] or in the incredibly beautiful version of General Relativity found by Regge, known as "Regge Calculus".[5] Much more appears when differential forms are at work and we shall see some of it in Chapter 7. Let us finish with a few words on *cubic homology*. In many applications it is easier to use squares and cubes instead of the triangles and tetrahedra we have being employing. Mathematicians are used to calculate homotopic properties (next chapter is devoted to applying complexes to obtain the fundamental group) in cubic complexes. For physicists they are still more important, as cubic homology is largely used in the lattice approach to gauge theories.[6] Now: cubes cannot, actually, be used as simplexes in the simple way presented above. The resulting topological numbers are different: for instance, a space with a single point would have all of the $b_j = 1$. It is possible, however, to proceed to a transformation on the complexes (called "normalization"), in such a way that the theory becomes equivalent to the simplex homology.[7] Triangulations are more fundamental because (i) any compact space (think on the square) can be triangulated, but not every compact space can be "squared" (think on the triangle), and (ii) the topological numbers coming up from triangulations coincide with those obtained from differential forms. In a rather permissive language, it is the triangles and tetrahedra which have the correct continuum limits.

---

[4] See Hwa & Teplitz 1966.
[5] Regge 1961; see also Misner, Thorne & Wheeler 1973, § 42.
[6] Becher & Joos 1982; their appendix contains a résumé of cubic homology.
[7] See Hilton & Wylie 1967.

Chapter 3

Homotopy

Some special deformations of a space can be of interest, even if
they only preserve part of its topological attributes.

3.1 General homotopy

We have said that, intuitively, two topological spaces are equivalent if one
can be continuously deformed into the other. Instead of the complete equivalence given by homeomorphisms — in general difficult to uncover — we
can more modestly look for some special deformations preserving only a
part of the topological characteristics. We shall in this section examine
one-parameter continuous deformations.

Roughly speaking, homotopies are function deformations regulated by
a continuous parameter, which may eventually be translated into space
deformations. The topological characterization thus obtained, though far
from complete, is highly significant.

In this section we shall state many results without any proof. This is
not to be argued as evidence for a neurotic dogmatic trend in our psychism.
In reality, some of them are intuitive and the proofs are analogous to those
given later in the particular case of homotopy between paths.

§ 3.1.1 **Homotopy between functions.** Let f and g be two continuous
functions between the topological spaces X and Y:

$$f, g : X \to Y.$$

Let again I designate the interval $[0, 1]$ included in $\mathbb{E}^1$. Then

f is *homotopic* to g ($f \approx g$) if there exists a continuous function
$F : X \times I \to Y$ such that $F(p, 0) = f(p)$ and $F(p, 1) = g(p)$ for every

$p \in X$. The function F is then a *homotopy* between f and g.

Two functions are homotopic when they can be continuously deformed into each other, in such a way that the intermediate deformations constitute a family of continuous functions between the same spaces: $F(p,t)$ is a one-parameter family of continuous functions interpolating between f and g. For fixed p, it gives a curve linking their values.

§ **3.1.2** Two constant functions $f, g : X \to Y$ with

$$f(p) = a \quad \text{and} \quad g(q) = b \quad \text{for all} \quad p, q \in X$$

are homotopic if Y is path-connected. In this case, a and b can be linked by a path $\gamma : \boldsymbol{I} \to Y$, with $\gamma(0) = a$ and $\gamma(1) = b$. Consequently,

$$F(p,t) = \gamma(t) \quad \text{for all} \quad (p,t) \in X \times \boldsymbol{I}$$

is a homotopy between f and g (Figure 3.1).

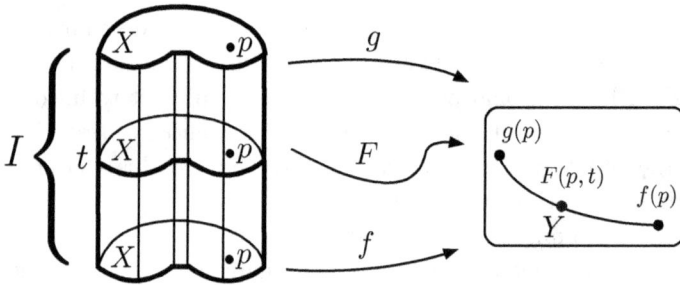

Fig. 3.1   On the left, the cartesian product $X \times \boldsymbol{I}$; on the right, the space Y where the images of f and g lie and, for each value of t, the images of the mediating function $F(p,t)$.

§ **3.1.3** We have mentioned in § 1.4.15 that the function set Y^X can be made into a topological space, for instance via the compact-open topology. We can then define paths on Y^X. Given two points in this space, like the above functions f and g, a homotopy F is precisely a continuous path on Y^X connecting them. As a consequence, two homotopic functions lie necessarily on the same path-component of Y^X.

§ **3.1.4** Homotopy is an equivalence relation between continuous functions, which consequently decomposes the set Y^X of continuous functions from X to Y into disconnected subsets, the equivalence classes. These classes are the *homotopy classes*, and the class to which a function f belongs is denoted by $[f]$.

The homotopy classes correspond precisely to the path-components of Y^X. The set of all classes of functions between X and Y is denoted by $\{X, Y\}$. Figures 3.8 and 3.9 below illustrate the decomposition in the particular case of curves on X.

§ 3.1.5 Composition preserves homotopy: if $f, g : X \to Y$ are homotopic and $h, j : Y \to Z$ are homotopic, then

$$h \circ f \approx j \circ g \approx h \circ g \approx j \circ f.$$

Consequently, composition does not individualize the members of a homotopy class: it is an operation between classes,

$$[f \circ g] = [f] \circ [g].$$

§ 3.1.6 Homotopy between spaces. The notion of homotopy may be used to establish an equivalence between spaces. Given any space Z, let $\mathrm{id}_Z : Z \to Z$ be the identity mapping on Z:

$$\mathrm{id}_Z(p) = p \quad \text{for every } p \text{ in } Z.$$

A continuous function $f : X \to Y$ is a homotopic equivalence between X and Y if there exists a continuous function $g : Y \to X$ such that

$$g \circ f \approx \mathrm{id}_X \quad \text{and} \quad f \circ g \approx \mathrm{id}_Y.$$

The function g is a kind of "homotopic inverse" to f. When such a homotopic equivalence exists, spaces X and Y are said to be of the *same homotopy type*.

Homotopy is a *necessary* condition for topological equivalence, though not a sufficient one. Every homeomorphism is a homotopic equivalence but not every homotopic equivalence is a homeomorphism. This means that homeomorphic spaces will have identical properties in what concerns homotopy (which is consequently a purely topological characteristic), but two spaces with the same homotopical properties are not necessarily homeomorphic — they may have other topological characteristics which are quite different. It will be seen in examples below that even spaces of different dimension can be homotopically equivalent.

§ 3.1.7 Contractibility is the first homotopic quality we shall meet. Suppose that the identity mapping id_X is homotopic to a constant function.

Putting it more precisely, there must be a continuous function $h : X \times \mathbf{I} \to \mathbf{X}$ and a constant function

$$f : X \to X, \quad f(p) = c$$

(a fixed point) for all $p \in X$, such that

$$h(p, 0) = p = \mathrm{id}_X(p) \quad \text{and} \quad h(p, 1) = f(p) = c.$$

When this happens, the space is homotopically equivalent to a point and said to be *contractible* (Figure 3.2).   The identity mapping id_X simply

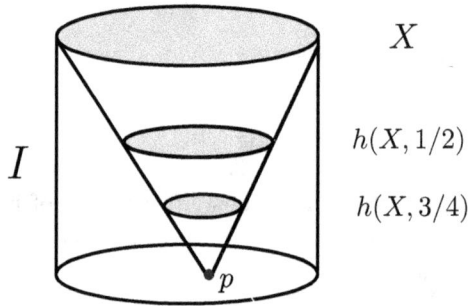

Fig. 3.2   There is a homotopy between X and a point: X is contractible.

leaves X as it is, while f concentrates X into one of its points. Contractibility means consequently that to leave X alone is homotopically equivalent to letting it shrink to a point. The interval $[0, 2\pi)$ with the induced topology (see § 1.4.13) is contractible, but the circle S^1 is not. With the quotient topology of § 1.5.3 the same interval $[0, 2\pi)$ is no more contractible — it becomes equivalent to S^1.

§ **3.1.8** A special, important result is:

Every vector space is contractible.

Let us see a special example. Take $X = \mathbb{E}^n$ and $Y = \{0\}$ (Figure 3.3, left side, shows the plane $\mathbb{E}^2$). Let f be the constant function

$$f : \mathbb{E}^n \to \{0\}, \quad f(x) = 0 \quad \forall x \in \mathbb{E}^n.$$

Let $g : \{0\} \to \mathbb{E}^n$ be the "canonical injection" of $\{0\}$ into $\mathbb{E}^n$, $g(0) = 0$. Then,

$$(f \circ g)(0) = f[g(0)] = f(0) = \mathrm{id}_Y(0).$$

As "0" is the only point in Y, we have shown that $f \circ g = \mathrm{id}_Y$, and so, that $f \circ g \approx \mathrm{id}_Y$. Also

$$(g \circ f)(x) = g[f(x)] = g(0) = 0.$$

Now, let

$$h : \mathbb{E}^n \times I \to \mathbb{E}^n, \quad h(x,t) = tx \in \mathbb{E}^n$$

(because of its vector space structure). Clearly

$$h(x,0) = 0 = (g \circ f)(x) \quad \text{and} \quad h(x,1) = x = \mathrm{id}_X(x).$$

So, h is a homotopy $\mathrm{id}_X \approx g \circ f$. The space $\mathbb{E}^n$ is consequently homotopic to a point, that is, contractible. The same is true of any open ball of $\mathbb{E}^n$.

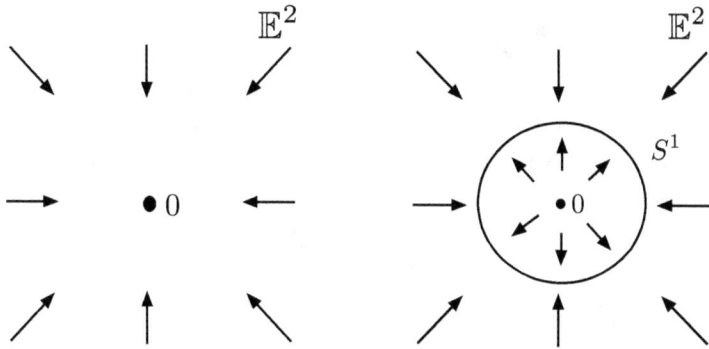

Fig. 3.3   Left: the plane is homotopically equivalent to a point (here, the origin), that is, it is contractible; Right: the punctured plane $\mathbb{E}^2/\{0\}$ is homotopically equivalent to the circle.

§ **3.1.9** Take $X = \mathbb{E}^2 - \{0\}$ and $Y = S^1$ (as in Figure 3.3, right side). Let

$$f : X \to Y, \quad f(x) = x/|x|,$$

and let $g : S^1 \to \mathbb{E}^2 - \{0\}$ be the canonical injection

$$g : e^{i\varphi} \in S^1 \to e^{i\varphi} \in \mathbb{E}^2 - \{0\}.$$

Then,

$$(f \circ g)(e^{i\varphi}) = f(e^{i\varphi}) = e^{i\varphi} = \mathrm{id}_{S^1}(e^{i\varphi}).$$

On the other hand,

$$(g \circ f)(x) = g(x/|x|) = x/|x|.$$

Now, the function $h : X \times \mathbf{I} \to X$ given by

$$h(x, t) = (1 - t)x/|x| + tx$$

is such that

$$h(x, 0) = (g \circ f)(x) \quad \text{and} \quad h(x, 1) = x = \mathrm{id}_X(x).$$

The conclusion is that $\mathbb{E}^2 - \{0\}$ and S^1 are homotopically equivalent.

This example is of significance in the study of the Aharonov–Bohm effect (§ 4.2.17). It has higher dimensional analogues. The following result, on the other hand, is relevant to the Dirac monopole: the space

$$\mathbb{E}^4 \backslash \{\text{points on the line } x^1 = x^2 = x^3 = 0\}$$

is homotopically equivalent to the sphere S^2. It can also be shown that

$$\mathbb{E}^3 \backslash \{\text{points on an infinite line}\} \approx S^1.$$

All these examples are cases of homotopical equivalences that are not complete topological equivalences, as the spaces involved do not even have the same dimensions. Notice also the relationship with the convexity of Section 2.3.2 in the last two examples.

§ **3.1.10** If either X or Y is contractible, every continuous function

$$f : X \to Y$$

is homotopic to a constant mapping (use $f = \mathrm{id}_Y \circ f$ or $f = f \circ \mathrm{id}_X$ at convenience). As said above, every vector space is contractible. Contractibility has important consequences in vector analysis. As we shall see later, some of the frequently used properties of vector analysis are valid only on contractible spaces — for example, the facts that divergenceless fluxes are rotationals and irrotational fluxes are potential — and will not hold if, for example, points are extracted from a vector space.

§ **3.1.11** We have repeatedly said that homotopies preserve part of the topological aspects. Contractibility, when extant, is one of them. We shall in the following examine some other, of special relevance because they appear as algebraic structures — the homotopy groups. The homology groups presented in § 2.3.26 are also invariant under homotopic transformations.

§ **3.1.12** A *retraction* of a topological space X onto a subspace Y is a continuous mapping $r : X \longrightarrow Y$ such that, for any $p \in Y$, $r(p) = p$. When such a retraction exists, Y is a *retract* of X.

In larger generality: a *deformation* of a topological space X is a family $\{h_s\}$ of mappings $h_s : X \to X$, with parameter $s \in I \equiv [0,1]$ such that h_0 is the identity mapping and the function

$$H : I \times X \to I \times X$$

defined by $H(s,p) = h_s(p)$ is continuous. When the mappings h_s are homeomorphisms, the deformation is an *isotopy*, or *isotopic deformation*. A deformation into or onto a subspace Y is a deformation of X such that the image $h_1 X$ is contained in (or is equal to) Y. A subspace Y of a topological space X is a *deformation retract* of X if there is a retraction $r : X \to Y$ which is a deformation. Deformation retractions preserve homotopy type.

3.2 Path homotopy

Paths defined on a space provide essential information about its topology.

3.2.1 *Homotopy of curves*

§ **3.2.1** Let us recall the previously given definition of a path on a topological space X: it is a continuous function $f : I \to X$. This is not the kind of path which will be most useful for us. Notice that I is contractible, so that every path is homotopic to a constant and, in a sense, trivial.

We shall rather consider, instead of such free-ended paths, *paths with fixed ends*. The value $x_0 = f(0)$ will be the *initial end-point*, $x_1 = f(1)$ the *final end-point* and $f(t)$ the *path* from x_0 to x_1. Given two paths f and g with the same end-points, they are *homotopic paths* (which will be indicated $f \approx g$) if there exists a continuous mapping

$$F : I \times I \to X$$

such that, for every $s, t \in I$,

$$F(s,0) = f(s) \qquad F(s,1) = g(s)$$

and

$$F(0,t) = x_0 \qquad F(1,t) = x_1.$$

§ **3.2.2** The function F is a *path homotopy* between f and g and represents, intuitively, a continuous deformation of the curve f into the curve g. For each fixed t, $F(s,t)$ is a curve from x_0 to x_1, intermediate between f and g, an interpolation between them (Figure 3.4). For each fixed s, $F(s,t)$ is

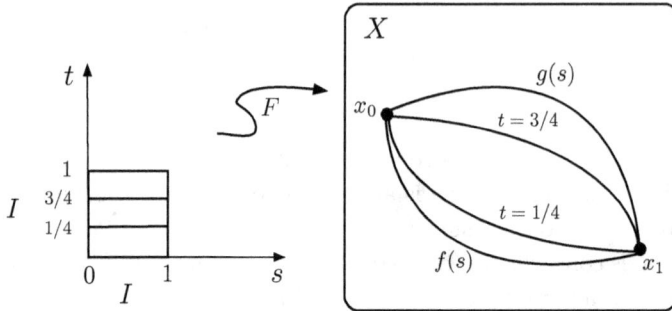

Fig. 3.4   For each value $t \in \mathbf{I}$, $F(s,t)$ gives a path, intermediate between f and g.

a curve from $f(s)$ to $g(s)$ (Figure 3.5). These transversal curves are called "variations" (see Chapter 19).

§ **3.2.3** Path homotopy is an equivalence relation:

(i) Given f, trivially $f \approx f$, as the mapping $F(p,t) = f(p)$ is a homotopy; by the way, it is the identity function on X^I.

(ii) Suppose F is a homotopy between f and g. Then,

$$G(p,t) = F(p, 1-t)$$

is a homotopy between g and f.

(iii) Suppose F is a homotopy $f \approx g$ and G a homotopy $g \approx h$ (see Figure 3.6). Define $H : \mathbf{I} \times \mathbf{I} \to X$ by the equations

$$H(s,t) = \begin{cases} F(s, 2t) & \text{for } t \in [0, 1/2] \\ G(s, 2t-1) & \text{for } t \in [1/2, 1] \end{cases}$$

H is well defined because, when $t = 1/2$,

$$[F(s, 2t)]_{t=1/2} = F(s,1) = g(s) = G(s,0) = [G(s, 2t-1)]_{t=1/2}.$$

As H is continuous on the two closed subsets

$$\mathbf{I} \times [0, 1/2] \quad \text{and} \quad \mathbf{I} \times [1/2, 1]$$

of $\mathbf{I} \times \mathbf{I}$, it is continuous on $\mathbf{I} \times \mathbf{I}$. It is a homotopy $f \approx h$.

§ **3.2.4** By definition, every pair of points inside each *path-component* of a given space X can be linked by some path. Let $\{C_\alpha\}$ be the set of path-components of X, which is denoted by $\pi_0(X)$:

$$\pi_0(X) = \{C_\alpha\}.$$

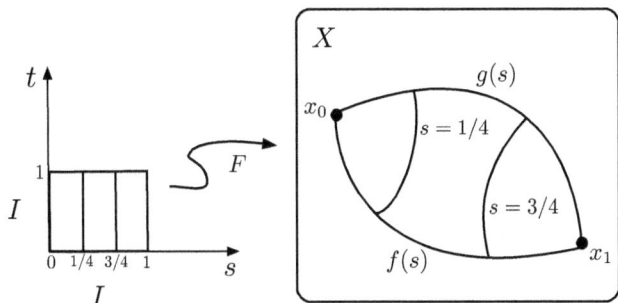

Fig. 3.5   For each fixed value s, $F(s, t)$ gives a path going from $f(s)$ to $g(s)$.

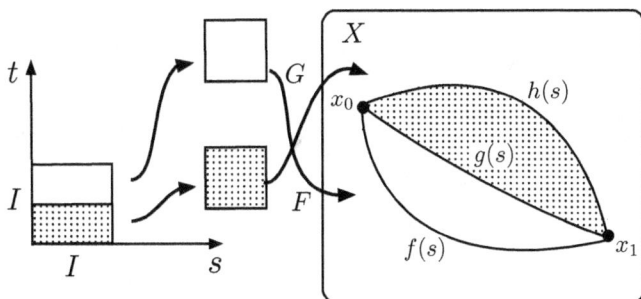

Fig. 3.6   If F is a homotopy $f \approx g$ and G is a homotopy $g \approx h$, then there exists also a homotopy $H : f \approx h$.

This notation is a matter of convenience: π_0 will be included later in the family of homotopy groups and each one of them is indicated by π_n for some integer n. The relation between the path homotopy classes on a space X and the path components of the space X^I of all paths on X is not difficult to understand (Figure 3.7). A homotopy F between two paths on X is a path on X^I and two homotopic paths f and g can be thought of as two points of X^I linked by the homotopy F.

§ 3.2.5 The intermediate curves representing the successive stages of the continuous deformation must, of course, lie entirely on the space X. Suppose X to have a hole as in Figure 3.8. Paths f and g are homotopic to each other and so are j and h. But f is homotopic neither to j nor to h. The space X^I is so divided in path-components, which are distinct for f

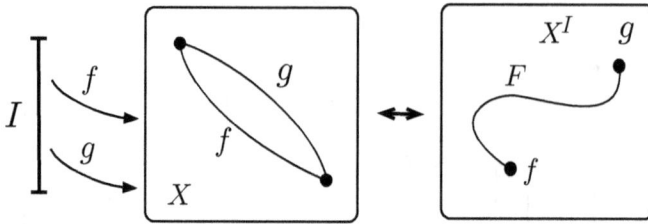

Fig. 3.7 Relation between path homotopy classes on a space X and the path components of the space X^I of all paths on X.

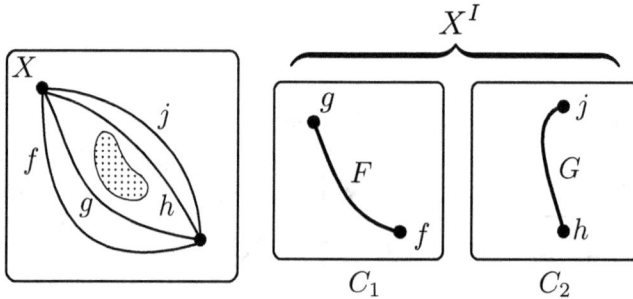

Fig. 3.8 Effect of the presence of a hole.

and j.

There are actually infinite components for X^I when X has a hole: a curve which turns once around the hole can only be homotopic to curves turning once around the hole. A curve which turns twice around the hole will be continuously deformable only into curves turning twice around the hole; and the same will be true for curves turning n times around the hole, suggesting a relation between the homotopy classes and the counting of the number of turns the curves perform around the hole. Another point is that a curve turning in clockwise sense cannot be homotopic to a curve turning anticlockwise even if they turn the same number of times (like the curves k and r in Figure 3.9). This suggests an *algebraic* counting of the turns.

We shall in what follows give special emphasis to some algebraic characteristics of paths and proceed to classify them into groups. An operation of path composition is defined which, once restricted to classes of closed paths, brings forth such a structure.

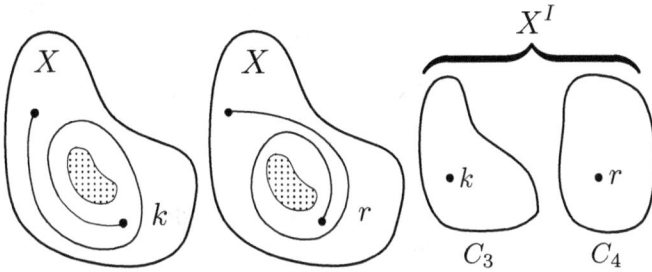

Fig. 3.9   Effect of curve orientations.

§ 3.2.6 Towards a group structure. Let us start by introducing the announced operation between paths. Take again the space X and f a path between x_0 and x_1. Let g be a path between x_1 and x_2. We define the *composition*, or *product* $f \bullet g$ of f and g as that path (Figure 3.10) given by

$$h(s) = (f \bullet g)(s) = \begin{cases} f(2s) & \text{for } s \in [0, 1/2] \\ g(2s - 1) & \text{for } s \in [1/2, 1]. \end{cases}$$

The composition h is a well defined curve from x_0 to x_2. It can be seen as a path with first half "f" and second half "g". The order in reading these products is from left to right, opposite to the usual composition of functions.

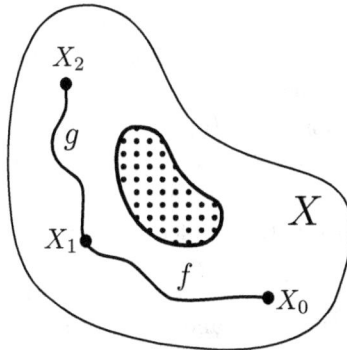

Fig. 3.10   Path composition.

§ 3.2.7 We shall show below that the operation "•" is well defined on the path-homotopy classes. It has the following "group-like" properties:

(i) Associativity: *if* $[f] \bullet ([g] \bullet [h])$ *and* $([f] \bullet [g]) \bullet [h]$ *are defined, then*

$$[f] \bullet ([g] \bullet [h]) = ([f] \bullet [g]) \bullet [h] \ .$$

Thus, path-homotopy classes with the operation of composition constitute a semigroup.

(ii) There are "identity" elements: for any $x \in X$, let $e_x \colon I \to X$ be such that $e_x(I) = x$. Thus, e_x is a constant path on X, with in particular both end-points equal to x. Given a path f between x_0 and x_1, we have

$$[f] \bullet [e_{x_1}] = [f] \qquad \text{and} \qquad [e_{x_0}] \bullet [f] = [f].$$

Notice however that the "identity" elements are different in different points (consequently, path-homotopy classes with the operation of composition constitute neither a monoid nor a groupoid — see Chapter 13).

(iii) Existence of "inverse" elements: the inverse of a path f from x_0 to x_1 is a path $f^{<-1>}$ from x_1 to x_0 given by $f^{<-1>}(s) = f(1-s)$. Thus,

$$[f] \bullet [f^{<-1>}] = [e_{x_0}] \qquad \text{and} \qquad [f^{<-1>}] \bullet [f] = [e_{x_1}].$$

§ 3.2.8 Before we start showing that operation • is well defined on homotopy classes, as well as that it is an associative product, let us repeat that, despite a certain similitude, the above properties do not actually define a group structure on the set of classes. The "identity" elements are distinct in different points, and the product $[f] \bullet [g]$ is not defined for any pair of equivalence classes: for that, $f(1) = g(0)$ would be required. The set of path-homotopy classes on X is not a group with the operation •. We shall see in the next section that the classes involving only *closed* paths do constitute a group, which will be called the *fundamental group*.

§ 3.2.9 In order to show that • is well defined on the homotopy classes, we must establish that, if $f \approx f'$ and $g \approx g'$, then

$$f \bullet g \approx f' \bullet g'.$$

Given the first two homotopies, we shall exhibit a homotopy between the latter. Supposing F and G to be path-homotopies respectively between

f, f' and g, g', a homotopy between $f \bullet g$ and $f' \bullet g'$ is (see Figure 3.11)

$$H(s,t) = \begin{cases} F(2s,t) & \text{for } s \in [0, 1/2], \\ G(2s-1,t) & \text{for } s \in [1/2, 1]. \end{cases}$$

For $s = 1/2$, we get $F(1,t) = x_1 = G(0,t)$ for any t. The function H is, thus, well defined and continuous. On the other hand,

$$H(s,0) = \begin{cases} f(s) & \text{for } s \in [0, 1/2], \\ g(2s-1) & \text{for } s \in [1/2, 1]. \end{cases}$$

defines $f \bullet g$, and

$$H(s,1) = \begin{cases} f'(s) & \text{for } s \in [0, 1/2], \\ g'(2s-1) & \text{for } s \in [1/2, 1]. \end{cases}$$

defines $f' \bullet g'$. As $H(0,t) = x_0$ and $H(1,t) = x_2$, the mapping H is indeed a path-homotopy between $f \bullet g$ and $f' \bullet g'$. To obtain the associativity

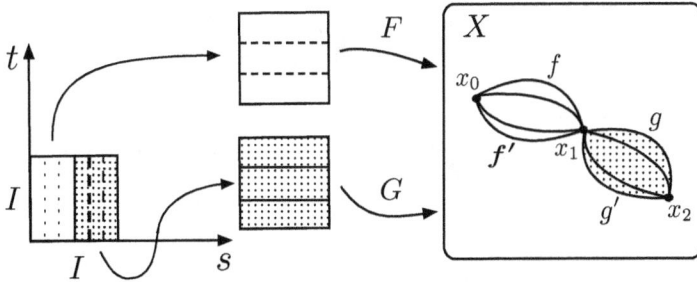

Fig. 3.11   Composition of homotopy classes, which leads to a group-like structure.

property, we should show that

$$f \bullet (g \bullet h) \approx (f \bullet g) \bullet h.$$

Through the mapping $f \bullet (g \bullet h)$, the image of s goes through the values of f when s goes from 0 to $1/2$, the values of g when s lies between $1/2$ and $3/4$, and the values of h when s goes from $3/4$ to 1:

$$(f \bullet (g \bullet h))(s) = \begin{cases} f(2s) & \text{for } s \in [0, 1/2], \\ (g \bullet h)(2s-1) & \text{for } s \in [1/2, 1]. \end{cases}$$

In more detail,

$$(f \bullet (g \bullet h))(s) = \begin{cases} f(2s) & \text{for } s \in [0, 1/2], \\ g[2(2s-1)] & \text{for } s \in [1/2, \frac{3}{4}], \\ h[2(2s-1)] & \text{for } s \in [3/4, 1]. \end{cases}$$

On the other hand,

$$((f \bullet g) \bullet h)(s) = \begin{cases} (f \bullet g)(2s) \text{ for } s \in [0, 1/2], \\ h(2s - 1) \text{ for } s \in [1/2, 1]. \end{cases}$$

or

$$((f \bullet g) \bullet h)(s) = \begin{cases} f(4s) \quad \text{ for } \quad s \in [0, 1/4], \\ g(4s - 1) \text{ for } s \in [1/4, 1/2], \\ h(2s - 1) \text{ for } \quad s \in [1/2, 1]. \end{cases}$$

Through the mapping $(f \bullet g) \bullet h$ the image of s goes through the values of f when s goes from 0 to 1/4, the values of g when s lies between 1/4 and 1/2, and the values of h when s goes from 1/2 to 1. The paths $f \bullet (g \bullet h)$ and $(f \bullet g) \bullet h$ have the same image, traversed nevertheless at different "rates". The desired homotopy is formally given by

$$F(s, t) = \begin{cases} f\left(\frac{4s}{t+1}\right) \quad \text{ for } \quad s \in [0, \frac{t+1}{4}], \\ g\left(4s - t - 1\right) \text{ for } s \in [\frac{t+1}{4}, \frac{t+2}{4}], \\ h\left(\frac{4s-t-2}{2-t}\right) \text{ for } \quad s \in [\frac{t+2}{4}, 1]. \end{cases}$$

The arguments of f, g and h are all in $[0, 1]$. It is easily checked that $F(s, 1)$ is the same as $(f \bullet (g \bullet h))(s)$, and that $F(s, 0)$ is the same as $((f \bullet g) \bullet h)(s)$.

3.2.2   *The fundamental group*

Classes of closed curves do constitute a discrete group, which is one of the main topological characteristics of a space. Here we shall examine the general properties of this group. Important examples of discrete groups, in particular the braid and knot groups, are described in the Chapter 14. Next section will be devoted to a method to obtain the fundamental group for some simple spaces in the form of word groups.

§ **3.2.10** We have seen in the previous section that the path product "$\bullet$" does not define a group structure on the whole set of homotopy classes. We shall now restrict that set so as to obtain a group — the fundamental group — which turns out to be a topological characteristic of the underlying space.

§ **3.2.11** Let X be a topological space and x_0 a point in X. A path whose both end-points are x_0 is called a *loop* with base point x_0. It is a continuous function $f : I \to X$, with

$$f(0) = f(1) = x_0.$$

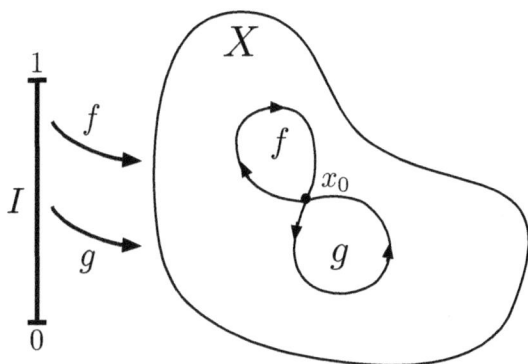

Fig. 3.12   Loops with a fixed base point.

§ 3.2.12 The set of homotopy classes of loops with a fixed base point does constitute a group with the operation "•". More precisely:

(i) Given two loops f and g with the same base point x_0, their composition $f \bullet g$ is well defined and is a loop with base point x_0 (Figure 3.12).
(ii) The properties of associativity, existence of an identity element $[e_{x_0}]$, and existence of an inverse $[f^{<-1>}]$ for each $[f]$ hold evidently .

§ 3.2.13 The group formed by the homotopy classes of loops with base point x_0 is called "the *fundamental group* of space X relative to the base point x_0" and is denoted $\pi_1(X, x_0)$. It is also known as the Poincaré group (a name not used by physicists, to avoid confusion) and *first homotopy group* of X at x_0, because of the whole series of groups $\pi_n(X, x_0)$ which will be introduced in Section 3.4.

§ 3.2.14 A question arising naturally from the above definition is the following: how much does the group depend on the base point? Before we answer to this question, some preliminaries are needed. Let γ be a path on X from x_0 to x_1 (Figure 3.13). Define the mapping

$$\gamma^{\#} : \pi_1(X, x_0) \to \pi_1(X, x_1)$$

through the expression

$$\gamma^{\#}([f]) = [\gamma^{-1}] \bullet [f] \bullet [\gamma].$$

This is a well defined mapping. If f is a loop with base point x_0, then

$$\gamma^{-1} \bullet f \bullet \gamma$$

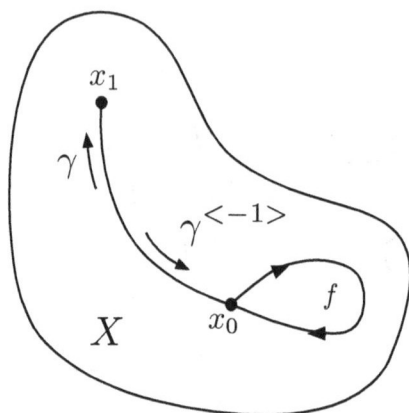

Fig. 3.13   Changing from one base point to another.

is a loop with base point x_1. Consequently, $\gamma^{\#}$ maps indeed $\pi_1(X, x_0)$ into $\pi_1(X, x_1)$. Finally, we have the following theorem:

The mapping $\gamma^{\#}$ is a group isomorphism.

Let us go by parts: a map φ of a group G with operation $(\cdot)$ into a group G' with operation $(\times)$ is a *group homomorphism* if, for any a and b in G,

$$\varphi(a \cdot b) = \varphi(a) \times \varphi(b).$$

In this case, φ is also called a *representation* (Chapter 18) of G in G'. The map φ will be a *group isomorphism*[1] if a homomorphism $\psi : G' \to G$ exists such that the compositions $\varphi \circ \psi$ and $\psi \circ \varphi$ are the identity mappings of G' and G respectively. In order to prove the theorem, we have first to show that $\gamma^{\#}$ is a homomorphism, that is, that it preserves the operation '•'. But

$$(\gamma^{\#}([f])) \bullet (\gamma^{\#}([g])) = ([\gamma^{-1}] \bullet [f] \bullet [\gamma]) \bullet ([\gamma^{-1}] \bullet [g] \bullet [\gamma])$$
$$= ([\gamma^{-1}] \bullet [f] \bullet [g] \bullet [\gamma]) = \gamma^{\#}([f] \bullet [g]).$$

So good for the homomorphism. Now, taking the inverse path γ^{-1}, let us see that $(\gamma^{-1})^{\#}$ is the inverse to $\gamma^{\#}$. Take $[f]$ in $\pi_1(X, x_0)$ and $[h]$ in $\pi_1(X, x_1)$. Then,

$$(\gamma^{-1})^{\#}[h] = [\gamma] \bullet [h] \bullet [\gamma^{-1}]$$

---

[1] Fraleigh 1974.

so that

$$\gamma^{\#}((\gamma^{-1})^{\#}[h]) = [\gamma^{-1}] \bullet ([\gamma] \bullet [h] \bullet [\gamma^{-1}]) \bullet [\gamma],$$

and finally

$$\gamma^{\#}((\gamma^{-1})^{\#}[h]) = [h].$$

It can be similarly verified that

$$(\gamma^{-1})^{\#}(\gamma^{\#}[f]) = [f].$$

§ 3.2.15 Wherever X is path-connected, there exist such mappings $\gamma^{\#}$ relating the fundamental group based on different points. A corollary is then:

> If X is path-connected and both x_0 and x_1 are in X, then $\pi_1(X, x_0)$ is isomorphic to $\pi_1(X, x_1)$.

§ 3.2.16 Suppose now that X is a topological space and C is a path-component of X to which x_0 belongs. As all the loops at x_0 belong to C and all the groups are isomorphic, we may write

$$\pi_1(C) = \pi_1(X, x_0).$$

The group depends only on the path-component and gives no information at all about the remaining of X. In reality, we should be a bit more careful, as the isomorphism $\gamma^{\#}$ above is not *natural* (or *canonical*): it depends on the path γ. This means that different paths between x_0 and x_1 take one same element of $\pi_1(X, x_0)$ into different elements of $\pi_1(X, x_1)$, although preserving the overall abstract group structure. The isomorphism is canonical only when the group is abelian. We shall be using this terminology without much ado in the following, though only in the abelian case is writing "$\pi_1(C)$" entirely justified. These considerations extend to the whole space X if it is path-connected, when we then talk of the group $\pi_1(X)$.

§ 3.2.17 A class $[f]$ contains its representative f and every other loop continuously deformable into f. Every space will contain a class $[c(t)]$ of trivial, or constant, closed curves, $c(t) = x_0$ for all values of t (Figure 3.14). If this class is the only homotopy class on X, then $\pi_1(X, x_0) = [c(t)]$, and we say that X is simply-connected. More precisely:

> X is *simply-connected* when it is path-connected
> and $\pi_1(X, x_0)$ is trivial for some $x_0 \in X$.

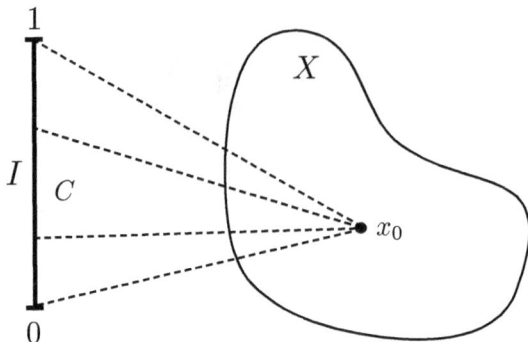

Fig. 3.14   A constant function.

§ **3.2.18** When X is simply-connected, every two loops on X can be continuously deformed into each other, and all loops can be continuously deformed to a point. When this is not the case, X is *multiply-connected*. This is a first evidence of how $\pi_1(X)$ reflects, to a certain degree, the qualitative structure of X. The fact that closed laces in a multiply-connected space cannot be caused to shrink to a point while remaining inside the space signals the presence of some defect, of a "hole" of some kind. The plane $\mathbb{E}^2$ is clearly simply-connected, but this changes completely if one of its points is excluded, as in $\mathbb{E}^2 - \{0\}$: loops not encircling the zero are trivial but those turning around it are not (Figure 3.15). There is a simple countable infinity of loops not deformable into each other. We shall see later (§ 3.2.33) that

$$\pi_1(\mathbb{E}^2 \backslash \{0\}) = \mathbb{Z},$$

the group of integer numbers. We shall also see (§ 3.2.34) that the exclusion of two points, say as in $\mathbb{E}^2 \backslash \{-1, +1\}$, leads to a non-abelian group π_1. The space $\mathbb{E}^3$ with an infinite line deleted (say, $x = 0, y = 0$) is also a clear non-trivial case. Some thought dedicated to the torus $T^2 = S^1 \times S^1$ will, however, convince the reader that such intuitive insights are yet too rough.

§ **3.2.19** The trivial homotopy class $[c(t)]$ acts as the identity element of π_1 and is sometimes denoted $[e_{x_0}]$. More frequently, faithful to the common usage of the algebrists, it is denoted by "0" when π_1 is abelian, and "1" when it is not (or when we do not know or care).

§ **3.2.20** All the euclidean spaces $\mathbb{E}^n$ are simply-connected:

$$\pi_1(\mathbb{E}^n) = 1.$$

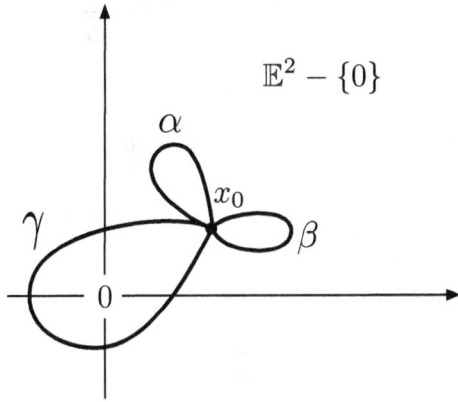

Fig. 3.15   Curves α and β are trivial. Curve γ, encircling the excluded zero, is not.

The n-dimensional spheres S^n for $n \geq 2$ are simply-connected:

$$\pi_1(S^n) = 1$$

for $n \geq 2$. We shall see below that the circle S^1 is multiply-connected and that

$$\pi_1(S^1) = \mathbb{Z}.$$

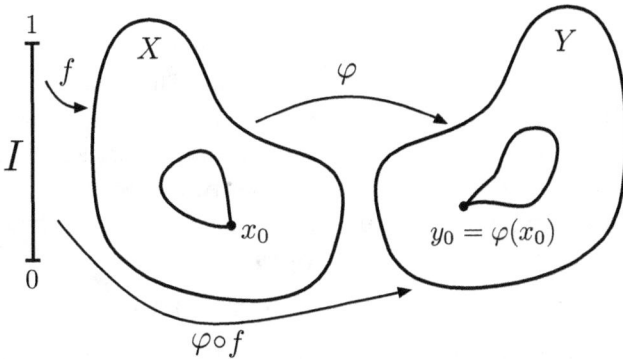

Fig. 3.16   A continuous mapping between spaces X and Y induces a homomorphism between their fundamental groups.

§ 3.2.21 Let us see now why the fundamental group is a topological invariant. The root of this property lies in the fact that continuous functions between spaces induce homomorphisms between their fundamental groups, even when they are not homeomorphisms (Figure 3.16). Indeed, let $\varphi : X \to Y$ be a continuous mapping with $\varphi(x_0) = y_0$. If f is a loop on X with base point x_0, then the composition $\varphi \circ f : I \to Y$ is a loop on Y with base point y_0. Consequently φ induces a mapping φ_* between $\pi_1(X, x_0)$ and $\pi_1(Y, y_0)$, defined by

$$\varphi_* : \pi_1(X, x_0) \to \pi_1(Y, y_0), \qquad \varphi_*([f]) = [\varphi \circ f].$$

This mapping φ_* is the "induced homomorphism", relative to the base point x_0. It can be shown that it is well defined: if f and f' are homotopic and $F : I \times I \to X$ is their homotopy, then $\varphi \circ F$ is a homotopy between the loops $\varphi \circ f$ and $\varphi \circ f'$. It can also be shown that φ_* is a homomorphism,

$$\varphi_*([f] \bullet [g]) = \varphi_*([f]) \bullet \varphi_*([g]).$$

Notice that φ_* depends not only on φ but also on the base point. This homomorphism has some properties of great importance, known in the mathematical literature under the name "functorial properties". We list them:

(i) If $\varphi : (X, x_0) \to (Y, y_0)$ and $\psi : (Y, y_0) \to (Z, z_0)$, then

$$(\psi \circ \varphi)_* = \psi_* \circ \varphi_*.$$

If $i : (X, x_0) \to (X, x_0)$ is the identity mapping, then i_* is the identity homomorphism.

(ii) If $\varphi : (X, x_0) \to (Y, y_0)$ is a homeomorphism, then φ_* is an isomorphism between $\pi_1(X, x_0)$ and $\pi_1(Y, y_0)$. This is the real characterization of the fundamental group as a topological invariant.

(iii) If (X, x_0) and (Y, y_0) are homotopically equivalent, then $\pi_1(X, x_0)$ and $\pi_1(Y, y_0)$ are isomorphic. Recall that two spaces are homotopically equivalent (or "of the same homotopic type") when two functions $j : X \to Y$ and h: $Y \to X$ exist satisfying

$$j \circ h \approx \mathrm{id}_X \quad \text{and} \quad h \circ j \approx \mathrm{id}_Y.$$

Here, these functions must further satisfy

$$j(x_0) = y_0 \quad \text{and} \quad h(y_0) = x_0.$$

Summing up: continuous functions induce homomorphisms between the fundamental groups; homeomorphisms induce isomorphisms.

§ **3.2.22** A contractible space is clearly simply-connected. The converse, however, is not true. We shall show later that the sphere S^2 is simply-connected, but it is clearly not contractible. The vector analysis properties mentioned in § 3.1.10 require real contractibility. They are consequently not valid on S^2, which is simply-connected.

§ **3.2.23** Before we finish this section let us mention an important and useful result. With the same notations above, let X and Y be topological spaces and consider the fundamental groups with base points x_0 and y_0. Then, the fundamental group of $(X \times Y, x_0 \times y_0)$ is isomorphic to the direct product of the fundamental groups of (X, x_0) and (Y, y_0):

$$\pi_1(X \times Y, x_0 \times y_0) \approx \pi_1(X, x_0) \otimes \pi_1(Y, y_0).$$

§ **3.2.24** As mentioned above, $\pi_1(S^1) = \mathbb{Z}$, the additive group of the integer numbers. This will be shown in Section 3.3.2 through the use of covering spaces. Although it does provide more insight on the meaning of the fundamental group, that method is rather clumsy. Another technique, much simpler, will be seen in Section 3.2.3. As an illustration of the last result quoted above, the torus $T^2 = S^1 \times S^1$ will have

$$\pi_1(T^2) = \mathbb{Z} \otimes \mathbb{Z}.$$

§ **3.2.25** For the reader who may be wondering about possible relations between homology and homotopy groups, let it be said that H_1 is the abelianized subgroup of the fundamental group. It contains consequently less information than π_1.

§ **3.2.26** The fundamental group, trivial in simply-connected spaces, may be very complicated on multiply-connected ones, as in the last examples. Quantum Mechanics on multiply-connected spaces[2] makes explicit use of π_1 (for a simple example, the Young double-slit experiment, see § 4.2.16). Feynman's picture, based on trajectories, is of course a very convenient formulation to look at the question from this point of view. The total propagator from a point A to a point B becomes a sum of contributions[3] of the different homotopy classes,

$$K(B, A) = \sum_\alpha \eta([\alpha]) K^\alpha(B, A).$$

---

[2] Schulman 1968.
[3] DeWitt-Morette 1969; Laidlaw & DeWitt-Morette 1971; DeWitt-Morette 1972.

K^α is the propagator for trajectories in the class $[\alpha]$, and η is a phase providing a one-dimensional representation of π_1(configuration space). It would be interesting to examine the case of many-component wavefunctions, for which such representation could perhaps become less trivial and even give some not yet known effects for non-abelian π_1.

3.2.3   *Some examples*

§ 3.2.27 Triangulations provide a simple means of obtaining the fundamental group. The method is based on a theorem[4] whose statement requires some preliminaries.

(i) Take a path-connected complex with vertices $v_1, v_2, v_3, \ldots$, and edges (v_1v_2), (v_1v_3), (v_2v_3), etc. To each oriented edge we attribute a symbol,

$$(v_iv_k) \to g_{ik}.$$

We then make the set of such symbols into a group, defining

$$g_{ik}^{-1} = g_{ki}$$

and imposing, for each triangle $(v_iv_jv_k)$, the rule

$$g_{ij}g_{jk}g_{ki} = 1,$$

the identity element. Roughly speaking, the element g_{ik} corresponds to "going along" the edge (v_iv_k), the product $g_{ij}g_{jk}$ to "going along" (v_iv_j) and then along (v_jv_k), and so on. The last rule above would say simply that going around a triangle, one simply comes back to the original point. We emphasize that one such rule must be imposed for each triangle (2-simplex).

(ii) It can be proven that on every path-connected polyhedron there exists at least one Euler path, a path through all the vertices which is contractible (that is, it is simple but not a loop).

§ 3.2.28 Once this is said, we enunciate the *calculating theorem*:

> *Take a vertex* v_0 *in a polyhedron K and a Euler path P starting from* v_0.
> *In the group defined above, put further* $g_{jk} = 1$ *for each edge* (v_jv_k)
> *belonging to P. Then, the remaining group G is isomorphic to the*
> *fundamental group of K with base point* v_0, $\pi_1(K, v_0)$.

---

[4]   Hu 1959; see also Nash & Sen 1983.

The proof is involved and we shall not even sketch it here. The theorem is a golden road to arrive at the fundamental group. The best way to see how it works is to examine some examples.

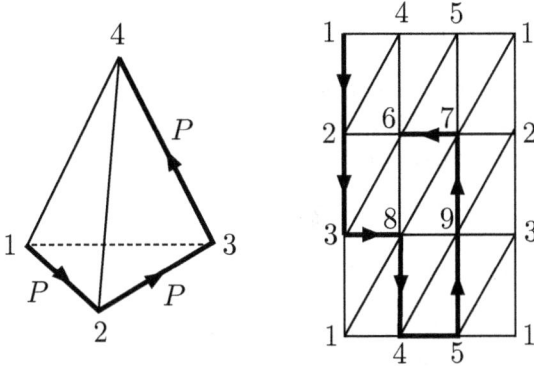

Fig. 3.17   Triangulations for the tetrahedron and the torus, with connected paths.

§ 3.2.29 Take again the (surface) tetrahedron as in the left diagram of Figure 3.17 and, for P, the connected path $1 \to 2 \to 3 \to 4$. For each triangle there is a condition:

$$g_{12}g_{24}g_{41} = 1, \quad g_{41}g_{13}g_{34} = 1, \quad g_{23}g_{34}g_{42} = 1, \quad g_{12}g_{23}g_{31} = 1.$$

We now impose also

$$g_{12} = g_{23} = g_{34} = 1,$$

because the corresponding edges belong to the chosen path. As a consequence, all the group elements reduce to the identity. The fundamental group is then $\pi_1(\text{tetrahedron}) = \{1\}$. The tetrahedron is simply-connected, and so are the spaces homeomorphic to it: the sphere S^2, the ellipsoid, etc.

§ 3.2.30 The torus. With the triangulation and path of Figure 3.17, right side, the 18 conditions are easily reduced to the forms

$$g_{49} = g_{89} = g_{78} = g_{68} = g_{36} = g_{26} = 1,$$
$$g_{27} = g_{39} = g_{29} = g_{17} = g_{35} =: g',$$
$$g_{24} = g_{64} = g_{14} = g_{65} = g_{75} = g_{13} = g_{18} =: g'',$$
$$g_{13}g_{35}g_{51} = 1,$$
$$g_{75}g_{51}g_{17} = 1.$$

The last two conditions can be worked out to give

$$g_{51} = g'g'' = g''g',$$

so that g' and g'' commute. The group has two independent generators commuting with each other. Consequently,

$$\pi_1(T^2) = \mathbb{Z} \times \mathbb{Z}.$$

§ **3.2.31** The *disc* in $\mathbb{E}^2$, that is, the circle S^1 and the region it circumscribes (Figure 3.18, left) is perhaps the simplest of all cases. The triangulation in the figure makes it evident that

$$g_{12} = g_{23} = g_{41} = 1,$$

because of the chosen path. The two conditions then reduce to

$$g_{24} = g_{34} = 1.$$

Consequently $\pi_1 = \{1\}$.

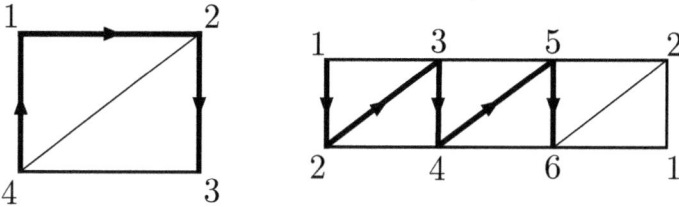

Fig. 3.18   Disk on the plane, and the Möbius band.

§ **3.2.32 Möbius band.** With the triangulation and path given in Figure 3.18, right diagram, it becomes quickly obvious that

$$g_{13} = g_{24} = g_{35} = g_{46} = 1$$

and

$$g_{62} = g_{52} = g_{61} = \text{ some independent element } g.$$

This means that we have the discrete infinite group with one generator, which is $\mathbb{Z}$: $\pi_1 = \mathbb{Z}$. Consequently, π_1 does not distinguish the Möbius band from the cylinder.

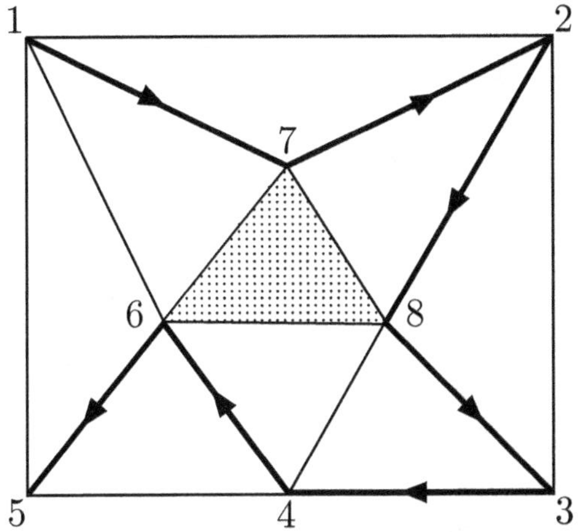

Fig. 3.19   A triangulating for the punctured disk.

§ **3.2.33 Disc with one hollow.** A triangulating complex for the once-punctured disk is given in Figure 3.19, as well as a chosen path. When two sides of a triangle are on the path, the third is necessarily in correspondence with the identity element. From this, it comes out immediately that

$$g_{12} = g_{23} = g_{45} = g_{78} = g_{48} = 1.$$

It follows also that $g_{68} = 1$. We remain with two conditions,

$$g_{51}g_{16} = 1 \quad \text{and} \quad g_{16}g_{67} = 1,$$

from which

$$g_{51} = g_{67} = g_{61} =: g.$$

One independent generator: $\pi_1 = \mathbb{Z}$. In fact, the hollowed disk is homotopically equivalent to S^1. Notice that the dark, "absent" triangle was not used as a simplex: it is just the hollow. If it were used, one more condition would be at our disposal, which would enforce $g = 1$. Of course, this would be the disk with no hollow at all, for which $\pi_1 = \{1\}$.

§ **3.2.34 The twice-punctured disk.** Figure 3.20 shows, in its left part, a triangulation for the disk with two hollows and a chosen path, the upper part being a repetition of the previous case. We find again

$$g_{51} = g_{67} = g_{61} =: g.$$

The same technique, applied to the lower part gives another, independent generator:

$$g_{4,13} = g_{3,13} = g_{3,11} =: g'.$$

The novelty here is that the group is non-commutative. Of course, we do not know *a priori* the relative behaviour of g and g'. To see what happens, let us take (right part of Figure 3.20) the three loops α, β and γ starting at point (10), and examine their representatives in the triangulation:

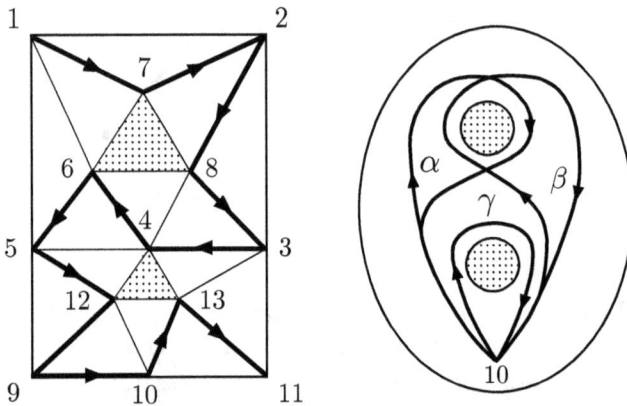

Fig. 3.20   A triangulating for the twice-punctured disk.

$$[\alpha] = g_{10,12}\, g_{12,5}g_{56}g_{67}g_{78}g_{84}g_{4,12}g_{12,10} = g_{67} = g$$

$$[\gamma] = g_{10,12}g_{12,4}g_{4}g_{13}g_{13,10} = g'$$

$$[\beta] = g_{10,13}g_{13,4}g_{46}g_{67}g_{78}g_{84}g_{4,13}g_{13,10} = (g')^{-1}gg'.$$

As α and β are not homotopic, their classes are different: $[\alpha] \neq [\beta]$. But

$$[\beta] = [\gamma^{-1}][\alpha][\gamma],$$

so that

$$[\alpha][\gamma] \neq [\gamma][\alpha], \quad \text{or} \quad gg' \neq g'g.$$

The group π_1 is non-abelian with two generators and no specific name. It is an unnamed group given by its presentation (see Section 14.2). It might be interesting to examine the Bohm–Aharonov effect corresponding to this case, with particles described by wavefunctions with two or more components to (possibly) avoid the loss of information on the group in 1-dimensional representations.

§ 3.2.35 The projective line RP^1. As the circle is homotopically equivalent to the hollowed disk, we adapt the triangulation of § 3.2.33 for S^1 (see Figure 3.21, compared with Figure 3.19) with identified antipodes:

$$\text{``1''} = \text{``}\hat{1}\text{''}, \quad \text{``2''} = \text{``}\hat{2}\text{''}, \text{ etc.}$$

Notice that this corresponds exactly to a cone with extracted vertex. The path is $1 \to 2 \to 3 \to 4 \to 5 \to 6$ and, of course, its antipode. An independent generator remains,

$$g = g_{14} = g_{26} = g_{16} = g_{36} = g_{56} = g_{15},$$

so that

$$\pi_1(RP^1) = \mathbb{Z}.$$

This is to be expected, as $RP^1 \approx S^1$.

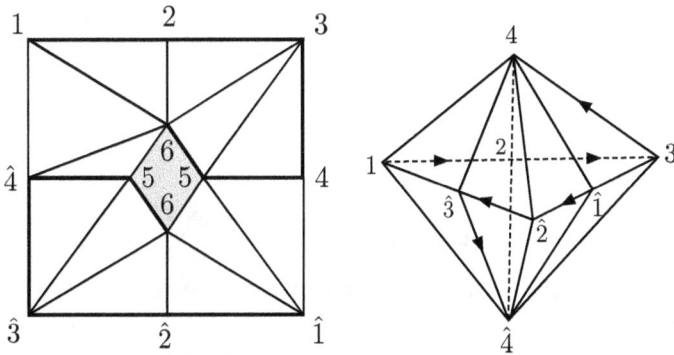

Fig. 3.21   Projective line RP^1 and projective plane RP^2.

§ 3.2.36 The projective plane RP^2. The sphere S^2 with identified antipodes may be described as in Figure 3.21 (§ 1.5.21). From the equations

$$g_{13}g_{34} = 1, \quad g_{34}g_{42} = 1, \quad g_{24}g_{41} = 1, \quad g_{31}g_{14} = 1,$$

we obtain the identifications

$$g := g_{13} = g_{43} = g_{42} = g_{41} = g_{31} = g^{-1},$$

so that $g^2 = 1$. The group has one cyclic generator of order 2. Consequently (see Section 14.3),

$$\pi_1(RP^2) = \mathbb{Z}_2.$$

This group may be represented by the multiplicative group $\{1, -1\}$. This example of a non-trivial finite group is of special interest. Compare the following three different kinds of loops at (say) vertex "1". Loop (1341) is trivial: it can be continuously contracted to a point. It corresponds to the identity element of π_1. Another loop is obtained by going from "1" to "4" and then to the antipode "$\hat{1}$" of "1" (which is identified to it). This is non-trivial, as such a loop cannot be deformed to a point and corresponds to the element "-1" of the group representation. Now, take this same loop twice:

$$\text{"1"} \rightarrow \text{"4"} \rightarrow \text{"}\hat{1}\text{"} \rightarrow \text{"}\hat{4}\text{"} \rightarrow \text{"1"}.$$

We see in the figure that such a loop can be progressively deformed into a point; this effect corresponds to the property $(-1)^2 = 1$ in the representation. The projective plane RP^2 is thus doubly-connected. This case has led to one of the oldest topological numbers in Physics, the Franck index turning up in the study of nematic crystals. See Section 28.3.3 and the figures therein.

3.3   Covering spaces

3.3.1   *Multiply-connected spaces*

§ **3.3.1** A remarkable characteristic of multiply-connected spaces is that functions defined on them are naturally multivalued. We have been using the word "function" for single-valued mappings but in this paragraph we shall be more flexible. To get some insight about this point, let us consider a simple case, like that illustrated in Figure 3.22. Suppose in some physical situation we have a "box" in which a function Ψ, obeying some simple differential equation, describes the state of a system. Boundary conditions — say, the values of Ψ on L_1 and L_2 — are prescribed by some physical reason. Under very ordinary conditions, Ψ will have a unique value at each point inside the "box" (the "configuration space"), found by solving the equation. In frequent cases, we could even replace the boundaries by other surfaces inside the "box", using the values on them as alternative boundary conditions.

Let us now deform the box so that it becomes an annular region, and then $L_1 = L_2$. Unless we had carefully chosen the boundary values at the start, Ψ will become multivalued on $L_1 = L_2$. Of course, this operation is a violence against the space: it changes radically its topology from a

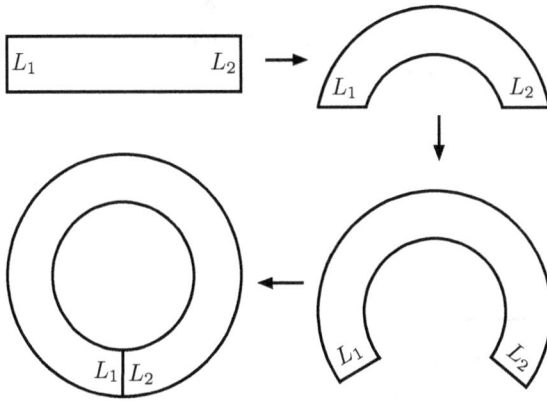

Fig. 3.22   Building up an annulus, and the raise of multiple–valuedness.

topology analogous to that of § 1.4.13 to another, akin to that of § 1.5.3. The initial box is simply-connected, the final annulus is multiply-connected. But the point we wish to stress is that, unless the boundary conditions were previously prepared "by hand" so as to match each other, Ψ will become multivalued. Had we started with the annulus as configuration space from the beginning, no boundaries as L_1 and L_2 would be present.

§ **3.3.2** If we want Ψ to be single-valued, we have to impose it by hand (say, through periodic conditions). This happens in Quantum Mechanics when the wavefunction is supposed to be single-valued: recall the cases of fine quantum behaviour exhibited by some macroscopic systems, such as vortex quantization in superfluids,[5] and flux quantization in superconductors.[6] Such systems are good examples of the interplay between physical and topological characteristics. They are dominated by strong collective effects. So stiff correlations are at work between all their parts that they may be described by a single, collective wavefunction. On the other hand, they have multiply-connected configuration spaces, and the quantizations alluded to come from the imposition, by physical reasons, of single-valuedness on the wavefunction. This leads to topological-physical effects.[7]

§ **3.3.3** Physical situations are always complicated because many supposi-

[5] See, for example, Pathria 1972.
[6] For example, Feynman, Leighton & Sands 1965, vol. III.
[7] See Dowker 1979.

tions and approximations are involved. We can more easily examine ideal cases through mathematical models. Consider, to begin with, on the plane $\mathbb{E}^2$ included in $\mathbb{E}^3$ (see Figure 3.23, left), the function defined by

$$\alpha(a) = 0, \quad \alpha(x) = \int_a^x \boldsymbol{A} \cdot dl,$$

where $\boldsymbol{A}$ is some vector. The integration is to be performed along some

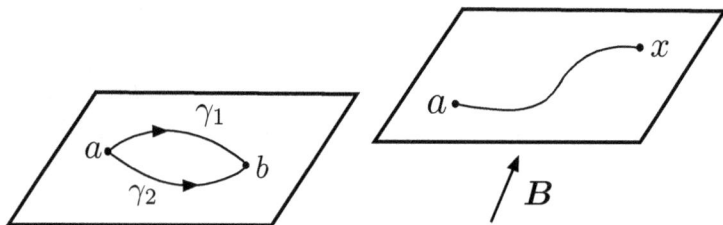

Fig. 3.23   Paths on the plane in absence (left) and presence (right) of a magnetic field.

curve, such as γ_1 or γ_2. The question of interest is: consider γ_1 and γ_2 to be curves linking points "a" and "b". Are the two integrals

$$\alpha_1(b) = \int_{\gamma_1} \boldsymbol{A} \cdot dl \quad \text{and} \quad \alpha_2(b) = \int_{\gamma_2} \boldsymbol{A} \cdot dl$$

equal to each other? It is clear that

$$\alpha_1(b) - \alpha_2(b) = \int_{\gamma_1 - \gamma_2} \boldsymbol{A} \cdot dl = \oint \boldsymbol{A} \cdot dl,$$

the last integration being around the closed loop starting at a, going through γ_1 up to b, then coming back to a through the inverse of γ_2, which is given by $(\gamma_2)^{-1} = -\gamma_2$. We see that $\alpha(b)$ will be single-valued iff

$$\oint \boldsymbol{A} \cdot dl = 0.$$

If the region S circumvented by $\gamma_1 - \gamma_2 = \gamma_1 + (\gamma_2)^{-1}$ is simply-connected, then $\gamma_1 - \gamma_2$ is just the boundary ∂S of S, and Green's theorem implies

$$\int_{\gamma_1 - \gamma_2} \boldsymbol{A} \cdot dl = \int_S \boldsymbol{rot}\, \boldsymbol{A} \cdot d\boldsymbol{\sigma}.$$

For general x, the single-valuedness condition for $\alpha(x)$ is consequently given by $\boldsymbol{rot}\, \boldsymbol{A} = 0$. In a contractible domain this means that some $\varphi(x)$ exists such that $\boldsymbol{A} = \boldsymbol{grad}\, \varphi$. In this case,

$$\alpha(x) = \int_a^x \boldsymbol{grad}\, \varphi \cdot dl = \varphi(x) - \varphi(a),$$

so that we can choose $\varphi(a) = 0$ and $\alpha(x) = \varphi(x)$. When this is the case, $\boldsymbol{A}$ is said to be *integrable* or *of potential type* (φ is its *integral*, or *potential*), a nomenclature commonly extended to $\alpha(x)$ itself.

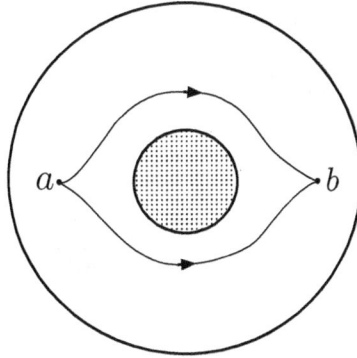

Fig. 3.24   The magnetic field through a hole in the plane.

§ **3.3.4** Now to some physics. Suppose $\psi(x)$ to be the wavefunction of an electron moving under the influence of a magnetic field $\boldsymbol{B} = \boldsymbol{rot}\,\boldsymbol{A}$, with $\boldsymbol{A}$ the vector potential. Forgetting about other effects, $(e/\hbar c)\,\alpha(x)$ is just the phase acquired by   (in the JWKB approximation of nonrelativistic Quantum Mechanics) when we go from a to x (Figure 3.23, right):

$$\psi(x) = \exp\left[i\frac{2\pi e}{hc}\int_a^x \boldsymbol{A}\cdot dl\right]\psi(a).$$

The monodromy condition,

$$\boldsymbol{B} = \boldsymbol{rot}\,\boldsymbol{A} = 0,$$

is the absence of field. Nonvanishing electromagnetic fields are, for this reason, *non-integrable phase factors*.[8]

§ **3.3.5** Consider now a multiply-connected configuration space, as the annulus in Figure 3.24. Because the central region is not part of the space, $\gamma_1 - \gamma_2$ is no more the boundary of a domain. We can still impose

$$\boldsymbol{A} = \boldsymbol{rot}\,\boldsymbol{v},$$

[8] Yang 1974.

for some $v(x)$, but then $v(x)$ is no longer single-valued.[9] We can still write

$$B = rot\,A,$$

but neither is A single-valued! In Quantum Mechanics there is no reason for the phases to be single-valued: only the states must be unique for the description to be physically acceptable. A state corresponds to a ray, that is, to a set of wavefunctions differing from each other only by phases. Starting at a and going through $\gamma_1 - \gamma_2$, the phase changes in proportion to the integral of A around the hole which, allowing multivalued animals, can be written

$$\Delta(phase) = \frac{2\pi e}{\hbar c} \oint A \cdot dl = \frac{2\pi e}{hc} \int_S rot\,A \cdot d\sigma$$
$$= \frac{2\pi e}{hc} \int_S B \cdot d\sigma = \frac{2\pi e}{hc} \Phi,$$

with Φ the flux of B through the surface S circumvented by $\gamma_1 - \gamma_2$ (which is not its boundary now !). In order to have

$$\psi(a) = \exp\left[i\frac{e}{\hbar c}\Phi\right]\psi(a)$$

single-valued, we must have

$$\frac{2\pi e}{hc}\Phi = 2\pi n,$$

that is, the flux is quantized:

$$\Phi = n\frac{hc}{e}.$$

It may happen that this condition does not hold, as in the Bohm–Aharonov effect (see § 4.2.17).[10]

§ **3.3.6** All these considerations (quite schematic, it is true) have been made only to emphasize that multi-connectedness can have very important physical consequences.

§ **3.3.7** We shall in what follows examine in some more detail the relation between the monodromy of a function and the eventual multiple-connectedness of the space on which it is defined. Summing it up, the following will happen. Let X be a multiply-connected space and Ψ a function on X. The function Ψ will be multivalued in general.

---

[9] Budak & Fomin 1973.
[10] Aharonov & Bohm 1959; 1961.

A covering space E will be an unfolding of X, another space on which Ψ becomes single-valued. Different functions will require different covering spaces to become single-valued, but X has a certain special covering, the universal covering $U(X)$, which is simply-connected and on which all functions become single-valued. This space is such that

$$X = U(X)/\pi_1(X).$$

The universal covering $U(X)$ may be roughly seen as that unfolding of X with one copy of X for each element of the fundamental group $\pi_1(X)$.

§ **3.3.8** Recall the considerations on the configuration space of a system of n identical particles (§ 1.5.8; see also § 3.3.28 and Section 14.2.5), which is $\mathbb{E}^{3n}/S_n$. In that case, $\mathbb{E}^{3n}$ is the universal covering and the fundamental group is isomorphic to the symmetric group, $\pi_1 \approx S_n$.

Consider, to fix the ideas, the case $n = 2$. Call x_1 and x_2 the positions of the first and the second particles. The covering space $\mathbb{E}^6$ is the set $\{(x_1, x_2)\}$. The physical configuration space X would be the same, but with the points (x_1, x_2) and (x_2, x_1) identified. Point (x_2, x_1) is obtained from (x_1, x_2) by the action of a permutation P_{12}, an element of the symmetric (or permutation) group S_2:

$$(x_2, x_1) = P_{12}(x_1, x_2).$$

A complex function $\Psi(x_1, x_2)$ (say, the wavefunction of the 2-particle system) will be single-valued on the covering space, but 2-valued on the configuration space. To make a drawing possible, consider instead of $\mathbb{E}^3$, the two particles on the plane $\mathbb{E}^2$. The scheme in Figure 3.25 shows how $\Psi(x_1, x_2)$ is single-valued on E, where $(x_1, x_2) \neq (x_2, x_1)$, and double-valued on X, where the two values correspond to the same point $(x_1, x_2) \equiv (x_2, x_1)$. There are two sheets because P_{12} applied twice is the identity.

Wavefunctions commonly used are taken on the covering space, where they are single-valued. The function $\Psi[P_{12}(x_1, x_2)]$ is obtained from $\Psi(x_1, x_2)$ by the action of an operator $U(P_{12})$ representing P_{12} on the Hilbert space of wavefunctions:

$$\Psi(x_2, x_1) = \Psi[P_{12}(x_1, x_2)] = U(P_{12})\Psi(x_1, x_2).$$

This is a general fact: the different values of a multivalued function are obtained by the action of a representation of a group, a distinct subgroup of $\pi_1(X)$ for each function. Above, the group S_2 is isomorphic to the cyclic group $\mathbb{Z}_2$ (notice the analogy of this situation with the covering related to the function $\sqrt{z}$ in § 3.3.13 below). The whole fundamental group will give all the values for any function. There are as many covering spaces of X as there are subgroups of $\pi_1(X)$.

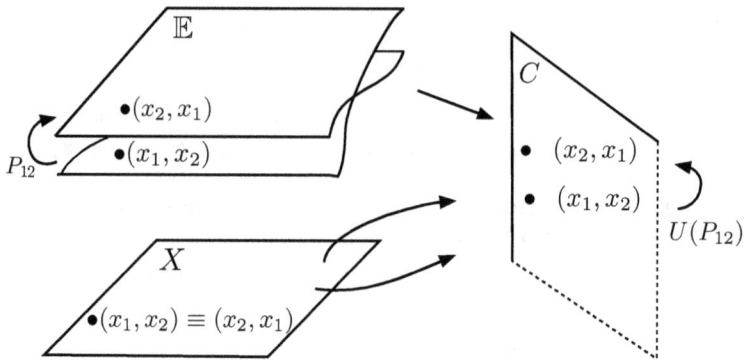

Fig. 3.25   Two-particle wavefunction: double–valued on X, single–valued on E.

§ **3.3.9 Percolation.** Phase transitions are more frequently signalled by singularities in physical quantities, as the specific heat. However, sometimes they show themselves as clear alterations in the topology of configuration space,[11] as in all the phenomena coming under the headname of percolation.[12] In its simplest form, it concerns the formation of longer and longer chains of (say, conducting) "guest" material in a different (say, isolating) "host" medium by the progressive addition of the former. The critical point (say, passage to conductivity) is attained when a first line of the "guest" material traverses completely the "host", by that means changing its original simply-connected character. As more and more complete lines are formed, the fundamental group becomes more and more complicated.

§ **3.3.10 Covering for braid statistics.** Instead of the above covering, braid statistics (see Section 14.2.5) requires, already for the 2-particle configuration space, a covering with infinite leaves (Figure 3.26).

§ **3.3.11 Poincaré conjecture.** Homotopy is a basic instrument in the "taxonomic" program of classifying topological spaces, that is, finding all

---

[11] Broadbent & Hammersley 1957; Essam 1972. Old references, but containing the qualitative aspects here referred to.
[12] An intuitive introduction is given in Efros 1986.

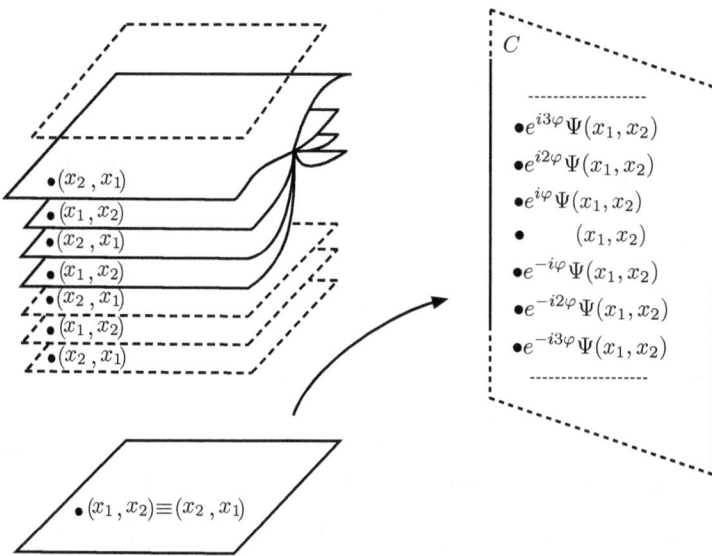

Fig. 3.26   The infinite unfolding of the 2-particle configuration space for braid statistics.

classes of homeomorphic spaces in a given dimension. This project has only been successful for 2-dimensional spaces. The difficulties are enormous, of course, but progress has been made in the 3-dimensional case. The main technique used is "surgery", through which pieces of a given space are cut and glued in a controlled way to get other spaces. The long debate concerning the Poincaré conjecture gives a good idea of how intricate these things are. We have seen that the sphere S^2 is simply-connected. Actually, it is the only simply-connected closed surface in $\mathbb{E}^3$. The conjecture was that the same holds in higher dimensions: S^n would be the only simply-connected closed surface in $\mathbb{E}^{n+1}$. It has been (progressively) proved for $n \geq 4$. The proof for Poincaré's original case, $n = 3$, has remained an open question,[13] and only recently seems to have been found.[14] It actually uses the differentiable structure. It has been found that 3-dimensional manifolds have a single differential structure, that is, they "become" a unique differentiable manifold once the additional differentiable structure is added. Thus, topological equivalence (homeomorphism) is the same as equivalence between differentiable manifolds (diffeomorphism). The conjecture has been

---

[13] For a popular exposition see Rourke & Stewart 1986.
[14] Perelman 2002, 2003.

rephrased as: *any closed 3-dimensional manifold is diffeomorphic to S^3*.[15]

3.3.2 Riemann surfaces

§ 3.3.12 The Riemann surface of a multivalued analytic function[16] is a covering space of its analyticity domain, a space on which the function becomes single-valued. It is the most usual example of a covering space. The considerations of the previous section hint at the interest of covering spaces to Quantum Mechanics. As said above, if X is the configuration space and $\pi_1(X, x_0)$ is non-trivial for some x_0 in X, there is no *a priori* reason for the wavefunction to be single-valued: this must be imposed by hand, as a physical principle. Some at least of the properties of their phases are measurable. What follows is a simplified description of the subject with the purpose of a "fascicule des résultats". For details, the reader is sent to the copious mathematical and physical literature.[17]

§ 3.3.13 Let us begin with the standard example (Figure 3.27). Consider on the complex plane $\mathbb{C}$ the function

$$f : \mathbb{C} \to \mathbb{C}, \quad f(z) = \sqrt{z}.$$

It is not analytic at $z = 0$, where its derivatives explode. If we insist on analyticity, the point $z = 0$ must be extracted from the domain of definition, which becomes (§ 3.1.9)

$$\mathbb{C} - \{0\} \approx S^1.$$

The function f is continuous, as its inverse takes two open sets (and so, their union) into an open set in $\mathbb{C} - \{0\}$. If we examine how a loop circumventing the zero is taken by $f(z)$, we discover that $\sqrt{z}$ simply takes one into another two values which are taken back to a same value by the inverse. Only by going twice around the loop in $\mathbb{C} - \{0\}$ can we obtain a closed curve in the image space. On the other hand, a loop not circumventing the zero is taken into two loops in the image space. The trouble, of course, comes from the two-valued character of $\sqrt{z}$ and the solution is well known: the function becomes monodromous if, instead of $\mathbb{C} - \{0\}$, we take as definition domain a space formed by two Riemann sheets with a half-infinite line (say, $\mathbb{E}_+^1$) in common. This new surface is a covering of $\mathbb{C} \backslash \{0\}$. The function $\sqrt{z}$ is $+|\sqrt{z}|$ on one sheet and $-|\sqrt{z}|$ on the other. They are related by a representation

---

[15] A recent review is Morgan 2004.
[16] See, for instance, Forsyth 1965.
[17] See, for example, Dowker 1979 and Morette–DeWitt 1969; 1972.

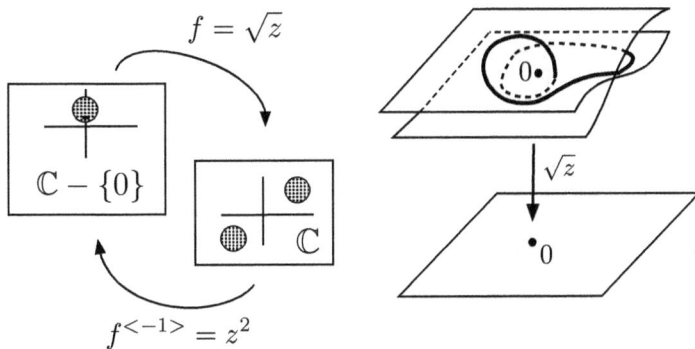

Fig. 3.27   Covering for the square–root function.

of the cyclic group $\mathbb{Z}_2$ given by $(+1, -1)$ and the multiplication. The group $\mathbb{Z}_2$ is a subgroup of

$$\pi_1(\mathbb{C} - \{0\}) = \mathbb{Z}.$$

The analogy with the statistical case of § 3.3.8 comes from the presence of the same group. The function $\sqrt[n]{z}$ would require a covering formed by n Riemann sheets.

§ **3.3.14** The idea behind the concept of covering space of a given multiply-connected space X is to find another space E on which the function is single-valued and a projection $p : E \to X$ bringing it back to the space X (Figure 3.28, left). Different functions require different covering spaces. A covering on which all continuous functions become single-valued is a *universal* covering space. Such a universal covering always exists and this concept provides furthermore a working tool to calculate fundamental groups.

§ **3.3.15** Let us be a bit more formal (that is, precise): consider two topological spaces X and E, and let

$$p : E \to X$$

be a continuous surjective mapping. Suppose that each point $x \in X$ has a neighbourhood U whose inverse image $p^{-1}(U)$ is the disjoint union of open sets V_a in E, with the property that each V_a is mapped homeomorphically onto U by p (see Figure 3.28, right). Then the set $\{V_a\}$ is a *partition* of $p^{-1}(U)$ into sheets. Map p is the *covering map*, or *projection*, and E is a

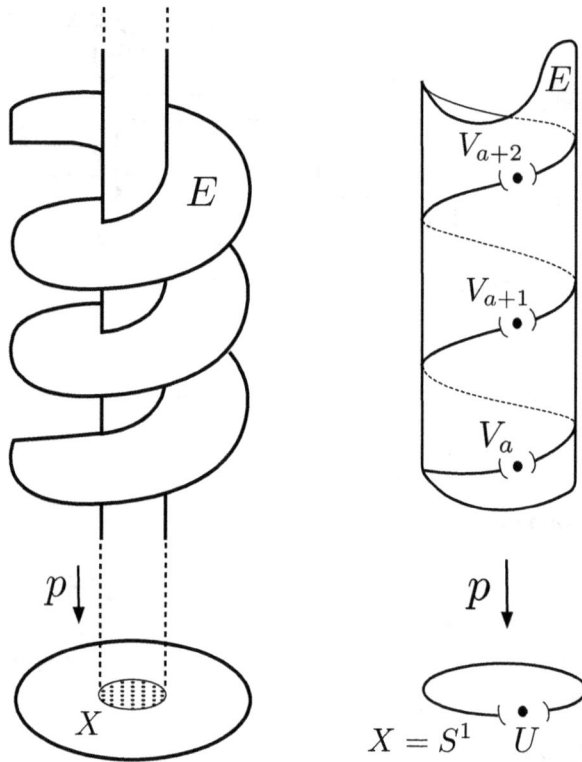

Fig. 3.28   A space X, its covering E and the partition related to a subset U.

covering space of X. Strictly speaking, the covering is given by the pair (E, p).

As a consequence of the above definition, we have:

(i) For each $x \in X$, the subset $p^{-1}(x)$ of E (called *fiber* over x) has a discrete topology.

(ii) p is a local homeomorphism.

(iii) X has a quotient topology obtained from E.

§ **3.3.16** If E is simply-connected and $p : E \to X$ is a covering map, then E is said to be the *universal covering space* of X. From this definition, the fundamental group of a universal covering space is

$$\pi_1(E) = \{1\}.$$

Up to homotopic equivalence, this covering with a simply-connected space is unique. The covering of § 3.3.13 is not, of course, the universal covering of $\mathbb{C}\backslash\{0\}$. As $\mathbb{C}\backslash\{0\}$ is homotopically equivalent to S^1, it is simpler to examine S^1.

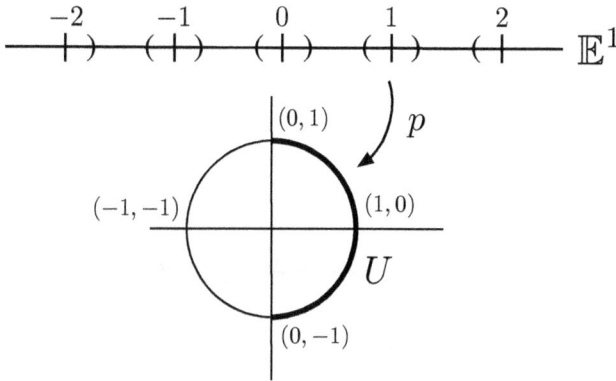

Fig. 3.29 $\mathbb{E}^1$ as the universal covering of S^1.

§ **3.3.17** The mapping $\mathbb{E}^1 \to S^1$ given by

$$p(x) = (\cos 2\pi x, \sin 2\pi x) = \exp[i2\pi x]$$

is a covering map. Take the point $(1,0) \in S^1$ and its neighbourhood U formed by those points in the right-half plane. Then,

$$p^{-1}(U) = \bigcup_{n=-\infty}^{\infty} \left(n - \tfrac{1}{4}, n + \tfrac{1}{4}\right).$$

The open intervals $V_n = (n - 1/4, n + 1/4)$ are (see Figure 3.29) homeomorphically mapped onto U by p. As $\mathbb{E}^1$ is simply-connected, it is by definition the universal covering space of S^1. Other covering spaces are, for instance, S^1 itself given as

$$S^1 = \{z \in \mathbb{C} \text{ such that } |z| = 1\}$$

with the mappings

$$p_n : S^1 \to S^1, \quad p_n(z) = z^n, \ n \in \mathbb{Z}_+ .$$

§ 3.3.18 Consider the torus $T^2 = S^1 \times S^1$. It can be shown that the product of two covering maps is a covering map. Then, the product
$$p \times p : \mathbb{E}^1 \times \mathbb{E}^1 \to S^1 \times S^1, \quad (p \times p)(x, y) = \big(p(x), p(y)\big),$$
with p the mapping of § 3.3.17, is a covering map and $\mathbb{E}^2$ (which is simply-connected) is the universal covering of T^2.

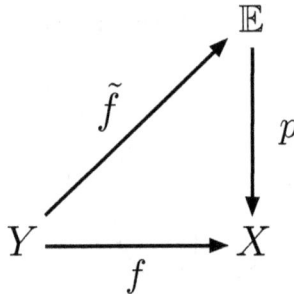

Fig. 3.30   Diagram for the lift of a function.

§ 3.3.19 There are several techniques to calculate the fundamental groups of topological spaces, all of them rather elaborate. One has been given in Section 3.2.3. We shall here describe another, which exploits the universal covering space. As our aim is only to show the ideas, we shall concentrate in obtaining the fundamental group of the circle, that is, $\pi_1(S^1)$.

§ 3.3.20 Let $p : E \to X$ be a covering map. If f is a continuous function of Y to X, the mapping $\tilde{f} : Y \to E$ such that
$$p \circ \tilde{f} = f$$
is the *lift*, or *covering*, of f. Pictorially, we say that the diagram of Fig. 3.30 is commutative. This is a very important definition. We shall be interested in lifts of two kinds of mappings: paths and homotopies between paths. In the following, some necessary results will simply be stated and, when possible, illustrated. They will be useful in our search for the fundamental group of S^1.

§ 3.3.21 Let (E, p) be a covering of X and $f : I \to X$ a path. The lift $\tilde{f}$ is the *path-covering* of f. If $F : I \times I \to X$ is a homotopy, then the homotopy
$$\tilde{F} : I \times I \to \mathbb{E}^1,$$
such that $p \circ \tilde{F} = F$, is the *covering homotopy* of F.

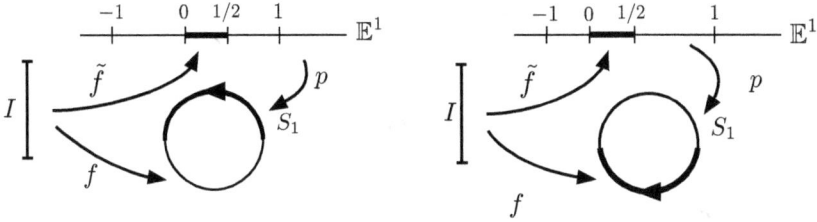

Fig. 3.31   Lifts for two paths on S^1.

§ **3.3.22** Take again the covering map of § 3.3.17, that is,

$$p(x) = (\cos 2\pi x, \sin 2\pi x) = \exp[i2\pi x].$$

Then, we have the following results:
(i) The path $f : I \to S^1$ given by

$$f(t) = (\cos \pi t, \sin \pi t),$$

with initial endpoint $(1,0)$, has the lift $\tilde{f}(t) = t/2$, with initial endpoint 0 and final endpoint $1/2$ (Fig. 3.31, left).
(ii) The path $f : I \to S^1$, given by

$$f(t) = (\cos \pi t, -\sin \pi t),$$

has the lift $\tilde{f}(t) = -t/2$ (Fig. 3.31, right).
(iii) The path given by

$$h(t) = (\cos 4\pi t, \sin 4\pi t)$$

traverses twice the circle S^1; it has the lift $\tilde{h}(t) = 2t$ (Fig. 3.32).

We enunciate now some theorems concerning the uniqueness of path- and homotopy-coverings.

Theorem 1. Let (E, p) be the universal covering of X, and $f : I \to X$ be a path with initial endpoint x_0. If $e_0 \in E$ is such that $p(e_0) = x_0$, then there is a unique covering path of f beginning at e_0 (Fig. 3.33).

Theorem 2. Let (E, p) be the universal covering of X, and $F : I \times I \to X$ be a homotopy with $F(0,0) = x_0$. If $e_0 \in E$ is such that $p(e_0) = x_0$, then there is a unique homotopy-covering

$$\tilde{F} : I \times I \to E$$

such that $\tilde{F}(0,0) = e_0$ (Fig. 3.34).

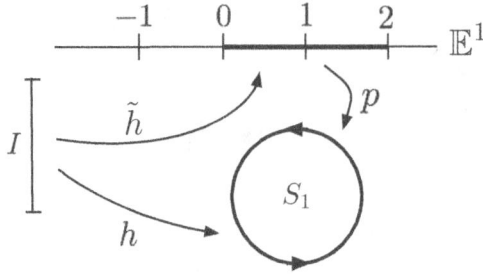

Fig. 3.32   Lift of the path $h(t) = (\cos 4\pi t, \sin 4\pi t)$ on the circle.

Theorem 3. Finally, the *monodromy theorem*, establishing the relation between covering spaces and the fundamental group. Let (E, p) be the universal covering of X, and let f and g be two paths on X from x_0 to x_1. Suppose $\tilde{f}$ and $\tilde{g}$ are their respective lifts starting at e_0. If f and g are homotopic, then $\tilde{f}$ and $\tilde{g}$ have also the same final endpoint,

$$\tilde{f}(1) = \tilde{g}(1) \ ,$$

and are themselves homotopic.

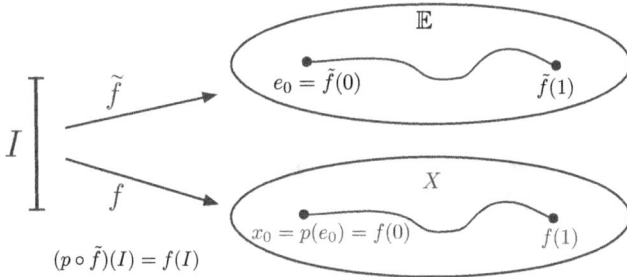

Fig. 3.33   Uniqueness of a lift starting at a given point on the covering.

§ 3.3.23 We shall now use these results to show that

$$\pi_1(S^1) = \mathbb{Z},$$

the additive integer group. To do it, we shall exhibit a group isomorphism between $\mathbb{Z}$ and $\pi_1(S^1, s_0)$, with the point $s_0 = (1, 0) \in S^1$ included in $\mathbb{C}$.

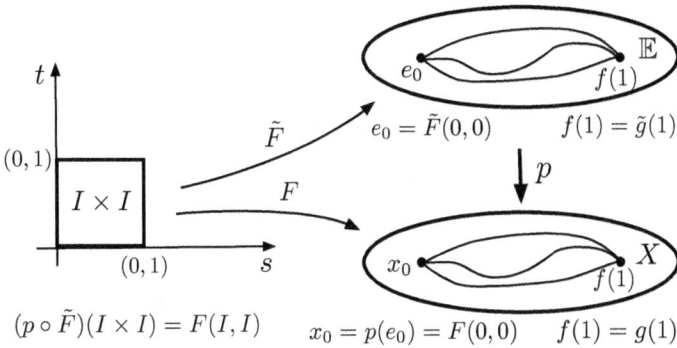

Fig. 3.34   Uniqueness of a homotopy lift starting at a given point on the covering.

Let $p : \mathbb{E}^1 \to S^1$ be defined by

$$p(t) = (\cos 2\pi t, \sin 2\pi t).$$

If f is a loop on S^1 with base point s_0, its lift $\tilde{f}$ is a path on $\mathbb{E}^1$ beginning at 0. The point $\tilde{f}(1)$ belongs to the set $p^{-1}(s_0)$, that is, $\tilde{f}(1) = $ some $n \in \mathbb{Z}$. The monodromy theorem tells us that the integer n depends only on the homotopy class of f. We may then define a mapping

$$\varphi : \pi_1(S^1, s_0) \to \mathbb{Z}, \quad \varphi[f] = \tilde{f}(1) = n \in \mathbb{Z}.$$

It remains to show that φ is a group isomorphism. To do that, we should show that it is onto, one-to-one and preserves the group structure.[18] We proceed through the following steps.

(i) φ is onto: let $n \in p^{-1}(s_0)$. Being $\mathbb{E}^1$ path-connected, we can choose the path $\tilde{f} : I \to \mathbb{E}^1$ from 0 to n. Then $f = p \circ \tilde{f}$ is a loop on S^1 with base point s_0, $\tilde{f}$ is its lift, and by definition

$$\varphi[f] = \tilde{f}(1) = n.$$

(ii) φ is injective: suppose

$$\varphi([f]) = \varphi([g]) = n.$$

Let us show that $[f] = [g]$. Take the respective lifts $\tilde{f}$ and $\tilde{g}$ from 0 to n. They are homotopic to each other, because $\mathbb{E}^1$ is simply-connected. If $\tilde{F}$ is the homotopy between $\tilde{f}$ and $\tilde{g}$, the mapping $F = p \circ \tilde{F}$ will be the

---

[18] Fraleigh 1974.

homotopy between f and g.

(iii) φ is a homomorphism: let f and g be two loops on S^1 at the point s_0, f and g their lifts on $\mathbb{E}^1$ at the point 0. Define a path on $\mathbb{E}^1$ by

$$\tilde{h}(t) = \begin{cases} \tilde{f}(2t) & \text{for } t \in [0, 1/2], \\ n + \tilde{g}(2t - 1) & \text{for } t \in [1/2, 1], \end{cases}$$

and suppose that $\tilde{f}(1) = n$ and $\tilde{g}(1) = m$. By construction, $\tilde{h}$ begins at 0. It is easy to see that $\tilde{h}$ is the lift of $f{\circ}g$, as the functions sine and cosine have periods $2\pi [p(n + t) = p(t)]$. Consequently $p \circ \tilde{h} = f{\circ}g$. On the other hand,

$$\varphi([f \circ g]) = \tilde{h}(1) = n + m = \varphi([f]) + \varphi([g]).$$

In simple words, a loop on S^1 will always be deformable into some loop of the form

$$f_n(t) = \exp[\tilde{f}_n(t)],$$

with $\tilde{f}_n(t) = nt$ and such that

$$\varphi[f_n] = \tilde{f}_n(1) = n.$$

Notice that, in order to show that φ is a homomorphism, we have used another structure present in $\mathbb{E}^1$, that of an additive group. Covering spaces in general do not present such a structure. Even so, some information on the fundamental group can be obtained through the following theorem:

§ 3.3.24 Theorem. *Let $p : (E, s_0) \to (X, x_0)$ be a covering map. If E is path-connected, there exists always a surjective mapping*

$$\varphi : \pi_1(X, x_0) \to p^{<-1>}(x_0).$$

If E is simply-connected, φ is bijective.

§ 3.3.25 Intuitively, $\pi_1(X)$ "counts" the number of sheets necessary to make of X a simply-connected space, or to obtain its universal covering. It is usual to write, for this reason,

$$X = E/\pi_1(X).$$

What about the other covering spaces, those which are not universal? We shall only state the general answer:

There is a covering space C for each subgroup K of π_1, obtained by factorizing the universal covering by that subgroup: $C = E/K$.

Another general fact is the following:

If a space C is locally homeomorphic to another space X,
then C is a covering space of X.

So, π_1 and its subgroups characterize all the covering spaces. Notice that the "counting" it provides comes between quotation marks because π_1 is not necessarily $\mathbb{Z}$, nor even an abelian group.

§ 3.3.26 The projective plane RP^2 (§ 1.5.21 and § 3.2.36) is obtained from S^2 by identifying each point $x \in S^2$ to its antipode $\hat{x} = -x$. This identification being an equivalence relation on S^2, RP^2 is a set of equivalence classes. A projection

$$p : S^2 \to RP^2$$

exists, given by $p(x) = [x] = |x|$. The topology is the quotient topology: U in RP^2 is open if $p^{-1}(U)$ is open in S^2. It is possible to show that p is a covering map. As S^2 is simply-connected, that is,

$$\pi_1(S^2) = \{1\},$$

the theorem of § 3.3.24 tells us that there is a bijection between $\pi_1(RP^2, r_0)$ and $p^{-1}(r_0)$. But then, $\pi_1(RP^2, r_0)$ is a group of rank 2, because $p^{-1}(r_0)$ is a set with 2 elements. Any group of rank 2 is isomorphic to the cyclic group $\mathbb{Z}_2$, so that $\pi_1(RP^2) \approx \mathbb{Z}_2$. We write $RP^2 = S^2/\mathbb{Z}_2$. It is a general result that, for any $n \geq 2$,

$$\pi_1(RP^n) \approx \mathbb{Z}_2 \quad \text{or} \quad RP^n = S^n/\mathbb{Z}_2.$$

As for the complex projective spaces CP^n, they are all simply-connected:

$$\pi_1(CP^n) = \{1\}, \text{ for any n.}$$

An alternative way to obtain π_1 has already been used (§ 3.2.23 and § 3.2.24) for cartesian products. Another example is the cylinder, which we consider next.

§ 3.3.27 As the cylinder is $S^1 \times I$, then

$$\pi_1(\text{cylinder}) \approx \pi_1(S^1) \times \pi_1(I) \approx \mathbb{Z} \times \{1\} \approx \mathbb{Z}.$$

Notice that, when calculating π_1 for cartesian products, one simply drops homotopically trivial spaces.

§ 3.3.28 Consider again (§ 3.3.8) the configuration space of a system of n identical particles, which is $\mathbb{E}^{3n}/S_n$. Recall that $\mathbb{E}^{3n}$ is the universal covering and $\pi_1 \approx S_n$ is a nice example of non-abelian fundamental group (when $n \geq 3$). When, in addition, the particles are impenetrable, the fundamental groups would be the braid groups B_n of Section 14.2.2. Such groups reduce to the symmetric group S_n in $\mathbb{E}^3$, but not on $\mathbb{E}^2$. As a consequence, quantum (and statistical) mechanics of identical impenetrable particles on $\mathbb{E}^2$ will be governed by braid groups. Instead of the usual permutation statistics given by the symmetric groups, which leads to the usual bosons and fermions, a braid statistics will be at work.[19] By the way, knots in $\mathbb{E}^3$ are characterized by the fundamental group of their complement in the host space (Section 14.3.3).

§ 3.3.29 Suppose a physical system with configuration space $\mathbb{E}^3$, described in usual spherical coordinates by the wavefunction $\Psi(r, \theta, \varphi)$. We shall consider only rotations around the $0z$ axis, and write simply $\Psi(\varphi)$. When we submit the system to a rotation of angle α around $0z$, the transformation will be represented by an operator $U(\alpha)$ acting on Ψ,

$$\Psi(\varphi + \alpha) = U(\alpha)\Psi(\varphi).$$

When $\alpha = 2\pi$, we would expect

$$\Psi(\varphi + 2\pi) = \Psi(\varphi),$$

that is, $U(2\pi) = 1$. The operator $U(\alpha)$ represents an element of $SO(3)$, the rotation group in $\mathbb{E}^3$ (a topological group). Roughly speaking, there is a mapping

$$U : [0, 2\pi) \to SO(3), \quad U : \alpha \to U(\alpha)$$

defining a curve on the $SO(3)$ space. Supposing $U(0) = U(2\pi)$ is the same as requiring this curve to be a closed loop. Now, it happens that $SO(3)$ is a doubly-connected topological space, so that $U(\alpha)$ is not necessarily single-valued. Actually,

$$SO(3) \approx RP^3 \approx S^3/\mathbb{Z}_2 .$$

By the way, this manifold is the configuration space for the spherical top, whose quantization[20] is consequently rather involved. In fact, this is why there are two kinds of wavefunctions: those for which

$$U(2\pi) = U(0) = 1,$$

---

[19] More details can be found in Aldrovandi 1992.
[20] Schulman 1968; Morette–DeWitt 1969, 1972; Morette–DeWitt, Masheshvari & Nelson 1979.

and those for which

$$U(2\pi) = -1.$$

The first type describes systems with integer angular momentum. Wavefunctions of the second type, called *spinors*, describe systems with half-integer angular momentum. The universal covering of $SO(3)$ is the group $SU(2)$ of the unitary complex matrices of determinant $+\,1$, whose manifold is the 3-sphere S^3. Consequently,

$$SO(3) = SU(2)/\mathbb{Z}_2.$$

The group $SU(2)$ stands with respect to $SO(3)$ in a way analogous as the square-root covering of § 3.3.13 stands to $\mathbb{C} - \{0\}$: in order to close a loop in $SU(2)$, we need to turn twice on $SO(3)$, so that only when $\alpha = 4\pi$ is the identity recovered. As $SU(2) = S^3$ is simply-connected, it is the universal covering of $SO(3)$.

§ **3.3.30** The simplest case of Quantum Mechanics on a multiply-connected space comes out in the well-known Young double-slit interference experiment. We shall postpone its discussion to § 4.2.16 and only state the result:

> *On a multiply-connected space, the wavefunction behaves as the superposition of its values on all the leaves of the covering space.*

3.4 Higher homotopy

Besides lassoing them with loops, which are 1-dimensional objects, we can try to capture holes in space with higher-dimensional closed objects. New groups then emerge, revealing new space properties.

§ **3.4.1** The attentive (and suspicious) reader will have frowned upon the notation $\pi_0(X)$ used for the set of path-connected components of space X in § 3.2.4. There, and in § 3.2.13, when the fundamental group π_1 was also introduced as the "first homotopy group", a whole series of groups π_n was announced. Indeed, both π_0 and π_1 are members of a family of groups involving classes of loops of dimensions $0, 1, 2, \ldots$. We shall now say a few words on the higher groups $\pi_n(X, x_0)$, for $n = 2, 3, 4$, etc.

§ **3.4.2** With loops, we try to detect space defects by lassoing them, by throwing loops around them. We have seen how it works in the case of points extracted from the plane, but the method is clearly inefficient to apprehend, say, a missing point in $\mathbb{E}^3$. In order to grasp it, we should tend something like a "net", a 2-dimensional "loop". Higher dimensional spaces and defects[21] require, in an analogous way, higher dimensional "loops". As happens for 1-loops, classes of n-loops constitute groups, just the π_n.

§ **3.4.3** The fundamental group was due to H. Poincaré. The groups π_n for general $n \in \mathbb{Z}$ have been introduced in the nineteen thirties by E. Cech and W. Hurewicz. The latter gave the most satisfactory definition and worked out the fundamental properties. His approach was restricted to metric spaces but was extended to general topological spaces by R. H. Fox in the nineteen forties. We shall here try to introduce the higher groups as natural extensions of the fundamental group.

§ **3.4.4** The group $\pi_1(X, x_0)$ is the set of homotopy classes of (1-dimensional) loops on X with base point x_0. With this in mind, our initial problem is to define 2-dimensional "loops". A 1-dimensional loop is a continuous mapping

$$f : I \to X, \quad \text{with } f(\partial I) = x_0,$$

where ∂I is a rough notation for the boundary of I, that is, the set of numbers $\{0, 1\}$. We can then define a 2-dimensional "loop" as a continuous mapping from $I^2 = I \times I$ into X, given by

$$f : I \times I \to X \quad \text{such that } f(\partial I^2) = x_0,$$

where ∂I^2 is the boundary of I^2.

§ **3.4.5** To extend all that to higher dimensions, we need beforehand an extension of the closed interval: denoted by I^n, it is an n-dimensional solid cube:

$$I^n = \{x = (x^1, x^2, \ldots, x^n) \in \mathbb{E}^n \quad \text{such that } 0 \leq x^i \leq 1, \forall i\}.$$

The set ∂I^n, the "boundary of I^n", is the cube surface, which can be defined in a compact way by

$$\partial I^n = \{x \in I^n \quad \text{such that } \prod_{i=1}^{n} x^i (1 - x^i) = 0\}.$$

§ **3.4.6** Let (X, x_0) be a topological space with a chosen point x_0. Denote by $\Omega_n(X, x_0)$ the set of continuous functions

$$f : I^n \to X \quad \text{such that } f(\partial I^n) = x_0.$$

---

[21] Applications of homotopy to defects in a medium are examined in Nash & Sen 1983.

Given two of such functions f and g, they are *homotopic* to each other if there exists a continuous mapping $F : \boldsymbol{I}^n \times \boldsymbol{I} \to X$ satisfying

$$F(x^1, x^2, \ldots, x^n; 0) = f(x^1, x^2, \ldots, x^n),$$
$$F(x^1, x^2, \ldots, x^n; 1) = g(x^1, x^2, \ldots, x^n),$$

where $(x^1, x^2, \ldots, x^n) \in \boldsymbol{I}^n$, and

$$F(x^1, x^2, \ldots, x^n, s) = x_0 \quad \text{when} \quad (x^1, x^2, \ldots, x^n) \in \partial \boldsymbol{I}^n \quad \text{for all} \quad s \in \boldsymbol{I}.$$

The function F is the homotopy between f and g. In shorthand notation,

$$F(\boldsymbol{I}^n, 0) = f \qquad F(\boldsymbol{I}^n, 1) = g \qquad F(\partial \boldsymbol{I}^n, \boldsymbol{I}) = x_0.$$

That this homotopy is an equivalence relation can be shown in a way analogous to the 1-dimensional case. The above set $\Omega_n(X, x_0)$ is consequently decomposed into disjoint subsets, the homotopy classes. The class to which f belongs will be once again indicated by $[f]$.

§ 3.4.7 Notice that, in the process of closing the curve to obtain a loop, the interval $\boldsymbol{I}$ itself becomes equivalent to a loop — from the homotopic point of view, we could take S^1 instead of $\boldsymbol{I}$ with identified endpoints. Actually, each loop could have been defined as a continuous mapping $f : S^1 \to X$. In the same way, we might consider the n-loops as mappings

$$f : S^n \to X.$$

This alternative definition will be formalized towards the end of this section.

§ 3.4.8 Let us introduce a certain algebraic structure by defining an operation "•" analogous to that of § 3.2.6. Given f and $g \in \Omega_n(X, x_0)$,

$$h(t_1, t_2, \ldots, t_n) = (f \bullet g)(t_1, t_2, \ldots, t_n)$$
$$= \begin{cases} f(2t_1, t_2, \ldots, t_n) & \text{for } t_1 \in [0, 1/2], \\ g(2t_1 - 1, t_2, \ldots, t_n) & \text{for } t_1 \in [1/2, 1]. \end{cases}$$

Operation • induces an operation "∘" on the set of homotopy classes of Ω_n:

$$[f] \circ [g] = [f \bullet g].$$

With the operation ∘, the set of homotopy classes of $\Omega_n(X, x_0)$ constitutes a group, the *n-th homotopy group* of space X with base point x_0, denoted by $\pi_n(X, x_0)$.

§ 3.4.9 Many of the results given for the fundamental group remain valid for $\pi_n(X, x_0)$ with $n \geq 2$. We shall now only list some general results of the theory of homotopy groups:

(i) If X is path-connected and $x_0, x_1 \in X$, then

$$\pi_n(X, x_0) \approx \pi_n(X, x_1) \quad \text{for all} \ n \geq 1.$$

(ii) If X is contractible, then $\pi_n(X) = 0 \ \ \forall \, n \in \mathbb{Z}_+$.

(iii) If (X, x_0) and (Y, y_0) are homotopically equivalent, then

$$\pi_n(X, x_0) \approx \pi_n(Y, y_0) \quad \text{for any} \ n \in \mathbb{Z}_+.$$

(iv) Take X and Y topological spaces and x_0, y_0 the respective chosen points; then for their topological product,

$$\pi_n(X \times Y, (x_0, y_0)) \approx \pi_n(X, x_0) \otimes \pi_n(Y, y_0).$$

From the property (iv), we can see that, for euclidean spaces,

$$\pi_n(\mathbb{E}^m) \approx \{0\}$$

for any n and m.

§ 3.4.10 The "functorial" properties discussed in § 3.2.21 keep their validity for $n \geq 2$. Let $\varphi \colon X \to Y$ be a continuous mapping with $\varphi(x_0) = y_0$. If $[f] \in \pi_n(X, x_0)$ for some n, then

$$\varphi \circ f : I^n \to Y$$

is a continuous mapping with base point y_0, that is,

$$(\varphi \circ f)(\partial I^n) = y_0.$$

Thus, $\varphi \circ f$ is an element of the class $[\varphi \circ f] \in \pi_n(Y, y_0)$. Consequently, φ induces a map

$$\varphi_* : \pi_n(X, x_0) \to \pi_n(Y, y_0), \quad \varphi_*([f]) = [\varphi \circ f]$$

for every $[f] \in \pi_n(X, x_0)$. This mapping, which can be shown to be well defined, is the "induced homomorphism" relative to the base point x_0. It has the following "functorial" properties:

(i) If $\varphi : (X, x_0) \to (Y, y_0)$ and $\psi : (Y, y_0) \to (Z, z_0)$, then $(\psi \circ \varphi)_* = \psi_* \circ \varphi_*$. Given the identity mapping $i : (X, x_0) \to (X, x_0)$, then i_* is the identity homomorphism.

(ii) If $\varphi \colon (X, x_0) \to (Y, y_0)$ is a homeomorphism, then φ_* is an isomorphism between $\pi_n(X, x_0)$ and $\pi_n(Y, y_0)$.

§ **3.4.11** In two important aspects the higher homotopy groups differ from the fundamental group:

(i) First, let (E, p) be the universal covering of X, and let $e_0 \in E$ such that $p(e_0) = x_0 \in X$. Then, the induced homomorphism

$$p_* : \pi_n(E, e_0) \to \pi_n(X, x_0)$$

is a group isomorphism for $n \geq 2$. This means that the universal covering, which for the fundamental group is trivial, keeps nevertheless all the higher homotopy groups of the space.

(ii) Second, for X any topological space, all the $\pi_n(X, x_0)$ for $n \geq 2$ are abelian (which again is not the case for the fundamental group).

§ **3.4.12** For any $n \in \mathbb{Z}_+$, the n-th homotopy group of the sphere S^n is isomorphic to $\mathbb{Z}$, that is, $\pi_n(S^n) \approx \mathbb{Z}$. In § 3.3.17 we have considered a particular case of the family of loops on S^1 given by

$$f_m(t) = \exp[i2\pi m t].$$

For each n, the corresponding lift is $\tilde{f}(t) = mt$. The group homomorphism $\varphi \colon \pi_1 \to \mathbb{Z}$ takes each class $[f_m]$ into m:

$$\varphi([f_m]) = m.$$

The parameter m is the number of times a member of the class $[f_m]$, say $f_m(t)$ itself, "covers" S^1. Or better, the image space of f_m "covers" S^1 a number m of times. In the same way, $\pi_n(S^n)$ contains the classes of functions whose image space "covers" S^n. The functions of a given class "covers" S^n a certain number of times, this number being precisely the labeling m given by φ. As the m-loops on a space X correspond to mappings

$$f : S^m \to X,$$

we have here maps $S^n \to S^n$ in which the values cover the target S^n a number m of times. This number m is known in the physical literature as *winding number* and turns up in the study of magnetic monopoles[22] and of the vacuum in gauge theories (see Chapter 32).[23] When the target space is a quotient as $S^m/\mathbb{Z}_2$, the winding number can assume half-integer values, as in the case of the Franck index in nematic systems (see Section 28.3.3).

---

[22] Arafune, Freund & Goebel 1975.
[23] See Coleman 1977; 1979.

§ **3.4.13** Consider the covering space $(\mathbb{E}^1, p)$ of S^1. As

$$p_* : \pi_n(\mathbb{E}^1, e_0) \to \pi_n(S^1, x_0)$$

is an isomorphism for all $n \geq 2$, then

$$\pi_n(S^1) \approx \pi_n(\mathbb{E}^1) \approx \{0\} \quad \forall n \geq 2.$$

Take the covering (S^n, p) of RP^n. As

$$p : \pi_n(S^n) \to \pi_n(RP^n)$$

is an isomorphism for $n \geq 2$, then

$$\pi_n(RP^n) \approx \pi_n(S^n) \approx \mathbb{Z}, \quad \forall n \geq 2.$$

§ **3.4.14** Let us now mention two alternative (and, of course, equivalent) definitions of $\pi_n(X, x_0)$. The first has just been alluded to, and said to be relevant in some physical applications. We have defined a 1-loop at x_0 as a continuous mapping $f : I \to X$ such that $f(\partial I) = x_0$. On the other hand, the quotient space obtained by the identification of the end-points of I is simply S^1. We can then consider a 1-loop on X as a continuous mapping

$$f : S^1 \to X, \quad \text{with} \quad f(1,0) = x_0.$$

In an analogous way, a 2-dimensional loop will be a continuous mapping

$$f : S^2 \to X.$$

The definition of a homotopy of functions $S^n \to X$, necessary to get $\pi_n(X, x_0)$, is the following:

Let (X, x_0) be a topological space with a chosen point, and call $\Omega = \cup_n \Omega_n$ the set of all continuous mappings

$$f : S^n \to X, \ n \in \mathbb{Z}_+$$

satisfying $f(1, 0, 0, \ldots, 0) = x_0$. Two functions f and g are homotopic if a continuous $F : S^n \times I \to X$ exists such that

$$F(\boldsymbol{x}, 0) = f(\boldsymbol{x}), \quad F(\mathbf{x}, 1) = g(\boldsymbol{x}),$$

$$F(1, 0, 0, \ldots, 0, s) = x_0, \ s \in \boldsymbol{I}.$$

Of course, F is a homotopy between f and g.

§ 3.4.15 Another definition, due to Hurewicz, involves the idea that a 2-dimensional loop is a "loop of loops". In other words: a 2-dimensional loop is a function $f : I \to X$ such that, for each $t \in I$, the image $f(t)$ is itself a loop on X, and $f(\partial I) = x_0$. With this in mind, we can endow the set $\Omega(X, x_0)$ of loops on X at x_0 with a topology (the compact-open topology of § 1.4.15), making it into a topological space. It turns out that $\pi_2(X, x_0)$ can be defined as the fundamental group of $\Omega(X, x_0)$.

More generally, we have the following: let X be a topological space and $\Omega(X, x_0)$ be the set of loops on X with base point x_0, itself considered as a topological space with the compact-open topology. If $n \geq 2$, the n-th homotopy group on X at x_0 is the $(n-1)$-th homotopy group of $\Omega(X, x_0)$ at c, where c is the constant loop at x_0:

$$\pi_n(X, x_0) \approx \pi_{n-1}(\Omega(X, x_0), c).$$

§ 3.4.16 A last word on π_0: 0-loops would be mappings from $\{0\} \subset I$ into X. Such loops can be deformed into each other when their images lay in the same path-component of X. It is natural to include their classes, which correspond to the components, in the family $\{\pi_n(X)\}$.

§ 3.4.17 Hard-sphere gas. A gas of impenetrable particles in $\mathbb{E}^3$ has, of course, non-trivial π_2. The classical problem of the hard-sphere gas is a good example of the difficulties appearing in such spaces. After some manipulation, the question reduces to the problem of calculating the excluded volume[24] left by impenetrable particles, which is as yet unsolved.

General references

Two excellent and readable books on homotopy are: Hilton 1953 and Hu 1959. A book emphasizing the geometrical approach, also easily readable, is Croom 1978. For beginners, giving clear introductions to the fundamental group: Munkres 1975, or Hocking & Young 1961. Very useful because they contain many detailed calculations and results, are: Greenberg 1967, and Godbillon 1971. An introduction specially devoted to physical applications, in particular to the problems of quantization on multiply-connected spaces, the Bohm–Aharonov effect, instantons, etc, is Dowker 1979. A good review on solitons, with plenty of homotopic arguments is Boya, Cariñena & Mateos 1978. A pioneering application to Gauge Theories can be found in Loos 1967.

---

[24] See, for instance, Pathria 1972.

Chapter 4

Manifolds and Charts

Topological manifolds can be, at least in a domain around each one of its points, approximated by euclidean spaces. They are, consequently, spaces on which coordinates make sense.

4.1 Manifolds

4.1.1 *Topological manifolds*

We have up to now spoken of topological spaces in great (and rough) generality. Spaces useful in the description of physical systems are most frequently endowed with much more structure, but not every topological space accepts a given additional structure. We have seen that metric, for instance, may be shunned by a topology. So, the very fact that one works with more complicated spaces means that some selection has been made. Amongst all the possible topological spaces, we shall from now on talk almost exclusively of those more receptive to additional structures of euclidean type, the topological manifolds.

§ 4.1.1 A topological manifold is a topological space S satisfying the following restrictive conditions:

(i) S is *locally euclidean*: for every point $p \in S$, there exists an open set U to which p belongs, which is homeomorphic to an open set in some $\mathbb{E}^n$. The number n is the *dimension of S at the point p*. Given a general topological space, it may have points in which this is not true: 1-dimensional examples are curves on the plane which cross themselves (crunodes), or are tangent to themselves (cusps) at certain points. At these "singular points" the neighbourhood above required fails to exist. Points in which they do exist are called "general points". Topological manifolds are thus entirely constituted by general points. Very important exceptions are the upper-half spaces $\mathbb{E}^n_+$ (§ 1.2.10). Actually, so important are they that we shall soften the condition to

(i') Around every $p \in S$ there exists an open U which is either homeomorphic to an open set in some $\mathbb{E}^n$ or to an open set in some $\mathbb{E}^n_+$. Dimension is still the number n. Points whose neighbourhoods are homeomorphic to open sets of $\mathbb{E}^n_+$ and not to open sets of $\mathbb{E}^n$ constitute the *boundary* ∂S of S. Manifolds including points of this kind are called *manifolds-with-boundary*. Those without such points are manifolds-without-boundary, or manifolds-with-null-boundary.

(ii) The space S has the same dimension n at all points. The number n is then the *dimension* of S, $n = \dim S$. The union of a surface and a line in $\mathbb{E}^3$ is not a manifold. This condition can be shown to be a consequence of the following one, frequently used in its stead:

(ii') S is connected. When necessary, a non-connected S can be decomposed into its connected components. For space-time, for instance, connectedness is supposed to hold because "we would have no knowledge of any disconnected component" not our own. Nowadays, with the belief in the existence of confined quarks and shielded gluons based on an ever increasing experimental evidence, one should perhaps qualify this statement.

(iii) S has a countable basis (i.e., it is second-countable). This is a pathology-exorcizing requirement — the Sorgenfrey line of § 1.3.18, for example, violates it.

(iv) S is a Hausdorff space. Again to avoid pathological behaviours of the types we have talked about in § 1.3.16.

§ **4.1.2** Not all the above conditions are really essential: some authors call *topological manifold* any locally-euclidean or locally-half-euclidean connected topological space. In this case, the four conditions above define a *countable Hausdorff topological manifold*. We have already said that Einstein's equations in General Relativity have solutions exhibiting non-Hausdorff behaviour in some regions of spacetime, and that some topologies proposed for Minkowski space are not second-countable. The fundamental property for all that follows is the local-euclidean character, which will allow the definition of coordinates and will have the role of a *complementarity principle*: in the local limit, the differentiable manifolds whose study is our main objective will be fairly euclidean. That is why we shall suppose from now on the knowledge of the usual results of Analysis on $\mathbb{E}^n$, which will be progressively adapted to manifolds in what follows.

It suffices that one point of S have no euclidean open neighbourhood to forbid S of being made into a manifold. And all non-euclidean opens sets are "lost" in a manifold.

4.1.2 *Dimensions, integer and other*

§ **4.1.3** The reader will have noticed our circumspection concerning the concept of dimension. It seems intuitively a very "topological" idea, because it is so fundamental. We have indeed used it as a kind of primitive concept. Just above, we have taken it for granted in euclidean spaces and defined dimensions of more general spaces in consequence. But only locally-euclidean spaces have been contemplated. The trouble is that a well established theory for dimension (sketched in § 4.1.4 below) only exists for metric second-countable spaces, of which euclidean spaces are a particular case.

The necessity of a "theory" to provide a well-defined meaning to the concept became evident at the end of the 19-th century, when intuition clearly showed itself a bad guide. Peano found a continuous surjective mapping from the interval $I = [0, 1]$ into its square $I \times I$, so denying that dimension could be the least number of continuous real parameters required to describe the space. Cantor exhibited a one-to-one correspondence between $\mathbb{E}^1$ and $\mathbb{E}^2$, so showing that the plane is not richer in points than the line, dismissing the idea of dimension as a measure of the "point content" of space and even casting some doubt on its topological nature. Mathematicians have since then tried to obtain a consistent and general definition.

Commentary 4.1 Notice that the definition of dimension used above in this text is actually sound: it simply transfers to a topological manifold, point by point, the dimension of $\mathbb{E}^n$, which is a vector space. And for vector spaces a well-defined notion does exist Morette–DeWitt— see Chapter 13, § 13.4.4. ◄

§ **4.1.4** A simplified rendering[1] of the *topological dimension* of a space X

---

[1] A classic on the subject is Hurewicz & Wallman 1941; it contains a historical introduction and a very commendable study of alternative definitions in the appendix.

is given by the following series of statements:

(i) the empty set $\emptyset$, and only $\emptyset$, has dimension equal to -1: dim $\emptyset = -1$;
(ii) dim $X \leq n$ if there is a basis of X whose members have boundaries of dimension $\leq n - 1$;
(iii) dim $X = n$ if dim $X \leq n$ is true, and dim $X \leq n - 1$ is false;
(iv) dim $X = \infty$ if dim $X \leq n$ is false for each n.

This rather baffling definition has at least two good properties: it does give n for the euclidean $\mathbb{E}^n$ and it is monotonous (in the sense that $X \subset Y$ implies dim $X \leq$ dim Y). But a close examination shows that it is rigorous only for separable metric (equivalently, metric second-countable) spaces. If we try to apply it to more general spaces, we get into trouble: for instance, one should expect that the dimension of a countable space be zero, but this does not happen with the above definition. Furthermore, there are many distinct definitions which coincide with that above for separable metric spaces but give different results for less structured spaces, and none of them is satisfactory in general.[2]

§ **4.1.5** In another direction, explicitly metric, we may try to define dimension by another procedure: given the space as a subset of some $\mathbb{E}^n$, we count the number $N(\varepsilon)$ of n-cubes of side ε necessary to cover it. We then make ε smaller and smaller, and calculate the Kolmogorov *capacity* (or *capacity dimension*)

$$d_c = \lim_{\varepsilon \to 0} \frac{\ln N(\varepsilon)}{\ln(1/\varepsilon)} .$$

Suppose a piece of line: divide it in k pieces and take $\varepsilon = 1/k$. Then, the number of pieces is $N(\varepsilon) = k$, and $d_c = 1$. A region of the plane $\mathbb{E}^2$ may be covered by $k^2 = N(\varepsilon)$ squares of side $\varepsilon = 1/k$, so that $d_c = 2$.

This capacity dimension gives the expected results for simple, usual spaces. It is the simplest case of a whole series of dimension concepts based on ideas of measure which, unlike the topological dimension, are not necessarily integer. Consider a most enthralling example: take the Cantor set of § 1.3.5. After the j-th step of its construction, 2^j intervals remain, each of length $1/3^j$. Thus, $N(\varepsilon) = 2^j$ intervals of length $\varepsilon = 1/3^j$ are needed to cover it, and

$$d_c = \frac{\ln 2}{\ln 3} \approx 0.6309 \ldots .$$

---

[2] A sound study of the subject is found in Alexandrov 1977.

Spaces with fractional dimension seem to peep out everywhere in Nature and have been christened *fractals* by their champion, B. B. Mandelbrot.[3] Notice that the fractal character depends on the chosen concept of dimension: the topological dimension of the Cantor set is zero. Fractals are of great importance in dynamical systems.[4]

§ **4.1.6** Spaces with non-integer dimensions have been introduced in the thirties (von Neumann algebras, see Section 17.5).

4.2 Charts and coordinates

§ **4.2.1** Let us go back to item (i) of § 4.1.1 on the definition of a topological manifold: every point p of the manifold has an euclidean open neighbourhood U, homeomorphic to an open set in some $\mathbb{E}^n$, and so to $\mathbb{E}^n$ itself. The homeomorphism

$$: U \to \text{open set in } \mathbb{E}^n$$

will give *local coordinates* around p. The neighbourhood U is called a *coordinate neighbourhood* of p. The pair (U, ψ) is a *chart*, or *local system of coordinates* (LSC) around p. To be more specific: consider the manifold $\mathbb{E}^n$ itself; an open neighbourhood V of a point $q \in \mathbb{E}^n$ is homeomorphic to another open set of $\mathbb{E}^n$. Each homeomorphism of this kind will define a *system of coordinate functions*, as u in Figure 4.1. For $\mathbb{E}^2$, for instance, we can use the system of cartesian coordinates

$$u^1(q) = x, \qquad u^2(q) = y,$$

or else the system of polar coordinates

$$u^1(q) = r \in (0, \infty), \qquad u^2(q) = \theta \in (0, 2\pi),$$

and so on.

§ **4.2.2** Take a homeomorphism $x : S \to \mathbb{E}^n$, given by

$$x(p) = (x^1, x^2, \ldots, x^n) = (u^1 \circ \psi(p), u^2 \circ \psi(p), \ldots, u^n \circ \psi(p)).$$

---

[3] Much material on dimensions, as well as beautiful illustrations on fractals and a whole account of the subject is found in Mandelbrot 1977.

[4] For a discussion of different concepts of dimension which are operational in dynamical systems, see Farmer, Ott & Yorke 1983.

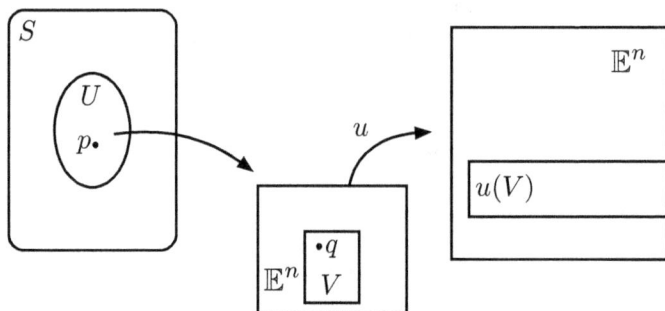

Fig. 4.1   Coordinates, and coordinate functions around a point p.

The functions

$$x^i = u^i \circ \psi : U \to \mathbb{E}^1$$

will be the *local coordinates* around p. We shall use frequently the simplified notation (U, x) for the chart. What people usually call coordinate systems (e.g. cartesian, polar, elliptic) are actually systems of coordinate functions, corresponding to the above u. This is a relevant distinction because distinct systems of coordinate functions require a different number of charts to plot a given space S. For $\mathbb{E}^2$ itself, one cartesian system is enough to chart the whole space: $U = \mathbb{E}^2$, with u the identity mapping. However, this is not true for the polar system: the coordinate θ is not related to a homeomorphism on the whole plane, as its inverse is not continuous — it takes points near to 0 and 2π back to neighbouring points in $\mathbb{E}^2$. There is always a half-line which is not charted (usually taken as $\mathbb{R}_+$); at least two charts are required by this system. One sometimes forgets this point, paying the price of afterwards finding some singularity. Some of the most popular singularities in Physics are not real, but of a purely coordinate origin.

A good example of a singular line which is not real, but only a manifestation of coordinate inadequacy, is the string escorting the Dirac magnetic monopole in its elementary formulation, which disappears when correct charts are introduced.[5] Another case is the Schwarzschild radius, not a singularity when convenient — albeit involved — coordinates are used.[6] The word "convenient" here may be a bit misleading: it means good for formal purposes, as to exhibit a particular property. The fact that singularities

---

[5] Wu & Yang 1975.
[6] Misner, Thorne & Wheeler 1973, Section 31.2.

are of coordinate origin does not mean that they will not be 'physically' observed, as measuring apparatuses can presuppose some coordinate function system.

§ **4.2.3** The coordinate homeomorphism could be defined in the inverse sense, from an open set of some $\mathbb{E}^n$ to some neighbourhood of the point $p \in S$. It is then called a *parameterization* .

§ **4.2.4** Of course, a given point $p \in S$ can in principle have many different coordinate neighbourhoods and charts. Remember the many ways used to plot the Earth in cartography. By the way, cartography was the birthplace of charts, whose use was pioneered by Hipparchos of Nicaea (to whom are also attributed the first trigonometric table and the discovery of the precession of the equinoxes) in the second century B.C.

§ **4.2.5** As the coordinate homeomorphism x of the chart (U, x) takes into a ball of $\mathbb{E}^n$, which is contractible, U itself must be contractible. This gives a simple criterium to have an idea on the minimum number of necessary coordinate neighbourhoods. We must at least be able to cover the space with charts with contractible neighbourhoods. Let us insist on this point: an LSC is ultimately $(U, x = u \circ \psi)$, and x has two pieces. The homeomorphism ψ takes U into some open V of $\mathbb{E}^n$; the coordinate function u chooses coordinates for $\mathbb{E}^n$ itself, taking it into some subspace — for instance, spherical coordinates (r, θ, φ) on $\mathbb{E}^3$ involve u:

$$\mathbb{E}^3 \rightarrow \mathbb{E}^1_+ \times (0, \pi) \times (0, 2\pi).$$

Recall that V is homeomorphic to $\mathbb{E}^n$. If a space N could be entirely covered by a single chart, ψ would be a homeomorphism between N and $\mathbb{E}^n$. If N is not homeomorphic to $\mathbb{E}^n$, it will necessarily require more than one chart. The minimum number of charts is the minimum number of open sets homeomorphic to $\mathbb{E}^n$ covering N, but the real number depends also of the function u. The sphere S^2 , for example, needs at least two charts (given by the stereographic projections from each pole into an $\mathbb{E}^2$ tangent in the opposite pole, see Section 23.3) but the imposition of cartesian coordinates raises this number to eight.[7]

Thus, summing up: the multiplicity of the necessary charts depends on the minimum number of euclidean open sets really needed to cover the space, and on the chosen system of coordinate functions.

---

[7] See Flanders 1963.

§ 4.2.6 The fact that a manifold may require more than one chart has a remarkable consequence. Transformations are frequently treated in two supposedly equivalent ways, the so called active (in which points are moved) and passive (in which their coordinates are changed) points of view. This can be done in euclidean spaces, and its generality in Physics comes from the euclidean "supremacy" among usual spaces. On general manifolds, only the active point of view remains satisfactory.

§ 4.2.7 Given any two charts (U, x) and (V, y) with $U \cap V \neq \emptyset$, to a given point $p \in U \cap V$ will correspond coordinates (see Figure 4.2)

$$x = x(p) \quad \text{and} \quad y = y(p).$$

These coordinates will be related by a homeomorphism between open sets of $\mathbb{E}^n$,

$$y \circ x^{<-1>} : \mathbb{E}^n \to \mathbb{E}^n,$$

which is a *coordinate transformation* and can be written as

$$y^i = y^i(x^1, x^2, \ldots, x^n). \tag{4.1}$$

Its inverse is $x \circ y^{<-1>}$, or

$$x^j = x^j(y^1, y^2, \ldots, y^n). \tag{4.2}$$

§ 4.2.8 Given two charts (U_α, ψ_α) and (U_β, ψ_β) around a point, the coordinate transformation between them is commonly indicated by a *transition function*

$$g_{\alpha\beta} : (U_\alpha, \psi_\alpha) \to (U_\beta, \psi_\beta)$$

and its inverse $g_{\alpha\beta}^{-1}$.

§ 4.2.9 Consider now the euclidean coordinate spaces as linear spaces, that is, taken with their vector structure. Coordinate transformations are relationships between points in linear spaces. If both

$$x \circ y^{<-1>} \quad \text{and} \quad y \circ x^{<-1>}$$

are C^∞ (that is, differentiable to any order) as functions in $\mathbb{E}^n$, the two local systems of coordinates (LSC) are said to be *differentially related*.

§ 4.2.10 An *atlas* on the manifold S is a collection of charts $\{(U_\alpha, \psi_\alpha)\}$ such that

$$\bigcup_\alpha U_\alpha = S.$$

The following theorem can be proven: *Any compact manifold can be covered by a finite atlas*, that is, an atlas with a finite number of charts.

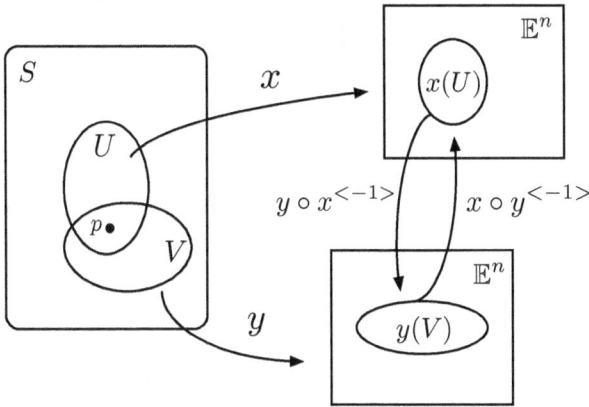

Fig. 4.2   Two distinct charts around p.

§ **4.2.11** If all the charts are related by linear transformations in their intersections, it will be a *linear atlas*.

§ **4.2.12** If all the charts are differentially related in their intersections, it will be a *differentiable atlas*. This requirement of infinite differentiability can be reduced to k-differentiability. In this case, the atlas will be a "C^k-atlas". Differentiating (4.1) and (4.2) and using the chain rule,

$$\delta^i_k = \frac{\partial y^i}{\partial x^j} \frac{\partial x^j}{\partial y^k} .$$

(4.3)

§ **4.2.13** This means that both jacobians are $\neq 0$. If some atlas exists on S whose jacobians are all positive, S is orientable. Roughly speaking, it has two faces. Most commonly found manifolds are orientable. The Möbius strip and the Klein bottle are examples of non-orientable manifolds.

§ **4.2.14** Suppose a linear atlas is given on S, as well as an extra chart not belonging to it. Take the intersections of the coordinate-neighbourhood of this chart with all the coordinate-neighbourhoods of the atlas. If in these intersections all the coordinate transformations from the atlas LSC's to the extra chart are linear, the chart is *admissible* to the atlas. If we add to a linear atlas all its admissible charts, we get a (linear) *complete atlas*, or (linear) *maximal atlas*. A topological manifold with a complete linear atlas is called a *piecewise-linear manifold* (usually, a "PL manifold").

§ **4.2.15** A topological manifold endowed with a certain differentiable atlas is a differentiable manifold (see Section 5.1). These are the most important manifolds for Physics and will deserve a lot of attention in the forthcoming chapters.

§ **4.2.16 Electron diffraction experiment.** A famous experiment involving a multiply–connected space is the well known Young double-slit interference experiment. In Quantum Mechanics, perhaps the simplest example appears in the similar 1927 Davisson & Germer electron diffraction experiment. We suppose the wave function to be represented by plane waves (corresponding to free particles of momentum $p = mv$) incident from the left (say, from an electron source S situated far enough to the left). A more complete scheme is shown in Figure 4.5, but let us begin by considering only the central part B of the future doubly-slitted obstacle (Figure 4.3). We are supposing the scene to be $\mathbb{E}^3$, so that B extends to infinity in the direction perpendicular to the drawing. Of course, the simple exclusion

Fig. 4.3   Barrier B extends to infinity in both directions perpendicular to the drawing.

represented by B makes the space multiply-connected, actually homeomorphic to $\mathbb{E}^3 \backslash \mathbb{E}^1$. A manifold being locally euclidean, around each point there is an open set homeomorphic to an euclidean space. When the space is not euclidean, it must be somehow divided into intersecting regions, each one euclidean and endowed with a system of coordinates. Here, it is already impossible to use a unique chart, at least two being necessary (as in Figure 4.4). We now add the parts A and C of the barrier (Figure 4.5). Diffraction sets up at the slits 1 and 2, which distort the wave. The slits act as new sources, from which waves arrive at a point P on the screen after propagating along straight paths γ_1 and γ_2 perpendicular to the wavefronts. If

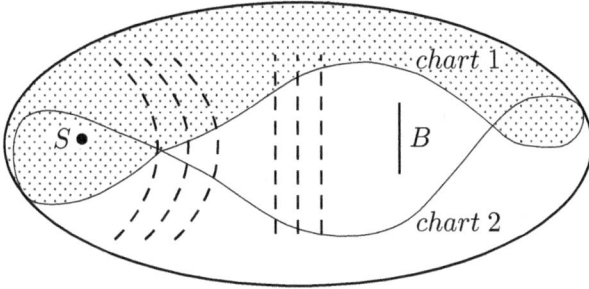

Fig. 4.4   At least two charts are necessary to cover $\mathbb{E}^3 \backslash \mathbb{E}^1$.

the path lengths are respectively l_1 and l_2, the wavefunction along γ_k will get the phase $2\pi l_k/\lambda$. The two waves will have at P a phase difference

$$\frac{2\pi|l_1 - l_2|}{\lambda} = \frac{2\pi m v |l_1 - l_2|}{h} .$$

Let us recall the usual treatment of the problem[8] and learn something

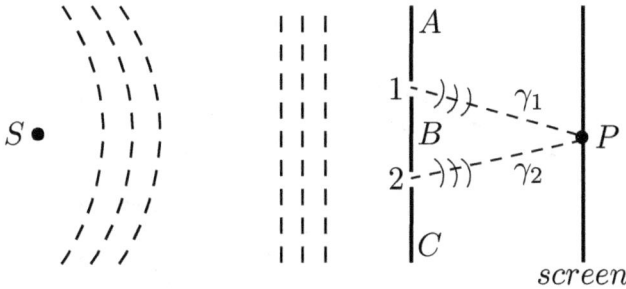

Fig. 4.5   Scheme or the double-slit diffraction nexperiment.

from Quantum Mechanics. At the two slit-sources, we have the same wave-function Ψ_0. The waves superpose and interfere all over the region at the right, in particular at P. The waves arrive at P as

$$\Psi_1 = \Psi_0 \exp\left[i\frac{2\pi l_1}{\lambda}\right] \quad \text{and} \quad \Psi_2 = \Psi_0 \exp\left[i\frac{2\pi l_2}{\lambda}\right].$$

---

[8] Furry 1963; Wootters & Zurek 1979.

Their superposition leads then to the relative probability density

$$\left| \exp\left[i\frac{2\pi l_1}{\lambda} \right] + \exp\left[i\frac{2\pi l_2}{\lambda} \right] \right|^2 = 2 + 2\cos\left[\frac{2\pi |l_1 - l_2|}{\lambda} \right].$$

This experimentally well verified result will teach us something important. The space is multiply-connected and the wavefunction is actually Ψ_1 on the first chart (on the first covering leaf) and Ψ_2 on the second chart (on the second covering leaf). We nevertheless obtain a single-valued Ψ at P by taking the superposition. Thus, we simply sum the contributions of the distinct leaves!

The morality is clear: wavefunctions on a multiply-connected manifold should be multiply-valued. Quantum Mechanics, totally supported by experiment, tells us that we can use as wavefunction the unique summation of the leaves contributions.

§ **4.2.17 Aharonov–Bohm effect.** We may go ahead[9] with the previous example. Drop parts A and C of the intermediate barrier, and replace part B by an impenetrable infinite solenoid orthogonal to the figure plane and carrying a magnetic field $\boldsymbol{B}$ (see Figure 4.6). The left side is replaced by a unique electron point source S. A single wavefunction Ψ_0 is prepared at

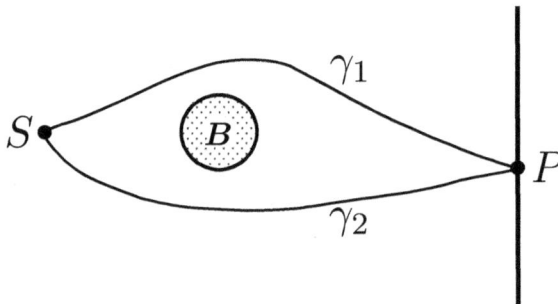

Fig. 4.6   A scheme for the Aharonov–Bohm effect.

point S but, the region of the solenoid being forbidden to the electrons, the domain remains multiply-connected. There is no magnetic field outside that region, so that the exterior (tri-)vector potential is a pure gauge:

$$\boldsymbol{A} = \boldsymbol{\nabla} f.$$

[9] See Furry 1963, Wu & Yang 1975 and Dowker 1979.

Waves going "north" and "south" from the forbidden region will belong to distinct leaves or charts. In the eikonal approximation, the wavefunction at P obtained by going along γ_1 will be

$$\Psi_1 = \Psi_0 \exp\left[\frac{i2\pi}{h}\int_{\gamma_1} pdq\right]$$

$$= \Psi_0 \exp\left[\frac{i2\pi}{h}\int_{\gamma_1}\left(m\boldsymbol{v} + \frac{e}{c}\boldsymbol{A}\right)\cdot d\boldsymbol{l}\right]$$

$$= \Psi_0 \exp\left[\frac{i2\pi}{h}\int_{\gamma_1} m\boldsymbol{v}\cdot d\boldsymbol{l} + \frac{i2\pi}{h}\frac{e}{c}\int_{\gamma_1}\boldsymbol{A}\cdot d\boldsymbol{l}\right]. \qquad (4.4)$$

The wavefunction at P obtained by going along γ_2 will be

$$\Psi_2 = \Psi_0 \exp\left[\frac{i2\pi}{h}\int_{\gamma_2} pdq\right]$$

$$= \Psi_0 \exp\left[\frac{i2\pi}{h}\int_{\gamma_2}\left(m\boldsymbol{v} + \frac{e}{c}\boldsymbol{A}\right)\cdot d\boldsymbol{l}\right]$$

$$= \Psi_0 \exp\left[\frac{i2\pi}{h}\int_{\gamma_2} m\boldsymbol{v}\cdot d\boldsymbol{l} + \frac{i2\pi}{h}\frac{e}{c}\int_{\gamma_2}\boldsymbol{A}\cdot d\boldsymbol{l}\right]. \qquad (4.5)$$

Taking a simplified view, analogous to the double slit case, of the kinematic part, the total phase difference at P will be given by

$$\Psi_2 = \Psi_1 \exp\left[i\left(\frac{2\pi|l_1 - l_2|}{\lambda} + \frac{2\pi e}{hc}\oint \boldsymbol{A}\cdot d\boldsymbol{l}\right)\right].$$

The closed integral, which is actually a line integral along the curve $\gamma_2 - \gamma_1$, is the magnetic flux. The effect, once the kinematical contribution is accounted for, shows that the vector potential, though not directly observable, has an observable circulation. The wavefunctions Ψ_1 and Ψ_2 are values of Ψ on distinct leaves of the covering space, and should be related by a representation of the fundamental group, which is here $\pi_1 = \mathbb{Z}$. A representation ρ of the group $\mathbb{Z}$, acting on any (one-dimensional) wavefunction, will be given by any phase factor like $\exp[i2\pi\alpha]$, for each real value of α,

$$\rho : n \to \exp[i2\pi\alpha\, n].$$

The value of α is obtained from above,

$$\alpha = \frac{|l_1 - l_2|}{\lambda} + \frac{e}{hc}\oint \boldsymbol{A}\cdot d\boldsymbol{l}.$$

Once this is fixed, one may compute the contribution of other leaves, such as those corresponding to paths going twice around the forbidden region. For each turn, the first factor will receive the extra contribution of the length of a circle around B, and the contribution of many-turns paths are negligible in usual conditions.

Chapter 5

Differentiable Manifolds

The relationship between a topological manifold and the coordinate systems it consents (or not!) to receive is very involved: some manifolds accept many distinct coordinate systems, others accept none. A particular manifold with a collection of good (that is, differentiable, admitting differential equations) systems of coordinates, and such that these can be exchanged always through differentiable transformations, is a differentiable (or "smooth") manifold.

5.1 Definition and overlook

§ 5.1.1 Suppose a differentiable atlas is given on a topological manifold S, as well as an extra chart not belonging to it. Take the intersections of the coordinate-neighbourhood of this chart with all the coordinate-neighbourhoods of the atlas. If in these intersections all the coordinate transformations from the atlas LSC's to the extra chart are C^∞, the chart is *admissible* to the atlas. If we add to a differentiable atlas all its admissible charts, we get a *complete atlas*, or *maximal atlas*, or C^∞-*structure*. The important point is that, given a differentiable atlas, its extension obtained in this way is unique.

A topological manifold with a complete differentiable atlas
is a differentiable manifold.

One might think that on a given topological manifold only one complete atlas can be defined — in other words, that it can "become" only one differentiable manifold. This is wrong: a fixed topological manifold can in principle accept many distinct C^∞-structures, each complete atlas with charts not admissible by the other atlases. This had been established for the first time in 1957, when Milnor showed that the sphere S^7 accepts 28 distinct complete atlases. The intuitive idea of identifying a differentiable manifold with its topological manifold, not to say with its point-set (when we say "a differentiable function on *the* sphere", "*the* space-time", etc), is actually dangerous (although correct for most of the usual cases, as the spheres S^n with $n \leq 6$) and, ultimately, false. That is why the mathematicians, who are scrupulous and careful people, denote a differentiable manifold by a pair (S, D), where D specifies which C^∞-structure they are using. Punctiliousness which pays well: it has been found recently, to general surprise, that $\mathbb{E}^4$ has infinite distinct differentiable structures!

Another point illustrating the pitfalls of intuition: not every topological manifold admits of a differentiable structure. In 1960, Kervaire had already found a 10-dimensional topological manifold which accepts no complete differentiable atlas at all. In the nineteen-eighties, a whole family of "non-smoothable" compact simply-connected 4-dimensional manifolds was found. And, on the other hand, it has been found that every non-compact manifold accepts at least one smooth structure.[1] Thus, compactness seems to act as a limiting factor to smoothness.

§ **5.1.2** The above definitions concerning C^∞-atlases and manifolds can be extended to C^k-atlases and C^k-manifolds in an obvious way. It is also possible to waive the Hausdorff conditions, in which case, as we have already said, the unicity of the solutions of differential equations holds only locally. Broadly speaking, one could define differentiable manifolds without imposing the Hausdorff condition, second-countability and the existence of a maximal atlas. These properties are, nevertheless, necessary to obtain some very powerful results, in particular the Whitney theorems concerning the imbedding of a manifold in other manifolds of higher dimension. We shall speak on these theorems later on, after the notion of imbedding has been made precise.

§ **5.1.3** A very important theorem by Whitney says that a complete C^k-atlas contains a C^{k+1}-sub-atlas for $k \geq 1$. Thus, a C^1-structure contains a C^∞-structure. But there is much more: it really contains an *analytic* sub-atlas. The meaning here is the following: in the definition of a differentiable atlas, replace the C^∞-condition by the requirement that the coordinate transformations be analytic. This will define an *analytic atlas*, and a manifold with an analytic complete atlas is an *analytic manifold*. The

---

[1] Quinn 1982.

most important examples of such are the Lie groups. Of course not all C^∞ functions are analytic, as the formal series formed with their derivatives as coefficients may diverge.

5.2 Smooth functions

§ **5.2.1** In order to avoid constant repetitions when talking about spaces and their dimensions, we shall as a rule use capital letters for manifolds and the corresponding small letters for their dimensions: dim $N = n$, dim $M = m$, etc.

§ **5.2.2** A function $f : N \to M$ is *differentiable* (or $\in C^k$, or still *smooth*) if, for any two charts (U, x) of N and (V, y) of M, the function

$$y \circ f \circ x^{<-1>} : x(U) \to y(V)$$

is differentiable ($\in C^k$) as a function between euclidean spaces.

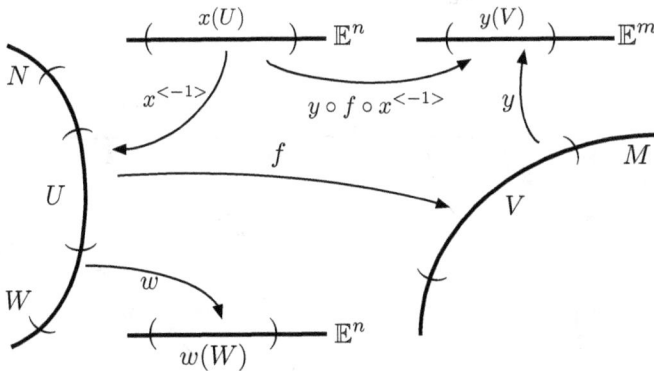

Fig. 5.1   Coordinate view of functions between manifolds.

§ **5.2.3** Recall that all the analytic notions in euclidean spaces are presupposed. This function $y \circ f \circ x^{<-1>}$, taking an open set of $\mathbb{E}^n$ into an open set of $\mathbb{E}^m$, is the *Expression of f in local coordinates*. We usually write simply $y = f(x)$, a very concise way of packing together a lot of things. We should keep in mind the complete meaning of this expression (see Figure 5.1): the point of N whose coordinates are $x = (x^1, x^2, \ldots, x^n)$ in chart (U, x) is taken by f into the point of M whose coordinates are $y = (y^1, y^2, \ldots, y^m)$ in chart (V, y).

§ **5.2.4** The composition of differentiable functions between euclidean spaces is differentiable. From this, it is not difficult to see that the same is true for functions between differentiable manifolds, because

$$z \circ (g \circ f) \circ x^{<-1>} = z \circ g \circ y^{<-1>} \circ y \circ f \circ x^{<-1>}.$$

If now a coordinate transformation is made, say $(U, x) \to (W, w)$ as in Figure 5.1, the new expression of f in local coordinates is $y \circ f \circ w^{<-1>}$. Thus, the function will remain differentiable, as this expression is the composition

$$y \circ f \circ x^{<-1>} \circ x \circ w^{<-1>}$$

of two differentiable functions: the local definition of differentiability given above is extended in this way to the whole manifold by the complete atlas. All this is easily extended to the composition of functions involving other manifolds (as $g \circ f$ in Figure 5.2).

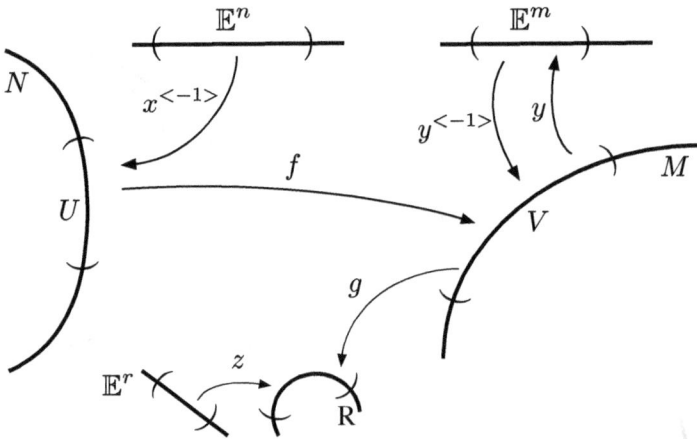

Fig. 5.2   A function composition.

§ **5.2.5** Each coordinate $x^i = u^i \circ \psi$ is a differentiable function

$$x^i : U \subset N \to \text{open set in } \mathbb{E}^1.$$

§ **5.2.6** A most important example of differentiable function is a *differentiable curve* on a manifold: it is simply a smooth function from an open set of $\mathbb{E}^1$ into the manifold. A *closed* differentiable curve is a smooth function from the circle S^1 into the manifold.

§ **5.2.7** We have seen that two spaces are equivalent from a purely topological point of view when related by a homeomorphism, a topology-preserving transformation. A similar role is played, for spaces with a differentiable structure, by a diffeomorphism:

A diffeomorphism is a differentiable homeomorphism whose inverse is also smooth.

§ **5.2.8** Two smooth manifolds are *diffeomorphic* when some diffeomorphism exists between them. In this case, besides being topologically the same, they have equivalent differentiable structures. The famous result by Milnor cited in the previous section can be put in the following terms: on the sphere S^7 one can define 28 distinct smooth structures, building in this way 28 differentiable manifolds. They are all distinct from each other because no diffeomorphism exists between them. The same holds for the infinite differentiable manifolds which can be defined on $\mathbb{E}^4$.

§ **5.2.9** The equivalence relation defined by diffeomorphisms was the starting point of an ambitious program: to find all the equivalence classes of smooth manifolds. For instance, it is possible to show that the only classes of 1-dimensional manifolds are two, represented by $\mathbb{E}^1$ and S^1. The complete classification has also been obtained for two-dimensional manifolds, but not for 3-dimensional ones, although many partial results have been found. The program as a whole was shown not to be realizable by Markov, who found 4-dimensional manifolds whose class could not be told by finite calculations.

5.3 Differentiable submanifolds

§ **5.3.1** Let N be a differentiable manifold and M a subset of N. Then M will be a (regular) *submanifold* of N if, for every point $p \in M$, there exists a chart (U, x) of the N atlas such that, for $p \in U$,

$$x(p) = 0 \in \mathbb{E}^n \quad \text{and} \quad x(U \cap M) = x(U) \cap \mathbb{E}^m \, ,$$

as in Figure 5.3. In this case M is a differentiable manifold by itself.

§ **5.3.2** This decomposition in coordinate space is a formalization of the intuitive idea of submanifold we get when considering smooth surfaces in $\mathbb{E}^3$. We usually take on these surfaces the same coordinates used in $\mathbb{E}^3$, adequately restricted. To be more precise, we implicitly use the inclusion

$$\text{i: Surface} \rightarrow \mathbb{E}^3$$

and suppose it to preserve the smooth structure. Let us make this procedure more general.

§ **5.3.3** A differentiable function $f : M \rightarrow N$ is an *imbedding* when
  (i) $f(M) \subset N$ is a submanifold of N;
  (ii) $f : M \rightarrow f(M)$ is a diffeomorphism.
  The above $f(M)$ is a differentiable *imbedded submanifold* of N. It corre-

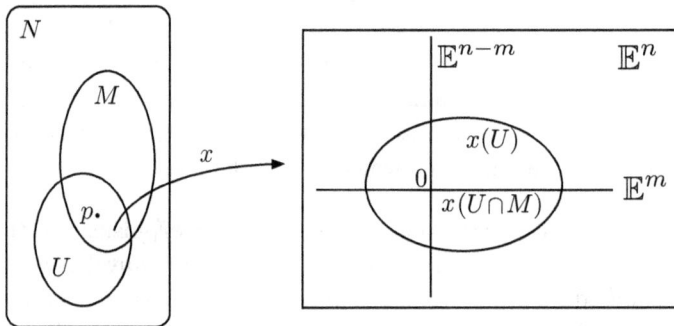

Fig. 5.3 M as a submanifold of N.

sponds precisely to our intuitive idea of submanifold, as it preserves globally all the differentiable structure.

§ **5.3.4** A weaker kind of inclusion is the following. A smooth function $f : M \rightarrow N$ is an *immersion* if, given any point $p \in M$, it has a neighbourhood U, with $p \in U \subset M$, such that f restricted to U is an imbedding. An immersion is thus a local imbedding and every imbedding is automatically an immersion. The set $f(M)$, when f is an immersion, is an *immersed* *submanifold*. Immersions are consequently much less stringent than imbeddings. We shall later (§ 6.4.33) give the notion of integral submanifold.

§ **5.3.5** These things can be put down in another (equivalent) way. Let us go back to the local expression of the function $f : M \to N$ (supposing $n \geq m$). It is a mapping between the euclidean spaces $\mathbb{E}^m$ and $\mathbb{E}^n$, of the type $y \circ f \circ x^{<-1>}$, to which corresponds a matrix $(\partial y^i / \partial x^j)$. The rank of this matrix is the maximum order of non-vanishing determinants, or the number of linearly independent rows. It is also (by definition) the rank of $y \circ f \circ x^{<-1>}$ and (once more by definition) the *rank* of f. Then, f is an immersion iff its rank is m at each point of M. It is an imbedding if it is an immersion and else an homeomorphism into $f(M)$. It can be shown that these definitions are quite equivalent to those given above.

§ **5.3.6** The mapping

$$f : \mathbb{E}^1 \to \mathbb{E}^2, \quad f(x) = (\cos 2\pi x, \sin 2\pi x)$$

is an immersion with $f(\mathbb{E}^1) = S^1 \subset \mathbb{E}^2$. It is clearly not one-to-one and so it is not an imbedding. The circle $f(\mathbb{E}^1)$ is an immersed submanifold but not an imbedded submanifold.

§ **5.3.7** The mapping

$$f : \mathbb{E}^1 \to \mathbb{E}^3, \quad f(x) = (\cos 2\pi x, \sin 2\pi x, x)$$

is an imbedding. The image space $f(\mathbb{E}^1)$, a helix (Figure 5.4), is an imbedded submanifold of $\mathbb{E}^3$. It is an inclusion of $\mathbb{E}^1$ in $\mathbb{E}^3$.

§ **5.3.8** We are used to think vaguely of manifolds as spaces imbedded in some $\mathbb{E}^n$. The question naturally arises of the validity of this purely intuitive way of thinking, so convenient for practical purposes. It was shown by Whitney that an n-dimensional differentiable manifold can always be immersed in $\mathbb{E}^{2n}$ and imbedded in $\mathbb{E}^{2n+1}$. The conditions of second-countability, completeness of the atlas and Hausdorff character are necessary to the demonstration. These results are used in connecting the modern treatment with the so-called "classical" approach to geometry (see Chapter 21). Notice that eventually a particular manifold N may be imbeddable in some euclidean manifold of lesser dimension. There is, however, no general result up to now fixing the minimum dimension of the imbedding euclidean space of a differentiable manifold. It is a theorem that 2-dimensional orientable surfaces are imbeddable in $\mathbb{E}^3$: spheres, hyperboloids, toruses are perfect imbedded submanifolds of our ambient space. On the other hand,

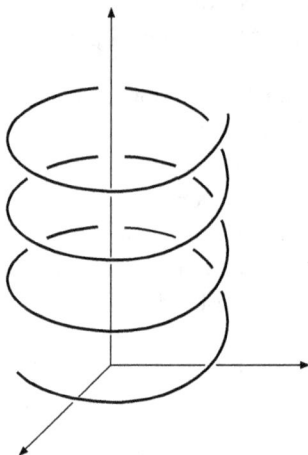

Fig. 5.4   A helix is an imbedded submanifold of $\mathbb{E}^3$.

it can be shown that non-orientable surfaces without boundary are not —
which accounts for our inability to visualize a Klein bottle. Non-orientable
surfaces are, nevertheless, imbeddable in $\mathbb{E}^4$.

General references

An excellent introduction is Boothby 1975. A short reference, full of illuminating comments, is the Chapter 5 of Arnold 1973. Nomizu 1956 is a very
good introduction to the very special geometrical properties of Lie groups.
The existence of many distinct differentiable structures on $\mathbb{E}^4$ was found
in 1983. It is an intricate subject, the proof requiring the whole volume
of Freed & Uhlenbeck 1984. Donaldson & Kronheimer 1991 summarize the
main results in a nearly readable way. The relationship between differential
aspects and topology is clearly discussed in Bott & Tu 1982. More recently,
such results (and new ones) have been re-obtained in a much simpler way,
in works stemming from those of Seiberg and Witten. For a review, see
Donaldson 1996.

Part 2: DIFFERENTIABLE STRUCTURE

Chapter 6

Tangent Structure

Vector fields will be defined on differentiable manifolds, as well as tensor fields and general reference frames (or bases). All these objects constitute the tangent structure of the manifold — metrics, for instance, are particular tensors. Vector fields provide the background to the modern approach to systems of differential equations, which we shall here ignore. They also mediate continuous transformations on manifolds, which we shall examine briefly.

6.1 Introduction

Think of an electron moving along the lines of force of an electric field created by some point charge. The lines of force are curves on the configuration space M (just $\mathbb{E}^3$ in the usual case, but it will help the mind to consider some curved space) and the electron velocity is, at each point, a vector tangent to one of them. Being tangent to a curve in M, it is tangent to M itself. A vector field is just that, a general object which reduces, at each point of the manifold, to a tangent vector. The first thing one must become aware of is that tangent vectors do not "belong" to M. At a fixed point p of M, they may be rescaled and added to each other, two characteristic properties of the members of a linear (or vector) space. M is not, in general, a linear space. Vectors belong to another space, the space tangent to M at p, which is indeed a linear space. The differentiable structure provides precise meanings to all these statements.

§ **6.1.1** Topological manifolds are spaces which, although generalizing their properties, still preserve most of the mild qualities of euclidean spaces. The differentiable structure has a very promising further consequence: it opens the possibility of (in a sense) approximating the space M by a certain $\mathbb{E}^m$ in the neighbourhood of each of its points.

Intuitively: the simplest euclidean space is the real line $\mathbb{E}^1$. A 1-dimensional smooth manifold will be, for instance, a smooth curve γ on the plane $\mathbb{E}^2$. In a neighbourhood of any point [say, point A in Figure 6.1(a)], the curve can be approximated by a copy of $\mathbb{E}^1$. Recall the high-school definition of a vector on the plane: given two points A and B, they determine the vector $V_{AB} = B - A$. If the tangent to the curve γ at A is the line

$$a(t) = A + t(B - A)$$

with parameter t, $a(t)$ is the best linear approximation to γ in a neighbourhood of $A = a(0)$. The vector

$$V_{AB} = \frac{da(t)}{dt}$$

can then be defined at the point A, since this derivative is a constant, and will be a vector tangent to the curve γ at A. This high-school view of a vector is purely euclidean: it suggests that point B lies in the same space as the curve. Actually, there is no mathematical reason to view the tangent space as "attached" to the point on the manifold (as in Figure 6.1(b)). This is done for purely intuitive, pictorial reasons.

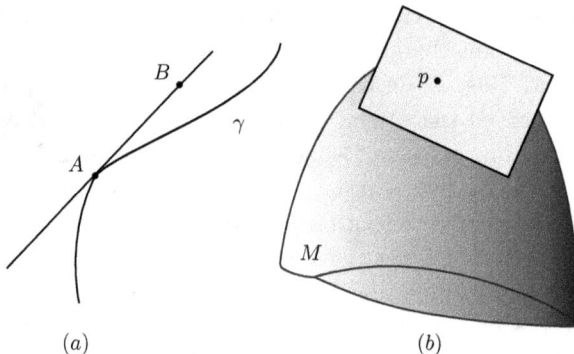

Fig. 6.1   (a) A vector as difference between points: $V_{AB} = B - A$; (b) Tangent space at $p \in M$ seen as "touching" M at p, a mere pictorial resource.

§ **6.1.2** Another characterization of this vector is, however, more convenient for the generalization we intend to use in the following. Let f be a smooth function on the curve γ, with real values: $f[\gamma(t)] \in \mathbb{E}^1$. Then, $\gamma(t) \approx a(t)$ in a neighbourhood of A and, just at A,

$$\frac{df}{dt} = \frac{df}{da}\frac{da}{dt} = V_{AB}\frac{df}{da}.$$

Thus, through the vector, to every function will correspond a real number. The vector appears in this case as a linear operator acting on functions.

§ 6.1.3 A vector $V = (v^1, v^2, \dots, v^n)$ in $\mathbb{E}^n$ can be viewed as a linear operator: take a point $p \in \mathbb{E}^n$ and let f be a function which is differentiable in a neighbourhood of p. Then, to this f the vector V will make to correspond the real number

$$V(f) = v^1\left[\frac{\partial f}{\partial x^1}\right]_p + v^2\left[\frac{\partial f}{\partial x^2}\right]_p + \dots + v^n\left[\frac{\partial f}{\partial x^n}\right]_p,$$

the directional derivative of f along V at p. Notice that this action of V on functions has two important properties:
(i) It is linear:

$$V(f + g) = V(f) + V(g).$$

(ii) It is a derivative, as it respects the Leibniz rule:

$$V(f \cdot g) = f \cdot V(g) + g \cdot V(f).$$

This notion of vector — a directional derivative — suits the best the generalization to general differential manifolds. The set of real functions on a manifold M constitutes — with the usual operations of addition, pointwise product and multiplication by real numbers — an algebra, which we shall indicate by $R(M)$. Vectors will act on functions, that is, they will extract numbers from elements of $R(M)$. This algebra will, as a consequence, play an important role in what follows.

6.2   Tangent spaces

§ 6.2.1 A differentiable curve *through a point* $p \in N$ is a differentiable curve $a : (-1, 1) \to N$ such that $a(0) = p$. It will be denoted by $a(t)$, with $t \in (-1, 1)$. When t varies in this interval, a 1-dimensional continuum of points is obtained on N. In a chart (U, ψ) around p these points will have coordinates

$$a^i(t) = u^i \circ \psi[a(t)].$$

§ 6.2.2 Let f be any differentiable real function on $U \ni p$,

$$f : U \to \mathbb{E}^1,$$

as in Figure 6.2. The vector V_p tangent to the curve $a(t)$ at point p is given by

$$V_p(f) = \frac{d}{dt}\left[(f \circ a)(t)\right]_{t=0} = \left[\frac{da^i}{dt}\right]_{t=0} \frac{\partial f}{\partial a^i}. \qquad (6.1)$$

Notice that V_p is quite independent of f, which is arbitrary. It is, as

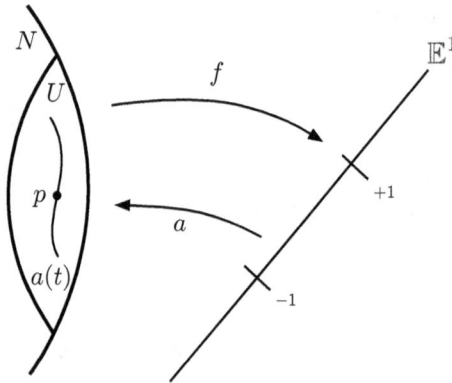

Fig. 6.2 A curve maps $\mathbb{E}^1$ into N, and a real function proceeds in the converse way. The definition of vector uses the notion of derivative on $\mathbb{E}^1$.

announced, an operator acting on the algebra $R(N)$ of real functions on N,

$$V_p : R(N) \to \mathbb{E}^1.$$

An alternative way of introducing a tangent vector is the following: suppose two curves through p, $a(t)$ and $b(t)$ with $a(0) = b(0) = p$. They are *equivalent* (intuitively: tangent to each other) at p if

$$\lim_{t \to 0}\left[\frac{a(t) - b(t)}{t}\right] = 0.$$

This is indeed an equivalence relation (tangency) and is chart-independent. The vector V_p is then defined as this equivalence class.

Now, the vector V_p, tangent at p to a curve on N, is a *tangent vector* to N at p. In the particular chart used in Eq.(6.1), (da^i/dt) are the components of V_p. Notice that, although the components are chart-dependent, the vector itself is quite independent.

§ 6.2.3 From its very definition, vector V_p satisfies:
$$V_p(\alpha f + \beta g) = \alpha V_p(f) + \beta V_p(g), \quad \forall \alpha, \beta \in \mathbb{E}^1, \quad \forall f, g \in R(N)$$
and
$$V_p(f \cdot g) = f \cdot V_p(g) + g \cdot V_p(f),$$
which is just the Leibniz rule. The formal definition of a *tangent vector* on the manifold N at the point p is then a mapping $V_p : R(N) \to \mathbb{E}^1$ satisfying conditions (i) and (ii). Multiplication of a vector by a real number gives another vector. The sum of two vectors gives a third one. So, the vectors tangent to N at a point p constitute a linear space, the *tangent space* T_pN of the manifold N at the point p.

§ 6.2.4 Given some local system of coordinate (LSC) around p, with $x(p) = (x^1, x^2, \ldots, x^n)$, the operators $\{\partial/\partial x^i\}$ satisfy the above conditions (i) and (ii) of § 6.1.3. They further span the whole space T_pN and are linearly independent. Consequently, any vector can be written in the form
$$V_p = V_p^i \frac{\partial}{\partial x^i}. \tag{6.2}$$
Notice that the coordinates are particular functions belonging to the algebra $R(N)$. The basis $\{\partial/\partial x^i\}$ is the *natural basis*, or *coordinate basis* associated to the given LSC, alternatively defined through the conditions
$$\frac{\partial x^j}{\partial x^i} = \delta_i^j. \tag{6.3}$$
The above V_p^i are, of course, the *components* of V_p in this basis. If $N = \mathbb{E}^3$, Eq.(6.2) reduces to the expression of the usual directional derivative following the vector $\boldsymbol{V}_p = (V_p^1, V_p^2, V_p^3)$, which is given by $\boldsymbol{V}_p \cdot \boldsymbol{\nabla}$. Notice further that T_pN and $\mathbb{E}^n$ are finite vector spaces of the same dimension and are consequently isomorphic. In particular, the tangent space to $\mathbb{E}^n$ at some point will be itself a copy of $\mathbb{E}^n$. Differently from the local euclidean character, which says that around each point there is an open set homeomorphic to the *topological space* $\mathbb{E}^n$, T_pN is isomorphic to the *vector space* $\mathbb{E}^n$.

§ 6.2.5 In reality, euclidean spaces are diffeomorphic to their own tangent spaces, and that explains part of their relative structural simplicity — in equations written on such spaces, one can treat indices related to the space itself and to the tangent spaces (and to the cotangent spaces defined below) on the same footing. This cannot be done on general manifolds, by reasons which will become clear later on. Still a remark: applying (6.2) to the coordinates x^i, one finds $V_p^i = V_p(x^i)$, so that
$$V_p = V_p(x^i) \frac{\partial}{\partial x^i}. \tag{6.4}$$

§ **6.2.6** The tangent vectors are commonly called simply *vectors*, or still *contravariant vectors*. As it happens to any vector space, the linear mappings

$$\omega_p : T_p N \to \mathbb{E}^1$$

constitute another vector space, denoted here $T_p^* N$, the dual space of $T_p N$. It is the *cotangent space* to N at p. Its members are *covectors*, or *covariant vectors*, or still *1-forms*. Given an arbitrary basis $\{e_i\}$ of $T_p N$, there exists a unique basis $\{\alpha^j\}$ of $T_p^* N$, called its *dual basis*, with the property

$$\alpha^j(e_i) = \delta_i^j.$$

Then, for any $V_p \in T_p N$,

$$V_p = \alpha^i(V_p)e_i. \tag{6.5}$$

Any $\omega_p \in T_p^* N$ will have, in basis $\{\alpha^j\}$, the expression

$$\omega_p = \omega_p(e_i)\,\alpha^i. \tag{6.6}$$

§ **6.2.7** The dual space is again an n-dimensional vector space. Nevertheless, its isomorphism to the tangent space is not canonical (i.e, basis – independent), and no internal product is defined on $T_p N$. An internal product will be present if a canonical isomorphism exists between a vector space and its dual. A metric (see Section 6.6) defined on a manifold establishes one such canonical isomorphism between its tangent spaces and their duals. And so does a symplectic form defined on a phase space (see Section 25.2).

§ **6.2.8** Let $f : M \to N$ be a C^∞ function between the differentiable manifolds M and N. Such a function induces a mapping

$$f_* : T_p M \to T_{f(p)} N$$

between tangent spaces (see Figure 6.3). If g is an arbitrary real function on N, $g \in R(N)$, this mapping is defined by

$$[f_*(X_p)](g) = X_p(g \circ f) \tag{6.7}$$

for every $X_p \in T_p M$ and all $g \in R(N)$. This mapping is a homomorphism of vector spaces. It is called the *differential* of f, by extension of the euclidean case: when $M = \mathbb{E}^m$ and $N = \mathbb{E}^n$, f_* is precisely the jacobian matrix. In effect, we can in this case use the identity map-

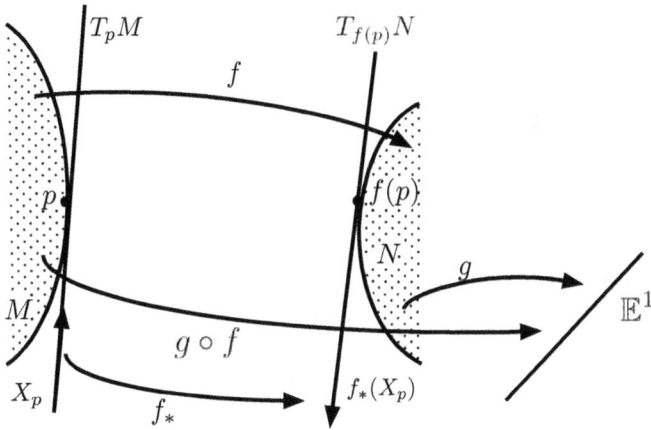

Fig. 6.3   A function f between two manifolds induces a function f_* between their tangent spaces.

pings as coordinate-homeomorphisms and write $p = (x^1, x^2, \ldots, x^m)$ and $f(p) = y = (y^1, y^2, \ldots, y^n)$. Take the natural basis,

$$X_p(g \circ f) = X_p^i \frac{\partial}{\partial x^i}[g(y)] = X_p^i \frac{\partial g}{\partial y^j} \frac{\partial y^j}{\partial x^i}.$$

Thus, the vector $f[_*(X_p)]$ is obtained from X_p by the (left-)product with the jacobian matrix:

$$[f_*(X_p)]^j = \frac{\partial y^j}{\partial x^i} X_p^i. \tag{6.8}$$

The differential f_* is also frequently written df. Let us take in the above definition $N = \mathbb{E}^1$, so that f is a real function on M (see Figure 6.4). As g in (6.7) is arbitrary, we can take for it the identity mapping. Then,

$$f_*(X_p) = df(X_p) = X_p(f). \tag{6.9}$$

§ **6.2.9** In a natural basis,

$$df(X_p) = X_p^i \frac{\partial f}{\partial x^i}. \tag{6.10}$$

Take in particular for f the coordinate function $x^j : M \to \mathbb{E}^1$. Then,

$$dx^j(X_p) = X_p(x^j). \tag{6.11}$$

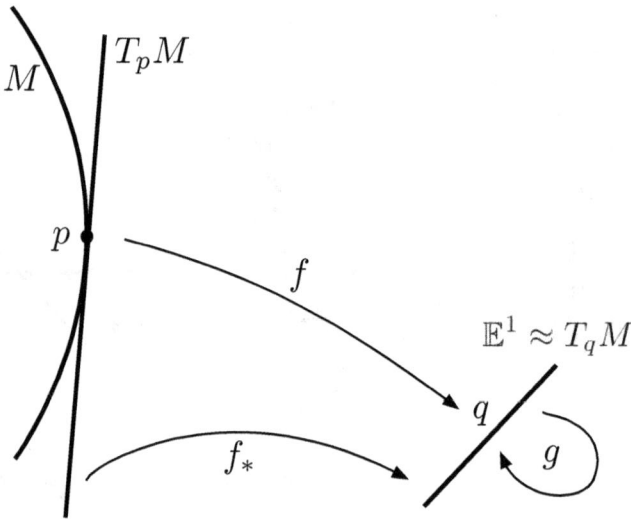

Fig. 6.4   Case of a real function on M.

For vectors belonging to the basis,

$$dx^j \left(\frac{\partial}{\partial x^i} \right) = \delta^j_i. \tag{6.12}$$

Consequently, the mappings $\{dx^j\}$ form a basis for the cotangent space, dual to the natural basis $\{\partial/\partial x^i\}$. Using (6.4), (6.10) and (6.11), one finds

$$df(X_p) = \frac{\partial f}{\partial x^i} \, dx^i(X_p).$$

As this is true for any vector X_p, we can write down the operator equation

$$df = \frac{\partial f}{\partial x^i} \, dx^i, \tag{6.13}$$

which is the usual expression for the differential of a real function. This is the reason why the members of the cotangent space are also called 1-forms: the above equation is the expression of a differential form in the natural basis $\{dx^j\}$, so that $df \in T^*_p M$. One should however keep in mind that, on a general differentiable manifold, the dx^j are not simple differentials, but linear operators: as stated in Eq. (6.11), they take a member of $T_p M$ into a number.

§ **6.2.10** As $T_p\mathbb{E}^1 \approx \mathbb{E}^1$, it follows that

$$df : T_p M \to \mathbb{E}^1.$$

If f is a real function on the real line, that is, if also $M = \mathbb{E}^1$, then all points, vectors and covectors reduce to real numbers. Taking in (6.6) $e_i = \partial/\partial x^i$ and $\alpha^i = dx^i$, we find the expression of the covector in a natural basis,

$$\omega_p = \omega_p \left(\frac{\partial}{\partial x^i} \right) dx^i. \tag{6.14}$$

As we have already stressed, if f is a function between general differentiable manifolds, $f_* = df$ will take vectors into vectors: it is a *vector-valued form*. We shall come back to such forms later on.

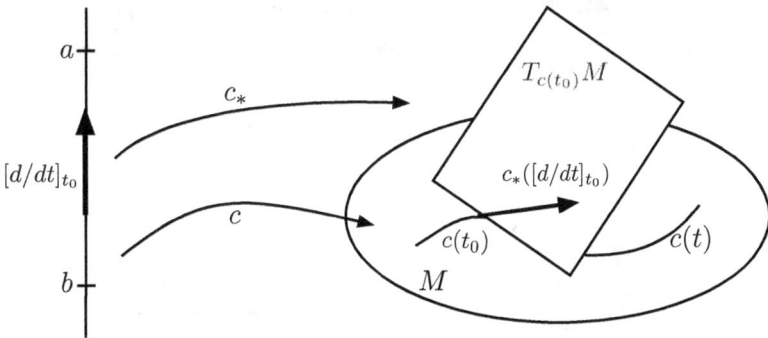

Fig. 6.5 $c_* \left([d/dt]_{t_0} \right)$ is a vector tangent to M at $c(t_0)$.

§ **6.2.11 Contact with usual notation.** Let $c : (a, b) \to M$ be a differentiable curve on M. Given the point $t_0 \in (a, b)$, then $[d/dt]_{t_0}$ is a tangent vector to (a, b) at t_0. It constitutes by itself a basis for the tangent space $T_{t_0}(a, b)$. Consequently, $c_* \left([d/dt]_{t_0} \right)$ is a vector tangent to M at $c(t_0)$ (see Figure 6.5). Given an arbitrary function $f : M \to \mathbb{E}^1$,

$$c_* \left[\frac{d}{dt} \right]_{t_0} (f) = \left[\frac{d}{dt} \right]_{t_0} (f \circ c).$$

Let us take a chart (U, x) around $c(t_0)$, as in Figure 6.6, in which the points $c(t)$ will have coordinates given by

$$x \circ c(t) = (c^1(t), c^2(t), \ldots, c^m(t)).$$

Then,

$$f \circ c(t) = [f \circ x^{<-1>}] \circ [x \circ c(t)]$$
$$= (f \circ x^{<-1>})(c^1(t), c^2(t), \ldots, c^m(t)).$$

Notice that $(f \circ x^{<-1>})$ is just the local expression of f, with the identity coordinate mapping in $\mathbb{E}^1$. In the usual notation of differential calculus, it is simply written f. In that notation, the tangent vector $c_*\,[d/dt]_{t_0}$ is fixed by

$$\left[\frac{d}{dt}(f \circ c)(t)\right]_{t_0} = \frac{\partial f}{\partial c^j}\frac{dc^j}{dt}\bigg|_{t_0} = \dot{c}^j(t_0)\,\frac{\partial f}{\partial c^j}\ .$$

The tangent vector is in this case the "velocity" vector of the curve at the point $c(t_0)$. Notice the practical way of taking the derivatives: one goes first, by inserting identity maps like $x \circ x^{<-1>}$, to a local coordinate representation of the function, and then derive it in the usual way.

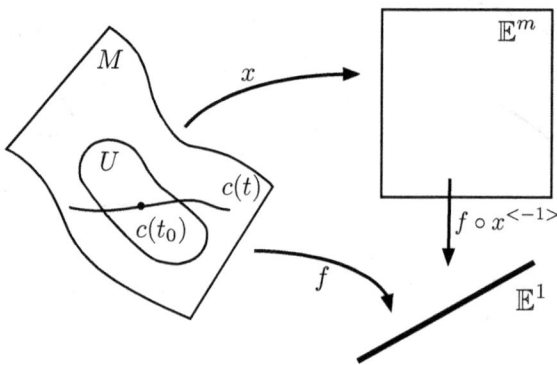

Fig. 6.6   Using a chart to show the relationship to usual notation.

§ **6.2.12** Up to isomorphisms, f_* takes $\mathbb{E}^m$ into $\mathbb{E}^n$. Suppose p is a critical point of f. Then, at p, $f_* = 0$, as the jacobian vanishes (see Section 21.5). Consider the simple case in which

$$M = S^1 = \{(x, y) \text{ such that } x^2 + y^2 = 1\}.$$

Let f be the projection

$$f(x, y) = y = (1 - x^2)^{1/2},$$

as in Figure 6.7. The jacobian reduces to

$$\frac{dy}{dx} = -x(1-x^2)^{-1/2},$$

whose critical points are $(0,1)$ and $(0,-1)$. All the remaining points are regular points of f. The points $f(x,y)$ which are images of regular points are the *regular values* of f. To the regular value P in Figure 6.7 correspond two regular points: p and p'. At p, the jacobian is positive; at p', it is negative. If we go along S^1 starting (say) from $(1,0)$ and follow the concomitant motion of the image by f, we see that the different signs of the jacobian indicate the distinct senses in which P is attained in each point of the inverse image. The sign of the jacobian at a point p is called the "degree of f at p":

$$\deg_p f = +1 \quad \text{and} \quad \deg_{p'} f = -1.$$

This is a general definition: the degree of a function at a regular point is the jacobian sign at the point. Now, the *degree of f at a regular value* is the sum of the degrees of all its counterimages:

$$\deg {}_P f = \text{sum of degrees at } f^{<-1>}(P).$$

In the example above, we have that $\deg_P f = 0$.

§ 6.2.13 An important theorem says the following: given a C^∞ function $f : M \to N$, then if M and N are connected and compact:

(i) f has regular values.
(ii) the number of counterimages of each regular value is finite.
(iii) the degree of f at a regular value is independent of that regular value.

The degree so defined depends solely on f. It is the Brouwer degree of f, denoted deg f.

§ 6.2.14 Consider the case of two circles S^1, represented for convenience on the complex plane, and the function $f : S^1 \to S^1$, $f(z) = z^n$. All points are regular in this case. All counterimages contain n points, each one of them with the same degree: $+1$ if $n > 0$, and -1 if $n < 0$. As a consequence, deg $f = n$. If we go along the domain S^1 and follow the corresponding motion in the image space, we verify that the degree counts the number of times the closed curve defined by f winds around the image space when the domain is covered once. The Brouwer degree is also known as the *winding number*. It gives the number "n" labelling the homotopy classes in

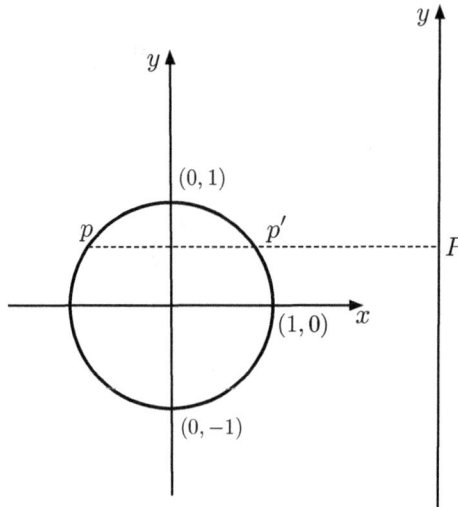

Fig. 6.7 Two regular points corresponding to one regular value by a horizontal projection of the circle.

$\pi_1(S^1)$. There is some variation in this nomenclature: the winding number is sometimes defined as a purely topological number — as we have done in section 3.4. As given above, it requires the differentiable structure and is more general, because the function f is not necessarily a loop. Including the differentiability requirement, however, does add something, because of the following important result (a special case of a theorem by Hopf):

Two differentiable maps are homotopic if and only if they have the same Brouwer degree.

Thus, the winding number characterizes the homotopy class of the mapping. In the higher dimensional case, we have mappings $S^n \to S^n$, related to the higher-order homotopy groups described in Section 3.4. As said in § 3.4.12, it is used to classify magnetic monopoles and the vacua of gauge theories.

6.3 Tensors on manifolds

§ 6.3.1 Tensors at a point p on a differentiable manifold are defined as tensors on the tangent space to the manifold at p. Let us begin by recalling the invariant (that is, basis-independent) definition of a tensor in linear algebra. Take a number r of vector spaces $V_1, V_2, \ldots, V_r$ and an extra vector space W. Take the cartesian product $V_1 \times V_2 \times \ldots \times V_r$. Then, the set $L^r(V_1, V_2, \ldots, V_r; W)$ of all the multilinear mappings of the cartesian product of r-th order into W is itself another vector space.[1] The special case $L^1(V; W)$, with W the field over which V is defined, is the dual space V^* of V; for a real vector space, $V^* = L^1(V; \mathbb{R})$.

§ **6.3.2** Consider now the cartesian product of a vector space V by itself s times, and suppose for simplicity V to be a real space. A real covariant tensor of order s on V is a member of the space $L(V \times V \times \ldots \times V; \mathbb{R})$, that is, a multilinear mapping

$$T_s^0 : \underbrace{V \times V \times \ldots \times V}_{s \text{ times}} \to \mathbb{R}.$$

By multilinear we mean, of course, that the mapping is linear on each copy of V. On the other hand, given two such tensors, say T and S, the linear (vector) structure of their own space is manifested by

$$(a\,T + b\,S)(v_1, v_2, \ldots, v_s) = a\,T(v_1, v_2, \ldots, v_s) + b\,S(v_1, v_2, \ldots, v_s).$$

§ **6.3.3 Tensor product.** The *tensor product* $T \otimes S$ is defined by

$$T \otimes S(v_1, v_2, \ldots, v_s, v_{s+1}, v_{s+2}, \ldots v_{s+q}) =$$
$$= T(v_1, v_2, \ldots, v_s)\, S(v_{s+1}, v_{s+2}, \ldots v_{s+q}), \qquad (6.15)$$

if T and S are respectively of orders s and q. The product $T \otimes S$ is, thus, a tensor of order $(s + q)$. The product $\otimes$ is clearly noncommutative in general. We have already said that the space of all tensors on a vector space is itself another linear space. With the operation $\otimes$, this overall tensor space constitutes a *noncommutative algebra* (Chapter 13).

§ **6.3.4** Let $\{\alpha^i\}$ be a basis for V^*, dual to some basis $\{e_i\}$ for V. A basis for the space of covariant s-tensors can then be built as

$$\{\alpha^{i_1} \otimes \alpha^{i_2} \otimes \alpha^{i_3} \otimes \ldots \otimes \alpha^{i_s}\}, 1 \le i_1, i_2, \ldots, i_s \le \dim V.$$

In this basis, an s-tensor T is written

$$T = T_{i_1 i_2 i_3 \cdots i_s}\, \alpha^{i_1} \otimes \alpha^{i_2} \otimes \alpha^{i_3} \otimes \ldots \otimes \alpha^{i_s}, \qquad (6.16)$$

where summation over repeated indices is implied. The *components* $T_{i_1 i_2 i_3 \cdots i_s}$ of the covariant s-tensor T are sometimes still presented as *the*

---

[1] "Vector space" here, of course, in the formal sense of linear space.

tensor, a practice mathematicians have abandoned a long while ago. Tensors as we have defined above are invariant objects, while the components are basis-dependent. On general manifolds, as we shall see later, the differentiable structure allows the introduction of tensor fields, tensors at different points being related by differentiable properties. Basis are extended in the same way. On general manifolds, distinct basis of covector fields are necessary to cover distinct subregions of the manifold. In this case, the components must be changed from base to base when we travel throughout the manifold, while the tensor itself, defined in the invariant way, remains the same. Consequently, it is much more convenient to use the tensor as long as possible, using local components only when they may help in understanding some particular feature, or when comparison is desired with results known of old. In euclidean spaces, where one coordinate system is sufficient, it is always possible to work with the natural basis globally, with the same components on the whole space. The same is true only for a few very special kinds of space (the toruses, for instance), a necessary condition for it being the vanishing of the Euler characteristic — technically, they must be *parallelizable*, a notion to be examined below (see § 6.4.13).

§ 6.3.5 In an analogous way, we define a *contravariant tensor* of order r: it is a multilinear mapping

$$T_0^r : \underbrace{V^* \times V^* \times \ldots \times V^*}_{r \text{ times}} \to \mathbb{R}.$$

The space dual to the dual of a (finite dimensional) vector space is the space itself: $(V^*)^* = V$. Given a basis $\{e_i\}$ for V, a basis for the contravariant r-tensors is

$$\{e_{i_1} \otimes e_{i_2} \otimes e_{i_3} \otimes \ldots \otimes e_{i_r}\}, 1 \leq i_1, i_2, \ldots i_r \leq \dim V.$$

A contravariant r-tensor T can then be written

$$T = T^{i_1 i_2 i_3 \cdots i_r} e_{i_1} \otimes e_{i_2} \otimes e_{i_3} \otimes \ldots \otimes e_{i_r},$$

the $T^{i_1 i_2 i_3 \cdots i_r}$ being its components in the given basis. Of course, the same considerations made for covariant tensors and their components hold here.

Commentary 6.1   Unlike the isomorphism between V and V^*, the isomorphism between $(V^*)^*$ and V is canonical (or natural), that is, basis–independent. $(V^*)^*$ and V can consequently be identified. This isomorphism $(V^*)^* = V$, however, only exists for finite-dimensional linear spaces. ◄

§ 6.3.6 A *mixed tensor*, covariant of order s and contravariant of order r, is a multilinear mapping

$$T_s^r : \underbrace{V \times V \times \ldots \times V}_{s \text{ times}} \times \underbrace{V^* \times V^* \times \ldots \times V^*}_{r \text{ times}} \to \mathbb{R}.$$

Given a basis $\{e_i\}$ for V and its dual basis $\{\alpha^i\}$ for V^*, a general mixed tensor will be written

$$T = T_{j_1 j_2 j_3 \ldots j_s}^{i_1 i_2 i_3 \ldots i_r} e_{i_1} \otimes e_{i_2} \otimes e_{i_3} \otimes \ldots \otimes e_{i_r} \otimes \alpha^{j_1} \otimes \alpha^{j_2} \otimes \alpha^{j_3} \otimes \ldots \otimes \alpha^{j_s}. \quad (6.17)$$

It is easily verified that, just like in the cases for vectors and covectors, the components are the results of applying the tensor on the basis elements:

$$T_{j_1 j_2 j_3 \ldots j_s}^{i_1 i_2 i_3 \ldots i_r} = T(\alpha^{i_1}, \alpha^{i_2}, \alpha^{i_3}, \ldots, \alpha^{i_r}, e_{j_1}, e_{j_2}, e_{j_3}, \ldots, e_{j_s}). \quad (6.18)$$

Important particular cases are:

(i) The T_0^0, which are numbers.
(ii) The T_0^1, which are vectors.
(iii) the T_1^0, which are covectors.

We see that a tensor T_s^r belongs to

$$V_1 \times V_2 \times \ldots \times V_r \times V_1^* \times V_2^* \times \ldots \times V_s^*.$$

Each copy V_i of V acts as the dual of V^*, as the space of its linear real mappings, and vice-versa. A tensor *contraction* is a mapping of the space of tensors T_s^r into the space of tensors T_{s-1}^{r-1}, in which V_i is supposed to have acted on some V_k^*, the result belonging to

$$V_1 \times V_2 \times \ldots \times V_{i-1} \times V_{i+1} \times \ldots \times V_r \times V_1^* \times V_2^* \times \ldots V_{k-1}^* \times V_{k+1}^* \ldots \times V_s^*.$$

The components in the above basis will be (note the "contracted" index j)

$$T_{m_1 m_2 m_3 \ldots m_{k-1} j \, m_{k+1} \ldots m_s}^{n_1 n_2 n_3 \ldots n_{i-1} j \, n_{i+1} \ldots n_r}.$$

§ 6.3.7 Important special cases of covariant tensors are the *symmetric* ones, those satisfying

$$T(v_1, v_2, \ldots, v_k, \ldots, v_j, \ldots) = T(v_1, v_2, \ldots, v_j, \ldots, v_k, \ldots)$$

for every j, k. An analogous definition leads to symmetric contravariant tensors. The space of all the symmetric covariant tensors on a linear space V can be made into a commutative algebra in the following way. Call $S_k(V)$ the linear space of covariant symmetric tensors of order k. The total space of symmetric covariant tensors on V will be the direct sum

$$S(V) = \bigoplus_{k=0}^{\infty} S_k(V).$$

Now, given $T \in S_k(V)$ and $W \in S_j(V)$, define the product $T W$ by

$$T W(v_1, v_2, \ldots, v_{k+j}) =$$
$$= \frac{1}{(k+j)!} \sum_{\{P\}} T(v_{P(1)}, v_{P(2)}, \ldots, v_{P(k)}) \times$$
$$\times W(v_{P(k+1)}, v_{P(k+2)}, \ldots, v_{P(k+j)}),$$

the summation taking place over all the permutations P of the indices. Notice that $T W \in S_{k+j}(V)$. With this symmetrizing operation, the linear space $S(V)$ becomes a commutative algebra. The same can be made, of course, for contravariant tensors.

§ **6.3.8** An algebra like those above defined (see § 6.3.3 and § 6.3.7), which is a sum of vector spaces,

$$V = \bigoplus_{k=0}^{\infty} V_k,$$

with the binary operation taking

$$V_i \otimes V_j \to V_{i+j},$$

is a *graded algebra*.

§ **6.3.9** Let $\{\alpha^i\}$ be a basis for the dual space V^*. A mapping $p : V \to \mathbb{E}^1$ defined by first introducing

$$P = \sum_{j_i=1}^{\dim V} P_{j_1 j_2 j_3 \cdots j_k} \alpha^{j_1} \otimes \alpha^{j_2} \otimes \alpha^{j_3} \otimes \ldots \otimes \alpha^{j_k},$$

with $P_{j_1 j_2 j_3 \ldots j_k}$ symmetric in the indices, and then putting

$$p(v) = P(v, v, \ldots, v) = \sum_{j_i=1}^{\dim V} P_{j_1 j_2 j_3 \cdots j_k} v^{j_1} v^{j_2} v^{j_3} \ldots v^{j_k}$$

gives a polynomial in the components of the vector v. The definition is actually basis-independent, and p is called a *polynomial function* of degree k. The space of such functions constitutes a linear space $P_k(V)$. The sum of these spaces,

$$P(V) = \bigoplus_{k=0}^{\infty} P_k(V),$$

is an algebra which is isomorphic to the algebra S(V) of § 6.3.7.

§ **6.3.10** Of special interest are the *antisymmetric tensors*, which satisfy

$$T(v_1, v_2, \ldots, v_k, \ldots, v_j, \ldots) = - T(v_1, v_2, \ldots, v_j, \ldots, v_k, \ldots)$$

for every pair j, k of indices. Let us examine the case of the antisymmetric covariant tensors. At a fixed order, they constitute a vector space by themselves. The tensor product of two antisymmetric tensors of order p and q is a $(p+q)$-tensor which is no more antisymmetric, so that the antisymmetric tensors do not constitute an algebra with the tensor product. We can, however, introduce another product which redresses this situation. Before that, we need the notion of *alternation* $\mathrm{Alt}(T)$ of a covariant tensor T of order s, which is a tensor of the same order defined by

$$\mathrm{Alt}(T)(v_1, v_2, \ldots, v_s) = \frac{1}{s!} \sum_{(P)} (\mathrm{sign}\, P)\, T(v_{p_1}, v_{p_2}, \ldots, v_{p_s}), \tag{6.19}$$

where the summation takes place on all the permutations P of the numbers $(1, 2, \ldots, s)$. The symbol $(\mathrm{sign}\, P)$ represents the parity of P. The tensor $\mathrm{Alt}(T)$ is antisymmetric by construction. If n is the number of elementary transpositions (Chapter 14) necessary to take $(1, 2, \ldots, s)$ into $(p_1, p_2, \ldots, p_s)$, then the parity will be $\mathrm{sign}\, P = (-)^n$.

§ **6.3.11** Given two antisymmetric tensors, ω of order p and η of order q, their *exterior product* $\omega \wedge \eta$ is the $(p+q)$-antisymmetric tensor given by

$$\omega \wedge \eta = \frac{(p+q)!}{p!q!} \mathrm{Alt}(\omega \otimes \eta).$$

This operation does make the set of antisymmetric tensors into an associative graded algebra, the *exterior algebra*, or *Grassmann algebra*. Notice that only tensors of the same order can be added, so that this algebra includes in reality all the vector spaces of antisymmetric tensors. We shall here only list some properties of real tensors which follow from the definition above:

(i) $(\omega + \eta) \wedge \alpha = \omega \wedge \alpha + \eta \wedge \alpha$;

(ii) $\alpha \wedge (\omega + \eta) = \alpha \wedge \omega + \alpha \wedge \eta$;

(iii) $a\,(\omega \wedge \alpha) = (a\,\omega) \wedge \alpha = \omega \wedge (a\,\alpha), \quad \forall\, a \in \mathbb{R}$;

(iv) $(\omega \wedge \eta) \wedge \alpha = \omega \wedge (\eta \wedge \alpha)$;

(v) $\omega \wedge \eta = (-)^{\partial_\omega \partial_\eta} \eta \wedge \omega$.

In the last property, concerning the commutation, ∂_ω and ∂_η are the orders respectively of ω and η.

If $\{\alpha^i\}$ is a basis for the covectors, the space of s-order antisymmetric tensors has a basis

$$\{\alpha^{i_1} \wedge \alpha^{i_2} \wedge \alpha^{i_3} \wedge \ldots \wedge \alpha^{i_s}\}, \ 1 \le i_1, i_2, \ldots, i_s \le \dim V.$$

An antisymmetric s-tensor can then be written

$$\omega = \frac{1}{s!} \omega_{i_1 i_2 i_3 \ldots i_s} \alpha^{i_1} \wedge \alpha^{i_2} \wedge \alpha^{i_3} \wedge \ldots \wedge \alpha^{i_s}, \tag{6.20}$$

the $\omega_{i_1 i_2 i_3 \ldots i_s}$'s being the components of ω in this basis. The space of antisymmetric s-tensors reduces automatically to zero for $s > \dim V$. Notice further that the dimension of the vector space formed by the antisymmetric covariant s-tensors is

$$\binom{\dim V}{s}.$$

The dimension of the whole Grassmann algebra is $2^{\dim V}$.

§ **6.3.12** A very important property of the exterior product is that it is preserved by mappings between manifolds. Let $f : M \to N$ be such a mapping and consider the antisymmetric s-tensor $\omega_{f(p)}$ on the vector space $T_{f(p)}N$. The function f determines then a tensor on $T_p M$ through

$$(f^* \omega)_p(v_1, v_2, \ldots, v_s) = \omega_{f(p)}(f_* v_1, f_* v_2, \ldots, f_* v_s). \tag{6.21}$$

Thus, the mapping f induces a mapping f^* between the tensor spaces, working however in the inverse sense (see the scheme of Figure 6.8): f^* is suitably called a *pull-back* and f_* is sometimes called, by extension, *pushfoward*. To make (6.21) correct and well-defined, f must be C^1. The pull-back has the following properties:

$$f^* \text{ is linear.} \tag{6.22}$$

$$f^*(\omega \wedge \eta) = f^* \omega \wedge f^* \eta. \tag{6.23}$$

$$(f \circ g)^* = g^* \circ f^*. \tag{6.24}$$

The pull-back, consequently, preserves the exterior algebra. Notice, however, the order-inversion in the last property above.

§ **6.3.13** Antisymmetric covariant tensors on differential manifolds are called differential forms. In a natural basis $\{dx^j\}$,

$$\omega = \frac{1}{s!} \omega_{j_1 j_2 j_3 \ldots j_s} dx^{j_1} \wedge dx^{j_2} \wedge dx^{j_3} \wedge \ldots \wedge dx^{j_s}.$$

The well-defined behaviour when mapped between different manifolds renders the differential forms the most interesting of all tensors. But then, as announced, we shall come to them later on.

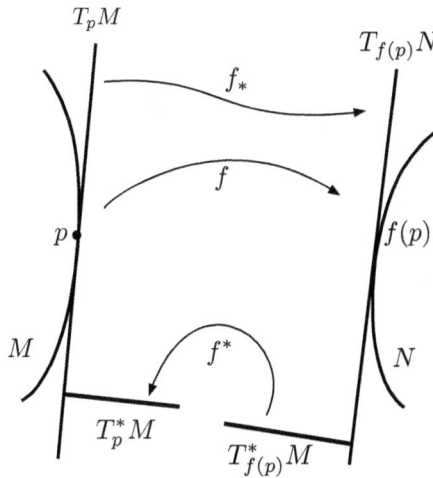

Fig. 6.8   A function f induces a push-foward f_* and a pull-back f^*.

§ 6.3.14 Let us now go back to differentiable manifolds. A tensor at a point $p \in M$ is a tensor defined on the tangent space T_pM. One can choose a chart around p and use for T_pM and T_p^*M the natural bases $\{\partial/\partial x^i\}$ and $\{dx^j\}$. A general tensor can then be written

$$T_s^r = T_{j_1 j_2 j_3 \ldots j_s}^{i_1 i_2 i_3 \ldots i_r} \frac{\partial}{\partial x^{i_1}} \otimes \frac{\partial}{\partial x^{i_2}} \otimes \frac{\partial}{\partial x^{i_3}} \otimes \ldots$$

$$\ldots \otimes \frac{\partial}{\partial x^{i_r}} \otimes dx^{j_1} \otimes dx^{j_2} \otimes dx^{j_3} \ldots \otimes dx^{j_s}. \tag{6.25}$$

In another chart, the natural basis will be $\{\partial/\partial x^{i'}\}$ and $\{dx^{j'}\}$, the same tensor being written

$$T_s^r = T_{j_1' j_2' j_3' \ldots j_s'}^{i_1' i_2' i_3' \ldots i_r'} \frac{\partial}{\partial x^{i_1'}} \otimes \frac{\partial}{\partial x^{i_2'}} \otimes \frac{\partial}{\partial x^{i_3'}} \otimes \frac{\partial}{\partial x^{i_r'}} \otimes \ldots dx^{j_1'} \otimes dx^{j_2'} \ldots dx^{j_3'} \otimes dx^{j_s'}$$

$$= T_{j_1' j_2' j_3' \ldots j_s'}^{i_1' i_2' i_3' \ldots i_r'} \frac{\partial x^{i_1}}{\partial x^{i_1'}} \otimes \frac{\partial x^{i_2}}{\partial x^{i_2'}} \otimes \frac{\partial x^{i_3}}{\partial x^{i_3'}} \otimes \frac{\partial x^{i_r}}{\partial x^{i_r'}} \otimes \ldots \frac{\partial x^{j_1'}}{\partial x^{j_1}} \frac{\partial x^{j_2'}}{\partial x^{j_2}} \cdots \frac{\partial x^{j_s'}}{\partial x^{j_s}}$$

$$\frac{\partial}{\partial x^{i_1}} \otimes \frac{\partial}{\partial x^{i_2}} \otimes \frac{\partial}{\partial x^{i_3}} \otimes \ldots \otimes \frac{\partial}{\partial x^{i_r}} \otimes \ldots dx^{j_1} \otimes dx^{j_2} \otimes dx^{j_3} \otimes \ldots \otimes x^{j_s},$$

$$\tag{6.26}$$

which provides the transformation rule for the components under changes of coordinates in the charts' intersection. Changes of basis unrelated to coordinate changes will be examined later on. We find frequently tensors defined by Eq. (6.26): they are introduced as those entities whose components transform in that way.

§ 6.3.15 It should be understood that a tensor is always a tensor with respect to a given group. In Eq. (6.26), the group of coordinate transformations is involved. General basis transformations (Section 6.5 below) constitute another group, and the general tensors above defined are related to that group. Usual tensors in $\mathbb{E}^3$ are actually tensors with respect to the group of rotations, $SO(3)$. Some confusion may arise because rotations may be represented by coordinate transformations in $\mathbb{E}^3$. But not every transformation is representable through coordinates, and it is advisable to keep this in mind.

6.4   Fields and transformations

Fields

§ 6.4.1 Let us begin with an intuitive view of vector fields. In the preceding sections, vectors and tensors have been defined at a fixed point p of a differentiable manifold M. Although we have been lazily negligent about this aspect, the natural bases we have used are actually $\{\,[\partial/\partial x^i]_p\,\}$. Suppose now that we extend these vectors throughout the whole chart's coordinate neighbourhood, and that the components are differentiable functions

$$f^i : M \to \mathbb{R}, \quad f^i(p) = X_p^i.$$

New vectors are then obtained, tangent to M at other points of the coordinate neighbourhood. Through changes of charts, vectors can eventually be got all over the manifold.

Now, consider a fixed vector at p, tangent to some smooth curve: it can be continued in the above way along the curve. This set of vectors, continuously and differentiably related along a differentiable curve, is a vector field. At point p,

$$X_p : R(M) \to \mathbb{R}.$$

At different points, X will map $R(M)$ into different points of $\mathbb{R}$, that is, a vector field is a mapping $X : R(M) \to R(M)$. In this way, generalizing that of a vector, one gets the formal definition of a vector field:

A *vector field* X on a smooth manifold M is a linear mapping
$X : R(M) \to R(M)$ obeying the Leibniz rule:
$X(f \cdot g) = f \cdot X(g) + g \cdot X(f)$, f, g $\in R(M)$.

§ 6.4.2 The tangent bundle. A vector field is so a differentiable choice of a member of T_pM for each p of M. It can also be seen as a mapping from M into the set of all the vectors on M, the union

$$TM = \bigcup_{p \in M} T_pM,$$

with the proviso that p is taken into T_pM:

$$X : M \to TM, \quad X : p \to X_p \in T_pM.$$

Given a function $f \in R(M)$, then $(Xf)(p) = X_p(f)$. In order to ensure the correctness of this second definition, one should establish a differentiable structure on the $2m$-dimensional space TM. Let π be a function

$$\pi : TM \to M, \quad \pi(X_p) = p,$$

to be called *projection* from now on.

§ 6.4.3 As for covering spaces (§ 3.3.15), open sets on TM are *defined* as those sets which can be obtained as unions of sets of the type $\pi^{-1}(U)$, with U an open set of M (so that π is automatically continuous). Given a chart (V, x) on M, such that $V \ni p$, we define a chart for TM as $(\tilde{V}, \tilde{x})$ with

$$\tilde{V} = \pi^{-1}(V) \tag{6.27}$$

$$\tilde{x} : \tilde{V} \to x(V) \times \mathbb{E}^m, \tag{6.28}$$

$$\tilde{x} : X_p \to (x^1(p), x^2(p), \ldots, x^m(p), X_p^1, X_p^2, \ldots, X_p^m), \tag{6.29}$$

where X_p^i are the components in

$$X_p = X_p^i \left[\frac{\partial}{\partial x^i} \right]_p.$$

Given another chart $(\tilde{W}, \tilde{y})$ on M, with $\tilde{W} \cap \tilde{V} \neq \emptyset$, the mapping $\tilde{y} \circ \tilde{x}^{<-1>}$, defined by

$$\tilde{y} \circ \tilde{x}^{<-1>} \left(x^1(p), x^2(p), \ldots, x^m(p), X_p^1, X_p^2, \ldots, X_p^m \right) =$$
$$= \left(y^1 \circ x^{<-1>}(x^1, x^2, \ldots, x^m), y^2 \circ x^{<-1>}(x^1, x^2, \ldots, x^m), \ldots, \right.$$
$$\left. y^m \circ x^{<-1>}(x^1, x^2, \ldots, x^m), Y_p^1, Y_p^2, \ldots, Y_p^m \right),$$

where (see § 6.2.8)

$$Y_p^I = \left(\text{jacobian of } y \circ x^{<-1>} \right)_j^i X_p^j$$

is differentiable. The two charts are, in this way, differentiably related. A complete atlas can in this way be defined on TM, making it into a differentiable manifold.

This differentiable manifold TM is the *tangent bundle*, the simplest example of a differentiable fiber bundle, or bundle space. The tangent space to a point p, T_pM, is called, in the bundle language, the *fiber* on p. The field X itself, defined by Eq.(6.27), is a *section* of the bundle. Notice that the bundle space in reality depends on the projection π for the definition of its topology.

§ **6.4.4 The commutator** ... Take the field X, given in some coordinate neighbourhood as $X = X^i(\partial/\partial x^i)$. As $X(f) \in R(M)$, one could consider the action of another field $Y = Y^j(\partial/\partial x^j)$ on $X(f)$:

$$YXf = Y^j \frac{\partial}{\partial x^j}(X^i)\frac{\partial f}{\partial x^i} + Y^j X^i \frac{\partial^2 f}{\partial x^j \partial x^i}.$$

This expression tells us that the operator YX, defined by

$$(YX)f = Y(Xf),$$

does not belong to the tangent space, due to the presence of the last term. This annoying term is symmetric in XY, and would disappear under anti-symmetrization. Indeed, as easily verified, the commutator of two fields

$$[X,Y] := (XY - YX) = \left(X^i \frac{\partial Y^j}{\partial x^i} - Y^i \frac{\partial X^j}{\partial x^i}\right)\frac{\partial}{\partial x^j} \qquad (6.30)$$

does belong to the tangent space and is another vector field.

§ **6.4.5 ... and its algebra.** The operation of commutation defines on the space TM a structure of linear algebra. It is easy to check that

$$[X,X] = 0$$

and

$$[[X,Y],Z] + [[Z,X],Y] + [[Y,Z],X] = 0,$$

the latter being the Jacobi identity. An algebra satisfying these two conditions is a *Lie algebra*. Thus, the vector fields on a manifold constitute, with the operation of commutation, a Lie algebra.

§ **6.4.6** Notice that a diffeomorphism f preserves the commutator:

$$f_*[X,Y] = [f_*X, f_*Y].$$

Furthermore, given two diffeomorphisms f and g, $(f \circ g)_*X = f_* \circ g_*X$.

§ **6.4.7 The cotangent bundle.** Analogous definitions lead to general tensor bundles. In particular, consider the union

$$T^*M = \bigcup_{p \in M} T_p^*M.$$

A covariant vector field, cofield or 1-*form* ω is a mapping

$$\omega : M \to T^*M$$

such that

$$\omega(p) = \omega_p \in T_p^*M, \ \forall \ p \in M.$$

§ **6.4.8** This corresponds, in just the same way as has been seen for the vectors, to a differentiable choice of a covector on each $p \in M$. In general, the action of a form on a vector field X is denoted

$$\omega(X) = <\omega, X>,$$

so that

$$\omega : TM \to R(M). \tag{6.31}$$

§ **6.4.9** In the dual natural basis (in other words, locally),

$$\omega = \omega_j dx^j. \tag{6.32}$$

Fields and cofields can be written respectively

$$X = dx^i(X)\frac{\partial}{\partial x^i} \ = \ <dx^i, X>\frac{\partial}{\partial x^i}, \tag{6.33}$$

$$\omega = \omega\left(\frac{\partial}{\partial x^i}\right)dx^i \ = \ <\omega, f\frac{\partial}{\partial x^i}>dx^i. \tag{6.34}$$

§ **6.4.10** The cofield bundle above defined is the cotangent bundle, or the *bundle of forms*. We shall see later (Chapter 7) that not every 1-form is the differential of a function. Those who are differentials of functions are called *exact* forms.

§ **6.4.11** We have obtained vector and covector fields. An analogous procedure leads to tensor fields. We first consider the tensor algebra over a point $p \in M$, consider their union for all p, topologize and smoothen the resultant set, then define a general tensor field as a section of this tensor bundle.

§ **6.4.12** At each $p \in M$, we have two m-dimensional vector spaces, T_pM and T_p^*M, of course isomorphic. Nevertheless, their isomorphism is not *natural* (or *canonical*). It depends on the chosen basis. Different bases fix isomorphisms taking the same vector into different covectors. Only the presence of an internal product on T_pM (due for instance to the presence of a metric, a case which will be examined later) can turn the isomorphism into a natural one. By now, it is important to keep in mind the total distinction between vectors and covectors.

§ **6.4.13** Think of $\mathbb{E}^n$ as a vector space: its tangent vector bundle is a Cartesian product. A tangent vector to $\mathbb{E}^n$ at a point p can be completely specified by a pair (p, V), where also V is a vector in $\mathbb{E}^n$. This comes from the isomorphism between each $T_p\mathbb{E}^n$ and $\mathbb{E}^n$ itself. Forcing a bit upon the trivial, we say that $\mathbb{E}^n$ is parallelizable, the same vector V being defined at all the different points of the manifold $\mathbb{E}^n$.

Given a general manifold M, it is said to be *parallelizable* if its tangent bundle is *trivial*, that is, a mere Cartesian product $TM = M \times \mathbb{E}^m$. In this case, a vector field V can be globally (that is, everywhere on M) given by (p, V). Recalling the definition of a vector field as a section on the tangent bundle, this means that there exists a global section. Actually, the existence of a global section implies the triviality of the bundle. This holds for any bundle: if some global section exists, the bundle is a Cartesian product.

Here there are some unforeseen results. All toruses are parallelizable. On the other hand, of all the spheres S^n, only S^1, S^3 and S^7 are parallelizable. The sphere S^2 is not — a result sometimes called *the hedgehog theorem*: you cannot comb a hairy hedgehog so that all its prickles stay flat. There will be always at least one point like the crown of the head.

The simplest way to find out whether M is parallelizable or not is based on the simple idea that follows: consider a vector $V \neq 0$. Then, the vector field $(\,, V)$ will not vanish at any point of M. Suppose that we are able to show that no vector field on M is everywhere nonvanishing. This would imply that TM is not trivial.

A necessary condition for parallelizability is the vanishing of the Euler–Poincaré characteristic of M. A very important point is:

All Lie groups are parallelizable differentiable manifolds.

§ **6.4.14 Dynamical systems.** Dynamical systems are described in Classical Physics by vector fields in the "phase" space $(q, \dot{q})$. Consider the free fall of a particle of unit mass under the action of gravity: call x the height and y the velocity, $y = \dot{x}$. From

$$\dot{y} = - g \text{ (constant)},$$

one gets the velocity in "phase" space $(v_x, v_y) = (y, -g)$. A scheme of this vector field is depicted in Figure 6.9.

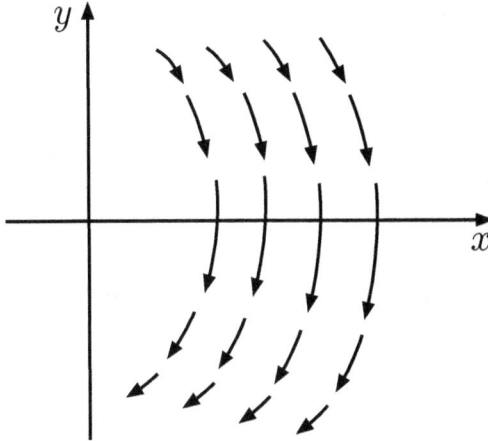

Fig. 6.9   Vector field scheme for $\dot{y} = -g$.

A classical system is completely specified by its velocity field in "phase" space, which fixes its time evolution (Chapter 25). Initial conditions simply choose one of the lines in the "flow diagram". Well, we should perhaps qualify such optimistic statements. In general, this perfect knowledge does not imply complete predictability. Small indeterminations in the initial conditions may be so amplified during the system evolution that after some time they cover the whole configuration space (see Section 15.2). This happens even with a simple system like the double oscillator with non-commensurate frequencies. The above example is precisely the field vector characterization of the system of differential equations

$$\dot{x} = y; \ \dot{y} = -g.$$

The modern approach to systems of differential equations is based on the idea of vector field.[2]

---

[2] A detailed treatment of the subject, with plenty of examples, is given in the little masterpiece Arnold 1973.

§ 6.4.15 Dynamical systems: maps.   Dynamical systems are also described, mainly in model building, by iterating maps like

$$x_{n+1} = f(x_n),$$

where x is a vector describing the state of some system. To help visualization, we may consider n as a discrete time. The state at the n-th stage is given by a function of the $(n-1)$-th stage, and so by its n-th iterate applied on the initial seed state x_0. The set of points $\{x_n\}$ by which the system evolves is the *orbit* of the system.

   An important concept in both the flow and the map pictures is the following: suppose there is a compact set A to which the sequence x_n (or, in the flow case, the state when t becomes larger and larger) converges for a given subset of the set of initial conditions. It may consist of one, many or infinite points and is called an *attractor*. It may also happen that A is a fractal, in which case it is a *strange* (or *chaotic*) *attractor*.[3] This is the case of the simple mapping

$$f : \boldsymbol{I} \to \boldsymbol{I}, \ \boldsymbol{I} = [0,1], f(x) = 4\,\lambda\,x\,(1-x),$$

popularly known as the "logistic map", which for certain values of $\lambda \in \boldsymbol{I}$ tends to a strange attractor akin to a Cantor set. Strange attractors are fundamental in the recent developments in the study of chaotic behaviour in non-linear dynamics.[4]

§ 6.4.16 Let us go back to the beginning of this chapter, where a vector at $p \in M$ was defined as the tangent to a curve $a(t)$ on M, with $a(0) = p$. It is interesting to associate a vector to each point of the curve by liberating the variation of the parameter t in Eq.(6.1):

$$X_{a(t)}(f) = \frac{d}{dt}\,(f \circ a)(t). \tag{6.35}$$

   This being so, $X_{a(t)}$ is the *tangent field to* $a(t)$, while $a(t)$ is the *integral curve* of X through p. In general, this is possible only locally, in a neighbourhood of p. When X is tangent to a curve globally, the above definition being extendable all over M, X is said to be a *complete field*. Let us, for the sake of simplicity, take a neighbourhood U around p and suppose that

---

   [3] See Farmer, Ott & Yorke 1983, where a good discussion of dimensions is also given.
   [4] For a short review, see Grebogi, Ott & Yorke 1987. A fairly complete discussion is given in Ott 1994. A matrix approach to continuous interpolation of iterative maps can be found in Aldrovandi 2001, mainly in Section 15.2; a generalization can be found in Aldrovandi 2014.

$a(t) \in U$, with coordinates $(a^1(t), a^2(t), \ldots, a^m(t))$. It follows then from (6.35) that

$$X_{a(t)} = \frac{da^i}{dt} \frac{\partial}{\partial a^i} \, . \tag{6.36}$$

In this case,

$$X_{a(t)}(a^i) = \frac{da^i}{dt} \tag{6.37}$$

is the component $X^i_{a(t)}$. In this sense, the field whose integral curve is $a(t)$ is given by da/dt. In particular, $X_p = [da/dt]_{t=0}$. Conversely, if a field is given by its components $X^k(x^1(t), x^2(t), \ldots, x^m(t))$ in some natural basis, its integral curve $x(t)$ is obtained by solving the system of differential equations $X^k = dx^k/dt$. The existence and unicity of solutions for such systems is in general valid only locally.

Transformations

Let us now address ourselves to what happens to differentiable manifolds under infinitesimal transformations to which, in a way, vector fields preside. More precisely, we examine the behaviour of general tensors under continuous transformations. The basic tool is the Lie derivative, which measures the variation of a tensor when small displacements take place on the manifold. We shall here mainly consider 1-dimensional displacements along a field (local) integral curve. The general multi-dimensional case will be seen later (Section 8.2).

§ **6.4.17** The *action of the group* $\mathbb{R}$ of the real numbers on the manifold M is defined as a differentiable mapping

$$\lambda : \mathbb{R} \times M \to M$$
$$\lambda : (t, p) \to \lambda(t, p)$$

satisfying

$$\lambda(0, p) = p$$
$$\lambda(t + s, p) = \lambda(t, \lambda(s, p)) = \lambda(s, \lambda(t, p)),$$

for all $p \in M$ and all $s, t \in \mathbb{R}$.

§ **6.4.18** At a fixed value of t, $\lambda(t, p)$ is a mapping

$$\lambda_t : M \to M, \quad \lambda_t : p \to \lambda(t, p),$$

a collective displacement of all the points of M. At fixed p, it is a mapping

$$\lambda_p : \mathbb{R} \to M, \quad \lambda_p : t \to \lambda(t, p),$$

which for each $p \in M$ describes a curve

$$\gamma(t) = \lambda_p(t),$$

the "*orbit* of p generated by the action of the group $\mathbb{R}$". The mapping λ is a 1-*parameter group* on M.

§ **6.4.19** The action so defined is a particular example of actions of Lie groups (of which $\mathbb{R}$ is a case) on manifolds. We shall see later (section 8.2) the general case. Notice that, being 1-dimensional, group $\mathbb{R}$ is abelian. Mathematicians use to call, by a mechanical analogy, M the *phase space*, λ the *flow*, and $\mathbb{R} \times M$ the *enlarged phase space*. Due to the special kind of group, it can be shown that only one orbit goes through each point p of M.

§ **6.4.20** Take a classical mechanical system and let its phase space M (see Chapter 25) be specified as usual by the points

$$(q, p) = (q^1, q^2, \ldots, q^n, p_1, p_2, \ldots, p_n).$$

The time evolution of the system, if the hamiltonian function is $H(q, p)$, is governed by the *hamiltonian flow*, which for a conservative system is

$$\lambda_{(q_0, p_0)}(t) = e^{tH(q_0, p_0)}; \qquad (q_0, p_0) \to (q_t, p_t).$$

Given a domain $U \subset M$, the Liouville theorem says that the above flow preserves its volume:

$$\text{vol } [\lambda(t)U] = \text{vol } [\lambda(0)U].$$

Suppose now that M itself has a finite volume. Then, after a large enough time interval, forcibly

$$(\lambda(t)U) \cap U \neq \emptyset.$$

In simple words: given any neighbourhood U of a state point (q, p), it contains at least one point which comes back to U for $t >$ some t_r. For large enough periods of time, a system comes back as near as one may wish to its initial state. In a more sophisticated language: for finite M, the set of points (q_t, p_t) (that is, the orbit) is *everywhere dense* in M. This is Poincaré's "théorème du rétour".

§ 6.4.21 Let $M = \mathbb{E}^3$, and $\bar{x} = (\bar{x}^1, \bar{x}^2, \bar{x}^3)$ a fixed point different from zero. Then,

$$\lambda_t(x) = (x^1 + \bar{x}^1 t, x^2 + \bar{x}^2 t, x^3 + \bar{x}^3 t)$$

defines a C^∞ action of $\mathbb{R}$ on M. For each $t \in \mathbb{R}$,

$$\lambda_t : \mathbb{E}^3 \to \mathbb{E}^3$$

is a translation taking x into $x + \bar{x}t$. Each vector $\bar{x}$ determines a translation. The orbits are the straight lines parallel to $\bar{x}$.

§ 6.4.22 To each flow λ corresponds a vector field: the *infinitesimal operator* (or *generator*) of λ is the field X defined by

$$X_p f = \lim_{\Delta t \to 0} \left\{ \frac{1}{\Delta t} \left[f(\lambda_p(\Delta t)) - f(p) \right] \right\}, \tag{6.38}$$

on each $p \in M$ and arbitrary $f \in R(M)$. A field X is thus a derivation along the differentiable curve $\gamma(t) = \lambda_p(t)$, which is its integral curve.

With $q = \lambda_p(t_0)$, we have the following:

$$\dot{\lambda}_p(t_0) f = \lambda_{p*} \left([d/dt]_{t_0} \right) f = [d/dt]_{t_0} (f \circ \lambda_p)$$

$$= \lim_{\Delta t \to 0} \frac{1}{\Delta t} \left\{ [f \circ \lambda_p(t_0 + \Delta t) - f \circ \lambda_p(t_0)] \right\}$$

$$= \lim_{\Delta t \to 0} \frac{1}{\Delta t} \left\{ f[\lambda(q + \Delta t)] - f(q) \right\}$$

$$= \lim_{\Delta t \to 0} \frac{1}{\Delta t} \left\{ f[\lambda_q(\Delta t)] - f(q) \right\} = X_{\lambda_p(t_0)} f.$$

As f is any element of $R(M)$, we have indeed

$$\dot{\lambda}_p(t_0) = X_{\lambda_p(t_0)}. \tag{6.39}$$

The above definition generalizes to manifolds, though only locally, the well known case of matrix transformations engendered by an invertible matrix $g(t) = \exp[tX]$, of which the matrix X is the generator,

$$X = [dg/dt]_{t=0} = Xe^{tX}|_{t=0}.$$

A matrix Y will transform according to

$$Y' = g(t)Yg^{-1}(t) = e^{tX}Ye^{-tX} \approx (1 + tX)Y(1 - tX) \approx Y + t[X,Y]$$

to first order in t, and we find the commutator in the role of the "first derivative":

$$[X,Y] = \lim_{t \to 0} \frac{1}{t} \left[g(t)\, Y\, g^{-1}(t) - Y \right].$$

§ **6.4.23** Take $M = \mathbb{E}^2$ and $\lambda \colon \mathbb{R} \times M \to M$ given by

$$\lambda(t, (x, y)) = (x + t, y),$$

that is, translations along the x-axis. The infinitesimal operator is then $X = d/dx$.

§ **6.4.24** We have seen that, given the action λ, we can determine the field X which is its infinitesimal generator. The inverse is not true in general, but holds locally: every field X generates locally a 1-parameter group. The restriction is related to the fact that to find out the integral curve we have to integrate differential equations (§ 6.4.16), for which the existence and unicity of solutions is in general only locally granted.

§ **6.4.25 Lie derivative.** In section 6.2 we have introduced the derivative of a differentiable function f along the direction of a vector

$$X : df(X_p) = X_p f.$$

It was a generalization to a manifold M of the directional derivative of a function on $\mathbb{E}^m$. Things are a bit more complicated when we try to derive more general objects. We face, to begin with, the problem of finding the variation rate of a vector field Y at $p \in M$ with respect to X_p. This can be done by using the fact that X generates locally a 1-parameter group, which induces an isomorphism

$$\lambda_{t*} : T_p M \to T_{\lambda_t(p)} M,$$

as well as its inverse λ_{-t*}. It becomes then possible to compare values of vector fields. We shall just state three different definitions, which can be shown to be equivalent.

The Lie derivative of a vector field Y on M with respect to the vector field X on M, at a point $p \in M$, is given by any of the three expressions:

$$\begin{aligned}
(L_X Y)_p &= \lim_{t \to 0} \frac{1}{t} \left[\lambda_{-t*}(Y_{\lambda(t,p)}) - Y_p \right] \\
&= \lim_{t \to 0} \frac{1}{t} \left[Y_p - \lambda_{t*}(Y_{\lambda(-t,p)}) \right] \\
&= -\left[\frac{d}{dt} \{ \lambda_{t*}(Y) \} \right]_{t=0}.
\end{aligned} \tag{6.40}$$

Each expression is more convenient for some different purpose. Notice that the vector character of Y is preserved by the Lie derivative: $L_X Y$ is a vector field. Let us examine the definition given in the first equality of Eqs.(6.40) (see Figure 6.10). The action $\lambda_p(t)$ induces an isomorphism between the

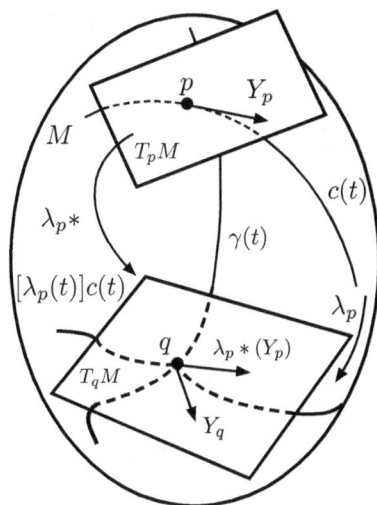

Fig. 6.10   Scheme for the Lie derivative.

tangent spaces T_pM and T_qM, with $q = \lambda_p(t)$. By this isomorphism, Y_p is taken into $\lambda_p(t)_*(Y_p)$, which is in general different from Y_q, the value of Y at q. By using the inverse isomorphism $\lambda_p(-t)_*$ we bring Y_q back to T_pM. In this last vector space we compare and take the limit. As it might be expected, the same definition can also be shown to reduce to

$$L_X Y = [X, Y].$$

One should observe that the concept of Lie derivative does not require any extra structure on the differentiable manifold M. Given the differentiable structure, Lie derivatives are automatically present.

§ **6.4.26** Let us consider, as shown in Figure 6.11, a little planar model for the water flow in a river of breadth $2a$: take the velocity field

$$X = (a^2 - y^2)e_1,$$

with

$$e_1 = \frac{\partial}{\partial x} \quad \text{and} \quad e_2 = \frac{\partial}{\partial y}.$$

It generates the 1-parameter group

$$\lambda_X(t, p) = (a^2 - y^2)te_1 + p,$$

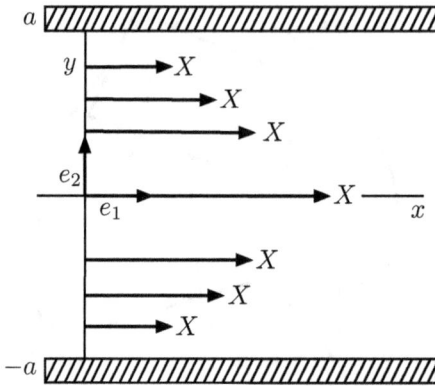

Fig. 6.11 A planar model for a river flow.

or
$$\lambda_X(t,p) = [x + (a^2 - y^2)t]e_1 + ye_2,$$
with $p = (x, y)$. The flow leaves the border points invariant. Consider now a constant transversal field, $Y = e_2$. It generates the group
$$\lambda_Y(s,p) = se_2 + p = xe_1 + (s+y)e_2.$$
A direct calculation shows that
$$(\lambda_Y\lambda_X - \lambda_X\lambda_Y)(p) = st\,(s+2y)e_1$$
or, to the lowest order, $(2sty)e_1$. The commutator $[X, Y]$ is precisely $2ye_1$, with the group
$$\lambda_{[X,Y]}(r,p) = 2\,y\,r\,e_1 + p = (x + 2\,y\,r)e_1 + ye_2.$$
From another point of view: examine the effect of
$$\lambda_{X*} : T_pM \to T_{\lambda_X(t,p)}M,$$
which is
$$\lambda_{X*}(Y_p)(f(x,y)) = Y_p(f \circ \lambda_X) = \frac{\partial}{\partial y}\left[f(x + (a^2 - y^2)t, y)\right]$$
$$= \left[-2yt\frac{\partial}{\partial x} + \frac{\partial}{\partial y}\right]f = -t\,[X,Y]f + Yf.$$
In this case,
$$-\frac{1}{t}\left\{\lambda_{X*}(t,p)(Y_p) - Y_p\right\}(f) = [X,Y]f,$$
which is the expression of the third definition in (6.40). Thus, *the Lie derivative turns up when we try to find out how a field Y experiences a small transformation generated by another field X.*

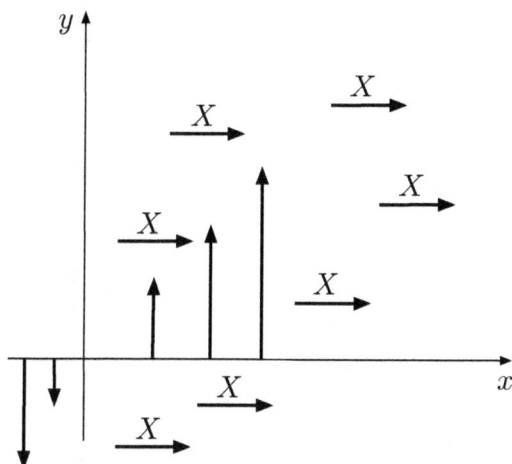

Fig. 6.12   Scheme for fields $X = e_1$ and $Y = xe_2$.

§ **6.4.27** Consider on the plane (Figure 6.12) the two fields

$$X = e_1 \text{ (constant)} \quad \text{and} \quad Y = xe_2,$$

again with $e_1 = \partial/\partial x$ and $e_2 = \partial/\partial y$. The integral curves are:

$$\gamma_X(s) = s$$

along e_1 and

$$\gamma_Y(t) = xt$$

along e_2. The groups generated by X and Y are:

$$\lambda_X(s,p) = p + se_1 = (x + s)e_1 + ye_2,$$
$$\lambda_Y(t,p) = p + xte_2 = xe_1 + (y + xt)e_2.$$

The Lie derivative measures the non-commutativity of the corresponding groups. We check easily that

$$\lambda_X\left[s, \lambda_Y(t,p)\right] - \lambda_Y\left[t, \lambda_X(s,p)\right] = -st\, e_2.$$

On the other side,

$$\lambda_{[X,Y]}(r,p) = re_2 + p.$$

We have drawn these transformations in Figure 6.13, starting at point $p = (2,1)$, and using $s = 2$ and $t = 1$. The difference is precisely that generated by the above commutator.

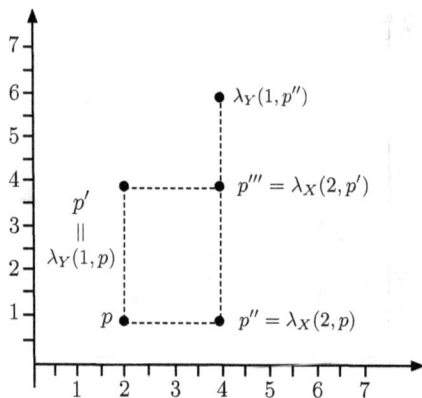

Fig. 6.13   Particular transformations for the previous scheme.

§ **6.4.28** Lie derivatives are a vast subject.[5] We can here only list some of their properties:

(i) Commutator:

$$[L_X, L_Y] = L_{[X,Y]}. \tag{6.41}$$

(ii) Jacobi identity:

$$[[L_X, L_Y], L_Z] + [[L_Z, L_X], L_Y] + [[L_Y, L_Z], L_X] = 0. \tag{6.42}$$

(iii) Function multiplying field:

$$L_X(fY) = (L_X f)Y + f L_X Y. \tag{6.43}$$

(iv) In a natural basis $\{\partial_i = \partial/\partial x^i\}$, they satisfy

$$L_{\partial_j}(Y) = \frac{\partial Y^i}{\partial x^j} \frac{\partial}{\partial x^i}. \tag{6.44}$$

This expression shows that the Lie derivative appears as a coordinate-independent version of the partial derivative.

(v) Take a basis $\{e_i\}$, in which $X = X^i e_i$ and $Y = Y^j e_j$. Then,

$$L_X Y = X(Y^i)e_i - Y(X^i)e_i + X^i Y^j L_{e_j} e_i. \tag{6.45}$$

§ **6.4.29** The Lie derivative of a covector field ω is defined by

$$(L_X \omega)Y = L_X(\omega(Y)) - \omega(L_X Y). \tag{6.46}$$

Thus,

$$(L_X \omega)Y = X(\omega(Y)) - \omega([X, Y]). \tag{6.47}$$

---

[5] For more details, see for instance Schutz 1985.

§ 6.4.30 This comes out as a consequence of the general definition for the Lie derivative of a tensor field of any kind, which is [cf. Eq.(6.40)]

$$(L_X T)_p = \lim_{t \to 0} \frac{1}{t} \left\{ T_p - \lambda_{t*}(T_{\lambda(-t,p)}) \right\}. \tag{6.48}$$

The maps induced by the 1-parameter group are to be taken as push-forward and/or pull-backs, according to the contravariant and/or covariant character of the tensor. Applied to a function f, this gives simply $X(f)$. Once the action of the Lie derivative is known on fields and cofields, the general definition is given as

$$
\begin{aligned}
(L_X T)(Y_1, Y_2, \ldots Y_s, &\omega^1, \omega^2, \ldots \omega^r) = \\
&= L_X[T(Y_1, Y_2, \ldots Y_s, \omega^1, \omega^2, \ldots \omega^r)] \\
&- T(L_X Y_1, Y_2, \ldots, Y_s, \omega^1, \omega^2, \ldots \omega^r) \\
&- T(Y_1, L_X Y_2, \ldots Y_s, \omega^1, \omega^2, \ldots \omega^r) - \ldots \\
&- T(Y_1, Y_2, \ldots, L_X Y_s, \omega^1, \omega^2, \ldots \omega^r) \\
&- T(Y_1, Y_2, \ldots, Y_s, L_X \omega^1, \omega^2, \ldots \omega^r) \\
&- T(Y_1, Y_2, \ldots Y_s, \omega^1, L_X \omega^2, \ldots \omega^r) - \ldots \\
&- T(Y_1, Y_2, \ldots, Y_s, \omega^1, \omega^2, \ldots, L_X \omega^r).
\end{aligned}
\tag{6.49}
$$

§ 6.4.31 Notice that L_X preserves the tensor character: it takes an $\binom{s}{r}$ tensor into another tensor of the same type. In terms of the components: in a natural basis $\{\partial_i\}$, the components of $L_X T$ are

$$
\begin{aligned}
(L_X T)^{ab\ldots r}_{ef\ldots s} = {}& X(T^{ab\ldots r}_{ef\ldots s}) - (\partial_i X^a) T^{ib\ldots r}_{ef\ldots s} - (\partial_i X^b) T^{ai\ldots r}_{ef\ldots s} - \ldots \\
& - (\partial_i X^r) T^{ab\ldots i}_{ef\ldots s} + (\partial_e X^i) T^{ab\ldots r}_{if\ldots s} + (\partial_f X^i) T^{ab\ldots r}_{ei\ldots s} + \ldots \\
& + (\partial_s X^i) T^{ab\ldots r}_{ei\ldots i}.
\end{aligned}
\tag{6.50}
$$

§ 6.4.32 The Lie derivative L_X provides the infinitesimal changes of tensorial objects under the 1-parameter group of transformations of which X is the generator. For this reason, Lie derivatives are basic instruments in the study of transformations imposed on differentiable manifolds (automorphisms) and, *a fortiori*, in the study of symmetries (see Section 8.2).

§ 6.4.33 We have seen in § 6.4.16 that, given a vector field X on a manifold M, there exists *locally* a curve on M which integrates it. Thus, there is a 1-dimensional manifold which is tangent to X. Unless the field is complete, the curve can only be an immersed submanifold (remember what has been said in section 5.3). We may consider many fields at a time, and ask for the general condition to relate fields to an imbedded submanifold N of M.

To begin with, a submanifold is a manifold, so that, if X and Y are tangent to N, so must be $[X, Y]$ (§ 6.4.4). Consider then a set of $n(\leq m)$ fields tangent to M. At each point p, they will generate some subspace of $T_p M$. If they are linearly independent, they generate a subspace of dimension n. Suppose this linear independence holds for all $p \in M$. Such an assignment of an n-dimensional subspace $D_p M$ of $T_p M$ for each $p \in M$ is called a *distribution* (not to be mistaken by singular functions !). If around each p there is an open set U and fields X_1, $X_2, \ldots, X_n$ forming a basis for $D_q M$ for all $q \in U$, then the distribution is said to be a *differentiable distribution*. The distribution is said to be *involutive* if it contains the commutator of every pair of its fields. Suppose now that N is an imbedded submanifold of M, with $i : N \to M$ being the imbedding. Then N is an *integral manifold* of the distribution if

$$i_*(T_p N) = D_p M \ \text{ for all } p \in M.$$

When there is no other integral manifold containing N, N is a "maximal" integral manifold. This gives the complete story of the relationships between the spaces tangent to a manifold and the spaces tangent to a submanifold.

Now there is a related and very strong result, concerning integrability around each point, the *Frobenius theorem*:

> *Given an involutive differentiable distribution on a manifold M, then through every point $p \in M$ there passes a (unique) maximal integral manifold $N(p)$, such that any other integral manifold through p is a submanifold of $N(p)$.*

Thus, the main condition for local integrability is the involutive character of the field set: all things being differentiable, there is a local submanifold around the point p whenever at p the fields do close a Lie algebra.

6.5 Frames

§ **6.5.1** Given a differentiable manifold M and an open set $U \subset M$, a set $\{X_i\}$ of m vector fields is a local *basis* of fields (or local *frame*) if, for any $p \in U$, $\{X_{(p)i}\}$ is a basis for $T_p M$. This means that each $X_{(p)i}$ is a tangent vector to M at p and that the $X_{(p)i}$'s are linearly independent. In principle, any set of m linearly independent fields can be used as a local basis. For some manifolds there exists a global basis. For most, only local bases exist.

§ **6.5.2** In particular, around every $p \in M$ there is a chart (U, x) and the set of fields

$$\left\{ \frac{\partial}{\partial x^i} \right\} : U \to TU, \quad p \to \left\{ \left[\frac{\partial}{\partial x^i} \right]_p \right\}$$

constitutes a basis. Field bases of this kind, directly related to local coordinates, are called *holonomous* (or holonomic) bases, or *coordinate* bases. A condition for a basis $\{X_i\}$ to be holonomous is that, for any two of its members, say X_j and X_k,

$$[X_j, X_k](f) = 0$$

for all $f \in R(M)$. Of course, this happens for $\{\partial/\partial x^i\}$ but it should be clear that this property is exceptional: most bases do not consist of all-commuting fields, and are called *anholonomic*, or *non-coordinate* bases.

§ **6.5.3** Take for example the ordinary spherical coordinates (r, θ, φ) in $\mathbb{E}^3$. The related holonomous basis is $(\partial_r, \partial_\theta, \partial_\varphi)$. We have seen that in $\mathbb{E}^3$ a vector is precisely the directional derivative; nevertheless, this basis does not give the usual form of the gradient. The velocity, for example, would be

$$V = V^r \partial_r + V^\theta \partial_\theta + V^\varphi \partial_\varphi$$

with components

$$V^r = \frac{dr}{dt} ; \quad V^\theta = \frac{d\theta}{dt} ; \quad V^\varphi = \frac{d\varphi}{dt}.$$

Usually, however, the velocity components are taken to be

$$V^r = \frac{dr}{dt} ; \quad V^\theta = r \frac{d\theta}{dt} ; \quad V^\varphi = r \sin\theta \frac{d\varphi}{dt},$$

which correspond to the anholonomous basis

$$X_r = \frac{\partial}{\partial r} ; \quad X_\theta = \frac{1}{r} \frac{\partial}{\partial \theta} ; \quad X_\varphi = \frac{1}{r \sin\theta} \frac{\partial}{\partial \varphi}. \tag{6.51}$$

These fields do not all commute with each other.

§ **6.5.4** We have seen that the commutator of two fields is another field. We can expand the commutator of two members of an anholonomic basis in that same basis,

$$[X_i, X_j] = C^k{}_{ij} X_k, \tag{6.52}$$

where the $C^k{}_{ij}$'s are called the *structure coefficients* of the basis algebra. For the above spherical basis the non-vanishing coefficients are

$$C^\theta{}_{r\theta} = C^\varphi{}_{r\varphi} = -\frac{1}{r}; \quad C^\varphi{}_{\theta\varphi} = -\frac{1}{r\sin\theta}$$

and their permutations in the lower indices (in which the coefficients are clearly antisymmetric). Notice: the coefficients are not necessarily constant and depend on the chosen basis. Clearly, a necessary condition for the basis to be holonomic is that $C^k{}_{ij} = 0$ for all commutators of the basis members. This condition, $C^k{}_{ij} = 0$ for all basis members, may be shown to be also sufficient for holonomy. The Jacobi identity, required by the Lie algebra, implies

$$C^n{}_{kl}C^i{}_{jn} + C^n{}_{jk}C^i{}_{ln} + C^n{}_{lj}C^i{}_{kn} = 0. \tag{6.53}$$

§ **6.5.5** Let us re-examine the question of frame transformations. Given two natural basis on the intersection of two charts, a field X will be written

$$X = X^i \frac{\partial}{\partial x^i} = X^{i'} \frac{\partial}{\partial x^{i'}}.$$

The action of X on the function $x^{j'}$ leads to

$$X^{j'} = \frac{\partial x^{j'}}{\partial x^i} X^i. \tag{6.54}$$

This expression gives the way by which field components in natural bases change when these bases are themselves changed. Here, basis transformations are intimately related to coordinate transformations. However, other basis transformations are possible: for example, going from the holonomic basis $(\partial_r, \partial_\theta, \partial_\varphi)$ to the basis $(X_r, X_\theta, X_\varphi)$ of Eq. (6.51) in the spherical case above is a basis transformation unrelated to a change of coordinates.

§ **6.5.6** Given an anholonomous basis $\{X_i\}$, it will always be possible to write locally each one of its members in some coordinate basis as

$$X_i = X_i^j \frac{\partial}{\partial x^j} .$$

By using a differentiable atlas, the components can be in principle obtained all over the manifold. Each change of natural basis will give new components according to

$$X_i^{k'} = X_i^j \frac{\partial x^{k'}}{\partial x^j}. \tag{6.55}$$

Notice that basis $\{X_i\}$ would be holonomous only if $X_i^j = \partial x^j/\partial y^i$, where $\{y^i\}$ is some other coordinate system. In that case, $\{X_i = \partial/\partial y^i\}$. General matrices (X_i^j) are not of this form, and an holonomous basis is more of an exception than a rule. More generally, a basis transformation will be given by

$$X_i^{k'} = X_i^j A_j^{k'}, \qquad (6.56)$$

where A is some matrix. Notice that each basis is characterized by the matrix (X_i^k) of its components in some previously chosen basis. Just above, a natural basis was chosen. The tangent spaces, being isomorphic to $\mathbb{E}^m$, possess each one a "canonical basis" of the type

$$v_1 = (1, 0, 0, \ldots, 0);$$
$$v_2 = (0, 1, 0, \ldots, 0);$$
$$\vdots \qquad (6.57)$$
$$v_m = (0, 0, 0, \ldots, 1).$$

The important point is that we can choose some starting basis from which all the other basis are determined by the matrices of their components. Such $m \times m$ matrices belong to the general linear space of $m \times m$ real matrices. As they are forcibly non-singular (otherwise the linear independence would fail and we would have no basis) and consequently invertible, they constitute the linear group $GL(m, \mathbb{R})$. Starting from one basis we obtain every other basis in this way, one basis for each transformation, one basis for each element of the group. The set of all basis at each point $p \in M$ is thus isomorphic to the linear group. But the transformation matrices A of Eq.(6.56) also belong to the group, so that we have a case of a group acting on itself.

Due to the peculiar form of the action shown in (6.56), we say that the transformations *act on the right* on the field basis, or that we have a *right-action* of the group. The frequent use of natural basis (in general more convenient for calculations) is responsible for some confusion between coordinate transformations and basis transformations, which are actually quite distinct.

§ **6.5.7** The case of covector field basis is analogous. Two natural basis are related by

$$dx^{j'} = \frac{\partial x^{j'}}{\partial x^i} dx^i. \qquad (6.58)$$

The elements of another basis $\{\alpha^i\}$ can be written as $\alpha^i = \alpha_j^i dx^j$ and will transform according to

$$\alpha_{j'}^i = \frac{\partial x^k}{\partial x^{j'}} \alpha_k^i. \qquad (6.59)$$

Under a general transformation,

$$\alpha^i_{j'} = A^k_{j'} \alpha^i_k, \tag{6.60}$$

so that the group of transformations acts *on the left* on the 1-form basis. Dual basis transform inversely to each other, so that, under the action, the value $< \omega, X >$ is invariant. That is to say that $< \omega, X >$ is basis-independent.

§ 6.5.8 The bundle of linear frames. Let B_pM be the set of all linear basis for T_pM. As we have said, it is a vector space and a group, just $GL(m, \mathbb{R})$. In a way similar to that used to build up TM as a manifold, the set

$$BM = \bigcup_{p \in M} B_pM$$

of all the basis on the manifold M can be viewed as a manifold. To begin with, we define a projection $\pi : BM \to M$, with

$$\pi\big(\{X_{(p)i}\} \in B_pM\big) = p.$$

A topology is defined on BM by taking as open sets the sets $\pi^{-1}(U)$, for each U an open set of M. Given a chart (U, x) of M, a basis at $p \in U$ is given by

$$(x^1, x^2, \ldots, x^m, X^1_1, X^2_1 \ldots, X^m_1, X^1_2 \ldots X^m_2, \ldots, X^1_m \ldots X^m_m),$$

where X^j_i is the j-th component of the i-th basis member in the natural basis. This gives the $(m + m^2)$ coordinates of a "point" on BM. It is possible to show that the mapping

$$U \times GL(m, \mathbb{R}) \to \mathbb{E}^{m+m^2}$$

is a diffeomorphism. Consequently, BM becomes a smooth manifold, the *bundle of linear frames* on M. We arrive thus to another fundamental fiber bundle. Let us list some of its characteristics:

(i) The group $GL(m, \mathbb{R})$ acts on each B_pM on the right [see Eq.(6.56)]; B_pM is here the *fiber* on p; this group of transformations is called the *structure group* of the bundle.

(ii) BM is locally trivial in the sense that every point $p \in M$ has a neighbourhood U such that $\pi^{-1}(U)$ is diffeomorphic to $U \times GL(m, \mathbb{R})$.

(iii) Concerning dimension:

$$\dim BM = \dim M + \dim GL(m, \mathbb{R}) = m + m^2.$$

§ 6.5.9 The fiber itself is $GL(m, \mathbb{R})$. A fiber bundle whose fiber coincides with the structure group is a *principal fiber bundle*. A more detailed study of bundles will be presented later on. Let us here only advance another concept. The tangent bundle has the spaces $T_p M$ as fibers. The action of $GL(m, \mathbb{R})$ on the basis can be thought of as an action on $T_p M$ itself: it is the group of linear transformations, taking a vector into some other. A bundle of this kind, on whose fibers (as vector spaces) the same group acts, is said to be an *associated bundle* to the principal bundle. Most common bundles are vector bundles on which some group acts. The main interest of principal bundles comes from the fact that properties of associated bundles are deducible from those of the principal bundle.

Coordinates, which are in general local characterizations of points on a manifold, are usually related to a local frame. One first chooses a frame at a certain point, consider the euclidean tangent space supposing it as "glued" to the manifold at the point, make its origin as a vector space (that is, the zero vector) to coincide with the point, then introduce cartesian coordinates, and finally move to any other coordinate system one may wish. By a change of frame, the set of coordinates will transform according to $x' = Ax$, or $x^{j'} = A_i^{j'} x^i$, as any contravariant vector. This leads to the differential

$$dx^{j'} = dA_i^{j'} x^i + A_i^{j'} dx^i.$$

Many physical problems involve comparison of rates of change of vector quantities in two different frames (recall for example the case of the "body" and the "space" frames in the rigid body motion, see Section 26.3.5 and on). Consider a general vector u, with

$$du^{j'} = dA_i^{j'} u^i + A_i^{j'} du^i.$$

The rate of change with a parameter t (usually time) will be

$$\frac{du^{j'}}{dt} = \frac{dA_i^{j'}}{dt} u^i + A_i^{j'} \frac{du^i}{dt}.$$

A velocity, for example, as seen from two frames, will have its components related by

$$v^{j'} = A_i^{j'} v^i + \frac{dA_i^{j'}}{dt} v^i.$$

Of course, we are here supposing that also the frames are in relative motion. We shall have more to say about such "moving frames" later (§ 9.3.5).

6.6   Metric and riemannian manifolds

The usual 3-dimensional euclidean space $\mathbb{E}^3$ consists of the set $\mathbb{R}^3$ of ordered triples plus the topology defined by the 3-dimensional balls. Such balls are defined through the use of the euclidean metric, a tensor whose components are, in the global cartesian coordinates, constant and given by $g_{ij} = \delta_{ij}$. We may thus say that $\mathbb{E}^3$ is $\mathbb{R}^3$ plus the euclidean metric. We use precisely this metric to measure lengths in our everyday life. It happens frequently that another metric is simultaneously at work on the same $\mathbb{R}^3$.

Suppose, for example, that the space is permeated by a medium endowed with a point-dependent refractive index (that is, a point-dependent electric and/or magnetic permeability) $n(p)$. Light rays (see Chapter 28) will in this case "feel" another metric, which will be

$$g'_{ij} = n^2(p)\delta_{ij}$$

if $n(p)$ is isotropic. To "feel" means that they will bend, acquire a "curved" aspect if looked at by euclidean eyes (like ours). Light rays will become geodesics of the new metric, the "straightest" possible curve if measurements are made using g'_{ij} instead of g_{ij}. As long as we proceed to measurements using only light rays, distances will be different from those given by the euclidean metric. Suppose further that the medium is some compressible fluid, with temperature gradients and all which is necessary to render point-dependent the derivative of the pressure p with respect to the fluid mass–density ρ at a fixed entropy S. The sound velocity will be given by

$$c_s^2 = \left(\frac{\partial p}{\partial \rho}\right)_S ,$$

and the sound propagation will be governed by geodesics of still another metric,

$$g''_{ij} = \frac{1}{c_S^2}\,\delta_{ij}.$$

Nevertheless, in both cases we use also the euclidean metric to make measurements, and much of geometrical optics and acoustics comes from comparing the results in both metrics involved. This is only to call attention to the fact that there is no such a thing like *the* metric of a space. It happens frequently that more than one is important in a given situation (for an example in elasticity, see Section 28.3). Let us approach the subject a little more formally.

§ **6.6.1** In the space of differential forms, a basis dual to the basis $\{X_i\}$ for fields in TM is given by those ω^j such that

$$\omega^j(X_i) \;=\; <\omega^j, X_i> \;=\; \delta_i^j, \tag{6.61}$$

so that $\omega = <\omega, X_j> \omega^j$. Given a field $Y = Y^i X_i$ and a form $z = z_j \omega^j$,

$$<z, Y> \;=\; z_i Y^i. \tag{6.62}$$

§ **6.6.2** Bilinear forms are covariant tensors of second order, taking $TM \times TM$ into $R(M)$. Recall that the tensor product of two linear forms w and z is defined by

$$(w \otimes z)(X, Y) = w(X) \cdot z(Y). \tag{6.63}$$

Given a basis $\{\omega^j\}$ for the space of 1-forms, the products $\omega^i \otimes \omega^j$, with $i, j = 1, 2, \ldots, m$, form a basis for the space of covariant 2-tensors, in terms of which a bilinear form g is written

$$g = g_{ij}\, \omega^i \otimes \omega^j. \tag{6.64}$$

Of course, in a natural basis,

$$g = g_{ij}\, dx^i \otimes dx^j. \tag{6.65}$$

The most fundamental bilinear form appearing in Physics is the Lorentz metric on $\mathbb{R}^4$, which defines Minkowski space and whose main role is to endow it with a partial ordering, that is, causality.[6]

§ **6.6.3** A *metric* on a smooth manifold is a bilinear form, denoted $g(X, Y)$, $X \cdot Y$ or $< X, Y >$, satisfying the following conditions:

(i) Of course, it is *bilinear*:

$$X \cdot (Y + Z) = X \cdot Y + X \cdot Z,$$

$$(X + Y) \cdot Z = X \cdot Z + Y \cdot Z.$$

(ii) It is *symmetric*:

$$X \cdot Y = Y \cdot X.$$

(iii) It is *non-singular*:

$$\text{If } X \cdot Y = 0 \text{ for every field } Y, \text{ then } X = 0.$$

§ **6.6.4** In the basis introduced in § 6.6.2, we have

---

[6] See Zeeman 1964.

$$g(X_i, X_j) = X_i \cdot X_j = g_{mn}\, \omega^m(X_i)\, \omega^n(X_j),$$

so that

$$g_{ij} = g(X_i, X_j) = X_i \cdot X_j. \tag{6.66}$$

The relationship between metrics and general frames (in particular, bases on 4-dimensional spacetimes, called four-legs, vierbeine or, more commonly, tetrads) will be seen in Chapter 9. As $g_{ij} = g_{ji}$ and we commonly write simply $\omega^i \omega^j$ for the symmetric part of the bilinear basis, then

$$\omega^i \omega^j = \omega^{(i} \otimes \omega^{j)} = \tfrac{1}{2}\left(\omega^i \otimes \omega^j + \omega^j \otimes \omega^i\right),$$

and we have

$$g = g_{ij}\omega^i \omega^j \tag{6.67}$$

or, in a natural basis,

$$g = g_{ij}dx^i dx^j. \tag{6.68}$$

§ 6.6.5 This is the usual notation for a metric. Notice also the useful symmetrizing notation (ij) for indices: in that notation, all indices $(ijk\ldots)$ inside the parenthesis are to be symmetrized. For antisymmetrization the usual notation is $[ijk\ldots]$, with square brackets, and means that all the indices inside the brackets are to be antisymmetrized.

Knowledge of the diagonal terms is enough for a metric: the off-diagonal may be obtained by *polarization*, that is, by using the identity

$$g(X, Y) = \tfrac{1}{2}\left[g(X + Y, X + Y) - g(X, X) - g(Y, Y)\right].$$

§ 6.6.6 A metric establishes a relation between vector and covector fields: Y is said to be the *contravariant image* of a form z if, for every X,

$$g(X, Y) = z(X).$$

If, in the dual bases $\{X_i\}$ and $\{\omega^j\}$, $Y = Y^i X_i$ and $z = z_j \omega^j$, then $g_{ij}Y^j = z_i$. In this case, we write simply $z_j = Y_j$. That is the usual role of the covariant metric, to *lower* indices, taking a vector into the corresponding covector. If the mapping $Y \rightarrow z$ so defined is onto, the metric is *non-degenerate*. This is equivalent to saying that the matrix (g_{ij}) is invertible. A contravariant metric $\hat{g}$ can then be introduced whose components are the elements of the matrix inverse to (g_{ij}). If w and z are the covariant images of X and Y, defined in a way inverse to the image given above, then

$$\hat{g}(w, z) = g(X, Y). \tag{6.69}$$

§ 6.6.7 All this defines on each T_pM and T_p^*M an internal product

$$(X, Y) := (w, z) := g(X, Y) = \hat{g}(w, z). \tag{6.70}$$

A beautiful case of the field-form duality created by a metric is found in hamiltonian optics, in which the momentum (eikonal gradient) is related to the velocity by the refractive index metric (see Section 30.3). There are many other in Physics. Let us illustrate by a howlingly simple example not only the relation of 1-forms to fields, but also that of both to linear partial differential equations. Consider on the plane the function (x, y are cartesian coordinates, a and b real constants)

$$f(x, y) = \frac{x^2}{a^2} + \frac{y^2}{b^2}.$$

Each case $f(x, y) = C$ (constant) represents an ellipse. The complete family of ellipses is represented by the gradient form df; that family is just the set of solutions of the differential equation $df = 0$. But f is also solution of the set of differential equations $X(f) = 0$, where X is the field

$$X = \frac{a^2}{x} \partial_x - \frac{b^2}{y} \partial_y.$$

Thus, a differential equation is given either by $df = 0$ or by a vector field. In the first case the form is the gradient of the solution, which vanishes at each value C. In the second case the solution must be tangent to the given field. The form is "orthogonal" to the solution curve, that is, it vanishes when applied to any tangent vector: $df(X) = X(f) = 0$. Thus, a curve is the integral of a field through tangency, and of a cofield through "gradiency". The word "orthogonal" was given quotation marks because no metric connotation is given to $df(X) = 0$. Of course, multiplying f by a constant will change nothing. The same idea is trivially extended to higher dimensions. In the example, we have started from a solution. We may start at a region around a point (x, y) and eventually obtain from the form

$$df = \frac{2x}{a^2} dx + \frac{2y}{b^2} dy$$

some local solution $f = Tdf$ (see § 7.2.11 for a systematic method to get it). This solution can be extended to the whole space, giving the whole ellipse. This is a special case, as of course not every field or cofield is integrable. In most cases they are only locally integrable, or non-integrable at all.

Suppose now that a metric g_{ij} is present, which relates fields and cofields. In the case above, the metric given by the matrix

$$(g_{ij}) = \text{diag} \left(1/a^2, 1/b^2 \right)$$

is of evident interest, as $f(v) = g(v,v)$, with v the position vector (x,y). To the vector v of components (x^j) will correspond the covector of components $(p_k = g_{kj}x^j)$ and the action of this covector on v will give simply

$$p(v) = p_k x^k = g_{ij}x^i x^j.$$

As we are also in a euclidean space, the euclidean metric $m_{ij} = \delta_{ij}$ may be used to help intuition. We may consider p and v as two euclidean vectors of components (p_k) and (x^k). Comparison of the two metrics is made by using $g(v,v) = m(p,v)$. Consider the curve

$$p(v) = g(v,v) = m(p,v) = C,$$

which is an ellipse. The vector v gives a point on the ellipse and the covector p, now assimilated to an euclidean vector, is orthogonal to the curve at each point, or to its tangent at the point. This construction, allowing one to relate a 1-form to a field in the presence of a non-trivial metric, is very much used in Physics. For rigid bodies, the metric m is the inertia tensor, the vector v is the angular velocity and its covector is the angular momentum. The ellipsoid is the inertia ellipsoid, the whole construction going under the name of Poinsot (more details can be found in section 26.3.10). In crystal optics, the Fresnel ellipsoid $\varepsilon_{ij}x^i x^j = C$ regulates the relationship between the electric field $\boldsymbol{E}$ and the electric displacement

$$\boldsymbol{D} = \varepsilon \boldsymbol{E},$$

where the metric is the electric permeability (or dielectric) tensor. In this case, another ellipsoid is important, given by the inverse metric ε^{-1}: it is the index, or Fletcher's ellipsoid (Section 30.7). In all the cases, the ellipsoid is defined by equating some hamiltonian to a constant.

§ **6.6.8** An important property of a space V endowed with an internal product is the following: given *any linear function* $f \in R(V)$, there is a unique $v_f \in V$ such that, for every $u \in V$,

$$f(u) = (u, v_f).$$

So, the forms include all the real linear functions on $T_p M$ (which is expected, they constituting its dual space), and the vectors include all the real linear functions on $T_p^* M$ (equally not unexpected, the dual of the dual being the space itself). The presence of a metric establishes a natural (or canonical) isomorphism between a vector space (here, $T_p M$) and its dual.

§ 6.6.9 The above definition has used fixed bases. As in general no base covers the whole manifold, convenient transformations are to be performed in the intersections of the definition domains of every pair of bases. If some of the above metric-defining conditions are violated at a point p, it can eventually come from something wrong with the basis: for instance, it may happen that two of the X_i are degenerate at p. A real singularity in the metric should be basis-independent. Non-degenerate metrics are called *semi-riemannian*. Although physicists usually call them just *riemannian*, mathematicians more frequently reserve this denomination to *non-degenerate positive-definite* metrics,

$$g : TM \times TM \to \mathbb{R}_+.$$

As it is not definite positive, the Lorentz metric — as repeatedly said — does not define balls and is consequently unable to provide for a topology on Minkowski spacetime.

§ 6.6.10 A *riemannian manifold* is a smooth manifold on which a riemannian metric is defined. A theorem (see Section 15.1.6) due to Whitney states that:

> *It is always possible to define at least one riemannian metric on an arbitrary differentiable manifold.*

§ 6.6.11 A metric is presupposed in any measurement: lengths, angles, volumes, etc. We may begin by introducing the *length of a vector field* X as

$$||X|| = (X, X)^{1/2}. \tag{6.71}$$

The length of a curve $\gamma : (a, b) \to M$ is then defined as

$$L_\gamma = \int_a^b \left|\left|\frac{d\gamma}{dt}\right|\right| dt. \tag{6.72}$$

§ 6.6.12 Given two points $p, q \in M$, a riemannian manifold, we consider all the piecewise differentiable curves γ with $\gamma(a) = p$ and $\gamma(b) = q$. The *distance* between p and q is the infimum of the lengths of all such curves between them:

$$d(p, q) = \inf_{\{\gamma(t)\}} \int_a^b \left|\left|\frac{d\gamma}{dt}\right|\right| dt. \tag{6.73}$$

In this way, a metric tensor defines a distance function on M.

Commentary 6.2   A metric is *indefinite* when $||X|| = 0$ does not imply $X = 0$. It is the case of the Lorentz metric for vectors on the light cone. ◀

§ **6.6.13 Motions** are transformations of a manifold into itself which preserve a metric given *a priori*. They are also called *isometries* in modern texts, but this term in general includes also transformations between different spaces. When represented by field vectors on the manifold, Eq.(6.50) will give the components of the Lie derivative:

$$(L_X g)_{\mu\nu} = X^\alpha \partial_\alpha g_{\mu\nu} + (\partial_\mu X^\alpha) g_{\alpha\nu} + (\partial_\nu X^\alpha) g_{\mu\alpha}.$$

Using the properties

$$(\partial_\mu X^\alpha) g_{\alpha\nu} = \partial_\mu X_\nu - X^\alpha \partial_\mu g_{\alpha\nu}$$

and

$$(\partial_\nu X^\alpha) g_{\alpha\mu} = \partial_\nu X_\mu - X^\alpha \partial_\nu g_{\alpha\mu},$$

it becomes

$$(L_X g)_{\mu\nu} = X^\alpha (\partial_\alpha g_{\mu\nu} - \partial_\mu g_{\alpha\nu} - \partial_\nu g_{\alpha\mu}) + \partial_\mu X_\nu + \partial_\nu X_\mu.$$

If we define the Christoffel symbol

$$\Gamma^\alpha{}_{\mu\nu} = \Gamma^\alpha{}_{\nu\mu} = \tfrac{1}{2} g^{\alpha\beta} \left[\partial_\mu g_{\beta\nu} + \partial_\nu g_{\beta\mu} - \partial_\beta g_{\mu\nu} \right], \tag{6.74}$$

whose meaning will become clear later (§ 9.4.23), then the Lie derivative acquires the form

$$(L_X g)_{\mu\nu} = \partial_\mu X_\nu - \Gamma^\alpha{}_{\mu\nu} X_\alpha + \partial_\nu X_\mu - \Gamma^\alpha{}_{\nu\mu} X_\alpha.$$

Introducing the "covariant derivative"

$$X_{\mu;\nu} = \partial_\nu X_\mu - \Gamma^\alpha{}_{\mu\nu} X_\alpha, \tag{6.75}$$

it can be written as

$$(L_X g)_{\mu\nu} = X_{\mu;\nu} + X_{\nu;\mu}. \tag{6.76}$$

The condition for isometry, $L_X g = 0$, then becomes

$$X_{\mu;\nu} + X_{\nu;\mu} = 0, \tag{6.77}$$

which is the *Killing equation*.[7] A field X satisfying it is a *Killing field* (the name Killing vector is more usual). There are powerful results concerning Killing fields.[8] For example, on a manifold M, the maximum number of

---

[7] See Davis & Katzins 1962.
[8] See Eisenhart 1949, chap.VI.

Killing fields is $m(m+1)/2$ and this number is attained only on spaces with constant curvature. It is a good exercise to find that the generators of the motions on Minkowski space are of two types:

$$J_{(\alpha)} = \partial_\alpha,$$

which generate translations, and

$$J_{(\alpha\beta)} = x_\alpha \partial_\beta - x_\beta \partial_\alpha,$$

generators of Lorentz transformations. Together, these operators generate the Poincaré group. Invariance under translations bespeaks spacetime homogeneity. Invariance under Lorentz transformations signifies spacetime isotropy. Such properties are seldom present in other manifolds: they may have analogues only on constant curvature spacetimes.[9]

The metrics concerned with light rays and sound waves, referred to in the introduction of this section, are both obtained by multiplying all the components of the euclidean metric by a given function. A transformation like

$$g_{ij} \to g'_{ij} = f(p)\, g_{ij}$$

is called a *conformal transformation*. Because in the measurements of angles the metric appears in a numerator and in a denominator, both metrics will give the same angle measurements. We say that conformal transformations preserve the angles, or the cones.

§ **6.6.14** Geometry, the very word witnesses it, has had a very strong historical relation to metric. Speaking of "geometries" has been, for a long time, synonymous to speaking of "kinds of metric manifolds". Such was, for instance, the case of the 19th century's discussions on non-euclidean "geometries" (see further in § 23.1 and on). The first statement of the first book of Descartes *Geometry* is that *every problem in geometry can easily be reduced to such terms that a knowledge of the lengths of certain straight lines is enough for its construction.* This comes from the impression, cogent to cartesian systems of coordinates, that we "measure" something (say, distance from the origin) when attributing coordinates to a point. Of course, we do not. Only homeomorphisms are needed in the attribution, and they are not necessarily isometric.

Nowadays, *geometry* — both the word and the concept behind it — has gained a much enlarged connotation. We hope to have made it clear that

---

[9] For applications in gravitation and cosmology, see Weinberg 1972, Chapter 13.

a metric on a differentiable manifold is an additional structure, chosen and introduced at convenience. As said, many different metrics can in principle be defined on the same manifold. Take the usual surfaces in $\mathbb{E}^3$: we always think of a hyperboloid, for instance, as naturally endowed with the (in the case, indefinite) metric induced by the imbedding in $\mathbb{E}^3$. Nevertheless, it has also at least one positive-definite metric, as ensured by Whitney's theorem. This character of metric, independence from more primitive structures on a manifold, is not very easy to reckon with. It was, according to Einstein, a difficulty responsible for his delay in building General Relativity:

> *Why were more seven years required for the construction of the general theory of relativity? The main reason lies in the fact that it is not so easy to free oneself from the idea that coordinates must have an immediate metrical meaning.*[10]

---

[10] Misner, Thorne &Wheeler 1973, page 5.

Chapter 7

Differential Forms

Forms provide a most economic language to describe physical and mathematical results. They allow for compact versions of practically every fundamental equation. This is illustrated below via examples, ranging from the laws of thermodynamics to hypersurfaces, from electrodynamics to homology.

7.1 Introduction

§ **7.1.1** Exterior differential forms are antisymmetric covariant tensor fields on smooth manifolds (§ 6.3.10 and on). Roughly speaking, they are those objects occurring under the integral sign. Besides being the central objects of integration on manifolds, however, these integrands have a lot of interest by themselves. They have been introduced by Cartan mainly because of the great operational simplicity they provide: they allow a concise shorthand formulation of the whole subject of vector analysis on smooth manifolds of arbitrary kind and dimension.

We are used to seeing, in the euclidean 3-dimensional space, line integrals written as

$$\int (A dx + B dy + C dz),$$

surface integrals as

$$\iint (P dx dy + Q dy dz + R dz dx),$$

and volume integrals as

$$\iiint T dx dy dz.$$

The differential forms appearing in these expressions exhibit a common
and remarkable characteristic: terms which would imply redundant integration, such as $dxdx$, are conspicuously absent. Intuition might seem enough
to eliminate redundancy, but there is a deeper reason for that: integrals
are invariant under basis transformations and the corresponding jacobian
determinants are already included in the integration measures, which are
henceby antisymmetric. We could almost say that, as soon as one thinks
of integration, only antisymmetric objects are of interest. This is a bit too
strong as, for instance, a metric may be involved in the integration measure.
However, differential calculus at least is basically concerned with antisymmetric objects with a well-defined behaviour under transformations, that
is, antisymmetric tensors.

§ 7.1.2 In the case of 1-forms (frequently called *Pfaffian forms*) antisymmetry is, of course, of no interest. We have seen, however, that they provide
basis for higher-order forms, obtained by exterior products (§ 6.3.11). Recall that the exterior product of two 1-forms (say, two members of a basis
$\{\omega^i\}$) is an antisymmetric mapping

$$\wedge : T_1^0(M) \times T_1^0(M) \to T_2^0(M),$$

where $T_s^r(M)$ is the space of (r, s)-tensors on M. In the basis so obtained,
a 2-form F, for instance, will be written

$$F = \tfrac{1}{2} F_{ij}\, \omega^i \wedge \omega^j.$$

§ 7.1.3 We shall denote by $\Omega^k(M)$ the space of the antisymmetric covariant tensors of order k on the space M, henceforth simply called *k-forms*.
Recall that they are tensor fields, so that in reality the space of the k-forms
on the manifold M is the union

$$\Omega^k(M) = \bigcup_{p \in M} \Omega^k(T_p M).$$

In a way quite similar to the previously defined bundles, the above space can
be topologized and made into another fiber bundle, the bundle of k−forms
on M. A particular k-form ω is then a section

$$\omega : M \to \Omega^k(M), \quad \omega : p \to \omega_p \in \Omega^k(T_p M).$$

It is a universally accepted abuse of language to call k-forms "differential
forms" of order k. We say "abuse", of course, because not every "differential
form" is the differential of something else.

§ 7.1.4 The exterior product — also called *wedge product* — is the generalization of the vector product in $\mathbb{E}^3$ to spaces of any dimension and thus, through their tangent spaces, to general manifolds. It is a mapping

$$\wedge : \Omega^p(M) \times \Omega^q(M) \to \Omega^{p+q}(M),$$

which makes the whole space of forms into a graded associative algebra. Recall that

$$\dim \Omega^p(M) = \binom{m}{p},$$

and that spaces of order $p > m$ reduce to zero. Thus, if α^p is a p-form and β^q is a q-form,

$$\alpha^p \wedge \beta^q = 0 \quad \text{whenever } p + q > m.$$

The space of 0-forms has as elements the real functions on M whose compositions, by the way, exhibit trivially the pull-back property.

§ 7.1.5 A basis for the maximal-order space $\Omega^m(M)$ is a single m-form

$$\omega^1 \wedge \omega^2 \wedge \omega^3 \ldots \wedge \omega^m.$$

In other words, $\Omega^m(M)$ is a 1-dimensional space. The nonvanishing elements of $\Omega^m(M)$ are called *volume elements,* or *volume forms.* Two volume elements v_1 and v_2 are said to be *equivalent* if a number $c > 0$ exists such that $v_1 = cv_2$. This equivalence divides the volume forms into two classes, each one called an *orientation.* This definition of orientation can be shown to be equivalent to that given in § 4.2.13. We shall come back to volume forms later.

Some naïve considerations in euclidean spaces provide a more pictorial view of Pfaffian forms. Let us proceed to them.

§ 7.1.6 Perhaps the most elementary and best known 1-form in Physics is the mechanical work, a Pfaffian form in $\mathbb{E}^3$. In a natural basis, it is written

$$W = F_k \, dx^k,$$

with the components F_k representing the force. The total work realized in taking a particle from a point "a" to point "b" along a line γ is

$$W_{ab}[\gamma] = \int_\gamma F_k \, dx^k,$$

and in general depends on the chosen line. It will be path-independent only when the force comes from a potential, as a gradient $F_k = -(\operatorname{grad} U)_k$. In this case W is $W = -dU$, truly the differential of a function, and

$$W_{ab} = U(a) - U(b).$$

A much used criterion for this integrability is to see whether $W[\gamma] = 0$ when γ is any closed curve. However, the work related to displacements in a non-potential force field is a typical "non-differential" 1-form: its integral around a closed curve does not vanish, and its integral between two points will depend on the path. Thus, the simplest example of a form which is not a differential is the mechanical work of a non-potential force. We shall later find another simple example, the heat exchange (§ 7.2.9).

Of a more geometrical kind, also the form appearing in the integrand in Eq.(6.72) is not a differential, as the arc length depends on the chosen curve. That is why the distance has been defined in Eq.(6.73) as an infimum.

§ **7.1.7** The gradient of a function like the potential $U(x, y, z)$ may be pictured as follows: consider the equipotential surfaces $U(x, y, z) = c$ (constant). The gradient field is, at each point $p \in \mathbb{E}^3$, orthogonal to the equipotential surface going through p, its modulus being proportional to the growth rate along this orthogonal direction. The differential form dU can be seen as this field (it is a cofield, but the trivial metric of $\mathbb{E}^3$ identifies field and cofields). For a central potential, these surfaces will be spheres of radii

$$r = \sqrt{x^2 + y^2 + z^2},$$

which are characterized by the form "dr". That is to say, the spheres are the integral surfaces of the differential equation $dr = 0$.

Despite the simplicity of the above view, it is better to see a gradient in $\mathbb{E}^3$ as the field of planes tangent to the equipotential surfaces and regard 1-forms in general as *fields of planes*. A first reason for this preference is that we may then imagine fields of planes that are not locally tangent to any surface: they are non-integrable forms. They "vary too quickly", in a non-differentiable way (as suggested by the right-up corner in the scheme of Figure 7.1). A second reason is that this notion is generalizable to higher order forms, which are fields of oriented continuum trellis of hyperplanes, a rather unintuitive thing. For instance, 1-forms on a space of dimension m are fields of $(m - 1)$-dimensional hyperplanes. Integrable forms are those trellis locally tangent to submanifolds. The final reason for the preference is, of course, that it is a correct view. The lack of intuition for the higher order case is the reason for which we shall not insist too much on this line[1]

---

[1] A beautiful treatment, with luscious illustrations, is given in Misner, Thorne & Wheeler 1973.

and take forms simply as tensor fields, which is equivalent. Let us only say a few more words on Pfaffian forms.

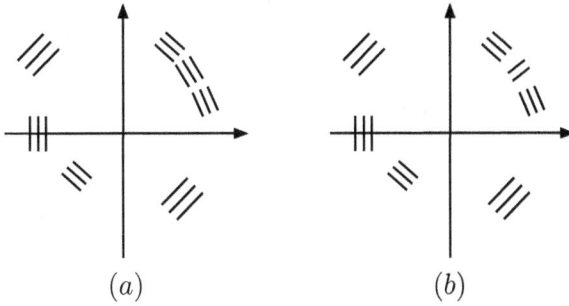

Fig. 7.1 (a) Integrable field of "planes"; (b) Non-integrable field of "planes".

§ **7.1.8** A 1-form is *exact* if it is a gradient, like $\omega = dU$. Being exact is not the same as being integrable. Exact forms are integrable, but non-exact forms may also be integrable if they are of the form $\alpha\, dU$. The same spheres "dr" of the previous paragraph will be solutions of $\alpha dr = 0$, where $\alpha = \alpha(x, y, z)$ is any well-behaved function. The field of planes is the same, the gradients have the same directions, only their modulus change from point to point (see Figure 7.2). Of course, this is related to the fact that fields of planes are simply fields of directions. The general condition for that is

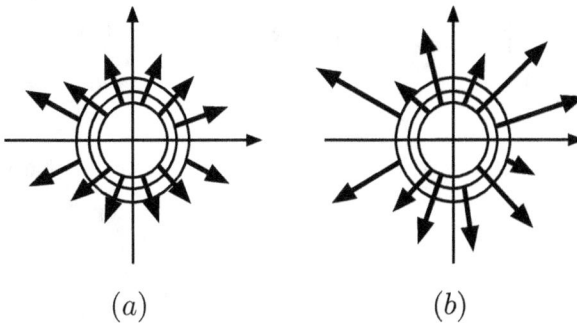

Fig. 7.2 (a) The "field" dr; (b) The "field" αdr: the moduli change from point to point, but the directions remain radial.

given by the Frobenius theorem (§ 6.4.33 and § 7.3.14).

Given a Pfaffian form ω, the differential equation $\omega = 0$ is the corresponding *Pfaffian equation*. It will have solutions if ω may be put into the form $\omega = \alpha\, df$. Otherwise, ω will be a field of planes which are not (even locally) tangent to families of surfaces, as happens with non-potential forces.

Consider Pfaffian forms on $\mathbb{E}^2$ [with its usual global cartesian coordinates (x, y)], on which fields of straight lines will replace those of planes. The line field formed by the axis Ox and all its parallels is fixed by dy, or $\alpha(x, y)dy$ for any α, as the solutions of $\alpha dy = 0$ are $y = $ constant. The fact that in αdy the modulus change from point to point (see Figure 7.3) does not change the line field, which is only a direction field. The line field of vertical lines, $x = $ constant, is $\alpha(x, y)dx$. The form

$$\omega = -\, adx + bdy,$$

where a and b are constants, will give straight lines (Figure 7.4 a)

$$y = (a/b)x + c\,,$$

whose tangent vectors are

$$v = (\dot{x}, \dot{y}) = \dot{x}\, \partial_x + \dot{y}\, \partial_y = \dot{x}\, \partial_x + (a/b)\, \dot{x}\, \partial_y.$$

The form ω is orthogonal to all such tangent vectors: $\omega(v) = 0$.

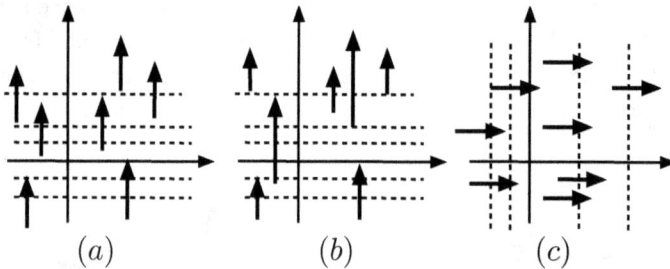

Fig. 7.3 (a) The line field dy; (b) The line field αdy; (c) The line field dx.

Next in complication would be the form

$$\omega = A(x, y)dx + B(x, y)dy.$$

The equation $\omega = 0$ will be always integrable, as the ordinary differential equation

$$\frac{dy}{dx} = -\, A/B$$

will have as solution a one-parameter family of curves $f(x,y) = c$. There will always exist an $\alpha(x,y,z)$ such that $\omega = \alpha\, df$. The form ω/α is exact and α is consequently called an *integrating denominator*. Every field of straight lines on the plane will find locally a family of curves to which it is tangent. It follows that

$$d\omega = (d\alpha\, \alpha) \wedge \omega.$$

The particular case of

$$\omega = -2x dx + dy,$$

depicted in Figure 7.4 (b), has for the Pfaffian equation solutions

$$y = x^2 + C,$$

with tangent vectors $\dot{x}\partial_x + 2x\dot{x}\partial_y$. All this holds no more in higher dimensions: fields of (hyper-)planes are not necessarily locally tangent to surfaces. When generalized to manifolds, all such line and plane fields are to be considered in the euclidean tangent spaces.

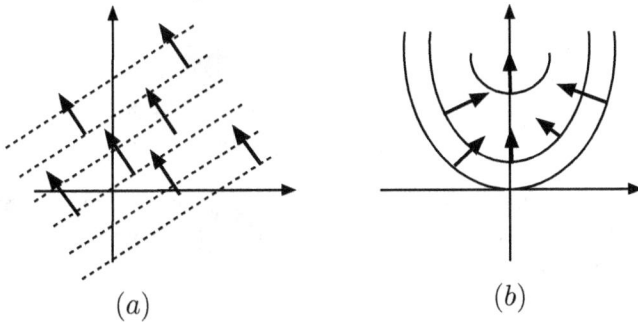

(a)   (b)

Fig. 7.4   (a) The line field $w = bdy - adx$; (b) The line field $w = -2x dx + dy$.

§ **7.1.9** A very useful object is the *Kronecker symbol*, defined by

$$\varepsilon^{k_1 k_2 k_3 \cdots k_p}_{j_1 j_2 j_3 \cdots j_p} = \begin{cases} +1 & \text{if the } j\text{'s are an even permutation of the } k\text{'s;} \\ -1 & \text{if the } j\text{'s are an odd permutation of the } k\text{'s;} \\ 0 & \text{in any other case.} \end{cases}$$

This symbol is a born antisymmetrizer. It may be seen as the determinant

$$\varepsilon^{k_1 k_2 k_3 \cdots k_p}_{j_1 j_2 j_3 \cdots j_p} = \begin{vmatrix} \delta^{k_1}_{j_1} & \delta^{k_2}_{j_1} & \cdots & \delta^{k_p}_{j_1} \\ \delta^{k_1}_{j_2} & \delta^{k_2}_{j_2} & \cdots & \delta^{k_p}_{j_2} \\ \cdots & \cdots & \cdots & \cdots \\ \delta^{k_1}_{j_p} & \delta^{k_2}_{j_p} & \cdots & \delta^{k_p}_{j_p} \end{vmatrix}. \tag{7.1}$$

It satisfies the relation

$$\varepsilon^{k_1 k_2 k_3 \cdots k_q}_{j_1 j_2 j_3 \cdots j_q} \, \varepsilon^{j_1 j_2 j_3 \cdots j_q n_1 n_2 n_3 \cdots n_p}_{m_1 m_2 m_3 \cdots m_{q+p}} = q! \, \varepsilon^{k_1 k_2 k_3 \cdots k_q n_1 n_2 n_3 \cdots n_p}_{m_1 m_2 m_3 \cdots\cdots m_{q+p}}. \tag{7.2}$$

When no doubt arises, we may write simply

$$\varepsilon_{j_1 j_2 j_3 \cdots j_p} = \varepsilon^{1\ 2\ 3\ \cdots\ p}_{j_1 j_2 j_3 \cdots j_p}. \tag{7.3}$$

When $p = m = \dim M$, then

$$\varepsilon_{j_1 j_2 j_3 \cdots j_m} \varepsilon^{j_1 j_2 j_3 \cdots j_m} = m! \tag{7.4}$$

§ 7.1.10 In form (7.3), the Kronecker symbol is of interest in the treatment of determinants. Given an $n \times n$ matrix $A = (A^{ij})$, some useful formulae involving its determinant are:

$$\det A = \varepsilon_{i_1 \cdots i_n} A^{1 i_1} A^{2 i_2} \dots A^{n i_n}; \tag{7.5}$$

$$\varepsilon_{i_1 \cdots i_n} A^{i_1 j_1} A^{i_2 j_2} \dots A^{i_n j_n} = \varepsilon^{j_1 j_2 j_3 \cdots j_n} \det A; \tag{7.6}$$

$$\varepsilon_{j_1 \cdots j_n} \varepsilon_{i_1 \cdots i_n} A^{i_1 j_1} A^{i_2 j_2} \dots A^{i_n j_n} = n! \det A. \tag{7.7}$$

Notice that we are using here upper indices only for notational convenience. We shall later meet Kronecker symbols of type (7.3) with indices raised by the action of a metric.

§ 7.1.11 Kronecker symbols are instrumental in calculations involving components of forms. Given a p-form

$$\alpha = \frac{1}{p!} \, \alpha_{j_1 j_2 \cdots j_p} \, \omega^{j_1} \wedge \omega^{j_2} \wedge \dots \wedge \omega^{j_p}, \tag{7.8}$$

each particular component holds a relationship with the set of all components:

$$\alpha_{j_1 j_2 \cdots j_p} = \frac{1}{p!} \, \varepsilon^{k_1 k_2 k_3 \cdots k_p}_{j_1 j_2 j_3 \cdots j_p} \, \alpha_{k_1 k_2 k_3 \cdots k_p}. \tag{7.9}$$

The basis for the space of p-forms can be written as

$$\omega^{j_1 j_2 \cdots j_p} = \omega^{j_1} \wedge \omega^{j_2} \wedge \dots \wedge \omega^{j_p} = \varepsilon^{j_1 j_2 j_3 \cdots j_p}_{k_1 k_2 k_3 \cdots k_p} \, \omega^{k_1} \otimes \omega^{k_2} \otimes \dots \otimes \omega^{k_p}. \tag{7.10}$$

§ 7.1.12 Given the p-form α and the q-form β, the components of the wedge product $\alpha \wedge \beta$ are, in terms of the components of α and β,

$$(\alpha \wedge \beta)_{i_1 i_2 \cdots i_{p+q}} = \frac{1}{p!} \, \varepsilon^{k_1 k_2 k_3 \cdots k_p j_1 j_2 j_3 \cdots j_q}_{i_1 i_2 i_3 \cdots \quad \cdots \quad i_{q+p}} \, \alpha_{k_1 k_2 k_3 \cdots k_p} \, \beta_{j_1 j_2 j_3 \cdots j_q}. \tag{7.11}$$

§ 7.1.13 A practical comment: in Eq. (7.8), one is supposed to sum over the whole range of all the indices. Many authors prefer to use only the independent elements of the basis: for example, as

$$\omega^1 \wedge \omega^2 = - \, \omega^2 \wedge \omega^1,$$

they are of course not independent. Instead of (7.8), those authors would write

$$\alpha = \alpha_{j_1 j_2 \cdots j_p} \, \omega^{j_1} \wedge \omega^{j_2} \wedge \ldots \wedge \omega^{j_p}, \tag{7.12}$$

without the factor $1/p!$ but respecting $j_1 < j_2 < j_3 < \ldots < j_p$ in the summation. We shall use one or another of the conventions, according to convenience.

§ 7.1.14 The main properties of the exterior product have been outlined in Chapter 6. Let us here only restate the rule concerning commutation:

$$\alpha^p \wedge \beta^q = (-)^{pq} \, \beta^q \wedge \alpha^p. \tag{7.13}$$

If p is odd,

$$\alpha^p \wedge \alpha^p = - \, \alpha^p \wedge \alpha^p = 0.$$

In particular, this holds for the elements ω^j of the basis. For a natural basis,

$$dx^i \wedge x^j = - \, dx^j \wedge dx^i,$$

so that $dx \wedge dx = 0$, $dy \wedge dy = 0$, etc. When no other product is present and no confusion is possible, we may omit the exterior product sign "$\wedge$" and write simply $dx^i dx^j = - \, dx^j dx^i$ or, using anticommutators, $\{dx^i, dx^j\} = 0$. A function f is a 0-form and

$$(f\alpha) \wedge \beta = \alpha \wedge (f\beta) = f(\alpha \wedge \beta). \tag{7.14}$$

Of course,

$$f \wedge \alpha = f\alpha = \alpha f. \tag{7.15}$$

Given any p-form α, we define the *operation of exterior product* by a 1-form ω through the application

$$\varepsilon(\omega) : \Omega^p(M) \to \Omega^{p+1}(M)$$

whose explicit form is

$$\varepsilon(\omega)\alpha = \omega \wedge \alpha, \quad p < m. \tag{7.16}$$

It is easy to check that the vanishing of the wedge product of two Pfaffian forms is a necessary and sufficient condition for their being linearly dependent.

7.2   Exterior derivative

§ **7.2.1** The 0-form f has the differential

$$df = \frac{\partial f}{\partial x^i}\, dx^i = \frac{\partial f}{\partial x^i} \wedge dx^i, \qquad (7.17)$$

which is a 1-form. The generalization of differentials to forms of any order is the *exterior differential*, an operation "d" with the following properties:

(i) The exterior differencial of a (k-form) is a certain $(k+1)$-form:
$$d : \Omega^k(M) \rightarrow \Omega^{k+1}(M).$$

(ii) It is distributive:
$$d(\alpha + \beta) = d\alpha + d\beta.$$

(iii) Chain rule:
$$d(\alpha \wedge \beta) = (d\alpha) \wedge \beta + (-)^{\partial_\alpha}\alpha \wedge \beta,$$
with ∂_α being the order of α.

(iv) It satisfies the so-called Poincaré lemma:
$$dd\alpha = 0,$$
which holds for any form α.

These properties define one and only one operation.

§ **7.2.2** To grasp something about condition (iv), let us examine the simplest case, a 1-form α in a natural basis $\{dx^k\}$: $\alpha = \alpha_i\, dx^i$. Its exterior differencial is

$$d\alpha = (d\alpha_i) \wedge dx^i + \alpha_i \wedge d(dx^i) = \frac{\partial \alpha_i}{\partial x^j}\, dx^j \wedge dx^i. \qquad (7.18)$$

If α is exact, $\alpha = df$ or in components, $\alpha_i = \partial_i f$, then

$$d\alpha = d^2 f = \frac{1}{2}\left[\frac{\partial^2 f}{\partial x^i \partial x^j} - \frac{\partial^2 f}{\partial x^j \partial x^i}\right] dx^i \wedge dx^j \qquad (7.19)$$

and the property $d^2 f = 0$ is just the symmetry of the mixed second derivatives of a function. Along the same lines, if α is not exact, we can consider

$$d^2 \alpha = \frac{\partial^2 \alpha_i}{\partial x^j \partial x^k} dx^j \wedge dx^k \wedge dx^i$$

$$= \frac{1}{2!}\left[\frac{\partial^2 \alpha_i}{\partial x^j \partial x^k} - \frac{\partial^2 \alpha_i}{\partial x^k \partial x^j}\right] dx^j \wedge dx^k \wedge dx^i = 0. \qquad (7.20)$$

Thus, the condition $d^2 \equiv 0$ comes from the equality of mixed second derivatives of the functions α_i, and is consequently related to integrability conditions. It is usually called the *Poincaré lemma*. We shall see later its relation to the homonym of § 2.3.16.

§ 7.2.3 It is natural to ask whether the converse holds: is every form α satisfying $d\alpha = 0$ of the type $\alpha = d\beta$? A form α such that $d\alpha = 0$ is said to be *closed*. A form α which can be written as a derivative, $\alpha = d\beta$ for some β, is said to be *exact*. In these terms, the question becomes: is every closed form exact? The answer, given below as the Poincaré inverse lemma, is *yes, but only locally*. It is true in euclidean spaces, and differentiable manifolds are locally euclidean. More precisely, if α is closed in some open set U, then there is an open set V contained in U where there exists a form β (the "local integral" of α) such that $\alpha = d\beta$. In words, every closed form is *locally* exact. But attention: if γ is another form of the same order of β and satisfying $d\gamma = 0$, then also

$$\alpha = d(\beta + \gamma).$$

There are, therefore, infinite forms β of which α is the differential. The condition for a closed form to be exact on the open set V is that V be contractible (say, a coordinate neighbourhood). On a smooth manifold, every point has an euclidean (consequently contractible) neighbourhood — and the property holds at least locally. When the whole manifold M is contractible, closed forms are exact all over M. When M is not contractible, a closed form may be non-exact, a property which would be missed from a purely coordinate point of view. Before addressing this subject, let us examine the use of the rules above in some simple cases.

Notice that, after the considerations of § 7.1.8, a form may be integrable without being closed. The general problem of integrability is dealt with by the Frobenius theorem (§ 6.4.33), whose version in terms of forms will be seen later (§ 7.3.14).

Commentary 7.1 By what we have said in § 7.1.6, the elementary length "ds" is a prototype of form which is not an exact differential, despite its appearance. Obviously the integral

$$\int_a^x ds$$

depends on the trajectory, leading thus to a multi-valued function of "x" (see Chapter 19 for more). ◂

§ 7.2.4 Take again the 2-form

$$F = \tfrac{1}{2} F_{ij}\, \omega^i \wedge \omega^j. \tag{7.21}$$

Its differential is the 3-form

$$dF = \tfrac{1}{2} \left[dF_{ij} \wedge \omega^i \wedge \omega^j + F_{ij}(d\omega^i) \wedge \omega^j - F_{ij}\, \omega^i \wedge d\omega^j \right]. \tag{7.22}$$

The computation is done by repeated use of properties (iii) and (ii) of § 7.2.1. One sees immediately the great advantage of using the natural basis $\omega^i = dx^i$. In this case, from property (iv), only one term remains:

$$dF = \tfrac{1}{2} dF_{ij} \wedge dx^i \wedge dx^j.$$

The component is a function, so its differential is just as given by Eq.(7.17):

$$dF = \tfrac{1}{2} \frac{\partial F_{ij}}{\partial x^k} dx^k \wedge dx^i \wedge dx^j. \tag{7.23}$$

We would like to have this 3-form put into the canonical form (7.8), with the components fully symmetrized. If we antisymmetrize now in pairs of indices (k, i) and (k, j), we in reality get 3 equal terms,

$$dF = \tfrac{1}{3!} \left[\partial_k F_{ij} + \partial_j F_{ki} + \partial_i F_{jk} \right] dx^k \wedge dx^i \wedge dx^j \tag{7.24}$$

or equivalently

$$dF = \tfrac{1}{3!} \left[\tfrac{1}{2!} \varepsilon_{kij}^{pqr} \partial_p F_{qr} \right] dx^k \wedge dx^i \wedge dx^j. \tag{7.25}$$

For a general q-form

$$\alpha = \frac{1}{q!} \alpha_{j_1 j_2 \ldots j_q} dx^{j_1} \wedge dx^{j_2} \wedge \ldots \wedge dx^{j_q},$$

the differential will be

$$
\begin{aligned}
d\alpha &= \frac{1}{q!} d(\alpha_{j_1 j_2 \ldots j_q}) \wedge dx^{j_1} \wedge dx^{j_2} \wedge \ldots \wedge dx^{j_q} \\
&= \frac{1}{q!} \frac{\partial \alpha_{j_1 j_2 \ldots j_q}}{\partial x^{j_0}} dx^{j_0} \wedge dx^{j_1} \wedge dx^{j_2} \wedge \ldots \wedge dx^{j_q} \\
&= \frac{1}{(q+1)!} \left(\frac{1}{q!} \varepsilon_{i_0 i_1 i_2 \ldots i_q}^{j_0 j_1 j_2 \ldots j_q} \frac{\partial \alpha_{j_1 j_2 \ldots j_q}}{\partial x^0} \right) dx^{i_0} \wedge dx^{i_1} \wedge dx^{i_2} \wedge \ldots \wedge dx^{i_q},
\end{aligned}
$$

which gives the components

$$(d\alpha)_{i_0 i_1 i_2 \ldots i_q} = \frac{1}{q!} \varepsilon_{i_0 i_1 i_2 \ldots i_q}^{j_0 j_1 j_2 \ldots j_q} \frac{\partial \alpha_{j_1 j_2 \ldots j_q}}{\partial x^0}. \tag{7.26}$$

§ 7.2.5 It is convenient to define the partial exterior derivative of α with respect to the local coordinate x^{j_0} by

$$\frac{\partial \alpha}{\partial x^{j_0}} = \frac{1}{q!} \frac{\partial \alpha_{j_1 j_2 \ldots j_q}}{\partial x^{j_0}} \wedge dx^{j_1} \wedge dx^{j_2} \wedge \ldots \wedge dx^{j_q} \tag{7.27}$$

so that

$$d\alpha = dx^{j_0} \wedge \frac{\partial \alpha}{\partial x^{j_0}}. \tag{7.28}$$

The expression for the exterior derivative in an arbitrary basis will be found below (see Eq. (7.90)). We shall see later (in Eq. (7.185) the real meaning of Eq. (7.27), and give in consequence still another closed expression for $d\alpha$.

§ 7.2.6 The invariant, basis-independent definition of the differential of a k-form is given in terms of its effect when applied to fields:

$$(k+1)d\alpha^{(k)}(X_0, X_1, \ldots, X_k) =$$

$$= \sum_{i=0}^{k}(-)^i \, X_i[\alpha(X_0, X_1, X_2, \ldots, X_{i-1}, \widehat{X}_i, X_{i+1}, X_{i+2}, \ldots X_k)]$$

$$+ \sum_{i<j}^{k}(-)^{i+j} \, \alpha\left([X_i, X_j], X_0, X_1, \ldots, \widehat{X}_i, \ldots, \widehat{X}_j, \ldots, X_k\right), \quad (7.29)$$

where, wherever it appears, the notation $\widehat{X}_n$ means that X_n is absent.

§ 7.2.7 Let us examine some facts in $\mathbb{E}^3$, where things are specially simple. There exists a global basis, the cartesian basis consisting of

$$e_1 = \frac{\partial}{\partial x^1} = \frac{\partial}{\partial x}; \quad e_2 = \frac{\partial}{\partial x^2} = \frac{\partial}{\partial y}; \quad e_3 = \frac{\partial}{\partial x^3} = \frac{\partial}{\partial z}.$$

Its dual is $\{dx^1, dx^2, dx^3\} = \{dx, dy, dz\}$. The euclidean metric will be, in this basis, written

$$g = \delta_{ij}dx^i dx^j.$$

Given a vector $V = V^i \partial/\partial x^i$, its covariant image Z is a form such that, for any vector U,

$$Z(U) = Z_i U^i = g(U, V) = \delta_{ij} V^j U^i,$$

so that $Z_i = \delta_{ij}V^j$. One uses the same names for a vector and for its covariant image, writing $Z_i = V_i$. So, to the vector V corresponds the form $V = V_i dx^i$. Its differential is

$$dV = \frac{\partial V_i}{\partial x^k} \, dx^k \wedge dx^i = \tfrac{1}{2} \left(\partial_k V_i - \partial_i V_k \right) dx^k \wedge dx^i,$$

or

$$dV = \tfrac{1}{2} \left(\mathrm{rot} V \right)_{ki} dx^k \wedge dx^i.$$

Think of electromagnetism in $\mathbb{E}^3$: the vector potential is the 1-form $A = A_i dx^i$, and the magnetic field is

$$H = dA = \mathrm{rot} A.$$

The derivative of a 0-form f is

$$df = (\mathrm{grad} f)_i \, dx^i.$$

Suppose the form V above to be just this gradient form. Then,

$$d^2 f = \mathrm{rot}\,\mathrm{grad} f$$

and the Poincaré lemma is here the well known property

$$\text{rot grad } f \equiv 0.$$

When the vector potential is the gradient of a function, $A = df$, the magnetic field vanishes:

$$H = d^2 f \equiv 0.$$

Consider now the second-order tensor

$$T = \tfrac{1}{2} T_{ij}\, dx^i \wedge dx^j.$$

In $\mathbb{E}^3$, to this tensor will correspond a unique vector (or 1-form) U, fixed by

$$T_{ij} = \varepsilon_{ijk} U_k.$$

Its differential is

$$dT = \tfrac{1}{2}\, \partial_k T_{ij}\, dx^k \wedge dx^i \wedge dx^j$$
$$= \tfrac{1}{2}\, \varepsilon_{ijk} \partial_k U_k\, dx^k \wedge dx^i \wedge dx^j$$
$$= (\boldsymbol{div}\, U)\, dx^1 \wedge dx^2 \wedge dx^3.$$

Taking

$$U_i = \tfrac{1}{2}\, \varepsilon_{ijk} (\boldsymbol{rot}\, V)_{jk},$$

the Poincaré lemma assumes still another well known avatar, namely, div rot $V \equiv 0$. The expression for the laplacian of a 0-form f,

$$\Delta f = \boldsymbol{div}\, \boldsymbol{grad} f = \partial_i \partial_i f,$$

is easily obtained.

A criterion to see the difference between a true vector and a second order tensor is the behaviour under parity $(x^i \to -x^i)$ transformation. A true vector changes sign, while a second order tensor does not. The magnetic field is such a tensor, and Maxwell's equation div $H = 0$ is $dH = 0$, actually the identity

$$d^2 A \equiv 0 \quad \text{if} \quad H = dA.$$

§ 7.2.8 Maxwell's equations, first pair. Consider the electromagnetic field strength in vacuum ($\mu_0 = \varepsilon_0 = c = 1$). It is a second-order antisymmetric tensor in Minkowski space, with the components

$$(F_{\mu\nu}) = \begin{pmatrix} 0 & H_3 & -H_2 & E_1 \\ -H_3 & 0 & H_1 & E_2 \\ H_2 & -H_1 & 0 & E_3 \\ -E_1 & -E_2 & -E_3 & 0 \end{pmatrix}. \tag{7.30}$$

The fourth row and column in the matrix correspond to the zeroth, or time components. The field strength can be written as a 2-form

$$F = \tfrac{1}{2} F_{\mu\nu}\, dx^\mu \wedge dx^\nu. \tag{7.31}$$

In detail,

$$F = H_1 dx^2 \wedge dx^3 + H_2 dx^3 \wedge dx^1 + H_3 dx^1 \wedge dx^2$$
$$+ E_1 dx^1 \wedge dx^0 + E_2 dx^2 \wedge dx^0 + E_3 dx^3 \wedge dx^0, \tag{7.32}$$

or

$$F = \tfrac{1}{2}\, \varepsilon_{ijk} H_i dx^j \wedge dx^k + E_j dx^j \wedge dx^0. \tag{7.33}$$

Using (7.25), we get

$$dF = \tfrac{1}{3!}\, \{\partial_\lambda F_{\mu\nu} + \partial_\nu F_{\lambda\mu} + \partial_\mu F_{\nu\lambda}\}\, dx^\lambda \wedge dx^\mu \wedge dx^\nu. \tag{7.34}$$

From (7.33),

$$dF = \boldsymbol{div} \cdot \boldsymbol{H}\, dx^1 \wedge dx^2 \wedge dx^3$$
$$+ \left[\frac{\partial H_1}{\partial x^0} - \left(\frac{\partial E_2}{\partial x^3} - \frac{\partial E_3}{\partial x^2}\right)\right] dx^0 \wedge dx^2 \wedge dx^3$$
$$+ \left[\frac{\partial H_3}{\partial x^0} - \left(\frac{\partial E_2}{\partial x^1} - \frac{\partial E_1}{\partial x^2}\right)\right] dx^0 \wedge dx^1 \wedge dx^2$$
$$+ \left[\frac{\partial H_2}{\partial x^0} - \left(\frac{\partial E_1}{\partial x^3} - \frac{\partial E_3}{\partial x^1}\right)\right] dx^0 \wedge dx^3 \wedge x^1. \tag{7.35}$$

Thus, the equation

$$dF = 0 \tag{7.36}$$

is the same as

$$\boldsymbol{div} \cdot \boldsymbol{H} = 0 \quad \text{and} \quad \partial_0 \boldsymbol{H} = -\,\boldsymbol{rot}\ \boldsymbol{E}. \tag{7.37}$$

This is the first pair of Maxwell's equations. Of course, this could have been seen already in Eq.(7.34), which gives them directly in the usual covariant expression

$$\partial_\lambda F_{\mu\nu} + \partial_\nu F_{\lambda\mu} + \partial_\mu F_{\nu\lambda} = 0. \tag{7.38}$$

Equation (7.36) says that the *electromagnetic form* F is closed. In Minkowski pseudo-euclidean space (supposedly contractible; recall that we do not know much about its real topology, § 1.2.14, § 1.2.18 and § 1.4.7), there exists then a 1-form

$$A = A_\mu dx^\mu$$

such that

$$F = dA = \tfrac{1}{2}\, [\partial_\mu A_\nu - \partial_\nu A_\mu]\, dx^\mu \wedge dx^\nu, \tag{7.39}$$

or in components,

$$F_{\mu\nu} = \partial_\mu A_\nu - \partial_\nu A_\mu.$$

The potential form A is not unique: given any 0-form f, we can also write $F = d(A + df)$. The potentials A and $A' = A + df$ give both the same field F. The transformation

$$A'_\mu = A_\mu + \partial_\mu f \tag{7.40}$$

is called a gauge transformation. The gauge invariance of F is thus related to its closedness and to the arbitrariness born from the Poincaré lemma.

We could formally define F as dA. In that case, the first pair of Maxwell's are not really equations, but constitute an identity. This point of view is justified in the general framework of gauge theories. From the quantum point of view, the fundamental field is the potential A, and not the field strength F. Although itself not measurable, its integral along a closed line is measurable (Aharonov–Bohm effect, seen in § 4.2.17). Furthermore, it is the field whose quanta are the photons. Even classically, there is a hint of its more fundamental character, coming from the lagrangian formalism: interactions of the electromagnetic field with a current j^μ are given by $A_\mu j^\mu$. There is also a further suggestion of the special character of $dF = 0$: unlike the second pair of Maxwell's equations (see below, § 7.4.16), the first pair does not follow from variations of the electromagnetic lagrangian.

§ **7.2.9 Thermodynamics of very simple systems.** We call "very simple systems" those whose states are described by points on a two-dimensional manifold with boundary, usually taken as diffeomorphic to the upper right quadrant of the plane $\mathbb{E}^2$. Thermodynamical coordinates are conveniently chosen so as to represent measurable physical variables. We shall use the entropy S and the volume V. The remaining physical quantities are then functions of these two variables (this is sometimes called the "entropy-volume representation"). The internal energy, for example, is $U = U(S, V)$. With obvious notation, the first principle of thermodynamics reads

$$dU(S, V) = T(S, V)dS - P(S, V)dV.$$

The heat "variation" TdS is usually denoted in textbooks by δQ or some other notation which already indicates that something is amiss. It is in reality another simple physical example of a 1-form which is not a differential: it is not an exact form, there exists no such a function as "Q" that makes this form into $TdS = dQ$. Though the same is true of the work PdV, the first principle says that the difference dU is an exact form. Taking the derivative,

$$d^2U = 0 = dT \wedge dS - dP \wedge dV.$$

But

$$dT = \left(\frac{\partial T}{\partial S}\right)_V dS + \left(\frac{\partial T}{\partial V}\right)_S dV,$$

and

$$dP = \left(\frac{\partial P}{\partial S}\right)_V dS + \left(\frac{\partial P}{\partial V}\right)_S dV.$$

Thus,

$$\left(\frac{\partial T}{\partial V}\right)_S dV \wedge dS = \left(\frac{\partial P}{\partial S}\right)_V dS \wedge dV.$$

Consequently,

$$\left(\frac{\partial T}{\partial V}\right)_S = -\left(\frac{\partial P}{\partial S}\right)_V,$$

which is one of Maxwell's *reciprocal relations*. The other relations are obtained in the same way, using, however, different independent variables at the start. All of them are integrability conditions, here embodied in the Poincaré lemma. A mathematically well founded formulation of Thermodynamics was initiated by Carathéodory[2] and is nowadays advantageously spelt in terms of differential forms, but we shall not proceed to it here.[3]

§ 7.2.10 We have introduced 1-forms, to start with, as differentials of functions (or 0-forms). We have afterwards said that not every 1-form is the differential of some function, and have found some examples of such non-differential forms: mechanical work (§ 7.1.6), thermodynamical heat and work exchanges (§ 7.2.9). This happens also for forms of higher order: not every p-form is the differential of some $(p-1)$-form. This is obviously related to integrability: given an exact form

$$\alpha = d\beta ,$$

β is its *integral*. The expression stating the closedness of α,

$$d\alpha = 0,$$

---

[2] Very nice résumés are found in Chandrasekhar 1939 and Born 1964.
[3] See for instance Mrugala 1978 and references therein.

when written in components, becomes a system of differential equations whose integrability (i.e., the existence of a unique integral β) is only granted locally.

§ 7.2.11 Inverse Poincaré lemma. The inverse Poincaré lemma says that every closed form α is *locally* exact and gives an expression for the integral of α. "Locally" has a precise meaning: if $d\alpha = 0$ at the point $p \in M$, then there exists a contractible neighbourhood of p in which a β exists such that $\alpha = d\beta$. To be more precise, we have to introduce still another operation on forms: given, in a natural basis, the p-form

$$\alpha(x) = \alpha_{i_1 i_2 \ldots i_p}(x)\, dx^{i_1} \wedge dx^{i_2} \wedge \ldots \wedge dx^{i_p},$$

the *transgression* of α is the $(p-1)$-form given by

$$T\alpha = \sum_{j=1}^{p}(-)^{j-1} \int_0^1 dt\, t^{p-1}\, x^{i_j}\, \alpha_{i_1 i_2 \ldots i_p}(tx)\, dx^{i_1} \wedge dx^{i_2} \wedge \ldots$$

$$\wedge dx^{i_{j-1}} \wedge dx^{i_{j+1}} \wedge \ldots dx^{i_p}. \quad (7.41)$$

Notice that, in the x-dependence of α, x is replaced by (tx) in the argument. As t ranges from 0 to 1, the variables are taken from the origin to x. In each term of the summation, labelled by the subindex j, the j-th differential dx^{i_j} is replaced by its integral x^{i_j}. In reality, the T operation involves a certain homotopy, and the above expression is frequently referred to as the *homotopy formula*. The operation is clearly only meaningful in a starshaped region, as x is linked to the origin by the straight line "tx", but can be generalized to a contractible region. The limitation of the result to be given below comes from this strictly local property. Well, the lemma then says that, *locally*, any form α can be written in the form

$$\alpha = d(T\alpha) + T(d\alpha). \quad (7.42)$$

The proof of this fundamental formula is rather involved and will not be given here.[4] It can nevertheless be directly verified from Eq.(7.41), by using the identity

$$\alpha(tx) = \frac{d}{dt}\,[t\alpha(tx)] - t\frac{d}{dt}\,[\alpha(tx)].$$

§ 7.2.12 The expression (7.42) tells us that, when $d\alpha = 0$,

$$\alpha = d(T\alpha), \quad (7.43)$$

---

[4] A constructive proof for general manifolds is found in Nash & Sen 1983. On $\mathbb{E}^n$, proofs are given in most textbooks, as for example Goldberg 1962 and Burke 1985.

so that α is indeed exact and the β looked for above is just $\beta = T\alpha$ (up to γ's such that $d\gamma = 0$). Of course, the formulae above hold globally on euclidean spaces, which are contractible.

§ **7.2.13** Take a constant magnetic field, $\boldsymbol{B} = \text{constant}$. It is closed by Maxwell's equation $\boldsymbol{\nabla} \cdot \boldsymbol{B} = 0$. It is the rotational of

$$\boldsymbol{A} = T\boldsymbol{B} = \boldsymbol{B} \wedge \boldsymbol{r},$$

as comes directly from Eq. (7.41).

§ **7.2.14** A simple test: take in $\mathbb{E}^2$ the form

$$v = dr = d\sqrt{x^2 + y^2} = \frac{x\,dx + y\,dy}{\sqrt{x^2 + y^2}}.$$

Then, as expected,

$$Tv = \int_0^1 dt\, \frac{tx^2 + ty^2}{\sqrt{t^2 x^2 + t^2 y^2}} = \sqrt{x^2 + y^2} = r.$$

§ **7.2.15** A fundamental property of the exterior derivative is its preservation under mappings: if $f : M \to N$, then the derivative of the pull-back is the pull-back of the derivative,

$$f^*(dw) = d(f^*w). \tag{7.44}$$

The way to demonstrate it consists in first showing it for 0-forms and then using induction. Let f be given by its local coordinate representation

$$y \circ f \circ x^{-1}, \quad \text{or} \quad y^i = f^i(x^1, x^2, \ldots, x^m),$$

with $i = 1, 2, \ldots, n$ (see the scheme of Figure 7.5). A field X on M, locally given by

$$X = X^i\, \frac{\partial}{\partial x^i},$$

is "pushed forward" to a field $f_* X$ on N, such that

$$(f_* X)(g) = X(g \circ f) = X^i \frac{\partial}{\partial x^i} g[f(x)]$$

$$= X^i \frac{\partial}{\partial x^i} g[y^1, y^2, \ldots, y^n]$$

$$= X^i \frac{\partial g}{\partial y^r}\frac{\partial y^r}{\partial x^i} = X^i \frac{\partial f^r}{\partial x^i}\frac{\partial g}{\partial y^r}.$$

This holds for any real function g, so that we may write

$$(f_* X) = X^i \frac{\partial f^r}{\partial x^i}\frac{\partial}{\partial f^r} = X(f^r)\frac{\partial}{\partial f^r}.$$

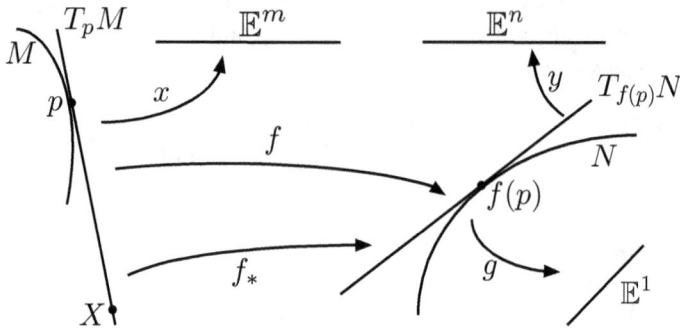

Fig. 7.5   Preservation of forms under mappings.

The pull-back of the 1-form ω on N is that 1-form $f^*\omega$ on M satisfying

$$(f^*\omega)(X) = \omega(f_*X) = (\omega \circ f_*)(X).$$

In a local basis, where $\omega = \omega_j \, dy^j$, we have

$$\omega(f_*X) = \omega_j \, dy^j \, [X^i \frac{\partial f^r}{\partial x^i} \frac{\partial}{\partial y^r}] = \omega_j \frac{\partial f^j}{\partial x^i} X^i.$$

For the specific case in which

$$\omega = dg = \frac{\partial g}{\partial y^r} \, dy^r,$$

we get

$$(f^*dg)(X) = dg(f_*X) = \frac{\partial g}{\partial y^r} \frac{\partial f^r}{\partial x^i} X^i.$$

A function g is pulled back to the composition, $f^*g = g \circ f$. Then,

$$\begin{aligned}(d[f^*g])(X) &= (d[g \circ f])(X) \\ &= \left(\frac{\partial g}{\partial y^j} \frac{\partial f^j}{\partial x^i} \, dx^i \right) \left(X^k \frac{\partial}{\partial x^k} \right) \\ &= \frac{\partial g}{\partial y^j} \frac{\partial f^j}{\partial x^i} X^i, \end{aligned}$$

so that

$$f^*dg = d[f^*g].$$

This is Eq. (7.44) for 0-forms. As already said, we now proceed by induction. Suppose the general result holds for $(p-1)$-forms. Take a p-form ω. Its pull-back is

$$f^*\omega = f^*[\frac{1}{p!}\,\omega_{i_1 i_2 i_3 \ldots i_p} dy^{i_1} \wedge dy^{i_2} \wedge dy^{i_3} \wedge \ldots \wedge y^{i_p}]$$

$$= \frac{1}{p!}\,[f^*\omega_{i_1 i_2 i_3 \ldots i_p} dy^{i_1} \wedge dy^{i_2} \wedge y^{i_3} \wedge \ldots \wedge y^{i_{p-1}}] \wedge f^* dy^{i_p}.$$

Therefore,

$$d[f^*\omega] = \frac{1}{p!}\,\{d[f^*\omega_{i_1 i_2 i_3 \ldots i_p} dy^{i_1} \wedge dy^{i_2} \wedge dy^{i_3} \wedge \ldots \wedge y^{i_{p-1}}] \wedge f^* dy^{i_p}$$

$$+ (-)^{p-1}[f^*\omega_{i_1 i_2 i_3 \ldots i_p} dy^{i_1} \wedge dy^{i_2} \wedge dy^{i_3} \wedge \ldots \wedge dy^{i_{p-1}}] \wedge d[f^* dy^{i_p}]\}.$$

But,

$$d[f^* dy^{i_p}](X) = d[dy^{i_p}(f_*X)],$$

which means that

$$d[f^* dy^{i_p}] = d[dy^{i_p} \circ f] = 0,$$

and the second term above vanishes. Now, using induction,

$$d[f^*\omega] = \frac{1}{p!}\,f^* d[\omega_{i_1 i_2 i_3 \ldots i_p} dy^{i_1} \wedge y^{i_2} \wedge dy^{i_3} \wedge \ldots \wedge dy^{i_{p-1}}] \wedge f^* dy^{i_p}$$

$$= \frac{1}{p!}\,f^*[d\omega_{i_1 i_2 i_3 \ldots i_p} dy^{i_1} \wedge y^{i_2} \wedge dy^{i_3} \wedge \ldots \wedge dy^{i_{p-1}} \wedge dy^{i_p}],$$

which is just

$$d[f^*\omega] = f^*[d\omega].$$

§ 7.2.16 **General Basis.** As far as derivations are involved, calculations are simpler in natural bases, but other bases may be more convenient when some symmetry is present. Let us go back to general basis and reexamine the question of derivation. Suppose $\{e_\mu\}$ is a general basis for the vector fields in an open coordinate neighbourhood U of M, and $\{\theta^\nu\}$ its dual basis. They can be related to the natural basis of a chart (U, x). Using the notation $(\partial/\partial x^\alpha) = \partial_\alpha$, the relations are

$$e_\mu = e_\mu{}^\alpha \partial_\alpha \tag{7.45}$$

and

$$\theta^\nu = \theta^\nu{}_\beta\, dx^\beta. \tag{7.46}$$

Conversely,

$$\partial_\alpha = e^\mu{}_\alpha\, e_\mu \qquad\qquad (7.47)$$

and

$$dx^\beta = \theta_\nu{}^\beta\, \theta^\nu, \qquad\qquad (7.48)$$

where we have used the orthogonality properties

$$e_\mu{}^\alpha e^\mu{}_\beta = \delta^\alpha_\beta \quad \text{and} \quad e_\nu{}^\alpha e^\mu{}_\alpha = \delta^\mu_\nu. \qquad\qquad (7.49)$$

The duality relations show that

$$e^\mu{}_\alpha = \theta^\mu{}_\alpha \quad \text{and} \quad e_\nu{}^\beta = \theta_\nu{}^\beta. \qquad\qquad (7.50)$$

Using (7.45) one can easily show that

$$[e_\mu, e_\nu] = C^\lambda{}_{\mu\nu}\, e_\lambda \qquad\qquad (7.51)$$

with the structure coefficients given by

$$C^\lambda{}_{\mu\nu} = [e_\mu(e_\nu{}^\beta) - e_\nu(e_\mu{}^\beta)]\, e^\lambda{}_\beta. \qquad\qquad (7.52)$$

We can now calculate

$$
\begin{aligned}
d\theta^\lambda = d[\theta^\lambda{}_\beta dx^\beta] &= \tfrac{1}{2}\,[\partial_\alpha\theta^\lambda{}_\beta - \partial_\beta\theta^\lambda{}_\alpha]\, dx^\alpha \wedge dx^\beta \\
&= \tfrac{1}{2}\,[e^\rho{}_\alpha e_\rho(e^\lambda{}_\beta) - e^\sigma{}_\beta e_\sigma(e^\lambda{}_\alpha)]\, e^\alpha{}_\mu e^\beta{}_\nu\, \theta^\mu \wedge \theta^\nu \\
&= \tfrac{1}{2}\,[e^\beta{}_\nu e_\mu(e^\lambda{}_\beta) - e^\beta{}_\mu e_\nu(e^\lambda{}_\beta)]\, \theta^\mu \wedge \theta^\nu.
\end{aligned}
\qquad (7.53)
$$

Using the relations (7.49) and the identities

$$e^\beta{}_\nu e_\mu(e^\lambda{}_\beta) = -\, e^\lambda{}_\beta e_\mu(e^\beta{}_\nu), \qquad\qquad (7.54)$$

we finally get

$$d\theta^\lambda = -\,\tfrac{1}{2}\, C^\lambda{}_{\mu\nu}\, \theta^\mu \wedge \theta^\nu. \qquad\qquad (7.55)$$

This equation is a "translation" of the commutation relations (7.51) to the space of forms. It tells us in particular that

$$d\theta^\lambda(e_\mu, e_\nu) = -\, C^\lambda_{\mu\nu}\,. \qquad\qquad (7.56)$$

From $df = (\partial f/\partial x^\mu)\, dx^\mu$ and using Eqs. (7.47) and (7.48), the differential of a 0-form in an anholonomic basis is

$$df = e_\mu(f)\, \theta^\mu\,. \qquad\qquad (7.57)$$

Suppose now the 1-form $A = A_\mu\theta^\mu$. By using (7.52), one easily finds

$$dA = \tfrac{1}{2}\,[e_\mu(A_\nu) - e_\nu A_\mu) - C^\lambda{}_{\mu\nu}A_\lambda]\, \theta^\mu \wedge \theta^\nu. \qquad (7.58)$$

Commentary 7.2 We shall see later that, on Lie groups, there is always a basis in which the structure coefficients are constant (the "structure constants"). In that case, Eq.(7.55) bears the name of "Maurer–Cartan equation". ◀

Basis $\{\theta^\mu\}$ will be *holonomic* (or natural, or coordinate) if some coordinate system $\{y^\mu\}$ exists in which

$$\theta^\mu = dy^\mu = \frac{\partial y^\mu}{\partial x^\alpha}\, dx^\alpha. \tag{7.59}$$

This means that θ^μ is an exact form, and a necessary condition is $d\theta^\mu = 0$. Conversely, this condition means that a coordinate system such as $\{y^\mu\}$ above exists, at least locally. From (7.55) comes the equivalent condition

$$C^\lambda{}_{\mu\nu} = 0.$$

§ 7.2.17 We can go back to the anholonomic spherical basis of § 6.5.3 for $\mathbb{E}^3$, and find the dual forms to the fields

$$X_r, \quad X_\theta, \quad X_\varphi.$$

They are given respectively by:

$$\omega^r = dr, \quad \omega^\theta = r\, d\theta, \quad \omega^\varphi = r\sin\theta\, d\varphi.$$

We may then write the gradient in this basis,

$$df = e_k(f)\, \omega^k = X_r(f)\, \omega^r + X_\theta(f)\, \omega^\theta + X_\varphi(f)\, \omega^\varphi \tag{7.60}$$

and check that this is the same as

$$df = (\partial_r f)\, dr + (\partial_\theta f)\, d\theta + (\partial_\varphi f)\, d\varphi, \tag{7.61}$$

which would be the expression in the natural basis related to the coordinates (r, θ, φ). The invariance of the exterior derivative is clear.

7.3 Vector-valued forms

§ 7.3.1 Up to now, we have been considering "ordinary" q-forms, antisymmetric linear mappings taking q vector fields into the real line. A vector-valued q-form will take the same fields into a vector space. If V is the vector space and $\omega^q(M)$ is the space of ordinary q-forms on the manifold M, a q-form ω with values in V is an element of the direct product $V \otimes \Omega^q(M)$. If $\{V_a\}$ is a basis for V, ω is written as

$$\omega = V_a\, \omega^a, \tag{7.62}$$

where ω^a are ordinary forms. Thus, a vector-valued form may be seen as a vector whose components are ordinary forms, or as a column of forms.

§ 7.3.2 Vector-valued forms are of fundamental importance in the theory of fiber bundles, where they appear as representatives of connections, curvatures, soldering, etc. They turn up everywhere in gravitational and gauge theories: gauge potentials and field strengths are in reality connections and curvatures respectively. They have been defined as direct products, and in consequence the operations on $\omega^q(M)$ ignore eventual operations occurring in V and vice-versa. The exterior derivative of the above form ω, for example, is defined as

$$dw = V_a \, dw^a. \tag{7.63}$$

Notice that the above definitions yield objects independent of the basis chosen for V. Under a basis change to $V_{a'} = U^a{}_{a'} V_a$, the component forms change to

$$\omega^{a'} = (U^{-1})^{a'}{}_a \omega^a,$$

so that

$$\omega = V_a \, \omega^a = V_{a'} \, \omega^{a'} \tag{7.64}$$

remains invariant. Usual forms ω^a have already been introduced as basis-independent objects in $\Omega^q(M)$, so that the whole object ω is basis-independent in both spaces.

§ 7.3.3 Algebra-valued forms. Of special interest is the case in which the vector space V has an additional structure of Lie algebra. The generators will satisfy commutation relations

$$[J_a, J_b] = f^c{}_{ab} J_c$$

and may be used as a basis for the linear space V. It is precisely what happens in the examples quoted above: gauge fields and potentials are forms with values in the Lie algebra of the gauge group.

There are two possible operations on such forms, the exterior product and the Lie algebra operation. Due to the possible anticommutation properties of exterior products, one must be careful when handling operations with the complete vector-valued forms. A bracket can be defined which dutifully accounts for everything: given two forms

$$\omega = J_a \, \omega^a \quad \text{and} \quad \alpha = J_b \, \alpha^b$$

of any orders, the bracket is[5]

$$[\omega, \alpha] := [J_a, J_b] \otimes \omega^a \wedge \alpha^b. \tag{7.65}$$

---

[5] Lichnerowicz 1955.

§ **7.3.4** Because we wish to stand by usual practice, we are employing above the same symbol [,] with two different meanings: in the left-hand side, the defined bracket; in the right-hand side, the commutator of the Lie algebra. The general definition of the bracket is consequently the following:

$$|[A, B]| = A \wedge B - (-)^{\partial_A \partial_B} B \wedge A, \tag{7.66}$$

with ∂_C the order of the algebra–valued form C. This is a *graded commutator*. It is also usual to left implicit the direct product character of the complete bracket, by omitting the symbol $\otimes$ in Eq. (7.65). When at least one of the involved forms is of even degree, no signs will come from the exterior product and the bracket reduces to a simple commutator. Otherwise, an anticommutator comes out.

§ **7.3.5** For example, a 1-form will be

$$A = J_a A^a{}_\mu dx^\mu. \tag{7.67}$$

Start by calculating

$$
\begin{aligned}
A \wedge A &= J_a J_b A^a{}_\mu A^b{}_\nu dx^\mu \wedge x^\nu \\
&= \tfrac{1}{2}[J_a, J_b] A^a{}_\mu A^b{}_\nu dx^\mu \wedge x^\nu \\
&= \tfrac{1}{2} J_c f^c{}_{ab} A^a{}_\mu A^b{}_\nu dx^\mu \wedge x^\nu,
\end{aligned}
\tag{7.68}
$$

the $f^c{}_{ab}$'s being the Lie algebra structure constants. If we compare with (7.65), it comes out that

$$A \wedge A = \tfrac{1}{2}[A, A]. \tag{7.69}$$

As announced, this is actually a graded commutator, though we use for it the usual commutator symbol.

§ **7.3.6** A particular case is given by the algebra $su(2)$ of the Lie group $SU(2)$ of special (that is, with determinant $= +1$) unitary complex matrices. The lowest-dimensional representation has generators $J_i = \tfrac{1}{2}\sigma_i$, where for the σ_i's we may take the Pauli matrices in the forms

$$
\sigma_1 = \begin{pmatrix} 0 & 1 \\ 1 & 0 \end{pmatrix}, \quad
\sigma_2 = \begin{pmatrix} 0 & -i \\ i & 0 \end{pmatrix}, \quad
\sigma_3 = \begin{pmatrix} 1 & 0 \\ 0 & -1 \end{pmatrix}.
$$

The 1-form (7.67) is then the matrix

$$
A = \frac{1}{2} \begin{pmatrix} A^3{}_\mu dx^\mu & A^1{}_\mu dx^\mu - iA^2{}_\mu dx^\mu \\ A^1{}_\mu dx^\mu + iA^2{}_\mu dx^\mu & -A^3{}_\mu dx^\mu \end{pmatrix}.
$$

It is also an example of a matrix whose elements are noncommutative.

§ **7.3.7** The differential of the 1-form (7.67) is easily obtained:

$$dA = J_a(\partial_\lambda A^a{}_\mu)\, dx^\lambda \wedge dx^\mu$$
$$= \tfrac{1}{2}\, J_a(\partial_\lambda A^a{}_\mu - \partial_\mu A^a{}_\lambda)\, dx^\lambda \wedge x^\mu. \tag{7.70}$$

In a non-holonomic basis, we would have found (using (7.55))

$$dA = \tfrac{1}{2}\, J_a(e_\lambda A^a{}_\mu - e_\mu A^a{}_\lambda - C^\nu{}_{\lambda\mu} A^a{}_\nu)\, \theta^\lambda \wedge \theta^\mu. \tag{7.71}$$

§ **7.3.8** In gauge theories, the gauge field (strength) F is a 2-form on a 4-dimensional space, given in terms of the (1-form) gauge potential A by

$$F = dA + \tfrac{1}{2}\,[A, A] = dA + A \wedge A. \tag{7.72}$$

The J_a's are the generators of the gauge group Lie algebra. To obtain the relations between the components, use Eqs. (7.72), (7.70) and (7.68) to write

$$F = \tfrac{1}{2}\, J_a(\partial_\mu A^a{}_\nu - \partial_\nu A^a{}_\mu + f^a{}_{bc} A^b{}_\mu A^c{}_\nu) dx^\mu \wedge dx^\nu. \tag{7.73}$$

Then, defining the components of F through

$$F = \tfrac{1}{2}\, J_a F^a{}_{\lambda\mu} dx^\lambda \wedge x^\mu, \tag{7.74}$$

we find the expression

$$F^a{}_{\mu\nu} = \partial_\mu A^a{}_\nu - \partial_\nu A^a{}_\mu + f^a{}_{bc} A^b{}_\mu A^c{}_\nu. \tag{7.75}$$

§ **7.3.9** As already said, even-order forms behave under commutation just as normal elements in the algebra: the bracket defined in Eq. (7.65) reduces to the algebra commutator when a 2-form, for example, is involved. For instance, $[A \wedge A, A] = 0$. Take the differential of (7.72),

$$dF = 0 + dA \wedge A - A \wedge dA = [dA, A] = [F - A \wedge A, A],$$

so that

$$dF + [A, F] = 0. \tag{7.76}$$

This relation, an automatic consequence of the definition (7.72) of F, is the *Bianchi identity*. Notice that it has taken us just one line to derive it.

§ **7.3.10** We shall see now that the expression

$$D_A = d + [A, \ ]$$

can be interpreted as a covariant derivative of a 2-form according to the connection A, just as

$$D_A = d + A \wedge$$

is the covariant derivative of a 1-form. This will be a bit more formalized below, in § 7.3.11. As we stand, such names are mere analogies to the riemannian case. Within this interpretation, the field is the covariant derivative of the connection proper, and the Bianchi identity establishes the vanishing of the covariant derivative of the field. By the same analogy, the field F is the curvature of the connection A. In components, Eq. (7.76) reads

$$0 = dF + [A, F] = \tfrac{1}{3!} J_a \left(\partial_{[\lambda} F^a{}_{\mu\nu]} + f^a{}_{bc} A^b{}_{[\lambda} F^c{}_{\mu\nu]} \right) dx^\lambda \wedge dx^\mu \wedge dx^\nu,$$

where the symbol $[\lambda\mu\nu]$ indicates that complete antisymmetrization is to be performed on the enclosed indices. It follows the vanishing of each component,

$$\partial_{[\lambda} F^a{}_{\mu\nu]} + f^a{}_{bc} A^b{}_{[\lambda} F^c{}_{\mu\nu]} = 0.$$

If we define the *dual* tensor

$$\widetilde{F}^{a\rho\lambda} = \tfrac{1}{2} \varepsilon^{\rho\lambda\mu\nu} F^a{}_{\mu\nu}, \tag{7.77}$$

the above expression can be rewritten as

$$\partial^\mu \widetilde{F}^a{}_{\mu\nu} + f^a{}_{bc} A^{b\mu} \widetilde{F}^c{}_{\mu\nu} = 0. \tag{7.78}$$

§ 7.3.11 Covariant derivatives. In order to understand the meaning of all that, we have to start by qualifying (7.63) and (7.64). The form has been supposed to take values on some unique vector space V which is quite independent of the manifold. Transformations in that vector space do not affect objects on the manifold. Consider now the case in which transformations in V, defined by matrices g, depend on the point on the manifold. If

$$W = J_a W^a$$

is a form of order ∂_W and the J_a's are matrices as in § 7.3.6, transformations in V will lead to

$$W' = g J_a g^{-1} W^a = g W g^{-1}.$$

This is the usual way transformations act on matrices, and will be seen later (Section 8.4) to be called an *adjoint action*. The matrices g are now supposed to be point-dependent, $g = g(x)$. We say that they are *gaugefied*.

Everything goes as before for W itself, but there is a novelty in dW: the derivative of the transformed form will now be

$$dW' = d(gWg^{-1}) = dg \wedge Wg^{-1} + gdWg^{-1} + (-)^{\partial W}gW \wedge dg^{-1}. \quad (7.79)$$

Only the second term of the r.h.s. has a "good" behaviour under the transformation, just the same as the original form W. We call covariant derivative of a tensor W a derivative DW which has the same behaviour as W under the transformation. Obviously, this is not the case of the exterior derivative dW. Let us examine how much it violates the covariance requirement. The 1-form $\omega = g^{-1}dg$ will be of special interest. By introducing at convenient places the expression $I = gg^{-1}$ for the identity, as well as its consequences

$$dgg^{-1} + gdg^{-1} = 0 \quad \text{and} \quad dg^{-1} = -g^{-1}dgg^{-1},$$

in terms of the graded commutator (7.66) dW' may be written as

$$\begin{aligned} dW' &= g\,dWg^{-1} + g\left(\omega \wedge W - (-)^{\partial W \partial \omega}W \wedge \omega\right)g^{-1} \\ &= g\left(dW + |[\omega, W]|\right)g^{-1}. \end{aligned} \quad (7.80)$$

We look now for a *compensating form*, a 1-form A transforming according to

$$A' = gAg^{-1} + gdg^{-1} = g\{A - \omega\}g^{-1},$$

and we verify that

$$dW' + |[A', W']| = g\{dW + |[A, W]|\}g^{-1}.$$

In geometrical language, an A transforming as above is a *connection* on the manifold. It follows from the last expression that

$$DW = dW + |[A, W]|$$

is the covariant derivative for any form transforming according to the expression

$$W' = gWg^{-1}.$$

This reduces to the previous expressions in the case of gauge fields. On the other hand, if W is a column vector of forms, and the matrices act as usual on column vectors,

$$W' = gJ_a W^a = gW,$$

then the covariant derivative is

$$DW = dW + AW,$$

with A the same connection as above. Of this type are the covariant derivatives of the source fields in gauge theories. In general, the g's belong to a group G. Quantities transforming as $W' = gWg^{-1}$ are said to belong to the adjoint representation of G, and quantities transforming as gW belong to linear representations. In gauge theories, the potentials A and their curvatures (field strengths) F belong to the adjoint representation. Source fields usually belong to linear representations. Let us retain that the expression of the covariant derivative depends both on the order of the form and on the representation to which it belongs in the transformation group.

Commentary 7.3 It is good to have in mind that some of the so-called "covariant" derivatives found in many texts are actually covariant "coderivatives", to be seen later (§ 7.4.19). ◄

§ 7.3.12 Moving frames. Some remarkable simplifying properties show up in euclidean spaces $\mathbb{E}^n$. The metric can be taken simply as $g_{ij} = \delta_{ij}$. There is a global "canonical" basis of column vectors K^α, and also a dual basis of rows

$$K_\alpha^T = (0, 0, \dots, 1, 0, \dots, 0)^T$$

with "1" only in the α-th entry ("T" stands for the transpose). $\mathbb{E}^n$ itself is diffeomorphic to both $T_p\mathbb{E}^n$ and $T_p^*\mathbb{E}^n$ for each $p \in \mathbb{E}^n$. Given a basis $\{e_i\}$, each member is a section in the tangent bundle, but can be seen here as a mapping

$$e_i : \mathbb{E}^n \to \mathbb{E}^n, \quad e_i : p \to e_i(p)$$

with $e_i(p)$ the vector e_i at the point p. Consequently, de_i is a 1-form taking $\mathbb{E}^n$ into $\mathbb{E}^n$, a vector-valued form. We write it

$$de_i = \omega_i{}^j e_j, \tag{7.81}$$

with $\omega_i{}^j$ some usual 1-forms. Differentiating this expression and using it again, one arrives immediately at

$$d\omega_i{}^j = \omega_i{}^k \wedge \omega_k{}^j. \tag{7.82}$$

Writing e_i in the canonical basis,

$$e_i = e_i{}^\alpha K_\alpha, \tag{7.83}$$

the elements of the dual basis $\{\omega^j\}$ will be

$$\omega^j = \omega^j{}_\beta (K^T)^\beta \tag{7.84}$$

with $\omega^j{}_\beta e_i{}^\beta = \delta_i^j$. Differentiating (7.83) and comparing with (7.81), we find

$$\omega_i{}^j = \omega^j{}_\alpha de_i{}^\alpha. \tag{7.85}$$

From $< e_i, e_j > = \delta_{ij}$, we get

$$< de_i, e_j > + < e_i, de_j > = 0$$

with the consequence

$$\omega_{ij} = -\,\omega_{ji}. \tag{7.86}$$

This antisymmetry justifies the rather awkward notation we have been using, with a careful but unusual index positioning: it is important to know where lowered and raised indexes go. Some more strict authors use a pointed notation — for example, $\omega^{j}{}_{\cdot\alpha}$ instead of the above $\omega^{j}{}_{\alpha}$. Notice that, if matrix notation is used, a condition like

$$\omega^{j}{}_{\beta} e_i{}^{\beta} = \delta_i^{j} \tag{7.87}$$

says that the matrix with entries $\omega^{j}{}_{\beta}$ is the inverse *to the transpose* of that with entries $e_i{}^{\beta}$.

Always keeping faith with this convention, we can define matrices inverse to those appearing in (7.83) and (7.84) so that

$$K_\alpha = e^{j}{}_{\alpha} e_j \quad \text{and} \quad (K^T)^{\beta} = \omega_i{}^{\beta} \omega^{i}.$$

Basis duality enforces $e^{j}{}_{\alpha} = \omega^{j}{}_{\alpha}$ and $\omega_i{}^{\beta} = e_i{}^{\beta}$. Differentiating (7.87) yields

$$d\omega^{j}{}_{\beta} = -\,\omega^{j}{}_{\alpha} \, de_i{}^{\alpha} \, e^{i}{}_{\beta} = \omega^{j}{}_{i} \, e^{i}{}_{\beta}.$$

Consequently,

$$d\omega^{j} = d\omega^{j}{}_{\beta}\,(K^T)^{\beta} = \omega^{j}{}_{i}\, e^{i}{}_{\alpha}\,(K^T)^{\alpha}$$

and

$$d\omega^{j} = \omega^{j}{}_{i} \wedge \omega^{i}. \tag{7.88}$$

The forms $\omega^{j}{}_{i}$ are the *Cartan connection forms* of space $\mathbb{E}^n$. Basis like $\{e_i\}$, which can be defined everywhere on $\mathbb{E}^n$, are called *moving frames* (*repères mobiles*). Equations (7.82) and (7.88) are the (Cartan) *structure equations* of $\mathbb{E}^n$. We have said that, according to a theorem by Whitney, every differentiable manifold can be locally imbedded (immersed) in some $\mathbb{E}^n$, for n large enough. Cartan has used moving frames to analyze the geometry of general smooth manifolds immersed in large enough euclidean spaces. More about this subject can be found in Section 22.2. For an example of moving frames in elasticity, see Section 28.3.2.

§ **7.3.13** Let us go back to the expressions of the exterior derivative. They have been given either in the rather involved basis-independent form (7.29) or, more simply, in the natural basis (7.28). When a form is given in a general basis,

$$\alpha = \tfrac{1}{p!}\, \alpha_{j_1 j_2 \cdots j_p}\, \omega^{j_1} \wedge \omega^{j_2} \wedge \ldots \wedge \omega^{j_p},$$

things get more complicated, as the derivatives of each basis element ω^{j_k} must be taken into account also. Consider a basis $\{\omega^i\}$ and its dual $\{e_j\}$. Writing the Cartan 1-form in the basis $\{\omega^i\}$ as

$$\omega_i^j = \Gamma^j{}_{ik}\omega^k,$$

where the $\Gamma^j{}_{ik}$'s are the connection components, Eqs.(7.88) and (7.52) tell us that the structure coefficients are the antisymmetric parts of these components, $C^j{}_{ik} = \Gamma^j{}_{[ik]}$, and we can choose

$$\Gamma^j{}_{ik} = e_m^j\, e_k(e_i^m). \tag{7.89}$$

This is the Cartan connection related to frame $\{e_i\}$ — also known as Weitzenböck connection. In terms of this connection, a covariant derivative $\nabla_j \alpha$ is defined whose components just appear in the expression for the exterior derivative in a general basis:

$$da = (-)^p \frac{1}{(p+1)!} \left\{ e_{[j_{p+1}} \alpha_{j_1 j_2 \cdots j_p]} + \Gamma^k{}_{[j_{p+1} j_r} \alpha_{j_1 j_2 \cdots j_{r-1} k j_{r+1} \cdots j_p]} \right\}$$
$$\omega^{j_1} \wedge \omega^{j_2} \wedge \ldots \wedge \omega^{j_p} \wedge \omega^{j_{p+1}}$$
$$= (-)^p \frac{1}{(p+1)!} \left\{ \nabla_{[j_{p+1}} \alpha_{j_1 j_2 \cdots j_p]} \right\} \omega^{j_1} \wedge \omega^{j_2} \wedge \ldots \wedge \omega^{j_p} \wedge \omega^{j_{p+1}}, \tag{7.90}$$

so that finally, Eq.(7.28) generalizes to

$$da = \omega^{j_0} \wedge \nabla_{j_0}\alpha = \varepsilon(\omega^{j_0})\nabla_{j_0}\alpha. \tag{7.91}$$

This expression is a bit more general than the well-known formulae giving the differential in general coordinate systems, and reduce to them for natural basis.

§ **7.3.14 Frobenius theorem, alternative version.** We have said that a set of linearly independent tangent vector fields

$$X_1, X_2, \ldots, X_n$$

on a manifold M are locally tangent to a submanifold N (of dimension $n < m$) around a point $p \in M$ if they are in involution,

$$[X_j, X_k] = c^i{}_{jk} X_i.$$

This means that, if we take such fields as members of a local basis $\{X_a\}$ on M, with $a = 1, 2, \ldots, n, n+1, \ldots, m$, the structure coefficients $c^a{}_{jk}$ vanish whenever $a \geq n+1$.

The dual version is the following: consider a set of Pfaffian forms $\theta^1, \ldots, \theta^n$. Linear independence means that their exterior product is nonvanishing (Section 22.2.2). If such forms are to be cotangent to a submanifold, they must close the algebra which is dual to the involution condition, and we must have Eq. (7.55) with structure coefficients $c^i{}_{jk}$ restricted to the indices $i, j, k = 1, 2, \ldots, n$, with the remaining vanishing. This is to say that $d\theta^i$ has only contributions "along" the θ^k's, that is,

$$d\theta^i \wedge \theta^1 \wedge \theta^2 \wedge \ldots \wedge \theta^n = 0.$$

This may be shown to be equivalent to the existence of a system of functions $f^1, \ldots, f^n$ such that

$$\theta^j = a^j_k \, df^k.$$

The set $\{df^k\}$ constitutes a local coordinate basis cotangent to the submanifold N, which is locally fixed by the system of equations $f^k = c^k$ (constants). The characterization of (local) submanifolds by forms is a most convenient method. Global cases have been previously seen: the (x, y)-plane in the euclidean $\mathbb{E}^3$ with cartesian coordinates may be characterized by $dz = 0$. The sphere S^2 given by

$$r = (x^2 + y^2 + z^2)^{1/2},$$

simply by $dr = 0$. Recall further that the form characterizing a surface is "orthogonal" to it. In $\mathbb{E}^3$, the straight line given by the x-axis will be given by $dy = 0$ and $dz = 0$.

7.4 Duality and coderivation

§ **7.4.1** In some previous examples, we have been using relationships between forms like

$$U_k = \tfrac{1}{2!} \varepsilon_{kij} T_{ij}$$

in $\mathbb{E}^3$, and

$$\widetilde{F}_{\mu\nu} = \tfrac{1}{2!} \varepsilon_{\mu\nu\rho\sigma} F^{\rho\sigma}$$

in Minkowski spacetime. These are particular cases of a general relation between p-forms and $(n-p)$-forms on a manifold N. Recall that the dimension of the space Ω^p of p-forms on an n-dimensional manifold is

$$\binom{n}{p} = \binom{n}{n-p}.$$

Of course, the space of $(n-p)$-forms is a vector space of the same dimension and so both spaces are isomorphic. The presence of a metric makes of this isomorphism a canonical one.

§ **7.4.2** Given the Kronecker symbol $\varepsilon_{i_1 i_2 \ldots i_n}$ and a metric $g^{j_1 k_1}$, we can define mixed-index symbols by raising some of the indices with the help of the contravariant metric,

$$\varepsilon^{j_1 j_2 \ldots j_p}_{i_{p+1} \ldots i_n} = g^{j_1 k_1} g^{j_2 k_2} \ldots g^{j_p k_p} \varepsilon_{k_1 k_2 \ldots k_p i_{p+1} \ldots i_n}.$$

A detailed calculation will show that

$$\varepsilon^{j_1 j_2 \ldots j_p i_{p+1} \ldots i_n} \varepsilon_{i_1 \ldots i_p i_{p+1} \ldots i_n} = \frac{(n-p)!}{g} \varepsilon^{j_1 j_2 \ldots j_p}_{i_1 \ldots i_p},$$

where $g = \det(g_{ij})$ is the determinant of the covariant metric.

§ **7.4.3 The dual of a form.** We shall give first a definition in terms of components, which is more appealing and operational. For a basis

$$\omega^{i_1} \wedge \omega^{i_2} \wedge \ldots \wedge \omega^{i_p}$$

for the space Ω^p of p-forms, we shall define a duality operation "$*$", called the *Hodge star-operation*, by an application of the form

$$* : \omega^p(N) \to \omega^{n-p}(N), \tag{7.92}$$

given explicitly by

$$*[\omega^{j_1} \wedge \omega^{j_2} \wedge \ldots \wedge \omega^{j_p}] = \frac{\sqrt{|g|}}{(n-p)!} \, \varepsilon^{j_1 j_2 \ldots j_p}{}_{j_{p+1} \ldots j_n} \omega^{j_{p+1}} \wedge \ldots \wedge \omega^{j_n}. \tag{7.93}$$

Here $|g|$ is the modulus of $g = \det(g_{ij})$. A p-form α will be taken into its dual, the $(n-p)$-form

$$*\alpha = *\left[\frac{1}{p!} \alpha_{j_1 j_2 \ldots j_p} \omega^{j_1} \wedge \omega^{j_2} \wedge \ldots \wedge \omega^{j_p} \right]$$

$$= \frac{\sqrt{|g|}}{(n-p)! \, p!} \, \varepsilon^{j_1 j_2 \ldots j_p}{}_{j_{p+1} \ldots j_n} \alpha_{j_1 j_2 \ldots j_p} \omega^{j_{p+1}} \wedge \ldots \wedge \omega^{j_n},$$

or equivalently,

$$*\alpha = \frac{\sqrt{|g|}}{(n-p)! \, p!} \, \varepsilon_{j_1 j_2 \ldots j_n} \alpha^{j_1 j_2 \ldots j_p} \omega^{j_{p+1}} \wedge \ldots \wedge \omega^{j_n}. \tag{7.94}$$

Notice that the components of $*\alpha$ are

$$(*\alpha)_{j_{p+1} \ldots j_n} = \frac{\sqrt{|g|}}{p!} \, \varepsilon_{j_1 j_2 \ldots j_p j_{p+1} \ldots j_n} \alpha^{j_1 j_2 \ldots j_p}. \tag{7.95}$$

The examples referred to in § 7.4.1 are precisely of this form, with the euclidean metric of $\mathbb{E}^3$ and the Lorentz metric, respectively. Although we have used a basis in the definition, the operation is in reality independent of any choice of basis. The invariant definition will be given in § 7.4.5.

§ 7.4.4 Consider the 0-form which is constant and equal to 1. Its dual will be the n-form

$$v = *1 = \frac{\sqrt{|g|}}{n!}\, \varepsilon_{j_1 j_2 \ldots j_n} \omega^{j_1} \wedge \omega^{j_2} \wedge \ldots \wedge \omega^{j_n} \tag{7.96}$$

$$= \sqrt{|g|}\; \omega^1 \wedge \omega^2 \wedge \ldots \wedge \omega^n, \tag{7.97}$$

which is an especial volume form (§ 7.1.5) called the *canonical volume form* corresponding to the metric g. Given the basis $\{e_j\}$, dual to $\{\omega^j\}$,

$$v(e_1, e_2, \ldots, e_n) = \frac{\sqrt{|g|}}{n!}\, \varepsilon_{j_1 j_2 \ldots j_n} \begin{vmatrix} \omega^{j_1}(e_1) & \omega^{j_1}(e_2) & \ldots & \omega^{j_1}(e_n) \\ \omega^{j_2}(e_1) & \omega^{j_2}(e_2) & \ldots & \omega^{j_2}(e_n) \\ \ldots & \ldots & \ldots & \ldots \\ \omega^{j_n}(e_1) & \omega^{j_n}(e_2) & \ldots & \omega^{j_n}(e_n) \end{vmatrix}$$

or

$$v(e_1, e_2, \ldots, e_n) = \frac{\sqrt{|g|}}{n!}\, \varepsilon_{j_1 j_2 \ldots j_n} \varepsilon_{j_1 j_2 \ldots j_n} = \sqrt{|g|}. \tag{7.98}$$

by Eqs. (7.1) and (7.4). This could of course have been obtained directly from (7.97). Given an arbitrary set of n fields $X_1, X_2, \ldots, X_n$,

$$v(X_1, X_2, \ldots, X_n) = \frac{\sqrt{|g|}}{n!}\, \varepsilon_{j_1 j_2 \ldots j_n} \begin{vmatrix} X_1^{j_1} & X_2^{j_1} & \ldots & X_n^{j_1} \\ X_1^{j_2} & X_2^{j_2} & \ldots & X_n^{j_2} \\ \ldots & \ldots & \ldots & \ldots \\ X_1^{j_n} & X_2^{j_n} & \ldots & X_n^{j_n} \end{vmatrix}$$

or

$$v(X_1, X_2, \ldots, X_n) = \sqrt{|g|}\, \det\left(X_i^j\right) \tag{7.99}$$

§ 7.4.5 Dual of a form: invariant definition. Take p fields $X_1, X_2, \ldots, X_p$ and call $X_1', X_2', \ldots, X_p'$ their respective covariant images. Then, the form $*\alpha$, dual to the p-form α, is defined as that unique $(n - p)$-form satisfying

$$\alpha(X_1, X_2, \ldots, X_p)\, v = (*\alpha) \wedge X_1' \wedge X_2' \wedge \ldots \wedge X_p' \tag{7.100}$$

for all sets of p fields $\{X_1, X_2, \ldots, X_p\}$. This is the invariant, basis independent, definition of operator $*$. It coincides with (7.94), and this tells us that what we have done there (and might be not quite evident), supposing that the operator $*$ ignores the components, is correct.

§ 7.4.6 Let us go back to Eq. (7.93) and check what comes out when we apply twice the star operator:

$$** \left[\omega^{j_1} \wedge \omega^{j_2} \wedge \ldots \wedge \omega^{j_p} \right] =$$

$$= \frac{|g|}{(n-p)!\, p!}\, \varepsilon^{j_1 j_2 \ldots j_p}{}_{j_{p+1} j_{p+2} \ldots j_n} \varepsilon^{j_{p+1} j_{p+2} \ldots j_n}{}_{i_1 i_2 \ldots i_p} \omega^{i_1} \wedge \omega^{i_2} \wedge \ldots \wedge \omega^{i_p}$$

$$= \frac{|g|}{(n-p)!\, p!}\, g^{j_1 k_1} g^{j_2 k_2} \ldots g^{j_p k_p} \varepsilon_{k_1 k_2 \ldots k_p j_{p+1} j_{p+2} \ldots j_n} g^{j_{p+1} i_{p+1}} g^{j_{p+2} i_{p+2}}$$

$$\ldots g^{j_n i_n} \varepsilon_{i_{p+1} \ldots i_n i_1 \ldots i_p} \omega^{i_1} \wedge \ldots \wedge \omega^{i_p}.$$

Now,

$$\varepsilon_{k_1 k_2 \ldots k_p j_{p+1} j_{p+2} \ldots j_n} g^{j_{p+1} i_{p+1}} g^{j_{p+2} i_{p+2}} \ldots g^{j_n i_n} g^{j_1 k_1} g^{j_2 k_2} \ldots g^{j_p k_p}$$

$$= \varepsilon^{j_1 j_2 \ldots j_p i_{p+1} i_{p+2} \ldots i_n} \det \left(g^{ij} \right) = \varepsilon^{j_1 j_2 \ldots j_p i_{p+1} i_{p+2} \ldots i_n} g^{-1}$$

so that

$$** \left[\omega^{j_1} \wedge \omega^{j_2} \wedge \ldots \wedge \omega^{j_p} \right] =$$

$$= \frac{|g|}{g} \frac{1}{(n-p)!\, p!}\, (-)^{p(n-p)} \varepsilon^{j_1 j_2 \ldots j_p}_{i_1 i_2 \ldots i_p} (n-p)!\, \omega^{i_1} \wedge \ldots \wedge \omega^{i_p}$$

$$= \frac{|g|}{g} \frac{1}{(n-p)!\, p!}\, (-)^{p(n-p)} (n-p)!\, p!\, \omega^{j_1} \wedge \ldots \wedge \omega^{j_p}$$

$$= \frac{|g|}{g} (-)^{p(n-p)} \omega^{j_1} \wedge \ldots \wedge \omega^{j_p}.$$

Thus, taking twice the dual of a p-form yields back the original form up to a sign and the factor $|g|/g$. The components of a metric constitute always a symmetric matrix, which can always be put in diagonal form in a convenient basis. The number of positive diagonal terms (P) minus the number of negative diagonal terms (N), given by

$$s = P - N = (n - N) - N = n - 2N,$$

is an invariant property of the metric (a theorem due to Sylvester), called its *signature*.[6] Minkowski metric, for instance, has signature $s = 2$. The factor $|g|/g$ is simply a sign

$$(-)^N = (-)^{(n-s)/2}.$$

We find thus that, for any p-form,

$$** \alpha^p = (-)^{p(n-p)+(n-s)/2} \alpha^p, \tag{7.101}$$

---

[6] We could have defined $s = N - P$ instead, without any change for our purposes.

so that the operator inverse to $*$ is

$$*^{-1} = (-)^{p(n-p)+(n-s)/2} *$$ (7.102)

when applied to a p-form and the metric has signature s. Of course, $|g|/g = 1$ for a strictly riemannian metric. For this reason the signature dependence of $*^{-1}$ is usually ignored in texts confined to strictly riemannian manifolds.

§ 7.4.7 Let $\{e_j\}$ be a basis in which the metric components are g_{ij}. The metric volume element introduced in § 7.4.4 could have been alternatively defined as the n-form v such that

$$v(e_1, e_2, \ldots, e_n)\, v(e_1, e_2, \ldots, e_n) = \det (g_{ij}) = g.$$ (7.103)

In reality, this only fixes v up to a sign. The manifold N has been supposed to be orientable and the choice of the sign in this case corresponds to a choice of orientation.

§ 7.4.8 The following property is true for forms α and β of the same order:

$$\alpha \wedge (*\beta) = \beta \wedge (*\alpha).$$ (7.104)

An inner product (α, β) between two forms of the same order can then be introduced: it is such that

$$\alpha \wedge (*\beta) = (\alpha, \beta)\, v.$$ (7.105)

§ 7.4.9 This inner product generalizes the inner product generated by the metric on the space of the 1-forms. In $\mathbb{E}^3$, for example, it comes out immediately that

$$*\, dx^1 = dx^2 \wedge dx^3 \qquad *\, dx^2 = dx^3 \wedge dx^1 \qquad *\, dx^3 = dx^1 \wedge dx^2. \quad (7.106)$$

Given two 1-forms $\alpha_i dx^i$ and $\beta_j dx^j$,

$$\alpha \wedge *\beta = (\alpha_i \beta_i)\, dx^1 \wedge dx^2 \wedge dx^3.$$

§ 7.4.10 We have already used the star operator in § 7.2.7. The trivial character of the euclidean metric has hidden it somewhat, but the correspondence between vectors and second-order antisymmetric tensors in $\mathbb{E}^3$ is given precisely by the star operator. It is essential to the definition of the laplacian of a function f:

$$\Delta f = \boldsymbol{div\ grad}\, f.$$ (7.107)

It is a simple exercise to check that

$$d * d\, f = (\Delta f)\, dx \wedge dy \wedge dz.$$ (7.108)

§ **7.4.11** Take the Minkowski space, with $g_{00} = -1$ and $g_{ii} = 1$, in the cartesian basis $\{dx^\alpha\}$, and with the convention $\varepsilon_{0123} = +1$. In this case,

$$* dx^1 = \tfrac{1}{3!} \sqrt{|g|}\, \varepsilon^1{}_{\alpha\beta\gamma}\, dx^\alpha \wedge x^\beta \wedge dx^\gamma = \varepsilon^1{}_{230}\, dx^2 \wedge dx^3 \wedge dx^0$$
$$= - dx^2 \wedge dx^3 \wedge dx^0 = - dx^0 \wedge dx^2 \wedge dx^3,$$
$$* dx^2 = - dx^0 \wedge dx^3 \wedge dx^1,$$
$$* dx^3 = - dx^0 \wedge dx^1 \wedge dx^2,$$
$$* (dx^1 \wedge dx^2) = dx^0 \wedge dx^3, \text{ etc.}$$
$$* (dx^0 \wedge dx^1) = - dx^2 \wedge dx^3,$$
$$* (dx^1 \wedge dx^2 \wedge dx^3) = - dx^0,$$
$$* (dx^0 \wedge x^1 \wedge dx^2) = dx^3, \text{ etc.}$$

The dual to the 1-form $A = A_\mu dx^\mu$ will be

$$* A = \tfrac{1}{3!} \sqrt{|g|}\, A^\mu \varepsilon_{\mu\lambda\rho\sigma}\, dx^\lambda \wedge dx^\rho \wedge dx^\sigma.$$

For the 2-form

$$F = \tfrac{1}{2!} F_{\mu\nu} dx^\mu \wedge dx^\nu,$$

we have

$$* F = \tfrac{1}{2!} \left(\tfrac{1}{2!} F^{\mu\nu} \varepsilon_{\mu\nu\rho\sigma} \right) dx^\rho \wedge dx^\sigma = \tfrac{1}{2!} \widetilde{F}_{\rho\sigma} dx^\rho \wedge dx^\sigma.$$

§ **7.4.12** In a 4-dimensional space, the dual of a 2-form is another 2-form. One could ask in which circumstances a 2-form can be self-dual (or antiself-dual), that is,

$$F = \pm * F.$$

This would require, from Eq.(7.101), that

$$F = \pm * F = \pm * [\pm * F] = * * F = (-)^{(4-s)/2} F = (-)^{s/2} F. \qquad (7.109)$$

In pseudo-euclidean spaces, self-duality of F implies the vanishing of F. In a euclidean 4-dimensional space, non-trivial self-duality is quite possible. In gauge theories, self-dual euclidean fields are related to *instantons*.

§ **7.4.13** The coderivative $\widetilde{d}$ of a p-form α is defined by

$$\widetilde{d}\alpha := (-)^p *^{-1} d * \alpha = - (-)^{n(p-1)+(n-s)/2} * d * \alpha. \qquad (7.110)$$

Perhaps *codifferential* would be a more appropriate name, but *coderivative* is more usual. A quick counting will tell that this additional exterior differentiation takes a p-form into a $(p - 1)$-form. There is more:

$$\widetilde{d}\widetilde{d} = (-)^{p-1}(-)^p *^{-1} d * *^{-1} d * = - *^{-1} d\, d * \equiv 0. \qquad (7.111)$$

§ **7.4.14** A form ω such that $\widetilde{d}\omega = 0$ is said to be *coclosed*. A p-form ω such that a $(p+1)$-form α exists satisfying $\omega = \widetilde{d}\alpha$ is *coexact*. In terms of components in a natural basis,

$$\widetilde{d}\alpha^{p+1} = -\frac{1}{p!}\left(\partial^j \alpha_{ji_1 \ldots i_p}\right) dx^{i_1} \wedge dx^{i_2} \wedge \ldots \wedge dx^{i_p}. \qquad (7.112)$$

Notice what happens with the components: each one will consist of a sum of derivatives by all those basis elements whose duals are not in the form-basis at the right. The coderivative is a generalization of the divergence, and is sometimes also called divergence. In a general basis, corresponding to Eq.(7.90) for the exterior derivative, a lengthy calculation gives

$$\widetilde{d}\alpha^{p+1} = -\frac{1}{p!}\left(\nabla^j \alpha_{ji_1 \ldots i_p}\right) \omega^{i_1} \wedge \omega^{i_2} \wedge \ldots \wedge \omega^{i_p}. \qquad (7.113)$$

Still another expression will be given in § 7.6.12. Only after that an expression will be found for the coderivative $\widetilde{d}(\alpha^p \wedge \beta^q)$ of the wedge product of two forms.

§ **7.4.15** Now, a *laplacian* operator can be defined which acts on forms of any order:

$$\Delta := \left(d + \widetilde{d}\right)^2 = d\widetilde{d} + \widetilde{d}d. \qquad (7.114)$$

On 0-forms, Δ reduces (up to a sign!) to the usual Laplace–Beltrami operator acting on functions.

Notice that the laplacian of a p-form is a p-form. Harmonic analysis can be extended to antisymmetric tensors of any order. A p-form ω such that

$$\Delta\omega = 0$$

is said to be *harmonic*. The harmonic p-forms constitute a vector space by themselves. From the very definition of Δ, a form simultaneously closed and coclosed is harmonic. The laplacian has the "commutation" properties:

$$d\,\Delta = \Delta\,d \qquad *\Delta = \Delta\,* \qquad \widetilde{d}\Delta = \Delta\,\widetilde{d}.$$

If A is a 1-form in $\mathbb{E}^3$, in which the trivial metric allows identification of 1-forms and vectors, $\Delta = d\widetilde{d} + \widetilde{d}d$ is the usual formula of vector calculus

$$\Delta A = \boldsymbol{grad\ div}\ A - \boldsymbol{rot\ rot}\ A.$$

§ **7.4.16 Maxwell's equations, second pair.** Using the results listed in § 7.4.11, Eq.(7.33) gives easily

$$
\begin{aligned}
*F = & -\ H_1 dx^1 \wedge dx^0 - H_2 dx^2 \wedge dx^0 - H_3 dx^3 \wedge dx^0 \\
& +\ E_1 dx^2 \wedge dx^3 + E_2 dx^3 \wedge dx^1 + E_3 dx^1 \wedge dx^2,
\end{aligned} \qquad (7.115)
$$

or equivalently

$$* F = H_i \, dx^0 \wedge x^i + \tfrac{1}{2} \varepsilon_{ijk} E_i \, dx^j \wedge dx^k. \qquad (7.116)$$

Comparison with Eq. (7.33) shows that the Hodge operator takes

$$\boldsymbol{H} \to \boldsymbol{E} \quad \text{and} \quad \boldsymbol{E} \to -\boldsymbol{H}.$$

It corresponds, consequently, to the usual dual transformation in electromagnetism, a symmetry of the theory in the sourceless case. If we calculate the coderivative of the electromagnetic form F, we find

$$\widetilde{d} F = \boldsymbol{\nabla} \cdot \boldsymbol{E} \, dx^0 + (\partial_0 \boldsymbol{E} - \boldsymbol{rot}\,\boldsymbol{H}) \cdot d\boldsymbol{x}. \qquad (7.117)$$

We have seen in § 7.2.8 that the first pair of Maxwell's equations is summarized in $dF = 0$. Now we see that, in the absence of sources, the second pair

$$\boldsymbol{\nabla} \cdot \boldsymbol{E} = 0 \quad \text{and} \quad \partial_0 \boldsymbol{E} = \boldsymbol{rot}\,\boldsymbol{H} \qquad (7.118)$$

is equivalent to $\widetilde{d} F = 0$. The first pair is metric-independent. The coderivative is strongly metric-dependent, and so is the second pair of Maxwell's equations. Equations (7.36) and (7.118) tell us that, in the absence of sources, F is a harmonic form. The first pair does not change when charges and currents are present, but the the second does: the first equation in (7.118) acquires a term $4\pi\rho$ in the right-hand side, and the second a term $-4\pi\boldsymbol{J}$. If we define on Minkowski space the current 1-form

$$j := 4\pi \left[\rho \, dx^0 - \boldsymbol{J} \cdot d\boldsymbol{x} \right], \qquad (7.119)$$

Maxwell's equations become

$$dF = 0 \quad \text{and} \quad \widetilde{d} F = j. \qquad (7.120)$$

The current form is coexact, as the last equation implies

$$\widetilde{d} j = 0, \qquad (7.121)$$

or in components,

$$\partial_0 \rho + \boldsymbol{\nabla} \cdot \boldsymbol{J} = 0, \qquad (7.122)$$

which is the continuity equation: *charge conservation is a consequence of the coexactness of the current form.* Notice that this is metric dependent. In the presence of charges, the electromagnetic form is no more harmonic, but remains closed. Consequently, in every contractible region of Minkowski space there exists a 1-form $A = A_\mu dx^\mu$ such that $F = dA$. From Eq. (7.120) we see that this potential form obeys the wave equation

$$\widetilde{d} d A = j, \qquad (7.123)$$

or in components,

$$\partial^\mu \partial_\mu A_\nu - \partial_\nu \partial^\mu A_\mu = j_\nu. \qquad (7.124)$$

The Lorenz gauge is the choice

$$\widetilde{d} A = \partial^\mu A_\mu = 0. \qquad (7.125)$$

§ 7.4.17 Here is a list of relations valid in Minkowski space, for forms of degrees 0 to 4. We use the compact notations

$$dx^{\mu\nu} = dx^\mu \wedge dx^\nu,$$

$$dx^{\mu\nu\sigma} = dx^\mu \wedge dx^\nu \wedge dx^\sigma,$$

$$dx^{\lambda\mu\nu\sigma} = dx^\lambda \wedge dx^\mu \wedge dx^\nu \wedge dx^\sigma.$$

We recall that the bracket $[\lambda, \mu, \dots]$ means a complete antisymmetrization in the included indices.

form	$*$	d	$\tilde{d}$
f	$f\,dx^{1230}$	$(\partial_\mu f)dx^\mu$	0
$A_\mu dx^\mu$	$\frac{1}{3!}A^\mu \varepsilon_{\mu\nu\rho\sigma}dx^{\nu\rho\sigma}$	$\frac{1}{2!}\partial_{[\mu}A_{\nu]}dx^{\mu\nu}$	$-\partial^\mu A_\mu$
$\frac{1}{2!}F_{\mu\nu}dx^{\mu\nu}$	$\frac{1}{2!}[\frac{1}{2!}F^{\mu\nu}\varepsilon_{\mu\nu\rho\sigma}]dx^{\rho\sigma}$	$\frac{1}{3!}\partial_{[\lambda}F_{\mu\nu]}dx^{\lambda\mu\nu}$	$-\partial^\mu F_{\mu\nu}dx^\nu$
$\frac{1}{3!}W_{\lambda\mu\nu}dx^{\lambda\mu\nu}$	$W^{[\lambda\mu\nu}dx^{\sigma]}$	$\frac{1}{4!}\partial_{[\lambda}W_{\mu\nu\sigma]}dx^{\lambda\mu\nu\sigma}$	$-\frac{1}{2!}\partial^\nu W_{\lambda\mu\nu}dx^{\lambda\mu}$
$\frac{1}{4!}V_{\lambda\mu\nu\rho}dx^{\lambda\mu\nu\rho}$	V_{1230}	0	$\frac{1}{3!}\varepsilon_{\mu\nu\rho\sigma}\partial^\mu V_{1230}dx^{\nu\rho\sigma}$

§ 7.4.18 With the notation of § 7.3.8, the field equations for gauge theories are the Yang–Mills equations

$$\partial^\lambda F^a{}_{\lambda\nu} + f^a{}_{bc}A^{b\lambda}F^c{}_{\lambda\nu} = J^a{}_\nu, \tag{7.126}$$

where $J^a{}_\nu$ is the source current. This is equivalent to

$$\partial_{[\lambda}\tilde{F}^a{}_{\mu\nu]} + f^a{}_{bc}A^b_{[\lambda}\tilde{F}^c{}_{\mu\nu]} = \tilde{J}^a{}_{[\lambda\mu\nu]}. \tag{7.127}$$

In invariant notation, (7.126) reads

$$\tilde{d}F + *^{-1}[A, *F] = J. \tag{7.128}$$

Thus, *in the sourceless case, the field equations are just the Bianchi identities written for the dual of F.* A point of interest of self-dual fields (§ 7.4.12) is that, for them, the sourceless Yang–Mills equations coincide with the Bianchi identities and any F of the form

$$F = dA + A \wedge A$$

will solve the field equations. Recall the discussion of § 7.3.11 on types of covariant derivatives, depending on the form degrees. The expression at the left-hand side of the equations above is the *covariant coderivative* of the 2-form F according to the connection A.

§ 7.4.19 Covariant coderivative. To adapt the discussion on covariant derivatives (§ 7.3.11) to the case of coderivatives, we simply notice that the transformations in the value (vector) space ignore the tensor content of the form, so that

$$(*W)' = *W' = g*Wg^{-1}.$$

We can therefore apply the same reasoning to obtain

$$d*W' + |[A',W']| = g\{d*W + |[A,*W]|\}g^{-1}.$$

Since

$$\widetilde{d}W = (-)^{\partial W}*^{-1}d*W,$$

we apply $(-)^{\partial W}*^{-1}$ to this expression to find the covariant coderivative

$$\widetilde{D}W = \widetilde{d}W + (-)^{\partial W}*^{-1}|[A,*W]|. \tag{7.129}$$

For W belonging to a linear representation, this is

$$\widetilde{D}W = \widetilde{d}W + (-)^{\partial W}*^{-1}A*W. \tag{7.130}$$

7.5   Integration and homology

We present here a quick, simplified view of an involved subject. The excuse for quickness is precisely simplicity. The aim is to introduce the crucial ideas and those simple notions which are of established physical relevance.

7.5.1   *Integration*

§ 7.5.1 Let us go back to our small talk of Section 7.1, where it was said that exterior differential forms are entities living under the shadow of the integral sign. The fundamental idea is as simple as follows: suppose we know how to integrate a form on a domain D of an euclidean space $\mathbb{E}^{m+n}$. Suppose besides that a differentiable mapping f is given, which maps D into some subset of the differentiable manifold M,

$$f : D \longrightarrow f(D) \subset M.$$

A form ω defined on M will be pulled back to a form $f^*\omega$ on D . The integral of ω on $f(D)$ is then *defined* as

$$\int_{f(D)} \omega = \int_D f^*\omega. \tag{7.131}$$

§ 7.5.2 Notice the importance of the "pulling-back" behaviour of forms. It is possible to show (this unforgivable phrase) that, given two diffeomorphisms between the interiors of D and $f(D)$, then they both lead to the same result: the definitions would then differ by a change of coordinate systems. "Interior", let us recall, means the set minus its boundary. Boundaries have been defined in § 2.3.16 for chains, which directs us to another point of relevance: *the integration domains will be chains.*

§ 7.5.3 We have in § 2.3.4 introduced polyhedra, which are the sets of points of simplicial complexes in euclidean spaces. We went further in § 2.3.5, defining *curvilinear polyhedra* and *curvilinear simplexes* on a general topological space as subsets of these spaces which are homeomorphic to polyhedra and simplexes in some euclidean space.

The line of thought adopted here is the following: first, define integration on euclidean simplexes; second, choose the mapping f to be a differentiable homeomorphism, fixing curvilinear simplexes on the manifold M; third, by using Eq.(7.131), define integration on these simplexes; finally, extend the definition to general p-chains on M. These last chains are defined in the same way, as sets homeomorphic to euclidean chains: they are usually called *singular p-chains.*

Commentary 7.4 The qualification "singular" is added because the homeomorphism f is only one-way differentiable and is not a diffeomorphism (see § 7.5.14 and § 7.5.20). ◄

§ 7.5.4 The integration of a p-form α on a simplex or polyhedron P on $\mathbb{E}^p$, in which it has the expression

$$\alpha = \alpha(x^1, x^2, \ldots, x^p)dx^1 \wedge dx^2 \wedge \ldots \wedge x^p \ ,$$

is defined simply as the usual Riemann integral

$$\int_P \alpha := \int_P \alpha(x^1, x^2, \ldots, x^p)dx^1 dx^2 \ldots dx^p, \tag{7.132}$$

as a limit of a summation.

§ 7.5.5 The definition of integration on a domain in a differentiable manifold M will require:

(i) The choice of an orientation in $\mathbb{E}^p$.
(ii) The choice of a differentiable homeomorphism f, defining on M a corresponding curvilinear polyhedron or simplex $f(P)$.

Then, the integral of a p-form ω will be

$$\int_{f(P)\subset M} \omega := \int_{P\subset\mathbb{E}^p} f^*\omega, \qquad (7.133)$$

$$\int_{f(S_p)} \omega := \int_{S_p} f^*\omega. \qquad (7.134)$$

§ 7.5.6 A *singular p-chain* σ_p on M is simply obtained: given on $\mathbb{E}^p$ a p-chain

$$c_p = m_1 S_p^1 + m_2 S_p^2 + \ldots + m_j S_p^j,$$

then

$$\sigma_p = m_1 \sigma_p^1 + m_2 \sigma_p^2 + \ldots + m_j \sigma_p^j,$$

where $\sigma_p^j = f(S_p^j)$ is a singular p-simplex. The coefficients m_j are the multiplicities of the corresponding curvilinear simplexes σ_p^j. Intuitively, they give the number of times one integrates over that simplex.

§ 7.5.7 Integration on the p-chain σ_p is now

$$\int_{\sigma_p} \omega := \sum_{i=1}^{j} m_i \int_{\sigma_p^i} \omega = \sum_{i=1}^{j} m_i \int_{S_p^i} f^*\omega. \qquad (7.135)$$

The boundary of a singular chain is defined in a natural way: first, the boundary of a curvilinear simplex is the image of the restriction of the mapping f to the boundary of the corresponding euclidean simplex,

$$\partial\sigma_p^j = f(\partial S_p^j).$$

Thus,

$$\partial\sigma_p = \sum_{i=1}^{j} m_i \partial\sigma_p^i. \qquad (7.136)$$

As the mapping f is a homeomorphism, it will take the empty set into the empty set. Thus,

$$\partial\partial\sigma_p^j = f(\partial\partial S_p^j) = f(\emptyset) = 0,$$

and

$$\partial\partial\sigma_p \equiv 0. \qquad (7.137)$$

The above definition of singular chains carries over with them all the algebraic properties of euclidean chains. It is then possible to define closed singular chains, or cycles, as well as exact singular chains, and transpose the same homology groups to chains on differentiable manifolds.

§ 7.5.8 We are now ripe to state one of the most important theorems of all Mathematics:

The integral of the exterior derivative of a form, taken on a singular chain, equals the integral of the form itself taken on the boundary of that chain:

$$\int_\sigma d\alpha = \int_{\partial\sigma} \alpha. \tag{7.138}$$

It is called by Arnold the *Newton–Leibniz–Gauss–Ostrogradski–Green–Stokes–Poincaré theorem*, a name tracing its historical evolution. Its fantastic generality was recognized by the last member of the illustrious team, and includes as particular cases all the vector analysis theorems associated to those eminent names. The first two patriarchs belong because, if f is a real function and σ is the real oriented interval (a, b), then we have that $\partial\sigma = b - a$, and

$$\int_\sigma df = \int_a^b df = \int_{\partial\sigma} f = f(b) - f(a). \tag{7.139}$$

Most authors call the theorem after the last-but-one member of the list.

§ 7.5.9 Although we shall not intend to demonstrate it, let us give only one indication: it is enough to suppose it valid for the euclidean case:

$$\int_{\sigma=f(S)} d\alpha = \int_S f^* d\alpha = \int_S d(f^* \alpha);$$

if it holds for the euclidean chain S, we can proceed:

$$= \int_{\partial S} f^* \alpha = \int_{f(\partial S)} \alpha = \int_{\partial\sigma} \alpha. \tag{7.140}$$

As to the demonstration for euclidean chains, it follows the general lines of the demonstrations of (say) Gauss theorem in vector analysis, with the necessary adaptations to general dimension and order.

Commentary 7.5 An immediate consequence of (7.138) is that the integral of any closed p-form on any p-boundary is zero. ◀

§ 7.5.10 A few examples in electromagnetism on $\mathbb{E}^3$. We shall alternate the short-hand notation we have been using with more explicit and usual ones.

(i) Take a closed curve γ in $\mathbb{E}^3$, and let S be some piece of surface bounded by γ: $\gamma = \partial S$. The circulation of the vector potential A along γ will be

$$\int_\gamma A = \int_{\gamma=\partial S} A \cdot dl = \int_S dA = \int_S (\boldsymbol{rot}\, A) \cdot d\sigma = \int_S H \cdot d\sigma = \int_S H,$$

that is to say, the flux of H through S.

(ii) Take now the flux of H through a closed surface (say, a sphere S^2 enclosing a ball B^3), which is

$$\int_{S^2} H = \int_{S^2} H \cdot d\sigma = \int_{\partial B^3} H \cdot d\sigma = \int_{B^3} dH = \int_{B^3} \boldsymbol{div}\, H\,.$$

On the other hand,

$$\int_{S^2} H = \int_{S^2} dA = \int_{S^2} (\boldsymbol{rot}\, A) \cdot d\sigma = \int_{\partial S^2 = 0} A = 0.$$

As the ball is of arbitrary size, this implies div $H = 0$.

(iii) Faraday law of induction: the circulation of E along a closed line is

$$\int_\gamma E = \oint E \cdot dl = \int_{\gamma=\partial S} E = \int_S dE = \int_S (\boldsymbol{rot}\, E) \cdot d\sigma = -\int_S \partial_0 H \cdot d\sigma.$$

As already said, H is a 2-tensor; by using the Hodge $*$ operation, it can be taken as an (axial) vector. Then,

$$dH = \boldsymbol{div}\, H \quad \text{and} \quad \tilde{d}H = \boldsymbol{rot}\, H.$$

§ 7.5.11 Total electric charge in a closed universe.

In the Friedmann–Robertson–Walker–Lemaître model for the universe,[7] it may happen that the space section of the universal spacetime is an open (infinitely extended, infinite-volume) or a closed (finite volume) manifold. The model reduces the possibilities to only these two.

Let us consider the closed case, in which the manifold is a 3-dimensional (expanding) sphere S^3. The total electric charge is given by Gauss law,

$$Q = 4\pi \int_{S^3} \rho = \int_{S^3} \boldsymbol{\nabla} \cdot \boldsymbol{E} = \int_{S^3} (\partial_i E_i)\, d^3 x.$$

But $\partial_i E_i = d*E$, with $E = E_i dx^i$. Consequently,

$$Q = \int_{S^3} d*E = \int_{\partial S^3} *E = 0$$

because $\partial S^3 = 0$. The argument can be adapted to an open universe, but then convenient conditions at large distances are required.

---

[7] See for example Weinberg 1972, Chap.15.

§ 7.5.12 Consider a force field on $\mathbb{E}^3$: suppose that the work necessary for a displacement between two points a and b is independent of the path. This is the same as saying that the integral along any closed path through a and b is independent of that closed path, and thus vanishes. For a conservative

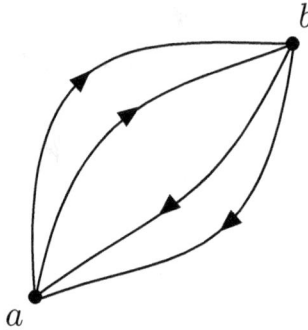

Fig. 7.6   Paths between two points in a force field.

system, the work

$$W = F_i \, dx^i$$

is a closed form. From

$$\oint F_i \, dx^i = 0$$

we deduce that the force is of potential origin, $F = -\, dU$ for some U, or

$$F_i = -\, (\partial U / \partial x^i).$$

§ 7.5.13 This is a particular case of a very important theorem by de Rham. Let us, before enunciating it, introduce a bit more of language: the integral of a p-form ω on a p-cycle σ is called the *period* of ω on σ. Clearly, the Stokes theorem implies that all the periods of an exact form are zero. The theorem ("first theorem of") de Rham has proved says that

a closed form whose periods are all vanishing is necessarily exact.

Notice that this holds true for any smooth manifold, however complicated its topology may happen to be.

§ 7.5.14 Let us make a few comments on the simplified approach adopted above. Integration is a fairly general subject. It is not necessarily related to topology, still less to differentiability. It requires a measure space, and for that the presupposed division of space is not a topology, but a σ-algebra (see Chapter 15). We are therefore, to start with, supposing also a compatibilization of this division with that provided by the topology. In other words, we are supposing a covering of the underlying set by Borel sets constituting the smallest σ-algebra generated by the topology. It is not excluded that two similar but different topologies generate the same σ-algebra. Only up to these points do the above statements involve (and eventually probe) *the* topology. But there are restrictions. For example, to be sure of the existence of a Borel measure, the topology must be Haussdorf and locally compact. The mapping f of § 7.5.1 is clearly a bit too strong to be assumed to exist for any domain D. We have later assumed the existence of a differentiable homeomorphism from euclidean spaces to the differentiable manifolds to the extent of taking chains from one into the other, etc. It should however be said, in favour of the above exposition, that it allows a fair view of the main relationships of integration to the "global" characteristics of the space. The interested reader is urged to consult more complete treatments.[8]

§ 7.5.15 On hypersurfaces.[9] A hypersurface is an $(n-1)$-dimensional space immersed in $\mathbb{E}^n$. It may be an imbedded manifold, or be only locally differentiable. We shall here suppose it an imbedding for the sake of simplicity. Suppose in $\mathbb{E}^n$ a hypersurface Γ given by the equation

$$\psi(x) = \psi(x^1, x^2, \ldots, x^n) = 0.$$

Then to Γ will correspond a special $(n-1)$-form, its volume form ω_Γ, in the following way. A requirement will be that the surface be nonsingular, that is,

$$\boldsymbol{grad}\ \psi = d\psi \neq 0,$$

at least locally. Let v be the $\mathbb{E}^n$ volume form,

$$v = dx^1 \wedge dx^2 \wedge \cdots \wedge dx^n.$$

Then ω_Γ is defined by

$$v = d\psi \wedge \omega_\Gamma.$$

Around a point p one is in general able to define new coordinates $\{u^k\}$ with positive jacobian $\|\partial x/\partial u\|$ and such that one of them, say u^j, is $\psi(x)$.

---

[8] Such as Choquet–Bruhat, DeWitt–Morette & Dillard–Bleick 1977.
[9] Gelfand & Shilov 1964.

Then,

$$v = \left\| \frac{\partial x}{\partial u} \right\| du^1 \wedge du^2 \wedge \cdots \wedge du^{j-1} \wedge d\psi \wedge du^{j+1} \wedge \cdots du^n. \qquad (7.141)$$

If around a point p it so happens that $\partial_j \psi \neq 0$, we may simply choose $u^{i \neq j} = x^i$ and $u^j = \psi(x)$, in which case

$$\omega_\Gamma = (-)^{j-1} \frac{dx^1 \wedge dx^2 \wedge \cdots dx^{j-1} \wedge dx^{j-1} \cdots \wedge dx^n}{\partial_j \psi}. \qquad (7.142)$$

A trivial example is the surface $x^1 = 0$, for which

$$\omega_\Gamma = dx^2 \wedge dx^3 \wedge \cdots \wedge dx^n.$$

These notions lead to the definition of hypersurface-concentrated distributions (generalized Dirac δ functions), given through test functions f on $\mathbb{E}^n$ by

$$(\delta(\psi), f) = \int_{\mathbb{E}^n} \delta(\psi) f = \int_\Gamma f = \int_\Gamma f(x) \omega_\Gamma.$$

Also the generalized step-function

$$\theta(\psi) = 1 \text{ for } \psi(x) \geq 0 \quad \text{and} \quad \theta(\psi) = 0 \text{ for } \psi(x) < 0,$$

can be defined by $\theta'(\psi) = \delta(\psi)$, with the meaning

$$\partial_j \theta = (\partial_j \psi)\, \delta(\psi).$$

It is sometimes more convenient to use an inverted notation. If Γ is the boundary of a domain D, we may want to use the characteristic function of D, a function which is 1 inside D and zero outside it. If D is the set of points x such that $\psi(x) \leq 0$, we define

$$\theta(\psi) = 1 \text{ for } \psi(x) \leq 0 \quad \text{and} \quad \theta(\psi) = 0 \text{ for } \psi(x) > 0,$$

so that $\theta(\psi)$ is the characteristic function of D. In this case we have that $\theta'(\psi) = -\delta(\psi)$, and

$$\mathbf{grad}\, \theta = -\delta(\psi)\, \mathbf{grad}\, \psi.$$

This corresponds to the usual practice of using, on a surface, the normal directed outward, and leads to general expressions for well known relations of vector analysis. Such distributions are of use, for example, in getting Maxwell's equations in integral form (Section 29.3).

7.5.2 *Cohomology of differential forms*

We begin by recalling — and in the meantime rephrasing — some of the comments about chains and their homologies, made in Chapter 2. Then we describe the dual structures which are found among the forms, cohomologies. Although cohomology is an algebraic structure with a very wide range of applications, differential forms are a subject of choice to introduce the main ideas involved.[10] Here we shall suppose all the chains already displayed on a general smooth manifold M.

§ **7.5.16 Homology, revisited.** A chain σ is a *cycle* (or is *closed*) if it has null boundary:

$$\partial\sigma = 0.$$

It is a *boundary* if another chain ρ exists such that

$$\sigma = \partial\rho.$$

Closed p-chains form a vector space (consequently, an additive group) Z_p. Boundaries likewise form a vector space B_p for each order p. Two closed p-chains σ and θ whose difference is a boundary are *homologous*. In particular, a chain which is itself a boundary is *homologous to zero*. Now, homology is an equivalence relation between closed forms. The quotient of the group Z_p by this relation is the *homology group*

$$H_p = Z_p/B_p.$$

For compact manifolds all these vector spaces have finite dimensions. The Betti numbers $b_p = \dim H_p$ are topological invariants, that is, characteristics of the topology defined on M.

§ **7.5.17 Cohomology.** Now for forms. A form ω is *closed* (or is a *cocycle*) if its exterior derivative is zero,

$$d\omega = 0.$$

It is *exact* (or a *coboundary*) if another form α exists such that

$$\omega = d\alpha.$$

---

[10] An excellent short introduction to this subject can be found in Godbillon 1971. Another excellent text, with an involved approach to many physical problems, including a rather detailed treatment of the decomposition theorems (and much more) is Marsden 1974. Finally, a treatise whose reading requires some dedication, and deserves it, is de Rham 1960. For some pioneering applications of cohomological ideas to Physics, see Misner & Wheeler 1957.

Closed p-forms constitute a vector space, denoted Z^p. The same happens to exact forms, which are in a vector space B^p. Two closed p-forms are said to be *cohomologous* when their difference is an exact form. In particular, an exact form is *cohomologous to zero*. Cohomology is an equivalence relation. The quotient of the cocycle group Z^p by this relation is another group, de Rham's *cohomology group* $H^p = Z^p/B^p$. Again for compact manifolds, all these vector spaces are of finite dimension, $b^p = \dim H^p$. Another fundamental result by de Rham is the following:

> *For compact manifolds, the homology group H_p and the cohomology group H^p are isomorphic.*

So, the Betti numbers are also the dimension of H^p: $b^p = b_p$. Roughly speaking, the number of independent p-forms which are closed but not exact is a topological invariant. This establishes a strong relation between forms and the topology of the manifold. Differential forms play just the role of the cochains announced in § 2.3.18.

§ 7.5.18 The above results, and the two identities,

$$\partial^2 \sigma \equiv 0 \quad \text{and} \quad d^2 \alpha \equiv 0,$$

show the deep parallelism between forms (integrands) and chains (integration domains). For compact manifolds without boundary, this parallelism is actually complete and assumes the characteristics of a *duality*: given a closed p-form ω^p and a closed p-chain σ^p, we can define a linear mapping

$$\omega^p, \sigma^p \to \; < \sigma^p, \omega^p > \; := \int_{\sigma^p} \omega^p \in \mathbb{R}, \qquad (7.143)$$

which has all the properties of a scalar product, just as in the case of a vector space and its dual. This product is in reality an action between homology and cohomology classes, and not between individual forms and chains.

This is a consequence of the two following properties:

(i) The integral of a closed form ω over a cycle σ depends only on the homology class of the cycle σ: if $\sigma - \theta = \partial\rho$, then

$$\int_\sigma \omega = \int_{\theta + \partial\rho} \omega = \int_\theta \omega + \int_{\partial\rho} \omega = \int_\theta \omega + \int_\rho d\omega = \int_\theta \omega.$$

(ii) The integral of a closed form ω over a cycle σ depends only on the cohomology class of the form ω: if $\omega - \alpha = d\beta$, then

$$\int_\sigma \omega = \int_\sigma (\alpha + d\beta) = \int_\sigma \alpha + \int_\sigma d\beta = \int_\sigma \alpha + \int_{\partial\sigma} d\beta = \int_\sigma \alpha.$$

§ 7.5.19 On a compact metric manifold, the star operator allows the definition of a *global inner product* between p-forms through an integration over the whole manifold M:

$$(\alpha, \beta) := \int_M \alpha \wedge *\beta. \tag{7.144}$$

This is a symmetric bilinear form. A one-to-one relation between forms and chains is then obtained by using (7.143) and (7.144): the chain σ can be "identified" with the form α_σ if, for every form ω,

$$< \sigma, \omega >= \int_\sigma \omega = (\alpha_\sigma, \omega) = \int_M \alpha_\sigma \wedge * \omega. \tag{7.145}$$

§ 7.5.20 A few additional comments. First, the compactness requirement is made to ensure the existence of the integrals. It can be softened to the exigency that at least one of the forms involved has a compact carrier: that is, it is different from zero only on a compact domain. Second, the complete duality between forms and chains really requires something else: chains on general smooth manifolds have been introduced through mappings which are only locally homeomorphisms, and nothing else has been required concerning the differentiability of their inverses. That is why they are called *singular* chains. On the other hand forms, as we have introduced them, are fairly differentiable objects. The above relation between forms and chains only exists when forms are enlarged so as to admit components which are distributions. Then, a general theory can be built with forms and chains as the same objects — this has been done by Schwartz and de Rham, who used for the new general objects, including both forms and chains, the physically rather embarrassing name *current* (proposed by Schwartz).

§ 7.5.21 We have seen that the homology groups on a topological manifold can be calculated in principle by the methods of algebraic topology. For compact manifolds, those results can be translated into results concerning the cohomology groups, that is, into properties of the forms defined on them. The simplest compact (bounded) manifolds are the balls B^n imbedded in $\mathbb{E}^n$. Their Betti numbers are

$$b^0(B^n) = b^n(B^n) = 1 \tag{7.146}$$

and

$$b^p(B^n) = 0 \quad \text{for } p = 1, 2, \ldots, n-1. \tag{7.147}$$

Here, another caveat: we have mainly talked about a particular kind of homology, the so-called *integer homology*, for chains with integer multiplicities. The parallelism between forms and chains is valid for *real homology*:

chains with real multiplicities are to be used. The vector spaces are then related to the real numbers, and the line $\mathbb{R}$ takes the place of the previously used set of integer numbers $\mathbb{Z}$. In the above example, the space $H^0(B^n)$ is isomorphic to the real line $\mathbb{R}$. This means that 0-forms (that is, functions) on the balls can be closed but not exact, although in this case they will be constants: their set is isomorphic to $\mathbb{R}$. This trivial result is no more valid for $p \neq 0$: in these cases, every closed form is exact. This is simply a pedantic rephrasing of what has been said before, since B^n is contractible.

§ **7.5.22** The next simplest compact manifolds are the n-dimensional spheres S^n imbedded in $\mathbb{E}^{n+1}$. For them,

$$H^0(S^0) \approx \mathbb{R}^2$$

$$H^0(S^n) \approx H^n(S^n) \approx \mathbb{R}$$

$$H^p(S^n) \approx 0 \quad \text{when} \quad p \neq 0, n. \tag{7.148}$$

On S^4, for instance, closed 1-, 2-, and 3-forms are exact. All 4-forms (which are of course necessarily closed) and closed functions are constants. On the sphere S^2, every closed 1-form (irrotational covariant vector field) is exact (that is, a gradient). All 2-forms and closed functions are constant.

§ **7.5.23** Spheres are the simplest examples of *compact manifolds*. Life is much simpler on such manifolds, on which the internal product (7.144) has many important properties. Take for instance a p-form β and a $(p-1)$-form α. Then,

$$
\begin{aligned}
(d\alpha, \beta) &= \int_M d\alpha \wedge *\beta \\
&= \int_M \left[d(\alpha \wedge *\beta) - (-)^{p-1} \alpha \wedge d*\beta \right] \\
&= \int_M d(\alpha \wedge *\beta) + (-)^p \int_M \alpha \wedge d*\beta \\
&= \int_{\partial M} \alpha \wedge *\beta + \int_M \alpha \wedge * \left[(-)^p *^{-1} d*\beta \right] \\
&= \int_M \alpha \wedge *\tilde{d}\beta.
\end{aligned}
$$

Consequently,

$$(d\alpha, \beta) = (\alpha, \tilde{d}\beta). \tag{7.149}$$

The operators d and $\tilde{d}$ are, thus, adjoint to each other in this internal product. It is also easily found that $*^{-1}$ is adjoint to $*$ and that the laplacian

$$\Delta = d\tilde{d} + \tilde{d}d$$

is self-adjoint:

$$(\Delta\omega, \gamma) = (\omega, \Delta\gamma). \tag{7.150}$$

There is more: on compact-without-boundary strictly riemannian manifolds, the internal product can be shown to be positive-definite. As a consequence, each term is positive or null in the right hand side of

$$(\Delta\omega, \omega) = (d\tilde{d}\omega, \omega) + (\tilde{d}d\omega, \omega) = (\tilde{d}\omega, \tilde{d}\omega) + (d\omega, d\omega).$$

Hence, in order to be harmonic, ω as to be both closed and coclosed. This condition is, on such manifolds, necessary as well as sufficient.

§ 7.5.24 Kodaira–Hodge–de Rham decomposition theorem. This theorem, in its present-day form, is the grown-up form of a well known result, of which primitive particular versions were known to Stokes and Helmholtz. Called *a fundamental theorem of vector analysis* by Sommerfeld,[11] it says that a differentiable enough vector field V in $\mathbb{E}^3$, with a good enough behaviour at infinity, may be written in the form

$$V = grad\, f + rot\, T + c,$$

where c is a constant vector. In its modern form, it is perhaps the deepest result of the above general harmonic analysis. It says that the inner product divides the space of p-forms into three orthogonal sub-spaces. In a rather weak version,

on a compact-without-boundary manifold, every form can be
decomposed in a unique way into the sum of one exact,
one co-exact and one harmonic form:

$$\omega = d\,\alpha + \tilde{d}\,\beta + h, \quad \text{with } \Delta h = 0. \tag{7.151}$$

The authors the theorem is named after have shown that, in reality, with the above notation, a form γ exists such that

$$\alpha = \tilde{d}\gamma \quad \text{and} \quad \beta = d\gamma,$$

which puts ω as the sum of a laplacian plus a harmonic form:

$$\omega = \Delta\gamma + h. \tag{7.152}$$

In consequence, no exact form is harmonic unless it is also coexact and belongs to the harmonic subspace; no harmonic form is purely exact, and so on. In particular, no harmonic form can be written down as $h = d\eta$.

---

[11] Sommerfeld 1964a.

Commentary 7.6   All these properties have been found when studying the solutions of the general Poisson problem

$$\Delta\omega = \rho\,.$$

It has solutions only when ρ belongs exclusively to the laplacian sector, or when ρ is not harmonic. ◄

§ 7.5.25 It is not difficult to verify that the harmonic forms constitute, by themselves, still another vector space. Another fundamental theorem by de Rham says the following:

> *On a compact-without-boundary manifold, the space of harmonic p-forms is isomorphic to the cohomology space $H^p(M)$.*

Thus, if Δ_p is the laplacian on p-forms,

$$b^p = \ \dim\ \ker\ \Delta_p = \dim\ H^p(M).$$

This is useful because it fixes the number of independent harmonic forms on the manifold. This number is determined by its topology.

§ 7.5.26 No electromagnetism on S^4. In order to fix the ideas and check the power of the above results, let us examine a specially simple case: abelian gauge theories on the sphere S^4. Instantons are usually defined as solutions of the free Yang–Mills equations on S^4, for a given gauge theory on Minkowski spacetime. The abelian case includes electromagnetism, for which the group is the 1-dimensional $U(1)$. The Bianchi identity and the Yang–Mills equations are simply

$$dF^a = 0 \ \text{ and } \ \ \widetilde{d}F^a = 0, \tag{7.153}$$

with an index a for each generator of the gauge group. Each F^a is a 2-form and the above equations require that F^a be simultaneously closed and coclosed. This is equivalent to require F^a to be harmonic. How is the space of harmonic 2-forms on S^4? De Rham's fundamental theorem tells us that it is isomorphic to $H^2(S^4)$. From Eq. (7.148), this is a space of zero dimension. We arrive to the conclusion that the only solution for (7.153) is the trivial one, the vacuum

$$F^a = 0.$$

We might say then that no instantons exist for abelian theories or, in particular, that no nontrivial electromagnetism exists on such a "spacetime" as S^4. Notice that, the result coming ultimately from purely topological reasons, it keeps holding for any space homeomorphic to S^4.

§ 7.5.27 Extensions of the decomposition theorem for the physically more useful cases of compact manifolds with boundary have been obtained. The question is far more complicated, because the different possible kinds of boundary conditions have to be analyzed separately. The operators d and $\tilde{d}$ are no more adjoint to each other. The boundary term now survives in the steps leading to Eq. (7.149):

$$(d\alpha, \beta) = (\alpha, \tilde{d}\beta) + \int_{\partial M} \alpha \wedge * \beta. \qquad (7.154)$$

The boundary conditions must be stated in an invariant way. For that, let us introduce some notation. Let

$$i : \partial M \to M$$

be the inclusion mapping attaching the boundary to the manifold. Let us introduce the following metric-dependent notions: the normal part of the form

$$\alpha : \alpha_n = i^*(*\alpha) \qquad (7.155)$$

and the tangent part of the form

$$\alpha : \alpha_t = i^*(\alpha). \qquad (7.156)$$

The form α will be parallel (or tangent) to ∂M if $\alpha_n = 0$. It will be perpendicular to ∂M if $\alpha_t = 0$. Adaptation to a field X is got by recalling that X is in relation to two distinct forms:

(i) its covariant image, a 1-form, and
(ii) the $(m-1)$-form $i_X v$, obtained from the volume form v through the interior product (see § 7.6.5).

Then, X is tangent to ∂M iff its covariant image is tangent to ∂M, or iff $i_X v$ is normal to ∂M. On the other hand, X is normal to ∂M iff $i_X v$ is tangent to ∂M. Also a stronger definition of harmonic forms is needed now:
   A form h is harmonic iff $d\,h = 0$ and $\tilde{d}\,h = 0$ hold simultaneously.
Then, a version[12] of the decomposition theorem valid for manifolds-with-boundary is

$$\omega = d\,\alpha_t + \tilde{d}\,\beta_n + h. \qquad (7.157)$$

Other versions are

$$\omega = d\alpha + \beta_t \quad \text{with } \beta_t \text{ satisfying } \tilde{d}\,\beta_t = 0. \qquad (7.158)$$

$$\omega = \tilde{d}\,\beta + \alpha_n \quad \text{with } \alpha_n \text{ satisfying } d\alpha_n = 0. \qquad (7.159)$$

---

[12] See Marsden 1974.

§ **7.5.28** A last remark: the inner product is used to obtain invariants. The action for electromagnetism, for example, is the functional of the basic fields A_μ given by

$$S[A] = (F, F) = \int F \wedge *F, \qquad (7.160)$$

where F is given by Eq. (7.39). For more general gauge theories, there is one such field for each generator in the Lie algebra of the gauge group. In order to obtain the action, an invariant metric has to be found also on the algebra. Once such a metric $K_{ab} = K(J_a, J_b)$ is given (see § 8.4.11), the action functional is taken to be

$$S[A] = K(J_a, J_b)(F^a, F^b) = K_{ab} \int F^a \wedge *F^b, \qquad (7.161)$$

with the F^a's now given as in (7.75).

Due to their intuitive content, a chapter on differential forms is a place of choice to introduce cohomology. It should be kept in mind, however, that cohomology is a very general and powerful algebraic concept with applications far beyond the above case, which is to be seen as a particular though important example. We shall meet cohomology again, in different contexts.

7.6 Algebras, endomorphisms and derivatives

While physicists have been striving to learn some geometry, geometers were progressively converting their subject into algebra. We have been leaning heavily on analytical terminology and way-of-thinking in the last chapters. In this paragraph we shall rephrase some of the previous results and proceed in a more algebraic tune.

§ **7.6.1** Let us start by stating (or restating) briefly some known facts. We refer to Chapter 13 for more detailed descriptions of algebraic concepts. An *algebra* is a vector space V with a binary operation $V \otimes V \to V$, submitted to certain general conditions. It may be associative or not, and commutative or not. A given algebra is a *Lie algebra* if its operation is anticommutative and satisfies the Jacobi identity. When the algebra is associative, but not a Lie algebra, it can be made into one: starting from any binary operation a Lie bracket can be defined as the commutator

$$[\alpha, \beta] = \alpha\beta - \beta\alpha,$$

which makes of any associative algebra a Lie algebra.

§ 7.6.2 An *endomorphism* (or linear operator) on a vector space V is a mapping $V \to V$ preserving its linear structure. If V is any algebra (not necessarily associative), then

$$\text{End}\, V = \{\text{set of endomorphisms on } V\}$$

is an associative algebra. And then the set $\{\text{End } V\}$ of its commutators is a Lie algebra. The generic name *derivation* is given to any endomorphism $D : V \to V$ satisfying Leibniz law:

$$D(\alpha\beta) = (D\alpha)\beta + \alpha(D\beta).$$

The Lie algebra $\{\text{End } V\}$ contains $D(V)$, the vector subspace of all the derivations of V, and the Lie bracket makes of $D(V)$ a Lie subalgebra of $\{\text{End } V\}$. This means that the commutator of two derivations is a derivation. Given an element $a \in V$, it defines an endomorphism $ad(a) = ad_a$, called the "adjoint action of a", by

$$ad_a(b) = [a, b].$$

This is a derivative because

$$ad_a(bc) = [a, bc] = b[a, c] + [a, b]c = ad_a(b)c + b\, ad_a(c).$$

The set

$$ad(A) = \{ad_a \text{ for all } a \in A\}$$

contains all the internal derivations of a Lie algebra A, and is itself a Lie algebra homomorphic to A.

§ 7.6.3 A *graded algebra* is a direct sum of vector spaces, $V = \oplus_k V_k$, with a binary operation taking

$$V_i \otimes V_j \to V_{i+j}.$$

If $\alpha \in V_k$, we say that k is the *degree* (or order) of α, and write $\partial_\alpha = k$. The standard example of graded algebra is that formed by the differential forms of every order on a manifold M,

$$\Omega(M) = \oplus_k \Omega^k(M),$$

with the exterior product

$$\wedge : \Omega^p(M) \times \Omega^q(M) \to \Omega^{p+q}(M)$$

as the binary operation. Let us go back to forms and introduce another endomorphism.

§ **7.6.4 Exterior product.** Let M be a metric manifold, $X = X^i \partial_i$ be a vector field on M written in some natural basis, and X' its covariant image $X' = X_i dx^i$. As said in § 7.1.14, the operation of *exterior product* by X' on any form α is defined by the application

$$\varepsilon(X') : \Omega^k(M) \to \Omega^{k+1}(M), \tag{7.162}$$

whose explicit form is

$$\varepsilon(X')\alpha = X' \wedge \alpha, \quad \text{with } k < m. \tag{7.163}$$

§ **7.6.5 Interior product.** On the other hand, given a vector field X, we define the operation of *interior product* by X, denoted $i(X)$ or i_X, acting on p-forms, is defined by the application

$$i_X : \Omega^p(M) \to \Omega^{p-1}(M),$$

whose explicit form is

$$\alpha \to i_X\alpha \tag{7.164}$$

The image $i(X)\alpha = i_X\alpha$ is that $(p-1)$-form which, for any set of fields $\{X_1, X_2, \dots, X_{p-1}\}$, satisfies

$$(i_X\alpha)(X_1, X_2, \dots, X_{p-1}) = \alpha(X, X_1, X_2, \dots, X_{p-1}). \tag{7.165}$$

If α is a 1-form,

$$i_X\alpha = i(X)\alpha = <\alpha, X> = \alpha(X). \tag{7.166}$$

The interior product of X by a 2-form ω is that 1-form satisfying

$$i_X\omega(Y) = \omega(X, Y)$$

for any field Y. For a form of general degree, it is enough to know that, for a basis element,

$$i_X \left[\alpha^1 \wedge \alpha^2 \wedge \alpha^3 \dots \wedge \alpha^p \right] =$$
$$= [i(X)\alpha^1] \wedge \alpha^2 \wedge \alpha^3 \dots \wedge \alpha^p - \alpha^1 \wedge [i(X)\alpha^2] \wedge \alpha^3 \dots$$
$$\dots \wedge \alpha^p + \alpha^1 \wedge \alpha^2 \wedge [i(X)\alpha^3] \wedge \dots \alpha^p + \dots$$
$$= \sum_{j=1}^{p} (-)^{j-1} \alpha^1 \wedge \alpha^2 \wedge \dots [i(X)\alpha^j] \wedge \dots \wedge \alpha^p$$
$$= \sum_{j=1}^{p} (-)^{j-1} \alpha^1 \wedge \alpha^2 \wedge \dots [\alpha^j(X)] \wedge \dots \wedge \alpha^p. \tag{7.167}$$

§ 7.6.6 If the manifold is a metric manifold, an alternative definition is

$$i_X = *^{-1}\varepsilon(X')* = (-)^{n(p-1)+(n-s)/2} *\varepsilon(X')*. \tag{7.168}$$

The interior product is adjoint to the exterior product defined just above:

$$(\varepsilon(X')\,\alpha^{p-1},\beta^p) = (X'\wedge\alpha^{p-1},\beta^p) = \int X'\wedge\alpha^{p-1}\wedge *\beta^p$$

$$= (-)^{p-1}\int \alpha^{p-1}\wedge X'\wedge *\beta^p = (-)^{p-1}\int \alpha^{p-1}\wedge *[*^{-1}X'\wedge *\beta^p]$$

$$= (-)^{p-1+(n-p+1)(p-1)+(n-s)/2}\int \alpha^{p-1}\wedge *[*X'\wedge *\beta^p]$$

$$= (-)^{n(p-1)+(n-s)/2}\int \alpha^{p-1}\wedge *[*\varepsilon(X')*\beta^p]$$

$$= \int \alpha^{p-1}\wedge *[i(X)\beta^p] = (\alpha^{p-1},i(X)\beta^p).$$

§ 7.6.7 Some properties of interest are (f being a real function and g a mapping between manifolds):

$$i_X f = 0; \tag{7.169}$$

$$i_X\,i_X\alpha \equiv 0; \tag{7.170}$$

$$i_{fX}\,\alpha = f i_X\,\alpha = i_X(f\alpha); \tag{7.171}$$

$$g^*(i_{g_*}X\alpha) = i_X(g^*\alpha); \tag{7.172}$$

$$i_X(\alpha\wedge\beta) = (i_X\alpha)\wedge\beta + (-)^{\partial\alpha}\alpha\wedge i_X\beta. \tag{7.173}$$

§ 7.6.8 We have already introduced a good many endomorphisms on the graded algebra

$$\omega(M) = \bigcup_{k=0}^{m} \Omega^k(M).$$

They are: d, $\tilde{d}$, $\varepsilon(\omega)$, ad_a, L_X, Δ and $i(X)$. An endomorphism E has *degree* r if

$$E:\Omega^p(M)\to\Omega^{p+r}(M).$$

Thus, L_X and Δ have degree $r=0$, d and $\varepsilon(\omega)$ have degree $r=+1$, $\tilde{d}$ and $i(X)$ have degrees $r=-1$. We can be more specific about derivations. For forms α^p and β^q, an endomorphism E in Ω is a *derivation* if its degree is *even* and

$$E(\alpha\wedge\beta) = E(\alpha)\wedge\beta + \alpha\wedge E(\beta). \tag{7.174}$$

It is an *antiderivation* if its degree is *odd* and

$$E(\alpha\wedge\beta) = E(\alpha)\wedge\beta + (-)^{\partial\alpha}\alpha\wedge E(\beta). \tag{7.175}$$

§ 7.6.9 Consequently, L_X is a derivation, and d and $i(X)$ are antiderivations. In reality, these three endomorphisms are not independent. Let us recall the expression for the Lie derivative of a p-form: if

$$\alpha = \alpha_{j_1 j_2 j_3 \ldots j_p} x^{j_1} \wedge dx^{j_2} \wedge \ldots dx^{j_p},$$

then

$$(L_X \alpha) = X(\alpha_{j_1 j_2 j_3 \ldots j_p}) dx^{j_1} \wedge dx^{j_2} \wedge \ldots dx^{j_p}$$
$$+ (\partial_{j_1} X^k)\alpha_{k j_2 j_3 \ldots j_p} dx^{j_1} \wedge dx^{j_2} \wedge \ldots dx^{j_p}$$
$$+ (\partial_{j_2} X^k)\alpha_{j_1 k j_3 \ldots j_p} dx^{j_1} \wedge dx^{j_2} \wedge \ldots \wedge x^{j_p} +$$
$$\ldots + (\partial_{j_p} X^k)\alpha_{j_1 j_2 j_3 \ldots j_{p-1} k} dx^{j_1} \wedge dx^{j_2} \wedge \ldots dx^{j_p}. \quad (7.176)$$

Take $p = 1$ and $\alpha = \alpha_j dx^j$:

$$L_X \alpha = X^i(\partial_i \alpha_j) dx^j + (\partial_j X^i)\alpha_i dx^j.$$

On the other hand,

$$i(X)d\alpha = i(X)[(\partial_j \alpha_i) dx^j \wedge dx^i]$$
$$= <X, \partial_j \alpha_i dx^j> dx^i - (\partial_j \alpha_i) dx^j <X, dx^i>$$
$$= X^j(\partial_j \alpha_i) dx^i - X^i(\partial_j \alpha_i) dx^j. \quad (7.177)$$

Therefore,

$$d[i_X \alpha] = d[X^i \alpha_i] = (\partial_i X^j)\alpha_j dx^i + X^i(\partial_j \alpha_i)dx^j. \quad (7.178)$$

We see that

$$L_X \alpha = i_X d\alpha + d i_X \alpha = \{i_X, d\}\alpha. \quad (7.179)$$

This is actually a general result, valid for α's of any order and extremely useful in calculations. It also illustrates another general property: the anticommutator of two antiderivations is a derivation. There is more in this line. Consider derivatives generically indicated by D, D', etc, and antiderivatives A, A', etc. Using the definitions, one finds easily that the square of an antiderivative is a derivative,

$$AA = D.$$

In the same token, one finds for the anticommutator

$$\{A, A'\} = D',$$

as well as the following relations for the commutator:

$$[D, D'] = D'' \quad \text{and} \quad [D, A] = A.$$

§ 7.6.10 Consequences of Eqs.(7.179) and (7.170) are the commutation properties

$$L_X i_X = i_X L_X \, ; \tag{7.180}$$

$$d(L_X \alpha) = L_X(d\alpha) \, . \tag{7.181}$$

The Lie derivative commutes both with the interior product and the exterior derivative. Other interesting properties are:

$$L_{fX} \alpha = f \, L_X \alpha + df \wedge i_X \alpha \, ; \tag{7.182}$$

$$[L_X, i_Y] = i_{[X,Y]} \, ; \tag{7.183}$$

$$L_{[X,Y]} \alpha = [L_X, L_Y] \alpha \, . \tag{7.184}$$

§ 7.6.11 By the way, Eq.(7.176) gives us the real meaning of $\partial \alpha / \partial x^{j_0}$ in Eq.(7.28), and provides a new version of it:

$$d\alpha = dx^j \wedge L_{\partial_j} \alpha \, . \tag{7.185}$$

In a general basis $\{e_k\}$, with $\{\omega^j\}$ its dual,

$$d\alpha = \omega^j \wedge L_{e_j} \alpha = \varepsilon(\omega^j) L_{e_j} \alpha. \tag{7.186}$$

This equation, which generalizes the formula

$$df = dx^i \frac{\partial f}{\partial x^i}$$

valid for 0-forms, is called the *Koszul formula* and shows how Lie derivatives generalize partial derivatives. We may use it to check the coherence between the Lie derivative and the exterior derivative. As the Lie derivative is a derivation, then

$$\begin{aligned}
d(\alpha \wedge \beta) &= dx^j \wedge L_{\partial_j}(\alpha \wedge \beta) \\
&= dx^j \wedge (L_{\partial_j} \alpha) \wedge \beta + \alpha \wedge (L_{\partial_j} \beta)] \\
&= (dx^j \wedge L_{\partial_j} \alpha) \wedge \beta + (-)^{\partial_\alpha} \alpha \wedge (dx^j \wedge L_{\partial_j} \beta) \\
&= (d\alpha) \wedge \beta + (-)^{\partial_\alpha} \alpha \wedge d\beta).
\end{aligned}$$

§ 7.6.12 It is also possible to establish a new expression for the codifferential. From (7.110) and (7.28),

$$\begin{aligned}
\widetilde{d} \alpha &= -(-)^{n(p-1)+(n-s)/2} {*} d {*} \, \alpha \\
&= -(-)^{n(p-1)+(n-s)/2} {*} \, \varepsilon(dx^j) {*} {*}^{-1} \frac{\partial}{\partial x^j} {*} \, \alpha.
\end{aligned}$$

Therefore,

$$\tilde{d}\alpha = -i(\partial_j)\left[*^{-1}\frac{\partial}{\partial x^j}*\right]\alpha$$

$$= -(-)^{n(n-p)+(n-s)/2}i(\partial_j)\left[*\frac{\partial}{\partial x^j}*\right]\alpha.$$

In an euclidean space with cartesian basis,

$$\left[*\frac{\partial}{\partial x^j}*\right]\alpha = (-)^{p(n-p)}\frac{\partial\alpha}{\partial x^j}\,, \tag{7.187}$$

so that we have

$$\tilde{d}\alpha = -i(\partial_j)\frac{\partial\alpha}{\partial x^j} = -i(\partial_j)L_{\partial^j}\alpha. \tag{7.188}$$

In a general basis, the latter is written as

$$\tilde{d}\alpha = -i(e_j)L_{e_j}\alpha. \tag{7.189}$$

This is not unexpected if we look at (7.185), and remember the given adjoint relation between $\varepsilon(dx^j)$ and $i(\partial_j)$. With (7.179), it leads directly to

$$\tilde{d}\alpha = -i(\partial_j)\circ d\circ i(\partial_j)\alpha. \tag{7.190}$$

§ **7.6.13** Using (7.188), the derivation character of the Lie derivative and (7.169)–(7.173), we find that

$$\tilde{d}(\alpha\wedge\beta) = \tilde{d}\alpha\wedge\beta + (-)^{\partial\alpha}\alpha\wedge\tilde{d}\beta$$
$$- (-)^{\partial\alpha}(L_{\partial^j}\alpha)\wedge[i(\partial_j)\beta] - [i(\partial_j)\alpha]\wedge(L_{\partial^j}\beta). \tag{7.191}$$

§ **7.6.14** From Eq.(7.167) it turns out that, if $\{X_j\}$ is the basis dual to $\{\alpha^k\}$, then

$$i(X_j)[\alpha^i\wedge\omega] = \delta^i_j\,\omega - \alpha^i\wedge i(X_j)\omega\,,$$

so that

$$\left[\alpha^i\wedge i(X_j) + i(X_j)\circ\alpha^i\wedge\right]\omega =$$
$$= \left[\varepsilon(\alpha^i)\circ i(X_j) + i(X_j)\circ\varepsilon(\alpha^i)\right]\omega$$
$$= \{\varepsilon(\alpha^i), i(X_j)\}\,\omega$$
$$= \delta^i_j\,\omega \tag{7.192}$$

for any ω. Consequently, we find the anticommutator

$$\{\varepsilon(\alpha^i), i(X_j)\} = \delta^i_j. \tag{7.193}$$

Using this and again Eq.(7.167), we find that, applied to any form ω of degree p, the operator

$$\sum_{j=1}^{n} \varepsilon(\alpha^j) i(X_j)$$

behaves like a "number operator":

$$\sum_{j=1}^{n} \varepsilon(\alpha^j) i(X_j)\omega^p = p\omega^p. \tag{7.194}$$

From Eq.(7.165) it follows that

$$\{i(X_i), i(X_j)\} = 0. \tag{7.195}$$

It is evident from the very definition of $\varepsilon(\omega)$ that

$$\{\varepsilon(\alpha^i), \varepsilon(\alpha^j)\} = 0. \tag{7.196}$$

Commentary 7.7 The last four equations are reminiscent of those respected by fermion creators $a_j^\dagger \leftrightarrow \varepsilon(\alpha^j)$, annihilators $a_i \leftrightarrow i(X_i)$, and the corresponding fermion number operator $\sum_j a_j^\dagger a_j \leftrightarrow \sum_{j=1}^{n} \varepsilon(\alpha^j) i(X_j)$. ◀

Again, in an euclidean space with the cartesian basis, we may use (7.188) and

$$d\alpha = \varepsilon(dx^j)\frac{\partial \alpha}{\partial x^j}$$

to find a curious relation of the laplacian to the anticommutator. Of course

$$\Delta = \{d, \tilde{d}\},$$

but the above expressions give also

$$\Delta = -\{\varepsilon(\alpha^i), i(X_j)\}\,\partial_i\partial^j = -\delta_j^i\,\partial_i\partial^j, \tag{7.197}$$

as they should. These formulas are the starting point of what some people call "supersymmetric" quantum mechanics and have been beautifully applied[13] to Morse theory and in the study of instantons.

§ 7.6.15 The Lie algebra $\chi(M)$ of fields on a smooth manifold M acts on the space C^∞ of differentiable functions on M. We have an algebra, and another space on which it acts as endomorphisms. The second space is a module and we say that C^∞ is a $\chi(M)$-module. With the action of the Lie derivatives, which due to Eq.(7.184) represent the Lie algebra, also the space of p-forms is a module.

---

13 Witten 1982a, b.

General references

A very good introduction to differential forms, addressed to engineers and physicists, but written by a mathematician, is Flanders 1963. A book containing a huge amount of material, written by a physicist, is Westenholz 1978. Slebodzinski 1970 is a mathematical classic, containing an extensive account with applications to differential equations and Lie groups. Also commendable are Burke 1985, Warner 1983 and Lovelock & Rund 1975.

Chapter 8

Symmetries

A physical system can be submitted to transformations. Some of these will make another system of it, but some other — its symmetries — leave it the same. The latter are of extreme importance — a system can be characterised, classified, described by them. In the age-old battle between things and the names trying to capture them, its symmetries provide the best photograph depicting a physical system.

8.1 Lie groups

The study of a topological group is much easier when the group operation is analytic. Even the algebraic structure becomes more accessible. This, however, requires that the underlying topological space have an analytic structure, more precisely: that it be an analytic manifold. We have already said (§ 5.1.3) that every C^∞ structure has an analytic substructure. That is why most authors prefer to define a Lie group as a C^∞ manifold, ready however to make good use of the analytic sub-atlas when necessary. A topological group has been defined in section 1.5.9 as a topological space G endowed with a group structure and such that both the mappings

$$m : G \times G \to G, \quad (g, h) \to g \cdot h,$$

and

$$\text{inv} : G \to G, \quad g \to g^{-1}$$

are continuous.

§ **8.1.1** A Lie group is a C^∞ manifold G on which a structure of algebraic group is defined, in such a way that the mappings

$$G \times G \to G, \quad (g, h) \to g \cdot h$$

and

$$G \to G, \quad g \to g^{-1}$$

are all C^∞.

§ **8.1.2** It follows from this definition that the mappings

$$L_g : G \to G, \quad h \to g \cdot h$$

and

$$R_g : G \to G, \quad h \to h \cdot g$$

are diffeomorphisms for every $g \in G$. These mappings are called respectively left-translation and right-translation induced by the element g.

§ **8.1.3** All the continuous examples previously given as topological groups are also Lie groups. Some manifolds (for example, the sphere S^3) accept more than one group structure, others (for example, the sphere S^2) accept none.

§ **8.1.4** An observation: there is a one-to-one correspondence

$$f : GL(n, \mathbb{R}) \times \mathbb{R}^n \to AL(n, \mathbb{R})$$

between the linear and the affine groups, given by

$$(L, t) \leftrightarrow \begin{bmatrix} L & t \\ 0 & 1 \end{bmatrix},$$

so that it is possible to introduce, on the affine group, a structure of differentiable manifold. This makes of f a diffeomorphism. With this structure, the group operation is C^∞.

Other examples may be obtained as *direct products*. Let G_1 and G_2 be two Lie groups. The product $G = G_1 \times G_2$ can be endowed with the C^∞ structure of the cartesian product, which makes of G the direct product group $G_1 \otimes G_2$ if the following operation is defined

$$G \times G \to G$$

whose explicit form is

$$((g_1, g_2), (g_1', g_2')) \to (g_1 \cdot g_1', g_2 \cdot g_2') .$$

It is important to notice that the affine group is not to be considered as the Lie group $GL(n, \mathbb{R}) \otimes \mathbb{R}^n$, because the product for $AL(n, \mathbb{R})$ is

$$G \times G \to G$$

with explicit form given by

$$((L,t),(L',t')) \to (LL', Lt' + t) .$$

This is an example of *semi-direct product*, denoted usually by

$$GL(n,\mathbb{R}) \oslash \mathbb{R}^n.$$

As S^1 is a Lie group, the product of n copies of S^1 is a Lie group, the "toral group" or n-torus

$$T^n = S^1 \otimes S^1 \otimes \ldots \otimes S^1.$$

§ 8.1.5 A subgroup of a topological group is also a topological group. An analogous property holds for Lie groups, under the conditions given by the following theorem:

> *Let G be a Lie group, and H an algebraic subgroup which is also an imbedded submanifold. Then, with its smooth structure of submanifold, H is a Lie group.*

§ 8.1.6 It is possible to show that the following subgroups of the linear groups (§ 1.5.18) are Lie groups:

(1) The orthogonal group $O(n)$, of dimension $n(n-1)/2$,

$$O(n) = \{X \in GL(n,\mathbb{R}) \text{ such that } X \cdot X^T = I\} ,$$

where X^T is the transposed of X. The special orthogonal group,

$$SO(n) = \{X \in O(n) \text{ such that } \det X = +1\} ,$$

is the group of rotations in $\mathbb{E}^n$. For the special value $n = 1$, it is trivial: $SO(1) = I$. Consecutive quotients of such groups are spheres:

$$SO(n)/SO(n-1) = S^{n-1} .$$

Thus, $SO(2) = S^1$. The groups $SO(n)$ are homotopically non-trivial: for $n = 2$,

$$\pi_1[SO(2)] = Z.$$

For the other cases, they are doubly-connected:

$$\pi_1[SO(n)] = Z_2 , \quad \text{for } n \geq 3 .$$

(2) The unitary group $U(n)$, of dimension n^2,

$$U(n) = \{X \in GL(n, \mathbb{C}) \text{ such that } X \cdot X^\dagger = I\},$$

where $X^\dagger$ is the adjoint (complex-conjugate transpose) of X. The groups $U(n)$ are multiply connected,

$$\pi_1[U(n)] = \mathbb{Z}, \text{ for } n \geq 1.$$

The "special" cases $SU(n)$, those for which furthermore $\det X = +1$, are of enormous importance in the classification and dynamics of elementary particles. With the exception of $SU(1) = SO(2)$, they are simply-connected:

$$\pi_1[SU(n)] = 1 \text{ for } n > 1.$$

The group $SU(2)$ is isomorphic to the sphere S^3 (see § 3.3.29) and describes the spin (see § 8.1.10 below). It is the universal covering of the rotation group in our ambient $\mathbb{E}^3$,

$$SU(2)/Z_2 = SO(3).$$

Such relationships are a bit more difficult to predict for higher dimensional cases, as shown by the example

$$SU(4)/Z_2 = SO(6).$$

Consecutive quotients are spheres:

$$U(n)/U(n-1) = SU(n)/SU(n-1) = S^{2n-1}.$$

(3) The symplectic group

$$Sp(n) = \{X \in GL(n, \mathbb{Q}) \text{ such that } X \cdot X^\dagger = I\}$$

is the group of linear canonical transformations for a system with n degrees of freedom. It is simply-connected for each value of n. Again, spheres come out from consecutive quotients

$$Sp(n)/Sp(n-1) = S^{4n-1}.$$

In particular, $Sp(1) = S^3$, which is the same manifold as $SU(2)$. This is, by the way, a good example of two group structures defined on the same manifold.

(4) The special linear group, of dimension $n^2 - 1$:

$$SL(n, \mathbb{R}) = X \in GL(n, \mathbb{R}) \text{ such that } \det X = 1.$$

(5) The special complex linear group, of dimension $4n^2 - 2$:

$$SL(n, \mathbb{C}) = X \in GL(n, \mathbb{C}) \text{ such that } \det X = 1.$$

§ **8.1.7** We know that the orthogonal group $O(n)$ preserves the euclidean scalar product: if $x, y \in \mathbb{E}^n$, then

$$\langle Tx, Ty \rangle = \langle x, y \rangle$$

for all $T \in O(n)$. This can be generalized to obtain a group preserving the non-definite ("pseudo-euclidean") scalar product

$$\langle x, y \rangle = \sum_{i=1}^{p} x^i y^i - \sum_{j=p+1}^{n} x^j y^j \ .$$

This group is the set of matrices

$$O(p, q) = \{ X \in GL(n, \mathbb{R}) \text{ such that } \mathbf{I}_{p,q} X^T \mathbf{I}_{p,q}^{-1} = X^{-1} \},$$

with $p + q = n$ and $\mathbf{I}_{p,q}$ the diagonal matrix with the first p elements equal to $(+1)$, and the q remaining ones equal to (-1). A case of particular importance is

$$SO(p, q) = \{ X \in O(p, q) \text{ such that } \det X = 1 \}.$$

Such "pseudo-orthogonal" groups are non-compact, their noblest example in Nature being the Lorentz group $SO(3, 1)$. More about these groups will be said in § 9.3.3.

§ **8.1.8** One more definition: let ϕ be an algebraic homomorphism between the Lie groups G_1 and G_2. Then, ϕ will be a *homomorphism of Lie groups* iff it is C^∞. Given the toral group T^n, then

$$\phi : \mathbb{E}^n \to T^n$$

with

$$(t_1, t_2, \ldots, t_n) \to (e^{2\pi t_1}, e^{2\pi t_2}, \ldots, e^{2\pi t_n})$$

is a homomorphism of Lie groups.

§ **8.1.9** A topological group is *locally compact* if around each of its points there exists an open set whose closure is compact (§ 1.3.14). This is a very important notion, and that by a technical reason. We are used to applying Fourier analysis without giving too much thought to its fundamentals. Putting it in simple words, the fact is that *Fourier analysis is always defined on a group* (or on a quotient of groups, or still on the ring of a group). Standard spaces are the circle S^1 (the 1-torus group T^1) for periodic functions, the real line additive group $(\mathbb{R}, +)$ and their products. The technical point is the following: the summations and integrals involved in Fourier series and integrals presuppose a measure, and this measure must be group-invariant (the same over all the group-space). And only on locally compact groups is the existence of such an invariant measure assured. These measures are called Haar measures (see § 8.2.20). Classical Fourier analysis is restricted to abelian locally compact groups like the above mentioned ones, but local compactness allows in principle extension of harmonic analysis to non-abelian groups, such as the special unitary group $SU(2) \approx S^3$ (Section 18.4).

§ 8.1.10 We have said (§ 3.3.29 and § 8.1.6) that the rotation group $SO(3)$, related to the angular momentum, is doubly-connected, with the special unitary group $SU(2)$ as covering space. This is very important for Physics, as only $SU(2)$ possesses the half-integer representations necessary to accommodate the fermions. The group $SO(3)$ is a subgroup of the Lorentz group $SO(3,1)$, which is also unable to take half-integer spins into account. Every elementary particle must "belong" to some representation of the Lorentz group in order to have a well-defined relativistic behaviour. The same must hold for the relativistic fields which represent them in Quantum Theory. To accommodate both fermions and bosons, it is necessary to go to the covering group of $SO(3,1)$, which is the complex special linear group $SL(2,C)$ (see Chapter 31).

§ 8.1.11 Grassmann manifolds. The Grassmann spaces G_{nd} of § 1.5.22, whose "points" are d-dimensional planes in euclidean n-dimensional spaces, can be topologized and endowed with a smooth structure, becoming manifolds. They can be obtained (or, if we wish, defined) as double quotients,

$$G_{nd} = G_d(\mathbb{E}^n) = O(n)/(O(d) \times O(n-d)).$$

If we consider oriented d-planes, we obtain a space $G_{nd}^{\#}$, which is a double covering of G_{nd}. In the complex case, the manifolds are

$$G_{nd}^C = G_d(\mathbb{C}^n) = U(n)/(U(d) \times U(n-d)).$$

As quotients of compact spaces, they are themselves compact.

§ 8.1.12 Stiefel manifolds. Denoted S_{nd} or $S_d(\mathbb{E}^n)$, these are spaces whose members are d-dimensional orthogonal frames in $\mathbb{E}^n$. They are found to be (or can be alternatively defined as)

$$S_{nd} = O(n)/O(n-d)$$

and their dimensions are dim $S_{nd} = d$. Stiefel manifolds have curious homotopic properties: *their lower homotopy groups vanish.* More precisely,

$$\pi_r(S_{nd}) = 0 \text{ for } (n-d-1) \geq r \geq 0.$$

As with Grassmann manifolds, we may consider complex Stiefel manifolds: $S_{nd}^C = S_d(\mathbb{C}^n)$ is the space of unitary d-dimensional frames in $\mathbb{C}^n$. For them,

$$\pi_r(S_{nd}^C) = 0 \ \text{ for } (2n - 2d - 1) \geq r \geq 0.$$

Because of these peculiar homotopic properties, Stiefel manifolds are of basic importance for the general classification of fiber bundles (Section 9.7).

Commentary 8.1 After what we have seen in § 6.4.3 and § 6.5.8, the very matrix elements can be used as coordinates on a matrix group. ◄

8.2 Transformations on manifolds

We proceed now to a short analysis of the action of groups on differentiable manifolds. In particular, continuous transformations on manifolds, in general, constitute continuous groups which are themselves manifolds, topological or differentiable. The literature on the subject is very extensive — we shall only occupy ourselves of some topics, mainly those essentially necessary to the discussion of bundle spaces. We have seen in section 6.4, when the concept of Lie derivative was introduced, the meaning of the action of $\mathbb{R}$ on a manifold M, $\mathbb{R}$ being considered as the additive (Lie) group of the real numbers. We shall now generalize that idea in a way which applies to both topological and Lie groups.

§ 8.2.1 Action of a group on a set Let G be a group, and M a set. The group G is said to act on M when there exists a mapping

$$\lambda : M \times G \to M, \quad (p, g) \to \lambda(p, g)$$

satisfying:
(i) $\lambda(p, e) = p$, $p \in M$, where e is the identity element of G.
(ii) $\lambda(\lambda(p, g), h) = \lambda(p, gh)$, $p \in M$ and $g, h \in G$.

The mapping λ is called the *right action* of G on M, and is generally denoted by R: we indicate the right action of g by $R_g x = x' \in M$, or more simply by $xg = x'$. Left action is introduced in an analogous way. Once such an action is defined on a set M, M is said to be a *G-space* and G a *transformation group* on M. A subset M' of M is *invariant* if $xg \in M'$ whenever $x \in M'$. Every subset M'' of M is contained in some invariant subset M', which is said to be the *generator* of M''.

§ 8.2.2 When G is a topological group and M a topological space, the action R is required to be continuous. When G is a Lie group and M a C^∞ manifold, R is required to be C^∞.

§ 8.2.3 As already said, we shall frequently use the abbreviated notation pg for $R(p, g) = \lambda(p, g)$, so that condition (ii) of § 8.2.1 above becomes

$$(pg)h = p(gh), \text{ for } p \in M \text{ and } g, h \in G.$$

In the C^∞ case, the mapping

$$R_g : M \to M, \quad p \to R_g(p) = pg$$

is a diffeomorphism and allows one to rewrite (ii) as

$$R_h R_g p = R_{gh} p.$$

§ 8.2.4 Effective action. An action is said to be *effective* when the identity element e is the only element of G preserving all the points of M, *i.e.*, when $R_g p = p$, $\forall p$, implies $g = e$. This means that at least some change comes out through the action of each group element.

§ 8.2.5 Transitive action. The action is said to be *transitive* when, given any two points p and q of M, there exists a g in G such that $pg = q$. We can go from a point of M to any other point by some transformation of the group. The space $\mathbf{E}^3$ is transitive under the group T_3 of translations. If g is unique for every pair (p, q), the action is *simply transitive*.

§ 8.2.6 Given the point p of a manifold M, the set of group members

$$H_p = \{h \in G \text{ such that } ph = p\}$$

is a subgroup of G, called the *isotropy group* (or *stability group*) of the point p.

§ 8.2.7 Homogeneous spaces. If G acts transitively on M, M is *homogeneous* under G. A homogeneous space has no invariant subspace, except itself and the empty set. Simple groups are themselves homogeneous by the action defined by the group multiplication. Space $\mathbf{E}^3$ is homogeneous under the translation group T_3 but not under the rotation group $SO(3)$. On the other hand, a sphere S^2 is homogeneous under $SO(3)$. Rotations in $\mathbb{E}^3$ leave fixed a certain point (usually taken as the origin around which the rotations take place). If p is a point of a homogeneous manifold M, another point q of M will be given by $q = pg$ for some g, or $q = phg$ for any $h \in H_p$. Thus,

$$q(g^{-1}hg) = phg = q,$$

so that $g^{-1}hg \in H_q$. Given any member of the isotropy group H_p, g will give a member of H_q and g^{-1} will do the same in the inverse way. As a consequence, the isotropy groups of all points on a homogeneous manifold are isomorphic. The Lie algebra has then a canonical decomposition of the form $G' = H' + T$, for some T.

A homogeneous space G/H has always a Riemann metric which is invariant under the action of G, and is also said to be a homogeneous riemannian space. We might in larger generality define a homogeneous riemannian space as a riemannian space M whose group of motions acts transitively on M.

The fact that a space may be obtained as a quotient of two Lie groups as G/H has very deep consequences, particularly in the case of homogeneous "symmetric" spaces. This happens when G has an involutive automorphism,

$$\sigma : G \to G, \quad \sigma^2 = 1.$$

The Lie algebra canonical decomposition $G' = H' + T$, in the presence of such an involution, has the form

$$[H', H'] \subset H', \qquad [H', T] \subset T, \qquad [T, T] \subset H'.$$

The bundle $G = (G/H, H)$ admits an invariant connection, called its canonical connection, which is determined by the space T. It has vanishing torsion and a very special form for the curvature. This special canonical connection is a restriction, to the quotient space, of the Maurer–Cartan form of the group G.

§ **8.2.8** The action of G is said to be *free* when no element but the identity e of G preserves *any* point of M, that is, $R_g p = p$ for some $p \in M$ implies $g = e$. Thus *a free action admits no fixed point*.

§ **8.2.9** The set of points of M which is obtained from a given point p by the action of G is the *orbit* of p:

$$\text{orbit}(p) = \{q = pg, \; g \in G\}.$$

Thus, one obtains the orbit of p by acting on p with all the elements of G. An orbit may be seen as the invariant subset generated by a single point:

$$\text{orbit}(p) = p \, G.$$

Every orbit is a transitive G-space by itself. A transitive space is clearly an orbit by one of its points.

§ **8.2.10** Everything which was said above about "right" action can be repeated for "left" action, with the notation L, and

$$L_g(p) = gp.$$

No confusion should arise by the use of the same notation for the (left or right) action on a manifold M and the (left or right) translation on the group G itself. For a physical example illustrating the difference between right- and left-actions, see Section 26.3.9.

§ **8.2.11** As $\mathbb{R}$ is a Lie group, a one-parameter group (§ 6.4.17) on a manifold M is an example of action, as it satisfies conditions (i) and (ii) of § 8.2.1, and is C^∞.

§ **8.2.12** Let G_1 and G_2 be two Lie groups and $\phi: G_1 \to G_2$ a group homomorphism. It is possible to show that

$$L : G_1 \times G_2 \to G_2, \quad (g_1, g_2) \to \phi(g_1)g_2$$

is a left-action.

§ **8.2.13** The best known case is the usual action of matrices on column vectors. We can rephrase it in a pedantic and precise way and call it the natural action (on the left) of $GL(n, \mathbb{R})$ on $\mathbb{R}^n$:

$$L : GL(n, \mathbb{R}) \times \mathbb{R}^n \to \mathbb{R}^n, \quad (A, x) \to Ax.$$

In an analogous way, we have the action on the right

$$R : \mathbb{R}^n \times GL(n, \mathbb{R})) \to \mathbb{R}^n, \quad (x, A) \to x^T A.$$

§ **8.2.14** The left-action of the affine group (§ 8.1.4) on $\mathbb{R}^n$ is given by

$$\left(\begin{bmatrix} L & t \\ 0 & 1 \end{bmatrix}, \begin{bmatrix} x^1 \\ x^2 \\ \vdots \\ x^n \\ 1 \end{bmatrix} \right) \to \begin{bmatrix} L & t \\ 0 & 1 \end{bmatrix} \cdot \begin{bmatrix} x^1 \\ x^2 \\ \vdots \\ x^n \\ 1 \end{bmatrix}$$

where $L \in GL(n, \mathbb{R})$ and $t \in \mathbb{R}^n$. If $L \in O(n)$, then we have action of a subgroup of $A(n, \mathbb{R})$ called the *group of rigid motions*, or *euclidean group* on $\mathbb{R}^n$.

§ **8.2.15 The Poincaré group** $PO(4, \mathbb{R})$. Also known as inhomogeneous Lorentz group, is a subgroup of the four-dimensional affine group $A(4, \mathbb{R})$, defined by

$$PO(4, \mathbb{R}) = \left\{ \begin{bmatrix} L & t \\ 0 & 1 \end{bmatrix} \text{ such that } L \in SO(3,1) \text{ and } t \in \text{Minkowski space} \right\}.$$

§ **8.2.16 Linear and Affine basis.** Let $B(\mathbb{R}^n)$ be the set of linear basis (see § 6.5.6) of $\mathbb{R}^n$ (here taken as synonym of $\mathbb{E}^n$ with the structure of vector space):

$$B(\mathbb{R}^n) = \{f := \{f_i\}, i = 1, 2, \ldots, n \text{ , the } f_i \text{ being linearly independent}\}.$$

Let us define the action

$$B(\mathbb{R}^n) \times GL(n, \mathbb{R}) \to B(\mathbb{R}^n)$$

defined by

$$(f, L) \to fL := \left\{ \sum_j f_j L_{j1}, \sum_j f_j L_{j2}, \ldots, \sum_j f_j L_{jn} \right\}.$$

Given two basis f and $\tilde{f}$ in $B(\mathbb{R}^n)$, there exists a unique $L \in GL(n, \mathbb{R})$ such that

$$\tilde{f} = f\, L.$$

Thus, the action is simply transitive. This means that there is a one-to-one correspondence between $GL(n, \mathbb{R})$ and $B(\mathbb{R}^n)$ given by

$$L \longleftrightarrow e\, L,$$

where $e = \{e_1, e_2, \ldots, e_n\}$ is the canonical basis for $\mathbb{R}^n$:

$$e_1 = (1, 0, 0, \ldots, 0), \quad e_2 = (0, 1, 0, \ldots, 0), \quad \text{etc.}$$

Once this correspondence is established, it is possible to endow the set $B(\mathbb{R}^n)$ with a topology and a C^∞ structure so as to make it diffeomorphic to $GL(n, \mathbb{R})$. $B(\mathbb{R}^n)$ is called the *basis space* of $\mathbb{R}^n$.

§ **8.2.17** What was done above can be repeated with a general vector space V instead of $\mathbb{R}^n$. Let us examine now the set of the affine basis on a vector space V:

$$A(V) = \{[f_1, f_2, \ldots, f_n; x] =: [f, x] \text{ such that } f \in B(V) \text{ and } x \in V\}.$$

The action of the affine group on the affine basis is given by

$$A(V) \times A(n, \mathbb{R}) \to A(V)$$

whose explicit form is

$$\left([f,x],\begin{bmatrix}L & t\\0 & 1\end{bmatrix}\right) \;\to\; [fL,ft+x] := [f,x]\begin{bmatrix}L & t\\0 & 1\end{bmatrix}$$

where

$$ft := \sum_{i=1}^{n} f_i t^i.$$

Given any two basis f and $\tilde{f} \in A(V)$, there is a unique element of $A(n,\mathbb{R})$ such that

$$\tilde{f} = f\begin{bmatrix}L & t\\0 & 1\end{bmatrix}.$$

Thus, there is a one-to-one correspondence between $A(n,\mathbb{R})$ and $A(V)$, given by

$$\begin{bmatrix}L & t\\0 & 1\end{bmatrix} \longleftrightarrow a\begin{bmatrix}L & t\\0 & 1\end{bmatrix},$$

where $a = \{e_1, e_2, \ldots, e_n, 0\}$ is the canonical basis for $A(V)$. With this correspondence, we can endow the set $A(V)$ with a topology and a C^∞ structure making it diffeomorphic to $A(n,\mathbb{R})$. As a manifold, $A(V)$ is the affine basis space on V. If, instead of a, another basis a' were used, the same structure would result, up to a diffeomorphism.

§ 8.2.18 We have learned (Section 1.5.1) how to obtain new manifolds from a given one, as quotient spaces by an equivalence relation. Suppose the relation between two points p and $q \in M$ defined by

$$p \approx q \longleftrightarrow \text{ there exists some } g \in G \text{ such that } q = R_g p.$$

It is an equivalence. The set $[p] = \{q \in M \text{ such that } q \approx p\}$, which is the orbit$(p)$, is the equivalence class with representative p. The set of all these classes is denoted by M/G: it is said to be the *quotient space* of M by G. The canonical projection is defined by

$$\pi : M \to M/G, \quad p \to [p].$$

A quotient topology can then be introduced on M/G, as well as a C^∞ structure.

An important particular case appears when M is itself a Lie group and G a subgroup of M. Then the quotient space M/G is an analytic manifold. The action $G \times M/G \to M/G$ is transitive and M/G is, consequently, a homogeneous manifold under the action of G. Manifolds of this type (group/subgroup) have many interesting special characteristics, as also the action is an analytic mapping, as well as the projection $M \to M/G$.

§ 8.2.19 Suppose the group G acts simultaneously on two manifolds M and N, with actions denoted m_g and n_g (see Figure 8.1). A mapping ϕ: $M \to N$ is *equivariant* (or an *intertwining map*) when $\phi \circ m_g = n_g \circ \phi$. The diagram at the right of Figure 8.1 is commutative.

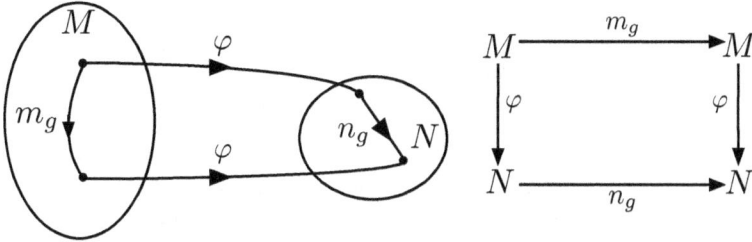

Fig. 8.1   Group acting on two manifolds.

§ 8.2.20 Invariant measure. Consider a set M on which an action of the group G is defined. Suppose there is a Borel measure μ on M (see Chapter 15). The measure μ is said to be an *invariant measure* when, for each measurable subset A of M and for all $g \in G$, we have

$$\mu(gA) = \mu(A).$$

Here "gA" means the set resulting from whatever action of G on the set A. In particular, the measure is left-invariant if $\mu(L_g A) = \mu(A)$ and right-invariant if $\mu(R_g A) = \mu(A)$. The Lebesgue measure on an euclidean space is invariant under translations and rotations; the Lebesgue measure on the sphere is invariant under the action of the rotation group on the sphere. Such Haar *measures* are sure to exist only on locally compact groups (§ 1.3.14, § 8.1.9). On such groups, they are unique up to positive factors. For the groups T^n and $(\mathbb{R}, +)$, the Haar measures are simply the (normalized) Lebesgue measures. Haar measures provide a characterization of compactness: the Haar measure on G is finite *iff* G is compact. This property stays behind the well known fact that Fourier expansions on compact spaces (groups!) are series while Fourier expansions on non-compact spaces are integrals (see Section 18.4).

§ 8.2.21 Invariant integration Given an invariant measure μ on M, the corresponding integral is invariant in the sense that

$$\int_M f(gx)d\mu(x) = \int_M f(x)d\mu(x).$$

An integral may be left-invariant, right-invariant, or both.

§ 8.2.22 Function spaces, revisited. Let us go back to § 8.2.7 and consider the space $C(M)$ of complex functions on a homogeneous space M. The space $C(M)$ may be made into a G-space by introducing the action

$$(R_g f)(m) = f(mg^{-1}).$$

Now, $C(M)$ is a vector space, which is fair, but it is not necessarily homogeneous. It has, in general, invariant subsets which are themselves linear spaces. Take for example a simple group G and $M = G$, the action being given by the group multiplication, say $Rg : x \to xg$. Take $C^h(G)$, the set of all homomorphisms of G into the non-vanishing complex numbers. Then if $h \in C^h(G)$, we have that $h \in C(G)$ and the set of all constant multiples of h constitutes an invariant (one-dimensional) subspace. Such subspaces are independent for distinct h's. When G is finite and commutative, each member of $C(G)$ is a unique sum of members, one from each invariant subspace. When G is infinite, such sums become infinite and some extra requirements are necessary. In general, one restricts $C(G)$ to a subspace of measurable (square integrable) functions with some (say, Lebesgue) Haar measure. Take for instance G as the circle S^1. The only measurable homomorphisms are

$$h_n(x) = e^{inx},$$

with $n = 0, \pm 1, \pm 2, \dots$. Then, any square integrable function f on G may be written in a unique way as

$$f(x) = \sum_n f_n e^{inx},$$

which is the Fourier theorem. To each irreducible representation $h_n(x)$ corresponds a "harmonic". As long as G is compact, the sums as above are discrete. Things are more involved when G is not compact. We have said that only locally compact groups have Haar measures. In the non-compact but still abelian cases, the number n above becomes continuous and the sums are converted into integrals. The best known case is $G = \mathbb{R} = \{$real numbers with addition $\}$, when

$$f(x) = \int_{\mathbb{R}} f(s)e^{isx}.$$

Let us say a few words on the compact non-commutative case. Take G as the rotation group in $\mathbb{E}^3$, and $M = S^2$, the spherical surface in $\mathbb{E}^3$. The space of all square-integrable functions on M divides itself into invariant subspaces M_j, one for each odd number $(2j+1)$ and such that

$$\dim M_j = (2j+1).$$

They are formed by 3-variable homogeneous polynomials of degree j which are harmonic, that is, satisfy the Laplace equation on M. Each M_j provides an irreducible representation, the $f_j \in M_j$ being the surface harmonics, and any function $f \in C(S^2)$ is uniquely expanded as

$$f = \sum_n f_j$$

with $f_j \in M_j$ (see also Section 18.4).

8.3 Lie algebra of a Lie group

We have said in section § 6.4.5 that the vector fields on a smooth manifold constitute a Lie algebra. Lie groups are differentiable manifolds of a very special type. We describe now (very superficially) the general properties of fields and forms on Lie groups.

§ **8.3.1** Consider the left action of a Lie group G on itself:

$$L_g : G \to G, \quad h \to L_g(h) = gh. \tag{8.1}$$

A first thing which is peculiar to the present case is that this action is a diffeomorphism. It induces of course the differential mapping between the respective tangent spaces,

$$L_{g*} = dL_g : T_hG \to T_{gh}G. \tag{8.2}$$

An arbitrary field X on G is a differentiable attribution of a vector X_g at each point g of G. Under the action of L_{g*}, its value X_h at the point h will be taken into some other field $X'_{gh} = L_{g*}(X_h)$ at the point gh (Figure 8.2, left). Suppose now that the field X is taken into itself by the left action of G:

$$L_{g*}(X_h) = X_{gh}. \tag{8.3}$$

In this case, X is said to be a *left-invariant field* of G and one writes

$$L_{g*}X = X. \tag{8.4}$$

This means that, for any function $f \in R(G)$,

$$(L_{g*}X_h)(f) = X_h(f \circ L_g) = X_{gh}(f). \tag{8.5}$$

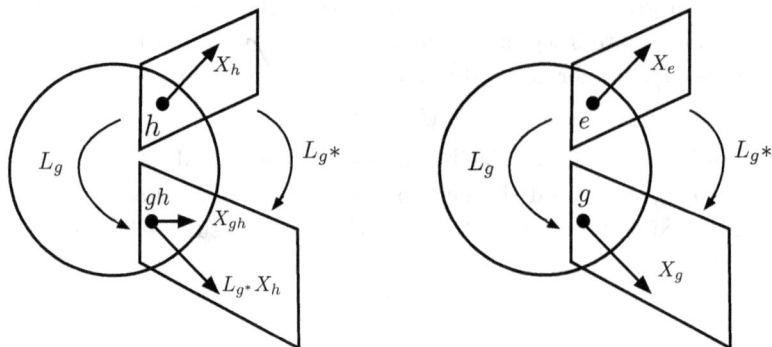

Fig. 8.2   Left-action and left-invariance.

§ **8.3.2** Notice that, in particular,

$$L_{g*}(X_e) = X_g \tag{8.6}$$

when X is left-invariant (Figure 8.2, right). Consequently, left-invariant fields are completely determined by their value at the group identity e. But not only the fields: their algebras are also completely determined, as diffeomorphisms preserve commutators (§ 6.4.6). Thus,

$$L_{g*}[X_e, Y_e] = [X_g, Y_g] \tag{8.7}$$

for any left-invariant fields X, Y. This is to say that the Lie algebra of left-invariant fields at any point on a Lie group G is determined by the Lie algebra of such fields at the identity point of G.

§ **8.3.3** This algebra of invariant fields, a subalgebra of the general Lie algebra of all the fields on G, is *the Lie algebra of the Lie group G*. It is usually denoted by $L(G)$, or simply G'. The vector space of G' is $T_e G$. A basis for G' will be given by d ($=$ dim G) linearly independent left-invariant fields X_α, which will satisfy

$$[X_\alpha, X_\beta] = C^\gamma{}_{\alpha\beta} X_\gamma. \tag{8.8}$$

§ **8.3.4** According to (8.7), this relation must hold (with the same $C^\gamma{}_{\alpha\beta}$'s) at any point of G, so that the structure coefficients are now point-independent. They are, for this reason, called the structure constants of G. The Lie algebra of G is thus a "small" (as compared with the infinite algebra of all fields) Lie algebra of d fields fixed by their values at one point of G.

Commentary 8.2 Right-invariant fields can be defined in an analogous way. They constitute a Lie algebra isomorphic to that of the left-invariant fields. ◀

§ 8.3.5 A p-form w on G is left-invariant if

$$L_g^* w = w. \tag{8.9}$$

Let us see how things work for 1-forms: given a form w_{gh} at gh, its pull-back is defined by

$$\langle L_{g^{-1}}^* w, X \rangle_h = \langle w, L_{g^{-1}*} X \rangle_{gh}. \tag{8.10}$$

If w is invariant,

$$\langle w, X \rangle_h = \langle w, L_{g^{-1}*} X \rangle_{gh}. \tag{8.11}$$

If also X is invariant,

$$\langle w, X \rangle_h = \langle w, X \rangle_{gh}. \tag{8.12}$$

Therefore, an invariant form, when applied to an invariant field, gives a constant.

§ 8.3.6 Invariant Pfaffian forms on Lie groups are commonly called *Maurer–Cartan forms*. They constitute a basis $\{w^\alpha\}$ for $L^*(G)$, dual to that of invariant fields satisfying Eq.(8.8). As a consequence, Eq.(7.55) tells us that they obey

$$dw^\gamma = \tfrac{1}{2} C^\gamma{}_{\alpha\beta} w^\alpha \wedge w^\beta. \tag{8.13}$$

This is the *Maurer–Cartan equation*, which can be put in a basis-independent form: define the vector-valued *canonical form*

$$w = X_\alpha w^\alpha. \tag{8.14}$$

When applied on a field Z, the canonical form simply gives it back:

$$w(Z) = X_\alpha w^\alpha(Z) = X_\alpha Z^\alpha = Z. \tag{8.15}$$

Then, a direct calculation puts the Maurer–Cartan equation in the form

$$dw + w \wedge w = 0. \tag{8.16}$$

§ 8.3.7 For a matrix group with elements g, it is easy to check that

$$w = g^{-1} dg \tag{8.17}$$

are matrices in the Lie algebra satisfying Eq.(8.16) (not forgetting that $dg^{-1} = -g^{-1} dg g^{-1}$).

§ **8.3.8** Given any $n \times n$ matrix A, its exponential is defined by the (convergent) series

$$e^A = I + A + \tfrac{1}{2}A^2 + \ldots = \sum_{j=0}^{\infty} \frac{1}{j!} A^j. \tag{8.18}$$

The set of matrices of type $\exp(tA)$, with $t \in \mathbb{R}$, constitutes an abelian group:

$$e^{tA}e^{sA} = e^{(t+s)A}, \quad e^{-tA}e^{tA} = I, \quad \text{etc.}$$

The mapping

$$a : \mathbb{R} \to GL(n, \mathbb{R}), \quad a(t) = \exp(tA),$$

is a curve on $GL(n, \mathbb{R})$ whose tangent at $t = 0$ is

$$\left. \frac{de^{tA}}{dt} \right|_{t=0} = A\,e^{tA}\big|_{t=0} = A. \tag{8.19}$$

So, $A \in T_I GL(n, \mathbb{R})$, or $A \in G'L(n, \mathbb{R})$. The set of matrices $\exp(tA)$ is the group generated by A. As A is arbitrary, we have shown that any $n \times n$ matrix belongs to $G'L(n, \mathbb{R})$. Thus, $G'L(n, \mathbb{R})$ is formed by all the $n \times n$ matrices, while $GL(n, \mathbb{R})$ is formed by those which are invertible.

§ **8.3.9** A very important result is *Ado's theorem*:

Every Lie algebra of a Lie group is a subalgebra
of $G'L(n, \mathbb{R})$, for some value of n.

For Lie groups, an analogous statement holds, but *only locally*: every Lie group is locally isomorphic to a subgroup of some $GL(n, \mathbb{R})$.

Concerning matrix notation and the use of a basis: a general matrix in $GL(n, \mathbb{R})$ will be written as $g = \exp(X_\alpha p^\alpha)$, where the X_α's constitute a basis in $G'L(n, \mathbb{R})$. The "components" p^α are the "group parameters" of g. The vector-valued form

$$w = X_\alpha w^\alpha = g^{-1}dg = g^{-1}(X_\beta dp^\beta)g = g^{-1}X_\beta\, g\, dp^\beta$$

will be a matrix of forms, with entries

$$w^i{}_k = (X_\alpha)^i{}_k w^\alpha = [g^{-1}X_\beta\, g]^i{}_k\, dp^\beta.$$

§ **8.3.10 Exponential mapping.** We have seen in § 6.4.17 how the group $\mathbb{R}$ acts on a manifold. Let us apply what was said there to the case in which the manifold is itself a Lie group:

$$\lambda : \mathbb{R} \times G \to G, \quad \lambda : (t, h) \to \lambda(t, h). \tag{8.20}$$

Take the orbits through the identity,

$$\lambda(0,e) = e \quad \text{and} \quad \lambda(t,e) = \lambda_e(t).$$

The theory of ordinary differential equations tells us that, in this case, there is an open $U \subset G$ around e in which the solution of Eq.(6.39),

$$\frac{d\lambda_e(t)}{dt} = X_{\lambda_e(t)} \tag{8.21}$$

is unique, for any X. Then, $a(t) = \lambda_e(t)$ is the integral curve of X through the identity and $X_e \in G'$. Now, when the manifold is a Lie group, this is a global result: the field X is *complete*, that is, $a(t)$ is defined for every $t \in \mathbb{R}$. Still more, the set $\{a_X(t)\}$, for all $t \in \mathbb{R}$ is a *one-parameter subgroup* of G generated by X. We can then introduce the *exponential mapping*, generalizing the case of $GL(n,\mathbb{R})$, as

$$\exp : G' \to G, \quad \exp(X) = a_X(0), \tag{8.22}$$

so that the subgroup is given by

$$a_X(t) = \exp(tX). \tag{8.23}$$

§ 8.3.11 **Normal coordinates.** This mapping is globally C^∞ and, in a neighbourhood around e, a diffeomorphism. In such a neighbourhood, it allows the introduction of a special LSC. Take a basis $\{J_\alpha\}$ of G' and

$$X = X^\alpha J_\alpha.$$

The algebra G' can be identified with $\mathbb{R}^d$ by $X \to (X^1, X^2, \ldots, X^d)$. As

$$a_X(1) = \exp(X^\alpha J_\alpha),$$

we can ascribe these coordinates to $a_X(1)$ itself. By Eq.(8.23), $a_X(t)$ would then have coordinates $\{tX^\alpha\}$:

$$[a(t)]^\alpha = tX^\alpha \quad \text{and} \quad [a(s)]^\alpha = sX^\alpha.$$

But $a(s)a(t) = a(s+t)$, so that

$$[a(s)a(t)]^\alpha = tX^\alpha + sX^\alpha = [a(t)]^\alpha + [a(s)]^\alpha.$$

Such local coordinates, for which the coordinates of a product are the sum of the factor coordinates, are called the *canonical*, or *normal coordinates*.

§ 8.3.12 **The Heisenberg Algebra and Group.** The usual Poisson bracket relation of classical mechanics for n degrees of freedom q^k,

$$\{p_i, p_j\} = \{q^k, q^l\} = 0, \qquad \{p_i, q^j\} = \delta_i^j,$$

and the commutation relations for their quantum correspondent operators,

$$[\hat{p}_i, \hat{p}_j] = [\hat{q}^k, \hat{q}^l] = 0, \qquad [\hat{p}_i, \hat{q}^j] = -i\hbar\delta_i^j I,$$

are formally the same. They constitute a Lie algebra going under the name of Heisenberg algebra. The corresponding Lie group is the Heisenberg group $\mathbb{H}_n$. The algebra may be characterized by parameters (p, q, s) in

$$\mathbb{R}^{2n+1} = \mathbb{R}^n \otimes \mathbb{R}^n \otimes \mathbb{R}^1.$$

Consider the $(n+2) \times (n+2)$ matrix

$$h(p, q, s) = \begin{bmatrix} 0 & p_1 & p_2 & \dots & p_n & s \\ 0 & 0 & 0 & \dots & 0 & q^1 \\ 0 & 0 & 0 & \dots & 0 & q^2 \\ \vdots & \vdots & \vdots & \vdots & \vdots & \vdots \\ 0 & 0 & 0 & \dots & 0 & q^n \\ 0 & 0 & 0 & \dots & 0 & 0 \end{bmatrix}.$$

It is immediate that the products of two matrices of this kind are given by

$$h(p, q, s)\, h(p', q', s') = h(0, 0, pq'),$$

$$[h(p, q, s)]^2 = h(0, 0, pq),$$

$$[h(p, q, s)]^m = 0 \text{ for } m > 2.$$

The commutator

$$[h(p, q, s), h(p', q', s')] = h(0, 0, pq' - p'q)$$

will define the Lie algebra. The group $\mathbb{H}_n$ is arrived at by the exponential map

$$h(p, q, s) \to H(p, q, s) = \exp[h(p, q, s)],$$

which is

$$H(p, q, s) = \begin{bmatrix} 1 & p_1 & p_2 & \dots & p_n & s + \frac{1}{2}pq \\ 0 & 1 & 0 & \dots & 0 & q^1 \\ 0 & 0 & 1 & \dots & 0 & q^2 \\ \vdots & \vdots & \vdots & \vdots & \vdots & \vdots \\ 0 & 0 & 0 & \dots & 1 & q^n \\ 0 & 0 & 0 & \dots & 0 & 1 \end{bmatrix}.$$

The group law may be expressed as

$$H(p, q, s)H(p', q', s') = H[p + p', q + q', s + s' + \tfrac{1}{2}(pq' - p'q)].$$

The centre, which here coincides with the commutator subgroup, is given by $C = \{H(0, 0, s)\}$. The Lebesgue measure on $\mathbb{R}^{2n+1}$ is a bi-invariant Haar measure on $\mathbb{H}_n$. Notice that other matrix realizations are possible, but the above one has the advantage that the inverse to $H(p, q, s)$ is simply

$$H(p, q, s)^{-1} = H(-p, -q, -s).$$

§ **8.3.13** A first hint of what happens in a noncommutative geometry is the following. Taking the Lie group algebraic structure into account, a field defines both a left derivative,

$$X_L f(g) = \left. \frac{d}{dt} f(e^{tX} g) \right|_{t=0}$$

and a right derivative

$$X_R f(g) = \left. \frac{d}{dt} f(g e^{tX}) \right|_{t=0} .$$

There is no reason for them to coincide when the group is non-abelian.

8.4 The adjoint representation

A representation of a group G is a homomorphism of G into some other group H, and a representation of a Lie algebra G' is a homomorphism of G' into some other algebra H'. The simplest cases are the linear group representations, those for which H is the group AutV of the linear invertible transformations of some vector space V, which is the "carrier space", or "representation space" of the representation. More details can be found in Section 18.2. We shall here consider some representations in terms of fields defined on the group manifold itself. They are essential to the understanding of Lie groups and of their action on other spaces.

The *adjoint representation* is a representation of a Lie group on the vector space of its own Lie algebra. Its differential is a representation of this Lie algebra on itself or, more precisely, on its derived algebra (§ 13.5.8).

§ **8.4.1** Isomorphisms of a Lie group G into itself and of a Lie algebra G' into itself are *automorphisms*. The set $\mathrm{Aut}(G)$ of all such automorphisms is itself a Lie group. For every $j \in \mathrm{Aut}(G)$, the differential $dj = j_*$ is an automorphism of G', such that

$$j(\exp X) = \exp(j_* X).$$

The diagram

$$
\begin{array}{ccc}
G' \to j_* = dj \to & G' \\
\downarrow & \downarrow \\
\exp & \exp \\
\downarrow & \downarrow \\
G \longrightarrow j \longrightarrow & G
\end{array}
$$

is then commutative.

§ 8.4.2 When working with matrices, we are used to seeing a matrix h be transformed by another matrix g in the form ghg^{-1}. This is a special case of a certain very special representation which is rooted in the very nature of a Lie group. The automorphisms j_* above mentioned belong to the group $\text{Aut}(G')$ of the linear transformations of G' (seen as a vector space). The differential operation

$$j \to j_* = dj$$

takes $\text{Aut}(G)$ into $\text{Aut}(G')$. It is a homomorphism, since

$$d(j \circ k) = dj \circ dk.$$

Consequently, it is a representation of $\text{Aut}(G)$ on G'. An important subgroup of $\text{Aut}(G)$ is formed by the *inner automorphisms* of G, which are combinations of left-and right-translations (§ 8.1.2) induced by an element g and its inverse g^{-1}:

$$j_g = L_g \circ R_{g-1} = R_{g-1} \circ L_g$$

$$j_g(h) = ghg^{-1}. \tag{8.24}$$

Each j_g is in reality a diffeomorphism, and $j_g(hk) = j_g(h) \cdot j_g(k)$. Thus, the mapping $g \to j_g$ is a group homomorphism. The mapping

$$dj_g = j_{g*} = L_{g*} \circ R_{(g^{-1})*} = R_{(g^{-1})*} \circ L_{g*}$$

belongs to $\text{Aut}(G')$.

§ 8.4.3 Now we arrive at our objective: the mapping

$$Ad : G \to Aut\ G', \quad \text{Ad}(g) = \text{Ad}_g = dj_g \tag{8.25}$$

is the *adjoint representation* of G. Given a field $X \in G'$, the effect of $\text{Ad}(g)$ on X is described by

$$\text{Ad}_g X = (R_{(g^{-1})*} \circ L_{g*})X. \tag{8.26}$$

Being X left-invariant, then

$$\text{Ad}_g X = R_{(g^{-1})*} X. \tag{8.27}$$

§ 8.4.4 Using (8.23), expression (8.26) may be written as

$$e^{t\text{Ad}_g X} = g\, e^{tX} g^{-1}. \tag{8.28}$$

Thus: take the curve $\exp(tX)$; transform it by j_g; then, $\text{Ad}_g X$ is the tangent to the transformed curve at the identity (Figure 8.3). This representation is of fundamental importance in modern gauge field theory, as both gauge potentials and field strengths belong to it (see Chapter 32).

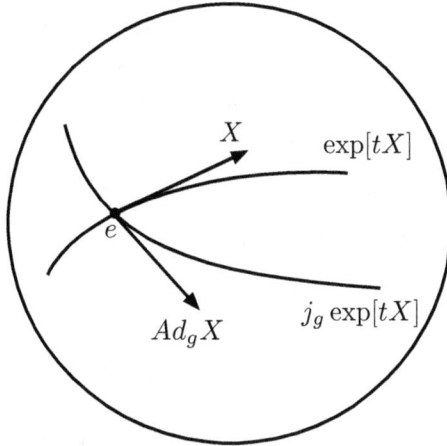

Fig. 8.3 Illustration of $\text{Ad}_g X$.

§ 8.4.5 The mapping $\text{ad} := d(\text{Ad})$ given by

$$\text{ad} : G' \to (\text{Aut}G)', \quad X \to \text{ad}_X \tag{8.29}$$

is the *adjoint representation* of the Lie algebra G'. To each field X in G' corresponds, by this representation, a transformation on the fields belonging to G', of which X will be the generator. We know that a field generates transformations on its fellow fields through the Lie derivative, so that

$$\text{ad}_X Y = L_X Y = [X, Y]. \tag{8.30}$$

In a basis $\{X_i\}$,

$$\text{ad}_{X_i} X_j = [X_i, X_j] = C^k{}_{ij} X_k. \tag{8.31}$$

Thus, the adjoint representation is realized by the matrices C_i whose elements $(C_i)^k{}_j$ are the structure constants,

$$(C_i)^k{}_j = [\text{ad}(X_i)]^k{}_j = C^k{}_{ij}. \tag{8.32}$$

§ 8.4.6 Notice that if g has its actions given by the matrices $U(g)$, and X is also a matrix (case of $GL(n, \mathbb{R})$), Eq.(8.26) gives simply the usual rule for matrix transformation UXU^{-1}. From a purely algebraic point of view, the adjoint representation is *defined* by (8.32): it is that representation by matrices whose entries are the structure constants. It is sometimes[1] called

---
[1] See, for instance, Gilmore 1974.

the *regular* representation, but we shall use this name for another kind of representation (see Section 18.3).

We may consider also the representations given by the action on the forms, through the pull-backs $L_{(g^{-1})*}$ and $R_{(g^{-1})*}$. Such representations on the covectors are called *coadjoint representations*. In the matrix notation of § 8.3.9, the adjoint representation will be given by a matrix A, such that

$$X'_\beta = g^{-1}(X_\beta)\, g = A_\beta{}^\alpha X_\alpha.$$

For $g = e^{X_\alpha p^\alpha}$, the vector-valued form

$$w = X_\alpha w^\alpha = g^{-1}(X_\beta)\, g\, dp^\beta$$

will be $A_\beta{}^\alpha dp^\beta$, so that $w^\alpha = A_\beta{}^\alpha dp^\beta$.

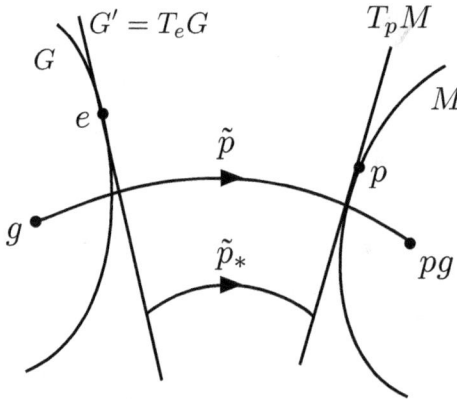

Fig. 8.4   Action R_g as a mapping of G into M.

§ 8.4.7 Let us go back to the action of groups on manifolds, studied on Section 8.2. Consider the right-action:

$$R_g : M \to M, \quad R_g(p) = pg \ \text{with}\ R_e(p) = p.$$

Let us change a bit the point of view (see Figure 8.4): take p fixed and change $g \in G$. The action R_g becomes a mapping of G into M, which we shall denote $\tilde{p}$:

$$\tilde{p} : G \to M, \quad \tilde{p}(g) = R_g(p) = pg. \tag{8.33}$$

The set of points $\tilde{p}(g)$ is, as said in § 8.2.9, the orbit of p by G.

§ **8.4.8** Then, the differential mapping

$$\widetilde{p}_* : T_e G \to T_p M$$

will take a field X of G' into some field $\bar{X}$ of $T_p M$,

$$\bar{X} = \widetilde{p}_*(X). \tag{8.34}$$

This mapping is an algebra homomorphism of G' into the Lie algebra of fields on some $U \ni p$. The following results are very important:

(1) If the action is effective, $\widetilde{p}_*$ is one-to-one; $\bar{X}$ is then a *fundamental field* on M, corresponding to X; taking all the $X \in G'$, the set $\widetilde{G}'$ of the corresponding fundamental fields is a representation of G'.
(2) If G acts also freely, $\widetilde{p}_*$ is an isomorphism: $G' \approx \widetilde{G}'$.

Summing up, the action of a group G on a manifold M around one of its points is thus fulfilled in the following way:

(i) For each group generator X there will be a "deputy" field $\bar{X}$ on M, its "nomination" being made through the mapping $\widetilde{p}_*$.
(ii) Each fundamental field will be the infinitesimal operator of transformations (§ 6.4.22) of a one-parameter group.
(iii) The set of fundamental fields will engender the group transformations on M.
(iv) The representation in the general case will be non-linear (more about that in Chapter 27).

§ **8.4.9** Let us try to put it all in simple words: given a group of transformations on a manifold under certain conditions, its generators are represented, in a neighbourhood of each point of the manifold, by fields on the manifold. Each generator appoints a field as its representative. This is what happens, for instance, when we represent a rotation around the axis $0z$ in $\mathbb{E}^3$ by the infinitesimal operator $x\partial_y - y\partial_x$, which is a field on $\mathbb{E}^3$. This operator acts on functions defined on $\mathbb{E}^3$ which carry a representation of the rotation group.

§ **8.4.10** We may ask now: under a group transformation, what happens to the fundamental fields themselves? On M,

$$(\overline{R_{g*} X})_p = \widetilde{p}_* \circ R_{g*} X = \widetilde{p}_*[Ad_{g^{-1}} X] = \overline{Ad_{g^{-1}} X},$$

where use was made of Eq.(8.27). We have been using the same notation R_g for the actions on G and M, so that we shall drop the bars when not strictly necessary:

$$R_{g*}\overline{X} = \overline{Ad_{g^{-1}}X}. \tag{8.35}$$

If we examine the left-action, we find

$$L_{g*}\overline{X} = \overline{Ad_g X}. \tag{8.36}$$

Notice the interchange $g \leftrightarrow g^{-1}$ between the two cases. The process is

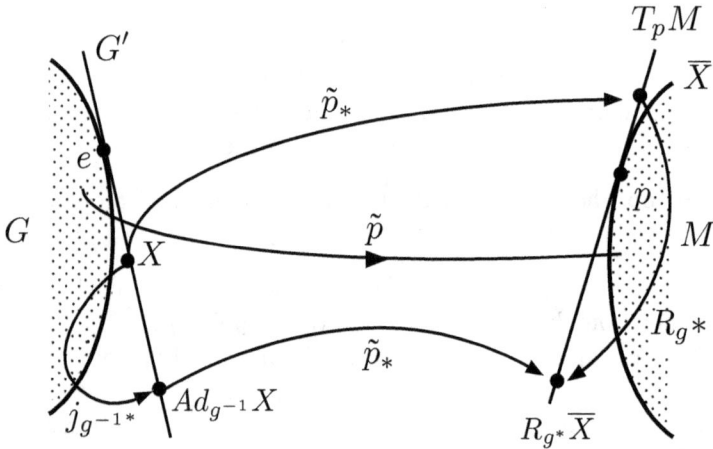

Fig. 8.5   Change of a fundamental field.

rather knotty. When we want to know how a fundamental field $\overline{X}$ changes in some point p of M, we begin by going back to the field X in G' which $\overline{X}$ represents. Transform X by the adjoint representation and then bring the result back by the mapping $\tilde{p}_*$ (Figure 8.5). This process is pictorially represented in the commutative diagram

$$
\begin{array}{ccc}
G' & \longrightarrow \tilde{p}_* \longrightarrow & T_pM \\
\downarrow & & \downarrow \\
Ad_{g^{-1}} & & R_{g*} \\
\downarrow & & \downarrow \\
G' & \longrightarrow \tilde{p}_* \longrightarrow & T_pM
\end{array}
$$

All this discussion is strictly local. It shows where the importance of the adjoint representation comes from and will be instrumental in the study of fiber bundles.

§ **8.4.11** The *Killing form* is a bilinear form γ on G' defined by

$$\gamma_{ij} = \gamma(X_i, X_j) = \text{tr} \, [\text{ad}(X_i) \cdot \text{ad}(X_j)], \tag{8.37}$$

or

$$\gamma_{ij} = C^k{}_{im} C^m{}_{jk}. \tag{8.38}$$

A theorem by Cartan says that

$$\det{(\gamma_{ij})} \neq 0$$

is a necessary and sufficient condition for G' to be a *semisimple* algebra (and G be a semisimple group, that is, without abelian invariant subgroups). Examples of non-semisimple groups are the linear groups, the affine group and the Poincaré group. On the other hand, the orthogonal (or pseudo-orthogonal, like the Lorentz group) and the unitary groups are semisimple. In the semisimple case, as γ is non-degenerate, it can be used as an invariant metric on G (invariant under the action of G itself), the *Cartan metric*. It is used in gauge theories, being the K metric of the basic lagrangian given by Eq.(7.161). Of course, it only makes sense when the gauge group (the structure group) is semisimple. Usual gauge theories use orthogonal or unitary groups.

Suppose G is a compact and semisimple group. A form on G is said to be invariant (not to be confused with a left-invariant form as defined in § 8.3.5 !) if it is a zero of the Lie derivatives with respect to all the generators X_α. As these constitute a vector basis, we can recall Eq.(7.186) to conclude that an invariant form is closed. With the Cartan metric, the coderivative and the laplacian are defined. From Eq.(7.189), every invariant form on G is also co-closed, and consequently harmonic. From these considerations follow very restrictive results on the topology of such groups.[2] For instance, the Betti numbers b_1 and b_2 are zero and $b_3 \geq 1$. For simple groups, $b_3 = 1$.

General references

Lie groups, and their relations to Lie algebras, in particular those of fields on manifolds, are widely treated in texts on differential geometry, such as Kobayashi & Nomizu 1963 and Spivak 1970. A more specific text is Nomizu 1956. A purely algebraic approach to Lie algebras is given in Jacobson 1962.

---

[2] Goldberg 1962.

Chapter 9

Fiber Bundles

A basic manifold to each point of which is attached a copy of another space, under very strict conditions ensuring good mathematical behaviour — that's the intuitive picture of a fibre bundle. This notion provides a compact, unified view of practically everything geometric. What follows is a purely heuristic introduction.

9.1 Introduction

We have already met a few examples of fiber bundles: the tangent bundle, the cotangent bundle, the bundle of linear frames. Also fibered spaces of another kind have been seen: the covering spaces, whose fibers are discrete spaces acted upon by the fundamental group. We shall in the following consider only bundles with differentiable fibers — differential fiber bundles. Of this type, we have glimpsed tensorial bundles in general, which include the tangent and the cotangent bundles as particular cases. Locally, bundles are direct-product manifolds, but globally they are nothing of the sort. In reality, their importance comes out mainly when global effects, or effects "in the large", are at stake. As Physics is largely based on (local) differential equations, such effects are usually taken into account in the boundary conditions.

We start with an intuitive presentation of the bundle idea, then proceed to examine the simplest cases, vector bundles. In vector bundles, whose prototypes are the tangent bundles, the fibers are vector spaces. We shall then proceed to the more involved bundle of frames, on which linear connections play their game. The frame bundle is the prototype of principal bundles, whose fibers are groups and which are the natural setting summing up all differential geometry. The frame bundle provides the natural background for General Relativity, and "abstract" principal bundles do the same for Gauge Theory.

§ 9.1.1 Intuitively, a fiber bundle is a manifold (the "base") to every point of which one "glues" another manifold (the "fiber"). For example, the sphere S^2 and all the planes ($\approx \mathbb{E}^2$) tangent to it (the sphere tangent bundle); or the same sphere S^2 and all the straight half-lines ($\approx \mathbb{E}^1_+$ normal to it (the "normal bundle"). The classical phase space of a free particle is a combination of the configuration space (base) and its momentum space (fiber), but it is a simple cartesian product and, as such, a trivial bundle (Section 25.1).

Notice however that the base and the fiber do not by themselves determine the bundle: it is necessary to specify *how* they are "glued" together. This is the role of the projection mapping. For instance, with the circle S^1 and the straight line $\mathbb{E}^1$ we can construct two different bundles: a trivial one — the cylinder — and the Möbius band, which is nontrivial. They cannot be distinguished by purely local considerations. By the way, the word "trivial" is here a technical term: a bundle is trivial when it is globally a cartesian product.

§ 9.1.2 To illustrate the possible import to Physics, the simplest example is probably the following[1]: suppose we have a scalar field on S^1, evolving in time according to the 2-dimensional Klein–Gordon equation

$$(\Box + m^2)\,\varphi = 0.$$

The d'Alembertian $\Box$ must be defined on the curved space (it is a Laplace–Beltrami operator Eq. (7.114) formed by S^1 and the time in $\mathbb{E}^1$. Now, how can we account for the two different possible spaces alluded to above? The equation is local and will have the same form in both cases. The answer lies in the boundary conditions: in the cylinder, it is forcible to use periodic conditions, but on the Möbius band, which is "twisted", one is forced to use antiperiodic conditions! Avis and Isham have performed the second-quantized calculations and found a striking result: the vacuum energy (lowest energy level) is different in the two cases! Fields on nontrivial bundles ("twisted fields") behave quite differently from usual fields.

We usually start doing Physics with some differential equations and *suppose* "reasonable" boundary conditions. If the comparison with experiments afterwards show that something is wrong, it may be that only these conditions, and not the equations, should be changed. Purely local effects

---

[1] Avis & Isham 1978, 1979; Isham 1978.

(such as the values of fields around a point at which the values are known) are relatively independent of topological (boundary) conditions. We say "relatively" because quantization is, as a rule, a global procedure. It assumes well-defined boundary conditions, which are incorporated in the very definition of the Hilbert space of wavefunctions. Energy levels, for example, are clearly global characteristics.

§ **9.1.3** Let us go back to the beginning: to constitute a bundle one appends a fiber to each point of the base. This is of course a pictorial point of view. One does not really need to attach to each point of the base its tangent space, for example, but it is true that it helps conceiving the whole thing. The fiber is, on each point, a copy of one same space, say $\mathbb{E}^m$ for the tangent space, $GL(m, \mathbb{R})$ for the frame bundle, etc. This abstract space, of which every fiber is but a copy, is called the *typical fiber*.

9.2 Vector bundles

§ **9.2.1** Given a differentiable manifold M, a vector space F and an open set $U \subset M$, the cartesian product $U \times F$ is a local vectorial bundle, for which U is the base space. If $x \in U$, the product $\{x\} \times F$ is the fiber on x. The mapping $\pi \colon U \times F \to U$ such that

$$\pi(x, f) = x$$

for every f in F is the bundle projection. Of course, the fiber on x is also $\pi^{-1}(x)$. As F is open, $U \times F$ is also open in $M \times F$.

§ **9.2.2** A *local fibered chart* is a pair (U, φ), where φ is a bijection

$$\varphi \colon U \times F \to U' \times F' \subset \mathbb{E}^n,$$

with n large enough. Such a chart provides a local system of coordinates (LSC) on the local bundle. A further condition is necessary: when we define local bundles as above for two open sets U_i and U_j in M, the coordinate transformation φ_{ij} in the intersection $U_i \cap U_j$, of the form $\varphi_{ij} = \varphi_j \circ \varphi_i^{-1}$, must be a diffeomorphism and obey

$$\varphi(x, f) = \big(\psi_1(x), \psi_2(x)f\big), \tag{9.1}$$

where x and f are coordinates of a point in the intersection and of a point in F, ψ_1 is a coordinate transformation in the intersection, and ψ_2 is a mapping taking the point represented by x into the set of linear mappings of F into F'; the result, $\psi_2(x)$, is an x-dependent mapping taking f into some f'. A set of charts (U_i, φ_i) satisfying the conditions of a complete atlas is a *vector fibered atlas*. As usual with their kin, such an atlas allows one to extend the local definitions given above to the whole M.

§ **9.2.3** A vector bundle is thus built up with a base space M, a typical fiber F which is a vector space, and an atlas. Suppose further that a Lie group G acts *transitively* on F: it will be called the *structure group* of the bundle. The bundle itself, sometimes called the *complete space*, is denoted

$$P = (M, F, G, \pi). \tag{9.2}$$

An important existence theorem[2] states that:

> *Given a Lie group G acting through a representation on a vector space F, and a differentiable manifold M, there exists at least one bundle (M, F, G, π).*

§ **9.2.4** A *section* σ is any C^∞ mapping

$$\sigma : U \to U \times F \tag{9.3}$$

such that, for every $p \in U \subset M$,

$$\pi(\sigma(p)) = p.$$

Such sections constitute by themselves an infinite-dimensional linear function space.

§ **9.2.5** We have already met the standard example of vector bundle, the *tangent bundle*

$$TM = (M, \mathbb{E}^m, GL(m, \mathbb{R}), \pi_T). \tag{9.4}$$

A vector field is a section $X : U \subset M \to TU$ such that $X(p) = X_p$, that is,

$$\pi_T \circ X(p) = p.$$

§ **9.2.6** This is a typical procedure: in general, points on the bundle are specified by sections. In physical applications, F is frequently a Hilbert space, wave-functions playing the role of sections. Why are bundles and all that almost never mentioned ? Simply because in most usual cases the underlying bundles are trivial, simple cartesian products. In wave mechanics, it is the product of the configuration space by a Hilbert space. Bundles provide the geometrical backstage for gauge theories (see Chapter 32) and it was precisely the flourishing of these theories that called attention to the

---

[2] Steenrod 1970.

importance of non-trivial bundles to Physics, mainly after Trautman[3] and Yang[4] uncovered their deeply geometrical character.

§ 9.2.7 We have above defined *local* sections. The reason is fundamental: it can be shown that only on trivial bundles there are global sections. The simple existence of a section defined everywhere on the base space (such as the usual wavefunctions) of a bundle ensures its direct-product character. Every bundle is *locally* trivial, only cartesian products are *globally* trivial. For the tangent bundle TM, this would mean that a field X can be defined everywhere on M by a single section. As we have said, M is, in this case, *parallelizable*. Lie groups are parallelizable manifolds. The sphere S^2 is not and, consequently, accepts no Lie group structure. In reality, only a few spheres can be Lie groups (S^1, S^3 and S^7), for this and other reasons.

§ 9.2.8 In general, many different fibers F constitute bundles like (9.2), with the same group G. They are called *associated bundles*. Bundles with a given base space and a given structure group can be classified. The classification depends, fundamentally, only on the topologies of the base and the group. It does not depend on the fiber. Consequently, the classification can be realized by taking into account only the principal bundles, in which the fiber is replaced by the group itself (see Section 9.7).

§ 9.2.9 Fibration. In gauge theories, the source fields belong usually to (associated) vector bundles. All the fibers are isomorphic to the typical fiber in a fiber bundle. The notion may be generalized to that of a fibration, in which the fibers are not necessarily the same on each point of the base space. Let us briefly describe the idea. Given two spaces E and M, a fibration is a mapping $\pi : E \to M$ possessing a property called "homotopy lifting", which consists in the following. Consider another space K, and a map $f : K \to E$. This leads to the composition

$$g = \pi \circ f : K \to M.$$

The map f is the "lift" of g. Consider now the homotopy class of g, a set of homotopic maps g_t with $g_0 = g$. If this homotopy lifts to a homotopy f_t of f, with $f_0 = f$, then π is said to have the homotopy property (a fiber bundle is a particular case, the bundle projection being a locally trivial fibration). Thus, the only requirement now is that all the fibers $\pi^{-1}(x)$, for $x \in M$, be of the same homotopy type.

---

[3] Trautman 1970.
[4] Yang 1974.

An example of recent interest is the *loop space*: start by taking as total space the set M_0^I of curves $x(t)$ with initial endpoint $x_0 = x(0)$. The final endpoint of each curve will be $x_1 = x(1)$. Take as fibration the mapping $\pi : x(t) \to x_1$ taking each path into its final endpoint. The mapping $\pi^{-1}(\mathrm{x})$ is the set of all curves $c_0(x)$ from x_0 to x. Now, choose some fixed path $c_0(\mathrm{x})$ from x_0 to x. Any other path of $\pi^{-1}(x)$ will differ from $c_0(x)$ by a loop through x_0, so that each new $c_0(x)$ determines an element of the loop space "LM". The mapping $\pi^{-1}(x)$ is homotopically equivalent to LM. Thus, in the fiber above, the initial endpoint is LM, and π satisfies the fibration requirements.

9.3   The bundle of linear frames

There are many reasons to dedicate a few special pages to the bundle of linear frames BM. On one hand, it epitomizes the notion of principal bundle and is of fundamental importance for geometry in general. This makes it of basic significance for General Relativity, as well as for Teleparallel Gravity (see Chapters 34 and 35). On the other hand, precisely because of this particular link to the geometry of the base underlying space, it has some particular properties not shared by other principal bundles. The intent here is to make clear its general aspects — which have provided the historical prototype for the general bundle scheme — while stressing those aspects well known to gravitation physicists which find no counterpart in the bundles which underlie gauge theories.

§ **9.3.1** Each point b of the bundle of linear frames, denoted BM, on the smooth manifold M is a frame at the point $p = \pi(b) \in M$, that is, a set of linearly independent vectors at p. The structural group $GL(m, \mathbb{R})$ acts *on the right* on BM as follows: given $a = (a_{ij}) \in GL(m, \mathbb{R})$ and a frame $b = (b_1, b_2, b_3, \ldots, b_m)$, then

$$b' = ba = (b'_1, b'_2, b'_3, \ldots, b'_m)$$

with

$$b'_i = a^j{}_i b_j = b_j a^j{}_i. \tag{9.5}$$

In the natural basis of a chart (U, x) around $\pi(b)$, b_i will be written

$$b_i = b_i{}^j \frac{\partial}{\partial x^j}$$

and $\{x^i, b_k{}^l\}$ provides a chart on BM around b. Take on $\mathbb{E}^m$ the canonical basis $\{K_i\}$, formed by columns K_i whose j-th element is δ_{ij}:

$$K_1 = (1,0,0,\ldots,0), \quad K_2 = (0,1,0,0,\ldots,0), \text{ etc.}$$

The frame $b \in BM$ can be seen as a mapping taking the canonical basis $\{K_k\}$ into $\{b_k\}$ (see Figure 9.1). More precisely: the frame b given by $b = (b_1, b_2, \ldots, b_m)$ is the linear mapping

$$b : \mathbb{E}^m \to T_{\pi(b)}M, \quad b(K_k) = b_k. \tag{9.6}$$

Being a linear mapping between two vector spaces of the same dimension, b is an isomorphism. It "appoints" the base member b_j as the "representative" of the canonical vector K_j belonging to the typical fiber $\mathbb{E}^m$. Consequently, two vectors X and Y of $T_{\pi(b)}M$ are images of two vectors r and s on $\mathbb{E}^m$:

$$X = b(r) = b(r^i K_i) = r^i b(K_i) = r^i b_i$$

and

$$Y = b(s) = b(s^i K_i) = s^i b(K_i) = s^i b_i.$$

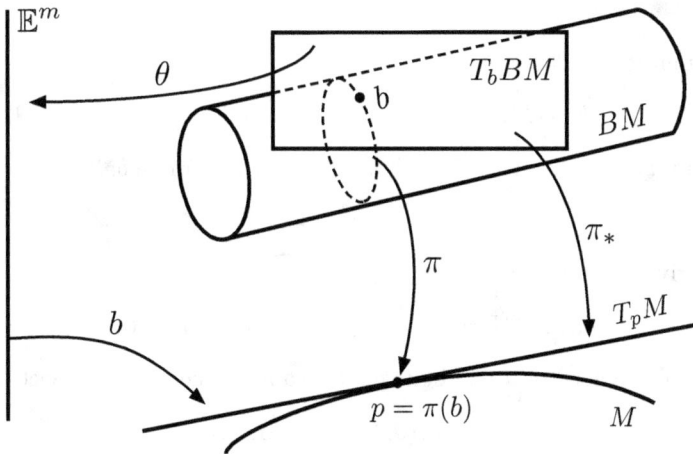

Fig. 9.1   Scheme of the frame bundle, with soldering.

9.3.1 Structure group

To see more of the action of the structure group on BM, notice that $GL(m, \mathbb{R})$ acts on the space $\mathbb{E}^m$ on the left (actually, we might build up an associated vector bundle with fiber $\mathbb{E}^m$): if $V \in \mathbb{E}^m$, $V = V^k K_k$, and $a = (a^j{}_i) \in GL(m, \mathbb{R})$,

$$(aV)^j = a^j{}_i V^i \quad \text{or} \quad aV = K_j a^j{}_i V^i.$$

The mapping b will give

$$b(aV) = b(K_j) a^j{}_i V^i = b_j a^j{}_i V^i$$

and

$$(ba)(V) = (ba)(V^i K_i) = V^i (ba)(K_i) = V^i u_i = b_j a^j{}_i V^i,$$

where use has been made of Eq. (9.5). Consequently,

$$b(aV) = (ba)(V). \tag{9.7}$$

It is instructive to interpret through this equation the action of $GL(m, \mathbb{R})$ on BM: it says that the diagram

$$
\begin{array}{ccc}
\mathbb{E}^m & \xrightarrow{a} & \mathbb{E}^m \\
{}_{ba}\searrow & & \downarrow b \\
& T_{\pi(b)}M &
\end{array}
$$

is commutative. The mapping

$$\tilde{b} : GL(m, \mathbb{R}) \to BM, \quad \tilde{b}(a) := R_a(b) = ba, \tag{9.8}$$

when applied to the group identity, gives just the frame b:

$$\tilde{b}(e) = be = b.$$

The derivative mapping will be

$$\tilde{b}_* : T_a GL(m, \mathbb{R}) \to T_{ba} BM, \quad \tilde{b}_*(J_a) = X_{ba}. \tag{9.9}$$

A group generator J will be taken into a fundamental field (Section 8.4):

$$J^* = \tilde{b}_*(J_e) = \text{a certain } X_b.$$

We may choose on the algebra $G'L(m, \mathbb{R})$ a convenient, canonical basis given by the $m \times m$ matrices $\Delta_i{}^j$ whose elements are

$$(\Delta_i{}^j)_a{}^b = \delta_i^b \delta_a^j. \tag{9.10}$$

In this basis, an element $J_e \in G'L(m, \mathbb{R})$ will be written $J = J^i{}_j \Delta_i{}^j$. The Lie algebra will be defined by the matrix commutator, with the commutator table

$$[\Delta_i{}^l, \Delta_j{}^k] = \delta_i^k \Delta_j{}^l - \delta_j^l \Delta_i{}^k. \tag{9.11}$$

The mapping $\tilde{b}_*$ can be shown to be an algebra homomorphism between $G'L(m, \mathbb{R})$ and the Lie algebra of fields on BM at b. This allows us to introduce a basis $\{E_i{}^j\}$ for the fundamental fields through

$$E_i{}^j = \tilde{b}_*(\Delta_i{}^j) = (\Delta_i{}^j)^*. \tag{9.12}$$

Thus, $J^* = J_j{}^i E_i{}^j$. The homomorphism leads to

$$[E_i{}^l, E_j{}^k] = \delta_i^k E_j{}^l - \delta_j^l E_i{}^k. \tag{9.13}$$

Now, $\pi \circ \tilde{b}$ is a constant mapping $G \to \pi(b)$, so that $\pi_* \circ \tilde{b}_* = 0$. Fields X such that $\pi_*(X) = 0$ are called *vertical* . The fundamental fields are clearly of this kind,

$$\pi_*(E_i{}^j) = 0. \tag{9.14}$$

They also obey Eq. (8.35),

$$R_{a^*}(J^*) = \tilde{b}_*[(Ad_{a^{-1}}J)_e]. \tag{9.15}$$

There is a one-to-one correspondence between the structure group and the space of frames. The fiber coincides with the group, so that the bundle is a principal bundle.

9.3.2 *Soldering*

Soldering is a very special characteristic of the bundle of linear frames, not found in other bundles (see again Figure 9.1). It is due to the existence of a peculiar vector-valued 1-form on BM, defined by

$$\theta : T_b BM \to \mathbb{E}^m, \quad \theta := b^{-1} \circ \pi_*. \tag{9.16}$$

This composition of two linear mappings will be an $\mathbb{E}^m$-valued form. Called *canonical form*, or *solder form*, θ can be written, in the canonical basis $\{K_i\}$ of $\mathbb{E}^m$, as

$$\theta = K_i \theta^i. \tag{9.17}$$

Each form θ^k is called a *soldering form*. It is possible to show that, under the right action of the structure group,

$$R_g^* \theta = g^{-1} \theta. \tag{9.18}$$

The name is not gratuitous. The presence of the solder form signals a coupling between the tangent spaces, to the bundle and to the base manifold, which is much stronger for BM than for other principal bundles. A consequence of utmost importance is that a connection on BM will have, besides the curvature it shares with all connections, another related form, torsion. Soldering is absent in bundles with "internal" spaces, which consequently exhibit no torsion.

9.3.3 *Orthogonal groups*

Let us say a few more words on the relation between the orthogonal groups and bilinear forms, introduced in § 8.1.7. A group of continuous transformations preserving a symmetric bilinear form η is an orthogonal group or, if the form is not positive-definite, a pseudo-orthogonal group. If an element of the group is represented by Λ, this defining property takes the matrix form

$$\Lambda^T \eta \Lambda = \eta, \tag{9.19}$$

where "T" indicates the transpose. As a consequence, any member A of the algebra, for which $\Lambda = e^A$ for some Λ, will satisfy

$$A^T = -\eta A \eta^{-1} \tag{9.20}$$

and will have $\mathrm{tr} A = 0$.

Let us go back to the real linear group $GL(m, \mathbb{R})$, and the basis (9.10). Given the bilinear form η, both basis and entry indices can be lowered and raised, as in $(\Delta^a{}_b)^i{}_j = \eta^{ai} \eta_{bj}$. One finds, for instance,

$$[\Delta_{ab}, \Delta_{cd}] = \eta_{bc} \Delta_{ad} - \eta_{da} \Delta_{cb}. \tag{9.21}$$

In this basis a member $K = K^{ab} \Delta_{ab}$ of the algebra will have as components its own matrix elements: $(K)^{ij} = K^{ij}$. The use of double-indexed basis for the algebra generators is the origin of the double-indexed notation (peculiar to the linear and orthogonal algebras) for the algebra-valued forms as, for example, the connection

$$\Gamma = J_a{}^b \Gamma^a{}_{b\mu} \, dx^\mu.$$

A special basis for the (pseudo) orthogonal group $SO(\eta)$ corresponding to η is given by the generators

$$J_{ab} = \Delta_{ab} - \Delta_{ba} = -J_{ba}. \tag{9.22}$$

All this leads to the usual commutation relations for the generators of orthogonal and pseudo-orthogonal groups,

$$[J^{ab}, J^{cd}] = \eta^{bc} J^{ad} + \eta^{ad} J^{bc} - \eta^{bd} J^{ac} - \eta^{ac} J^{bd}. \tag{9.23}$$

Let us insist on some widely used nomenclature: the usual group of rotations in 3-dimensional euclidean space is the special orthogonal group, indicated $SO(3)$. Being "special" means connected to the identity, that is, represented by 3×3 matrices of determinant $= +1$. Orthogonal and pseudo-orthogonal groups are usually indicated by $SO(\eta) = SO(p, q)$, with (p, q) fixed by the signs in the diagonalized form of η. The group of rotations in n-dimensional euclidean space will be $SO(n)$, the Lorentz group will be $SO(3, 1)$, etc.

9.3.4 *Reduction*

We may in many cases replace $GL(m, \mathbb{R})$ by some subgroup in such a way as to obtain a sub-bundle. The procedure is called "bundle reduction". For example, $GL(m, \mathbb{R})$ can be reduced to the orthogonal subgroup $O(m)$ (or to its pseudo-orthogonal subgroups). The bundle BM of the linear frames reduces to the sub-bundle

$$OM = (M, O_p M, O(m), \pi),$$

where $O_p M$ is the set of orthogonal frames on $T_p M$. Now, $T_p M$ is isomorphic to $\mathbb{E}^m$, the typical fiber of the associated tangent bundle and on which there exists an internal product which is just invariant under the action of $O(m)$. Let us consider now a consequence of the reduction to $SO(\eta)$.

9.3.5 *Tetrads*

The most interesting point of reduction is that, in the process, each basis of BM defines on M a riemannian metric. Suppose on $\mathbb{E}^m$ the invariant internal product to be given by the euclidean (or a pseudo-euclidean) metric η, with

$$(r, s) = \eta_{\alpha\beta} r^\alpha s^\beta. \tag{9.24}$$

Given

$$X = b(r) = r^i b_i \quad \text{and} \quad Y = b(s) = s^i b_i,$$

a riemannian (or pseudo-riemannian) metric on M can be defined by

$$g(X, Y) = (b^{-1} X, b^{-1} Y) = (r, s). \tag{9.25}$$

It is possible to show that g is indeed riemannian (or pseudo-riemannian, if $O(m)$ is replaced by some pseudo-orthogonal group).

The procedure can be viewed the other way round: given a riemannian g on M, one takes the subset in BM formed by the $b = (b_1, b_2, \ldots, b_m)$ which are orthogonal according to g. The resulting bundle, OM, is the *bundle of orthogonal frames* on M. In the case of interest for General Relativity, it is the pseudo-orthogonal Lorentz group $SO(3,1)$ which is at work in the tangent space, and we must take for η the Lorentz metric of Minkowski $\mathbb{E}^{3,1}$ space. Of course, $SO(3,1)$ is also a subgroup of $GL(4,\mathbb{R})$. Given a natural basis $\{\partial_\mu\}$, a general basis $\{h_\alpha\}$ of BM has elements

$$h_\alpha = h_\alpha{}^\mu \partial_\mu, \tag{9.26}$$

and its dual basis is $\{h^\beta\}$,

$$h^\beta = h^\beta{}_\mu dx^\mu \tag{9.27}$$

with

$$h^\beta(h_\alpha) = h^\beta{}_\mu h_\alpha{}^\mu = \delta^\beta_\alpha . \tag{9.28}$$

To reduce to the bundle OM, we impose "orthogonality" by some g:

$$g(h_\alpha, h_\beta) = g_{\mu\nu} h_\alpha{}^\mu h_\beta{}^\nu = \eta_{\alpha\beta} . \tag{9.29}$$

We can calculate, for $X = X^\mu \partial_\mu$ and $Y = Y^\sigma \partial_\sigma$,

$$g(X,Y) = g_{\mu\nu} dx^\mu(X) dx^\nu(Y) = g_{\mu\nu} h_\alpha{}^\mu h^\alpha(X) h_\beta{}^\nu h^\beta(Y)$$

$$= \eta_{\alpha\beta} h^\alpha(X) h^\beta(Y) = \eta_{\alpha\beta} h^\alpha{}_\mu h^\beta{}_\nu X^\mu Y^\nu.$$

Thus, X and Y being arbitrary,

$$g_{\mu\nu} = \eta_{\alpha\beta} h^\alpha{}_\mu h^\beta{}_\nu. \tag{9.30}$$

From this expression and (9.29) we recognize in these pseudo-orthogonal frames the *tetrad fields* $h^\alpha{}_\mu$ (or *four-legs*, or still *vierbeine*). Each base $\{h_\alpha\}$ determines a metric by Eq. (9.30): it "translates" the Minkowski metric into another, riemannian metric. It is important to notice that the tetrads belong to the differentiable structure of the manifold. They are there as soon as some internal product is supposed on the typical tangent space. Unlike connections — to be introduced in next section — they represent no new, additional structure. Their presence will be at the origin of torsion. Only the metric turns up in the Laplace–Beltrami operator appearing in the Klein–Gordon equation of § 9.1.2, which governs the behaviour of boson fields (see Section 31.2.2). Tetrads only come up explicitly in the Dirac equation. This fact makes of fermions privileged objects to probe into tetrad fields. In particular, they exhibit a direct coupling to torsion.

9.4 Linear connections

Connections materialize in the notion of parallel transport. Consider a (piecewise differentiable) curve γ on a manifold. A vector X undergoes parallel transport if it is displaced along γ in such a way that its angle with the curve (i.e. with the tangent to γ) remains constant (Figure 9.2). This intuitive view supposes a metric (see § 9.4.23), but actually a connection suffices. A connection determines a covariant derivative of the type described in § 7.3.11, and the vector is parallel-transported when its covariant derivative vanishes. The covariant derivative can be extended to any tensor, and a tensor is parallel-transported when its covariant derivative vanishes. The notion of parallel transport for general principal bundles will be introduced (in § 9.6.20) in an analogous way.

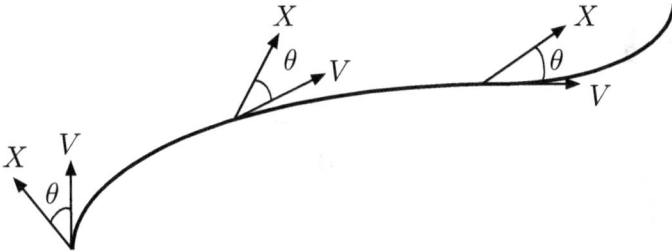

Fig. 9.2 In a metric space, a vector X is parallel–transported if its angle with the curve remains constant

§ **9.4.1** Ask a friend to collaborate in the following experiment.[5] He must stand before you, with his right arm straight against his side but with his thumb pointing at you. He must keep rigidly the relative position of the thumb with respect to the arm: no rotation of thumb around arm axis allowed. He will then (i) lift the arm sideways up to the horizontal position; (ii) rotate the arm horizontally so that it (the arm) points at you at the end; you will be seeing his fist, with the thumb towards your right; (iii) finally he will drop the arm back to his side. The thumb will still be pointing to your right. The net result will be a 90° rotation of the thumb in the horizontal plane. Notice that his hand will have been moving along three great arcs of a sphere S^2 of radius L (the arm's length). Its is just as if you looked

---

[5] This simple example is adapted from Levi 1993.

at the behaviour of a vector on "earth": (a) initially at the south pole S and pointing at you (see Figure 9.3); (b) transported along the rim from S to H, all the time pointing at you; (c) taken along the equator to the meridian just facing you (the vector becomes progressively visible and, at F, it will be entirely toward your right; (d) transported southwards back to S. The net rotation is a measure of earth's curvature. For a small parallel

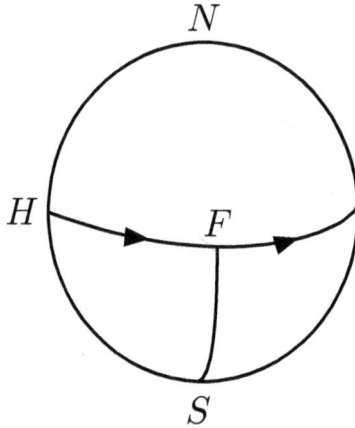

Fig. 9.3   The arms experiment.

displacement dx^k, the variation of a vector X will be given by

$$\delta X^i = -\Gamma^i{}_{jk} X^j dx^k,$$

where $\Gamma^i{}_{jk}$ represents precisely the connection. Along a curve, one must integrate. It so happens that the curvature is the rotational of Γ, and that, in the case above, the curve bounds one-eigth of the whole earth's surface, $4\pi L^2$ (see details in Section 22.4.1). Stokes theorem transforms the line integral into the surface integral of the curvature over the octant. The curvature is constant, and equal to $1/L^2$, so that what remains is just an octant's area divided by L^2, which gives the rotation angle, $\pi/2$. The same procedure, if followed on the plane, which is a flat (zero curvature) space, would take the vector back to its initial position quite unmodified.

§ 9.4.2 There is no such a thing as "curvature of space". This is perhaps still more evident when general connections are introduced (section 9.6). Curvature is a property of a connection, and a great many connections may be defined on the same space. Different particles feel different connections and different curvatures. There might be a point for taking the Levi–Civita connection as part of the very definition of spacetime, as is frequently done. It seems far wiser, however, to take space simply as a manifold, and connection (with its curvature) as an additional structure. We shall here be concerned with the general notion of linear connection and its formal developments, such as the relations involving curvature, frames and torsion. Though nowadays presented in a quite intrinsic way, all this has evolved from the study of subspaces of euclidean spaces. The historical approach, embodied in the so-called "classical differential geometry" is very inspiring and a short account of it is given from Section 22.3.1 on.

§ **9.4.3** A *linear connection* is a G'-valued form Γ leading fundamental fields back to their corresponding generators:

$$\Gamma(J^*) = J.$$

In particular,

$$\Gamma(E_i{}^j) = \Delta_i{}^j. \tag{9.31}$$

As Γ is a form with values on G', it may be written in basis $\{\Delta_i{}^j\}$ as $\Gamma = \Delta_i{}^j \, \Gamma^i{}_j$, with $\Gamma^i{}_j$ usual 1-forms. Then,

$$\Gamma^k{}_l(E_i{}^j) = \delta_i^k \, \delta_l^j. \tag{9.32}$$

A field X on BM such that $\Gamma(X) = 0$ is said to be *horizontal*.

§ **9.4.4** Notice that the definition of vertical fields as in Eq. (9.14) is inherent to the smooth structure and quite independent of additional structure. But horizontal fields are only defined once a connection is given. Distinct connections will define distinct fields as horizontal.

§ **9.4.5 5.** A connection can be proven to satisfy, besides the defining relation given by $\Gamma(J^*) = J$, the covariance condition

$$(R_a^*\Gamma)(X) = Ad_{a^{-1}}\Gamma(X).$$

Conversely, any 1-form on the complete space satisfying both these conditions is a connection. Because of the form of the covariance condition, we say that the connection "belongs" to the adjoint representation.

§ **9.4.6** The presence of a connection has an important consequence on the solder form θ. Once a connection is given, θ becomes an isomorphism of vector spaces (though not of algebras — see § 9.4.8 below). There will be a *unique* set $\{E_i\}$ of horizontal vectors such that

$$\theta(E_i) = K_i \text{ or } \theta^j(E_i) = \delta_i^j. \tag{9.33}$$

Notice that the fields E_i are exactly dual to the solder forms. Also

$$\pi_*(E_i) = b \circ \theta(E_i) = b_i.$$

Thus, on the base manifold and in a given basis $\{b_i\}$, the soldering forms are represented by that base of forms which is dual to $\{b_i\}$.

Given any vector $V = V^j K_j$ in $\mathbb{E}^m$, the horizontal vector $V^j E_j$ on BM is the *basic* or *standard* vector field associated to V. For each frame b, the vectors E_i and $E_i{}^j$ can be shown to be linearly independent, so that the set $\{E_i, E_i{}^j\}$ constitute a basis for $T_b BM$. Although connection-dependent, this basis is independent of the charts. The "supertangent" fiber bundle TBM obtained in this way has a structure of direct product, and the complete space of the bundle of frames is consequently a parallelizable manifold. Any vector at b may be decomposed into a vertical and a horizontal part,

$$X = VX + HX = X^i{}_j E_i{}^j + X^i E_i. \tag{9.34}$$

§ 9.4.7 Actually, the connection might be defined as a form vanishing on the "horizontal" space H_b spanned by the E_i. Using the horizontal projection $X \to HX$, the covariant differential of a p-form ω is that $(p+1)$-form $D\omega$ which, acting on $(p+1)$ vectors $X_1, X_2, \ldots, X_{p+1}$, gives

$$D\omega(X_1, X_2, \ldots, X_{p+1}) = d\omega(HX_1, HX_2, \ldots, HX_{p+1}). \tag{9.35}$$

It is consequently the "horizontalized" version of the exterior derivative. The covariant derivative is thus defined on the bundle complete manifold. In order to "bring it down" to the base manifold, use must be made of a section-induced pull-back.

§ 9.4.8 We can further show that

$$[E_i{}^j, E_k] = \delta^j_k E_i. \tag{9.36}$$

Nevertheless, unlike $\tilde{b}_*$, θ is not an algebra homomorphism. The algebra basis $\{E_i, E_i^j\}$ can be completed by putting

$$[E_i, E_j] = - F_m{}^n{}_{ij} E_n{}^m + T^k{}_{ij} E_k. \tag{9.37}$$

The notation is not without a purpose. The detailed calculations give for the coefficients $F_m{}^n{}_{ij}$ and $T^k{}_{ij}$ the values of the curvature and the torsion, whose meaning we shall examine in the following. Let us retain here that the curvature appears as the vertical part of the commutator of the basic fields, and the torsion as its part along themselves.

The projection π_* is a linear mapping from the start, which becomes an isomorphism of vector spaces (between the horizontal subspace H_b and

$T_{\pi(b)}M$) when a connection is added. The decomposition of $T_b BM$ into horizontal and vertical is, therefore, complete in what concerns the vector space, but not in what concerns the algebra.

§ 9.4.9 Torsion. Given a connection Γ, its *torsion* form T is the covariant differential of the canonical form θ,

$$T = D\theta. \tag{9.38}$$

It is consequently given by

$$T(X,Y) = d\theta(HX, HY) = K_i d\theta^i (X^j E_j, X^k E_k). \tag{9.39}$$

We find that

$$T = -\tfrac{1}{2} K_k T^k{}_{ij}\theta^i \wedge \theta^j. \tag{9.40}$$

In detail, the invariant expression of T is

$$T = d\theta + \Gamma \wedge \theta + \theta \wedge \Gamma. \tag{9.41}$$

If the solder form is written as

$$K_i\theta^i = K_a h^a{}_\mu dx^\mu$$

in a natural basis, the components of the torsion tensor are

$$T^a{}_{\mu\nu} = \partial_\mu h^a{}_\nu - \partial_\nu h^a{}_\mu + \Gamma^a{}_{b\mu}h^b{}_\nu - \Gamma^a{}_{b\nu}h^b{}_\mu. \tag{9.42}$$

Using the property

$$\Gamma^a{}_{b\mu} = h^a{}_\rho \Gamma^\rho{}_{\lambda\mu} h_b{}^\lambda + h^a{}_\rho \partial_\mu h_b{}^\rho, \tag{9.43}$$

the spacetime-indexed torsion is found to be

$$T^\rho{}_{\mu\nu} \equiv h_a{}^\rho T^a{}_{\mu\nu} = \Gamma^\rho{}_{\nu\mu} - \Gamma^\rho{}_{\mu\nu}. \tag{9.44}$$

§ 9.4.10 Curvature. The *curvature* form F of the connection Γ is its own covariant differential:

$$F = D\Gamma = d\Gamma + \Gamma \wedge \Gamma. \tag{9.45}$$

Being G'-valued, its components are given by

$$F = \tfrac{1}{2} \Delta_a{}^b R^a{}_{b\mu\nu}\theta^\mu \wedge \theta^\nu. \tag{9.46}$$

In a natural basis,

$$R^a{}_{b\mu\nu} = \partial_\mu \Gamma^a{}_{b\nu} - \partial_\nu \Gamma^a{}_{b\mu} + \Gamma^a{}_{e\mu}\Gamma^e{}_{b\nu} - \Gamma^a{}_{e\nu}\Gamma^e{}_{b\mu}. \tag{9.47}$$

Using property (9.43), the spacetime-index components of the curvature tensor

$$R^\rho{}_{\lambda\mu\nu} \equiv h_a{}^\lambda h^b{}_\lambda R^a{}_{b\mu\nu}$$

are found to be

$$R^\rho{}_{\lambda\mu\nu} = \partial_\mu \Gamma^\rho{}_{\lambda\nu} - \partial_\nu \Gamma^\rho{}_{\lambda\mu} + \Gamma^\rho{}_{\kappa\mu}\Gamma^\kappa{}_{\lambda\nu} - \Gamma^\rho{}_{\kappa\nu}\Gamma^\kappa{}_{\lambda\mu}. \tag{9.48}$$

§ 9.4.11 From the above expressions, by taking derivatives and reshuffling the terms, we can find the first Bianchi identity

$$DF = dF + [\Gamma, F] = 0. \tag{9.49}$$

We can also find the second Bianchi identity

$$dT + [\Gamma, T] + [\theta, F] = 0. \tag{9.50}$$

When $T = 0$, it reduces to the simple form

$$[\theta, F] = 0. \tag{9.51}$$

§ 9.4.12 A vector field X is parallel-transported along a curve $\gamma(s)$ if it is the projection of a horizontal field all along γ. Its covariant derivative must then vanish. This means that, when displaced of $d\gamma$ along γ, it satisfies

$$\frac{dX^k}{ds} + \Gamma^k{}_{ij} X^i \frac{d\gamma^j}{ds} = 0, \tag{9.52}$$

or

$$dX^k = -\Gamma^k{}_{ij} X^i d\gamma^j. \tag{9.53}$$

A geodesic curve is a self-parallel curve, that is, a curve along which its own tangent vector (velocity) $d\gamma^i/ds$ is parallel-transported. It obeys consequently the geodesic equation

$$\frac{d^2\gamma^k}{ds} + \Gamma^k{}_{ij} \frac{d\gamma^i}{ds} \frac{d\gamma^j}{ds} = 0.$$

§ 9.4.13 More about geodesics will be said in Chapter 24. Here, only a few general aspects of them will interest us. Geodesics provide an easy view of the meaning of curvature. Consider a vector X^k at the point "o", as in Figure 9.4, which is parallel-transported by a connection as given by Eq.(9.53). First, parallel transport X^k from "o" to "a" along the small bit of geodesic $d\alpha'$, and then parallel transport it from "a" to "c" along the small bit of geodesic $d\beta'$. The resulting vector is denoted X'^k. Second, parallel transport X^k from "o" to "b" along the small bit of geodesic $d\alpha''$, and then parallel transport it from "b" to the same point "c" along the small bit of geodesic $d\beta''$, forming in this way an infinitesimal parallelogram The resulting vector is denoted X''^k. The *angle deficit* between vectors X'^k and X''^k, denoted δX^k, is a measure of the curvature:

$$\delta X^k = -R^k{}_{ijl} X^i d\alpha^j d\beta^l. \tag{9.54}$$

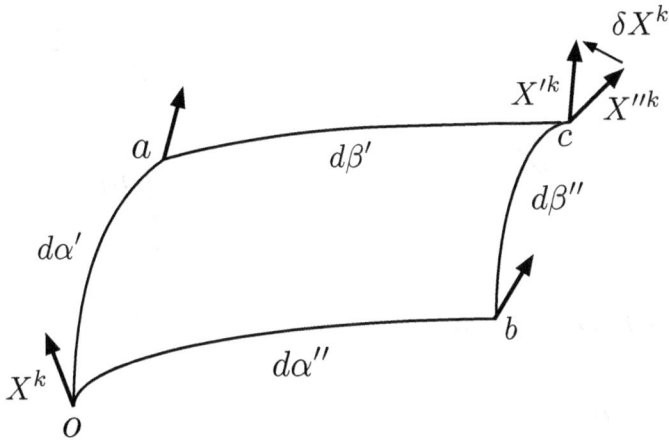

Fig. 9.4   How curvature can be found through the behavior of a vector field.

§ **9.4.14** Geodesics allow also an intuitive view of the effect of torsion, and this in a rather abrupt way: torsion disrupts the above infinitesimal geodesic parallelograms. In effect, if we transport as above the very geodesic bits in the presence of torsion (as in Figure 9.5), we find that a gap between the extremities shows up. This gap, characterized by a *distance deficit* and denoted ΔX^k, is a measure of torsion:

$$\Delta X^k = T^k{}_{ij}\, d\beta^i d\alpha^j. \tag{9.55}$$

Commentary 9.1   When the connection has both curvature and torsion, the parallel transport of a vector X^k will give rise to both an angle and a distance deficit.
◄

§ **9.4.15** Given two fixed fields X and Y, the curvature R can be seen as a family of mappings $R(X, Y)$ taking a field into another field according to

$$R(X,Y)Z = [\nabla_X \nabla_Y - \nabla_Y \nabla_X - \nabla_{[X,Y]}]Z$$
$$= \nabla_X(\nabla_Y Z) - \nabla_Y(\nabla_X Z) - \nabla_{[X,Y]}Z, \tag{9.56}$$

where ∇ is the covariant derivative and ∇_X its projection along X (see Eq. 9.63 below). In the same token, the torsion can be seen as a field-valued 2-tensor T such that

$$T(X,Y) = \nabla_X Y - \nabla_Y X - [X,Y]. \tag{9.57}$$

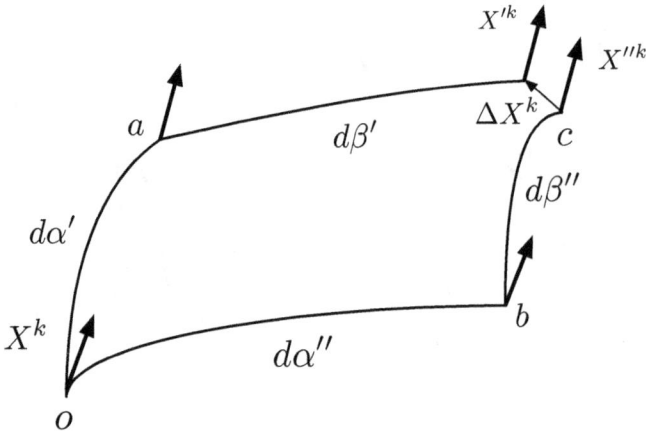

Fig. 9.5   Torsion detected by comparing geodesics.

The Bianchi identities (9.50) or (9.49 become, in this language,

$$(\nabla_X R)(Y, Z) + R(T(X, Y), Z) + \text{(cyclic permutations)} = 0, \qquad (9.58)$$

and

$$(\nabla_Z T)(X, Y) - R(X, Y)Z + T(T(X, Y), Z) + \text{(cyclic permutations)} = 0. \qquad (9.59)$$

The covariant derivative has the properties

$$\nabla_{fX} Y = f \nabla_X Y, \qquad (9.60)$$

and

$$\nabla_{X+Z} Y = \nabla_X Y + \nabla_Z Y. \qquad (9.61)$$

§ 9.4.16 The relation of the above expressions to the more usual component forms is given, for the curvature, by

$$R(e_a, e_b) e_c = R^f{}_{cab} e_f. \qquad (9.62)$$

Let us profit to give some explicit expressions. Take a vector field basis $\{e_a\}$, with

$$[e_a, e_b] = f^c{}_{ab} e_c$$

and find first that

$$\nabla_X Y = X^a [e_a Y^c + Y^b \Gamma^c{}_{ba}] e_c, \qquad (9.63)$$

of which a particular case is

$$\nabla_{e_a} e_b = \Gamma^c{}_{ba} e_c. \tag{9.64}$$

Then, we calculate

$$\nabla_X Y = X^a \nabla_{e_a}(Y^b e_b) = X^a[e_a Y^c + Y^b \Gamma^c{}_{ba}]e_c. \tag{9.65}$$

We find next:

$$[X, Y] = [X^a e_a, Y^b e_b] = [X^a e_a(Y^c) - Y^a e_a(X^c) + X^a Y^b f^c{}_{ab}]e_c \tag{9.66}$$

and

$$\nabla_{[X,Y]} Z = [X(Y^d) - Y(X^d) + X^a Y^b f^d{}_{ab}]\nabla_{e_d} Z. \tag{9.67}$$

Using these expressions, we get

$$R(e_a, e_b)Z = \nabla_{e_a}(\nabla_{e_b} Z) - \nabla_{e_b}(\nabla_{e_a} Z) - f^c{}_{ab}\nabla_{e_c} Z. \tag{9.68}$$

Finally, using Eq. (9.62), we obtain

$$R^f{}_{cab} = e_a(\Gamma^f{}_{cb}) - e_b(\Gamma^f{}_{ca}) + \Gamma^f{}_{da}\Gamma^d{}_{cb} - \Gamma^f{}_{db}\Gamma^d{}_{ca} - f^g{}_{ab}\Gamma^f{}_{cg}. \tag{9.69}$$

In a holonomic basis, curvature assumes the form

$$R(\partial_\mu, \partial_\nu)\partial_\sigma = R^\rho{}_{\sigma\mu\nu}\partial_\rho.$$

Some further useful expressions are:

$$\begin{aligned} R(X, Y)Z &= [\nabla_X \nabla_Y - \nabla_Y \nabla_X - \nabla_{[X,Y]}]Z \\ &= X^a Y^b R(e_a, e_b)Z \\ &= X^a Y^b Z^d R(e_a, e_b)e_d \\ &= X^a Y^b Z^c R^f{}_{cab} e_f\,; \end{aligned} \tag{9.70}$$

$$\begin{aligned} T(e_a, e_b) &= \nabla_{e_a} e_b - \nabla_{e_b} e_a - f^g{}_{ab} e_g \\ &= (\Gamma^g{}_{ba} - \Gamma^g{}_{ab} - f^g{}_{ab})\, e_g\,, \end{aligned} \tag{9.71}$$

and

$$\begin{aligned} T(X, Y) &= X^a Y^b T(e_a, e_b) \\ &= X^a Y^b \{\Gamma^c{}_{ba} - \Gamma^c{}_{ab} - f^c{}_{ab}\}e_c. \end{aligned} \tag{9.72}$$

In a holonomic basis, torsion is written as

$$T(\partial_\mu, \partial_\nu) = T^\rho{}_{\mu\nu}\partial_\rho. \tag{9.73}$$

§ 9.4.17 The *horizontal lift* of a vector field X on the base manifold M is that (unique! see below) horizontal field $X^\#$ on the bundle space which is such that

$$\pi_*(X_b^\#) = X_{\pi(b)}$$

for all b on BM. We only state a few of the most important results concerning this notion:

    (i) For a fixed connection, the lift is unique;

    (ii) The lift of the sum of two vectors is the sum of the corresponding lifted vectors;

    (iii) The lift of a commutator of two fields is the horizontal part of the commutator of the corresponding lifted fields.

§ 9.4.18 A *horizontal curve* on BM is a curve whose tangent vectors are all horizontal. The horizontal lift of a smooth curve

$$\gamma : [0,1] \to M, \quad t \to \gamma_t$$

on the base manifold M is a horizontal curve $\gamma_t^\#$ on BM such that

$$\pi(\gamma_t^\#) = \gamma_t.$$

Given $\gamma(0)$, there is a unique lifted curve starting at a chosen point $b_0 = \gamma_0^\#$. This point is arbitrary in the sense that any other point on the same fiber will be projected on γ_0.

§ 9.4.19 Parallel transport (or *parallel displacement*) along a curve: take a point b_0 on BM as above, such that

$$\pi(b_0) = \gamma_0.$$

The unique horizontal lift going through b_0 will have an end point b_1, on another fiber $\pi^{-1}(\gamma_1)$, or, if we prefer, such that

$$\pi(b_1) = \gamma_1 = \gamma(1).$$

Now, if we vary the point b_0 on the *initial* fiber $\pi^{-1}(\gamma_0)$, we shall obtain other points on the *final* fiber $\pi^{-1}(\gamma_1)$. This defines a mapping $\gamma^\#$ between the two fibers. As any horizontal curve is mapped into another horizontal curve by the group action R_g,

$$\gamma^\# \circ R_g = R_g \circ \gamma^\#,$$

then the mapping

$$\gamma^\# : \pi^{-1}(\gamma_0) \to \pi^{-1}(\gamma_1)$$

is an isomorphism. This isomorphism, by which each point of the initial fiber is taken into a point of the final fiber, is the parallel displacement along the curve γ. In these considerations, it is only necessary that the curve be piecewise C^1.

§ **9.4.20 Formal characterization.** Recall the mapping (9.6): each frame b is seen as a map from $\mathbb{E}^m$ to $T_{\pi(b)}M$, which "appoints" the base member b_j as the representative of the j-th canonical vector K_j of $\mathbb{E}^m$. And it will take a general vector $X = X^j K_j$ of the typical fiber $\mathbb{E}^m$ into

$$b(X) = X^j b_j$$

on M. Also vice-versa: given any vector $V = V^j b_j$ tangent to M at $p = \pi(b)$, its inverse will provide a vector

$$b^{-1}(V) = V^j K_j$$

on $\mathbb{E}^m$. We may call $b^{-1}(V)$ the "paradigm" of the tangent vector V. Now, each frame b is a point on BM. Consider $b_0 \in \pi^{-1}(\gamma_0)$ as above. It will put any V_p of $T_{\pi(b)}M$ into correspondence with an euclidean vector

$$V_0 = b_0^{-1}(V_p).$$

The same will be true of any point $b(t) = b_t$ along the horizontal lift of the curve γ_t. The parallel-transported vector V_t at each point $\gamma(t)$ of the curve is defined as

$$b_t(V_0) = b_t[b_0^{-1}(V_p)].$$

Thus, at each point, one takes the corresponding nominee of the same euclidean vector one started with. We say then that "V is kept parallel to itself".

§ **9.4.21 Associated bundles.** (a more formal approach) The considerations on horizontal and vertical spaces, parallel displacements, etc, may be transferred to any associated bundle AM on M, with typical fiber F. Let us first give a formal definition of such an associated bundle. We start by defining a right-action of the structure group G on $BM \times F$ as follows: given $(b, v) \in BM \times F$, we define

$$R_g : (b, v) \to (bg, g^{-1}v),$$

for each $g \in G$. Then AM may be defined as the quotient of $BM \times F$ by this action. There is then a natural mapping of $BM \times F$ into AM, given by the quotient projection. Given a point u on AM, the vertical space is defined as the space tangent to F at u. Things are more involved for the horizontal space. Consider the above natural mapping of $BM \times F$ into AM and choose a point $(b, v) \in BM \times F$ which is mapped into u. Fix $v \in F$ and this will become a mapping of BM into AM. Then the horizontal subspace in AM is defined as the image of the horizontal subspace of BM by this new mapping.

These considerations allow one to define the *covariant derivative of a section* along a curve on an associated bundle AM, that is, of any tensor field, given a connection on BM. Lifts on AM are defined in the same way as those of BM. A section of AM on a curve γ_t will be given by a mapping σ such that

$$\pi_{AM} \circ \sigma(\gamma_t) = \gamma_t$$

all along the curve (Figure 9.6). For each fixed t, call $\gamma_t^{\#t+\varepsilon}$ the parallel displacement of the fiber $\pi_{AM}^{-1}(\gamma_{t+\varepsilon})$ from $\gamma_{t+\varepsilon}$ to γ_t. Then the covariant derivative, which measures how much the section deviates from horizontality in an infinitesimal displacement, is

$$D_{\gamma_t}\sigma = \lim_{\varepsilon \to 0} \frac{1}{\varepsilon} \{\gamma_t^{\#t+\varepsilon}[\sigma(\gamma_{t+\varepsilon})] - \sigma(\gamma_t)\}. \tag{9.74}$$

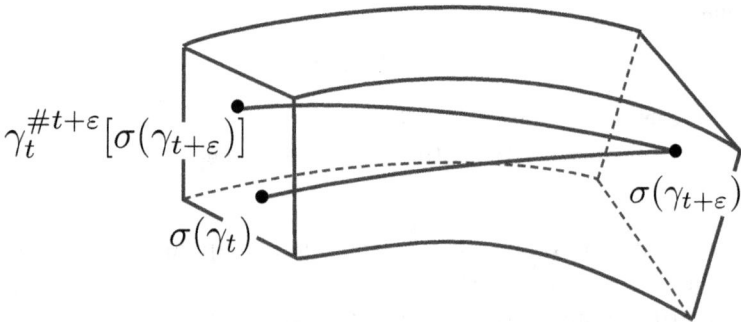

Fig. 9.6 Covariant derivative of a section along a curve.

The covariant derivative in the direction of a field X at a point p is the covariant derivative along a curve which is tangent to X at p. From the above definition of covariant derivative, a section (in general, a tensor) is said to be parallel-displaced along a curve iff the corresponding covariant derivative vanishes. When its covariant derivative is zero along any field, it will be parallel-transported along any curve. In this case, we say that the connection "preserves" the tensor.

The definition is actually an usual derivative, only taking into account the whole, invariant version of a tensor. This means that, besides derivating the components, it derivates also the basis members involved. Take a tensor like $T = T^\rho{}_\sigma e_\rho \otimes w^\sigma$. Its covariant derivative will be

$$DT = dT^\rho{}_\sigma e_\rho \otimes w^\sigma + T^\rho{}_\sigma de_\rho \otimes w^\sigma + T^\rho{}_\sigma e_\rho \otimes dw^\sigma.$$

Using Eqs.(7.81) and (7.88) for the adapted frames of Section 22.2.3, it becomes

$$DT = [e_\lambda(T^\rho{}_\sigma) + \Gamma^\rho{}_{\mu\lambda}T^\mu{}_\sigma - \Gamma^\nu{}_{\sigma\lambda}T^\rho{}_\nu]\, w^\lambda \otimes e_\rho \otimes w^\sigma.$$

The covariant derivative along a curve will be the contraction of this derivative with the vector tangent to the curve at each point (that is, if u is the tangent field, its index will be contracted with the derivative index). In the above example, it will be

$$D_u T = u^\lambda \left[e_\lambda(T^\rho{}_\sigma) + \Gamma^\rho{}_{\mu\lambda}T^\mu{}_\sigma - \Gamma^\nu{}_{\sigma\lambda}T^\rho{}_\nu \right] e_\rho \otimes w^\sigma.$$

§ **9.4.22** In this way, the above notion of covariant derivative applies to general tensors, sections of associated bundles of the frame bundle, and gives the usual expressions in terms of components (say) in a natural basis, duly projected along the curve, that is, contracted with its tangent vector. Say, for a covariant vector, we find the expressions used in § 6.6.13,

$$D_\nu X_\mu \equiv X_{\mu;\nu} = \partial_\nu X_\mu - \Gamma^\alpha{}_{\mu\nu}X_\alpha. \tag{9.75}$$

This semicolon notation for the covariant derivative, usual among physicists, does not include the antisymmetrization. In invariant language,

$$DX = \tfrac{1}{2}\, D_{[\nu}X_{\mu]}dx^\mu \wedge dx^\nu.$$

For a contravariant field,

$$D_\nu X^\mu = X^\mu_{;\nu} = \partial_\nu X^\mu + \Gamma^\mu{}_{\alpha\nu}X^\alpha. \tag{9.76}$$

§ **9.4.23 The Levi–Civita connection.** The covariant derivative of a metric tensor will have components

$$D_\lambda g_{\mu\nu} = g_{\mu\nu;\lambda} = \partial_\lambda g_{\mu\nu} - \Gamma^\alpha{}_{\mu\lambda}g_{\alpha\nu} - \Gamma^\alpha{}_{\nu\lambda}g_{\mu\alpha}. \tag{9.77}$$

This will vanish when the connection preserves the metric. The components of the torsion tensor in a natural basis are $T^\alpha{}_{\mu\lambda} = \Gamma^\alpha{}_{\lambda\mu} - \Gamma^\alpha{}_{\mu\lambda}$. In principle, there exists an infinity of connections preserving a given metric, but only one of them has vanishing torsion. In this case, the connection is symmetric in the lower indices and we can solve the above expression to find

$$\Gamma^\alpha{}_{\mu\nu} = \Gamma^\alpha{}_{\nu\mu} = \tfrac{1}{2}\, g^{\alpha\beta}[\partial_\mu g_{\beta\nu} + \partial_\nu g_{\beta\mu} - \partial_\beta g_{\mu\nu}], \tag{9.78}$$

just the Christoffel symbol of Eq. (6.74). Summing up: given a metric, there exists a unique torsionless connection which preserves it, whose components in a natural basis are the usual Christoffel symbols and is called "the Levi–Civita connection of the metric". Usual riemannian curvature is the curvature of this connection, which is the connection currently used in General Relativity. The hypothesis of gravitation universality gives priority to this connection, as it says that all particles respond to its presence in the same way.

It is with respect to this connection that parallel transport acquires the simple, intuitive meaning of the heuristic introduction: a vector is parallel-displaced along a curve if its modulus and its angle with the tangent to the curve remain constant. Of course, measuring modulus and angle presupposes the metric. And it is the curvature of this connection which is meant when one speaks of the "curvature of a (metric) space". The discovery of "curved spaces", or non-euclidean geometries (Chapter 23), has been historically the germ of modern geometry.

§ **9.4.24** Consider a manifold M and a point $p \in M$. We define the symmetry s_p at p as a diffeomorphism of a neighbourhood U of p into itself which sends $\exp(X)$ into $\exp(-X)$ for all $X \in T_pM$. This means in particular that normal coordinates change signs. When such a symmetry exists, the space is said to be "locally symmetric".

Suppose then that a linear connection Γ is defined on M. We denote M with this fixed connection by (M, Γ). A differentiable mapping f of M into itself will be an 'affine transformation' if the induced mapping

$$f_* : TM \to TM$$

maps horizontal curves into horizontal curves. This means that f_* maps each parallel vector field along each curve γ into a parallel vector field along the curve $f(\gamma)$.

The affine transformations on M constitute a Lie group. If the symmetry s_p above is an affine transformation, (M, Γ) is an 'affine locally symmetric manifold'. This only happens when Γ's torsion is $T = 0$ and its curvature satisfies $\nabla R = 0$.

On the other hand, (M, Γ) is said to be an 'affine symmetric manifold' if, for each $p \in M$, the symmetry s_p can be extended into a global affine transformation (compare with section 8.2.7). On every affine symmetric manifold M the group of affine transformations acts transitively. Thus, M may be seen as a homogeneous space, $M = G/H$. The connection on G will be the torsion-free connection above referred to.

9.5 Principal bundles

In a principal bundle the fiber is a group G. Other bundles with G as the structure group are "associated" to the principal. General properties are better established first in principal bundles and later transposed to the associated. The paradigm is the linear frame bundle.

§ 9.5.1 We have already met the standard example of principal fiber bundle, the bundle of linear frames

$$BM = (M, B_P M, GL(m, R), \pi_B), \qquad (9.79)$$

in which the fiber $B_p M$ is isomorphic to the structure group.

Let (M, F, G, π) be a vector bundle, with G acting on F on the left. We can obtain a new bundle by replacing F by G and considering the left action of G on itself. Such will be a *principal bundle*, indicated by (M, G, π), and the bundle (M, F, G, π) is said to be *associated* to it. We have already seen that the tangent bundle TM is associated to BM.

Recall the formal definition of an associated bundle for BM in § 9.4.21. It is a particular example of the general definition, in which, as there, we start by defining a right-action of the structure group G on $P \times F$ as follows: given $(b, v) \in P \times F$, we define

$$R_g : (b, v) \to (bg, g^{-1}v),$$

for each $g \in G$. Then the associated bundle AM, with the fiber F on which a representation of G is at work, is defined as the quotient

$$AM = (P \times F)/R_g.$$

There is then a natural mapping $\xi : P \times F \to AM$. Parametrizing $b = (p, g)$, ξ is the quotient projection

$$\xi(b, v) = \xi((p, g), v) = (p, v).$$

Or, if we prefer,

$$\xi(b, v) = (\text{class of } b \text{ on } (P, v)) = (\text{orbit of } b \text{ by the action of } G, v).$$

§ 9.5.2 Conversely, consider a principal bundle (M, G, π). Take a vector space F which carries a faithful (i.e, isomorphic) representation ρ of G:

$$\rho : G \to Aut(F), \quad \rho \to \rho(g). \qquad (9.80)$$

The space F may be, for example, a space of column vectors on which G is represented by $n \times n$ matrices $\rho(g)$:

$$gf := \rho(g)f. \qquad (9.81)$$

A bundle (M, F, G, π) is got in this way, which is associated to (M, G, π). Notice that different representations lead to different associated bundles. There are therefore infinite such bundles for each group.

§ **9.5.3** Locally, a point in (M, F, G, π) will be represented by coordinates

$$(p, f) = (x^1, x^2, \ldots, x^m, f^1, \ldots, f^n).$$

A LSC transformation will lead to some (p', f'). Both f and f' belong to F. If the action of G on F is (simply) transitive, there will be a (unique) matrix $\rho(g)$ such that

$$f' = \rho(g)f.$$

Thus, the group action accounts for the LSC transformations in the fiber.

§ **9.5.4** A point on the principal bundle will be "found" by a section σ,

$$(p, f) = \sigma(p) \tag{9.82}$$

with

$$\pi(p, f) = \pi \circ \sigma(p) = p. \tag{9.83}$$

§ **9.5.5** Let us now proceed to the formal definition. It requires a lot of things in order to ensure that the bundle as a whole is a differentiable manifold. A C^∞ principal fiber bundle is a triplet

$$P = (M, G, \pi)$$

such that P (the complete space) and M (the base space) are C^∞ differentiable manifolds, and G (the structure group) is a Lie group satisfying the following conditions:

(1) G acts freely and effectively on P on the right,

$$R_g : P \times G \to P.$$

This means that no point of P is fixed under the action, and no subgroup of G is the stability group of some point of P.

(2) M is the quotient space of P under the equivalence defined by G, $M = P/G$; the projection $\pi : P \to M$ is C^∞; for each p of M, G is simply transitive on the fiber $\pi^{-1}(p)$; so, the fiber is homogeneous, and we say that "the group preserves the fiber".

(3) P is locally trivial: for every p of M, there exists a neighbourhood $U \ni p$ and a C^∞ mapping

$$F_U : \pi^{-1}(U) \to G$$

such that F_U commutes with R_g for every g in G. The combined mapping

$$f_U : \pi^{-1}(U) \to U \times G, \quad f_U(b) = \big(\pi(b), F_U(b)\big) \qquad (9.84)$$

is a diffeomorphism, called a *trivialization*. Notice that there is a F_U for each U (see Figure 9.7).

With the first condition, we may generalize here the mapping b of Eq. (9.6). When the action is free and effective, there is a Lie algebra isomorphism between the group Lie algebra and the tangent to the fiber (§ 8.4.8). Take an associated bundle AM, with a typical fiber F which is usually a vector space and whose copies in the bundle we shall call "realized fibers". Choose a starting basis on F. A point b of the principal bundle may be seen as a mapping from F into the realized fiber on $\pi(b)$, with image $z_b = L_b z_0$, and with z_0 indicating the "zero-section", which will be defined in § 9.5.8. The group identity will deputize

$$z_e = L_e z_0 = z_0$$

as the set of representatives of the starting basis members. Each member of F will be thus translated into a point of the realized fiber, and each point z of the realized fiber will "come" from a member $b^{-1}(z)$ of F, its "paradigm". The typical fiber F is in this way "installed" on $\pi(b)$.

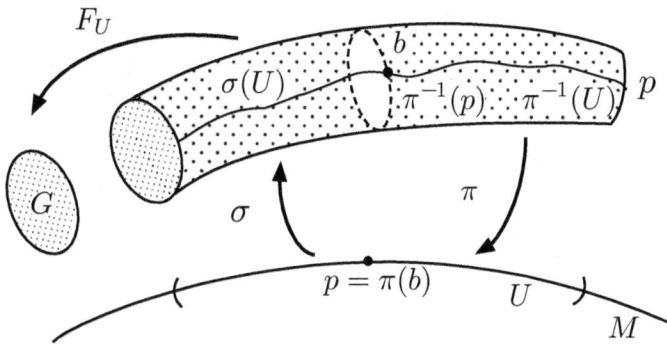

Fig. 9.7   A bundle trivialization.

§ **9.5.6** We have seen in sections 8.2 and 8.4 that, when a group G acts on a manifold M, each point p of M defines a mapping

$$\tilde{p} : G \to M, \quad \tilde{p}(g) = pg = R_g(p).$$

If the action is free and we restrict ourselves to orbits of p, this mapping is a diffeomorphism. Well, given a point b in a fiber, every other point in the fiber is on its orbit since G acts in a simply transitive way. Thus, $\tilde{b}$ is a diffeomorphism between the fiber and G,

$$\tilde{b} : G \to \pi^{-1}(\pi(b)) \subset P, \quad \tilde{b}(g) = R_g(b) = bg. \tag{9.85}$$

§ **9.5.7** To say that $M = P/G$ is to say that $\pi(bg) = \pi(b)$. To say that F_U commutes with R_g means that $F_U(bg) = F_U(b)g$, or

$$F_U \circ R_g(b) = R_g \circ F_U(b).$$

As a consequence (see Figure 9.8),

$$\pi_* \circ R_{g*}(X_b) = \pi_*(X_b) \tag{9.86}$$

and

$$F_{U*} \circ R_{g*} = R_{g*} \circ F_{U*} . \tag{9.87}$$

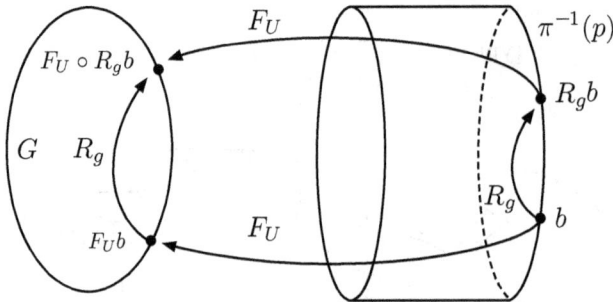

Fig. 9.8   The interplay of trivialization, projection and group action.

§ **9.5.8** An important theorem is

A bundle is trivial if and only if there exists a global C^∞ section.

That is, if there exists a C^∞ mapping $\sigma : M \to P$ with $\pi \circ \sigma = id_M$. In the general case, sections are only locally defined. Each trivialization defines a special local section: for $p \in M$ and $b \in P$ with $\pi(b) = p$, such a section is given by

$$\sigma_U : U \to \pi^{-1}(U), \quad \sigma_U(p) = b[F_U(b)]^{-1} \tag{9.88}$$

so that

$$b = \sigma_U(p) F_U(b). \tag{9.89}$$

Thus, if $F_U(b) = g$, then

$$\sigma_U(p) = bg^{-1}.$$

But then

$$F_U(\sigma_U(p)) = F_U(bg^{-1}) = F_U(b)[F_U(b)]^{-1} = e.$$

This section takes p into a point $\sigma_U(p)$ such that

$$f_U(\sigma_U(p)) = (p, e).$$

It is called the *zero section* of the trivialization f_U.

§ **9.5.9** F_U is a mapping of P into G. When restricted to a fiber, it is a diffeomorphism. Within this restriction, coordinate changes are given by the *transition functions*

$$g_{UV} : U \cap V \to G, \quad g_{UV}(p) = F_U(b)[F_V(b)]^{-1}. \tag{9.90}$$

They are C^∞ mappings satisfying

$$g_{UV}(p)\, g_{VW}(p) = g_{UW}(p). \tag{9.91}$$

Notice that something like $[F_U(b)]^{-1}$ will always take b into the point (p, e), in the respective trivialization, but the point corresponding to the identity "e" may be different in each trivialization. And, as each chart (U, x) around p will lead to a different trivialization, the point on the fiber corresponding to e will be different for each chart. To simplify the notation, it is usual to write $F_U(x)$, where $x \in \mathbb{E}^m$ is the coordinate of $p = \pi(b)$ in the chart (U, x).

The bundle commonly used in gauge theories has an atlas $(U_i, x_{(i)})$ with all the U_i identical. Changes of LSC reduce then to changes in the coordinate mappings, and the transition functions g_{UV} represent exactly the local gauge transformations,[6] which correspond here to changes of zero sections. From

$$b = \sigma_U(p) F_U(b) \quad \text{and} \quad \sigma_V(p) = b F_V^{-1}(b)$$

it follows that

$$\sigma_V(p) = \sigma_U(p) F_U(b) F_V^{-1}(b) = \sigma_U(p) g_{UV}(p), \tag{9.92}$$

---

[6] Wu & Yang 1975.

which shows precisely how sections change under LSC transformations. Equation (9.90) says furthermore that, given ξ in the fiber,

$$\xi_U = g_{UV}\xi_V.\tag{9.93}$$

If ξ is a column vector in fiber space, this is written in matrix form:

$$\xi'_i = g_{ij}\,\xi_j.\tag{9.94}$$

Gauge transformations are usually introduced in this way: source fields φ belonging to some Hilbert space are defined on Minkowski space. They transform according to (9.94),

$$\varphi'_i(x) = S_{ij}(x)\varphi_j(x).\tag{9.95}$$

This means that LSC transformations are at work only in the "internal" spaces, the fibers. The fields φ carry a representation of the gauge group in this way. In the fiber, a change of LSC can be looked at in two ways: either as a coordinate change as above or as a point transformation with fixed coordinates. The non-triviality appears only in the point dependence of the transition function.

§ **9.5.10** Giving the coordinate neighbourhoods and transition functions completely characterizes the bundle. It is possible to show that:

(i) if either G or M is contractible, the bundle is trivial;

(ii) if a principal bundle is trivial, every bundle associated to it is trivial.

§ **9.5.11 Sub-bundles.** Fiber bundles are differentiable manifolds. Can we introduce the notion of immersed submanifolds, while preserving the bundle characteristics? In other words, are there sub-bundles? The answer is yes, but under rather strict conditions. We have seen a particular case in § 9.3.4. As there is a lot of structure to preserve, including group actions, we must begin by defining homomorphisms between bundles.

Given two bundles P and P', with structure groups G and G', a bundle homomorphism between them includes a mapping $f : P \to P'$ and a group homomorphism $h : G \to G'$, with

$$f(bg) = f(b)h(g).\tag{9.96}$$

If f is an immersion (an injection) and h is a monomorphism (an injective homomorphism), then we have an immersion of P in P'. In this case, P is a sub-bundle of P'. If furthermore P and P' have the same base space, which remains untouched by the homomorphism, then we have a *group reduction*, and P is a *reduced bundle* of P'.

§ 9.5.12 Induced bundles. We may require less than that. Suppose now a bundle P with base space B. If both another manifold B' and a continuous mapping $f : B' \to B$ are given, then by a simple use of function compositions and pull-backs one may define a projection, as well as charts and transition functions defining a bundle P' over B'. It is usual to say that the map f between base spaces induces a mapping $f_* : P' \to P$ between complete spaces and call $P' = f^*P$ the induced bundle, or the "pull-back" bundle.

Suppose there is another base-space-to-be B'' and another analogous map $f' : B'' \to B$ leading to a bundle P'' over B'' in just the same way. If $B' = B''$ and the maps are homotopic, then P' and P'' are equivalent. Such maps are used to demonstrate the above quoted results on the triviality of bundles involving contractible bases and/or fibers. They are also used to obtain general bundles as induced bundles of Stiefel manifolds, which allows their classification (§ 9.7.2).

§ 9.5.13 It might seem that the above use of a general abstract group is far-fetched and that Physics is concerned only with transformation groups acting on "physical" spaces, such as spacetime and phase spaces. But the above scheme is just what appears in gauge theories (see Section 32.2 on). Gauge groups (usually of the type $SU(N)$) are actually abstract, acting on some "internal" spaces of wavefunctions defined on Minkowski base space. The first statement in § 9.5.10 would say that, if Minkowski space is contractible, the bundles involved in gauge theories are trivial if no additional constraints are imposed via boundary conditions.

9.6   General connections

A connection is a structure defined on a principal bundle. We have called "linear connections" those connections on the bundle of linear frames. Let us now quote the main results on connections in general.

§ 9.6.1 Consider the tangent structure of the complete bundle space P. At a given point b, $T_b P$ has a well defined decomposition into a vertical space V_b, tangent to the fiber at b and its linear complement, which we shall (quite prematurely) call the horizontal space H_b. The vertical space is defined as

$$V_b = \{X \in T_b P \text{ such that } \pi_* X = 0\}. \tag{9.97}$$

In words: π_* projects a vector on P into a vector on the base space M. A vertical vector lies along the fiber and projects into the zero of $T_{\pi(b)}M$. As to H_b, the mere fact that it is the linear complement to V_b fixes it at b, but is not sufficient to determine it in other points in a neighbourhood of b.

§ **9.6.2** The mapping $\tilde{b} : G \to P$ given in Eq. (9.85) induces the differential

$$d\tilde{b} = \tilde{b}_* : T_g G \to T_{bg} P$$

or

$$\tilde{b}_* : X_g \to \overline{X}_{bg},$$

taking fields on G into fields on P. The composition

$$\pi \circ \tilde{b} : G \to M$$

is a constant mapping, taking all the points of G into the same point $\pi(b)$ of M:

$$(\pi \circ \tilde{b})(g) = \pi(bg) = \pi(b).$$

Thus,

$$d((\pi \circ \tilde{b}) = \pi_* \circ \tilde{b}_* = 0.$$

Consequently, given any X on G, $\tilde{b}_*(X)$ is vertical. Recall what was said in § 8.4.8: applied to the present case, G is acting on P and $\tilde{b}_*(X)$ is a fundamental field. Given the generators $\{J_a\}$ in $G' = T_e G$, $\tilde{b}_*$ will take them into the fundamental fields

$$\bar{J}_a = \tilde{b}_*(J_a) \in T_{be} P = T_b P. \tag{9.98}$$

Each $\bar{J}_a$ will be a vertical field and the algebra generated by $\{\bar{J}_a\}$, which is tangent to the fiber, will be isomorphic to G' and will represent it on P. Given a vertical field $\overline{X}$, there is a unique X in G' such that

$$\overline{X} = \tilde{b}_*(X).$$

The field $\overline{X}$ is said to be "engendered" by X, which it "represents" on P. The set $\{\bar{J}_a\}$ may be used as a basis for V_b.

§ **9.6.3** A fundamental field $\overline{X}$, under a group transformation, will change according to Eq. (8.35):

$$R_{g*}(\overline{X}) = \overline{Ad_{g^{-1}}(X)} = \tilde{b}_*[Ad_{g^{-1}}(X)]. \tag{9.99}$$

Strong relations between the tangent spaces to P, M and G arise from the mappings between them. A trivialization around b will give

$$f_U(b) = (\pi, F_U)(b) = (\pi(b), F_U(b)) = (p, g).$$

It is convenient to parametrize b by writing simply $b = (p, g)$. The mapping $\tilde{b}_*$ will take a X_e in G' into its fundamental representative

$$\tilde{b}_*(X_e) = \overline{X}_b = \overline{X}_{(p,g)}. \tag{9.100}$$

If $b' = (p, e)$, then $\tilde{b}'_*(X_e) = \overline{X}_{(p,e)}$. But

$$\overline{X}_{(p,g)} = R_{g*}\overline{X}_{(p,e)} = R_{g*}\tilde{b}_*(X_e).$$

This is also

$$\overline{X}_{(p,g)} = \tilde{b}_*(X_g) = \tilde{b}_* \circ R_{g*}(X_e).$$

Recall that F_U commutes with R_g.

§ **9.6.4** A general vector X_b can be decomposed into vertical and horizontal components,

$$X_b = V_b X + H_b X. \tag{9.101}$$

Trouble comes out precisely when we try to separate these components. The local trivialization shows, as discussed above, that a part of $V_b X$ comes from G' as a fundamental field. We have obtained a purely vertical contribution to X_b, that coming from G'. But another contribution may come from M, the base space. Let us see how.

Take a field $X_p \in T_p M$. It can be lifted to $T_b M$ by a section, such as the zero section of Eq. (9.89):

$$\sigma_{U*} X_p = \widehat{X}_{(p,e)}. \tag{9.102}$$

To obtain $\widehat{X}_{(p,g)}$, it is enough to displace the point by the group action,

$$\widehat{X}_b = \widehat{X}_{(p,g)} = R_{g*} \circ \sigma_{U*}(X_p). \tag{9.103}$$

A field X_b on the complete space will thus have two contributions: this one from $T_p M$, and another, purely vertical, coming from G' as in Eq. (9.100):

$$X_b = R_{g*} \circ \sigma_{U*}(Y_p) + \tilde{b}_*(X_e) = \widehat{X}_b + \overline{X}_b. \tag{9.104}$$

And now comes the trouble: $\overline{X}_b$ is purely vertical by construction, but $\widehat{X}_b$ is not necessarily purely horizontal. It may have a vertical part (Figure 9.9). The total vertical component will be, consequently, $\overline{X}_b$ plus some contribution from $\widehat{X}_b$.

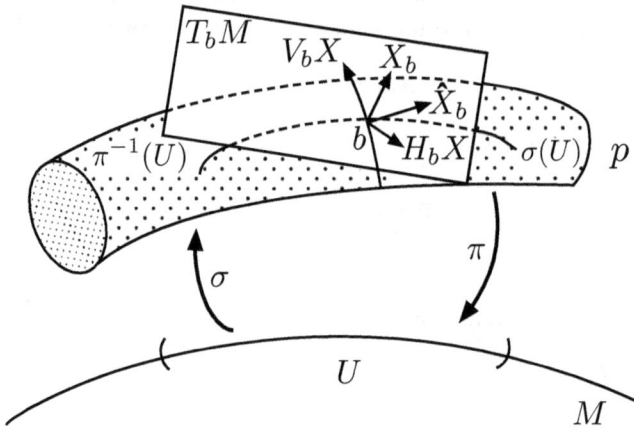

Fig. 9.9   The lift of any field can have a vertical part.

§ **9.6.5** Consequently, the decomposition (9.101) is, in the absence of any further structure, undefined. Putting it into another language: P is locally a direct product $U \times G$ and we are looking for subspaces of $T_b P$ which are tangent to the fiber ($\approx G$) and to an open set in the base space ($\approx U$). The fields tangent to a submanifold must constitute a Lie algebra by themselves, because a submanifold is a manifold. The horizontal fields do not form a Lie algebra: the commutator of two of them has vertical components. When acted upon (through the Lie derivative) by another horizontal field, a horizontal field $H_b X$ comes up with a vertical component. In a neighbouring point b', $H_{b'} X$ will not necessarily be that same field $H_b X$.

§ **9.6.6** Summing up: in a bundle space it is impossible to extricate vertical and horizontal subspaces without adding further structure. The additional structure needed is a connection. Strictly speaking, a connection is a differentiable distribution (§ 6.4.33), a field H of ("horizontal") subspaces H_b of the tangent spaces $T_b P$ satisfying the conditions:

(i) At each b, H_b is the linear complement of V_b, so that every vector may be decomposed as in Eq. (9.101).
(ii) H_b is invariant under the right action of G, that is, $R_{g*} H_b = H_{bg}$.

This field of horizontal subspaces is the kernel of a certain 1-form which completely characterizes it. It is indeed quite equivalent to this 1-form, and we shall prefer to define the connection as this very form.

§ **9.6.7** A *connection* is a 1-form on P with values on G' which spells out the isomorphism between G' and V_b:

$$\Gamma : V_b \to G'.$$

Its explicit form is

$$\Gamma(X_b) = Z \in G' \text{ such that } \tilde{b}_* Z = X_b, \text{ or } \overline{Z} = X_b. \tag{9.105}$$

It is a "vertical" form: when applied to any horizontal field, it vanishes:

$$\Gamma(H_b X) = 0 \; \forall X. \tag{9.106}$$

In particular, from (9.98),

$$\Gamma(\bar{J}_a) = J_a. \tag{9.107}$$

§ **9.6.8** Being a form with values in G', it can be written

$$\Gamma = J_a \, \Gamma^a \tag{9.108}$$

with

$$\Gamma^a(\bar{J}_b) = \delta_b^a. \tag{9.109}$$

It is, in a restricted sense, the inverse of $\tilde{b}_*$. It is *equivariant* (§ 8.2.19) because it transforms under the action of G in the following way: for any fundamental field $\overline{X}$,

$$(R_{g*}\Gamma)(\overline{X}) = \Gamma(R_{g*}\overline{X}) = \Gamma(\overline{Ad_{g^{-1}}X})$$
$$= \Gamma \circ \tilde{b}_*(Ad_{g^{-1}}X) = Ad_{g^{-1}}X = Ad_{g^{-1}}\Gamma(\overline{X}).$$

Therefore,

$$\Gamma \circ R_{g*} = R_{g*}\Gamma = Ad_{g^{-1}}\Gamma, \tag{9.110}$$

or equivalently

$$\Gamma(R_{g*}\overline{X}) = (R_{g*}\Gamma)(\overline{X}) = (g^{-1}J_a g)\,\Gamma^a(\overline{X}). \tag{9.111}$$

A scheme of the various mappings involved is given in Figure 9.10.

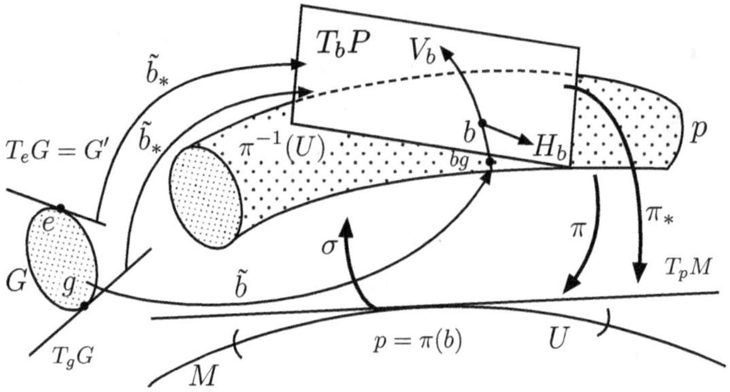

Fig. 9.10   General, local view of a differentiable bundle.

§ 9.6.9 A connection form defines, through (9.106), horizontal spaces at every point b of P. Notice that, in principle, there is an infinity of possible connections on P, and distinct connections will determine different horizontal spaces at each b.

Gauge potentials are connections on principal fiber bundles (a statement to be slightly corrected in § 9.6.13). From a purely geometrical point of view, they are quite arbitrary. Only under fixed additional *dynamical* conditions (in the case, they must be solutions of the dynamical Yang-Mills equations with some boundary conditions) do they become well determined.

In order to transfer the decomposition into vertical and horizontal spaces to an associated bundle, we again recall what happened in the frame bundle case (§ 9.4.21). Given a point u on an associated bundle AM with fiber F, the vertical space is simply defined as the space tangent to F at u. For the horizontal space, we take the natural mapping ξ of $P \times F$ into AM (§ 9.5.1) and chose a point $(b, v) \in P \times F$ such that

$$\xi(b, v) = u.$$

Fix v: this will become a mapping ξ_v of P into AM,

$$\xi_v(b) = u'.$$

Then, the horizontal subspace in AM is defined as the image of the horizontal subspace of P by ξ_{v*}.

§ 9.6.10 Well, all this may be very beautiful, but Γ is a form on P, it "inhabits" (the cotangent bundle of) the bundle space. We need a formulation in terms of forms on the base space M. In order to bring Γ "down" to M, we will be forced to resort to the use of sections — the only means to "pull Γ back to earth". Let us consider two zero sections related to two charts on M, with open sets U and V. They will be related by Eq. (9.92),

$$\sigma_V(p) = \sigma_U(p)g_{UV}(p),\tag{9.112}$$

in which the transition function g_{UV}, given by (9.90), mediates the transformation of section σ_U to section σ_V. It is a mapping

$$g_{UV} : U \cap V \to G.$$

Its differential will take a field X on M into a field on G,

$$g_{UV*} : T_pM \to T_{g_{UV}}G,$$
$$g_{UV*} : X \to g_{UV*}(X).\tag{9.113}$$

§ 9.6.11 Recall the behavior of the Maurer–Cartan form w on G: when applied to a field on G, it gives the same field at the identity point, that is, that same field on G' (Section 8.3.5 and on). It will be necessary to pull w back to M, which will be done by g_{UV}. We shall define the G'-valued form on M as

$$w_{UV}(X) = (g_{UV}^* w)(X) = w \circ g_{UV*}(X).\tag{9.114}$$

To the field $g_{UV*}(X)$ on G will correspond a fundamental field on P by the mapping of Eq. (8.34). As here $b = \sigma_U(p)$, we can write

$$\overline{g_{UV*}(X)} = \tilde{\sigma}_{U*}[g_{UV*}(X)].\tag{9.115}$$

As to the connection Γ, it will be pulled back to M by each section:

$$\Gamma_U = \sigma_U^*\Gamma \quad \text{and} \quad \Gamma_V = \sigma_V^*\Gamma\tag{9.116}$$

Notice that Γ_U and Γ_V are forms (only locally defined) on M with values in G'.

§ 9.6.12 Now, let us take the differential of (9.112) by using the Leibniz formula, and apply the result to a field X on M:

$$\sigma_V^*(X) = \sigma_U^*(X)g_{UV}(p) + \sigma_U(p)g_{UV*}(X),\tag{9.117}$$

whose detailed meaning is

$$\sigma_V^*(X) = R_{g_{UV}*}[\sigma_{U*}(X)] + \tilde{\sigma}_{U*}[g_{UV*}(X)].\tag{9.118}$$

When we apply Γ to this field on P, the first term on the right-hand side will be

$$\Gamma\{R_{g_{UV}*}[\sigma_{U*}(X)]\} = (R_{g_{UV}}^*)(\sigma_{U*}X) = (Ad_{g_{UV}^{-1}}\Gamma)\,\sigma_{U*}X$$

by (9.110), and thus also equal to

$$Ad_{g_{UV}^{-1}}(\sigma_U^*\Gamma)(X) = Ad_{g_{UV}^{-1}}\Gamma_U(X).$$

The second term, on the other hand, will be $\Gamma\{\overline{g_{UV^*}(X)}\}$ by Eq. (9.115). This will be a certain field of G', in reality the field $g_{UV^*}(X)$ on G brought to G', of which $\overline{g_{UV^*}(X)} = \sigma_{U^*}g_{UV^*}(X)$ is the fundamental representative field on P. A field on G is brought to G' by the Maurer–Cartan form w, so that

$$\Gamma\{\overline{g_{UV^*}(X)}\} = w[g_{UV^*}(X)] = (g_{UV}^*w)(X) = w_{UV}(X).$$

Consequently,

$$\Gamma(\sigma_{V^*}(X)) = (\sigma_V^*\Gamma)(X)$$

will be

$$\Gamma_V(X) = Ad_{g_{UV}^{-1}}\Gamma_U(X) + w_{UV}(X), \tag{9.119}$$

or

$$\Gamma_V = Ad_{g_{UV}^{-1}}\Gamma_U + w_{UV}. \tag{9.120}$$

This gives Γ pulled back to M, in terms of a change of section defined by g_{UV}. In general, one prefers to drop the (U,V) indices and write this equation in terms of the group-valued mapping $g = g_{UV}$. Using Eqs.(9.111) and (8.17), it becomes

$$\Gamma' = g^{-1}\Gamma g + g^{-1}dg. \tag{9.121}$$

In this notation, we repeat, Γ and Γ' are G'-valued 1-forms on M. In a natural basis, $\Gamma = \Gamma_\mu dx^\mu = J_a\Gamma^a{}_\mu dx^\mu$,

$$\Gamma'_\mu = g^{-1}\Gamma_\mu g + g^{-1}\partial_\mu g \tag{9.122}$$

and

$$J_a\Gamma'^a{}_\mu = g^{-1}J_a g\,\Gamma^a{}_\mu + g^{-1}\partial_\mu g. \tag{9.123}$$

§ **9.6.13** In these expressions the reader will have recognized the behavior of a gauge potential under the action of a gauge transformation, or the change in the Christoffel due to a change of basis. Here, the small correction we promised in § 9.6.9:

Gauge potentials are pulled-back connections on principal
fiber bundles with the gauge group as structure
group and spacetime as base space.[7]

They are defined on the base space and so they are section-dependent. A section is what is commonly called "a gauge". Changes of sections are gauge transformations. The geometrical interpretation of the underlying structure of gauge theories, pioneered by Trautman and a bit resented at first, is nowadays accepted by everybody.[8] It helps clarifying many important points. Let us only call attention to one of such. The "vacuum" term $g^{-1}dg$ in (9.121) is a base-space representative of the Maurer–Cartan form, Eq. (9.114). This form is a most important geometrical characteristic of the group, in reality connected to many of its topological properties. The vacuum of gauge theories is thereby strongly related to the basic properties of the gauge group.

§ 9.6.14 A 1-form satisfying condition (9.106) is said to be a *vertical form*. On the other hand, a form γ on P which vanishes when applied to any vertical field,

$$\gamma(V_p X) = 0, \quad \forall X, \tag{9.124}$$

is a *horizontal 1-form*. The canonical form on the frame bundle is horizontal. Vertical (and horizontal) forms of higher degrees are those which vanish when at least one horizontal (respectively, vertical) vector appears as their arguments.

§ 9.6.15 Given a connection Γ, horizontal spaces are defined at each point of P. Given a p-form ω on P with values in some vector space V, its *absolute derivative* (or *covariant derivative*) according to the connection Γ is the $(p+1)$-form

$$D\omega = H d\omega = d\omega \circ H, \tag{9.125}$$

where H is the projection to the horizontal space:

$$D\omega(X_1, X_2, \ldots, X_{p+1}) = d\omega(HX_1, HX_2, \ldots, HX_{p+1}). \tag{9.126}$$

The covariant derivative is clearly a horizontal form. An important property of D is that it preserves the representation: if ω belongs to a certain representation, so does $D\omega$. For example, if

$$R_g^* \omega = Ad_{g^{-1}}\omega,$$

---

[7] Trautman 1970.
[8] Daniel & Viallet 1980; Popov 1975; Cho 1975.

then also

$$R_g^* D\omega = Ad_{g^{-1}} D\omega. \tag{9.127}$$

This property justifies the name "covariant" derivative. A connection also defines horizontal spaces in associated bundles, and a consequent covariant derivative. As fibers are carrier spaces to representations of G, D will in that case take each element into another of the same representation.

§ 9.6.16 Going back to principal bundles, D is the horizontally projected exterior derivative on P. If we take $V = G'$ and $\omega = \Gamma$ itself, the resulting 2-form

$$F = D\Gamma \tag{9.128}$$

is the *curvature form* of Γ. From Eq. (9.127),

$$R_g^* F(X_1, X_2) = F(R_{g*}X_1, R_{g*}X_2) = Ad_{g^{-1}} F(X_1, X_2). \tag{9.129}$$

Being a form with values in G', it can be written

$$F = \tfrac{1}{2} J_a F^a{}_{\mu\nu}\omega^\mu \wedge \omega^\nu, \tag{9.130}$$

with $\{\omega^\mu\}$ a base of horizontal 1-forms. Equation (9.129) is then

$$R_g^* F = F' = \tfrac{1}{2} g^{-1} J_a\, g\, F^a{}_{\mu\nu}\, \omega^\mu \wedge \omega^\nu, \tag{9.131}$$

or

$$F' = g^{-1} F g. \tag{9.132}$$

§ 9.6.17 A closed global expression of F in terms of Γ is got from Eq. (7.29), which for the present case is

$$2d\Gamma(X,Y) = X[\Gamma(Y)] - Y[\Gamma(X)] - \Gamma([X,Y]). \tag{9.133}$$

A careful case study for the vertical and horizontal components of X and Y leads to[9]

$$F(X,Y) = d\Gamma(X,Y) + \tfrac{1}{2}[\Gamma(X),\Gamma(Y)], \tag{9.134}$$

which is the compact version of Cartan's *structure equations*. This is to be compared with the results of § 7.3.12: there, we had a zero-curvature connection (also called a *flat connection*) on $\mathbb{E}^n$. A better analogy is found in submanifolds (Chapter 22).

---

[9] Kobayashi & Nomizu 1963, vol. I.

§ **9.6.18** We can write more simply

$$F = d\Gamma + \Gamma \wedge \Gamma. \tag{9.135}$$

An immediate consequence is the *Bianchi identity*

$$DF = 0, \tag{9.136}$$

which follows from $DF = dF \circ H$ and $\Gamma \circ H = 0$.

§ **9.6.19 Curvature.** The curvature form can be pushed back to the base manifold by a local section σ, as was done for the connection form. It is customary to simplify the notation by writing $(\sigma^* F) = F$. In a natural basis, Eq. (9.135) becomes

$$F = \tfrac{1}{2} J_a F^a{}_{\mu\nu} \, dx^\mu \wedge dx^\nu, \tag{9.137}$$

where

$$F^a{}_{\mu\nu} = \partial_\mu \Gamma^a{}_\nu - \partial_\nu \Gamma^a{}_\mu + C^a{}_{bc} \Gamma^b{}_\mu \Gamma^c{}_\nu. \tag{9.138}$$

Here, we recognize the expression of the field strength in terms of gauge potentials. For linear connections, it is convenient to use the double-index notation for the generators $J_a{}^b$ of the linear group $GL(m, \mathbb{R})$ (or of one of its subgroups). In that case, the above $\Gamma^a{}_\mu$ becomes $\Gamma^a{}_{b\mu}$, $F^a{}_{\mu\nu}$ becomes $F^a{}_{b\mu\nu}$, and so on. Using the structure constants for $GL(m, \mathbb{R})$, Eq. (9.138) acquires its usal form for linear connections. With the Lorentz group generators, it becomes the usual expression of the curvature tensor in terms of the Christoffel symbols.

Notice here a very important point: curvature is a characteristic of a connection. What we have is always the *curvature of a connection*.

§ **9.6.20 Parallel transport.** The simplest way to introduce the notion of parallel transport for general principal bundles is through the covariant derivative. An object is parallel-transported along a curve when its covariant derivative is zero all along the curve. Parallelism on associated bundles may be introduced along the lines of § 9.6.19. A member of a realized fiber is taken in a parallel way along a curve when its paradigm in the typical, abstract fiber remains the same (Figure 9.11).

In such a case, G acts through some representation. A member of the fiber belongs to its carrier space. In the case of a gauge theory, a source field ψ will belong to some Hilbert space fiber on which G performs gauge transformations. Think it as a column vector and let $\{T_a\}$ be a basis for the corresponding matrix generators. The connection will be a matrix

$$\Gamma = T_a \Gamma^a$$

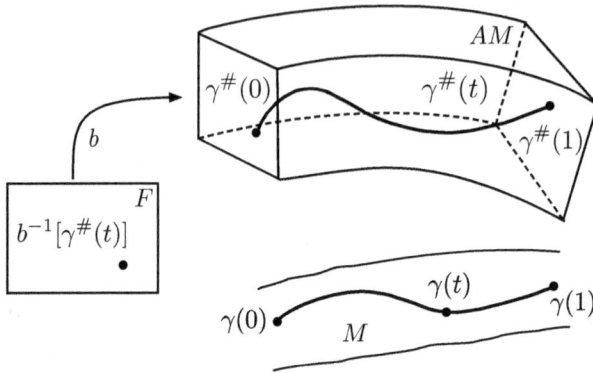

Fig. 9.11 Parallel transport of any object: its paradigm in the typical fiber remains the same all along.

or, in a local natural basis,

$$\Gamma = T_a \Gamma^a{}_\mu dx^\mu.$$

The covariant derivative will be

$$D\psi = d\psi + \Gamma\psi = (\partial_\mu \psi + \Gamma^a{}_\mu T_a \psi) dx^\mu. \qquad (9.139)$$

This derivative D allows one to introduce a notion of parallelism in neighbouring points: $\psi(x)$ is said to be parallel to $\psi(x + dx)$ when

$$\psi(x + dx) - \psi(x) = \Gamma_\mu \psi dx^\mu, \qquad (9.140)$$

that is, when $D\psi = 0$ at x. What is the meaning of it? For linear connections and tangent vectors, this generalizes the usual notion of parallelism in euclidean spaces. In gauge theories, in which the associated fibers contain the source fields, it represents a sort of "internal" state preservation.

Suppose a gauge model for the group $SU(2)$, as the original Yang–Mills case. The nucleon wavefields ψ will have two components, being isotopic-spin Pauli spinors. Suppose that at point x the field ψ is a pure "up" spinor, representing a proton. If $D\psi = 0$ at x, then $\psi(x + dx)$ given by Eq. (9.140) will also be a pure proton at $(x + dx)$.

All this is summed up again in the following: all over a parallel transport of an object, the corresponding element in typical fiber stays the same (Figure 9.12).

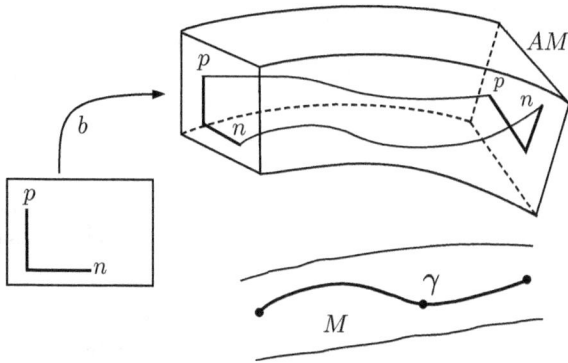

Fig. 9.12   Illustration of parallel transport.

§ 9.6.21 Holonomy groups. Consider again a principal fiber bundle

$$P = (M, G, \pi).$$

Fix a point p on M and consider the set of all closed curves (supposed piecewise-smooth) with endpoints p. The parallel displacement according to a connection Γ along each loop will define an isomorphism of the fiber $\pi^{-1}(p)$ into itself. The set of all such isomorphisms constitute a group $\mathrm{Hol}(p)$, the "holonomy group of Γ at the point p". The subgroup engendered by those loops which are homotopic to zero is the "restricted holonomy group of Γ at point p", denoted $\mathrm{Hol}_0(p)$. In both cases, the group elements take a member of the fiber into another, so that both groups may be thought of as subgroups of the structure group G. It is possible to show (when M is connected and paracompact) that:

(i) $\mathrm{Hol}(p)$ is a Lie subgroup of G;

(ii) $\mathrm{Hol}_0(p)$ is a Lie subgroup of $\mathrm{Hol}(p)$;

(iii) $\mathrm{Hol}_0(p)$ is a normal subgroup of $\mathrm{Hol}(p)$, and $\mathrm{Hol}(p)/\mathrm{Hol}_0(p)$ is countable.

Take an arbitrary point b belonging to the fiber $\pi^{-1}(p)$, and consider the parallel transport along some loop on M: it will lead b to some $b' = ba$, so that to each loop will correspond an element a of G. In this way one finds a subgroup of G isomorphic to $\mathrm{Hol}(p)$. We call this group $\mathrm{Hol}(b)$. Had we chosen another point c on the fiber, another isomorphic group $\mathrm{Hol}(c)$ would be found, related to $\mathrm{Hol}(b)$ by the adjoint action. Thus, on any point of the fiber we would find a representation of $\mathrm{Hol}(p)$.

Consider an arbitrary $b \in \pi^{-1}(p)$. Call $P(b)$ the set of points of P which can be attained from b by a horizontal curve. Then it follows that (i) $P(b)$ is a reduced bundle with structure group $\text{Hol}(b)$, and (ii) the connection Γ is reducible to a connection on this sub-bundle. $P(b)$ is called the "holonomy bundle" at b.

§ **9.6.22 The Ambrose–Singer holonomy theorem.** Take the holonomy bundle $P(b)$ and fix a point $h \in P(b)$. Consider the curvature form F of the connection Γ at h, denoted F_h. The theorem states that the Lie algebra of $\text{Hol}(b)$ is a subalgebra of G' spanned by all the elements of the form $F_h(X, Y)$, where X and Y are arbitrary horizontal vectors at h.

§ **9.6.23 Berry's phase.** Fields or wavefunctions respond to the presence of connections by modifying the way they feel the derivatives. We have seen that they can answer to linear connections and to gauge potentials. If a wavefunction ψ represents a given system, the effect can be seen by displacing ψ along a line. As a member of an associated bundle, ψ will remain in the same state if parallel transported. In the abelian case, integration of the condition $D\psi = 0$ along the line adds a phase to ψ (as in § 4.2.17). In the non-abelian case, ψ moves in the "internal" space in a well-defined way (as in § 9.6.20). There is, however, still another kind of connection to which a system may react: connections defined on parameter spaces.

The effect was brought to light in the adiabatic approximation.[10] Suppose we have a system dependent on two kinds of variables: the usual coordinates and some parameters which, in normal conditions, stay fixed. We may think of an electron in interaction with a proton. The variables describing the position of the proton, very heavy in comparison with the electron, can be taken as fixed in a first approximation to the electron wavefunction ψ. Consider now the proton in motion — its variables are "slow variables" and can be seen as parameters in the description of the electron. Suppose that it moves (follows some path on the electron parameter space) somehow and comes back to the initial position. This motion is a closed curve in the parameter space. Normally nothing happens — the electron just comes back to the original wavefunction ψ. But suppose something else: that the parameter space "is curved" (say, the proton is constrained to move on a sphere S^2). A non-flat connection is defined on the parameter space. The loop described by the proton will now capture the curvature flux in the surface it circumscribes (again as in § 4.2.17, or in

---

[10] Berry 1984.

the "experiment" of § 9.4.1). The wavefunction will acquire a phase which, due to the curvature, is no more vanishing. This is Berry's phase.[11] The connection in parameter space has the role of an effective vector potential. This is of course a very crude example, which is far from doing justice to a beautiful and vast subject.[12] Even in a fixed problem, what is a parameter and what is a variable frequently depends on the conditions. And it was found later[13] that this "geometrical phase" (sometimes also called "holonomy phase") can appear even in the absence of parameters, in the time evolution of some systems. There is now a large amount of experimental confirmation of its existence in many different physical situations.

9.7 Bundle classification

Let us finish the chapter with a few words on the classification of fiber bundles. Steenrod's theorem (§ 9.2.3) is a qualitative result, which says that there is always at least one bundle with a base space M and a group G. We have seen (§ 9.1.1) that with the line and the circle at least two bundles are possible: the trivial cylinder and the twisted Möbius band. Two questions[14] come immediately to the mind, and can in principle be answered:

(1) In how many ways can M and G be assembled to constitute a complete space P? The answer comes out from the universal bundle approach.
(2) Given P, is there a criterion to measure how far it stands from the trivial bundle? The answer is given by the theory of characteristic classes.

Notice to begin with that each associated bundle is trivial when the principal bundle P is trivial. Consequently, an eventual classification of vector bundles is induced by that of the corresponding principal and we can concentrate on the latter.

§ **9.7.1 Back to homogeneous spaces.** Let us recall something of what was said about homogeneous spaces in § 8.2.7. If M is homogeneous under the action of a group G, then we can go from any point of M to any other point of M by some transformation belonging to G. A homogeneous space

---

[11] Simon 1983.
[12] A very good review is Zwanziger, Koenig & Pines 1990.
[13] Aharonov & Anandan 1987 and 1988.
[14] Nash & Sen 1983.

has no invariant subspace but itself. Simple Lie groups are homogeneous by the actions (right and left) defined by the group multiplication. Other homogeneous spaces can be obtained as quotient spaces. The group action establishes an equivalence relation, $p \approx q$, if there exists some $g \in G$ such that $q = R_g p$. The set

$$[p] = \{\, q \in M \text{ such that } q \approx p \,\} = \mathrm{orbit}_G(p)$$

is the equivalence class with representative p. The set of all these classes is the quotient space of M by the group G, denoted by M/G and with dimension given by

$$\dim M/G = \dim M - \dim G.$$

The canonical projection is

$$\pi : M \to M/G, \quad p \to [p].$$

All the spheres are homogeneous spaces:

$$S^n = SO(n+1)/SO(n).$$

Consider in particular the hypersphere

$$S^4 = SO(5)/SO(4).$$

It so happens that $SO(5)$ is isomorphic to the bundle of orthogonal frames on S^4. We have thus a curious fact: the group $SO(5)$ can be seen as the principal bundle of the $SO(4)$-orthogonal frames on S^4. This property can be transferred to the de Sitter spacetimes: they are homogeneous spaces with the Lorentz group as stability subgroup, respectively,

$$dS(4,1) = SO(4,1)/SO(3,1) \quad \text{and} \quad dS(3,2) = SO(3,2)/SO(3,1).$$

And the de Sitter groups are the respective bundles of (pseudo-) orthogonal frames. Can we generalize these results? Are principal bundles always related to homogeneous spaces in this way? The answer is that it is *almost* so. We shall be more interested in special orthogonal groups $SO(n)$. In this case, we recall that the Stiefel manifolds (§ 8.1.12) are homogeneous spaces,

$$S_{nk} = SO(n)/SO(k).$$

They do provide a general classification of principal fiber bundles.

§ **9.7.2 Universal bundles.** Given a smooth manifold M and a Lie group G, it is in principle possible to build up a certain number of principal fiber bundles with M as base space and G as structure group. One of them is the direct product, $P = M \times G$. But recall that the general definition of a principal bundle includes the requirement that M be the quotient space of P under the equivalence defined by G, that is, $M = P/G$.

Let us take a bit seriously the following easy joke. Say, to start with, that principal bundles are basically product objects obeying the relation $M = P/G$, which are trivial when in this relation we can multiply both sides by G in the naïve arithmetic way to obtain $P = M \times G$. Thus, nontrivial bundles are objects $P = M \diamondsuit G$, where $\diamondsuit$ is a "twisted" product generalizing the cartesian product — as the Möbius band generalizes the cylinder. How many of such "twisted" products are possible? Answering this question would correspond to somehow classifying the principal fiber bundles. It is actually possible to obtain a homotopic classification of the latter, and this is done through a construct which is itself a very special fiber bundle. This special fiber bundle is the *universal bundle* for G.

> *The principal bundle* $P = (M, G, \pi)$ *is universal for the group* G
> *if the complete space* P *is contractible, that is, if all*
> *the homotopy groups* $\pi_k(P) = 0$.

When $\pi_k(P) = 0$ only for $k < m$, P is called "m-universal". We shall only consider the cases in which G is either an orthogonal or an unitary group. Given G, it would be enough to find some contractible bundle with G as structure group. We shall find it, the base space being a Grassmann manifold and the complete space being a Stiefel manifold.

Take the orthogonal group $O(m)$ and its subgroup $O(d)$, with $d \leq m$. Using the notation (see § 9.5.1)

> *Complete space = (base space, structure group, projection)*,

let us consider the bundle

$$O(m)/O(m-d) = (O(m)/O(d) \times O(m-d), O(d), \text{ quotient projection}).$$

If we recall what has been previously said on Grassmann spaces (§ 1.5.22, and § 8.1.11) and on Stiefel manifolds (§ 8.1.12), we see that this is

> *Stiefel = (Grassmann, $O(d)$, quotient projection)*

or equivalently

$$S_{md} = (G_{md}, O(d), \text{ quotient projection}).$$

The situation, of which a local scheme is given in Figure 9.13, is just that of a bundle of base space G_d, fiber $O(d)$ and complete space S_d. We may

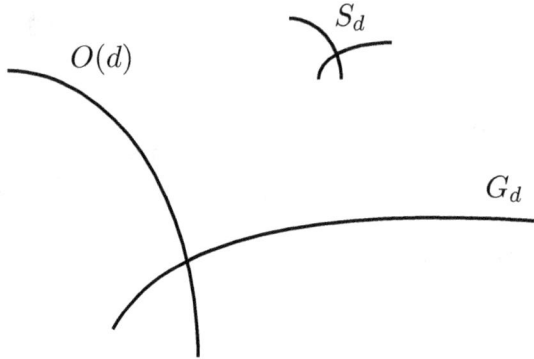

Fig. 9.13 Local view of the Stiefel bundle.

as well obtain

$$S_{md} = (G^{\#}_{md}, SO(d), projection)$$

for the covering of G_{md}. In the complex case, we can construct bundles

$$S^C_{md} = (G^C_{md}, U(d), \; projection).$$

As said previously,

$$\pi_r(S_{md}) = 0 \;\; \text{for} \; (m - d) > r \geq 0,$$

and

$$\pi_r(S^C_{md}) = 0 \;\; \text{for} \; 2(m - d) > r \geq 0.$$

As a consequence, we may take m to infinity while retaining d fixed: then $S_{\infty d}$ and $S^C_{\infty d}$ will have all the homotopy groups trivial and will be contractible. We have thus universal bundles whose base spaces are Grassmann manifolds of d-planes of infinite-dimensional euclidean (real or complex) spaces. We shall call these base spaces generically B.

Now, we state the Classification Theorem:

> *Consider the principal bundles of base M and group G; the set of bundles they form is the set of homotopy classes of the mappings $M \to B$.*

It may seem rather far-fetched to employ infinite-dimensional objects, but in practice we may use the base for the m-universal bundle. In this way the problem of classifying bundles reduces to that of classifying mappings between base manifolds. The number of distinct "twisted" products is thus the number of homotopy classes of the mappings $M \to B$. In reality there is another, very strong result:

> *Given P as above, then for a large enough value of m there*
> *exists a map $f : M \to G_{md}$ such that P is the*
> *pull-back bundle of S_{md}: $P = f^* S_{md}$.*

Thus, every bundle is an induced bundle (see § 9.5.12) of some Stiefel manifold. The Stiefel spaces appear in this way as bundle-classifying spaces.

Formally at least, it would be enough to consider real Stiefel spaces and orthogonal (or pseudo-orthogonal) groups, because any compact group is a subgroup of some $O(q)$ for a high enough q. $U(n)$, for instance, is a subgroup of $O(2n)$. For a general Lie group G, the Stiefel manifold as base space is replaced by $O(m)/G \times O(m-d)$. There will be one bundle on M for each homotopy class of the maps

$$M \to O(m)/G \times O(m-d).$$

The question of how many twisted products there exist can thus be answered. Unfortunately, all this is true only in principle, because the real calculation of all these classes is usually a loathsome, in practice non realizable, task.

§ **9.7.3 Characteristic classes.** These are members of the real cohomology classes $\{H^n(M)\}$ of differential forms on the base space M. They answer to the question concerning the "degree of twistness" of a given twisted product, or how far it stands from the direct product. To obtain them, one first finds them for the universal bundle, thereby estimating how far it is from triviality. Then one pulls the forms back to the induced bundles; as the pull-back preserves the cohomological properties, the same "twisting measure" holds for the bundle of interest. Thus, the first task is to determine the cohomology groups

$$H^n[O(m)/G \times O(m-d)].$$

For the most usual groups, such classes are named after some illustrious mathematicians. For $G = O(n)$ and $SO(n)$, they are the Pontryagin classes, though the double covering of special groups produce an extra series of classes, the Euler classes. For $U(n)$, they are the Chern classes. These are the main ones. Other may be arrived at by considering non-real cohomologies.

Part 3: NATURE'S EXTREME GEOMETRIES

Chapter 10

Quantum Geometry

Everything said up to now concerns the geometry involved in classical physics, commutative geometry. We shall in this chapter briefly broach the subject of noncommutative geometry,[1] and then proceed to some related aspects in Quantum Mechanics.

10.1 Quantum goups: a pedestrian outline

§ 10.1.1 Think of a group of usual matrices. Matrices in general do not commute. Their entries consist of real or complex numbers $a^i{}_j$. We are used to multiplying matrices (that is, to perform the group operation) and, in doing so, the entries themselves get multiplied according to the well-known rules. Complex numbers commute with each other and we do not trouble with their order when multiplying them:

$$a^i{}_j \, a^m{}_n = a^m{}_n \, a^i{}_j.$$

§ 10.1.2 Matrix groups are Lie groups, that is, smooth manifolds whose points are the group elements. To each group element, a point on the group manifold, corresponds a matrix. Thus, each matrix will have its coordinates. What are the coordinates of a matrix? Just the entries $a^m{}_n$ if they are real and, if they are complex, their real and imaginary parts (§ 6.4.3 and § 6.5.8). Thus, although themselves noncommuting, matrices are represented by sets of commuting real numbers. Their non-commutativity is embodied in the rules to obtain the entries of the product matrix from those of the matrices being multiplied.

Suppose now that, in some access of fantasy, we take the very entries $a^i{}_j$ as noncommutative. This happens to be not pure folly, provided some

---

[1] Connes 1990.

rules are fixed to ensure a minimum of respectability to the new structure
so obtained. With convenient restrictions, such as associativity, the new
structure is just a Hopf algebra (§ 13.6.2).

§ **10.1.3** Very roughly speaking, quantum groups (the physicists' unhappy
name for Hopf algebras)[2] are certain sets of matrices whose elements $a^i{}_j$
are themselves non-commutative. They are not groups at all (hence the
unhappiness of the name), but structures generalizing them. The non-
commutativity is parametrized in the form

$$R^{rs}{}_{mn} a^m{}_i a^n{}_j = R^{pq}{}_{ij} a^r{}_p a^s{}_q,$$

where the $R^{rs}{}_{mn}$ are complex coefficients and the respectability conditions
are encoded in constraints which are imposed on them. In particular, a
direct calculation shows that the imposition of associativity, starting from
the above general commutativity assumptions, leads to the Yang–Baxter
equation (§14.2.7; see also Section 27.5) in the form

$$R^{jk}{}_{ab} R^{ib}{}_{cr} R^{ca}{}_{mn} = R^{ij}{}_{ca} R^{ck}{}_{mb} R^{ab}{}_{nr}.$$

Tecnically, these R-matrices, satisfying the Yang–Baxter equation, belong
to a particular kind of Hopf algebras, the so-called quasi-triangular algebras.

§ **10.1.4** Well, the Hopf structure would probably remain an important
tool of interest in limited sectors of higher algebra if it did not happen
that the Yang–Baxter equation turned up in a surprisingly large number of
seemingly disparate topics of physical concern: lattice models in Statistical
Mechanics (about the relation with braids and knots, see Section 28.2.3),
integrability of some differential equations,[3] the inverse scattering method,[4]
the general problem of quantization (Section 10.2), etc.

§ **10.1.5** A first question comes immediately to the mind: if now the entries
themselves are no more commutative, what about the coordinate roles they
enjoyed? The coordinates themselves become noncommutative — and that
is where noncommutative geometry comes to the scene forefront. While
the new matrices of noncommutative entries constitute Hopf algebras, the
entries themselves constitute other spaces of mathematical predilection, von
Neumann algebras (Section 17.5).

---

[2] Woronowicz uses the term "pseudo-group". Concerning the name "quantum groups",
the physicists are not alone in their guilt: people looking for material on Hopf algebras
in the older mathematical literature will have to look for "annular groups" ... The best
name is probably "bialgebras".

[3] McGuire 1964.

[4] See different contributions in Yang & Ge 1989.

§ **10.1.6** A first idea would be to take the above noncommuting matrix elements also as matrices. Matrices of matrices then turn up. This is where the large use of direct-product matrices comes out. In diagonal block matrices, the blocks work independently from each other and will correspond to abelian bialgebras. Drop the diagonal character and you will have general block matrices. By the way, finite von Neumann algebras are precisely algebras of block matrices (well, actually almost everything a physicist bothers about is a von Neumann algebra. The novelty is that up to recent times most of these algebras were commutative).

§ **10.1.7** Groups are usually introduced in Physics as sets of transformations on some carrier space. Would quantum groups also preside over transformations? If so, the above matrices with non-commuting entries would be expected to act upon column vectors, these with non-commuting components. This is actually so,[5] and the study of the carrier spaces (called quantum spaces, or Manin spaces) seems simpler than that of the Hopf algebras themselves.[6] The general case is, however, very involved also.[7]

§ **10.1.8** There are other gates into the realm of bialgebras. Hopf introduced them originally in homology theory, but other simpler cases have been found since then. One may, for instance, proceed in a way similar to that used to introduce the classical groups as transformation groups (§ 8.1.7). These are sets of transformations preserving given sesquilinear forms.[8] This is probably the most apealling approach for physicists. In knot theory, they appear in presentations of groups and algebras. In physics, the original approach was related to Lie algebras deformations.

Classical Sine–Gordon equation is related to the Lie algebra $sl(2, R)$. Once quantized, instead of this algebra, this integrable equation exhibitted another structure, which was recognized by Kulish and Reshetikin as a deformation of $sl(2, R)$, and called it its "quantum" version. Drinfeld[9] has then given the new structure a general descripton, through the consideration of phase spaces defined on Lie groups (see Chapter 27).

§ **10.1.9** A most interesting approach is that pioneered by Woronowicz,[10] which we could call the "Fourier gate". It relates to harmonic analysis on

---

[5] Manin 1989.

[6] See for instance Fairlie, Fletcher & Zachos 1989, and references therein.

[7] See Ocneanu's postface to Enoch & Schwartz 1992.

[8] Dubois–Violette & Launer 1990.

[9] Drinfeld 1983.

[10] Woronowicz 1987.

groups. It goes from the Pontryagin duality for abelian groups, through the Tanaka–Krein duality for non-abelian groups, to still more general theories (see § 18.4 and those following it). The whole subject is very involved in its formal aspects, and still a research subject for physicists and mathematicians.[11] It involves deep points of the theory of Banach algebras (Section 16.6) and Hopf–Banach algebras. We retain here only the point that coordinates can become non-commutative and proceed to a heuristic discussion of a formalism well known to physicists.

10.2   Noncommutative geometry

§ **10.2.1** People may think that "it is evident" that Quantum Mechanics is concerned with a noncommutative geometry. Actually, that the *geometry* is noncommutative is not so evident. Despite the foresight of Dirac who, in his basic paper,[12] calls commutators "quantum derivations", the well known noncommutativity in Quantum Mechanics is of algebraic, not geometric, character. The difference rests, of course, in the absence of specifically geometric structures in the algebra of operators, such as differentiable structure, differential forms, connections, metrics and the like — in a word, in the absence of a *differential* geometry. On the other hand, some noncommutativity comes up even in Classical Mechanics: the Poisson bracket is the central example. But this is precisely the point — the Poisson bracket is a highly strange object.

§ **10.2.2** Let us consider (see Section 25.1 and on) the classical phase space $\mathbb{E}^{2n}$ of some mechanical system with generalized coordinates

$$q = (q^1, q^2, \ldots, q^n)$$

and momenta

$$p = (p_1, p_2, \ldots, p_n).$$

Dynamical quantities like $F(q, p)$ and $G(q, p)$ on $\mathbb{E}^{2n}$ constitute an associative algebra with the usual pointwise product $F \cdot G$ as operation. Given any associative algebra, one may get a Lie algebra with the commutator as operation. Of course, due to the commutativity, the classical Lie algebra of dynamical functions coming from the pointwise product is trivial. It is the

---

[11] A general idea on the whole subject can be obtained in Gerstenhaber & Stasheff 1992.

[12] Dirac 1926.

peculiar noncommutative Lie algebra defined by the Poisson bracket which is physically significant. This is a rather strange situation from the mathematical point of view, as the natural algebraic brackets are those coming as commutators. The Poisson bracket stands apart because it does not come from any evident associative algebra of functions. We know, however, that a powerful geometric background, the hamiltonian (or symplectic) structure, lies behind the Poisson bracket, giving to its algebra a meaningful and deep content.

On the other hand, in Quantum Mechanics, the product in the algebra of dynamical functions (the operators) is noncommutative and the consequent commutator is significant — but there seems to exist no structure of the symplectic type. Now, it is a general belief that the real mechanics of Nature is Quantum Mechanics, and that classical structures must come out as survivals of those quantal characteristics which are not completely "erased" in the semiclassical limit. It is consequently amazing that precisely those quantal structures — somehow leading to the basic hamiltonian formalism of Classical Mechanics, mainly the symplectic structure — be poorly known. The answer to this problem, however, is known. And it just requires the introduction of more geometric structure in the quantum realm. The geometry coming forth is, however, of a new kind: it is noncommutative.

§ 10.2.3 In rough words, the usual lore of noncommutative geometry[13] runs as follows.[14] Functions on a manifold M constitute an associative algebra $C(M)$ with the pointwise product (Section 17.4). This algebra is full of content, because it encodes the manifold topology and differentiable structure. It contains all the information about M. The differentiable structure of smooth manifolds, for instance, has its counterpart in terms of the derivatives acting on $C(M)$, the vector fields. On usual manifolds (point manifolds, as the phase space above), this algebra is commutative. The procedure consists then in going into that algebra and work out everything in it, but "forgetting" about commutativity (while retaining associativity). In the phase space above, this would mean that $F \cdot G$ is transformed into some non-abelian product $F \circ G$, with "∘" a new operation.[15] The resulting

---

[13] Dubois–Violette 1991.

[14] Coquereaux 1989.

[15] In this chapter, the symbol "∘" is taken for the so-called "star-product" of quantum Wigner functions. It has nothing to do with its previous use for the composition of mapppings. It is the Fourier transform of the convolution, for which the symbol "*" is used of old.

geometry of the underlying manifold M will thereby "become" noncommutative. Recall that a manifold is essentially a space on which coordinates (an ordered set of real, commutative point functions) can be defined. When we go to noncommutative manifolds, the coordinates, like the other functions, become noncommutative. Differentials come up in a very simple way through the (then nontrivial) commutator

$$[F, G] = F \circ G - G \circ F.$$

Associativity of the product $F \circ G$ implies the Jacobi identity for the commutator, *i.e.*, the character of Lie algebra. With fixed F, the commutator is a derivative "with respect to F": the Jacobi identity

$$[F, [G, H]] = [[F, G], H] + [G, [F, H]]$$

is just the Leibniz rule for the new "product" defined by $[\,,\,]$.

§ 10.2.4 In order to approach this question in Quantum Mechanics, the convenient formalism is the Weyl–Wigner picture. In that picture, quantum operators are obtained from classical dynamical functions via the Weyl prescription, and the quantum formalism is expressed in terms of Wigner functions, which are "c-number" functions.

Let us only recall in general lines how the Weyl prescription[16] works for a single degree of freedom, the coordinate-momentum case (Section 14.1.1). The "Wigner functions" $A_W(q, p)$ are written as Fourier transforms $F[A]$ of certain "Wigner densities" $A(a, b)$,[17]

$$A_W(q,p) = F[A] = \frac{2\pi}{h} \iint \exp[i2\pi(aq + ibp)/h]A(a,b). \qquad (10.1)$$

Then the Weyl operator $\boldsymbol{A}(\boldsymbol{q}, \boldsymbol{p})$, a function of operators $\boldsymbol{q}$ and $\boldsymbol{p}$ corresponding to the Wigner function A_W, is

$$\boldsymbol{A}(\boldsymbol{q},\boldsymbol{p}) = \frac{2\pi}{h} \iint \exp[i2\pi(a\boldsymbol{q} + ib\boldsymbol{p})/h]A(a,b). \qquad (10.2)$$

We may denote by $\widehat{F}$ this operator Fourier transform, so that

$$\boldsymbol{A} = \widehat{F}[F^{-1}[A_W]] \qquad (10.3)$$

and

$$A_W = F[\widehat{F}^{-1}[\boldsymbol{A}]]. \qquad (10.4)$$

---

[16] See for instance Galetti & Toledo Piza 1988.
[17] Baker 1958, and Agarwal & Wolf 1970.

The Wigner functions are, despite their c-number appearence, totally quantum objects. They are representatives of quantum quantities (they will include powers of $\hbar$, for example) and only become classical in the limit $\hbar \to 0$, when they give the corresponding classical quantities. They embody and materialize the correspondence principle. The densities

$$A = F^{-1}[A_W] = \widehat{F}^{-1}[\boldsymbol{A}] \tag{10.5}$$

include usually Dirac deltas and their derivatives. All this may seem rather strange, that c-number functions can describe Quantum Mechanics. Actually, this is a lie: the Wigner functions are "c-number" functions indeed, but they do not multiply each other by the usual pointwise product. In order to keep faith to the above correspondence rule with quantum operators, a new product "$\circ$" has to be introduced, as announced in § 10.2.3. This means that we can describe quantum phenomena through functions, *provided we change the operation in their algebra*. The product "$\circ$" related to quantization is called (rather unfortunately) the "star-product".[18] The simplest way to introduce it is by changing the usual multiplication rule of the Fourier basic functions

$$\varphi_{(a,b)}(q,p) = \exp[i2\pi(aq + ibp)/h]. \tag{10.6}$$

We impose

$$\varphi_{(a,b)}(q,p) \circ \varphi_{(c,d)}(q,p) = \exp[-i(ad - bc)h/4\pi]\,\varphi_{(a+c,b+d)}(q,p) \tag{10.7}$$

instead of the "classical"

$$\varphi_{(a,b)}(q,p) \cdot \varphi_{(c,d)}(q,p) = \varphi_{(a+c,b+d)}(q,p). \tag{10.8}$$

This last expression says that the functions provide a basis for a representation of the group R^2, which is self-dual under Fourier transformations. Imposing equation (10.7) instead, with the extra phase, means to change R^2 into the Heisenberg group (§ 8.3.12). This is enough to establish a complete correspondence between the functions and the quantum operators. The commutator of two functions, once the new product is defined, is no longer trivial:

$$\{A_W, B_W\}(q,p) = [A_W, B_W]_\circ(q,p) =$$
$$= \left(\frac{2\pi}{h}\right)^2 \int dx\,dy\,dz\,dw\,A^{(x,y)}B^{(z,w)}$$
$$\times \frac{i\,4\pi}{h}\sin\left[\frac{h}{4\pi}(yz - xw)\right]\varphi_{(x+z,y+w)}(q,p).$$

---

[18] Bayen, Flato, Fronsdal, Lichnerowicz & Sternheimer 1978.

This "quantum bracket", called Moyal bracket after its discoverer, allows in principle to look at quantum problems in a way analogous to hamiltonian mechanics. Derivatives are introduced through the bracket: for instance, $\{q, F\}_{\text{Moyal}}$ is the derivative of F with respect to the "function" q. A symplectic structure appears naturally,[19] whose differences with respect to that of Classical Mechanics (shown in Chapter 25) correspond exactly to the quantum effects. Calculations are of course more involved than those with the Poisson bracket. Due to the star product, for instance, the derivative of p with respect to q is no more zero, but just the expected

$$\{q, p\} = i\hbar.$$

If we want to attribute coordinates to the quantum phase space, the only way to keep sense is to use no longer the usual euclidean space with commutative numbers, but to consider coordinate functions with a "∘" product. In this picture, we repeat, quantization preserves the dynamical functions, but changes the product of their algebra.

§ **10.2.5** And here comes the crux: in the limit $\hbar \longrightarrow 0$, the Moyal bracket gives just the Poisson bracket. The strangeness of the Poisson bracket is thus explained: though not of purely algebraic content, it is the limit of an algebraic and nontrivial bracket coming from a non-commutative geometry. The central algebraic operation of Classical Mechanics is thus inherited from Quantum Mechanics, but this only is seen after the latter has been given its full geometrical content. There is much to be done as yet, mainly when we consider more general phase spaces, and a huge new field of research on noncommutative geometrical structures is open.

All this is to say that Physics has still much to receive from Geometry. It is an inspiring fact that an age-old structure of Classical Mechanics finds its explanation in the new geometry.

General References

Non-commutative geometry has deserved a wide attention in the last two decades and only a few examples will be referred to here: from more mathematical themes — its relation to matrix theory (Connes, Douglas & Schwarz 1998) — through the quantum Hall effect (Pasquier 2007) to the harmonic oscillator (Hassanabadi, Derakhsani & Zarrinkamar 2016, which includes a commendable list of other references) and to Cosmology (Monerat, Corrêa Silva, Neves, Oliveira–Neto, Rezende Rodrigues & Silva de Oliveira 2015).

---

[19] Dubois–Violette, Kerner & Madore 1990.

Chapter 11

Cosmology: the Standard Model

The largest geometry in nature endeavours to describe the entire Universe, as far as human capacity can get data to support it. The Cosmos, or large scale spacetime, is modeled as a riemannian four-space $\mathcal{S}$, governed by the strictures of General Relativity. This requires a source, which is modeled as a gas whose molecules are galaxies and/or their agglomerates.

11.1 The geometrical perspective

The standard version of Einstein's equation,

$$R^{\mu\nu} - \tfrac{1}{2}R\, g^{\mu\nu} = \frac{8\pi G}{c^4}\, T^{\mu\nu} \, , \tag{11.1}$$

comes from a rather simple physical argument.[1] Gravitation appears in the left-hand side, represented by a function of the curvature, whose existence is encapsulated in a non-trivial metric $g_{\mu\nu}$. By the minimal coupling prescription, in all expressions valid in Special Relativity the Lorentz metric $\eta_{\alpha\beta}$ is to be replaced by $g_{\mu\nu}$ and all derivatives are to be replaced by covariant ones, using the Levi–Civita connection related to that metric. At the right-hand side, energy appears as the source of gravitation. Relativistic energy is represented by the symmetric energy momentum tensor density $T^{\mu\nu}$ modified, however, due to the minimal coupling prescription, by the presence of curvature. Energy is to be conserved: translating again from Special Relativity, the covariant divergence of $T^{\mu\nu}$ must vanish. On the left-hand side, we must then have some expression involving the curvature which has also a vanishing divergence — and the simplest one is just the

---

[1] More detail can be found in Chapter 34 and those following it.

Einstein tensor

$$G^{\mu\nu} = R^{\mu\nu} - \tfrac{1}{2} R\, g^{\mu\nu}.$$

We consequently equate both conserved quantities up to a constant,

$$G^{\mu\nu} = \kappa\, T^{\mu\nu},$$

and that constant is fixed as $\kappa = 8\pi G/c^4$ by fitting with the newtonian limit.

There is here, however, a problem in this reasoning. The fundamental field would be the metric $g_{\mu\nu}$, but we have imposed a basic role to its related curvature by requiring the presence of $G^{\mu\nu}$, simply because it is the simplest curvature-dependent expression with vanishing covariant divergence. The fundamental field seems to be forced to appear only *through* the curvature which stems from it. Actually, it is the curvature who plays the role of fundamental field.

Aa a matter of fact, if the metric is to be the real fundamental field, things are different, because $g_{\mu\nu}$ itself is divergence-free: this stands in the very definition of the Levi–Civita connection, and means that a term $\Lambda g^{\mu\nu}$, with Λ a constant, can be added to $G^{\mu\nu}$. In that case, the Einstein equation acquires its extended form

$$R^{\mu\nu} - \tfrac{1}{2} R\, g^{\mu\nu} - \Lambda g^{\mu\nu} = \frac{8\pi G}{c^4}\, T^{\mu\nu}. \tag{11.2}$$

It turns out that this is the equation which is able to fit, in principle, the present-day cosmological data, with Λ the "cosmological constant". As a matter of fact, the "cosmological term" shows itself as essential, as it relates to the observationally dominant "dark energy". This equation stands as the foundation for the "Cosmological Standard Model".

This leads, however, to another, more fundamental conundrum. The Lorentz metric is a solution of Eq. (11.1) with $T_{\mu\nu} = 0$, but not of Eq.(11.2). Minkowski space $\mathcal{M}$ is no more a solution in the absence of gravitational source. In the presence of a cosmological term, absence of energy leads to a de Sitter space $d\mathcal{S}$, a hyperbolic spacetime with pseudo-radius l such that $\Lambda = 3l^{-2}$, solution of

$$R^{\mu\nu} - \tfrac{1}{2} R\, g^{\mu\nu} = \Lambda g^{\mu\nu} = \frac{3}{l^2}\, g^{\mu\nu}. \tag{11.3}$$

This would mean that, in the minimal coupling prescription, all quantities related to the Minkowski space should be replaced by the corresponding quantities related to de Sitter space. We should start from a "de Sitter

Special Relativity".[2] We shall leave this problem to Chapter 37, and here only examine the standard version of the cosmological model.

On the geometrical side, the Standard Model supposes for the large-scale spacetime S a very simple metric, subsumed in the Friedmann–Lemaître–Robertson–Walker (FLRW) interval[3]

$$ds^2 = g_{\mu\nu}dx^\mu dx^\nu = c^2 dt^2 - a^2(t)\left[\frac{dr^2}{1-kr^2} + r^2 d\theta^2 + r^2 \sin^2\theta d\phi^2\right]. \quad (11.4)$$

A spherical coordinate system is taken for the space sector, which is nevertheless globally inflated in time via the scale parameter $a(t)$. It is always convenient to consider the radial coordinate r as dimensionless, just as the angles θ and ϕ, and suppose the scale parameter $a(t)$ as carrying the length dimension. The expression between brackets corresponds then to the possible space sections Γ, depending on the parameter k: it can be a sphere S^3 (if $k = +1$), a hyperbolic hypersurface $\mathcal{H}^{2,1}$ (if $k = -1$) or an euclidean space $\mathbb{E}^3$ (if $k = 0$).

Expression (11.4) for the interval actually incorporates two very important simplifying principles:

- the *cosmological principle*, which postulates that the space-section Γ is homogeneous and isotropic — a hypothesis which can only be assumed for very large scales: the picture is that of an ideal gas whose constituent molecules are galaxies and/or their agglomerates;
- the *universal time principle*, which states[4] that time flows quite independently of space positioning, so that the topology of the four-dimensional manifold S is actually the direct product $\mathbb{E}^1 \times \Gamma$.

11.2   The physical content

The content of the universe, actual source of large-scale gravitation, is modeled by the energy-momentum of a perfect fluid,[5]

$$T^{\mu\nu} = (p + \epsilon)\, U^\mu U^\nu - p\, g^{\mu\nu}, \quad (11.5)$$

where $\epsilon = \rho c^2$ is the energy density, p is the pressure and U^μ is a streamline four-velocity. A fluid is said to be "perfect" if it is isotropic when looked at from a frame carried along its streamlines, that is, along the local integral

---

[2] For more detail, see Aldrovandi & Pereira 2009.
[3] Friedmann 1922 and 1924; Lemaître 1931; Robertson 1935; Walker 1936.
[4] See Dirac 1958b and 1959 and/or Narlikar 2002.
[5] Weinberg 1972, Kolb & Turner 1994.

curves of the four-velocities U^μ. Such frames are sometimes also called "comoving frames".

Dynamics is governed by Einstein's equation with a cosmological term (11.2). This equation determines, in principle, all acceptable large-scale spacetimes, or "universes". Minkowski space, with metric

$$\eta = \mathrm{diag}(1, -1, -1, -1) = \eta_{\alpha\beta} dx^\alpha dx^\beta$$

in cartesian coordinates, which is equivalent to Eq. (11.4) with $k = 0$ and $a(t) \equiv 1$, is a solution when both Λ and $T^{\mu\nu}$ are zero.

It will be of interest to introduce some particular combinations of the scale parameter and its derivatives. The first will be the Hubble function

$$H(t) \equiv \frac{\dot{a}}{a}. \tag{11.6}$$

Another will be the concavity

$$C(t) \equiv \frac{\ddot{a}}{a} = H^2 + \dot{H}, \tag{11.7}$$

and, finally, the function

$$E(t) \equiv \frac{\dot{a}^2}{a^2} + \frac{kc^2}{a^2} = H^2 + \frac{kc^2}{a^2}, \tag{11.8}$$

which will be found to be related to the total energy.

Once substituted into the extended Einstein's equation (11.2) with source (11.5), the metric shown in interval (11.4) leads to the Friedmann equations for the scale parameter $a\,(t)$:

$$E(t) = \frac{\Lambda c^2}{3} + \frac{8\pi G}{3c^2}\,\epsilon, \tag{11.9}$$

$$C(t) = \frac{\Lambda c^2}{3} - \frac{4\pi G}{3c^2}\,(\epsilon + 3p)\,. \tag{11.10}$$

It is a common practice in cosmology to indicate present-day values by the index "0". Today's value of the Hubble function, for example, is the Hubble constant H_0. And the red-shift parameter z is defined by

$$1 + z = \frac{a_0}{a(t)}, \tag{11.11}$$

in terms of the scale parameter and its present-day value a_0.

The "fluid" energy density ϵ and pressure p include those of usual matter, dark matter and radiation. It will be convenient to use for them the notations ϵ_m and p_m and introduce the "dark energy density"

$$\epsilon_\Lambda = \frac{\Lambda c^4}{8\pi G}. \tag{11.12}$$

The cosmological term is equivalent to an "exotic" fluid, with the equation of state

$$\epsilon_\Lambda = -p_\Lambda = \text{constant}.$$

We shall use a simplifying notation, defining

$$\epsilon_\Lambda = -p_\Lambda = -\tfrac{1}{2}\left(\epsilon_\Lambda + 3\,p_\Lambda\right).$$

It is then possible to write ϵ_T and p_T, respectively, for the total energy and pressure, and rewrite equations (11.9)–(11.10) as

$$E(t) = \frac{8\pi G}{3c^2}\left(\epsilon_m + \epsilon_\Lambda\right) = \frac{8\pi G}{3c^2}\,\epsilon_T \qquad (11.13)$$

and

$$C(t) = \frac{8\pi G}{3c^2}\left[\epsilon_\Lambda - \tfrac{1}{2}\left(\epsilon_m + 3\,p_m\right)\right] = -\frac{4\pi G}{3c^2}\left(\epsilon_T + 3p_T\right). \qquad (11.14)$$

As announced, $E(t)$ is a measure of the total energy. The above expression for $C(t)$ relates the scale parameter concavity to the total energy and pressure. By the way, the middle equation in (11.14) exhibits clearly the contrasting effects of a (positive) cosmological constant and matter: $C < 0$ indicates decelerating expansion, whereas $C > 0$ an accelerating expansion.

The energy conservation expression involves both $C(t)$ and $E(t)$:

$$\dot{\epsilon}_T = \frac{3c^2}{8\pi G}\,\dot{E}(t) = \frac{3c^2}{4\pi G}\,H\left[C(t) - E(t)\right] = -3H(t)\left[\epsilon_T + p_T\right]. \qquad (11.15)$$

11.3   The cosmos curvature

The Christoffel symbols of metric (11.4) are rather involved, but can be all put together by using Kronecker deltas:[6]

$$\Gamma^\alpha{}_{\mu\nu} = \delta_{\mu\nu}\left[\delta_{1\mu}\frac{\delta_0^\alpha a^2 H + \delta_1^\alpha kr}{1 - kr^2}\right.$$
$$+ r\left[\delta_0^\alpha a^2 Hr - \delta_1^\alpha\left(1 - kr^2\right)\right]\left(\delta_{2\mu} + \delta_{3\mu}\sin^2\theta\right) - \delta_2^\alpha\delta_{3\mu}\sin\theta\cos\theta\bigg]$$
$$+ \left(\delta_{\lambda\mu}\delta_\nu^\alpha + \delta_{\lambda\nu}\delta_\mu^\alpha\right)(1 - \delta_0^\alpha)\left(\delta_{0\lambda}H + (\delta_2^\alpha + \delta_3^\alpha)\delta_{1\lambda}\frac{1}{r} + \delta_3^\alpha\delta_{2\lambda}\cot\theta\right).$$

$$(11.16)$$

---

[6] We assume the summation convention restricted to upper-lower contractions. Repeated lower-lower and upper-upper indices are not summed over. The usual relativistic $(0, 1, 2, 3)$ notation will be extended in what follows to the indexing of matrix row and columns.

Anti-symmetries of the Riemann curvature tensor in both the first and the second pairs of indices can be obscured by the metric factors necessary to raise and/or lower them. The components $R^{\alpha\beta}{}_{\mu\nu}$, with two upper and two lower indices, are the simplest. Again through the use of Kronecker deltas, we find

$$
\begin{aligned}
R^{\alpha\beta}{}_{\mu\nu} &= \frac{1}{c^2} \left(\delta^\beta_\mu \delta^\alpha_\nu - \delta^\alpha_\mu \delta^\beta_\nu \right) \left[\left(H^2 + \dot{H} \right) \left(\delta_{0\mu} + \delta_{0\nu} \right) \right. \\
&\quad + \left. \left(H^2 + \frac{kc^2}{a^2} \right) \left(1 - \delta_{0\mu} - \delta_{0\nu} \right) \right] \\
&= \frac{1}{c^2} \left(\delta^\beta_\mu \delta^\alpha_\nu - \delta^\alpha_\mu \delta^\beta_\nu \right) \left[C \left(\delta_{0\nu} + \delta_{0\mu} \right) + E \left(1 - \delta_{0\mu} - \delta_{0\nu} \right) \right].
\end{aligned}
$$

The Einstein equation should give the gravitational field (that is, the Riemann tensor) in terms of the source characters. The Riemann tensor components given above depend on the scale parameter and its derivatives exactly through the expressions which appear in the Friedmann equations and can, consequently, be written directly in terms of the source contributions: equations (11.13, 11.14) lead immediately to the components

$$
R^{\alpha\beta}{}_{\mu\nu} = \frac{8\pi G}{3c^4} \left(\delta^\beta_\mu \delta^\alpha_\nu - \delta^\alpha_\mu \delta^\beta_\nu \right) \left\{ \epsilon_T - \tfrac{3}{2} (\epsilon_T + p_T)(\delta_{0\nu} + \delta_{0\mu}) \right\} . \tag{11.17}
$$

They depend, ultimatetly, only on the time coordinate. The Ricci tensor will have components

$$
\begin{aligned}
R^\alpha{}_\mu &= -\frac{1}{c^2} \delta^\alpha_\mu \left[\left(H^2 + \dot{H} \right) (1 + 2\delta_{0\mu}) + 2 \left(H^2 + \tfrac{k}{a^2} \right)(1 - \delta_{0\mu}) \right] \\
&= -\frac{1}{c^2} \delta^\alpha_\mu \left[C(1 + 2\delta_{0\mu}) + 2E(1 - \delta_{0\mu}) \right]. \tag{11.18}
\end{aligned}
$$

It can be presented alternatively as the matrix

$$
\begin{aligned}
(R^\alpha{}_\mu) &= -\frac{1}{c^2} \begin{pmatrix} 3C & 0 & 0 & 0 \\ 0 & 2E + C & 0 & 0 \\ 0 & 0 & 2E + C & 0 \\ 0 & 0 & 0 & 2E + C \end{pmatrix} \\
&= \frac{4\pi G}{c^4} \begin{pmatrix} \epsilon_T + 3p_T & 0 & 0 & 0 \\ 0 & p_T - \epsilon_T & 0 & 0 \\ 0 & 0 & p_T - \epsilon_T & 0 \\ 0 & 0 & 0 & p_T - \epsilon_T \end{pmatrix}, \tag{11.19}
\end{aligned}
$$

whose trace is the scalar curvature

$$
\begin{aligned}
R &= -\frac{6}{c^2} \left[2H^2 + \dot{H} + \frac{kc^2}{a^2} \right] \\
&= -\frac{6}{c^2} (E + C) = \frac{8\pi G}{c^4} (3\, p_T - \epsilon_T). \tag{11.20}
\end{aligned}
$$

Commentary 11.1 The Gaussian curvature will be

$$K_G = -\frac{R}{12c^2} = \frac{2\pi G}{3c^4}(\epsilon_T - 3p_T).$$

◄

By the way, from this comes a general "equation of state", which must hold for any homogeneous isotropic model with scalar curvature R:

$$p_T = \frac{\epsilon_T}{3} + \frac{Rc^4}{24\pi G}. \tag{11.21}$$

Finally, the Einstein tensor will be

$$(G^\alpha{}_\mu) = (R^\alpha{}_\mu - \tfrac{1}{2}\delta^\alpha_\mu R) = \begin{pmatrix} \frac{3E}{c^2} & 0 & 0 & 0 \\ 0 & \frac{2C+E}{c^2} & 0 & 0 \\ 0 & 0 & \frac{2C+E}{c^2} & 0 \\ 0 & 0 & 0 & \frac{2C+E}{c^2} \end{pmatrix}$$

$$= \frac{8\pi G}{c^4} \begin{pmatrix} \epsilon_T & 0 & 0 & 0 \\ 0 & -p_T & 0 & 0 \\ 0 & 0 & -p_T & 0 \\ 0 & 0 & 0 & -p_T \end{pmatrix}. \tag{11.22}$$

This is, by the way, an alternative version of equation (11.2).

The most convenient form of the Friedmann equations is written in terms of dimensionless variables. Indeed, define the parameters

$$\rho_{crit} = \frac{3H_0^2}{8\pi G}; \quad \Omega_m = \frac{\rho_m}{\rho_{crit}} = \frac{8\pi G\rho_m}{3H_0^2}; \tag{11.23}$$

$$\Omega_\Lambda = \frac{\Lambda c^2}{3H_0^2} = \frac{\rho_\Lambda}{\rho_{crit}}; \quad \Omega_\kappa(t) = -\frac{\kappa c^2}{a^2 H_0^2}; \quad \omega = \frac{p_m}{\epsilon_m}. \tag{11.24}$$

Parameter ω, in particular, is frequently used in the literature. With these parameters, we can write

$$\frac{E}{H_0^2} - \frac{\kappa c^2}{a^2 H_0^2} = \frac{H^2}{H_0^2} = \Omega_\Lambda + \Omega_\kappa + \Omega_m \tag{11.25}$$

and

$$\frac{C}{H_0^2} = \Omega_\Lambda - \frac{1+3\omega}{2}\Omega_m. \tag{11.26}$$

For present-day values, we get the standard normalization of observed parameters:

$$\Omega_\Lambda + \Omega_{\kappa 0} + \Omega_{m0} = 1.$$

Observations favor the values

$$\Omega_\kappa \approx 0, \qquad \Omega_\Lambda \approx 0.75, \qquad \Omega_m \approx 0.25,$$

the latter including all visible matter, radiation and dark matter.

General References

Very commendable general references are: Ludvigsen 1999, Harrison 2001, Rindler 2007 and Weinberg 2008.

Chapter 12

Planck Scale Kinematics

For millennia, knowledge of the underlying geometry has come from our everyday-life experience. Even rather sophisticated topological results have been arrived at through practice: we pave plane surfaces with regular hexagons, but spherical soccer balls with pentagons. In the last few centuries great advances in experimental physics have provided major advances in that knowledge. Astronomical observations in the last few decades are seemingly suggesting new changes, which can come to cause a complete revision of our basic notions, even of the space we live in.

Recent years have witnessed a growing interest in the possibility that Special Relativity may have to be modified at ultra-high energies.[1] Reasons have been advanced both from the theoretical and the experimental-observational points of view.

On the theoretical side, suggestions in that direction come from the physics at the Planck scale, where a fundamental length parameter, given by the Planck length

$$l_P = \left(\frac{G\hbar}{c^3}\right)^{1/2}$$

naturally shows up. Since a length contracts under a Lorentz transformation, the Lorentz symmetry is usually assumed to be somehow broken at that scale.[2] On the experimental side, intimations come basically from the

---

[1] See, for example, Magueijo & Smolin 2002, Amelino–Camelia 2002, Kowalski–Glikman 2005 and 2006, and Das & Kong 2006.

[2] Amelino–Camelia 2000, Protheroe & Meyer 2000, Ahluwalia 2002, Jacobson, Liberati & Mattingly 2002, Myers & Pospelov 2003 and Brandenberger & Marti 2002.

propagation of extragalactic very-high energy gamma-ray bursts. According to these observations, higher energy gamma-rays seem to travel slower than lower energy ones.

A possible explanation for such behaviour may be obtained from quantum gravity considerations, according to which high energies might cause small-scale fluctuations in the texture of spacetime. These fluctuations could, for example, act as small-scale lenses, interfering in the propagation of ultra-high energy photons. The higher the photon energy, the more it changes the spacetime structure, the larger the interference will be. This kind of mechanism could be the cause of the recently observed delay in high energy gamma-ray flares from the heart of the galaxy Markarian 501.[3] Those observations compared gamma rays in two energy ranges, from 1.2 to 10 TeV, and from 0.25 to 0.6 TeV. The first group arrived on Earth four minutes later than the second. If this comes to be confirmed, it will constitute a violation of Special Relativity.

Then comes the important point. Whenever one talks about possible violation of Special Relativity, one immediately thinks about violation of Lorentz symmetry. However, the kinematic group of Special Relativity is not Lorentz but Poincaré, which includes, in addition to Lorentz, also spacetime translations. This means that an eventual violation of Special Relativity does not necessarily mean that the Lorentz symmetry is violated. A violation of spacetime translations is another, quite reasonable possibility.

An example of such possibility is provided by the de Sitter group. As is well known, in stereographic coordinates this group is obtained from Poincaré simply by replacing spacetime translations by a combination of translations and *proper* conformal transformations.[4] In this case, the Lorentz symmetry is preserved, but the very notions of energy and momentum will change, which will then produce concomitant changes in the Special Relativity dispersion relation. These changes could eventually be able to account for any experimental discrepancy of ordinary Special Relativity. Furthermore, it seems to be much more sensible to think about changes in the definitions of energy and momentum than in the definition of spin, which would be the case if Lorentz symmetry were violated — not to mention the deep relation of Lorentz symmetry with causality.

Cosmology is perhaps the boldest enterprise of Physical Science — trying to describe the far out Universe with the laws it has found around the Earth. It has only recently become a phenomenology, and it is not surpris-

---

[3] Albert, Ellis, Mavromatos, Nanopoulos, Sakharov & Sarkisyan 2007.
[4] Gürsey 1962.

ing that some of its results are baffling. The very presence of a cosmological constant is bewildering enough. It is surely desirable that an explanation for it be found within the scheme of local physics, however modified. This may be the case, provided local kinematics is described by de Sitter invariant Special Relativity[5] instead of the ordinary Poincaré invariant Special Relativity.

Progress in Physics has been achieved through the use of a succession of correspondence principles. Special Relativity has for General Relativity the role classical physics has for quantum physics. Any change in the first will find its echoes in the second. The local kinematics of Special Relativity underpins that of General Relativity. Changing from Poincaré to de Sitter local kinematics impinges heavily on the more general kinematics.

We think of General Relativity as the dynamics of the gravitational field. In this sense, we can look at the presence of local *cosmological constants* as a kinematical effect. It changes spacetime itself, and consequently it redefines its kinematics. The other fundamental interactions exhibit no such character — they are, in this sense, more essentially dynamical. On the other hand, they are described within the gauge paradigm, which is more receptive to conformal symmetry. Einstein's sourceless equation is not conformally invariant, while the Yang–Mills equation is. It may well be that the passage from Poincaré to de Sitter relativities — encapsulated in the addition of the proper conformal transformations to usual translations — have far less traumatic consequences for electromagnetic, strong and weak interactions than it has for gravitation.

Our standard geometry was born from our intuitive notions of distance and space, notions which take their roots on our everyday experience. We have seen that, at distances very short compared with that experience, mechanics becomes quantal and geometry becomes non-commutative. Only quantum objects are able to probe very small pieces of space and geometry reveals itself as being quite different from that of our everyday experience. Nevertheless, for very large distances, we do maintain our standard geometry. We keep using that geometry to parametrise and fit the data we receive from very far away.

What if also for very large distances geometry should be quite another? Maybe dark-matter and dark-energy would turn up only because we are forcing the use of usual geometry for scales for which it is also no more valid. This is exactly what happens when the underlying kinematics is

---

[5] Aldrovandi, Beltrán Almeida & Pereira 2007.

assumed to be ruled by the de Sitter group. In fact, in this case spacetime will no longer reduce locally to a Minkowski spacetime, but to a de Sitter spacetime. This in turn means that spacetime will no longer present a riemannian structure, but will be described by a more general structure, known as Cartan geometry.[6] More precisely, it will be described by a special kind of Cartan geometry that reduces locally to a de Sitter geometry — and which is for this reason called de Sitter–Cartan geometry.[7]

In Chapter 37 the de Sitter invariant Special Relativity will be studied in some more detail, and a glimpse of the consequences for gravitation and cosmology will be presented. Among many other peculiarities, this geometry is consistent with Penroses's[8] cyclic view of the Universe.[9]

[6] Sharpe 1997.

[7] Wise 2010.

[8] Penrose 2011.

[9] Araujo, Jennen, Pereira, Sampson & Savi 2015.

Part 4: MATHEMATICAL TOPICS

Chapter 13

The Basic Algebraic Structures

Current mathematical literature takes for granted the notions given here, as a *minimum minimorum*, the threshold of algebraic literacy. As a rule, we shall only retain below those ideas which are necessary to give continuity to the presentation.

13.1 Some general concepts

Asking for the reader's forbearance let us only recall, to keep the language at hand, that the *cartesian set product* $U \times V$ of two sets U and V is the set of all ordered pairs (u, v) with $u \in U$ and $v \in V$.

A (binary) *relation* R on a set S: a subset of $S \times S$. Given a relation R between the points of a set, we write "pRq" to mean that p is in that relation with q. If p is not in relation R with q, we write this as a negative: "$\smallsmile pRq$".

Relation R is an *equivalence relation* when it is reflexive (each point p is in that relation with itself, pRp, $\forall p$), symmetric (pRq implies qRp) and transitive (pRq and qRr together imply pRr). Equality is the obvious example.

Loosely speaking, "near" is an equivalence, but "far" is only symmetric. On the plane with coordinates (x, y), the relation "has the same coordinate $x = x_0$" establishes an equivalence between all the points on the same vertical line, that is, for all values of y. An equivalence relation (for which notations $p \approx q$ and $p \equiv q$ are usual) divides a point set into *equivalence classes*, subsets of points having that relation between each other. In the above example on the plane, each vertical line is an equivalence class and can be labelled by the value of x_0.

Relation R is an *order relation* when it is reflexive, antisymmetric (pRq

plus qRp implies $p = q$) and transitive. An usual notation for pRq is, in this case, $p \le q$.

An *internal binary operation* "∘" on a set S is a rule assigning to each ordered pair of elements (a, b) of S another element "$a \circ b$" of S. More precisely, it is a mapping $\circ : S \times S \longrightarrow S$.

When $a \circ b = b \circ a$ for all $a, b \in S$, the operation is *commutative*. When $a \circ (b \circ c) = (a \circ b) \circ c$ for all $a, b, c \in S$, the operation is *associative*.

An operation establishes a structure on S, and a good notation would be $< S, \circ >$. A structured set is, however, most frequently denoted simply by the symbol of the set point itself, here S.

13.2   Groups and lesser structures

Groups are the most important of the algebraic structures. Humans enjoy their presence in some special way, through the bias of symmetry, and learning processes seem to make use of them. Since their unheeded introduction by an ill-fated young man, its fortune has been unsurpassed in Physics. We present next some relevant definitions related to groups and their lesser structures.

§ **13.2.1 Groups.** A group is a set point G on which is defined a binary operation

$$* : G \times G \longrightarrow G$$

taking the cartesian set product of G by itself into G, with the four following properties:

(i) For all $g, g' \in G$, the result $g * g'$ belongs to G (the operation is internal).

(ii) There exists in G an element e (the identity, or neutral element) such that, for all $g \in G$,

$$e * g = g * e = g.$$

(iii) To every $g \in G$ corresponds an inverse element g^{-1} which is such that

$$g^{-1} * g = g * g^{-1} = e .$$

(iv) The operation is associative: for all g, g', g'' in G,

$$(g * g') * g'' = g * (g' * g'').$$

The group $(G, *)$ is commutative (or abelian) when $g * g' = g' * g$ holds for all $g, g' \in G$. The operation symbol $*$ is usually omitted for multiplicative groups. The *center* of G is the set of the $g \in G$ which commute with all the elements of G. The *order* of a group G, written $|G|$, is the number of elements of G.

§ 13.2.2 Centralizer. The centralizer of a subset X of a group G is the set of elements of G which commute with every member of X; the centralizer of G itself is, of course, its center. If a is an element of G, its *conjugate class* $[a]$ is the set of all elements of G which can be put under the form xax^{-1} for some $x \in G$. Two conjugate classes $[a]$ and $[b]$ are either identical or disjoint. The element a belongs to the center of G if and only if $[a] = \{a\}$.

§ 13.2.3 Generators. Given a group G, if there is a set of $a_i \in G$ such that every $g \in G$ can be written in the monomial form $g = \prod_i (a_i)^{n_i}$ for some set of exponents $\{n_i\}$, we say that the set $\{a_i\}$ *generates* G, and call the a_i's *generators* of G. In the monomials, as the a_i's are not necessarily commutative, there may be repetitions of each a_i in different positions. When the set $\{a_i\}$ is finite, G is *finitely generated*. The number of generators of G is, in this case, called the *rank* of G.

§ 13.2.4 Transformation groups. As presented above, we have an abstract group, which is a concern of Algebra (as a mathematical discipline). A group formed by mappings from a set S on itself (that is, automorphisms) is a *transformation group*. The bijective mappings of S constitute the "largest" transformation group on S and the identity mapping constitutes the "smallest".

Commentary 13.1 The main interest of groups to Physics lies precisely on transformations preserving some important feature of a system. The word *symmetry*, when attributed to a physical system, is commonly reserved to the set of transformations preserving that system's hamiltonian — which presides over its time-evolution. ◄

§ 13.2.5 Homomorphism. The name *homomorphism* is given in general to any mapping preserving an algebraic structure. Thus, a mapping $h :$ $G \longrightarrow H$ of a group $(G, *)$ into another group $(H, \circ)$ is a *homomorphism* if it preserves the group structure, that is, if it satisfies

$$h(g * g') = h(g) \circ h(g').$$

The set $\ker h := \{\, g \in G \text{ such that } h(g) = id_H \}$ is the *kernel* of h and the set $\operatorname{im} h := \{ h \in H \text{ such that some } g \in G \text{ exists for which } h = h(g)\}$ is the *image* of h. When such a mapping exists, G and H are said to be *homomorphic*. A homomorphism of a set endowed with some algebraic structure (such as a group, a ring, a linear space, ...) into itself is an *endomorphism*. A *monomorphism* is an injective (one-to-one) homomorphism. A surjective homomorphism is an *epimorphism*.

§ **13.2.6 Isomorphism.** An *isomorphism* is a bijective (that is, one-to-one and onto) homomorphism. When such a mapping exists, the two algebraic objects are *isomorphic*.

§ **13.2.7 Canonical.** The word *canonical* is used for a homomorphism taking a given member of the starting set always into the same member of the target set. Thus, a canonical isomorphism between linear spaces is a basis-independent isomorphism. Generalizations of these concepts for other algebraic structures are straightforward.

§ **13.2.8 Representations.** A homomorphism of an abstract group into a transformation group is called a *representation* of the group. In the physicist usual language, abstract groups are frequently identified to some important homomorphic transformation group turning up in some physical system. We seldom talk about "the group $SO(3)$ of real 3×3 special orthogonal matrices"; we prefer to speak of "the group of rotations in the 3-dimensional euclidean space".

A homomorphism taking every element of a group into its own identity element is a *trivializer*. It leads to the so-called *trivial representation*. The theory of group representations is a vast chapter of Higher Mathematics, of which a very modest résumé is given in Chapter 18.

§ **13.2.9 Groupoids, monoids, semigroups.** We have seen that four requirements are made in the definition of a group. When some of them are not met, we have less stringent structures.[1] Thus, when only (b) and (c) hold, G is said to be a *groupoid*. If only (d) holds, G is a *semigroup* (some authors add condition (b) in this case). When only (b) and (d) are satisfied, G is a *monoid*. A monoid is seen as a unital semigroup, a groupoid appears as a non-associative group, and so on.

---

[1] There are large fluctuations concerning the definitions of monoid and semigroup. Some authors (such as Hilton & Wylie 1967) include the existence of neutral element for semigroups, others (Fraleigh 1974) only associativity. In consequence, the schemes here shown, concerned with these structures, are author-dependent. We follow, for simplicity, the glossary Maurer 1981.

In this way, groups being much more widely known, it is simpler to get at the other structures from them, by dropping some of their defining conditions. This is illustrated in the diagram

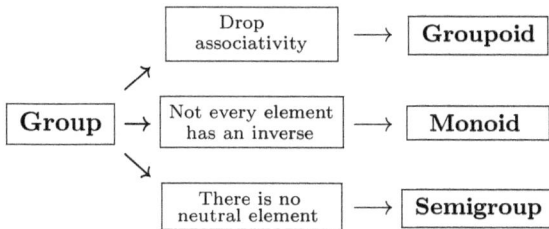

A general, more constructive scheme is

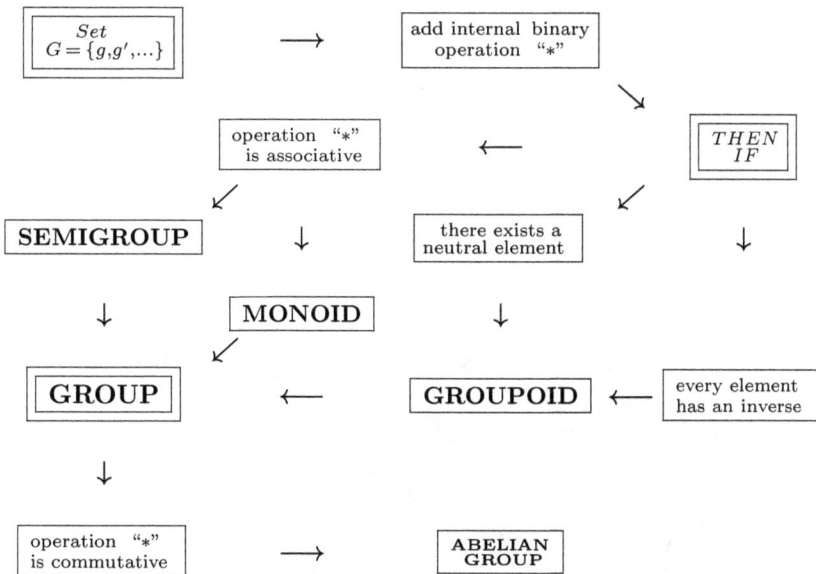

Commentary 13.2 Amongst the "lesser" structures, the term "semigroup" is widely used in physical literature to denote anything which falls short of being a group by failing to satisfy some condition. For example, the time evolution operators for the diffusion equation are defined only for positive time intervals and have no inverse. Also without inverses are the members of the so called "renormalization group" of Statistical Mechanics and Field Theory. But be careful. So strong is the force of established language, however defective, that a physicist referring to the "renormalization semigroup", or "monoid", runs the risk of being stoned to death by an outraged audience. ◀

We may further indulge ourselves with some Veblen-like diagrams (see Figure 13.1), with the structures in the intersections of the regions representing the properties.

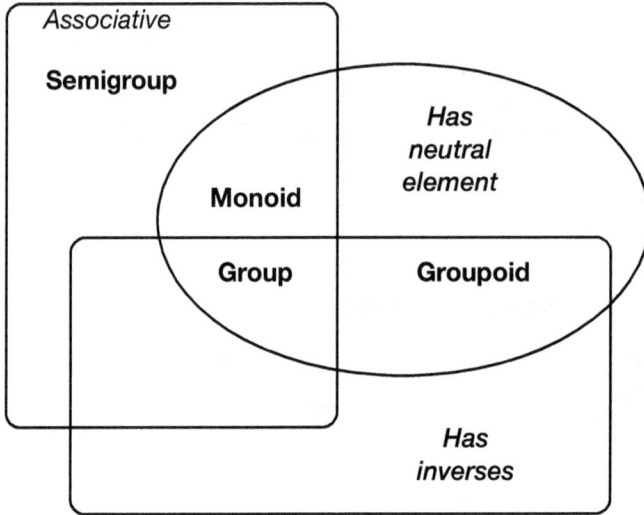

Fig. 13.1   The basic structures and their properties.

§ **13.2.10 Subgroups.** A subset H of a group G forming a group by itself with the same operation is a *subgroup*. The group G itself, as well as the group formed solely by its identity element, are *improper subgroups*. Any other subgroup is *proper*. A subgroup H of a group G is a *normal subgroup* (or *invariant subgroup, or else ideal*) if $ghg^{-1} \in H$ for all $g \in G$ and $h \in H$. A group is *simple* if it has no invariant proper subgroup. A group is *semisimple* if it has no *abelian* invariant subgroup. Every simple group is, of course, also semisimple.

If the order of G is finite, then the order of any subgroup H must divide it (Lagrange theorem) and the *index* of H in G is $|G|/|H|$. If $|G|$ is a finite prime number, the Lagrange theorem will say that G has no proper subgroups and, in particular, will be simple. Given a group G with elements a, b, c, etc, the set of all elements of the form $aba^{-1}b^{-1}$ constitutes a normal subgroup, the *commutator subgroup* indicated $[G, G]$. The quotient $G/[G, G]$ (the set of elements of G which are not in $[G, G]$) is an abelian group, the *abelianized* group of G. The classical example is given by the fundamental group $\pi_1(M)$ of a manifold M and the first homology group $H_1(M)$: the latter is the abelianized group of the former.

Commentary 13.3 The homomorphism $\alpha : G \longrightarrow G/[G,G]$ taking a group G into its abelianized subgroup is canonical (that is, independent of the choice of the original generators) and receives the name of *abelianizer*. ◀

§ 13.2.11 Solvable and nilpotent groups. Groups can be classified according to their degree of non-commutativity. Given two subsets A and B of a group G, we define their *commutator* as $[A, B] = \{$ all elements in G of the form $aba^{-1}b^{-1}$, with $a \in A$ and $b \in B\}$. We start with the abelian quotient $G/[G,G]$. We may then form two sequences of subgroups in the following way. Define $G_1 = [G, G]$, the commutator subgroup of G; proceed to $G_2 = [G_1, G_1]$ and so on to the n-th term of the series, $G_n = [G_{n-1}, G_{n-1}]$. Putting $G = G_0$, then $G_{n-1} \supset G_n$ for all n. If it so happens that there exists an integer k such that $G_n =$ identity for all $n \geq k$, the group G is *solvable* of class k. A second sequence is obtained by defining $G^0 = G$ and $G^n = [G, G^{n-1}]$. If $G^n =$ identity for $n \geq k$, the group G is *nilpotent* of class k. Of course, both sequences are trivial for abelian groups. And the less trivial they are, the farther G stands from an abelian group.

13.3 Rings, ideals and fields

These are structures mixing up two internal operations.

§ 13.3.1 Rings. A ring $< R, +, \cdot >$ is a set R on which two binary internal operations, "$+$" and "$\cdot$", are defined and satisfy the following axioms:

(i) $< R, + >$ is an abelian group.
(ii) "$\cdot$" is associative.
(iii) Both operations respect the distributive laws: for all $a, b, c \in R$,

$$a \cdot (b + c) = a \cdot b + a \cdot c \quad \text{and} \quad (a + b) \cdot c = a \cdot c + b \cdot c.$$

The multiplication symbol "$\cdot$" is frequently omitted: $a \cdot b = ab$. When the operation "$\cdot$" is commutative, so is the ring. When a multiplicative identity "1" exists, such that $1 \cdot a = a \cdot 1 = a$ for all $a \in R$, then $< R, +, \cdot >$ is a ring with unity (*unital*, or *unit ring*). In a unit ring, the multiplicative inverse

(not necessarily existent) to $a \in R$ is an element a^{-1} such that
$$a{\cdot}a^{-1} = a^{-1}{\cdot}a = 1.$$
If every nonzero element of R has such a multiplicative inverse, then $< R, +, \cdot >$ is a *division ring*. The subset R' of R is a *subring* if $a{\cdot}b \in R'$ when $a \in R'$ and $b \in R'$.

Commentary 13.4 Let G be an abelian group. The ring R is G-graded if, as a group, R is a direct sum of groups R_α, for $\alpha \in G$, such that $R_\alpha \times R_\beta$ is contained in $R_{\alpha+\beta}$. The notation "na" is frequently used to indicate "$a + a + \ldots + a$", with n summands. ◀

As illustrations, we list below a few examples of rings:

- The set $\mathbb{Z}$ of integer numbers with the usual operations of addition and multiplication is a commutative ring, though not a division ring;
- The set $\mathbb{Z}_n$ of integers modulo n is also a commutative (but not a division) ring, formed from the cyclic group $< \mathbb{Z}_n, + >$ with the multiplication modulo n then indicated simply by "$\cdot$" ; it is denoted $< \mathbb{Z}_n, +, \cdot >$;
- The set $R(t)$ of polynomials in the variable t with coefficients in a ring R is also a ring ;
- The set $M_n[R]$ of $n \times n$ matrices whose entries belong to a ring R is itself a ring with unity if R is a ring with unity; it is not a division ring;
- The set of real or complex functions on a topological space S, with addition and multiplication given by the pointwise addition and product:
$$(f + g)(x) = f(x) + g(x) \quad \text{and} \quad (fg)(x) = f(x)\,g(x).$$

Some relevant additional properties are:

- It may happen that positive integers n_i exist for which $(n_i a) = 0$ for all $a \in R$. Then $n = \text{minimum } \{n_i\}$ is the *characteristic* of R. When no such n_i's exist, R is of characteristic zero. The rings $\mathbb{Z}$, $\mathbb{R}$ and $\mathbb{C}$ are of this kind. $\mathbb{Z}_n$ is of characteristic n.

- Two elements l and r of a ring $< R, +, . >$ are (respectively left and right) divisors of zero if they are nonzero and such that $l.r = 0$. In a commutative ring, left (right) divisors of zero are right (left) *divisors of zero*. In the ring $< \mathbb{Z}_n, +, . >$, all numbers not relatively prime to n are divisors of zero (and only them). So, 2 and 3 are divisors of zero in $\mathbb{Z}_6$. In particular, $\mathbb{Z}_n$ has no divisors of zero when n is a prime number.

- An *integral domain* is a commutative ring with unity and with no divisors of zero. Every field is an integral domain. Every finite integral domain is a field. $\mathbb{Z}_n$ is a field when n is prime.

§ 13.3.2 Ideal.

The sub-ring R' of a ring R is a left-ideal if

$$a \times b \in R' \text{ for all } a \in R \text{ and } b \in R'.$$

It is a right-ideal if

$$a \times b \in R' \text{ for all } a \in R' \text{ and } b \in R.$$

And it is a *bilateral ideal* if

$$a \times b \in R' \text{ when either } a \in R' \text{ or } b \in R.$$

When nothing else is said, the word *ideal* is used for bilateral ideals. When R' is such a bilateral ideal, there is a natural multiplication defined on the group R/R'. The resulting ring is a quotient ring. The ring $< \mathbb{Z}_n, +, . >$ is the quotient of $\mathbb{Z}$ by the ideal formed by all the multiples of n: $\mathbb{Z}_n = \mathbb{Z}/n\mathbb{Z}$. The ring R' is a *maximal ideal* of R if, for any other ideal R'', $R' \subset R''$ implies $R'' = R'$. In the ring of complex functions on a topological space S, those functions vanishing at a certain point $p \in S$ form an ideal. If R is a ring with unity, and N is an ideal of R containing an element with a multiplicative inverse, then $N = R$. Because of the analogy, the name "ideal" is also used for a normal subgroup.

Two additional relevant notions involving operations on sets, and which are dual to each other, are:

- *Set ideal:* On a set S, an ideal is a family $\mathcal{I}$ of subsets of S such that (i) the whole set S does *not* belong to $\mathcal{I}$, (ii) the union of two members of $\mathcal{I}$ is also a member, and (iii) any subset of S contained in a member of $\mathcal{I}$ is also a member.

- *Filter:* A filter on the set S is a family $\mathcal{F}$ of subsets of S such that (i) the empty set $\emptyset$ does *not* belong to $\mathcal{F}$, (ii) the intersection of two members of $\mathcal{F}$ is also a member, and (iii) any subset of S containing a member of $\mathcal{F}$ is also a member. The simplest example is the set of all open neighbourhoods of a fixed point $p \in S$. An *ultrafilter* is a filter which is identical to any filter finer to it. Filters and ultrafilters are essential to the study of continuity and convergence in non-metric topological spaces.[2]

---

[2] A good reference on filters and their role in geometry is Elworthy, Le Jan & Li 2010.

§ 13.3.3 Field.

A field is a commutative division ring. The sets of real numbers ($\mathbb{R}$) and complex numbers ($\mathbb{C}$) constitute fields. Familiarity with real and complex numbers makes fields the best known of such structures. It is consequently more pedagogical to arrive at the concept of ring by lifting progressively their properties. Let us start with a Field: dropping "product commutativity" yields a Division Ring. Dropping "existence of product inverse" leads to a Unital Ring. Dropping then the "existence of product neutral" yields a Ring. The Figure below shows these properties in a pictorial form.

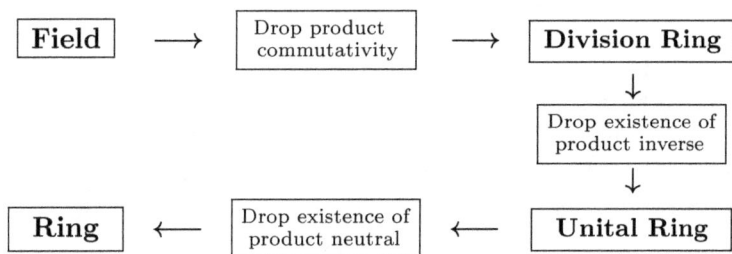

Fig. 13.2　The basic relations between Field and Ring.

§ 13.3.4 Ring of a group.
We can always "linearize" a group by passing into a ring. There are two rings directly related to a given group. A first definition is as follows. Take a ring R and consider the set of all formal summations of the form

$$a = \sum_{g \in G} a(g)\, g,$$

where only a finite number of the coefficients $a(g) \in R$ are non-vanishing. Assume then addition and multiplication in the natural way:

$$a + b = \sum_{g \in G} a(g)\, g + \sum_{g \in G} b(g)\, g = \sum_{g \in G} [a(g) + b(g)]\, g \tag{13.1}$$

and

$$ab = \sum_{g,h \in G} a(g)\, b(h)\, g\, h. \tag{13.2}$$

The summations constitute a new ring, the *group ring* $R(G)$ of G over R. If R is a unital ring, each $g \in G$ can be identified with that element "a" of

$R(G)$ whose single coefficient is $a(g) = 1$. The group is thereby extended to a ring. If G is non-abelian, so will be the ring. Now, to each $a \in R(G)$ corresponds an R-valued function on G, $f_a \colon G \to R$, such that

$$f_a(g) = a(g) \quad \text{and} \quad a = \sum_{g \in G} f_a(g)\, g.$$

Conversely, to any function $f \colon G \to R$ will correspond a ring member $\sum_{g \in G} f(g)g$. Notice that, given

$$a = \sum_{g \in G} a(g)\, g \quad \text{and} \quad b = \sum_{h \in G} b(h)h,$$

the product $ab = \sum_{g \in H} f_{ab}(g)g$ will be given by

$$ab = \sum_{g \in H} \left[\sum_{h \in H} f_a(h)\, f_b(gh^{-1}) \right] g.$$

To the product ab will thus correspond the convolution

$$f_{ab}(g) = \sum_{h \in H} f_a(h)\, f_b(gh^{-1}).$$

We arrive thus at another definition: given a group G, its group ring $R(G)$ over R is the set of mappings $f \colon G \to R$, with addition defined as

$$(f_1 + f_2)(g) = f_1(g) + f_2(g),$$

and multiplication in the ring given by the convolution product

$$(f_1 * f_2)(g) = \sum_{h \in G} f_1(h)\, f_2(h^{-1}g).$$

The condition concerning the finite number of coefficients in the first definition is necessary for the summations to be well-defined. Also the convolution requires a good definition of the sum $\sum_h$ over all members of the group. That is to say that it presupposes a measure on G. We shall briefly discuss measures in Section 15.1. If G is a multiplicative group, $R(G)$ is indicated by $< R, +, \cdot >$ and is "the ring of G over R". The restriction $< R, \cdot >$ contains G. If G is noncommutative, $R(G)$ is a noncommutative ring. $R(G)$ is sometimes called the "convolution ring" of G, because another ring would come out if the multiplication were given by the pointwise product

$$(f_1 f_2)(g) = f_1(g) f_2(g).$$

The analogy with Fourier analysis is *not* fortuitous (see Section 18.2 and those following). In order to track groups immersed in a ring R, we can use idempotents. An *idempotent* (or *projector*), as the name indicates, is an element p of the ring such that $p \cdot p = p$. A group has exactly one idempotent element.

13.4   Modules and vector spaces

These are structures obtained by associating two of the above ones.

§ **13.4.1 Modules.** An R-module is given by an abelian group M of elements $\alpha, \beta, \gamma, \ldots$ and a ring $R = \{a, b, c, \ldots\}$ with an operation of external multiplication of elements of M by elements of R satisfying the four axioms:

  (i) $a\,\alpha \in M$;
  (ii) $a\,(\alpha + \beta) = a\,\alpha + a\,\beta$;
  (iii) $(a + b)\,\alpha = a\,\alpha + b\,\alpha$;
  (iv) $(ab)\,\alpha = a\,(b\,\alpha)$.

It is frequently denoted simply by M. The external product has been defined in the order RM, with the element R at the left, so as to give things like $a\alpha$. The above module is called a "left-module". We can define right modules in an analogous way. A bilateral module (*bimodule*) is a left- and right-module. When we say simply "module", we always mean actually a bimodule.

Modules generalize vector spaces, which are modules for which the ring is a field. As vector spaces belong to the common lore, we might better grasp modules by starting with a vector space and going backwardly through the steps of Figure 13.3. But here we are trying to be constructive. We shall recall their main characteristics in a slightly pedantic way while profiting to introduce some convenient language.

§ **13.4.2 Vector spaces.** A linear space (on the field F, for us the real or complex numbers) is an abelian group V with the addition operation "$+$", on which is defined also an external operation of ("scalar") multiplication by the elements of F. For all vectors $u, v \in V$ and numbers $a, b \in F$, the following five conditions should be met:

  (i) $au \in V$;
  (ii) $a(bu) = (ab)u$;
  (iii) $(a + b)u = au + bu$;
  (iv) $a(u + v) = au + av$;
  (v) $1u = u$.

We say that V is a linear (or vector) space *over* F. The field F is a vector space over itself. If the field F is replaced by a ring, we get back a module.

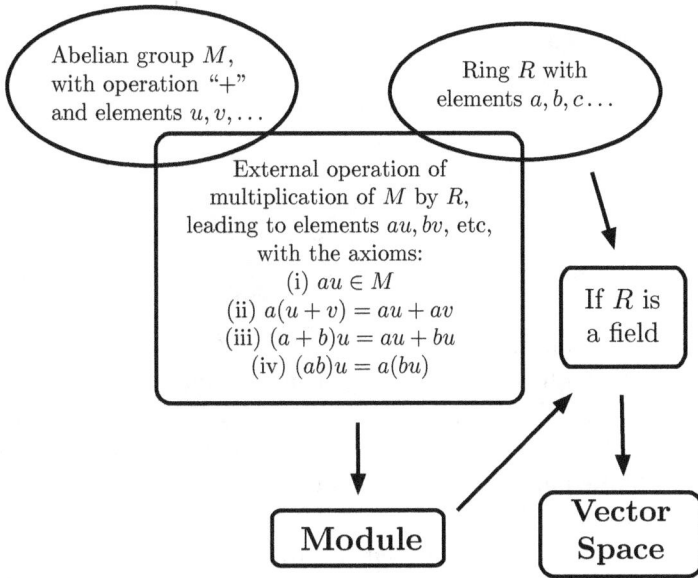

Fig. 13.3   Modules and vector spaces.

V is a *real* vector space if $F = \mathbb{R}$. The name *linear space* is to be preferred to the more current one of vector space, but as this pious statement seems useless, we use both of them indifferently.

§ **13.4.3 The notion of action.**   We repeat that the *cartesian set product* $U \times V$ of two sets U and V is the set of all pairs (u, v) with $u \in U$ and $v \in V$. A well-defined mapping $U \times V \to V$ goes under the general name of "action of U on V". The above axioms for vector spaces define an action of the field F on V. Linear representations of algebras are actions on modules and the classification of modules is intimately related to that of representations.

§ **13.4.4 Dimension.**   A family $\{v_k\}$ of vectors is said to be *linearly dependent* if there are scalars $a_k \in F$, not all equal to zero, such that $\sum_k a_k v_k = 0$. If, on the contrary, $\sum_k a_k v_k = 0$ implies that all $a_k = 0$, the v_k's are *linearly independent*. This notion is extended to infinite families: an infinite family is linearly independent if every one of its finite subfamilies is linearly independent. The number of members of a family is arbitrary.
The maximal number of members of a linearly independent family is, how-

ever, fixed: it is the *dimension* of V, indicated dim V. If $\{v_k\}$ is a family of linear independent vectors, a vector subspace W, with dim $W <$ dim V, will be engendered by a subfamily $\{w_k\}$ with dim W members. The set of vectors of V with zero coefficients a_k along the w_k's will be the linear complement of W.

§ **13.4.5 Dual space.** Vector spaces are essentially duplicated. This is so because the space V^* formed by linear mappings on a vector space is another vector space, its dual. The image of $v \in V$ by the linear mapping $k \in V^*$ is indicated by $< k, v >$.

When V is finite-dimensional, the dual of V is a twin space, isomorphic to V. In the infinite-dimensional case, V is in general only isomorphic to a subspace of V^{**}.

A mapping of V into its dual is an *involution*. The isomorphism $V \approx V^*$, even in the finite-dimensional case, is in general not canonical. This means that it depends of the basis chosen for V. The presence of a norm induces a canonical isomorphism between V and (at least part of) V^*, and involutions are frequently defined as a mapping $V \to V$ (see below, Section 13.5.1). The action of the dual on V may define an inner product $V \times V \to F$.

§ **13.4.6 Inner product.** An inner product is defined as a mapping from the cartesian set product $V \times V$ into $\mathbb{C}$, defined by

$$V \times V \to \mathbb{C}, \quad (v, u) \to < v, u >$$

with the following properties:

(i) $< v_1 + v_2, u > = < v_1, u > + < v_2, u >$;

(ii) $< a\,v, u > = a < v, u >$;

(iii) $< v, u > = < u, v >^*$ (so that $< v, v > \in \mathbb{R}$);

(iv) $< v, v > \geq 0$;

(v) $< v, v > = 0 \leftrightarrow v = 0$.

The action of the dual V^* on V defines an inner product if there exists an isomorphism between V and V^*, which is the case when V is of finite dimension. An inner product defines a topology on V. Vector spaces endowed with a topology will be examined in Chap. 16 and Chap. 17.

Infinite dimensional vector spaces differ deeply from finite dimensional ones. For example, closed bounded subsets are not compact neither in the norm nor in the weak topology.

§ 13.4.7 Endomorphisms and projectors.

An *endomorphism* (or linear operator) on a vector space V is a mapping $V \to V$ preserving its linear structure. Projectors, or idempotents, are here particular endomorphisms p satisfying $p^2 = p$. Given a vector space E and a subspace P of E, it will be possible to write any vector v of E as $v = v_p + v_q$, with $v_p \in P$. The set Q of vectors v_q will be another subspace, the supplement of P in E. Projectors p on P (which are such as $p(v_p) = v_p, \operatorname{Im} p = P$) are in canonical (that is, basis independent) one-to-one correspondence with the subspaces supplementary to P in E: to the projector p corresponds the subspace $Q = \ker p$.

§ 13.4.8 Tensor product.
Let A and B be two vector spaces (on the same field F) with respective basis $\{x_i\}$ and $\{y_j\}$. Consider the linear space C with basis $\{x_i y_j\}$, formed by those sums of formal products $\sum_{m,n} a_{mn} x_m y_n$ defined by

$$\sum_{m,n} a_{mn} x_m y_n = \sum_n \left(\sum_m a_{mn} x_m \right) y_n = \sum_m x_m \left(\sum_n a_{mn} y_n \right),$$

when only a finite number of the coefficients $a_{mn} \in F$ is different from zero. We obtain in this way a novel vector space, the tensor product of A and B, denoted $A \otimes B$. The (above made use of) alternative notations $x_i y_j$ for $x_i \otimes y_j$ and $\sum_{m,n} a_{mn} x_m y_n$ for $\sum_{m,n} a_{mn} x_m \otimes y_n$ are both usual. Given two elements

$$a = \sum_m a_m x_m \in A \quad \text{and} \quad b = \sum_n b_n y_n \in B,$$

then

$$a \otimes b = \sum_{m,n} a_m b_n x_m \otimes y_n.$$

The elements of $A \otimes B$ have the following three properties:

(i) $(a + a') \otimes b = a \otimes b + a' \otimes b$;
(ii) $a \otimes (b + b') = a \otimes b + a \otimes b'$;
(iii) $r\,(a \otimes b) = (r\,a) \otimes b = a \otimes (rb)$,

for all $a, a' \in A$, $b, b' \in B$ and $r \in F$. Conversely, these properties define C in a basis-independent way.

Commentary 13.5 Consider the space D of all the mixed bilinear mappings
$$\rho : A \otimes B \to F, \quad (a, b) \to \rho(a, b).$$
Then $A \otimes B$ is dual to D. In the finite dimensional case, they are canonically isomorphic if $< a \otimes b, \rho >= \rho(a, b)$. ◀

The product of a space by itself, $A \otimes A$, has special characteristics. It may be divided into a symmetric part, given by sums of formal products of the form $\sum_{m,n} a_{mn} x_m x_n$, with $a_{mn} = a_{nm}$, and an antisymmetric part, formed by those sums of formal products of type $\sum_{m,n} a_{mn} x_m x_n$ with $a_{mn} = -a_{nm}$.

§ 13.4.9 Affine space. An *affine space* A is a subspace of a linear space V whose elements may be written in the form $a = k + v_0$, with k in a linear subspace of V and v_0 a fixed point of V. Roughly speaking, it is a vector space V in which the origin (the zero vector) is not fixed, or not relevant. Or, if we so wish, it is the space of the points of V. Vectors are then differences between points in the affine space. Physics deals actually with affine spaces whenever invariance under translations holds, so that the true position of the origin is irrelevant.

13.5 Algebras

The word *algebra* denotes, of course, one of the great chapters of Mathematics. But, just as the word "topology", it has also a particular, restricted sense: it also denotes a very specific algebraic structure.

§ 13.5.1 Algebra. An algebra is a vector space A over a field F (for us, $\mathbb{R}$ or $\mathbb{C}$) on which is defined a binary operation (called *multiplication*)

$$m : A \otimes A \to A, \quad (\alpha, \beta) \to m(\alpha, \beta) = \alpha\beta$$

such that, for all $a \in F$ and $\alpha, \beta, \gamma \in A$, the following conditions hold:

(i) $(a\alpha)\beta = a(\alpha\beta) = \alpha(a\beta)$;
(ii) $(\alpha + \beta)\gamma = \alpha\gamma + \beta\gamma$;
(iii) $\alpha(\beta + \gamma) = \alpha\beta + \alpha\gamma$.

This defines an action of the vector space on itself. Once so defined, A is an "algebra on F". When $F = \mathbb{C}$, a mapping $\alpha \in A \to \alpha^* \in A$ is an *involution*, or adjoint operation, if it satisfies the postulates

(i) $\alpha^{**} = \alpha$;
(ii) $(\alpha\beta)^* = \beta^*\alpha^*$;
(iii) $(a\alpha + b\beta)^* = a^*\alpha^* + b^*\beta^*$.

In that case α^* is the *adjoint* of α.

§ 13.5.2 Kinds of algebras.

The algebra A is *associative* if further

(iv) $(\alpha\beta)\gamma = \alpha(\beta\gamma) \ \forall \alpha, \beta, \gamma \in A.$

This property can be rendered by saying that the diagram

$$
\begin{array}{ccc}
& A \otimes A & \\
\overset{m \otimes id}{\nearrow} & & \overset{m}{\searrow} \\
A \otimes A \otimes A & & A \\
\underset{id \otimes m}{\searrow} & & \underset{m}{\nearrow} \\
& A \otimes A &
\end{array}
$$

is commutative. We say that α and $\beta \in A$ commute when $m(\alpha, \beta) = m(\beta, \alpha)$. Algebra A is *commutative* when $m(\alpha, \beta) = m(\beta, \alpha) \ \forall \ \alpha, \beta \in A$. A module can be assimilated to a commutative algebra.

Commentary 13.6 The *center* of an algebra A is the subalgebra formed by those elements commuting with all elements of A. In more detail: the centralizer (the term "commutant" is more used for von Neumann algebras) X' of a subset X of A is the set of elements of A commuting with every member of X; the center is the centralizer of A itself. ◄

A is a *unit algebra* (or unital algebra) if it has a unit, that is, an element "e" such that

$$\alpha e = e \alpha = \alpha, \ \forall \alpha \in A.$$

Notice that each $\alpha \in A$ defines a mapping $h_\alpha \colon F \to A$ by $h_\alpha(a) = a\,\alpha$. Of course, $\alpha = h_\alpha(1)$. The existence of a unit can be stated as the existence of an element "e" such that $h_e(a) = a\,e = a$ for all $a \in F$. This is the same as the commutativity of the diagram

$$
\begin{array}{ccc}
& A \otimes A & \\
\overset{h \otimes id}{\nearrow} & & \overset{id \otimes h}{\nwarrow} \\
F \otimes A & \downarrow m & A \otimes F \\
\searrow & & \swarrow \\
& A &
\end{array}
$$

Homomorphisms of unital algebras take unit into unit. A unit subalgebra of A will contain the unit of A. We can put some of such structures in a scheme:

```
┌──────────────┐      ┌───────────────┐      ┌──────────┐
│ Vector Space │  →   │ Add associative│  →   │ Algebra  │
└──────────────┘      │ multiplication │      └──────────┘
                      └───────────────┘
       ↓                                            ↓
┌──────────────────┐   ┌──────────┐   ┌───────────────┐
│ Add inner product│   │ Add unit │   │ Add involution│
└──────────────────┘   └──────────┘   └───────────────┘
       ↓                    ↓      ↖         ↓
  ┌──────────────┐    ┌──────────────┐   ┌──────────────┐
  │ Inner Product│    │  Involutive  │   │  Involutive  │
  │    Space     │    │ Unit Algebra │   │   Algebra    │
  └──────────────┘    └──────────────┘   └──────────────┘
```

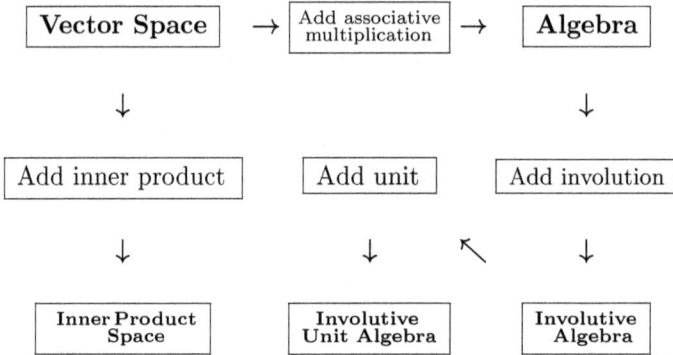

An element α of an unit-algebra A is an *invertible* element if there exists in A an element α^{-1} such that

$$\alpha\,\alpha^{-1} = \alpha^{-1}\alpha = e.$$

The set $G(A)$ of the invertible elements of A is a group with the multiplication, "the group of the algebra A".

A *graded algebra* is a direct sum of vector spaces, $A = \oplus_k A_k$, with the binary operation taking

$$A_i \otimes A_j \to A_{i+j}.$$

If $\alpha \in A_k$, we say that k is the *degree* (or *order*) of α, and write $\partial_\alpha = k$. An example is the space of differential forms of every order on a manifold M, which is actually a *differential graded algebra*, that is, a graded algebra on which is defined a *graded derivative*, a derivation D such that

$$D(\alpha\beta) = (D\alpha)\beta + (-)^{\partial_\alpha}\alpha D\beta.$$

The standard example of graded derivative is the exterior derivative. Differential graded algebras are of special interest because especially prone to cohomology.

§ 13.5.3 **Lie algebra.** An algebra is a Lie algebra if its multiplication (called the "Lie bracket") is anticommutative and satisfies the Jacobi identity

$$m(\alpha, m(\beta, \gamma)) + m(\gamma, m(\alpha, \beta)) + m(\beta, m(\gamma, \alpha)) = 0.$$

Starting from any binary operation, the Lie bracket can always be defined as the commutator

$$[\alpha, \beta] = \alpha\beta - \beta\alpha,$$

and builds, from any associative algebra A, a Lie algebra A_L. If A is any algebra (not necessarily associative, even merely a vector space),

$$\text{End } A = \{\text{set of endomorphisms on } A\}$$

is an associative algebra. Then, the set $[\text{End } A]$ of its commutators is a Lie algebra. A vector basis $\alpha_1, \alpha_2, \ldots, \alpha_n$ for the underlying vector space will be a basis for the Lie algebra.

Lie algebras have a classification analogous to groups. They may be solvable, nilpotent, simple, semisimple, etc, with definitions analogous to those given for groups.

Commentary 13.7 Drop in the algebra A the external multiplication by scalars of its underlying vector space. What remains is a ring. As long as this external multiplication is irrelevant, we can talk of a ring. The usual language is rather loose in this respect, though most people seem to prefer talking about "algebras". ◄

Commentary 13.8 Modules may be obtained by applying a projector (here, an element p of the algebra such that $p * p = p$) to an algebra. ◄

§ **13.5.4 Enveloping algebra.** To every finite Lie algebra A will correspond a certain unital associative algebra, denoted $\mathcal{U}$ and called the *universal enveloping algebra* of A. Given a basis $\{\alpha_1, \alpha_2, \ldots, \alpha_n\}$ for A, $\mathcal{U}$ will be generated by the elements $\{\alpha_1^{\nu_1}, \alpha_2^{\nu_2}, \ldots, \alpha_n^{\nu_n}\}$, with $\nu_j = 0, 1, 2, \ldots n$.

§ **13.5.5** Let us quote a few amongst the main properties of $\mathcal{U}$:

(i) $\mathcal{U}$ admits a unique anti-automorphism "†", called the "principal anti-automorphism of $\mathcal{U}$", which is $X^\dagger = -X$ for every $X \in \mathcal{U}$;
(ii) there exists a one-to-one homomorphism $\Delta : \mathcal{U} \to \mathcal{U} \otimes \mathcal{U}$ such that $\Delta(X) = X \otimes 1 + 1 \otimes X$ for all $X \in \mathcal{U}$; it is called the "diagonal mapping of $\mathcal{U}$";
(iii) each derivation of A (see § 13.5.8 below) admits a unique extension to a derivation of $\mathcal{U}$;
(iv) there exists a one-to-one correspondence between the representations of A and the representations of $\mathcal{U}$: every representation of A can be extended to a unique representation of $\mathcal{U}$, and the restriction to A of every representation of $\mathcal{U}$ defines a representation of A.

§ 13.5.6 Algebra of a group. The algebra of a group G comes out when the group ring of G is a field, usually $\mathbb{R}$ or $\mathbb{C}$. It is frequently called the "group convolution algebra", to distinguish it from the other algebra which would come if the pointwise product were used.

Commentary 13.9 This other algebra is the set of real or complex functions on the group G, with addition and multiplication given by the pointwise addition and product:

$$(f + g)(x) = f(x) + g(x) \quad \text{and} \quad (f\,g)(x) = f(x)\,g(x).$$

It is sometimes also called the group algebra. ◀

§ 13.5.7 Dual algebra. On algebras, the involution is required to submit to some conditions, by which it transfers the algebraic properties into the dual space, which thereby becomes a dual algebra.

    A norm, for example that coming from an inner product, can define a topology on a linear space. Addition of a topology, through an inner product or by other means, turns linear spaces into much more complex objects. These will be left to Chapter 17.

§ 13.5.8 Derivation. The generic name *derivation* is given to any endomorphism $D : A \to A$ for which Leibniz's rule

$$D(\alpha\beta) = (D\alpha)\beta + \alpha(D\beta)$$

holds. The Lie algebra End A contains $D(A)$, the vector subspace of all the derivations of A, and the Lie bracket makes of $D(A)$ a Lie subalgebra of [End A], called its "derived algebra". This means that the commutator of two derivations is a derivation. Each member $a \in A$ defines an endomorphism $ad(a) = ad_a$, called the "adjoint action of a", by

$$ad_a(b) = [a, b] \quad \forall\, b.$$

The set of differentiable real or complex functions on a differentiable manifold M constitutes an algebra. The vector fields (derivations on M) are derivations of this algebra, which consequently reflects the smooth structure of the space itself. If M is a Lie group, consequently endowed with additional structure, the algebra of functions will gain extra properties reflecting that fact.

13.6 Coalgebras

§ 13.6.1 General case. Suppose now that spaces A and B are algebras over a certain field, say $\mathbb{C}$. Then $A \otimes B$ is also an algebra (the tensor product of algebras A and B) with the product defined by

$$(a \otimes b)(a' \otimes b') = (a\,a') \otimes (b\,b').$$

If A and B are associative unit algebras, so will be their tensor product.

The product in A is, in reality, a mapping

$$m : A \otimes A \to A, \quad a \otimes a' \to a\,a'.$$

Its dual mapping

$$\Delta : A \to A \otimes A$$

is called the *coproduct*, or *comultiplication* (or still *diagonal mapping*). It is supposed to be associative. The analogue to associativity (the "coassociativity") for the comultiplication would be the property

$$(id \otimes \Delta)\,\Delta(x) = (\Delta \otimes id)\,\Delta(x)$$

as homomorphisms of A in $A \otimes A \otimes A$:

$$
\begin{array}{ccc}
 & A \otimes A & \\
{\scriptstyle \Delta}\nearrow & & \searrow{\scriptstyle \Delta \otimes id} \\
A & & A \otimes A \otimes A\,. \\
{\scriptstyle \Delta}\searrow & & \nearrow{\scriptstyle id \otimes \Delta} \\
 & A \otimes A &
\end{array}
$$

Once endowed with this additional structure, A is a coalgebra. The coalgebra is *commutative* if $\Delta(A)$ is included in the symmetric part of $A \otimes A$. Let us put it in other words: define a permutation map

$$\sigma : A \otimes A \to A \otimes A, \quad \sigma(x \otimes y) = y \otimes x.$$

Then the coalgebra is commutative if $\sigma \circ \Delta = \Delta$.

§ 13.6.2 Bialgebras, or Hopf algebras. An associative unit algebra A is a Hopf algebra (or *bialgebra*, or still (old name) *annular group*) if it is a coalgebra satisfying:

(i) The product is a homomorphism of unit coalgebras.

(ii) The coproduct is a homomorphism of unit algebras,

$$\Delta(xy) = \Delta(x)\Delta(y).$$

The general form is

$$\Delta(x) = I \otimes x + x \otimes I + \sum_j x_j \otimes y_j,$$

with x_j, $y_j \in A$. From (ii), $\Delta I = I \otimes I$. When

$$\Delta(x) = I \otimes x + x \otimes I,$$

x is said to be "primitive".

Let us present two more mappings:

1. A map $\varepsilon : A \to \mathbb{C}$ defining the *counit* of the coproduct Δ, which is given by

$$(\varepsilon \otimes id)\Delta(x) = (id \otimes \varepsilon)\Delta(x) = x,$$

$$
\begin{array}{ccc}
 & A \otimes A & \\
{}^{\Delta}\nearrow & & \searrow^{\varepsilon \otimes id} \\
x \in A & & x \in A\,. \\
{}^{\Delta}\searrow & & \nearrow_{id \otimes \varepsilon} \\
 & A \otimes A &
\end{array}
$$

It is an algebra homomorphism:

$$\varepsilon(xy) = \varepsilon(x)\varepsilon(y).$$

2. Consider the map $\gamma : A \to A$, given by an antihomomorphism $\gamma(xy) = \gamma(y)\gamma(x)$ such that

$$m(id \otimes \gamma)\,\Delta(x) = m(\gamma \otimes id)\Delta(x) = \varepsilon(x)I.$$

It is described in the diagram

$$
\begin{array}{ccc}
A \otimes A & \xrightarrow{id \otimes \gamma} & A \otimes A \\
{}^{\Delta}\nearrow & & \searrow^{m} \\
x \in A & & \varepsilon(x)\,\mathbf{I} \in A. \\
{}^{\Delta}\searrow & & \nearrow_{m} \\
A \otimes A & \xrightarrow[\gamma \otimes id]{} & A \otimes A
\end{array}
$$

The map $\gamma(x)$ is called the *antipode* (or co-inverse) of x. Given the permutation map $\sigma(x \otimes y) = y \otimes x$, then $\Delta' = \sigma \circ \Delta$ is another coproduct on A, whose antipode is $\gamma' = \gamma^{-1}$.

Commentary 13.10 Some people call bialgebras structures as the above up to the existence of the counit, and reserve the name Hopf algebras to those having further the antipode. ◄

Commentary 13.11 Write $\Delta x = (\Delta_1 x, \Delta_2 x)$. Then both

$$(\varepsilon \otimes id)\Delta x = \varepsilon(\Delta_1 x)\Delta_2 x \quad \text{and} \quad (id \otimes \varepsilon)\Delta x = \varepsilon(\Delta_2 x)\Delta_1 x$$

should be x, so that

$$\Delta_1 x = \frac{1}{\varepsilon(\Delta_2 x)}\, x \quad \text{and} \quad \Delta_2 x = \frac{1}{\varepsilon(\Delta_1 x)}\, x\,.$$

Furthermore, $x\cdot\gamma(x) = \varepsilon(x)\varepsilon(\Delta_1 x)\varepsilon(\Delta_2 x)\, \mathbf{I}$. ◄

Hopf algebras appear in the study of products of representations of unital algebras. A representation of the algebra A (see from Sec. 18.2 on) will be given on a linear space V by a linear homomorphic mapping ρ of A into the space of linear operators on V. The necessity of a Hopf algebra comes out when we try to compose representations, as we usually do with angular momenta. We take two representations (ρ_1, V_1) and (ρ_2, V_2) and ask for a representation fixed by ρ_1 and ρ_2, on the product $V_1 \otimes V_2$. In order to keep up with the requirements of linearity and homomorphism, it is unavoidable to add an extra mapping, the coproduct Δ. Once this is well established, the product representation will be $\rho = (\rho_1 \otimes \rho_2)\, \Delta$.

The universal enveloping algebra of any Lie algebra has a natural structure of Hopf algebra with the diagonal mapping (see § 13.5.5) as coproduct. But also algebras of functions on groups may lead to such a structure. Particular kinds of Hopf algebras, called quasi-triangular, are more commonly known to physicists under the name of "quantum groups".

§ 13.6.3 R-matrices.
To give an example of what has been just said above, it will be necessary to use the direct-product notation, to be detailed only later, in Section 14.2.6. The Hopf algebra is a "quasi-triangular algebra" if:

(i) Δ and $\Delta' = \sigma \circ \Delta$ are related by conjugation $\sigma \circ \Delta(x) = R\,\Delta(x)\,R^{-1}$, for some matrix $R \in A \otimes A$. This means that, for a commutative Hopf algebra $R = I \otimes I$, we have also that

(ii) $(id \otimes \Delta)(R) = R_{13}R_{12}$;

(iii) $(\Delta \otimes id)(R) = R_{13}R_{23}$;

(iv) $(\gamma \otimes id)(R) = R^{-1}$.

Then the Yang–Baxter equation of Section 14.2.7 follows:

$$R_{12}\, R_{13}\, R_{23} = R_{23}\, R_{13}\, R_{12}.$$

General references

Recommended general texts covering the topics of this chapter are Fraleigh 1974, Kirillov 1974, Warner 1972, Majid 1990 and Bratelli & Robinson 1979.

Chapter 14

Discrete Groups: Braids and Knots

These are the original groups, called into being by Galois. Physicists became so infatuated with Lie groups that it is necessary to say what we mean by discrete groups: those which are not continuous, on which the topology is either undefined or the discrete topology.

14.1   Discrete groups

Discrete groups can be of finite order (like the group of symmetries of a crystal, or that of permutations of members of a finite set) or of infinite order (like the braid groups). In comparison with continuous groups, their theory is very difficult: additional structures as topology and differentiability tend to provide extra information and make things easier. As a consequence, whereas Cartan had been able to classify the simple Lie groups at his time, the general classification of finite simple groups has only been terminated in the seventeen-eighties.

Commentary 14.1   As in all highly sophisticated subjects, the landscape here is full of wonders (results such as: "any group of 5 or less elements is abelian"; or "the order of a simple group is either even or prime"; or still "any two groups of order 3 are isomorphic" and "there are, up to isomorphisms, only two groups of order 4: the so called Klein 4-group and the cyclic group Z_4") and amazements (like the existence and order of the Monster group). ◄

Practically all the cases of discrete groups we shall meet here are fundamental groups of some topological spaces. These are always found in terms of some generators and relations between them. Let us say a few words on this way of treating discrete groups, taking for simplicity a finite rank.

405

14.1.1   *Words and free groups*

Consider a set A of n elements, $A = \{a_1, a_2, \ldots, a_n\}$. We shall use the names *letters* for the elements a_j and *alphabet* for A itself. An animal with p times the letter a_j will be written a_j^p and will be called a *syllable*. A finite string of syllables, with eventual repetitions, is (of course) a *word*. Notice that there is no reason to commute letters: changing their orders lead to different words. The *empty word* "1" has no syllables.

There are two types of transformations acting on words, called *elementary contractions*. They correspond to the usual manipulations of exponents: by a contraction of first type, a symbol like $a_i^p a_i^q$ becomes a_i^{p+q}; by a second type contraction, a symbol like a_j^0 is replaced by the empty word "1", or simply dropped from the word. With these contractions, each word can be reduced to its simplest expression, the *reduced word*. The set $F[A]$ of all the reduced words of the alphabet A can be made into a group: the product $u \cdot v$ of two words u and v is just the reduced form of the juxtaposition uv. It is possible to show that this operation is associative and ascribes an inverse to every reduced word. The resulting group $F[A]$ is the *word group* generated by the alphabet A. Each letter a_k is a *generator*.

Words may look at first as too abstract objects. They are actually extremely useful. Besides obvious applications in Linguistics and decoding,[1] the word groups are largely used in Mathematics, and have found at least one surprising application in Geometry: they classify the 2-dimensional manifolds.[2] In Physics, they are used without explicit mention in elementary Quantum Mechanics. Recall the Weyl prescription[3] (the "correspondence rule") to obtain the quantum operator Weyl

$$(p^m q^n) = \boldsymbol{W}(p^m q^n)$$

corresponding to a classical dynamical quantity like $p^m q^n$:

$$\boldsymbol{W}(p^m q^n) = \frac{1}{2^n} \sum_{k=0}^{n} \binom{n}{k} \boldsymbol{q}^k \boldsymbol{p}^m \boldsymbol{q}^{n-k}$$

$$= \frac{1}{2^m} \sum_{k=0}^{m} \binom{m}{k} \boldsymbol{p}^k \boldsymbol{q}^n \boldsymbol{p}^{m-k} \tag{14.1}$$

where bold-faced letters represent operators. The first few cases are

$$\boldsymbol{W}(pq) = \tfrac{1}{2}(\boldsymbol{pq} + \boldsymbol{qp})$$

---

[1] Schreider 1975.
[2] Doubrovine, Novikov & Fomenko 1979, vol. III.
[3] Weyl 1932.

$$W(pq^2) = \tfrac{1}{3}\left(pq^2 + qpq + q^2p\right),$$

and so on. The quantum operator corresponding to a polynomial $p^m q^n$ in the classical degree of freedom "q" and its conjugate momentum "p" is the (normalized) sum of all the words one can obtain with m times the letter p and n times the letter q.

Now, given a general discrete group G, it will be a *free group* if it has a set $A = \{a_1, a_2, \ldots, a_n\}$ of generators such that G is isomorphic to the word group $F[A]$. In this case, the a_j are the *free generators* of G. The number of letters, which is the rank of G, may eventually be infinite. The importance of free groups comes from the following theorem:

Every group G is a homomorphic image of some free group $F[A]$.

This means that a mapping $f : F[A] \to G$ exists, preserving the group operation. In a homomorphism, in general, something is "lost": many elements in $F[A]$ may be taken into a same element of G. $F[A]$ is in general too rich. Something else must be done in order to obtain an isomorphism. As a rule, a large $F[A]$ is taken and the "freedom" of its generators is narrowed by imposing some relationships between them.

14.1.2 *Presentations*

Consider a subset $\{r_j\}$ of $F[A]$. We build the minimum normal subgroup R with the r_j as generators. The quotient $F[A]/R$ will be a subgroup, corresponding to putting all the $r_j = 1$. An isomorphism of G onto $F[A]/R$ will be a *presentation* of G. The set A is the set of generators and each r_j is a *relator*. Each $r \in R$ is a *consequence* of $\{r_j\}$. Each equation $r_j = 1$ is a *relation*. Now, another version of the theorem of the previous section is:

Every group G is isomorphic to some quotient group of a free group.

Commentary 14.2 In this way, groups are introduced by giving generators and relations between them. Free groups have for discrete groups a role analogous to that of coordinate systems for surfaces: these are given, in a larger space, by the coordinates and relations between them. Of course, such "coordinates" being non-commutative, things are much more complicated than with usual coordinates and equations seldom lead to the elimination of variables. Related to this point, there is a difficulty with presentations: the same group can have many of them, and it is difficult to know whether or not two presentations refer to the same group. ◀

14.1.3 *Cyclic groups*

The simplest discrete groups are the cyclic groups, which are one-letter groups. A group G is a cyclic group if there is an element "a" such that any other element (including the identity) may be obtained as a^k for some k. It is of order n if the identity is a^n. Good examples are the n-th roots of 1 in the complex plane. They form a group isomorphic to the set $\{0, 1, 2, 3, \ldots, n - 1\}$ of integers with the operation of addition modulo n. This is the cyclic group $\mathbb{Z}_n$. There is consequently one such group $\mathbb{Z}_n$ for each integer $n = |\mathbb{Z}_n|$. The simplest case is $\mathbb{Z}_2$, which can be alternatively seen as a multiplicative group of generator

$$a = -1 : \mathbb{Z}_2 = \{1, -1\}.$$

The operation is the usual multiplication. Every cyclic group is abelian. Every subgroup of a cyclic group is a cyclic group. Any two cyclic groups of the the same finite order are isomorphic. Thus, the groups $\mathbb{Z}_n$ classify all cyclic groups and for this reason $\mathbb{Z}_n$ is frequently identified as *the* cyclic group of order n. Any infinite cyclic group is isomorphic to the group $\mathbb{Z}$ of integers under addition.

Given a group G and an element $a \in G$, then the cyclic subgroup of G generated by a,

$$\langle a \rangle = \{a^n : n \in \mathbb{Z}\}$$

is the smallest subgroup of G which contains a. If $\langle a \rangle = G$, then a generates G entirely, and G is itself a cyclic group.

Consider an element $g \in G$. If an integer n exists such that $g^n = e$, then n is the *order* of the element g, and g belongs to a cyclic subgroup. When no such integer exists, g is said to be of infinite order. If every element of G is of finite order, G is a *torsion group*. G is *torsion-free* if only its identity is of finite order. In an abelian group G, the set T of all elements of finite order is a subgroup of G, the *torsion subgroup* of G.

14.1.4 *The group of permutations*

Let A be an alphabet, $A = \{a_1, a_2, \ldots, a_n\}$. A *permutation* of A is a one-to-one function of A onto A (a bijection $A \to A$). The usual notation for a fixed permutation in which each a_j goes into some a_{p_j} is

$$\begin{pmatrix} a_1 & a_2 & \cdots & a_{n-1} & a_n \\ a_{p_1} & a_{p_2} & \cdots & a_{p_{n-1}} & a_{p_n} \end{pmatrix}. \tag{14.2}$$

The set of all permutations of an n-letter alphabet A constitutes a group under the operation of composition ("product"), the n-th *symmetric group*, denoted S_n. The order of S_n is $(n!)$.

Commentary 14.3 The expression "a permutation group" is used for any (proper or improper) subgroup of a symmetric group. This is very important because every finite group is isomorphic to some permutation group (Cayley's theorem). ◀

The permutation of the type

$$\begin{pmatrix} a_1 \ a_2 \ \cdots \ a_{j-1} \ a_j \\ a_2 \ a_3 \ \cdots \ \ a_j \ \ a_1 \end{pmatrix}$$

is a *cycle* of length j, usually denoted simply $(a_1, a_2, \ldots, a_j)$. A product of two cycles is not necessarily a cycle. A product of disjoint cycles is commutative. A cycle of length 2 is a *transposition* — for example: (a_1, a_5). Any permutation of S_n is a product of disjoint cycles. Any cycle is a product of transpositions,

$$(a_1, a_2, \ldots, a_n) = (a_1, a_2)(a_2, a_3)(a_3, a_4) \ldots (a_n, a_1).$$

Thus, any permutation of S_n is a product of transpositions.

Given a permutation s, the number of transpositions of which s is a product is either always even or always odd. The permutation s itself is, accordingly, called *even* or *odd*. The number of even permutations in S_n equals the number of odd permutations (and equals $n!/2$). The even permutations of S_n constitute a normal subgroup, the *alternating group* A_n.

The symmetric group can be introduced through a presentation. Define as generators the $(n-1)$ elementary transpositions $s_1, s_2 \ldots, s_{n-1}$ such that s_i exchanges only the i-th and the $(i+1)$-th entry:

$$s_i = \begin{pmatrix} 1 \ 2 \cdots \ \ i \ \ i+1 \cdots \ n-1 \ n \\ 1 \ 2 \cdots i+1 \ \ i \ \ \cdots n-1 \ n \end{pmatrix} \tag{14.3}$$

Each permutation will be a word with the alphabet $\{s_j\}$. The s_i's obey the relations

$$s_j s_{j+1} s_j = s_{j+1} s_j s_{j+1}, \tag{14.4}$$

$$s_i s_j = s_j s_i \quad \text{for } |i-j| > 1, \tag{14.5}$$

$$(s_i)^2 = 1, \tag{14.6}$$

which determine completely the symmetric group S_n. Any group with generators satisfying these relations is isomorphic to S_n. Condition (14.6) is of fundamental importance: it limits the number of different possible exchanges, and is responsible by the fact that S_n is of finite order. Its absence, as will be seen in §'s 14.2.4 and 14.2.5, leads to braid groups, which are of infinite order.

Commentary 14.4 Many groups are only attained through presentations. This is frequently the case with fundamental groups of spaces. A good question is the following: given two presentations, can we know whether or not they "present" the same group? This is a version of the so-called "word problem". It was shown by P. S. Novikov that there can exist no general procedure to answer this question. ◀

Suppose that in the permutation s there are n_1 1-cycles, n_2 2-cycles, etc. The *cycle type* of a permutation is given by the numbers $(n_1, n_2, \ldots)$. Different permutations can be of the same cycle type, with the same set $\{n_j\}$. The importance of the cycle type comes from the following property:

> *Permutations of the same cycle type go into each other under the adjoint action of any element of S_n: they constitute conjugate classes.*

Repeating: to each set $\{n_j\}$ corresponds a conjugate class of S_n. We can attribute a variable t_r to each cycle of length "r" and indicate the cycle structure of a permutation by the monomial $t_1^{n_1} t_2^{n_2} t_3^{n_3} \ldots t_r^{n_r}$. Then, to all permutations of a fixed class will be attributed the same monomial above. Such monomials are invariants under the action of the group S_n. The total number of permutations with such a fixed cycle configuration is

$$\frac{n!}{\prod_{j=1}^{n} n_j! j^{n_j}}.$$

The n-variable generating function for these numbers is the so-called cycle indicator polynomial[4]

$$C_n(t_1, t_2, t_3, \ldots t_n) = \sum_{\{n_j\}} \frac{n!}{\prod_{j=1}^{n} n_j! j^{n_j}} t_1^{n_1} t_2^{n_2} t_3^{n_3} \ldots t_r^{n_r}. \tag{14.7}$$

The summation takes place over the sets $\{n_i\}$ of non-negative integers for which

$$\sum_{i=1}^{n} i\, n_i = n.$$

Of course, such a summation of invariant objects is itself invariant. This is an example of a very important way of characterizing discrete groups: by invariant polynomials. Though not the case here, it is sometimes easier

---

[4] Comtet 1974.

to find the polynomials than to explicit the group itself. This happens for example for the knot groups (see Section 14.3.5 below).

The above invariant polynomial for the symmetric group turns up frequently in Statistical Mechanics of systems of identical particles.[5] For these systems they play the fundamental role of a partition function, from which all the basic physical variables are obtained, and which is, of course, invariant under particle permutations (Chapter 28).

14.2 Braids

14.2.1 *Geometrical braids*

A braid may be seen[6] as a family of non-intersecting curves $(\gamma_1, \gamma_2, \ldots \gamma_n)$ on the cartesian product $\mathbb{E}^2 \times I$ with

$$\gamma_j(0) = (P_j, 0) \quad \text{for} \quad j = 1, 2, \ldots, n,$$
$$\gamma_j(1) = (P_{\sigma(j)}, 1) \quad \text{for} \quad j = 1, 2, \ldots, n,$$

where σ is an index permutation (Figure 14.1). A braid is *tame* when its curves are differentiable, i.e., have continuous first-order derivatives (are of class C^1). Otherwise, it is said to be *wild*.

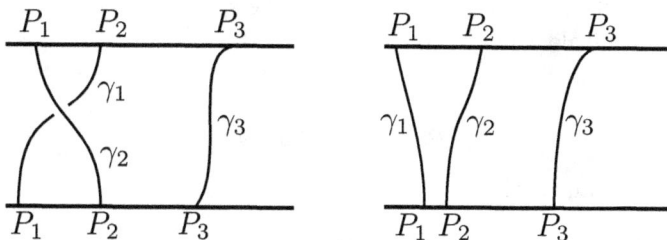

Fig. 14.1 Geometrical braids, seen as curves on $\mathbb{E}^2 \times I$.

---

[5] For a description of the symmetric and braid groups leading to braid statistics, see Aldrovandi 1992.

[6] Doubrovine, Novikov & Fomenko 1979, vol. II.

14.2.2 *Braid groups*

Braids constitute groups, which were first studied by Artin. There are many possible approaches to these groups. We shall first look at them as the fundamental groups of certain spaces. Consider n distinct particles on the euclidean plane $M = \mathbb{E}^2$. Their configuration space will be

$$M^n = \mathbb{E}^{2n} = \{x = (x_1, x_2, \ldots, x_n)\},$$

the n-th Cartesian product of manifold M. Suppose further that the particles are impenetrable, so that two of them cannot occupy the same position in $\mathbb{E}^2$. To take this into account, define the set

$$D_n = \{x_1, x_2, \ldots, x_n\}$$

such that $x_i = x_j$ for some i, j, and consider its complement in M^n,

$$F_n M = M^n \backslash D_n. \tag{14.8}$$

Then the *pure braid group* P_n is the fundamental group of this configuration space:

$$P_n = \pi_1[F_n M].$$

If the particles are identical, indistinguishable, the configuration space is still reduced: two points x and x' are "equivalent" if $(x_1, x_2, \ldots, x_n)$ and $(x'_1, x'_2, \ldots, x'_n)$ differ only by a permutation, a transformation belonging to the symmetric group S_n. Let $B_n M$ be the space obtained by identification of all equivalent points, the quotient of the configuration space by the symmetric group S_n:

$$B_n M = [F_n M]/S_n. \tag{14.9}$$

Then the fundamental group

$$\pi_1[B_n M]$$

is the *full braid group* B_n, or simply braid group. Artin's braid group is the full braid group for $M = \mathbb{E}^2$, but the above formal definition allows generalization to braid groups on any manifold M.

Of course, $F_n M$ is just the configuration space for a gas of n impenetrable particles, and $B_n M$ is the configuration space for a gas of n *impenetrable and identical* particles. Consequently, quantization of a system of n indistinguishable particles must start from such highly complicated, multiply-connected space.[7] As such a quantization employs just the fundamental group of the configuration space,[8] it must be involved with braid groups. And, of course, statistical mechanics will follow suit.

14.2.3 *Braids in everyday life*

In reality, braid groups[9] are concerned with real, usual braids. They count among the simplest examples of *experimental* groups: we can easily build their elements in practice, multiply them, invert them. They are related to the (still in progress) study of general weaving patterns, which also includes knots and links. Figure 14.2 depicts some simple braids of 3 strands: take two copies of the plane $\mathbb{E}^2$ with 3 chosen, "distinguished" points; link distinguished points of the two copies in any way with strings; you will have a braid.

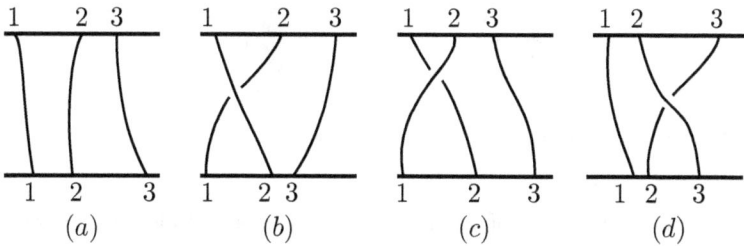

Fig. 14.2   Members of braid group B_n: first, basic steps into weaving.

Part (a) of Figure 14.2 shows the trivial 3-braid, with no interlacing of strands at all. Parts (b) and (d) show the basic, elementary steps of weaving, two of the simplest nontrivial braids. By historical convention, the strings are to be considered as going *from top to bottom*. Notice that in the drawing, the plane $\mathbb{E}^2$ is represented by a line just for facility. In part (b), the line going from 2 to 1 goes down behind that from 1 to 2. Just the opposite occurs in part (c). Braids (b) and (c) of Figure 14.2 are

---

[7] Leinaas & Myrheim 1977: a very good discussion of the configuration spaces of identical particles systems. Wavefunctions for bosons and fermions are found without resource to the summations of wavefunctions of distinguishable particles usually found in textbooks. Braid groups, although not given that name, are clearly at play.

[8] Schulman 1968; Laidlaw & DeWitt–Morette 1971; DeWitt–Morette 1972; DeWitt–Morette, Masheshwari & Nelson 1979.

[9] Birman 1975: the standard mathematical reference.

different because they are thought to be drawn between two planes, so that the extra dimension needed to make strings go behind or before each other is available. Braids are multiplied by composition: given two braids A and B, $A \times B$ is obtained by drawing B below A. Figure 14.3 shows the product of braid (b) of Figure 14.2 by itself. Figure 14.4 shows (b) × (d).

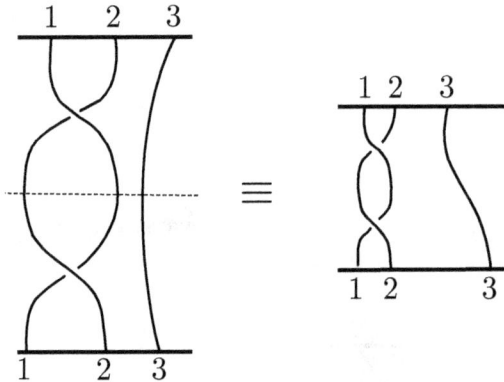

Fig. 14.3 Next steps into weaving, as products of the elementary first steps.

The trivial braid (a) of Figure 14.2 is the neutral element: it changes nothing when multiplied by any braid. It is easily verified that (b) and (c) are inverse to each other. The product is clearly non-commutative (compare (b) × (d) and (d) × (b)). In reality, any braid of 3 strands may be obtained by successive multiplications of the elementary braids (b) and (d) and their inverses. Such elementary braids are consequently said to *generate* the 3rd braid group which is, by the way, usually denoted B_3. The procedure of building by products from elementary braids may be used indefinitely. The braid group is consequently of infinite order. Of course, each braid may be seen as a mapping $\mathbb{E}^2 \to \mathbb{E}^2$, and (a) of Figure 14.2 is the identity map.

Commentary 14.5 All this can be easily generalized to the n-th braid group B_n, whose elements are braids with n strands. The reader is encouraged to proceed to real experiments with a few strings to get the feeling of it. ◀

A basic point is the following: consider the Figure 14.3. Each point on it is, ultimately, sent into itself. It would appear that it corresponds to the identity, but that is not the case! The identity is given by (a) of Figure 14.2, and Figure 14.3 cannot be reduced to it by any continuous change of point positions on $\mathbb{E}^2$. It cannot be unwoven! Nevertheless, a short experiment will show that it would be possible to disentangle it if the space were $\mathbb{E}^3$. As every braid is a composition of the elementary braids, that would mean that any braid on $\mathbb{E}^3$ may be unbraided ... as witnessed by millennia of practice with hair braids. Hair braids on $\mathbb{E}^2$ can be simulated by somehow gluing together their extremities, thereby eliminating one degree of freedom.

Because braids can be unwoven in $\mathbb{E}^3$, the braid group reduces to the symmetric group and Quantum and Statistical Mechanics in $\mathbb{E}^3$ remain what they are usually. Differences could however appear in the 2-dimensional case. Anyhow, from the point of view of maps on $\mathbb{E}^2$, we see that the "identity" exhibits infinite possibilities! Each particle sees the others as forbidden points, as holes. Repeated braid multiplication $(b) \times (b)$ of Figure 14.2 will lead to paths starting at 1 and turning 2, 3, ... times around a hole representing 2.

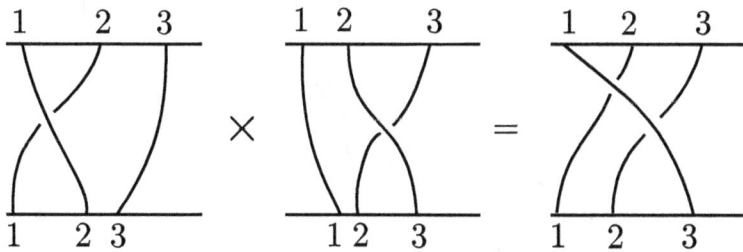

Fig. 14.4   Further example of braid product.

Of course, all this strongly suggests a relation to the fundamental group of $\mathbb{E}^2$ with holes. It is indeed through this relation that mathematicians approach braid groups, as seen below. The modified "identity" of Figure 14.3 would be simply twice the transposition of points 1 and 2. More generally, any permutation of points becomes multiform: the n-th braid group is an enlargement of the group of permutations S_n. Mathematicians have several definitions for B_n, the above "configuration space" definition allowing, as said, generalization to braid groups on any manifold M. But it can be, instead, introduced via a presentation.

14.2.4 *Braids presented*

The braid group B_n has also, like S_n, $(n-1)$ generators σ_j satisfying the relations

$$\sigma_j\sigma_{j+1}\sigma_j = \sigma_{j+1}\sigma_j\sigma_{j+1}, \tag{14.10}$$

$$\sigma_i\sigma_j = \sigma_j\sigma_i \quad \text{for } |i-j| > 1 . \tag{14.11}$$

They are the same as the two relations (14.4) and (14.5) for S_n. The absence of condition (14.6), however, makes of B_n a quite different group. It is sufficient to say that, while S_n is of finite order, B_n is infinite.

14.2.5 *Braid statistics*

The absence of the relation (14.6) has, as we have said, deep consequences. Unlike the elementary exchanges of the symmetric group, the square of an elementary braid is not the identity. In many important applications, however, it is found that σ_j^2 differs from the identity in a well-defined way. In the simplest case, σ_j^2 can be expressed in terms of the identity and σ_j, which means that it satisfies a second order equation like

$$(\sigma_j - x)(\sigma_j - y) = 0,$$

where x and y are numbers. In this case, the σ_j's belong to a subalgebra of the braid group algebra, called Hecke algebra. This is the origin of the so-called skein relations, which are helpful in the calculation of the invariant polynomials of knot theory.

In Quantum Mechanics, a basis for a representation of a braid group will be given by operators $U(\sigma_j)$ acting on wavefunctions according to

$$\psi'(x) = U(\sigma_j)\psi(x) = e^{i\varphi}\psi(x).$$

But now there is no constraint enforcing $U(\sigma_j^2) = 1$, so that

$$U^2(\sigma_j)\psi(x) \equiv U(\sigma_j^2)\psi(x) = e^{i2\varphi}\psi(x),$$
$$U(\sigma_j^3)\psi(x) = e^{i3\varphi}\psi(x),$$

and so on. The representation is now, like the group, infinite. It is from the condition $U(\sigma_j^2) = 1$ that the possibilities of phase values for the usual n-particle wavefunctions are reduced to two: as twice the same permutation leads to the same state, $U(\sigma_j^2)\psi(x) = \psi(x)$ so that $e^{i\varphi} = \pm 1$. The two signs correspond to wave-functions which are symmetric and antisymmetric under exchange of particles, that is, to bosons and fermions. When statistics is governed by the braid groups, as is the case for two-dimensional configuration spaces of impenetrable particles, the phase $e^{i\varphi}$ remains arbitrary and there is a different statistics for each value of φ. Such statistics are called braid statistics.

14.2.6 *Direct product representations*

Representations of the braid groups can be obtained with the use of direct products of matrix algebras. Suppose the direct product of two matrices A and B. By definition, the matrix elements of their direct product $A \otimes B$ are

$$< ij|A \otimes B|mn > \,= \,< i|A|m >< j|B|n > . \qquad (14.12)$$

On the same token, the direct product of 3 matrices is given by

$$< ijk|A \otimes B \otimes C|mnr > \,= \,< i|A|m >< j|B|n >< k|C|r > . \qquad (14.13)$$

And so on. The direct product notation compactifies expressions in the following way. Let $T = A \otimes B$, and E be the identity matrix. Then we write

$$T_{12} = A \otimes B \otimes E, \qquad (14.14)$$

$$T_{13} = A \otimes E \otimes B, \qquad (14.15)$$

$$T_{23} = E \otimes A \otimes B, \text{ etc.} \qquad (14.16)$$

A useful property of direct products is

$$(A \otimes B \otimes C)(G \otimes H \otimes J) = (AG) \otimes (BH) \otimes (CJ),$$

and analogously for higher order products. We may also use the notation

$$T^{ij}{}_{mn} =< ij|T|mn > .$$

Given a matrix $\widehat{R}$, an expression like

$$\widehat{R}^{kj}{}_{ab}\widehat{R}^{bi}{}_{cr}\widehat{R}^{ac}{}_{mn} = \widehat{R}^{ji}{}_{ca}\widehat{R}^{kc}{}_{mb}\widehat{R}^{ba}{}_{nr} \qquad (14.17)$$

is equivalent to

$$\widehat{R}_{12}\widehat{R}_{23}\widehat{R}_{12} = \widehat{R}_{23}\widehat{R}_{12}\widehat{R}_{23}, \qquad (14.18)$$

which is the "braid equation", name usually given to (14.10). To show it, look at s_1, s_2 as $s_1 = S_{12}$ and $s_2 = S_{23}$, S being some direct product as above. Then find

$$< ijk|s_1s_2s_1|mnr >= S^{ij}{}_{pq}S^{qk}{}_{vr}S^{pv}{}_{mn}$$

and

$$< kji|s_2s_1s_2|mnr >= S^{ji}{}_{qs}S^{kq}{}_{mv}S^{vs}{}_{nr},$$

so that the braid equation is

$$S^{kj}{}_{ab}S^{bi}{}_{cr}S^{ac}{}_{mn} = S^{ji}{}_{ca}S^{kc}{}_{mb}S^{ba}{}_{nr}.$$

We have found above conditions for representations of B_3. Higher order direct products of projectors will produce representations for higher order braid groups. In the general case, given a matrix $\widehat{R} \in Aut(V \otimes V)$ satisfying relations as above and the identity $E \in Aut(V)$, a representation of B_N on $V^{\otimes N}$ is obtained with generators

$$\sigma_i = E \otimes E \otimes E \otimes \ldots \widehat{R}_{i,i+1} \otimes \ldots \otimes E \otimes E$$
$$= (E\otimes)^{i-1}\widehat{R}_{i,i+1}(\otimes E)^{N-i}. \tag{14.19}$$

14.2.7 *The Yang–Baxter equation*

With the notation above, we may easily establish a direct connection of the braid relations to the Yang–Baxter equation, usually written

$$R_{12}R_{13}R_{23} = R_{23}R_{13}R_{12}, \tag{14.20}$$

which is the same as

$$R^{jk}{}_{ab}R^{ib}{}_{cr}R^{ca}{}_{mn} = R^{ij}{}_{ca}R^{ck}{}_{mb}R^{ab}{}_{nr}. \tag{14.21}$$

Define now another product matrix by the permutation $\widehat{R} = PR$, $\widehat{R}^{ij}{}_{mn} = R^{ji}{}_{mn}$. The above expressions are then equivalent to

$$\widehat{R}_{12}\widehat{R}_{23}\widehat{R}_{12} = \widehat{R}_{23}\widehat{R}_{12}\widehat{R}_{23}, \tag{14.22}$$

just the braid equation. The "permutation" relation is thus a very interesting tool to obtain representations of the braid groups from Yang–Baxter solutions and vice-versa. Notice that, due to this equivalence, many people give the name "Yang–Baxter equation" to the braid equation. An important point is that Yang–Baxter equations come out naturally from the representations of the Lie algebra of any Lie group. Thus, each such Lie algebra representation will provide a solution for the braid relations.[10]

The relation between this matrix formulation and our first informal representation of braids by their plane drawings leads to an instructive matrix-diagrammatic formulation. It is enough to notice the relationship

$$\begin{smallmatrix}a \\ \\ c\end{smallmatrix}\diagdown\begin{smallmatrix}b \\ \\ d\end{smallmatrix} \quad \Longleftrightarrow \quad \widehat{R}^{ab}{}_{cd}$$

and proceed to algebrize diagrams by replacing concatenation by matrix multiplication, paying due attention to the contracted indices. Looking at Figure 14.5, we see that the braid equation (14.10), becomes exactly the Yang–Baxter equation in its form (14.17).

---

[10] See Jimbo's contribution in Yang & Ge 1989.

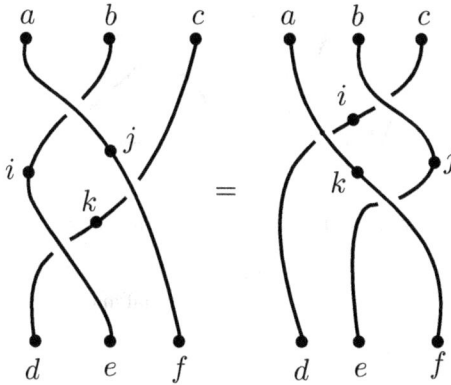

Fig. 14.5 $\quad R^{ab}{}_{ij} R^{jc}{}_{kf} R^{ik}{}_{de} = R^{bc}{}_{ij} R^{ai}{}_{dk} R^{kj}{}_{ef}$

14.3 Knots and links

The classification of knots has deserved a lot of attention from physicists like Tait and Kelvin at the end of the 19th century, when it was thought that the disposition of the chemical elements in the periodic table might be related to some kind of knotting in the ether. Motivated by the belief in the possibility of a fundamental role to be played by weaving patterns in the background of physical reality, they have been the pioneers in the (rather empirical) elaboration of tables[11] of "distinct" knots. Nowadays the most practical classification of knots and links is obtained via "invariant polynomials". Braids constitute highly intuitive groups of a more immediate physical interest, and there is a powerful theorem relating braid and knots.[12]

14.3.1 *Knots*

Consider the two knots in Figure 14.6. As anyone can find by experiencing with a string, they are "non-equivalent". This means that we cannot obtain one from the other without somehow tying or untying, that is, passing one of the ends through some loop. The mathematical formalization of

---

[11] See for instance Rolfsen 1976.

[12] On knots: for an introductory, intuitive view, see Neuwirth 1979. For a more qualitative appraisal, see Birman 1991; an involved treatment, but with a very readable first chapter, is Atiyah 1991.

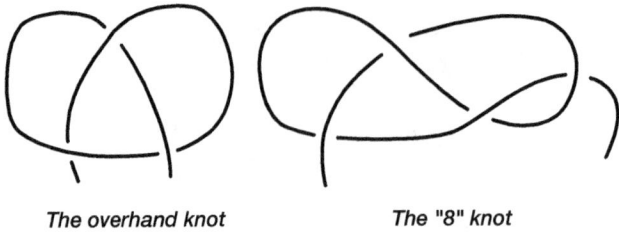

The overhand knot   **The "8" knot**

Fig. 14.6   Two of the simplest, non-trivial and distinct knots.

this intuitive notion of "different knots" is the so called "knot problem" and leads to an involved theory. The characterization is completely given by the notion of knot- (or link-) type, which is as sound as unpractical. Actually, there is no practical way to establish the distinction of every two given knots. There are however two methods allowing an imperfect solution of the problem. One of them attributes to every given knot (or link) a certain group, the other attributes a polynomial. They are imperfect because two different knots may have the same polynomial or group. The characterization by the knot groups is stronger: two knots with the same group have the same polynomials, but not vice-versa. On the other hand, polynomials are easier to find out.

We must somehow ensure the stability of the knot, and we do it by eliminating the possibility of untying. We can either extend the ends to infinity or simply connect them. We shall choose the latter, obtaining the closed versions drawn more symmetrically as in the Figure 14.7. The example to the right is equivalent to the circle, which is the trivial knot.

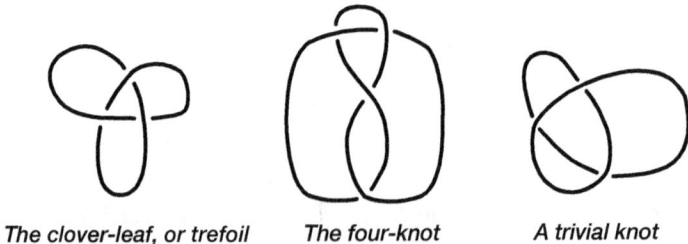

The clover-leaf, or trefoil   **The four-knot**   **A trivial knot**

Fig. 14.7   Examples of standard knots.

Now, the formal definition: a knot is any 1-dimensional subspace of $\mathbb{E}^3$ which is homeomorphic to the circle S^1. Notice that, as spaces, all knots are topologically equivalent. How to characterize the difference between the above knots, and between knots in general? The answer comes from noticing that tying and untying are performed in $\mathbb{E}^3$, and the equivalence or not is a consequence of the way in which the circle is plunged in $\mathbb{E}^3$. Two knots A and B are equivalent when there exists a continuous deformation of $\mathbb{E}^3$ into itself which takes A into B. This definition establishes an equivalence relation, whose classes are called knot-types. The trivial knot, equivalent to the circle itself, is called the *unknot*. The trefoil and the four-knot overleaf are of different and non-trivial types. We shall see below (§ 14.3.3) how to define certain groups characterizing knot-types.[13]

14.3.2 *Links*

Links are intertwined knots, consequently defined as spaces homeomorphic to the disjoint union of circles. The left example of Figure 14.8 shows a false link, whose component knots are actually independent. Such links are also called "unknots". The center and right examples of the Figure show two of the simplest collectors' favorites. Of course, knots are particular one-component links, so that we may use the word "link" to denote the general case.

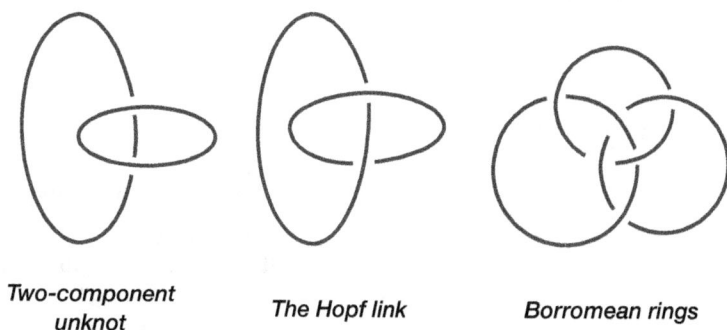

Two-component unknot   **The Hopf link**   **Borromean rings**

Fig. 14.8   Some standard links.

We have above talked loosely of "continuous deformation" of the host

---

[13] Crowell & Fox 1963; Doubrovine, Novikov & Fomenko 1979, vol. II.

space taking one knot into another. Let us make this idea more precise. The first step is the following: two knots A and B are *equivalent* when there exists a homeomorphism of $\mathbb{E}^3$ into itself which takes A into B. This is fine, but there is better. We would like to put arrows along the lines, to give knots an orientation. The equivalence just introduced would not take into account different orientations of the knots.

A more involved notion will allow A and B to be "equal" only if also their orientations coincide once A is deformed into B. An *isotopy* (or isotopic deformation) of a topological space M is a family of homeomorphisms h_t of M into itself, parametrized by $t \in [0, 1]$, and such that

(i) $h_0(p) = p$ for all $p \in M$, and,

(ii) the function $h_t(p)$ is continuous in both p and t.

Isotopies provide the finest notion of "continuous orientation-preserving deformations of M into itself". They constitute a special kind of homotopy, in which each member of the one-parameter family of deformations is invertible. When M is the host space of links, this definition establishes an equivalence relation, whose classes are called *link-types*. Link-types provide the complete characterization of links, but it has a serious drawback: given a link, it is a very difficult task to perform isotopies to verify whether or not it may be taken into another given link. That is why the experts content themselves with incomplete characterizations, such as the link group and the invariant polynomials.

14.3.3   *Knot groups*

Knots, as defined above, are all homeomorphic to the circle and consequently topologically equivalent as 1-dimensional spaces. We have seen that knot theory in reality is not concerned with such 1-dimensional spaces themselves, but with how $\mathbb{E}^3$ englobes these deformed "circles". Given the knot K, consider the complement $\mathbb{E}^3 \backslash K$. The knot group of K is the fundamental group of this complement, $\pi_1(\mathbb{E}^3 \backslash K)$. It is almost evident that the group of the trivial knot is $\mathbb{Z}$. Simple experiments with a rope will convince the reader that such groups may be very complicated. The trefoil group is the second braid group B_2. As already said, knot groups do not completely characterize knot types: two inequivalent knots may have the same group.

14.3.4 *Links and braids*

The relation between links and braids is given by Alexander's theorem, which requires a preliminary notion. Given a braid, we can obtain its *closure* simply by identifying corresponding initial and end points. Experiments with pieces of rope will once again be helpful. For instance, with the sole generator σ_1 of the two-strand-group B_2 we can build the Hopf link and the trefoil: they are respectively the braids σ_1^2 and σ_1^3 when their corresponding ends meet. Alexander's theorem says that

> *to every link-type corresponds a closed braid,*
> *provided both braid and link are tame .*

Given the braid β whose closure (denoted $\hat{\beta}$) corresponds to a link K, we write $\hat{\beta} = K$. This means that a link-type may be represented by a word in the generators of some braid group B_n. Experiments also show that this correspondence is not one-to-one: many braids may correspond to a given link-type. Thus, we obtain knots and links (their closures) if we connect corresponding points of a braid.

14.3.5 *Invariant polynomials.*

The relation between links and braids is the main gate to the most practical characterizations of links, the invariant polynomials.[14] Great progress has been made on this altogether fascinating subject in the last decades of the 20th century.

The idea of somehow fixing invariance properties through polynomials in dummy variables is an old one. Already Poincaré used polynomials, nowadays named after him, as a shorthand to describe the cohomological properties of Lie groups.[15] For a group G, such polynomials are

$$p_G(t) = b_0 + b_1 t + b_2 t^2 + \ldots + b_n t^n,$$

where the b_k's are the Betti numbers of G. They are, of course, invariant under homeomorphisms. Notice that each b_k is the dimension of a certain space, that of the harmonic k-forms on G. Or, if we prefer, of the spaces of cohomology equivalence classes. We have said in Section 14.1.4 that the cycle indicators (14.7) are invariant polynomials of the symmetric group, and that the coefficients of each monomial is the number of elements of the respective cycle configuration, or conjugate class.

---

[14] Commendable collections of papers are Yang & Ge 1989 and Kohno 1990.
[15] Goldberg 1962.

Invariant polynomials are a characterization of knots, which is weaker than the knot group: knots with distinct groups may have the same polynomial. But they are much easier to compute. And it may happen that a new polynomial be able to distinguish knots hitherto undifferentiated.

Actually, only rather recently, thanks to Conway, it became really easy to find some of them, because of his discovery of skein relations. There are at present a few different families of polynomials. To the oldest family belong the Alexander polynomials, found in the nineteen-thirties. The way to their computation was rather involved before a skein relation was found for them.

Skein relations, which are different for each family of polynomials, provide an inductive way to obtain the polynomial of a given link from those of simpler ones.

Suppose three links that only differ from each other at one crossing. There are three types of crossing: $\diagdown\!\!\!\diagup$, its inverse $\diagup\!\!\!\diagdown$, and the identity, or "uncrossing" $)($.

The polynomial of a knot K is indicated by a bracket $< K >$. If a knot K' differs from K only in one crossing, then their polynomials differ by the polynomial of a third knot in which the crossing is abolished. There are numerical factors in the relation, written in terms of the variable of the polynomial. Instead of drawing the entire knot inside the bracket, only that crossing which is different is indicated. For example, the Alexander polynomials of K and K' are related by

$$< \diagup\!\!\!\diagdown >_A - < \diagdown\!\!\!\diagup >_A + \tfrac{t-1}{\sqrt{t}} <)(>_A = 0, \qquad (14.23)$$

the index "A" indicating "Alexander". This relation says that the σ_j's are in a Hecke algebra: it is a graphic version of

$$\sigma_j^{-1} - \sigma_j + \tfrac{t-1}{\sqrt{t}}\, \boldsymbol{I} = 0. \qquad (14.24)$$

The skein relation must be supplemented by a general rule

$$< HL >=< H >< L >$$

if H and L are unconnected parts of HL, and by a normalization of the bubble (the polynomial of the unknot), which is different for each family of polynomials. For the Alexander polynomial, $< O > = 1$. A skein relation relates polynomials of different links, but is not in general enough for a full computation. It will be interesting for the knowledge of $< K >$ only if $< K' >$ is better known.

Kauffman extended the previous weaving patterns by introducing the so-called monoid diagrams, including objects like $\cup_\cap$. If we add then convenient relations like

$$< \! \diagdown \!\!\!\!\diagup \, > \; = t^{1/2} <\,)(\, > - (t^{1/2} + t^{-1/2}) < \! \overset{\cup}{\cap} \, > \; , \qquad (14.25)$$

and

$$< \! \diagup \!\!\!\!\diagdown \, > \; = t^{-1/2} <\,)(\, > - (t^{1/2} + t^{-1/2}) < \! \overset{\cup}{\cap} \, > \; , \qquad (14.26)$$

we can go down and down to simpler and simpler links, and at the end only the identity and simple blobs O remain.

The animal $\overset{\cup}{\cap}$ represents a projector. Kauffman's decomposition is justified by Jones discovery of representations of the braid group in some special von Neumann algebras, which are generated by projectors. Jones has thereby found other polynomials, and also clarified the meaning of the skein relations. The cycle indicator polynomial appears as the partition function of a system of identical particles (Section 28.1.2). Jones polynomials appear as the partition function of a lattice model (Section 17.6).

General references

Highly recommended general texts covering the topics of this chapter are: Adams 1994, Birman 1991, Crowell & Fox 1963, Fraleigh 1974, Kauffman 201, Neuwirth 1965 and Yang & Ge 1989.

Chapter 15

Sets and Measures

A topology is essential to the definition of continuity of functions defined on a set. Some additional restructuring, essential to allow for integration and for the notion of probability, are presented here.

15.1 Measure spaces

15.1.1 *The algebra of subsets*

A family of subsets of S is a topology if it includes S itself, the empty set $\emptyset$, all unions of subsets and all intersections of a finite number of them. We shall here describe collections of subsets of another kind, profiting in the while to introduce some notation and algebraic terminology. Given two subsets A and B of a set S,

$A - B = A\backslash B =$ difference of A and $B = \{p \in A$ such that $p \notin B\}$
$A \cup B =$ union of A and $B = \{$p $\in A$ or $p \in B\}$
$A \; \Delta \; B =$ symmetric difference of A and $B = (A\backslash B) \cup (B\backslash A)$.

15.1.2 *Measurable space*

Suppose that a family R of subsets is such that

(i) it contains the difference of every pair of its members
(ii) it contains the union of every pair of its members.

In this case it will contain also the empty set $\emptyset$, and all finite unions and intersections. More than that, a first algebraic structure emerges. The symmetric difference operation Δ is a binary internal operation, taking a pair (A, B) of subsets into another subset, $A \; \Delta \; B$. A pair such as (A, B)

belongs to twice R, that is, to the cartesian set product $R \times R$ of R by itself. The notation is $(A, B) \in R \times R$. An internal binary operation such as Δ is indicated by

$$\Delta : R \times R \to R, \quad (A, B) \to A \Delta B.$$

With this operation, R constitutes an abelian group. The neutral element is $\emptyset$ and each subset is its own inverse. Other binary internal operations are of course present, such as the difference $\setminus$ and the intersection

$$\cap : (A, B) \to A \cap B = A \setminus (A \setminus B).$$

The latter is associative,

$$A \cap (B \cap C) = (A \cap B) \cap C.$$

The relationship of $\cap$ and Δ is distributive:

$$A \cap (B \Delta C) = (A \cap B) \Delta (A \cap C)$$
$$(A \Delta B) \cap C = (A \cap C) \Delta (B \cap C).$$

The scheme is the same as that of the integer numbers, with Δ for addition and $\cap$ for multiplication. Such a structure, involving two binary internal operations obeying the distributive laws, one constituting an abelian group and the other being associative, is a ring (Section 13.3). A family R of subsets as above will be a *ring of subsets* of S. The power set of any S is a ring of subsets. Suppose now also that

(iii) $S \in R$.

S will work as a unit element for the "multiplication" $\cap$:

$$A \cap S = S \cap A = A.$$

In this case the whole structure is a "ring with unity". In the present case, R is more widely known as the Boolean algebra. Because of the historical prestige attached to the last name, the ring R is called an *algebra of subsets*, and indicated by A. Let us make one more assumption:

(iv) R contains all the countable unions of its members.

A family satisfying (i)–(iv) is called a σ-*algebra* (sometimes also a "σ-field"). A topology is essential to a clear and proper definition of the notion of continuity. A σ-algebra is the minimum structure required for the construction of measure and probability theories. The pair (S, A) formed by a space S and a particular σ-algebra is for this reason called a *measurable space*.

15.1.3 *Borel algebra*

It is possible, and frequently desirable, to make topology and measure compatible with each other. This is done as follows. Suppose some family C of subsets of S is given which does not satisfy (i) – (iv). It is then possible to show that there exists a smallest σ-algebra $A(C)$ of S including C, and that it is unique. $A(C)$ is said to be the σ-algebra *generated* by C. Consider now a topology T defined on S. The family T is, as in the case above, such that there will be a smallest σ-algebra $A(T)$ generated by T. This is the *Borel σ-algebra*, and every one of its members is a *Borel set*. The open intervals of $\mathbb{E}^1$ generate a Borel σ-algebra. If $T = $ indiscrete topology, little will remain of it in this procedure.

Commentary 15.1 As already said and repeated since the beginning of this text there are, besides topologies, many dissections of a set, each one convenient for a certain purpose. We have just seen σ-algebras. Other exemples are filters and ultrafilters (Section 13.3), which are instrumental in the study of continuity and convergence in non-metric topological spaces. ◄

15.1.4 *Measure and probability*

Given a set S and a family $A = \{A_i\}$ of its subsets, a real *set function* is given by

$$f : A \to \mathbb{R}, \quad f(A_i) = \text{some real number}.$$

Notice: this function takes a subset into a real number. We shall suppose that A contains the empty set and the finite unions of its members, and define a *positive set function* as a mapping $m : A \to \mathbb{R}_+$. Suppose further that, for every finite collection of disjoint sets $\{A_i \in A, i = 1, 2, \ldots, n\}$, the two following conditions hold:

(i) $m \left(\bigcup_{i=1}^{n} A_i \right) = \sum_{i=1}^{n} m(A_i)$.

(ii) $m(\emptyset) = 0$.

The function m is then said to be *finitely additive*. If the conditions hold even when n is infinite, m is *countably additive*. A *positive measure* is precisely such a countably additive set function on S, with the further proviso that A be a σ-algebra on S. The sets $A_i \in A$ are the *measurable subsets* of S and, for each set A_i, $m(A_i)$ is the "measure of A_i". The whole structure is denoted (S, A, m) and is called a *measure space*. If S is a countable union of subsets A_i with each $m(A_i)$ finite, the measure m is "σ-finite". Given any set algebra on S, it generates a σ-algebra, and any positive set function m is extended into a positive measure. If m is σ-finite, this extension is unique (Hahn extension theorem). The measure m is *finite* if $m(S)$ is finite.

A *probability space* is a measure space (S, A, m) such that $m(S) = 1$. In this case each set $A_i \in A$ is an *event* and $m(A_i)$ is the probability of event A_i. On locally compact topological spaces we may choose the closed compact subsets as Borel sets. A positive measure on a locally compact Hausdorff space is a *Borel measure*. A good example is the *Lebesgue measure* on $\mathbb{E}^1$: the Borel σ-algebra is that generated by the open intervals (a, b) with $b \geq a$ and the measure function is

$$m[(a, b)] = b - a.$$

The Lebesgue measure extends easily to $\mathbb{E}^n$.

15.1.5 *Partition of identity*

Consider a closed subset U of a differentiable manifold M. Then, there is a theorem which says that there exists a smooth function f_U (the characteristic function of U) such that $f_U(p) = 1$ for all $p \in U$, and $f_U(p) = 0$ for all $p \notin U$. Suppose further that M is paracompact. This means that M is Hausdorff and each covering has a locally finite sub-covering. Given a smooth atlas, there will be a locally finite coordinate covering $\{U_k\}$. Then, another theorem says that a family $\{f_k\}$ of smooth functions exists such that

(i) the support of $f_k \subset U_k$.
(ii) $0 \leq f_k(p) \leq 1$ for all $p \in M$.
(iii) $\sum_k f_k(p) = 1$ for all $p \in M$.

The family $\{f_k\}$ is a "partition of the identity". The existence of a partition of the identity can be used to extend a general local property to the whole space, as in the important examples below.

15.1.6 *Riemannian metric*

Once assured that a partition of the identity exists, we may show that a differentiable manifold has always a riemannian metric (§ 6.6.10). As each coordinate neighbourhood is euclidean, we may define on each U_k the euclidean metric

$$g^{(k)}_{\mu\nu} = \delta_{\mu\nu}.$$

A riemannian metric on M will then be given by

$$g_{\mu\nu}(p) = \sum_k f_k(p) g_{\mu\nu}^{(k)}(p).$$

15.1.7 *Measure and integration*

On the same token, as we know how to integrate over each euclidean U_k, the integral over M of any m-form ω is defined as

$$\int_M \omega = \sum_k \int_{U_k} \omega^{(k)} f_k(p),$$

where $\omega^{(k)}$ is the coordinate form of ω on U_k.[1]

15.2 Ergodism

Well-known examples of probability spaces are found in classical statistical mechanics (Chapter 28). Each one of the statistical ensembles uses a different Borel measure $F(q, p)$ and provides a different relationship between microscopic and macroscopic quantities. The Lebesgue measure

$$dqdp = dq^1 dq^2 \ldots dq^n dp_1 dp_2 \ldots dp_n$$

gives the volume of a domain U in phase space M as $\int_U dqdp$, that is, $F(q, p) = 1$. By the Liouville theorem, this volume is preserved by the microscopic dynamics. Systems in equilibrium are described by time-independent Borel measures. In this case we usually write

$$d\mu = F(q, p)dqdp$$

for the measure and the measure of U,

$$\mu(U) = \int_U F(q, p)dqdp,$$

is constant in time. When this happens for any U, we say that the microscopic hamiltonian flow is measure-preserving. The expected value of a macroscopic quantity A is

$$< A > = \int_M a(q, p)F(q, p)dqdp = \int_M a(q, p)d\mu,$$

where $a(q, p)$ is the corresponding microscopic quantity. Notice however that the only thing which is warranted to be preserved is the measure of any volume element. There is no information on anything else. This subject evolved into a sophisticated theory involving contributions from every chapter of Mathematics, the Ergodic Theory.

---

[1] Kolmogorov & Fomin 1977; Choquet-Bruhat, DeWitt-Morette & Dillard-Bleick 1977.

15.2.1 *Types of flow*

A particularly important question is the following: what is the flow of a volume element U in phase space M? There are three qualitatively different possibilities:

1. Non-ergodic flow: U moves without distortion and returns to its initial position after some finite interval of time; the total flow of U covers a small region of M. Consider a point p on M, and think of U as the initial uncertainty on its position; then the position at any time is perfectly determined, as well as the "error", which remains just U.

2. Ergodic flow: the shape of U is only slightly changed during the flow but the system never comes back to its initial configuration; the total flow of U sweeps a large region of M, possibly the whole of it; the points originally in U become a dense subset of M. If there is an initial "error" U in the position of the point p, then, after some time, p can be at any point of M. Previsibility is lost. This situation of overall sweeping of phase space by an initially small domain is generically called *ergodicity*.

3. Mixing flow: the shape of U is totally distorted; the distance between two initially neighbouring points diverges exponentially in time:

$$d(t) \approx e^{at}d(0).$$

The coefficient "a" in the exponent is a much used characterization of chaoticity, the "Lyapunov exponent". Because of the underlying deterministic dynamics, this case is frequently referred to as "deterministic chaos". Mixing implies ergodicity, but the converse is not true. Now, Sinaï has shown the "billiard theorem": a system of N balls in a box of hard walls is a mixing system. The evolution of even such a simple system as 2 balls enclosed in a box is, thus, very complicated.

15.2.2 *The ergodic problem*

All these considerations stand behind the famous ergodic problem. Suppose an isolated system with fixed energy E. Such a system is described by the microcanonical ensemble (Section 28.1.2) and the representative point travels on the hypersurface defined by the hamiltonian

$$H(q,p) = E$$

on the phase space. There are of course the integrals of motion, which reduce this hypersurface to a smaller subspace. We consider the average behaviour of the representative point on this reduced phase space. Any macroscopic observation of the system will last for a time interval T large in comparison with the microscopic times involved. Thus, what is really observed is a time-average over the microscopic processes, something like

$$\bar{a}_T = \frac{1}{2T} \int_{-T}^{T} dt \, a[q(t), p(t)]. \tag{15.1}$$

However, a basic notion of Statistical Mechanics is that the value of a macroscopic quantity is obtained as an ensemble average, that is,

$$<A> = \int_M a(q, p) \, d\mu. \tag{15.2}$$

Boltzmann's ergodic theorem says that this expectancy (average on phase space) equals the time average for large intervals of time: if you call

$$\bar{a} = \lim_{T \to \infty} \bar{a}_T \tag{15.3}$$

then

$$<A> = \bar{a}. \tag{15.4}$$

The interval T is supposed to be large not only with respect to the times involved in the detailed microscopic processes (like, for example, the scattering times of the constituent particles), but also as compared with those times relevant for the establishment of equilibrium (relaxation time, free flight between the walls, etc).

The ergodic problem is summarized in the question: is the ergodic theorem valid? Or, which is the same, can we replace one average by the other? If the answer is positive, we can replace statistics by a dynamical average.

Roughly speaking, the answer is that the theorem is true provided the measure on phase space has a certain property, so that statistics is actually never eliminated. To give an idea of the kind of questions involved, we shall briefly describe the basic results.

A first point refers to the very existence of the limit in (15.3). A second point is concerned with the independence of $\bar{a}$ on the particular flow (the particular hamiltonian).

Concerning the limit question, there are two main points of view, depending on the type of convergence assumed. One is that of Birkhoff, the other that of von Neumann. Suppose a finite-volume subset S of phase space. Then, we have two different theorems:

1. Birkhoff's theorem.

If the dynamical function $f(q,p)$ on S is such that

$$\int_S f[q(0), p(0)] d\mu < \infty,$$

then

$$\lim_{T \to \infty} \left[\frac{1}{2T} \int_{-T}^{T} dt f[q(t), p(t)] \right]$$

exists for all points (q, p) of S and is independent of the chosen origin of time. This limit will be identified with the average $\bar{f}$, but notice that this is a particular definition, assuming a particular type of convergence.

2. von Neumann's theorem.

Consider the Hilbert space of square-integrable dynamical functions on S. The inner product

$$(f, g) = \int_S f(q, p) g(q, p) d\mu_S$$

defines a norm $||f||$. Then there exists a function $\bar{f}$ such that

$$\lim_{T \to \infty} ||f - \bar{f}|| = 0.$$

If f is simultaneously integrable and square-integrable, Birkhoff's and von Neumann's limits coincide over S, except for functions defined on sets of zero measures. Of course, everything here holds only up to such sets.

This seems to settle the question of the limit, though it should be noticed that other topologies on the function spaces could be considered. Equation (15.4) is valid for both cases above, *provided* an additional hypothesis concerning the measure $d\mu$ is assumed. In simple words, the measure $d\mu$ should not divide the phase space into non-communicating sub-domains. Phase space must not be decomposed into flow-invariant sub-regions.

Let's be a bit more precise: a space is metric-indecomposable (or metrically transitive) if it cannot be separated into two (or more) regions whose measures are invariant under the dynamical flow and different from 0 or 1.

The condition for the ergodic theorem to be true is then that the phase space be metrically transitive. This means that there is no tendency for a point to abide in a sub-domain of phase space, or that no trajectory remains confined to some sub-region. In particular, there must be no hidden symmetries. It is in general very difficult to know whether or not this is the case for a given physical system, or even for realistic models.

General references

Highly recommended general texts covering the topics of this chapter are Arnold 1976, Balescu 1975, Jancel 1969 and Mackey 1978.

Chapter 16

Topological Linear Spaces

Adding topologies to vector spaces leads to more sophisticated algebraic structures. For infinite dimensional manifolds, the resulting topological linear spaces play the role analogous to that of euclidean spaces for finite dimensional manifolds, both as purveyors of coordinates and, in the differentiable cases, as tangent spaces. In general, topology is defined on a vector space through a norm.

16.1   Inner product space

A linear space endowed with an inner product is an "inner product space". Given the inner product

$$V \times V \to \mathbb{C}, \quad (v, u) \to < v, u >,$$

the number

$$||v|| = \sqrt{< v, v >}$$

is the norm of v induced by the inner product. This is a special norm, as general norms will be defined as in next section, independently of inner products. Some consequences, valid for this particular case, are:

1. The Cauchy–Schwarz inequality:

$$| < v, u > | \leq ||v|| \cdot ||u||.$$

2. The triangular inequality, or sub-additivity:

$$||v + u|| \leq ||v|| + ||u||.$$

3. The parallelogram rule:

$$||v + u||^2 + ||v - u||^2 = 2||v||^2 + 2||u||^2.$$

Let us add some further concepts. Two members u and v of a linear space (that is, of course, two vectors) are *orthogonal*, indicated $u \perp v$, if $< v, u > \, = 0$. For them will hold the Pythagoras theorem:

$$u \perp v \rightarrow ||v + u||^2 = ||v||^2 + ||u||^2.$$

16.2   Norm

A norm on a linear space V over the field $\mathbb{C}$ is a mapping

$$V \rightarrow \mathbb{R}, \quad v \rightarrow ||v||,$$

the following conditions holding for all $v, u \in V$, and $\lambda \in \mathbb{C}$:

(i) $||v + u|| \leq ||v|| + ||u||$.
(ii) $||\lambda v|| = |\lambda| \, ||v||$.
(iii) $||v|| \geq 0$.
(iv) $||v|| = 0 \leftrightarrow v = 0$.

It will be a *seminorm* if only the properties (i) and (ii) hold.

16.3   Normed vector spaces

Once endowed with a norm, V will be a normed vector space.[1] Internal product spaces are special cases, as we have seen that an internal product defines a norm. Norm is however a more general concept, as there are norms which are not induced by an internal product. The parallelogram law is a consequence of an internal product and does not necessarily hold for a general norm. A norm is a distance function, and defines the *norm topology* (also called the *strong topology*, and sometimes *uniform topology*). Normed spaces are metric topological spaces.

On normed vector spaces, the linear structure allows the introduction of one further concept: let V be such a space and a, b two of its points. Define the "straight line" between a and b by the curve

$$f : I \rightarrow V, \quad f(t) = (1 - t)a + tb.$$

A subset C of V is *convex* if, for every pair $a, b \in C$, all the points $f(t)$ also lie on C. The whole V is always convex, and so is also every vector subspace of V. Convex sets are sometimes called *starshaped* sets. Closed differential forms are always exact in a convex domain.

---

[1] Helmerg 1969.

16.4 Hilbert space

A Hilbert space[2] is an inner product space which is complete under the inner product norm topology. The standard case has for point set the set of sequences

$$\boldsymbol{v} = (v_1, v_2, v_3, \ldots) = \{v_i\}_{i=1}^{\infty} = \{v_i\}_{i \in \mathbb{N}}$$

of complex numbers such that

$$\sum_{i=1}^{\infty} |v_i|^2 < \infty.$$

The inner product is defined as

$$< \boldsymbol{v}, \boldsymbol{u} > = \sum_{i=1}^{\infty} v_i u_i^*.$$

A topological basis is prescribed by the open balls with the distance function

$$d(\boldsymbol{v}, \boldsymbol{u}) = | < \boldsymbol{v}, \boldsymbol{u} > |.$$

Hilbert spaces generalize euclidean spaces to the infinite dimensional case. They can be shown to be connected. It is possible to establish a one-to-one correspondence between the above sequences and bounded functions, by which such spaces become function spaces. The spaces of wavefunctions describing negative-energy states in Quantum Mechanics are Hilbert spaces of this kind. Positive-energy states constitute spaces far more complicated and are sometimes called "Dirac spaces" by physicists.

Let us now consider sequences of vectors (in a Hilbert space, sequences of the above sequences). The sequence of vectors $\{\boldsymbol{v}_n = (v_{n_1}, v_{n_2}, v_{n_3}, \ldots)\}$ is an *orthogonal sequence* if $\boldsymbol{v}_n \perp \boldsymbol{v}_m = 0$ for all pairs of distinct members, and is *orthonormal* if further it is true that $\|\boldsymbol{v}_n\| = 1$ for each member. In these cases we talk of an *orthogonal system* for the linear space. A theorem says that an orthogonal family of non-zero vectors is linearly independent.

The Hilbert space $\mathcal{H}$, defined as above, contains a countably infinite orthogonal family of vectors. Furthermore, this family is dense in $\mathcal{H}$, so that $\mathcal{H}$ is separable. In this case, consider a vector $\boldsymbol{u}$. The number $< \boldsymbol{u}, \boldsymbol{v}_n >$ is the *n-th coordinate*, or the *n-th Fourier coefficient*[3] of $\boldsymbol{u}$ with respect to the system $\{\boldsymbol{v}_m\}$.

---

[2] Halmos 1957.
[3] Dieudonné 1960.

An example of separable Hilbert space is the following: consider the complex-valued functions on the interval $[a,b] \in \mathbb{R}$. Then the space L^2 of all absolutely square integrable functions is a separable Hilbert space:

$$\mathcal{H} = L^2 = \left\{ f \text{ on } [a,b] \text{ with } \int_a^b |f(x)|^2 dx < \infty \right\}.$$

In greater generality, we may consider also non-separable Hilbert spaces. These would come out if, in the definition given above, instead of $v = \{v_i\}_{i \in \mathbb{N}}$, we had $v = \{v_\alpha\}_{\alpha \in \mathbb{R}}$: the family is not indexed by natural numbers, but by real numbers in the continuum. This definition would accommodate Dirac spaces. The energy eigenvalues, for the discrete or the continuum spectra, are precisely the indexes labeling the family elements, wavefunctions or kets. There are nevertheless new problems in this continuum-label case: the convergent summations $\sum_{i=1}^{\infty}$ used in the very definition of Hilbert space become integrals. In order to define integrals over a set, one needs additional structures: those of a σ-algebra of subsets, and that of a measure (see Chapter 15). Such Hilbert spaces will depend also on the choice of these structures.

16.5 Banach space

We have seen that a Hilbert space is an inner product space which is complete under the inner product norm topology. More general, a Banach space is a normed vector space which is complete under the norm topology. In this case, each one of its Cauchy sequences in the norm topology is convergent.

16.6 Topological vector spaces

A Banach space is a topological vector space when both the addition operation and the scalar multiplication are continuous in the norm topology. Although these rather abstract concepts hold in finite-dimensional spaces, they are actually fundamental in the study of infinite-dimensional spaces, which have quite distinct characteristics.

A general scheme is shown in Figure 16.1. Normed spaces have metric topologies. If they are also complete, they are Banach spaces. On the other hand, the norm may come from an inner product, or not. When it does and furthermore the space is complete, it is a Hilbert space. If the linear operations (addition and scalar multiplication) are continuous in the norm topology (inner product or not), we have topological vector spaces.

Commentary 16.1 The word *metrizable*, when applied to such spaces, means that its topology is given by a translation-invariant metric. ◄

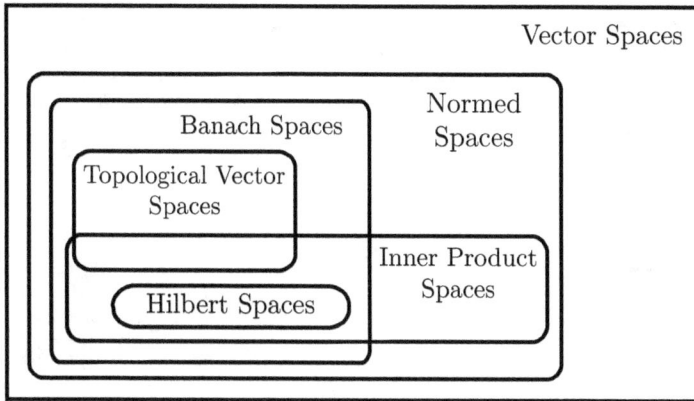

Fig. 16.1 A scheme of the relationships between the different kinds of topological vector space.

Let us recall that the dual space to a given linear space V is that linear space (usually denoted V^*) formed by all the linear mappings from V into its own field. When V is finite-dimensional, V^* is related to V by an isomorphism which in general is not canonical, but V^{**} is canonically isomorphic to V. In the infinite-dimensional case, V is in general only isomorphic to a subspace of V^{**}. The image of $v \in V$ by $k \in V^*$ is indicated by $< k, v >$.

On a topological vector space V, another topology is defined through the action of the V^*. It is called *the* weak topology and may be defined through convergence: a sequence $\{v_n\}$ converges weakly to $v \in$ V if, for every $k \in V^*$,

$$< k, v_n > \rightarrow < k, v >$$

as $n \rightarrow \infty$. As the names indicate, the norm topology is finer than the weak topology: a sequence may converge weakly and not converge in the norm topology. There are many other possible topologies, in effect. Figure 16.2 shows a scheme of linear spaces and some of their topologies.

A very useful notion is the following: a subset U of a topological vector space is a *bounded set* if it obeys the following condition of "archimedean" type: for any neighbourhood V of the origin there exists a number $n > 0$ such that $nV \supset U$.

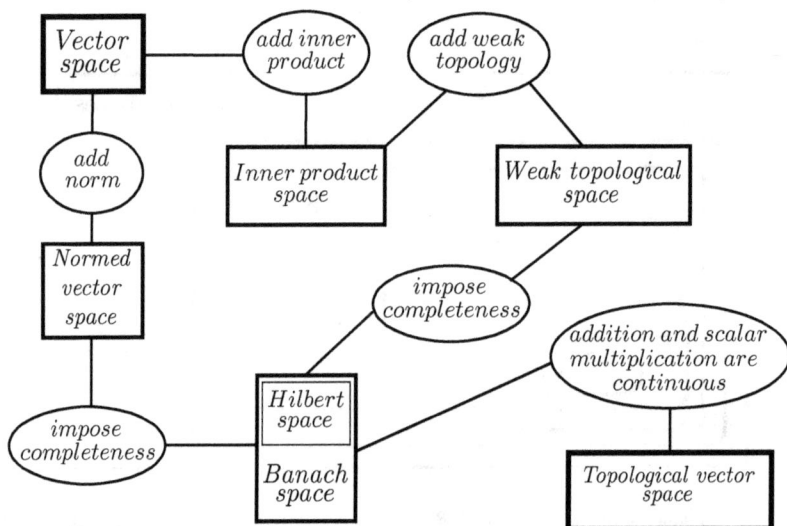

Fig. 16.2   Two ways to get a topological vector space.

16.7   Function spaces

Consider the space $C^\infty(M, \mathbb{C})$ of differentiable complex functions on a manifold M. It is a vector space to start with. Define on this space an internal operation of multiplication

$$C^\infty(M, \mathbb{C}) \otimes C^\infty(M, \mathbb{C}) \longrightarrow C^\infty(M, \mathbb{C}).$$

To help the mind, we may take the simplest multiplication, the pointwise product, defined by

$$(fg)(x) = f(x)g(x).$$

Then $C^\infty(M, \mathbb{C})$ becomes an algebra. Actually, it is as associative, commutative *-algebra (described in Section 17.3). We may in principle introduce other kinds of multiplication and obtain other algebras within the same function space.

As hinted in the discussion on the Hilbert space, there are very important cases in which the norm involves a measure. They combine in this way the above ideas with those given in Chapter 15, and the resulting structure is far more complicated. For this reason we leave Banach and *-algebras to Chapter 17. Differentiability on infinite-dimensional manifolds, which supposes topological vector spaces to provide a tangent structure, is discussed in Chapter 19.

General references

Good general texts covering the topics of this chapter are Kolmogorov & Fomin 1977 and Bratelli & Robinson 1979.

Chapter 17

Banach Algebras

We give here a sketchy account of what happens when the vector spaces are, furthermore, algebras. This includes spaces of operators, of particular interest to quantum physics. That is why we start by recalling some basic points behind the usual idea of quantization, and use some well-known aspects of quantum theory to announce some notions to be developed afterwards.

17.1 Quantization

Quantum observables related to a physical system are self-adjoint operators acting on a complex Hilbert space $\mathcal{H}$. Each state is represented by an operator, the "density matrix" ρ. In some special cases $\rho^2 = \rho$ and the state, called "pure", can be represented by a single projector $|\psi><\psi|$, where the ket $|\psi>$ is an element of $\mathcal{H}$. This exceptional situation is that supposed in wave mechanics, in which $|\psi>$ is said to be "the state" of the system and everything is described in terms of a wavefunction as, for example,

$$(x) = < x|\psi > .$$

All predictions are of a purely statistical character. Expectation values are attributed to an observable operator A as averages given by

$$< A > = \frac{\operatorname{tr}(\rho A)}{\operatorname{tr} \rho} .$$

If the system is in the pure state $|\psi>$, the value of A is

$$A_\psi = \frac{< \psi|A|\psi >}{< \psi|\psi >} .$$

Given an operator A and a function $f(z)$, then $f(A)$ represents (under conditions given below) another operator. For an isolated system, time evolution is fixed by the fact that $|\psi(t_1)>$ is related to the same ket at another time, $|\psi(t_2)>$, by a unitary evolution operator,

$$|\psi(t_2)> = \exp[-i(t_2 - t_1)H] |\psi(t_1)>,$$

with H the hamiltonian of the system.

Thus, ultimately, quantization deals with operators acting on Hilbert spaces. Such operators constitute by themselves other linear spaces and submit to some peculiar conditions, imposed by physical and/or coherence reasons. For example, in scattering problems the final state must be obtained for times very large as compared to any other time interval characteristic of the process, so that $t_2 = \infty$ for all purposes. For analogous reasons, also $t_1 = -\infty$. Whether or not this is a well-defined notion depends on the convergence of the evolution operator

$$U(t) = \exp[-i(t_2 - t_1)H],$$

and consequently on the topology defined on the space of operators. Some norm must be introduced to provide a good notion of convergence and boundedness of operators.

Summing up, we have a topological linear space of operators. Which leads to a Banach space. And, as operators compose between themselves by product, an algebra of operators is present, which is a Banach algebra. The algebra must contain the adjoint of each one of its members, so that what really appears is a special type of Banach algebra, called *-algebra. As expectation values are the only physically accessible results and are given by matrix elements, it is a weak topology which must be at work. This leads to a still more specialized kind of algebra, a W^*-algebra. Let us briefly describe such spaces, pointing whenever possible to the main relationships with quantum requirements.

Commentary 17.1 Quantum operators are preferably bounded, in the sense that its spectrum is somehow limited. Instead of using directly non-bounded operators, like the momentum in wave mechanics, one considers their exponentials. There must be a norm and, as suggested by the scattering example, a parameter-dependent operator must be able to be continued indefinitely in the parameter. The operator linking the initial and final states must belong to the algebra, wherefrom the completeness requirement. ◄

17.2 Banach algebras

A Banach space is a complete normed vector space, and a Banach algebra $\mathcal{B}$ (older name: "normed ring") is a Banach space with an associative internal multiplication. One can always consider it to be a unit algebra, with unity element I. If not, one is always able to make the "adjunction" of I. This is not as trivial as it may seem, but is guaranteed by a theorem.

The norm must be consistent with the algebra structure. It must satisfy, for α in the field of the vector space $\mathcal{B}$ and all A and $B \in \mathcal{B}$, the conditions:

(i) $\|\alpha A\| = |\alpha| \, \|A\|$;
(ii) $\|A + B\| \leq \|A\| + \|B\|$;
(iii) $\|AB\| \leq \|A\| \cdot \|B\|$;
(iv) $\|A^* A\| = \|A\|^2$.

Once endowed with a norm, the space becomes a metric space, with the balls $\{\mathbf{v}$ such that $\|\mathbf{v} - \mathbf{u}\| < \varepsilon\}$ around each $\mathbf{u}$. The algebra is *involutive* if, besides the involution postulates, the norm is preserved by the involution:

$$\|A^*\| = \|A\|.$$

The Banach space $\mathcal{B}$ is *symmetric* (or *self-adjoint*) if, for each element $A \in \mathcal{B}$, $\mathcal{B}$ contains also A^*.

Commentary 17.2 Thus, the space of quantum operators must belong to a symmetric involutive Banach algebra. The involution is the mapping taking each operator into its adjoint. If A is an acceptable operator, so is its adjoint, which furthermore has the same squared values. ◀

In quantum theory, we are always interested in functions of operators. In order to have them well-defined, we need some preliminary notions. The *spectrum* SpA of an element $A \in \mathcal{B}$ is the set of complex numbers λ for which the element $A - \lambda \mathbf{I}$ is *not* invertible.[1] An important theorem says that SpA is a closed set, a non-empty compact subset of $\mathbb{C}$. Consider the complement of SpA, that is, the set $\mathbb{C} \backslash$SpA of those complex numbers λ for which the element $A - \lambda \mathbf{I}$ *is* invertible. The function

$$R_A(\lambda) = [A - \lambda \mathbf{I}]^{-1}$$

is called the *resolvent* of A. This function is analytic in the complement $\mathbb{C} \backslash$SpA. If $\mathcal{B}$ happens to be a field over $\mathbb{C}$, it contains also $z \in \mathbb{C}$ and Sp $z = \emptyset$. Thus, the only Banach *field* over $\mathbb{C}$ is $\mathbb{C}$ itself (this is the Gelfand–Mazur theorem).

As SpA is compact, there exists a real number

$$\rho(A) = \sup_{\lambda \in \text{Sp}A} |\lambda| \,,$$

which is called the *spectral radius* of A. Actually,

$$\rho(A) = \lim_{n \to \infty} \|A^n\|^{1/n} \,.$$

---

[1] More about that can be seen in Aldrovandi 2001.

Given a complex polynomial with complex coefficients,

$$P(z) = a_0 + a_1 z + a_2 z^2 + \ldots + a_n z^n,$$

we can form a polynomial belonging to the Banach algebra,

$$P(A) = a_0 + a_1 A + a_2 A^2 + \ldots + a_n A^n$$

for each $A \in \mathcal{B}$. As powers of the same operator commute, the mapping $P(z) \to P(A)$ is an algebra homomorphism. The sets of polynomials in a fixed A will constitute a commutative algebra. Taking now all the elements of $\mathcal{B}$, the set of all polynomials constitute the *polynomial algebra* of $\mathcal{B}$, which is in general far from commutative.

This procedure of obtaining elements of $\mathcal{B}$ by "extension" of complex functions may be taken further. More precisely, it may be taken up to the following point: consider an open subset $o(f)$ of the set $\mathrm{Sp}A$ and let $AN(o(f))$ be the algebra of analytic functions on $o(f)$. The topology used is that of the uniform convergence on compact sets. Then, there exists a homomorphism of the algebra $AN(o(f))$ into $\mathcal{B}$, given by

$$f(A) = \frac{1}{2\pi} \int_\gamma f(\lambda)\, R_A(\lambda)\, d\lambda.$$

This homomorphism includes the above polynomial case, and the good subset $o(f)$, as the notation suggests, depends on the function f. The integration is along γ, which is any closed curve circumscribing the entire set $\mathrm{Sp}A$.

Commentary 17.3   We learn thus, by the way, how to get a function of a given quantum operator. ◀

Consider now the case of *commutative* Banach algebras, which are the natural setting for standard harmonic analysis[2] (§ 18.3.4). To each such algebra one associates $I(\mathcal{B})$, the set of its maximal proper (that is, $\ne \mathcal{B}$ itself) ideals. $I(\mathcal{B})$ is compact. An important point is that to each maximal ideal corresponds a character of $\mathcal{B}$, a homomorphic mapping $\chi : \mathcal{B} \to \mathbb{C}$. Given χ, then $\chi^{-1}(0)$ is a maximal ideal of $\mathcal{B}$. This interpretation of the characters as maximal ideals leads to the Gelfand transformation: it relates a function on $I(\mathcal{B})$ to each element of $\mathcal{B}$, given by $\widehat{A}(\chi) = \chi(A)$. The set of values of the function $\widehat{A}$ coincides with $\mathrm{Sp}A$.

Let $\mathcal{B} = R(X)$ be the algebra of real continuous functions on the compact X with the pointwise product $(fg)(x) = f(x)g(x)$ as operation. The set of functions vanishing at a point x is an ideal: $f(x) = 0$ implies

---

[2] Katznelson 1976.

$(fg)(x) = 0$ for any g. Each closed ideal is formed by those functions which vanish on some closed subset $Y \subset X$. There is a correspondence between the maximal ideals and the points of X: we can identify $x \in X$ to the maximal ideal $I(x)$ of functions f such that $f(x) = 0$ and the space X itself to the quotient $R(X)/I$, where I is the union of all $I(x)$.

17.3 *-algebras and C*-algebras

A *-algebra[3] is a complete normed algebra with involution. A C*-algebra is a Banach algebra over the field $\mathbb{C}$ of complex numbers, endowed with an antilinear involution $T \to T^*$ such that

$$(TS)^* = S^*T^* \quad \text{and} \quad ||T^*T|| = ||T||^2.$$

Only in the framework of *-algebras can we talk about self-adjointness: A is *self-adjoint* if $A = A^*$, and A is *normal* if it commutes with its adjoint:

$$A^*A = AA^*.$$

Also, A is *positive* if

$$A = A^* \quad \text{and} \quad \text{Sp}A \subset \mathbb{R}_+.$$

In a C*-algebra we can have square-roots. If A is positive, then there exists a B in the C*-algebra such that $A = B^*B$.

Commentary 17.4 The algebra of quantum observables is, under reasonable conditions, a C*-algebra. The density matrices, which represent the possible states, are positive operators. The space of states is contained in the space of positive operators. ◄

The frequent use of C*-algebras to treat operators on Hilbert spaces justifies a more specific definition. In this case, a C*-algebra is denoted by $L(\mathcal{H})$ and is an involutive Banach algebra of bounded (see Section 19.2.3) operators taking a complex Hilbert space $\mathcal{H} = \{\xi\}$ into itself, endowed with the norm

$$||T|| = \text{Sup}_{||\xi|| \leq 1} ||T\xi||$$

and the involution $T \to T^*$ defined by

$$< T^*\xi, \phi > = < \xi, T\phi >, \quad \forall\, \xi, \phi \in \mathcal{H}.$$

We can then prove that $||T^*T|| = ||T||^2$.

Let us list some of the main topologies defined on $L(\mathcal{H})$:

---

[3] Dixmier 1982.

(1) The norm topology, defined by the norm

$$||T|| = \text{Sup}_{\xi \in \mathcal{H}, ||\xi|| \leq 1} ||T\xi||.$$

For finite dimensions, T is a matrix and $||T||^2$ is the highest eigenvalue of TT^*. For infinite dimensions, $||T||^2$ is the spectral radius of TT^*.

(2) The strong topology, weaker than the norm topology, is defined in such a way that the sequence T_n converges to T iff $T_n\xi$ converges to $T\xi$ in $\mathcal{H}$ for all $T\xi \in \mathcal{H}$.

(3) The weak topology, weakest of the three, is defined by the statement that $T_n\xi$ converges to $T\xi$ in $\mathcal{H}$ iff, for all $\xi, \zeta \in \mathcal{H}$, $| < T_n\xi, \zeta > |$ converges to $| < T\xi, \zeta > |$.

(4) The "σ-weak" topology, defined by taking two sequences in $\mathcal{H}$, $\{\xi_i\}$ and $\{\eta_k\}$, with

$$\Sigma_i ||\xi_i||^2 < \infty \quad \text{and} \quad \Sigma_k ||\eta_k||^2 < \infty.$$

Then

$$|T| := \Sigma_n |(\xi_n, T\eta_n)|$$

is a seminorm on $L(\mathcal{H})$ and defines the σ-weak topology.

Let us recall that, as an algebra, $\mathcal{A}$ will have a "dual" space, formed by all the linear mappings of $\mathcal{A}$ into the real line. We write "dual", with quotation marks, because infinite dimensional vector spaces are deeply different from the finite dimensional vector spaces and one of the main differences concerns precisely the dual. For finite dimensional spaces, the dual of the dual is the space itself: $(V^*)^* = V$. This is no more true in the infinite dimensional case, the general result being that $(V^*)^* \supset V$. This is only to prepare for the "caveat dual" which is essential in the study of infinite dimensional vector spaces.

Even if $\mathcal{H}$ is of countable dimension, $L(\mathcal{H})$ is not. Consequently, $L(\mathcal{H})$ is not isomorphic to the dual $L^*(\mathcal{H})$. There is a (unique) subspace $L_*(\mathcal{H})$ of $L^*(\mathcal{H})$ which is isomorphic to $L(\mathcal{H})$, and this is usually called the predual of $L(\mathcal{H})$. The predual is a closed subspace of $L^*(\mathcal{H})$, of which $L(\mathcal{H})$ is the dual, and it has the σ-weak topology.

Let us see how this relates to properties of spaces. If X is any compact Hausdorff space, then the set $C(X)$ of continuous complex functions on X is a commutative unit algebra over C. It has a natural involution "*" given by

$$(f^*)(x) = f(x)^*,$$

and a norm

$$||f|| = \text{Sup}_{x \in X} |f(x)|,$$

which satisfies

$$||f^* f|| = ||f||^2.$$

With this norm, the space $C(X)$ is complete. It is consequently a C*-algebra. By the way, a general C*-algebra has just this structure, up to the commutative property. To every compact Hausdorff space corresponds a commutative C*-algebra. If X has only a single point p, $X = \{p\}$, then $C(X) = C$ and each mapping $F : X \to Y$ determines a functional

$$F^* : C(Y) \to C \quad \text{on} \quad C(Y) : (F^* f)(p) = f(F(p)).$$

This functional can be shown to be linear and multiplicative. Thus, to points of Y correspond functionals on $C(Y)$. The Gelfand–Naimark theorem states that this correspondence is one-to-one.

This has a deep consequence: each commutative unital C*-algebra $\mathcal{A}$ is the algebra of continuous complex functions on a compact space Y: $\mathcal{A} = C(Y)$. And Y can be identified with the set of linear multiplicative functionals on the algebra. A linear multiplicative functional on the algebra is a character. Thus, Y is the set of characters of $\mathcal{A}$. Finally: each continuous mapping $F : X \to Y$ between two compact Hausdorff spaces induces a C* homomorphism

$$F^* : C(Y) \to C(X), \quad \text{with} \quad (F^* f)(x) = f(F(x)),$$

so that these properties are, at least partially, carried over from compact to compact by continuous mappings.

17.4   From geometry to algebra

We have said in Section 17.2 that a compact space X can be seen as a subspace of the algebra $C(X)$ of continuous functions on X with the pointwise product as multiplication. Each point of X is an ideal formed by those functions which vanish at the point. Maximal ideals are identifiable to characters of the algebra and in section 17.3 we have indeed said that if X is a compact then it can be identified to the characters of $C(X)$. This has been the starting point of a process by which Geometry has been recast as a chapter of Algebra. We shall only say a few words on the subject, which provides one of the gates into non-commutative geometry and may become important to Physics in the near future.

Consider again the space $C^\infty(M, C)$ of differentiable complex functions on a manifold M. It is clearly a linear space. Define on this space an internal operation of multiplication

$$C^\infty(M, C) \times C^\infty(M, C) \to C^\infty(M, C).$$

The function space becomes an algebra. When there is no unit in this algebra, we can always add it. What results is a unital algebra. To help the mind, we can take the simplest, usual commutative pointwise product of complex functions: $(fg)(x) = f(x)g(x)$. Suppose we are able to introduce also some norm. With the pointwise product, $C^\infty(M, C)$ becomes an associative commutative *-algebra.

We can in principle introduce other kinds of multiplication and obtain other algebras with the same starting space. Inspired by phase spaces, for instance, we might think of introducing a Poisson bracket. From a purely algebraic point of view, a Poisson algebra is a commutative algebra A, as above provided with a map $\{\ ,\ \} : A \times A \to A$ such that:

(i) A is Lie algebra with the operation $\{\ ,\ \}$.
(ii) The bracket is a derivative in A: $\{a, bc\} = b\{a, c\} + \{a, b\}c$.

Once endowed with such a function algebra and bracket, a space M is said to have a Poisson structure. Notice that M is not necessarily a phase space: a Poisson structure can in principle be introduced on any differentiable manifold. But there is still more. The differentiable structure of M is encoded[4] in the *-algebra $C^\infty(M, C)$. Each property of M has a translation into a property of $C^\infty(M, C)$. If we restrict ourselves to the space $C^0(M, C)$ of functions which are only continuous, only the topological properties of M remain encoded, the information on the differentiable structure of M being "forgotten".

For instance, the above mentioned theorem by Gelfand and Naimark throws a further bridge between the two structures. It states, roughly, that any unit abelian *-algebra is isomorphic to some algebra like $C^0(M, C)$, with M some compact manifold. The difference between $C^\infty(M, C)$ and $C^0(M, C)$ lays in the fact that the former has a great number of derivations, the vector fields on M.

In the algebraic picture, such derivations[5] are replaced by the derivations in the *-algebra, that is, endomorphisms of $C^\infty(M, C)$ satisfying the

---

[4] Dubois–Violette 1991.
[5] Dubois–Violette 1988.

Leibniz rule,[6] like the above example with the Poisson bracket. Such derivations constitute a Lie algebra if $C^\infty(M, C)$ is associative. Vector fields on M have components which are differentiable up to a given order. Higher order differentiability translates itself into properties of derivations on the algebra. Summing up, the topological and geometrical properties of the manifold M are somehow taken into account in the algebra of functions $C^\infty(M, C)$.

Commentary 17.5 Also algebraic properties of M are reflected in $C^0(M, C)$. If M has also the structure of a topological group, for example, $C^0(M, C)$ will get some extra properties. ◄

Commentary 17.6 We can define "compact quantum spaces" (quantum groups) by dropping the commutativity requirement related to the pointwise product. In this case, each C*-algebra with unit can be seen as the algebra of "continuous functions" on a certain quantum space. ◄

The simplest way to pass into noncommutative geometry[7] is first to go to $C^\infty(M, C)$ and there dismiss the commutative character of the *-algebra. This means altering the pointwise product into some other, non-commutative product. The direct relation to the manifold M becomes fuzzy. The best example comes out in the Weyl–Wigner picture of Quantum Mechanics. As said in Section 10.2, you start with usual functions $F(q, p)$ on the most usual euclidean phase space of Classical Mechanics (§ 25.1). Instead of using the pointwise product, you deform it into the star product "∘". Thus

$$F(q, p) \cdot G(q, p)$$

is replaced by

$$F(q, p) \circ G(q, p).$$

But then the functions themselves become ambiguous — they are in general built up from more elementary objects (monomials, for example) and these are ambiguous. For example, from the classical

$$F(q, p) = qp$$

you have to choose either

$$F(q, p) = q \circ p \quad \text{or} \quad F(q, p) = p \circ q.$$

In building up functions, one usually uses a lot of function-of-function stuff, and now one is forced to go to the very beginning and redefine each function

---

[6] Connes 1980.
[7] Connes 1986.

from the most elementary redefined ones. The results are the Wigner functions, actually c-number representatives of quantum dynamical quantities which furthermore belong to an algebra with a deformed Poisson bracket, the Moyal bracket.

Non-commutative geometry comes out clearly. Either you go on using the coordinates q and p as ordinary euclidean-valued functions — but loose any connection with quantum reality; or you take also the coordinate functions q and p as belonging to the deformed function algebra — and then the coordinates are no more commutative. The Moyal bracket endows the algebra of Wigner functions with a Poisson structure.

17.5 von Neumann algebras

An involutive symmetric subalgebra M of $L(\mathcal{H})$ containing the unit and closed by the weak topology is a von Neumann algebra[8] (or W^*-algebra). The distinction between von Neumann algebras and general *-algebras is essential: von Neumann algebras are closed under the weak topology while *-algebras are closed under the norm topology. Of course, von Neumann algebras are particular cases of *-algebras, but they are not, in general, separable by the norm topology. Figure 17.1 is a scheme summarizing the main definitions. It should be compared with Figure 16.2 of the previous chapter.

Let us rephrase all this, while introducing some more ideas.[9] A von Neumann algebra M is a nondegenerate self-adjoint algebra of operators on a Hilbert space $\mathcal{H}$, which is closed under the weak operator topology. This is the locally convex topology in $L(\mathcal{H})$ induced by the family of seminorms

$$x \in L(\mathcal{H}) \rightarrow \quad |<x\zeta|\xi>| \text{ for all } \zeta,\xi \in \mathcal{H}.$$

The von Neumann *bicommutant theorem* says that $M = M''$: M is the commutant of its own commutant.

More precisely: if S is a subset of $L(\mathcal{H})$, its commutant will be $S' = \{ x \in L(\mathcal{H}) \text{ such that } xs = sx \text{ for all } s \in S \}$. Thus, $S'' = (S')'$. Call alg(S) the algebra generated by S. Suppose two things:
(i) S is symmetric: $x \in S$ implies $x^* \in S$.
(ii) $1 \in S$.

---

[8] Dixmier 1981.
[9] Takesaki 1978.

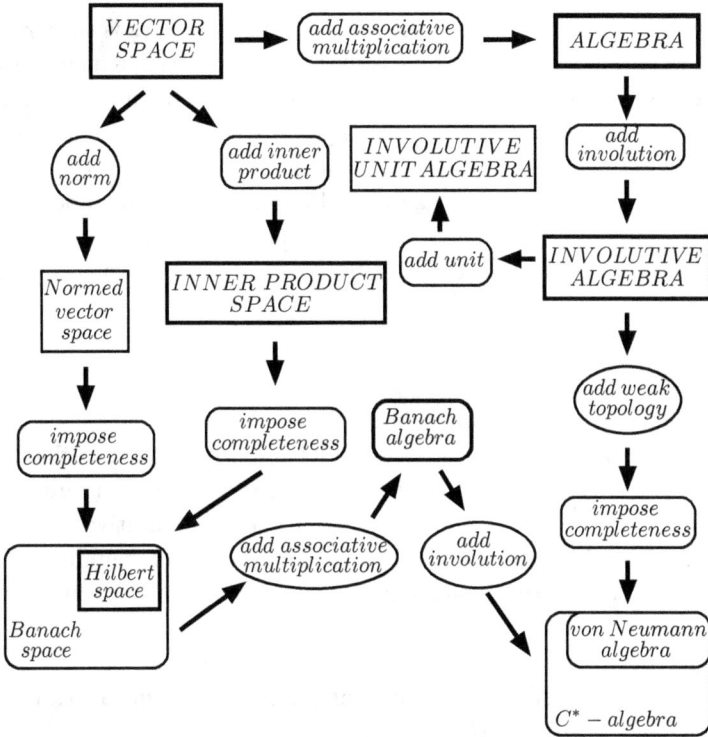

Fig. 17.1 A scheme summarizing Banach spaces and related algebras.

Then the theorem says that alg(S) is strongly dense in S'' (consequently, it is also weakly dense in S'').

An important fact is that the set of all projectors on a von Neumann algebra M, with the identity included, generates M. We recall that a subset U of an algebra M is said to generate M if the set of all the polynomials obtained with all the members of U is dense in M. As the projectors are idempotents, polynomials in projectors are simply linear combinations (see Section 17.6 for a finite dimensional example).

The classification of von Neumann algebras is based on the properties of its projectors. The center of a von Neumann algebra is abelian. A *factor* is a von Neumann algebra with trivial center,

$$M \cap M' = \mathbb{C}.$$

This means that the center is formed by the complex multiples of the identity.

One of the greatest qualities of von Neumann algebras is their receptivity to integration. In effect, it is possible to define on them a measure theory generalizing Lebesgue's, and that despite their noncommutativity. A factor, as said above, is a von Neumann algebra whose center is $\mathbb{C}$. The space of factors contained in a von Neumann algebra M is itself Borel-measurable (Section 15.1.2). Call the measure μ. Each factor can be labelled by an index t belonging to a borelian set, and denoted by $M(t)$. Then the whole algebra M (according to a theorem by von Neumann) is given by the decomposition

$$M = \int M(t)\, d\mu(t)$$

This property justifies the name "factor" and reduces the problem of finding all the von Neumann algebras to that of classifying all the possible factors. The last grand steps in this measure-algebraic program were given by A. Connes. They naturally opened the gates to noncommutative analysis and to noncommutative geometry. Another recent, astonishing finding by Jones (see Section 17.6) is that von Neumann algebras are intimately related to knot invariants.

Projectors are ordered as follows. Let p and q be two projectors in the von Neumann algebra M. We say that p and q are equivalent, and write $p \approx q$, if there exists $u \in M$ such that

$$p = uu^* \quad \text{and} \quad u^*u = q.$$

We say that $p \leq q$ (q dominates p) if there exists $u \in M$ such that

$$p = uu^* \quad \text{and} \quad u^*uq = u^*u.$$

This means that u^*u projects into a subspace of $q\mathcal{H}$, that is, that $u^*u \leq q$. If both $p \leq q$ and $q \leq p$, then $p \approx q$. We say further that $p \perp q$ if $pq = 0$.

Now, if M is a factor, then it is true that, given two projections p and q, either $p \leq q$ or $q \leq p$. The terminology is not without recalling that of transfinite numbers. The projector q is *finite* if the two conditions $p \leq q$ and $p \approx q$ together imply $p = q$. The projector q is *infinite* if the conditions

$$p \leq q, \quad p \approx q, \quad p \neq q$$

can hold simultaneously. The projector p is *minimal* if $p \neq 0$ and $q \leq p$ implies $q = 0$ (p only dominates 0). The factors are then classified in types, denoted I, II_1, II_∞, and III:

I: If there exists a minimal projector.

II_1: If there exists no minimal projector and all projectors are finite

II_∞: If there exists no minimal projector and there are finite and infinite projectors.

III: If the only finite projection is 0.

On a factor there exists a dimension function $d :$ {projections on M} $\to [0, \infty]$ with the suitable properties:

(i) $d(0) = 0$.

(ii) $d(\Sigma_k p_k) = \Sigma_k d(p_k)$ if $p_i \perp p_j$ for $i \neq j$.

(iii) $d(p) = d(q)$ if $p \approx q$.

It is possible then to show that, conversely, $d(p) = d(q)$ implies $p \approx q$. This "Murray–von Neumann dimension" d can then be normalized so that its values have the following ranges:

- Type $I : d(p) \in \{0, 1, 2, \ldots, n$, with possibly $n = \infty\}$;
  If n is finite, type I_n; if not, type I_∞;
- Type $II_1 : d(p) \in [0, 1]$;
- Type $II_\infty : d(p) \in [0, \infty]$;
- Type $III : d(p) \in \{0, \infty\}$.

We have been talking about a *-algebra $\mathcal{A}$ as a set of operators acting on some Hilbert space $\mathcal{H}$. For many purposes, it is interesting to make abstraction of the supposed carrier Hilbert space and consider the algebra by itself, taking into account as far as possible only its own properties, independent of any realization of its members as operators. From this point of view, we speak of the "abstract *-algebra". The realization as operators on a Hilbert space is then seen as a representation of the algebra, with $\mathcal{H}$ as the carrier space. In that case we speak of a "concrete *-algebra".

Commentary 17.7 The name "W*-algebra" is frequently reserved to the abstract von Neumann algebras. ◀

It so happens that the abstract algebra is rich enough to provide even an intrinsic realization on a certain Hilbert space. This comes out of the GNS (Gelfand–Naimark–Segal) construction. The GNS construction is a method to obtain a von Neumann algebra from a *-algebra $\mathcal{A}$. One starts by building a Hilbert space. The linear forms

$$\varphi : \mathcal{A} \to \mathbb{C}$$

constitute a vector space. A form φ is *positive* if $\varphi(x^*x) \geq 0$ for all $x \in \mathcal{A}$. Given a positive form φ, one defines an inner product by

$$< x, y > = \varphi(y^*x).$$

There may exist zeros of φ, elements $x \neq 0$ but with $\varphi(x^*x) = 0$. The set of zeros (kernel of φ) form an ideal I $=$ ker φ in $\mathcal{A}$. The Hilbert space $\mathcal{H}_\varphi$ is then the completion of the quotient of $\mathcal{A}$ by this ideal, $\mathcal{A}/$I. Then, if it is a C*-algebra, $\mathcal{A}$ will act on it by left multiplication and this action is the GNS representation. The von Neumann algebra is the completion of the image of this representation. Even if $\mathcal{A}$ is a factor, the GNS von Neumann algebra can have a non-trivial center, due to the process of completion.

In Quantum Statistical Mechanics, von Neumann algebras are traditionally attained in the following way. One starts by "preparing" the formalism for a finite system (finite volume and number of particles; or finite lattice and lattice parameter), with all operators being finite matrices. Each state is a density matrix ρ (see § 28.1) and the space of states is given by the set of such positive operators. The expectation value of an observable A in state ρ is $\mathrm{tr}(\rho A)$.

In such finite models, no phase transition is ever found. Then one proceeds to the thermodynamic limit, volumes and/or lattice going to infinity. And one supposes that all this is well-defined, though the limit procedure is very delicate. Phase transitions are eventually found. Actually, only a particular type of infinite algebras can be found as the limit of finite algebras, the so-called *hyperfinite* algebras. The enormous majority of operator algebras cannot be attained in this way. The direct study of infinite but non-hyperfinite algebras, which could describe physical systems "beyond the thermodynamic limit", is a major program of Constructive Field Theory.[10]

17.6   The Jones polynomials

Let us now examine some particular finite-dimensional von Neumann algebras, of special interest because of their relationship with braids and knots. A finite-dimensional von Neumann algebra is just a product of matrix algebras, and can be represented in the direct-product notation. In his work[11] dedicated to the classification of factors, Jones[12] was led to examine[13] certain complex von Neumann algebras[14] A_{n+1}, generated by the identity I

---

[10] Of which a remarkable presentation is Haag 1993.
[11] Jones 1983.
[12] Jones 1987.
[13] Jones 1985, and his contribution in Kohno 1990.
[14] Wenzl 1988.

plus n projectors $p_1, p_2, \ldots, p_n$ satisfying

$$p_i^2 = p_i = p_i^\dagger, \tag{17.1}$$

$$p_i p_{i \pm 1} p_i = \tau p_i, \tag{17.2}$$

$$p_i p_j = p_j p_i \quad \text{for} \quad |i - j| \geq 2. \tag{17.3}$$

The complex number τ, the inverse of which is called the Jones index, is usually written by Jones as

$$\tau = \frac{t}{(1 + t)^2},$$

where t is another complex number, more convenient for later purposes. For more involved von Neumann algebras, the Jones index is a kind of dimension (notice: in general a complex number) of subalgebras in terms of which the whole algebra can be decomposed in some sense. In lattice models of Statistical Mechanics, with a spin variable at each vertex, the Jones index is the dimension of the spin-space (Section 28.2). Conditions (17.2) and (17.3) involve clearly a "nearest neighbor" prescription, and are reminiscent of the braid relations. We shall see below that some linear combinations of the projectors and the identity do provide braid group generators.

Consider the sequence of algebras A_n. We can add the algebra $A_0 = \mathbb{C}$ to the club. If we impose that each algebra A_n embeds naturally in A_{n+1}, it turns out that this is possible for arbitrary n only if t takes on some special values: either t is real positive, or t is of the form

$$t = \exp[\pm 2\pi i / k]$$

with k = 3, 4, 5, ... in which case

$$\tau = \frac{1}{4 \cos^2(\pi / k)}.$$

For these values of t there exists a *trace* defined on the union of the A_n's, defined as a function into the complex numbers, tr: $\cup_n A_n \to \mathbb{C}$, entirely determined by the conditions

$$\mathrm{tr}(ab) = \mathrm{tr}(ba); \tag{17.4}$$

$$\mathrm{tr}(w\, p_{n+1}) = \tau\, \mathrm{tr}(w) \quad \text{for} \quad w \in A_n; \tag{17.5}$$

$$\mathrm{tr}(a^\dagger a) > 0 \quad \text{if} \quad a \neq 0; \tag{17.6}$$

$$\mathrm{tr}(I) = 1. \tag{17.7}$$

Conditions (17.1–17.7) determine the algebra A_n up to isomorphisms.

Such algebras were known to physicists, as A_n had essentially been used by Temperley and Lieb[15] in their demonstration of the equivalence between the ice-type and the Potts models,[16] the only difference being in the projector normalization. Define new projectors $E_i = dp_i$, where "d" is a number. If $\tau = d^2$, the conditions become

$$E_i^2 = dE_i; \tag{17.8}$$

$$E_i E_{i\pm1} E_i = E_i; \tag{17.9}$$

$$E_i E_j = E_j E_i \quad \text{for} \quad |i - j| \geq 2. \tag{17.10}$$

A fascinating thing about these algebras is that they lead to a family of invariant polynomials for knots (§ 14.3.5), the Jones polynomials. But to physicists, perhaps the main point is that the partition function of the Potts model is a Jones polynomial for a certain choice of the above variable "t". This entails a relationship between lattice models and knots.

We shall in what follows make large use of the terminology introduced in § 14.1 and the subsequent paragraphs. A representation of the braid group B_n in the above algebra is given as follows: to each generator corresponds a member of the algebra:

$$r(\sigma_i) = G_i = \sqrt{t}\,[tp_i - (I - p_i)]. \tag{17.11}$$

Actually, these operators G_i are just the invertible elements of the algebra, so that B_n appears here as the "group of the algebra" A_n. The inverse to the generators are

$$G_i^{-1} = \frac{1}{\sqrt{t}}\,[t^{-1}p_i - (I - p_i)]. \tag{17.12}$$

Each generator G will satisfy a condition of the type

$$(G - a)(G - b) = 0.$$

This means that the squared generators are linear functions of the generators, so that we have a Hecke algebra. One has

$$P = \frac{G - a}{b - a}$$

(normalized so that $P^2 = P$) as the projector on the eigenspace of b, in terms of which

$$G = (b - a)P + aI = a(I - P) + bP.$$

[15] Temperley & Lieb 1971.
[16] See Baxter 1982.

The projectors are

$$p_i = [G_i + \sqrt{t}]/[(1+t)\sqrt{t}],$$

and the condition $p_i^2 = p_i$ is equivalent to

$$G_i^2 = \sqrt{t}(t-1)\,G_i + t^2,$$

or

$$(G_i - t\sqrt{t})(G_i + \sqrt{t}) = 0,$$

or still

$$t\,G_i^{-1} - t^{-1}G_i + \frac{t-1}{\sqrt{t}}\,I = 0. \tag{17.13}$$

Let us introduce an inspiring notation. Indicate each braid generator by $G_i = \diagup\!\!\!\diagdown$, where it is implicit that the first line is the i-th, and all the unspecified lines are "identity" lines. The identity itself is indicated by $)($ and G_i^{-1} suitably by $\diagdown\!\!\!\diagup$. Relation (17.13) can then be drawn as

$$t\diagdown\!\!\!\diagup - t^{-1}\diagup\!\!\!\diagdown + \frac{t-1}{\sqrt{t}}\,)(= 0. \tag{17.14}$$

This is a skein relation, emerging here as the representative of an algebraic relation. The representation of the braid group B_n takes place actually in a Hecke sub-algebra of A_n, which this relation determines.

Commentary 17.8 For the Alexander polynomial, the relation is

$$(G - \sqrt{t})(G + 1/\sqrt{t}) = 0.$$

◄

In Kauffman's monoid diagrams, the projectors E_i are represented by $\cup\atop\cap$ and will give to the relation (17.11) the form

$$\diagup\!\!\!\diagdown = \frac{t^{3/2}+t^{1/2}}{d}\,{\cup\atop\cap} - t^{1/2}\,)(. \tag{17.15}$$

The first projectors are shown in Figure 17.2 (for the case $n = 4$). A simple blob gives just "d" as a number multiplier: the bubble is normalized by

$$\bigcirc = d.$$

Condition (17.10) is immediate. Conditions (17.9) and (17.8) are shown respectively in Figure 17.3 and Figure 17.4.

The reason for the name "monoid diagrams" is simple to understand here. Projectors, with the sole exception of the identity, are not invertible. Once we add projectors to the braid group generators, what we have is no

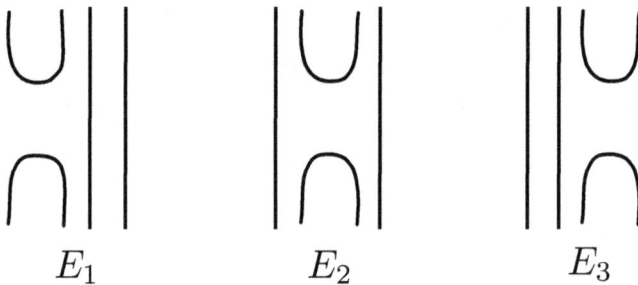

Fig. 17.2   Projectors for 4-strings Kauffman's monoid diagrams.

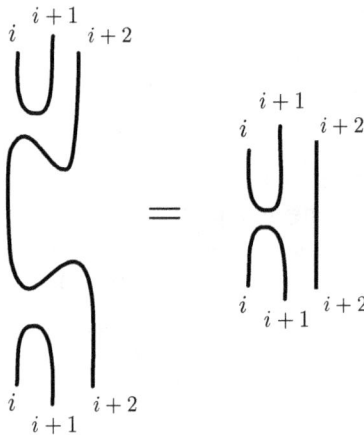

Fig. 17.3   Condition (17.9) for the projectors.

more a group, but a monoid (§ 13.2.9). The addition of projectors to the braid generators, or the passing into the group algebra, turns the matrix-diagram relationship into a very powerful technique.

The Jones polynomials are obtained as follows. Given a knot, obtain it as the closure $\hat{b}$ of a braid b. Then the polynomial is

$$V_{\hat{b}}(t) = \left[-\frac{1+t}{\sqrt{t}} \right]^{n-1} \operatorname{tr}\,[r(b)]. \tag{17.16}$$

Given a knot, draw it on the plane, with all the crossings well-defined. Choose a crossing and decompose it according to (17.14) and (17.15). Two

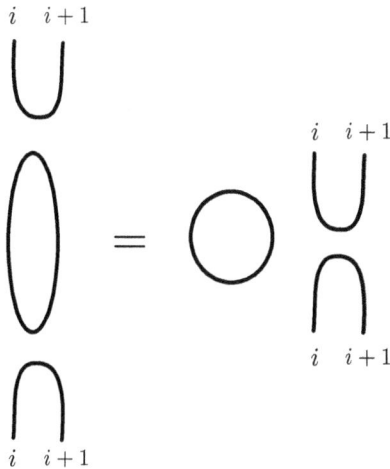

Fig. 17.4   Condition (17.8) for the projectors.

new knots come out, which are simpler than the first one. The polynomial of the starting knot is equal to the sum of the polynomials of these two new knots. Do it again for each new knot. In this way, the polynomial is related to the polynomials of progressively simpler knots. At the end, only the identity and the blob remain.

Jones has shown that his polynomials are isotopic invariants. The Jones polynomial is able to distinguish links which have the same Alexander polynomial. Perhaps the simplest example is the trefoil knot: there are actually two such knots, obtained from each other by inverting the three crossings. This inversion corresponds, in the Jones polynomial, to a transformation $t \rightarrow t^{-1}$, which leads to a different Laurent polynomial. This means that the two trefoils are not isotopic.

General references

Recommended general texts covering the topics of this chapter are: Kirillov 1974, Haag 1993, Bratelli & Robinson 1979, Dixmier 1981, Dixmier 1982, Kaufman 1991, Jones 1991 and Aldrovandi 2001.

Chapter 18

Representations

A representation is a way of making a group (or algebra) more "visible":
it is a homomorphism into some other group (or algebra), usually related
to transformations turning up in some physical system.

18.1   Introductory remarks

In general, a representation[1] of a group G is a homomorphism of G into
some other group H. A representation of a Lie algebra G' of a Lie group G
will be a homomorphism of G' into some other Lie algebra H'. All this is
rather abstract. Actually, representation theory is a way to look at groups
as sets of transformations: H is a transformation group, formed typically
by transformations on vector spaces. Thus, linear representations are those
for which H is the group Aut V of the linear invertible transformations of a
vector space V, or, roughly speaking, those for which H is a matrix group.
A current, though misleading, practice is to use the expression "representation" for V itself. For algebras, H' is a matrix algebra, or something
generalizing it. We shall here concentrate on group representations. We
shall make a passage through linear representations, finite and infinite,[2]
and come back to the group, to examine the case of regular representations, for which the carrier spaces are spaces of functions on the group
itself. We finish with a short incursion into Fourier analysis,[3] a chapter of
representation theory[4] of permanent interest to Physics.

---

[1] A very sound though compact text on representations is Kirillov 1974.
[2] Mackey 1955.
[3] Gel'fand, Graev & Vilenkin 1966.
[4] Mackey 1978.

18.2   Linear representations

Consider a vector space V and the set of transformations defined on it. Of all such transformations, those which are continuous and invertible constitute the linear group on V, indicated by $L(V)$. A linear representation of the group G is a continuous function T defined on G, taking values on $L(V)$ and satisfying the homomorphism requirement,

$$T(gh) = T(g)T(h). \tag{18.1}$$

We speak, in general, of the "representation $T(g)$". The space V is the *representation space*, or the *carrier space*. The representation is *faithful* if T is one-to-one. Call "e" the identity of G, and E the identity of $L(V)$. It follows from the requirement $T(g^{-1}) = T^{-1}(g)$ that $T(e) = E$. A very particular case is the *trivial representation*, which takes all the elements of G into E. The representation $T(g)$ is said to be *exact* when the identity of G is the only element taken by T into E. If $\{e_i\}$ is a basis for V, the matrix $T(g) = (T_{ij}(g))$ will have entries given by the transformation

$$T(g)e_i = \sum_j T_{ij}(g)e_j. \tag{18.2}$$

§ **18.2.1 Dimension.** Ado's theorem (see § 8.3.9) may be rephrased as follows: every finite-dimensional Lie algebra has a faithful linear representation. Or still: every finite-dimensional Lie algebra can be obtained as a matrix algebra. If V is a finite space, such matrices are finite and d_T (the dimension of V) is the *dimension of representation* $T(g)$. The homomorphism condition is then simply

$$T_{ij}(gh) = \sum_{k=1}^{d_T} T_{ik}(g)\, T_{kj}(h). \tag{18.3}$$

When V is infinite, on the other hand, we must worry about the convergence of the involved series. V can be, for instance, a Hilbert space. In the general case of a carrier space endowed with an inner product (u, v), the adjoint $A^\dagger$ of a matrix A satisfies $(Au, v) = (u, A^\dagger v)$ for all $u, v \in V$. It comes out that, if $T(g)$ is a representation, then $T^\dagger(g^{-1})$ is another representation, called the representation "adjoint" to the representation $T(g)$ (this is not to be mistaken by the adjoint representation of a Lie group).

§ **18.2.2 Unitary representations.** A representation $T(g)$ of a group G in the inner product space V is unitary if, for all u, v in V,

$$(T(g)u, T(g)v) = (u, v). \tag{18.4}$$

In this case,

$$T^\dagger(g)T(g) = E$$

and

$$T^\dagger(g) = T^{-1}(g) = T(g^{-1}),$$

so that a unitary representation coincides with its adjoint. The name comes from the matrix case: in a matrix unitary representation, all the representative matrices are unitary.

§ **18.2.3 Equivalent representations.** Let $T(g)$ be a representation on some carrier space V, and let us suppose that there exists a linear invertible mapping f into some other vector space U, that is, $f : V \to U$. Then,

$$T'(g) = f \circ T(g) \circ f^{-1}$$

is a representation with representation space U. The relation between such representations, obtained from each other by a linear invertible mapping, is an equivalence. Two such representations are thus *equivalent representations* and may be seen as different "realizations" of one same representation. A well known example is the rotation group in euclidean 3-dimensional space, $SO(3)$: the vector representation, corresponding to angular momentum 1, is that generated by the fields

$$\tfrac{1}{2}\varepsilon_{ijk}(x^j\partial^k - x^k\partial^j)$$

if seen as transformations on functions and fields on $\mathbb{E}^3$; it is that of the real orthogonal 3×3 matrices if seen as acting on column vectors.

§ **18.2.4 Characters.** Consider, as a simple and basic example, the case in which the carrier space V is a one-dimensional space. Matrices reduce to functions, so that $T(g) = \chi(g)I$, with I the identity operator and χ a non-vanishing complex function on G. As T is a homomorphism,

$$\chi(gh) = \chi(g)\chi(h)$$

for all $g, h \in G$. A function like χ, taking G homomorphically into $\mathbb{C}\backslash\{0\}$, is called a *character*. If G is a finite group, there exists some integer n such that

$$\chi(g^n) = \chi^n(g) = 1,$$

so that

$$|\chi(g)| = 1 \quad \text{and} \quad \chi^*(g) = \chi(g)^{-1}.$$

Every finite abelian group is the direct product of cyclic groups $\mathbb{Z}_N$. Such a cyclic group has a single generator g such that $g^N = I$. Hence,

$$\chi^N(g) = 1.$$

For $\mathbb{Z}_N$, consequently, the characters are the N-th roots of the identity, one for each group element. $\mathbb{Z}_N$ is, therefore, isomorphic to its own group of characters. The characters of the direct product of two groups are the respective product of characters, and it follows that every finite abelian group is isomorphic to its group of characters.

The characters of finite abelian groups have the following further properties:

(i) $\Sigma_{g \in G} \chi(g) = 0$.
(ii) $\Sigma_{g \in G} \chi_1(g) \chi_2^*(g) = 0$,   whenever $\chi_1 \neq \chi_2$.

The characters are thus orthogonal to each other. They are in consequence linearly independent, they span the space of functions on the group, and form a basis for the vector space of complex functions on G. If the group is unitary, so is the character representation:

$$\chi^*(g) = \chi^{-1}(g).$$

§ 18.2.5 Irreducible representations.
Suppose that the carrier space V has a subspace V' such that $T(g)v' \in V'$ for all $v' \in V'$ and $g \in G$. The space V' is an *invariant subspace*, and $T(g)$ is *reducible*. When no invariant subspace exists, the representation is *irreducible*. In the reducible case there are two new representations, a $T'(g)$ on V' and another, $T''(g)$ on the quotient space $V'' = V/V'$. There exists always a basis in which the matrices of a reducible representation acquire the bloc-triangular form

$$\begin{bmatrix} T'(g) & K(g) \\ 0 & T''(g) \end{bmatrix}.$$

There is in general no basis in which $K(g)$ vanishes. This happens, however, when V' admits a linear complement V'' which is also an invariant subspace, so that V is the direct sum $V = V' \oplus V''$ of both. Recall that this means that any $v \in V$ can be written as $v = v' + v''$, with $v' \in V'$ and $v'' \in V''$. In this case

$$T(g)v = T(g)v' + T(g)v'' = T'(g)v' + T''(g)v''.$$

The representations $T'(g)$ and $T''(g)$ are the restrictions of $T(g)$ to V' and V'' and we write

$$T(g) = T'(g) + T''(g).$$

Notice that we are incidentally defining an *addition* of representations. The best physical example is the usual addition of angular momenta.

If V' or V'' are themselves direct sums of other invariant spaces, the procedure may be continued up to a final stage, when no more invariant subspaces exist. When a representation $T(g)$ is such that it is possible to arrive at a final decomposition

$$T(g) = \Sigma_j T_j(g),$$

with each $T_j(g)$ an irreducible representation, $T(g)$ is said to be *completely reducible*. An important related result is the following:

All unitary finite-dimensional representations
are completely reducible.

This is not always true for infinite-dimensional representations. On the other hand, all the irreducible representations of a commutative group have dimension 1.

§ 18.2.6 Tensor products.
Let $T(g)$ and $S(g)$ be two representations of G, respectively on spaces V and U. They define another representation of G, the *tensor product* (or Kronecker product) $R = T \otimes S$ of T and S, acting on the direct product of V and U. Operators A, B acting on V, and C, D acting on U will satisfy

$$(A \otimes C)(B \otimes D) = (AB) \otimes (CD) .$$

With tensor products we go into higher dimensional representations, so that there is a higher chance of obtaining reducible representations.

Coming back to the characters, we have above defined them for one-dimensional representations. For general matrix representations, they are defined as the trace of $T(g)$,

$$\chi_T(g) = \text{ tr } T(g) = \sum_{k=1}^{d_T} T_{kk}(g), \qquad (18.5)$$

where d_T is the dimension of the representation. Due to the trace properties, the character depends only on the class of equivalent representations. Of course, when the group is not finite, the summations on the group used above have to be examined in detail. In particular, for continuum groups they become integrals and some measure must be previously defined. We shall come to this point later. Let us only state two properties which hold anyway: the character of an addition is the sum of characters, and the character of a tensor product is the product of the characters:

$$\chi_{T+S}(g) = \chi_T(g) + \chi_S(g), \tag{18.6}$$

$$\chi_{T\otimes S}(g) = \chi_T(g)\chi_S(g). \tag{18.7}$$

18.3 Regular representation

§ 18.3.1 **Invariant spaces.** Let M be a homogeneous space under G and consider $\mathcal{L}$ the set of functions $f : M \to V$, where V is some other space. Then $\mathcal{L}$ is said to be an *invariant space* of functions if $f(x) \in \mathcal{L}$ implies $f(gx) \in \mathcal{L}$. It may happen that some subspace of $\mathcal{L}$ be invariant by itself. To each such invariant subspace will correspond a representation, given by

$$T(g)f(x) = f(g^{-1}x). \tag{18.8}$$

Commentary 18.1 Notice to avoid confusion that, as already mentioned, some authors use the name "regular representation" in senses different from that used here; see for instance Hamermesh 1962, Gilmore 1974. ◄

§ 18.3.2 **Invariant measures.** As introduced above, $\mathcal{L}$ is not even a topological space. If there exists an invariant measure on M, we may instead take for $\mathcal{L}$ the space of complex, square-integrable functions. In this case, given the inner product

$$(f_1, f_2) = \int_M f_1^*(x)f_2(x)d\mu(x), \tag{18.9}$$

the representation [18.8] is unitary:

$$(T(g)f_1, T(g)f_2) = \int_M f_1^*(g^{-1}x)f_2(g^{-1}x)d\mu(x)$$

$$= \int_M f_1^*(x)f_2(x)d\mu(x) = (f_1, f_2). \tag{18.10}$$

Recall that on locally-compact groups, the existence of a left-invariant measure is guaranteed, as is the existence of a right-invariant measure. Such Haar measures, by the way, do not necessarily coincide (groups for which they coincide are called *unimodular*). Given the actions of a group G on itself, the functions defined on G will carry the left and right representations:

$$L(h)f(g) = (L_h f)(g) = f(h^{-1}g), \tag{18.11}$$

$$R(h)f(g) = (R_h f)(g) = f(gh). \tag{18.12}$$

§ 18.3.3 Generalities. If G has a left-invariant measure, we may take for function space the set of square-integrable functions. $L(g)$ is then called the *left-regular representation*. In an analogous way, $R(g)$ is the *right-regular representation*. Such representations, given by operators acting on functions defined on the group itself, are of the utmost importance. First, because it happens that

> *any irreducible representation of G is equivalent to some regular representation.*

And second, because their study is the starting point of generalized Fourier analysis, or Fourier analysis on general groups (or still, non-commutative harmonic analysis — see Section 18.4 below).

§ 18.3.4 Relation to von Neumann algebras. What we have here are representations in terms of operators on function spaces. We have been forced to restrict such function spaces through the measure requirements.

The inner product introduced above defines a topology, and we may go a step further. We require that the function space be complete in this topology, so that they become Hilbert spaces. The "$T(g)$" introduced above belong consequently to spaces of operators acting on Hilbert spaces. They may be added and multiplied by scalars, constituting consequently a linear space. Such operator spaces are themselves normed spaces, the norm being here the weak one, i.e., that provided by the internal product. Restricting again the whole scheme, we content ourselves with those subspaces of normed operators which are complete, so as to obtain a Banach space. A product of such operators is well-defined, so that they constitute Banach algebras. On the other hand, they are involutive spaces, and complete by the weak topology given by the internal product. Thus, finally, they actually make up von Neumann algebras.[5]

Summarizing: the regular representations are group homomorphisms from the group G into von Neumann algebras. Wherefrom comes a relationship[6] between the classification of irreducible group representations and the classification of von Neumann algebras (§ 17.5). Just as for the latter, the groups are said to be of type I, II, III according to their representations. Only for type I groups does the simple ordinary decomposition of a representation into a sum (in general, an integral) of irreducible representations hold. For types II and III, the space of representations is in general not

---

[5] Mackey 1968.
[6] Mackey 1978, chap. 8.

even (first-)separable.

18.4 Fourier expansions

We discuss in this section some small bits of Fourier analysis, starting with the main qualitative aspects of the elementary case.

§ **18.4.1 The standard cases.** Consider a periodic square-integrable complex function f on the line $\mathbb{E}^1$. Being periodic means that f is actually defined on the circle S^1, which is the manifold of the group $U(1) = SO(2)$. The Fourier expansion for f is then

$$f(x) = \sum_{n=-\infty}^{\infty} e^{inx} \tilde{f}(n), \qquad (18.13)$$

where the discrete coefficients

$$\tilde{f}(n) = \frac{1}{2\pi} \int_0^{2\pi} dx e^{-inx} f(x) \qquad (18.14)$$

are the values of the Fourier transform of f. Notice that:

(i) The original space $U(1)$ is compact and the series is discrete, $n \in \mathbb{Z}$.
(ii) For each n, e^{inx} is a character of $U(1)$.
(iii) The characters form irreducible, unitary representation of $U(1)$.
(iv) The Fourier series is consequently an expansion in terms of non-equivalent irreducible unitary representations of the original group $U(1)$.

Conversely,

(i') Also $\mathbb{Z}$ is a (non-compact) group with the addition operation, $(\mathbb{Z}, +)$, and $\tilde{f}(n)$ is its Fourier expansion.
(ii') For each x, e^{inx} can be seen as a character for $\mathbb{Z}$.
(iii') Such characters form irreducible, unitary representation of $\mathbb{Z}$.
(iv') The Fourier integral is consequently an expansion in terms of irreducible unitary representations of $\mathbb{Z}$.

Suppose now we dropped the periodicity condition: we would then have

$$f(x) = \int_{-\infty}^{\infty} dy e^{iyx} \tilde{f}(y) \qquad (18.15)$$

and

$$\tilde{f}(y) = \frac{1}{2\pi} \int_{-\infty}^{\infty} dx e^{-iyx} f(x) \qquad (18.16)$$

with both x and $y \in \mathbb{E}^1$. Statements analogous to those made above keep holding, with the difference that now both groups are $(\mathbb{E}^1, +)$, which is non-compact.

Suppose finally we started with the cyclic group $\mathbb{Z}_N$, which is a kind of "lattice circle". Then,

$$f(m) = \sum_{n=1}^{N} e^{i(2\pi/N)mn} \tilde{f}(n) \tag{18.17}$$

and

$$\tilde{f}(n) = \frac{1}{2\pi} \sum_{m=1}^{N} e^{-i(2\pi/N)mn} f(m). \tag{18.18}$$

Both groups are $\mathbb{Z}_N$, compact and discrete.

§ 18.4.2 Pontryagin duality. According to the discussion above, $\mathbb{Z}$ is "Fourier-dual" to S^1 and vice-versa. The line $\mathbb{E}^1$ is self-dual and so is $\mathbb{Z}_N$. These results keep valid for other cases, such as the euclidean 3-space $\mathbb{E}^3$: the Fourier transformations establish a duality between the space of functions on the original space (which may be seen as the translation group T_3 in $\mathbb{E}^3$) and the space of the Fourier transforms, which are functions on — in principle — another space. The latter is the space of (equivalence classes of) unitary irreducible representations of T_3, and constitutes another group. It is a general result that, when the original group G is commutative and locally compact, the Fourier-dual is also a commutative locally compact group. Furthermore, if they are compact or discrete, the duals are respectively discrete or compact. This duality appearing in the abelian case is the Pontryagin duality. Local compactness is required to ensure the existence of a Haar measure, which in the example above is just the (conveniently normalized) Lebesgue measure.

§ 18.4.3 Noncommutative harmonic analysis. Harmonic analysis may be extended to other groups, although with increasing difficulty. There is a complete theory for abelian groups. For non-abelian groups, only the compact case is well established and the subject has been christened "non-commutative harmonic analysis". We have said (Section 17.2) that commutative Banach algebras are the natural setting for "standard" harmonic analysis, that is, for Fourier analysis on abelian groups.

For non-commutative groups, non-commutative Banach algebras are the natural setting. As a good measure is necessary, the research has been concentrated on locally-compact groups. The reason for the special simplicity of abelian groups is that their unitary irreducible representations have dimension one and the tensor product of two such representations is another one of the same kind. Each such representation may be considered simply as a function $f \in \mathbb{C}(G)$,

$$f : G \to \mathbb{C}, \quad g \to f(g),$$

with

$$f(g_1 g_2) = f(g_1) f(g_2). \tag{18.19}$$

Tensor products are then reduced to simple pointwise products of functions. The set of inequivalent unitary irreducible representations is consequently itself a group. This is the property which does not generalize to the non-commutative case.

Consider a representation T_λ on a Hilbert space H_λ. A group representation extends to a ring representation. The general Fourier transform is

$$\tilde{f}(\lambda) = \sum_{h \in G} T_\lambda(h) \, df(h), \tag{18.20}$$

where $\tilde{f}(\lambda) \in \text{End} H_\lambda$. It takes an element $f \in \mathbb{R}(G)$ of the ring into $\tilde{f} \in \tilde{G}$.

§ **18.4.4 The Peter–Weyl theorem.**[7] A compact group G has a countably infinite set of representations, which we shall label by the index "α". Each representation T_α is acting on some vector space of dimension

$$d_\alpha = dim \, T_\alpha(G),$$

and each group element g will be represented by the matrix $T_\alpha(g)$ with elements

$$\left[T_\alpha(g) \right]_{ij}, \quad i, j = 1, 2, \ldots, d_\alpha.$$

The set $\tilde{G}$ of all the unitary irreducible representations of G is a sum of all such spaces, and is called "the unitary dual of G".

The Peter–Weyl theorem says that, with an invariant normalized measure dg on G, the set $\{[T_\alpha(g)]_{ij}\}$ of all matrix elements of all the representations is a complete orthogonal system for the square-integrable functions on G. More precisely, if $f : G \to \mathbb{C}$ and

$$\int |f|^2 dg < \infty,$$

---

[7] Vilenkin 1969.

then

$$f(g) = \sum_{\alpha} \sum_{i,j=1}^{d_\alpha} \sqrt{d_\alpha} \, [T_\alpha(g)]_{ij} \, f^\alpha{}_{ij} \qquad (18.21)$$

where

$$f^\alpha{}_{ij} = d_\alpha \int dg f(g) [T_\alpha(g)]^*_{ij} \qquad (18.22)$$

are the Fourier components of f. The matrix elements $[T_\alpha(g)]_{ij}$, in terms of which every square-integrable function can be expanded, are the "special functions" we are used to. Every time we have a spherically symmetric problem, for example, we obtain ultimately solutions in terms of Legendre polynomials. These polynomials are exactly the above matrix elements for the representations of the rotation group.

§ **18.4.5 Tanaka–Krein duality.** The Fourier transformation makes use of the unitary irreducible representations of the group, and that is why harmonic analysis is a chapter of representation theory. When G is a compact nonabelian group, the dual $\tilde{G}$ is in fact a category, that of the finite dimensional representations of G. It is a category of vector spaces (or an algebra of blocs). The representations of this category (representations of categories are called functors) constitute a group isomorphic to G. This new duality, between two different kinds of structures — a group and a category — is called the Tanaka–Krein duality.

§ **18.4.6 Quantum groups.**

Is it possible to enlarge the notion of group to another object, so that its dual comes to be an object of the same kind? The complete answer has been found in the case of finite groups: the more general objects required are *Hopf algebras* (§ 13.6.2), frequently called *quantum groups* in recent times. On infinite groups, the operators must stand in Hopf algebras which are also von Neumann algebras. Amongst such Hopf–von Neumann algebras, some can be chosen (called Kac algebras) that respect some kind of Fourier duality. But different notions of duality are possible when we go to the finest details, and the subject is still a province of mathematical research. Due to the clear relationship between quantization and Fourier analysis, it would be interesting to find quantum groups as those generalizations of groups allowing for a "good" notion of Fourier duality.

General references

Good general texts covering the topics above are Vilenkin 1969, Mackey 1978, Kirillov 1974, Katznelson 1976 and Enoch & Schwartz 1992.

Chapter 19

Variations and Functionals

A geometrical approach to variations is given that paves the way to the introduction of functionals and creates the opportunity for a glimpse at infinite-dimensional manifolds.

19.1 Curves

19.1.1 *Variation of a curve*

To study the variation of a curve we begin by "placing it in the middle of a homotopy". A curve on a topological space M is a continuous function

$$\gamma : I = [-1, +1] \to M .$$

We shall consider only paths with fixed ends, from the initial end-point $a = \gamma(-1)$ to the final end-point $b = \gamma(+1)$. Given paths γ, α, β going from point a to point b, they are homotopic (we write $\gamma \approx \alpha \approx \beta$) if there exists a continuous mapping

$$F : I \times I \to M,$$

such that, for every t, s in I,

$$F(t, 0) = \gamma(t) \quad F(t, 1) = \alpha(t) \quad F(t, -1) = \beta(t)$$

and

$$F(-1, s) = a, \ \ F(1, s) = b.$$

The family of curves F is a homotopy including γ, α and β, or, if we wish, a continuous deformation of the curve γ into the curves α and β. For each fixed s,

$$\gamma_s(t) = F(t, s)$$

is a curve from a to b, intermediate between α and β, with

$$\gamma(t) = \gamma_0(t)$$

somewhere in the middle. But also, for each fixed t,

$$\gamma_t(s) = F(t, s)$$

is a curve going from the point $\alpha(t)$ to the point $\beta(t)$ while s treads I and meets $\gamma(t)$ somewhere in between (see Figure 19.1). Let us fix this notation: $\gamma_s(t)$ is a curve with parameter t at fixed s; $\gamma_t(s)$ is a curve with parameter s at fixed t. The one-parameter family γ_s of curves is called a *variation* of γ.

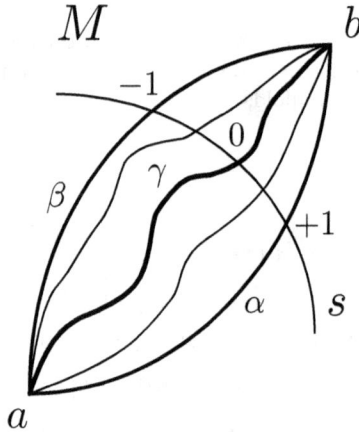

Fig. 19.1   Variation of curve γ: a homotopy of curves with the same end-points.

19.1.2   *Variation fields*

We now add a smooth structure: M is supposed to be a differentiable manifold and all curves are differentiable paths. Each $\gamma_s(t)$ will have tangent vectors

$$V_s(t) = \frac{d}{dt}\gamma_s(t) = \dot{\gamma}_s,$$

which are its "velocities" at the points $\gamma_s(t)$. But also the transversal curves $\gamma_t(s)$ will have their tangent fields, $d\gamma_t(s)/ds$. Consider now the curve $\gamma(t)$: besides its velocity

$$V(t) = V_0(t) = \dot{\gamma}(t) = \frac{d}{dt}\gamma_0(t),$$

it will display along itself another family of vectors, a vector field induced by the variation, (see Figure 19.2)

$$X(t) = \frac{d}{ds}\gamma_t(s)|_{s=0} = \frac{d}{ds}\gamma_t(0).$$

$X(t)$ will be a vector field on M, defined on each point of $\gamma(t)$. This vector field X is called an *infinitesimal variation* of γ. It is sometimes called a Jacobi field, though this designation is more usually reserved to the case in which γ is self–parallel.

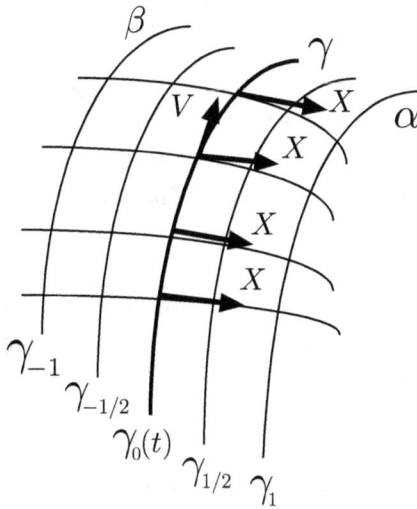

Fig. 19.2   Vector fields related to a variation.

19.1.3   *Path functionals*

Non-exact Pfaffian forms (§ 7.1.2), whose integrals depend on the curve along which it is taken, open the way to path functionals. A trivial example is the length itself: the integral $\int_\gamma ds$ depends naturally on the path. Another example is the work done by a non-potential force, which depends on the path γ along which the force is transported from point a to point b: the integral

$$W_{ab}[\gamma] = \int_\gamma F_k\, dx^k$$

is actually a "function" of the path, it will depend on the functional form of $\gamma(t)$. Paths belong to a space of functions, and mappings from function spaces into $\mathbb{R}$ or $\mathbb{C}$ are *functionals*. Thus, work is a functional of the trajectory.

But perhaps the most important example in mechanics is the action functional related to the motion between two points a and b along the path γ. Let the points on γ be given by coordinates γ^i and the corresponding velocity be given by $\dot{\gamma}^i$. We can use time as the parameter "t", so that the "velocity" above is the true speed. Let the lagrangian density at each point be given by $\mathcal{L}[\gamma^i(t), \dot{\gamma}^i(t), t]$. The action functional of the motion along γ will be

$$A[\gamma] = \int_\gamma \mathcal{L}[\gamma^i(t), \dot{\gamma}^i(t), t]\, dt. \tag{19.1}$$

To each path γ going from a to b will correspond such an action. Consider the space $H_{ab} = \{\gamma\}$ of all such paths. The functional A is a mapping

$$A : H_{ab} \to \mathbb{E}^1.$$

Notice that the underlying space M remains in the background: the coordinates of its points are parameters. If they appear in the potential function U in the usual way, through

$$\mathcal{L} = T(\dot{x}^i) - U(x^i),$$

only its values $U(\gamma^i)$ on the path will actually play a role. Clearly the integral A does not depend on the points, not even on the values of the sole acting parameter t. The action A depends actually only on how γ depends on t.

19.1.4 *Functional differentials*

Once we have established that A is a real-valued function on H_{ab}, we may think of differentiating it. But then, to start with, H_{ab} should be a manifold. The question is: H_{ab} being of infinite dimension, how to make of it a manifold? As repeatedly said, manifolds are spaces on which coordinates make sense. The euclidean spaces are essential to finite dimensional manifolds, as they appear both as coordinate spaces and as tangent spaces at each point of differentiable manifolds. Here, the place played by the euclidean spaces would be played by Banach spaces, infinite dimensional spaces endowed with topology and all that. We keep this point in mind, but actually use an expedient: we take as coordinates the γ^i above, understanding by this that, through the infinite possible values they may take, we are in reality covering the space H_{ab} with a chart. The high cardinality involved is hidden in the apparently innocuous "γ", which contains an arbitrary functional freedom. This assumption makes life much simpler than a detailed examination of Banach spaces characteristics, and we shall admit it. Let us only remark that, Banach spaces being vector spaces, we are justified in adding the coordinates γ^i as we shall presently do. There is, however, a clear problem in multiplying them by scalars, as then the end-points could change. This means that paths do not constitute a topological vector space. It is fortunate that we shall only need addition in what follows.

A differential of the action A (which does not always exist) can be introduced as follows. Take the path γ and one of its "neighbours" in the variation family, given by the coordinates $\gamma^i + \delta\gamma^i$. We represent collectively the coordinates of the path points by γ and $\gamma + \delta\gamma$. Suppose then that we may separate the difference between the values of their actions into two terms,

$$A[\gamma + \delta\gamma] - A[\gamma] = F_\gamma[\delta\gamma] + R, \tag{19.2}$$

in such a way that $F_\gamma[\delta\gamma]$ is a *linear* functional of $\delta\gamma$, and R is of second (or higher) order in $\delta\gamma$. When this happens, A is said to be *differentiable*, and F is its differential. By a smart use of Dirac deltas, we can factorize $\delta\gamma$ out of $F_\gamma[\delta\gamma]$, obtaining an expression like

$$F_\gamma[\delta\gamma] = \frac{\delta F}{\delta\gamma}\delta\gamma.$$

(examples can be found in Sec. 20.2). The quantity "$\delta\gamma$" is usually called *the variation*, and the factor $\delta F/\delta\gamma$ is the *functional derivative*. This derivative has actually many names. In functional calculus its is called Fréchet (or strong) derivative. In mechanics, due to the peculiar form it exhibits, it is called Lagrange derivative.

For the action functional (19.1), we have

$$A[\gamma + \delta\gamma] - A[\gamma] = \int_\gamma \mathcal{L}[\gamma^i(t) + \delta\gamma^i(t), \dot\gamma^i(t) + \delta\dot\gamma^i(t)]dt - \int_\gamma \mathcal{L}[\gamma^i(t), \dot\gamma^i(t)]dt$$

$$= \int_\gamma \left[\frac{\partial\mathcal{L}}{\partial\gamma^i}\delta\gamma^i + \frac{\partial\mathcal{L}}{\partial\dot\gamma^i}\delta\dot\gamma^i\right]dt + O(\delta\gamma^2),$$

so that

$$F_\gamma[\delta\gamma] = \int_\gamma \left[\frac{\partial\mathcal{L}}{\partial\gamma^i}\delta\gamma^i + \frac{\partial\mathcal{L}}{\partial\dot\gamma^i}\delta\dot\gamma^i\right]dt. \tag{19.3}$$

The second term in the bracket is

$$\int_\gamma \left[\frac{\partial \mathcal{L}}{\partial \dot\gamma^i}\frac{d}{dt}\delta\gamma^i\right] dt = \int_\gamma \frac{d}{dt}\left[\frac{\partial \mathcal{L}}{\partial \dot\gamma^i}\delta\gamma^i\right] dt - \int_\gamma \left[\frac{d}{dt}\frac{\partial \mathcal{L}}{\partial \dot\gamma^i}\right]\delta\gamma^i dt$$

$$= \left[\frac{\partial \mathcal{L}}{\partial \dot\gamma^i}\delta\gamma^i\right]_{t=1} - \left[\frac{\partial \mathcal{L}}{\partial \dot\gamma^i}\delta\gamma^i\right]_{t=0} - \int_\gamma \left[\frac{d}{dt}\frac{\partial \mathcal{L}}{\partial \dot\gamma^i}\right]\delta\gamma^i dt.$$

As the variations are null at the end-points, only the last term remains and we get

$$F_\gamma[\delta\gamma] = \int_\gamma \left[\frac{\partial \mathcal{L}}{\partial \gamma^i} - \frac{d}{dt}\frac{\partial \mathcal{L}}{\partial \dot\gamma^i}\right]\delta\gamma^i dt. \tag{19.4}$$

The integrand is now the Lagrange derivative of $\mathcal{L}$. The Euler–Lagrange equations

$$\frac{\partial \mathcal{L}}{\partial \gamma^i} - \frac{d}{dt}\frac{\partial \mathcal{L}}{\partial \dot\gamma^i} = 0 \tag{19.5}$$

are extremum conditions on the action: they fix those curves γ which extremize $A[\gamma]$.

19.1.5 *Second variation*

Going further, we may add to the differentiable structure of M a connection Γ. The covariant derivative along an arbitrary vector field X is generally indicated by ∇_X. Given a curve γ, whose tangent velocity field will be

$$V(t) = \dot\gamma(t) = \frac{d}{dt}\gamma(t),$$

the covariant derivative $V^i D_i W$ of a field or form W along the curve γ will be indicated by

$$\frac{DW}{Dt} = \nabla_V W.$$

The acceleration, for example, will be

$$a = \frac{DV}{Dt} = \nabla_V V.$$

The field (or form) W will be parallel-transported along γ if

$$\nabla_V W = 0.$$

The infinitesimal variation field X, or Jacobi field, will submit to a second order ordinary differential equation, the Jacobi equation (see Chapter 24). Curves are very special examples of functions. We have discussed functionals on spaces of curves. Let us now address ourselves to more general function spaces.

19.2 General functionals

A heuristic introduction to functionals which generalize the path functionals will be given in what follows. Attention will be called on the topologies necessarily involved. The approach is voluntarily repetitive. Functionals generalize to operators. Functionals are real- or complex-valued functions on some spaces of functions, while operators take such spaces into other spaces of the same kind. Only that minimum necessary to the presentation of the very useful notion of functional derivative will be considered.

19.2.1 *Functionals*

Let us gather all the courage and be a bit repetitive. A functional is a mapping from a function space into $\mathbb{R}$ or $\mathbb{C}$. Let us insist on the topic by giving a heuristic introduction[1] to the notion of functional as a generalization of a function of many variables. Consider the function

$$f : \mathbb{R}^N \to \mathbb{C},$$

written as $f(x^1, x^2, \ldots, x^N)$. It is actually a composite function, as the argument may be seen as another function

$$x : \{\text{index set}\} \to \mathbb{R}^N, \quad x : \{1, 2, 3, \ldots, N\} \to (x^1, x^2, \ldots, x^N),$$

which attributes to each integer i the value x^i. We have thus a function of function $f[x(k)]$, starting from a discrete set of indices.

Why do we still talk of f as "a function of x"? Because the index set is fixed through the process of calculating a value of f. Given this set, the function "x" will fix a value $(x^1, x^2, \ldots, x^N)$. To each value of x, f will then attribute a real or complex number. Or, if we prefer, to another function

$$y : \{\text{index set}\} \to \mathbb{R}^N, \quad y : \{1, 2, 3, \ldots, N\} \to (y^1, y^2, \ldots, y^N),$$

f will attribute another number. The function f is thus dependent only on the set $\{x\}$ of functions from $\{1, 2, 3, \ldots, N\}$ into $\mathbb{R}^N$, and not at all on the set $\{1, 2, 3, \ldots, N\}$. This set is actually fixed, so that f is only a function

$$f : \{x\} \to \mathbb{C}.$$

Take now, instead of the discrete set $\{1, 2, 3, \ldots, N\}$, a set of points in some continuum, say, the interval $I = [0, 1]$. The set $\{x\}$ will then be a set of functions on I, each one some $x(t)$. And $f[x(t)]$ will depend on *which* function $x(t)$ is considered, that is to say, f will be a functional on the space

---
[1] Inspired in Balescu 1975.

of functions defined on I. This space might, in principle, be anything, but functionals are usually introduced on topological spaces, such as "the set of square integrable functions on the line", or "the set of twice-differentiable functions on the sphere S^2".

Functionals appear in this way as functions depending on many (actually, infinite) variables, or on variables whose indices belong to a continuum. For instance, "t" may be time, $x(t)$ a trajectory between two fixed points in $\mathbb{E}^3$, and f will be a functional on trajectories [say, the classical mechanical action (19.1)].

We may, of course, take the converse point of view and consider usual few-variable functions as functionals on spaces of functions whose arguments belong to a discrete finite set. Or we may go further in the first direction, and consider many-dimensional sets of continuous indices. For instance, in relativistic field theory (see Chapter 31) such indices are the coordinates in Minkowski space, the role of the degrees of freedom being then played by fields $\varphi(x, y, z, ct)$. A splendid functional will be the action functional for the fields φ.

19.2.2 *Linear functionals*

The reasoning above makes clear the interest of functionals: they are the "functions" defined on infinite dimensional spaces, specially on spaces whose "infinity" has the power of the continuum or more. As function spaces can be always converted into linear spaces, a functional can be defined as any complex function defined on a linear space. Once we have established that a functional is a function from a function space $\{f\}$ to $\mathbb{C}$, we may think of applying to it the usual procedures of geometry and analysis. But then, to start with, the function space should be a good (topological) space. Better still, it should be a manifold. Banach spaces will play here the role played by the euclidean spaces in the finite case. Some authors[2] give another name to topological vector spaces: they call "euclidean spaces" any inner-product linear space, be it finite-dimensional or not. It is necessary to adapt euclidean properties to the infinite-dimensional case and then consider linear objects.[3] We shall below list the main results in words as near as possible to those describing the finite dimensional case, while stressing the notions, here more delicate, of continuity and boundedness.

---

[2] Kolmogorov & Fomin 1977.
[3] For a very good short introduction to analysis on infinite-dimensional spaces, see Marsden 1974.

In general a functional is a mapping from a topological linear space into $\mathbb{C}$. The mapping W is a *linear functional* when

$$W[f + g] = W[f] + W[g]$$

and

$$W[kf] = kW[f].$$

19.2.3 *Operators*

Let X and Y be two topological vector spaces. Then, an operator is any mapping $M : X \to Y$. It is a *linear operator* if

$$M[f + g] = M[f] + M[g]$$

and

$$M[kf] = kM[f].$$

It is *continuous at a point* f_0 of X if, for any open set $V \subset Y$ around $M[f_0]$, there exists an open set U around f_0 such that $M[f] \in V$ whenever $f \in U$. It is *continuous* on the space X if it is continuous at each one of its points. For normed spaces, this is equivalent to saying that, for any tolerance $\varepsilon > 0$, there is a spread $\delta > 0$ such that

$$\|f_1 - f_2\| < \varepsilon \quad \text{implies} \quad \|Mf_1 - Mf_2\| < \delta.$$

Recall (Section 16.6) that a subset U of a topological vector space is a *bounded set* if, for any neighbourhood V of the space origin, there exists a number $n > 0$ such that $nV \supset U$. The mapping-operator M is a *bounded operator* if it takes bounded sets into bounded sets. It so happens that every continuous operator is bounded. A linear functional is a particular case of operator, when $Y = \mathbb{C}$. We may thus pursue our quest by talking about operators in general.

19.2.4 *Derivatives: Fréchet and Gateaux*

Let X and Y now be two normed spaces and M an operator

$$M : X \to Y.$$

The operator M will be a (strongly) *differentiable operator* at $f \in X$ if there exists a linear bounded operator M'_f such that

$$M[f + g] = M[f] + M'_f[g] + R[f, g], \tag{19.6}$$

where the remainder $R[f, g]$ is of second order in g, that is to say,

$$\frac{||R[f, g]||}{||g||} \to 0$$

when $||g|| \to 0$. When this happens, $M_f'[g] \in Y$ is the *strong differential* (or *Fréchet differential*) of M at f The linear operator M_f' is the *strong derivative*, (or *Fréchet derivative*), of M at f. The theorem for the derivative of the function of a function holds for the strong derivative.

We may define another differential: the *weak differential* (or *Gateaux differential*) is

$$D_f M[g] = \left[\frac{d}{dt} M[f + tg] \right]_{t=0} = \lim_{t \to 0} \frac{M[f + tg] - M[f]}{t} , \qquad (19.7)$$

the convergence being understood in Y's norm topology. In general, $D_f M[g]$ is not linear in g. When $D_f M[g]$ happens to be linear in g, then the linear operator $D_f M$ is called the *weak derivative*, (or *Gateaux derivative*) at f. The theorem for the derivative of the function of a function does *not* hold in general for the weak derivative.

If M is strongly differentiable, then it is weakly differentiable and both differentials coincide. But the inverse is not true. The existence of the weak differential is a warrant of the existence of the strong differential *only if $D_f M$ is continuous* as a functional of f. Anyhow, when the strong differential exists, one may use for it the same expression

$$\left[\frac{d}{dt} M[f + tg] \right]_{t=0} ,$$

which is frequently more convenient for practical calculational purposes.

The practical use of all that will be illustrated in the next Chapter, with applications to Physics, or more precisely, to lagrangian field theory.

General references

Highly recommended general texts covering the topics of this chapter are Lanczos 1986, Kobayashi & Nomizu 1963, Kolmogorov & Fomin 1977, and Choquet–Bruhat, DeWitt–Morette & Dillard–Bleick 1977.

Chapter 20

Functional Forms

The lore of exterior differential calculus can be extended to spaces whose "points" are functions, provided very careful attention is dedicated to the far more involved topologies. It leads to exterior functional calculus — including many of well-known resources, such as lagrangian approaches in general. We insist on the word "calculus" because the approach will be purely descriptive, operational and practical. It is of particular relevance in field theory and specially helpful in clarifying the geometry of field spaces.

20.1 Introductory remarks

Exterior differential calculus is a very efficient means to compactify notation and reduce expressions of tensor analysis to their essentials. It has been for long the privileged language of the geometry of finite dimensional spaces, though it has become a matter of necessity for physicists in recent times. On the other hand, functional techniques and the closely related variational methods have long belonged to the common lore of Theoretical Physics. What follows is a mathematically naïve introduction to exterior variational calculus,[1] which involves differential forms on spaces of infinite dimension in close analogy with the differential calculus on finite dimensional manifolds. Only "local" aspects will be under consideration, meaning by that properties valid in some open set in the functional space of fields. As it has been the case with differential forms on finite-dimensional manifolds, functional forms may come to be of help also in the search of topological functional characteristics.

Special cases of exterior differential calculus have been diffusely applied

---

[1] Olver 1986; Aldrovandi & Kraenkel 1988.

to the study of some specific problems, such as the BRST symmetry,[2] and anomalies,[3] but its scope is far more general.

20.2   Exterior variational calculus

20.2.1   *Lagrangian density*

Let us consider a set of fields $\phi = \{\phi^a\} = (\phi^1, \phi^2, \ldots, \phi^N)$ defined on a space M which, to fix the ideas, we shall suppose to be the Minkowski spacetime, even though, as some examples will make clear, all that follows is easily adaptable to fields defined on other manifolds. The word "field" is employed here with its usual meaning in Physics: it may be a scalar, a vector, a spinor, a tensor, etc. What is important is that it represents an infinity of functions, vectors, etc, and describe a continuum infinity of degrees of freedom. Each kind of field defines a bundle with M as base space, and we shall take local coordinates (x^μ, ϕ^a) on the bundle. As in the bundles we have already met, the basic idea is to use the x's and the ϕ's as independent coordinates:[4] the base space coordinates x^μ and the functional coordinates ϕ^a. If $\mathcal{L}[\phi(x)]$ is a lagrangian density, its total variation under small changes of these extended (or bundle) coordinates will be

$$\delta_T \mathcal{L} = \delta \mathcal{L} + d\mathcal{L} = (\delta_a \mathcal{L})\,\delta\phi^a + (\partial_\mu \mathcal{L})\,dx^\mu, \qquad (20.1)$$

where $\delta_a \mathcal{L}$ is a shorthand for $\delta \mathcal{L}/\delta\phi^a$ (in analogy to $\partial_\mu f = \partial f/\partial x^\mu$) , and $\delta\phi^a$ is the purely functional variation of ϕ^a. We are thus using a natural basis for these 1-forms on the bundle. Vector functional fields (in the geometrical sense of the word) can be introduced, and the set of derivatives $\{\delta/\delta\phi^a\}$ can be used as a "natural" local basis for them, dual to $\{\delta\phi^a\}$. A general Field X (we shall call "Field", with capital F, geometrical fields on the functional space) will be written as

$$X = X^a \frac{\delta}{\delta\phi^a}.$$

In the spirit of field theory, all information is contained in the fields, which are the degrees of freedom. Consequently, $\mathcal{L}$ will be supposed to have no

---

[2] The original papers are: Becchi, Rouet & Stora 1974, 1975, 1976 and Tyutin 1975; Kugo & Ojima 1978 give a more "quantal" (rather than geometrical) approach; later good presentations are Stora 1984; Zumino, Wu & Zee 1984; Faddeev & Shatashvilli 1984.

[3] Bonora & Cotta–Ramusino 1983.

[4] See, for instance, Anderson & Duchamp 1980.

explicit dependence on x^μ, so that

$$\partial_\mu \mathcal{L} = (\delta_a \mathcal{L}) \, \partial_\mu \phi^a.$$

Of course, for the part concerned with variations in spacetime (*i.e.*, in the arguments of the fields), we have usual forms and

$$d^2 \mathcal{L} = \tfrac{1}{2} \left(\partial_\lambda \partial_\mu \mathcal{L} - \partial_\mu \partial_\lambda \mathcal{L} \right) dx^\lambda \wedge dx^\mu = 0. \tag{20.2}$$

20.2.2 *Variations and differentials*

This is precisely one of the results of exterior calculus which we wish to extend to the ϕ-space. This extension is natural for the δ operator:

$$\delta^2 \mathcal{L} = \tfrac{1}{2} \left(\delta_a \delta_b \mathcal{L} - \delta_b \delta_a \mathcal{L} \right) \delta \phi^a \wedge \delta \phi^b = 0. \tag{20.3}$$

Here $\delta\phi^a \wedge \delta\phi^b$ provides the antisymmetrization of the product $\delta\phi^a \delta\phi^b$, just the exterior product of the differentials of the coordinates $\delta\phi^a$ and $\delta\phi^b$. We proceed to strengthen the parallel with usual smooth forms. In order to enforce the boundary-has-no-boundary property for the total variation, we must impose

$$\delta_T^2 = (\delta + d)^2 = \delta d + d\delta = 0. \tag{20.4}$$

But

$$\begin{aligned} \delta_T^2 \mathcal{L} &= (\delta d + d\delta)\mathcal{L} = \delta\phi^a \wedge \delta_a(\partial_\mu \mathcal{L} dx^\mu) + dx^\mu \wedge \partial_\mu(\delta_a \mathcal{L} \delta\phi^a) \\ &= (\delta_a \partial_\mu \mathcal{L} - \partial_\mu \delta_a \mathcal{L}) \, dx^\mu \delta\phi^a \wedge dx^\mu, \end{aligned}$$

so that (20.4) requires

$$\delta_a \partial_\mu \mathcal{L} = \partial_\mu \delta_a \mathcal{L}. \tag{20.5}$$

The anticommutation of δ and d implies the commutation of the respective derivatives. This behaviour is no real novelty, as it appears normally in differential calculus when we separate a manifold into two subspaces.

The total variation δ_T does not commute with spacetime variations, as

$$[\partial_\mu, \delta_T] f = (\partial_\mu \delta x^\lambda) \partial_\lambda f, \tag{20.6}$$

but the purely functional variation δ does.

20.2.3 *The action functional*

We shall from now on consider only purely functional variations. Furthermore, instead of densities as the $\mathcal{L}$ above, we shall consider only objects integrated on spacetime, such as the action functional

$$S[\phi] = \int d^4x\, \mathcal{L}[\phi(x)]. \tag{20.7}$$

For simplicity of language, we shall sometimes interchange the terms "lagrangian" and "action". Differentiating the expression above,

$$\delta S[\phi] = \int d^4x\, \delta\mathcal{L}[\phi(x)] = \int d^4x \left(\delta_a \mathcal{L}[\phi(x)] \right) \delta\phi^a(x). \tag{20.8}$$

20.2.4 *Variational derivative*

We shall suppose that the conditions allowing to identify the strong and the weak derivatives are satisfied (see Chapter 19). Thus, the differential[5] of a functional $F[\phi]$ at a point ϕ of the function space along a direction $\eta(x)$ in that space will be defined by

$$F'_\phi[\eta] = \lim_{\epsilon \to 0} \frac{F[\phi + \epsilon\eta] - F[\phi]}{\epsilon} = \left[\frac{dF[\phi + \epsilon\eta]}{d\epsilon} \right]_{\epsilon=0}. \tag{20.9}$$

It is a linear operator on $\eta(x)$, and $F'[\eta] = 0$ is a linearized version of the equation $F[\phi] = 0$. The integrand in (20.8) is the Fréchet derivative of $\mathcal{L}[\phi]$ along $\eta = \delta\phi/\epsilon$. The presence of the integration, allied to property (20.5), justifies the usual procedures of naïve variational calculus, such as "taking variations (in reality, functional derivatives and not differentials) inside the common derivatives" which, allied to an indiscriminate use of integrations by parts (that is, assuming convenient boundary conditions), lends to it a great simplicity.

20.2.5 *Euler forms*

The vanishing of the expression inside the parentheses in (20.8) gives the field equations, $\delta_a \mathcal{L}[\phi(x)] = 0$. Given a set of field equations $E_a[\phi(x)] = 0$, we shall call its *Euler Form* the expression

$$E[\phi] = \int d^4x\, E_a[\phi(x)]\delta\phi^a(x). \tag{20.10}$$

---

[5] The expression given is actually that of the weak (Gateaux) differential. When the strong derivative exists, so does the weak and both coincide. We suppose it to be the case, and use the most practical expression.

The exterior functional (or variational) differential of such an expression will be defined as

$$\delta E[\phi] = \int d^4x\, \delta E_a[\phi(x)] \wedge \delta\phi^a(x)$$

$$= \tfrac{1}{2} \int d^4x \left(\delta_b E_a[\phi(x)] - \delta_a E_b[\phi(x)] \right) \delta\phi^b(x) \wedge \delta\phi^a(x). \quad (20.11)$$

The differential of (20.8) is immediately found to be zero: $\delta^2 S[\phi] = 0$.

20.2.6   *Higher order forms*

In analogy to the usual 1-forms, 2-forms, etc, of exterior calculus, we shall call 1-Forms, 2-Forms, etc, with capitals, the corresponding functional differentials such as (20.8) and (20.11). A p-Form will be an object like

$$Z[\phi] = \frac{1}{p!} \int d^4x\, Z_{a_1 a_2 \ldots a_p}[\phi(x)]\, \delta\phi^{a_1}(x) \wedge \delta\phi^{a_2}(x) \wedge \ldots \wedge \delta\phi^{a_p}(x), \quad (20.12)$$

the exterior product signs $\wedge$ indicating a total antisymmetrization quite analogous to that of differential calculus.

20.2.7   *Relation to operators*

A thing which is new in Forms is that their components in a natural "coframe" $\{\delta\phi^a\}$ as above may be operators, in reality acting on the first $\delta\phi^a$ at the right. Take, for instance, the Euler-Form for a free scalar field,

$$E[\phi] = \int d^4x \left[\Box_x + m^2 \right] \phi_a(x) \delta\phi^a(x). \quad (20.13)$$

Its differential will be

$$\delta E[\phi] = \int d^4x \{ \delta_{ab} \left[\Box_x + m^2 \right] \} \delta\phi_a(x) \wedge \delta\phi^b(x) = 0, \quad (20.14)$$

because the component $\delta_{ab} \left[\Box_x + m^2 \right]$ is a symmetric operator. The vanishing of the d'Alembertian term may be seen, after integration by parts, as a consequence of

$$\delta\partial_\mu \phi^a(x) \wedge \delta\partial^\mu \phi_a(x) = 0.$$

The use of operatorial components provides an automatic extension to the larger space containing also the field derivatives, avoiding the explicit use of jet bundles of rigorous variational calculus.[6]

---

[6] Anderson & Duchamp 1980.

20.2.8   *Continuum Einstein convention*

The indices $\{a_j\}$ in (20.12) are, of course, summed over, as they are repeated. To simplify notation, we shall from now on extend this Einstein convention to the spacetime variables x^μ and omit the integration sign, as well as the $(p!)$ factor. Its implicit presence should however be kept in mind, as integration by parts will be frequently used. In reality, to make expressions shorter, we shall frequently omit also the arguments. Equation (20.10), for example, will be written simply

$$E[\phi] = E_a \delta\phi^a. \tag{20.15}$$

Finally, we shall borrow freely from the language of differential calculus: a Form W satisfying $\delta W = 0$ will be said to be a *closed* Form, and a Form W which is a variational differential of another, $W = \delta Z$, will be an *exact* Form.

20.3   Existence of a lagrangian

20.3.1   *Inverse problem of variational calculus*

In rough terms, the fundamental problem of variational calculus is to find equations (the field equations) whose solutions lead some functional (the action functional) to attain its extremal values. The inverse problem of variational calculus is concerned with the question of the existence of a lagrangian for a given set of field equations. It can then be put in a simple way in terms of the Euler Form E: is there a 0-Form S, as in (20.7), such that $E = \delta S$? Or, when is E locally an exact Form?

20.3.2   *Helmholtz–Vainberg theorem*

Consider the expression (20.11). The $E_a[\phi(x)]$ are densities just as $\mathcal{L}[\phi(x)]$, and the differentials appearing are Fréchet differentials,

$$\delta E_a = \{\delta_b E_a[\phi]\}\, \delta\phi^b = E_a'[\delta\phi]. \tag{20.16}$$

As said, $E_a'[\eta] = 0$ is a linearized version of the equation $E_a[\phi] = 0$. The Helmholtz–Vainberg necessary and sufficient condition[7] for the existence of a local lagrangian[8] is that, in a ball around ϕ in the functional space,

$$\epsilon^a E_a'[\eta] = \eta^a E_a'[\epsilon] \tag{20.17}$$

[7] Vainberg 1964.
[8] Aldrovandi & Pereira 1986, 1988.

for any two increments η, ϵ. In our notation, with increments η^a along ϕ^a and ϵ^b along ϕ^b, (20.16) tells that this is equivalent to $\delta_b E_a = \delta_a E_b$ or, from (20.11),

$$\delta E = 0. \tag{20.18}$$

20.3.3 Equations with no lagrangian

We shall see in next section a variational analogue of the Poincaré inverse lemma of differential calculus: for a Form to be locally exact, it is necessary and sufficient that it be closed. In this case, $E_a = \delta_a \mathcal{L}$ for some $\mathcal{L}$: the equation does come from a Lagrangian. There are, however, equations of great physical interest which are not related to an action principle in terms of the fundamental physical fields involved.

Navier–Stokes equation

Let us look at the notorious case of the equation

$$\rho \, \partial_t v_i + \rho v_j \partial^j v_i + \partial_i p - \mu \partial^j \partial_j v_i = 0, \tag{20.19}$$

which, together with the incompressibility condition

$$\partial_i v^i = 0 \tag{20.20}$$

describes the behaviour of an incompressible fluid of density ρ and coefficient of viscosity μ. We shall, for simplicity, consider only the stationary case, in which the first term in (20.19) vanishes. The physical fields of interest are the fluid velocity components v^j and its pressure p. We learn from Fluid Mechanics that the pressure is the Lagrange multiplier for the incompressibility condition, so that we can write the Euler Form as

$$E = \left[\rho v_j \partial^j v_i + \partial_i p - \mu \partial^j \partial_j v_i \right] \delta v^i - (\partial_j v^j) \delta p \,, \tag{20.21}$$

with the relative sign conveniently chosen. After putting E under the more convenient form

$$E = \rho \left(v_j \partial^j v_i \right) \delta v^i + \delta \left[\tfrac{1}{2} \mu (\partial_j v_i \partial^j v^i) - p(\partial_j v^j) \right],$$

a direct calculation shows that

$$\delta E = \delta \left[\rho (v_j \partial^j v_i) \delta v^i \right] \neq 0. \tag{20.22}$$

The "offending" non-lagrangian term can be immediately identified as

$$\rho (v_j \partial^j v_i).$$

The power of exterior variational calculus is well illustrated in these few lines, which summarize the large amount of information necessary to arrive at this result.[9]

Korteweg–de Vries equation

This is another example of interest, for which the Euler Form is given by

$$E = (u_t + u u_x + u_{xxx})\, \delta u, \qquad (20.23)$$

the indices indicating derivatives with respect to t and x. That no lagrangian exists can be seen from the simple consideration, for instance, of the first term in δE, given by $\delta u_t \wedge \delta u$, which is nonvanishing and cannot be compensated by any other contribution.

This example, by the way, illustrates an important point: the existence or not of a lagrangian depends on which field is chosen for the role of fundamental physical field. Above, such field was supposed to be u. In terms of u no lagrangian exists. However, a lagrangian does exist in terms of a field variable ϕ if we put $u = \phi_x$, in which case E becomes the closed Form

$$E = (\phi_{tx} + \phi_x \phi_{xx} + \phi_{xxxx})\, \delta\phi. \qquad (20.24)$$

However, when the choice of the fundamental physical field is given by some other reason, as in quantum field theory, it is of no great help that a lagrangian may be found by some smart change of variable.

There is an obvious ambiguity in writing the Euler Form for a set of two or more field equations, as multiplying each equation by some factor leads to an equivalent set. Such a freedom may be used to choose an exact Euler Form and to give the lagrangian a correct sign, for example leading to a positive hamiltonian.

20.4   Building lagrangians

20.4.1   *The homotopy formula*

It has been seen in § 7.2.11 that every differential form is given locally by a very convenient expression in terms of a differential and a transgression, which embodies the Poincaré inverse lemma. We shall here adapt that

---

[9] Finlayson 1972.

expression to Forms. Let us begin by defining the operation T on the p-Form Z. If

$$Z[\phi] = Z_{a_1 a_2 \ldots a_p}[\phi] \, \delta\phi^{a_1} \wedge \delta\phi^{a_2} \wedge \ldots \wedge \delta\phi^{a_p}, \qquad (20.25)$$

then TZ is defined as the $(p-1)$-Form given by

$$TZ[\phi] = \sum_{j=1}^{p} (-)^{j-1} \int_0^1 dt \, t^{p-1} Z_{a_1 a_2 \ldots a_p}[t \, \phi] \, \phi^{a_j} \, \delta\phi^{a_1} \wedge \delta\phi^{a_2} \ldots$$
$$\delta\phi^{a_{j-1}} \wedge \delta\phi^{a_{j+1}} \wedge \ldots \wedge \delta\phi^{a_p}. \qquad (20.26)$$

The fields ϕ^a appearing in the argument of $Z_{a_1 a_2 \ldots a_p}$ are multiplied by the variable t before the integration is performed. As t runs from 0 to 1, the field values are continuously deformed from 0 to ϕ^a. This is a homotopy operation[10] in ϕ-space. A more general homotopy

$$\phi_t = t \, \phi + (1 - t) \, \phi_0$$

with $\phi_0 \neq 0$ can be used, but without real gain of generality.

The important point is that the ϕ-space is supposed to be a starshaped domain around some "zero" field (each point may be linked to zero by straight lines). Spaces of this kind are called "affine" spaces by some authors. Some important field spaces, however, are not affine. For example, the space of metrics used in General Relativity includes no zero, nor does the space of chiral fields with values on a Lie group. For such cases, the use of (20.26) is far from immediate (see Sec. 32.3).

The Poincaré inverse lemma says that, on affine functional spaces, Z can always be written locally as

$$Z[\phi] = \delta(TZ) + T(\delta Z). \qquad (20.27)$$

This result may be obtained from (20.26) by direct verification. A consequence is that a closed Z will be locally exact: $Z = \delta(TZ)$. For a closed Euler Form E, this gives immediately the Vainberg "homotopy formula", which provides the lagrangian as

$$\mathcal{L} = TE. \qquad (20.28)$$

As the operator T is the "transgression operator", this expression is called the "transgression formula". It provides a systematic procedure to find a lagrangian for a given equation, when it exists. Equation (20.27) allows furthermore a systematic identification of those pieces of a given E which are lagrangian-derivable and those which are not. This was done directly

---

[10] Nash & Sen 1983.

in (20.22), but (20.27) may be useful in more complicated cases. No term in (20.23) is lagrangian-derivable, since there it happens that

$$T\delta E = E \tag{20.29}$$

and

$$\delta TE = 0. \tag{20.30}$$

When $\mathcal{L}$ does exist, a trivial rule to obtain it from $E = \delta\mathcal{L} = E_a\delta\phi^a$ comes out when E_a is a polynomial in the fields and/or their derivatives: replace in E each factor $\delta\phi^a$ by ϕ^a and divide each monomial of the resulting polynomial by the respective number of fields (and/or their derivatives).

20.4.2 Examples

Helmholtz–Korteweg lagrangian

If the first term in (20.21) is dropped, the remaining terms would come from

$$\mathcal{L} = \tfrac{1}{2}\,\mu(\partial_j v_i \partial^j v^i) - (p\partial_j v^j). \tag{20.31}$$

As to Eq.(20.24), it comes immediately from the lagrangian

$$\mathcal{L} = \tfrac{1}{2}\,\varphi\varphi_{tx} + \tfrac{1}{3}\,\varphi\varphi_x\varphi_{xx} + \tfrac{1}{2}\,\varphi\varphi_{xxxx}. \tag{20.32}$$

Born–Infeld electrodynamics

A simple example of the use of (20.26) in a non-polynomial theory may be found in the Born–Infeld electrodynamics.[11] With $F_{\mu\nu} = \partial_\mu A_\nu - \partial_\nu A_\mu$ and $F^2 = F_{\mu\nu}F^{\mu\nu}$, its Euler Form is

$$E = \partial^\mu \left(\frac{F_{\mu\nu}}{\sqrt{1 - F^2/(2k)}} \right) \delta A^\nu. \tag{20.33}$$

In this case,

$$TE = A^\nu \partial^\mu \left(\int_0^1 dt \frac{t\,F_{\mu\nu}}{\sqrt{1 - t^2 F^2/(2k)}} \right) \tag{20.34}$$

gives, after an integration and a convenient antisymmetrization,

$$\mathcal{L} = k\left(\sqrt{1 - F^2/(2k)} - 1 \right). \tag{20.35}$$

---

[11] Born & Infeld 1934.

Einstein's equations

It is sometimes possible, by a clever picking-up of terms, to exhibit the Euler Form directly as an exact Form, thereby showing the existence and the explicit form of a lagrangian. Take Einstein's equations for the pure gravitational field. Its Euler Form is

$$E = \sqrt{-g}\left[R_{\mu\nu} - \tfrac{1}{2}g_{\mu\nu}(R+\Lambda)\right]\delta g^{\mu\nu}, \tag{20.36}$$

with Λ the cosmological constant. We can recognize

$$\delta\sqrt{-g} = -\tfrac{1}{2}\sqrt{-g}\,g_{\mu\nu}\delta g^{\mu\nu}$$

in the second term and separate

$$\delta R = \delta g_{\mu\nu}R^{\mu\nu}g_{\mu\nu}\delta R^{\mu\nu}$$

to write

$$E = \delta\big(\sqrt{-g}(R+\Lambda)\big) - \sqrt{-g}\,g_{\mu\nu}\delta R^{\mu\nu}.$$

Of these two terms, the latter is known to be a divergence[12] and the first exhibits the Hilbert–Einstein lagrangian. The factor $\sqrt{-g}$ is to be expected if we recall the implicit integration in (20.36). It plays the role of an integrating factor, as E would be neither invariant nor closed in its absence.

Spinor field

The Euler Form for a 1/2-spin field in interaction with the electromagnetic field is

$$E = \delta\overline{\psi}\left[i\gamma^{\mu}(\partial_{\mu} - ieA_{\mu})\psi - m\psi\right] - \left[i(\partial_{\mu} + ieA_{\mu})\overline{\psi}\gamma^{\mu} + m\overline{\psi}\right]\delta\psi$$
$$+ \left[\partial^{\mu}F_{\mu\nu} + e\overline{\psi}\gamma_{\nu}\psi\right]\delta A^{\nu}.$$

The corresponding lagrangian is

$$\mathcal{L} = \tfrac{1}{2}\left\{i\overline{\psi}\gamma^{\mu}(\partial_{\mu} - ieA_{\mu})\psi - i(\partial_{\mu} + ieA_{\mu})\overline{\psi}\gamma^{\mu}\psi\right\} - m\overline{\psi}\psi - \tfrac{1}{4}F^{\mu\nu}F_{\mu\nu}$$
$$= \tfrac{1}{2}\left\{i\overline{\psi}\gamma^{\mu}\partial_{\mu}\psi - i(\partial_{\mu}\overline{\psi})\gamma^{\mu}\psi\right\} - m\overline{\psi}\psi + eA_{\mu}\overline{\psi}\gamma^{\mu}\psi - \tfrac{1}{4}F^{\mu\nu}F_{\mu\nu}.$$

Complex scalar field

In this case, the Euler Form is given by

$$E = \delta\varphi^{*}\left[\Box\varphi - ie(\partial_{\mu}A^{\mu})\varphi - 2ieA^{\mu}\partial_{\mu}\varphi - e^{2}A^{\mu}A_{\mu}\varphi\right]$$
$$+ \left[\Box\varphi^{*} + ie(\partial_{\mu}A^{\mu})\varphi^{*} + 2ieA^{\mu}\partial_{\mu}\varphi^{*} - e^{2}A^{\mu}A_{\mu}\varphi^{*}\right]\delta\varphi,$$

and the lagrangian is

$$L = -\left[\partial_{\mu} + ieA_{\mu}\right]\varphi^{*}\left[\partial^{\mu} - ieA^{\mu}\right]\varphi + m\varphi\varphi^{*} + \lambda\varphi^{4} - \tfrac{1}{4}F^{\mu\nu}F_{\mu\nu}.$$

---

[12] Landau & Lifshitz 1975.

Second order fermion equation

Applying twice the Dirac operator, we obtain a second order equation for the fermion, which includes a spin-field coupling introduced by Fermi to account for the anomalous magnetic moment of the neutron. In this case,

$$
\begin{aligned}
E = \ & \delta\overline{\psi}\Big[\Box\psi - ie(\partial_\mu A^\mu)\psi - 2ieA^\mu\partial_\mu\psi - e^2 A^\mu A_\mu\psi - \frac{e}{2}\sigma^{\mu\nu}F_{\mu\nu}\psi\Big] \\
& + \Big[\Box\overline{\psi} + ie(\partial_\mu A^\mu)\overline{\psi} + 2ieA^\mu\partial_\mu\overline{\psi} - e^2 A^\mu A_\mu\overline{\psi} - \frac{e}{2}\overline{\sigma^{\mu\nu}F_{\mu\nu}}\Big]\delta\psi \\
& + \Big[\partial^\mu F_{\mu\nu} + e\overline{\psi}\gamma_\nu\psi + e\,\partial^\mu(\overline{\psi}\sigma_{\mu\nu}\psi)\Big]\delta A^\nu .
\end{aligned}
$$

The Fermi term

$$
\begin{aligned}
E_F &= -\frac{e}{2}\,\delta\big(\overline{\psi}\sigma^{\mu\nu}F_{\mu\nu}\psi\big) \\
&= -e\,\delta\big(\overline{\psi}\sigma^{\mu\nu}F_{\mu\nu}\psi\partial_\mu A_\nu\big) \\
&= e\,\delta\big[A_\nu\partial_\mu(\overline{\psi}\sigma^{\mu\nu}F_{\mu\nu}\psi)\big]
\end{aligned}
$$

is an example of non-minimal coupling. The interaction lagrangian is, of course,

$$
\mathcal{L}_F = -\frac{e}{2}\,\overline{\psi}\,\sigma^{\mu\nu}F_{\mu\nu}\,\psi.
$$

The electric current j_ν is given by the *complete* Lagrange derivative

$$
j_\nu = \frac{\delta L_F}{\delta A_\nu},
$$

as usual for the matter currents in gauge theories.

20.4.3 Symmetries of equations

Other notions from differential calculus can be implemented in the calculus of Forms. One such is that of a Lie derivative. Recall that a general field X will be written

$$
X = X^a\delta/\delta\varphi^a .
$$

Suppose that X represents a transformation generator on the φ-space. On Forms, the transformation will be given by the Lie derivative L_X. The Lie derivative L_X, acting on Forms, will have properties analogous to those found in differential calculus. In particular, it commutes with differentials, so that

$$
L_X E = L_X\delta\mathcal{L} = \delta L_X\mathcal{L} . \tag{20.37}
$$

Consequently, a symmetry of the lagrangian ($L_X\mathcal{L} = 0$) is a symmetry of the equation ($L_X E = 0$), but the equation can have symmetries which are not symmetries of the lagrangian. This is a well known fact, but here we find a necessary condition for that:

$$\delta L_X \mathcal{L} = 0.$$

Still other notions of differential calculus translate easily to Forms, keeping quite analogous properties. Such is the case, for example, of the interior product $i_X W$ of a field X by a Form W, which has the usual relation to the Lie derivative,

$$L_X W = i_X(\delta W) + \delta(i_X W).$$

20.4.4 *Final remarks*

We have shown some examples of the power of exterior variational calculus in treating in a very economic way some involved aspects of field theories. All examined cases were "local", valid in some open set of the field space. The last decades of the 20th century have witnessed a growing interest in the global, topological properties of such spaces. Anomalies, BRST symmetry and other peculiarities (see Section 32.3) are now firmly believed to be related to the cohomology of the field functional spaces, this belief coming precisely from results obtained through the use of some special variational differential techniques. Many global properties of finite dimensional manifolds are fairly understood and transparently presented in the language of exterior differential forms. The complete analogy of the infinite dimensional calculus suggests that, besides being of local interest, it is a natural language to examine also global properties of field spaces.

General references

Information of interest on the topics above can be found in Marsden 1974, Olver 1986, and Aldrovandi & Kraenkel 1988 .

Chapter 21

Singular Points

Singularities of any object defined on a manifold (like curves or functions) signal non-trivial aspects of its underlying topology. Such objects can be used to probe properties of the host manifold, which are actually independent of them. We give in this chapter a very sketchy presentation of some of their most remarkable aspects.

21.1   Index of a curve

Given a vector field X on a smooth manifold M, the point $p \in M$ will be a *singular point* of X if $X_p = 0$. Singular points[1] of a vector field give information on the underlying topology. For example, there may be fields without singular points on the euclidean plane, but not on the sphere S^2.

The point p will be a critical point of the function f if it is a singular point of the gradient of f,

$$X = (\partial^i f)\partial_i.$$

As the gradient is actually a 1-form,

$$df = (\partial_i f)dx^i,$$

this supposes a metric to introduce X as the contravariant image of the differential form df. Recall that the presence of singular points on M signals a non-trivial tangent bundle. We shall in what follows (except in Section 21.8 below) suppose that singular points, if existent, are always non-degenerate.

Let us begin with the simplest non-trivial case, which occurs when $M = \mathbb{E}^2$. Take a field X and a singular point $p \in \mathbb{E}^2$. Take another point $q \in \mathbb{E}^2$

---

[1] A very complete treatment of the subject is given in Doubrovine, Novikov & Fomenko 1979, Vol. II, § 13–15.

and suppose it to move around p, describing a closed curve α never touching p. Let us fix a point q_0 on the curve and follow the field X_q as q moves along α. It is X_{q_0} at the start, and is X_{q_0} again when q arrives back at q_0. As it travels along α, X will "turn around itself" a certain number of times, both in the clockwise sense (taken by convention as negative) and in the counterclockwise sense (positive by convention). The algebraic sum of this number of turns is the *index of the curve* with respect to the field X. For instance, the index equals $+1$ in the case pictured in Figure 21.1. If it

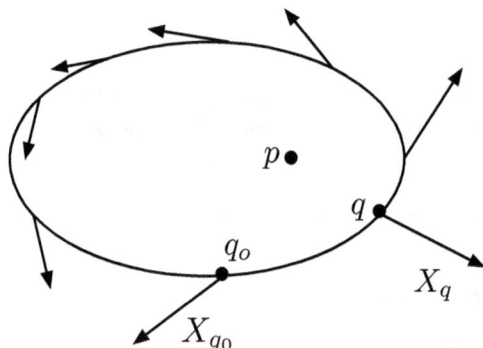

Fig. 21.1 Field X turns once (in the counterclockwise sense) around itself while traveling along the curve: index = +1.

so happens that p is not a singular point of the field X, we can always find, thanks to the continuity of X, a small enough neighbourhood of p inside which all curves have null index. In Figure 21.2 we show an example: in the complex plane version of $\mathbb{E}^2$, the behaviour of the field

$$X(z) = z^2 = (x + iy)^2$$
$$= x^2 - y^2 + i\,2xy$$
$$= (x^2 - y^2)\,\partial_x + 2xy\,\partial_y,$$

when it traverses the circle $|z| = 1$ around its singular point $z = 0$. The index is $+\,2$. It is easy to see that, had we taken a point outside the curve, the index would be zero. In general, the circle will have index $(+n)$ with respect to the field

$$X(z) = z^n$$

and $(-n)$ with relation to the field

$$X(z) = z^{-n}$$

(which illustrates the role of the orientation). A practical way to find the index is to draw the vectors directly at the origin and follow the angle φ it makes with $\mathbb{R}_+$.

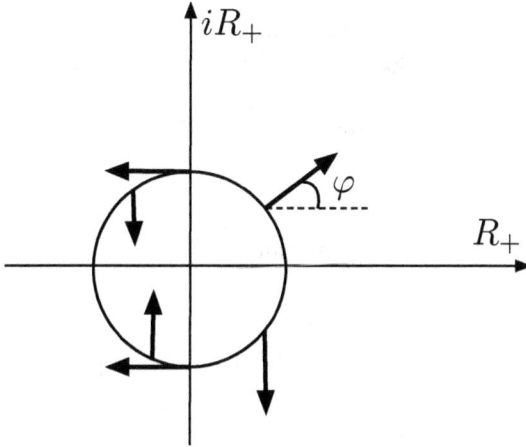

Fig. 21.2   Following the vector directions for $X(z) = z^2$.

Let us formalize what is depicted in the figure: suppose a chart (U, c) and a field

$$X = X^1 e_1 + X^2 e_2.$$

Let $U' = U - \{\text{singular points of } X\}$, and $p \in U'$. Define the mapping $f : U' \to S^1$, from U' into the unit circle given by

$$f(p) = \frac{X_p}{|X_p|}.$$

In the case $X(z) = z^2$, with $p = z$,

$$f(z) = \frac{x^2 - y^2}{x^2 + y^2} + i\,\frac{2xy}{x^2 + y^2} = X^1 + i\,X^2.$$

Then, in a neighbourhood of $f(p) \in S^1$, let us take a local angular coordinate φ, $\varphi \circ f(p)$ (Figure 21.3). Such a coordinate, as said in § 4.2.5, does not cover the whole plane including S^1: its inverse is discontinuous and φ leaves out the axis $\varphi = 0$. Two charts are actually necessary, each one covering the axis left out by the other. In the intersection, a coordinate transformation

$$\varphi = \varphi' + \alpha,$$

with constant α, is defined. As their difference is a constant, the coordinate differentials do coincide outside the origin: $d\varphi = d\varphi'$. Thus,

$$d\varphi = d\arctan\frac{x^2}{x^1} = \frac{x^2 dx^1 - x^1 dx^2}{|x|^2}$$

is a differential form well defined on the whole plane outside the singular points. The index of a closed curve $\gamma : S^1 \to U'$ is then defined as

$$\text{ind } \gamma = \frac{1}{2\pi} \oint d\varphi . \tag{21.1}$$

In the example $X(z) = z^2$,

$$f = \arctan\frac{2xy}{x^2 - y^2} .$$

It is easier to define

$$z = r\exp[i\varphi] \quad \text{and} \quad X(z) = r^2 \exp[2i\varphi],$$

so that

$$f = 2\varphi.$$

While φ goes from 0 to 2π, f goes from 0 to 4π. Consequently,

$$\text{ind } \gamma = \frac{1}{2\pi} \oint d(2\varphi) = 2.$$

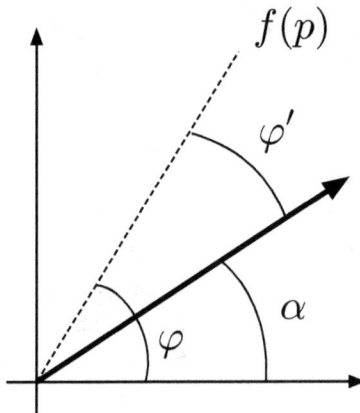

Fig. 21.3   Two charts are necessary for using angular coordinates.

Notice an important thing: the mapping $f(z)$, defined through the field X, takes two points of U' in one same point of S^1. The index is just the number of points from the domain of f taken into one point of its image. This is a general property, which will allow the generalization to higher dimensional manifolds. Let us list some results which can be proven about the indices:

(i) They do not change under continuous deformations of the curve γ, provided γ never touches any singular point.

(ii) They do not change under continuous deformations of the field, provided X never has a singular point on the curve.

(iii) Every disk whose contour has a non-vanishing index related to a field contains some singular point of that field.

(iv) The index of a curve contained in a small enough neighbourhood of a singular point is independent of the curve, that is, it is the same for any curve. The index so obtained is the *index of the singular point*. This index point is chart-independent.

(v) If a curve encloses many singular points, its index is the sum of the indices of each point. A good example is given by the field

$$V(z) = z^2 - (z/2).$$

It has two singular points in the unit disk. The index of $z = 0$ is 2, that of $z = 1/2$ is zero, and the index of the unit curve $|z|^2 = 1$ is 2.

(vi) The index of a singular point is invariant under homeomorphism; this allows the passage from the plane to any bidimensional manifold, as it is a purely local property.

21.2 Index of a singular point

In order to generalize all this to singular points of fields on general differentiable manifolds, let us start by recalling what was said in Section 5.3:

Any differentiable manifold M of dimension m can be imbedded in an euclidean manifold of high enough dimension.

Consider a singular point p of a field X on M. Around it, there will be a neighbourhood U diffeomorphic to $\mathbb{E}^m$,

$$f(U) \approx \mathbb{E}^m.$$

We may transfer X to $\mathbb{E}^m$ through the differential mapping f_* and consider a sphere $S^{(m-1)}$ around $f(p)$ with a radius so small that p is the only singular point inside it. To simplify matters, let us forget the diffeomorphism and write simply "p" for "$f(p)$", "X_p" for "$X_{f(p)}$", etc. With this notation, define then the mapping

$$h : S^{(m-1)} \to S^{(m-1)}$$

with the explicit form

$$h(p) = \frac{f_*[X_p]}{|f_*[X_p]|} =: r \, \frac{X_p}{|X_p|}. \tag{21.2}$$

The *index of the singular point* is the Brouwer degree (§ 6.2.14) of this mapping.

21.3 Relation to topology

On a compact manifold, the number of singular points of a fixed field X is finite. Still another beautiful result:

The sum of all the indices of a chosen vector field on a
compact differentiable manifold M equals the
Euler–Poincaré characteristic of M.

In this way we see that this sum is ultimately independent of the field which has been chosen — it depends only on the underlying topology of the manifold. As a consequence, on a manifold with $\chi \neq 0$, each field *must* have singular points! Information on the topology of a manifold can be obtained by endowing it with a smooth structure and analyzing the behaviour of vector fields. We had seen other, more direct means to detect defects, holes, etc by lassoing or englobing them. The present differential method (which points to *differential topology*) gives another way and can be pictured out by the image of throwing a fluid through the manifold and looking for sinks, whirlpools, sources, etc.

21.4 Basic two-dimensional singularities

In the two-dimensional case, drawing the local integral lines is of great help to get intuition on the corresponding fields. Figure 21.4 shows some simple singular points, with their names and corresponding indices. Notice that

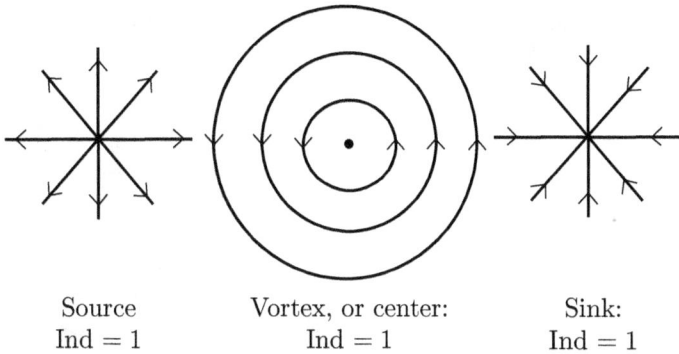

Source — Vortex, or center: — Sink:
Ind = 1 — Ind = 1 — Ind = 1

Fig. 21.4 Basic index = +1 two-dimensional singularities.

the index does not change if we invert the field orientation. Figure 21.5 shows two other singularities of great importance: the saddle point and the dipole. Sources, crosspoints and sinks may be taken as "elementary" singular points.

For 2-dimensional manifolds, we may cast a bridge towards the homological version of the Euler number by taking a triangulation and putting a source at each vertex, a crosspoint replacing each edge, and a sink at the center of each loop.

21.5 Critical points

Notice from the drawings that the index is not enough to characterize completely the kind of singular point. Such a characterization requires further analysis, involving a field "linearization": near the singular point, it is approximated so as to acquire a form $\dot{x} = Ax$, with x a set of coordinates and A a matrix. The eigenvalues of A provide then a complete classification.[2] On n-dimensional metric spaces, as said, connection may be made with the critical points of functions, which are singular points of their gradient fields.[3]

Let us go back to the differentiable function $f : M \to N$. We have defined its rank as the rank of the jacobian matrix of its local expression in

---

[2] Arnold 1973, chap.3.
[3] A huge amount of material, mainly concerned with dynamical systems, is found in Guckenheimer & Holmes 1986.

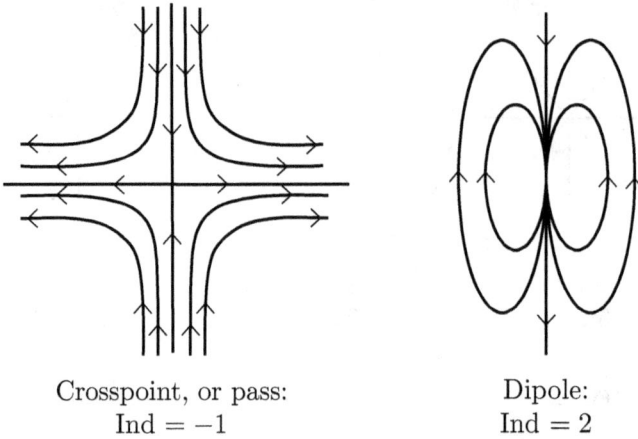

Crosspoint, or pass:	Dipole:
Ind = −1	Ind = 2

Fig. 21.5   Other basic two-dimensional singularities.

coordinates. This is clearly a local concept. The points of M in which this rank is maximal, that is, rank $f = \min(m, n)$ are called *regular points* of f. The points at which rank $f < \min(m, n)$ are *critical points* (or *extrema*) of f. The study of functions in what concerns their extrema is the object of *Morse theory*, of whose fascinating results we shall only say a few words.[4]

21.6   Morse lemma

Consider differentiable real functions $f : N \to \mathbb{E}^1$. Let $p \in N$ be a critical point of f. The critical point p is *non-degenerate* if the hessian matrix of the composition of functions $x^{<-1>}$, f and z,

$$\left[\frac{\partial^2 (z \circ f \circ x^{<-1>})}{\partial x^i \partial x^j} \right]_{x(p)} \tag{21.3}$$

is non-singular for some pair of charts (U, x) and (V, z) on N and $\mathbb{E}^1$ respectively. Differentiability conditions ensure that in this case the non-degeneracy is independent of the choice of the charts.

Here comes a first result, the Morse lemma: in the non-degenerate case, there exists a chart (W, y) around the critical point p such that $y(p) = 0$ and, for $y \in y(W)$, the function $f \circ y^{-1}(y_1, y_2, \ldots, y_n)$ can be written as a

---

[4] Milnor 1973.

quadratic form:

$$f \circ y^{-1}(y_1, y_2, \ldots, y_n) = f(p) - y_1^2 - y_2^2 - \cdots - y_k^2 + y_{k+1}^2 + y_{k+2}^2 + \cdots + y_n^2 \quad (21.4)$$

for some k, with $0 \leq k \leq n$. Notice that $f \circ y^{-1}$ is just the expression of f in local coordinates, as we are using the trivial chart ($\mathbb{E}^1$, identity mapping) for $\mathbb{E}^1$. The integer k, the number of negative signs in the quadratic form, is the *Morse index* of the critical point. When $k = n$, the point is a *maximum* of f, as in the quadratic form all the other points in the neighbourhood give lesser values to f. When $k = 0$, p is a *minimum*. Otherwise, it is a *saddle-point*. The integer k is independent of the choice of coordinates because it is the signature of a quadratic form. The relation with the (singular point) indices is as follows: taking the Morse quadratic form and studying its gradient, we find that minima have index $= +1$, maxima have index $= (-)^n$, and saddle points have alternate signs, ± 1.

The Lemma has a first important consequence: in the neighbourhood W in which the quadratic expression is valid, there are no other critical points. Thus,

each non-degenerate critical point is isolated.

Another important consequence, valid when N is compact and f has only non-degenerate critical points, is that

the number of critical points is finite.

This comes, roughly speaking, from taking a covering of N by including charts as the (W, y) above, one for each critical point, and recalling that any covering has a finite subcovering in a compact space.

Take $N = S^2$, the set of points of $\mathbb{E}^3$ satisfying

$$x^2 + y^2 + z^2 = 1.$$

The projection on the z axis is a real function,

$$z = \pm(1 - x^2 - y^2)^{1/2}.$$

It has a maximum at $z = 1$, around which

$$z \approx 1 - (x^2 + y^2)/2$$

(so, index 2), and a minimum at $z = -1$, around which

$$z \approx -1 + (x^2 + y^2)/2$$

(so, index 0).

21.7   Morse indices and topology

Another enthralling result links the critical points of *any* smooth function
to the topology of the space: if N is compact, and n_k denotes the number
of critical points with index k, then

$$\sum_{i=0}^{n}(-)^k n_k = \chi(N), \tag{21.5}$$

the Euler characteristic of N. This means in particular that the sum is
independent of the function f: every function will lead to the same result.
In order to know the Euler characteristic of a space, it is enough to examine
the critical points of any real function on it. Using the previous example,
one finds immediately

$$\chi(S^2) = 2.$$

The relationship of critical points to topology is still deeper. Again for
N compact, each number n_k of critical points with index k satisfies the
Morse inequality

$$n_k \geq b_k(N), \tag{21.6}$$

with $b_k(N)$ the k-th Betti number of N. All this is only meant to give a flavor of this amazing theory. There are stronger versions of these inequalities,
like the polynomial expression

$$\sum_{k=0}^{n}(n_k - b_k)t^k = (1+t)\sum_{i=0}^{n} q_k t^k$$

with each $n \geq 0$, from which the above expression for $\chi(N)$ comes out when
$t = -1$.

Summing up, all we want here is to call the attention to the strong connection between the topology of a differentiable manifold and the behaviour
of real functions defined on it. For example, take the torus imbedded in $\mathbb{E}^3$,
as in Figure 21.6, and consider the height function, given by the projection
on the z-axis. It has one maximum, one minimum and two saddle points.
Thus,

$$\chi(T^2) = (-)^2 \times 1 + (-)^1 \times 2 + (-)^0 \times 1 = 0.$$

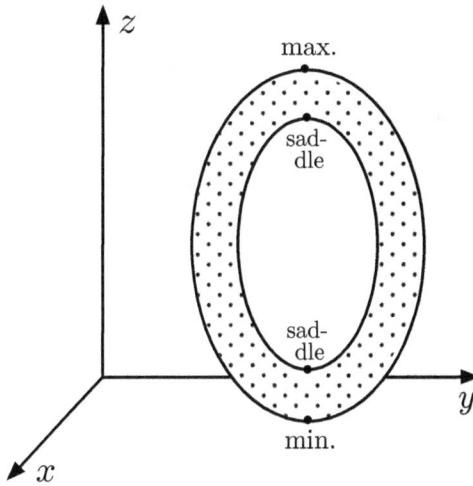

Fig. 21.6   Torus: critical points of the height function.

21.8   Catastrophes

Morse theory shows a kind of stability concerning isolated singularities. The index of each critical point is fixed, and immediately exhibited by Eq. (21.4) once the good system of coordinates is found. It was a fantastic discovery that much of this stability keeps holding when the non-degeneracy condition is waived. In this case the critical points are no more isolated — they constitute lines of singularity. After what we have said, we could expect lines of minima, or maxima, or saddle points. Any function is approximated by the unique Morse quadratic form around each critical point. It was found by R. Thom[5] that functions with lines of singularities ("catastrophes") are also described by elementary expressions, few-variables polynomials, around the singularities. The general expression is not unique, but it is always one of a few basic forms, dependent on the dimensions involved. These "elementary catastrophes" have been completely classified for dimensions ≤ 5.

The main stimulus for Thom came from optics and the idea of form evolution in biology. Zeeman has applied the theory to biology, medicine, social sciences, etc.[6] In Optics they appear as caustics, and their existence,

---

[5]   Thom 1972. A more readable text is Thom 1974.
[6]   His works are collected in Zeeman 1977.

limited number and standard forms have been beautifully confirmed.[7] We might expect their avatars in many other fields, as bifurcations in non-linear systems and in phase transitions related to the vacuum (minimum of some potential) degeneracy.

It should be noticed, however, that the theory is purely qualitative. It allows one say that, in a given physical system, the singularities must be there, and have such or such form, but it does not tell at which scale nor when they will show up. Though some initial successes have been achieved, the feeling remains that there is much as yet to be done if we want to make of it a practical tool for physical applications. A general appraisal of the theory, as well as of the controversy its applications have raised, is found in Woodcock & Davis 1980.

General references

A commendable reference about singular points is Doubrovine, Novikov & Fomenko 1979. Other general refrences covering the subjects of this chapter are Arnold 1973, Kobayashi & Nomizu l963, and Milnor 1973.

---

[7] Berry 1976.

Chapter 22

Euclidean Spaces and Subspaces

Euclidean spaces have been the cornerstones of all geometry — and their historical name does justice to that. They have kept a role of standards for all kinds of comparison. Other spaces are related to them by mappings and their properties are always somehow referred to their properties, at least locally. They surely deserve a particular presentation, albeit minimal like that given below.

22.1 Introductory remarks

Euclidean spaces are, as we have seen, the fundamental spaces to which manifolds are locally homeomorphic. In addition, differentiable manifolds can always be imbedded in some euclidean space of high enough dimension (Section 5.3). For a n-dimensional manifold N, the Whitney theorem only guarantees that this is always possible in a $(2n+1)$-dimensional space, but sometimes a smaller dimension is enough: for instance, S^2 may be imbedded in $\mathbb{E}^3$. On the other hand, one more dimension is not always sufficient: the 4-dimensional Schwarzschild space, solution of Einstein's equations, cannot be imbedded in $\mathbb{E}^5$. No general rule giving the minimal dimension for euclidean imbedding is known. We shall here consider, for simplicity, the manifold N immersed in some $\mathbb{E}^{n+d}$, with d large enough. This will allow us to touch on some results of what is nowadays called "classical" differential geometry.

22.2   Structure equations

22.2.1   *Moving frames*

We have seen in § 7.3.12 that, given a moving frame $\{e_i\}$ on $\mathbf{E}^{n+d}$ with dual basis $\{\omega^j\}$, its structure equations are

$$d\omega^j = \omega^j{}_i \wedge \omega^i \tag{22.1}$$

and

$$d\omega^j{}_i = \omega^k{}_i \wedge \omega^j{}_k . \tag{22.2}$$

22.2.2   *The Cartan lemma*

An important general result is the Cartan lemma: let

$$\{\alpha^i, \quad \text{with } i = 1, 2, \ldots, r \le m\}$$

be a set of r linearly independent 1-forms on $\mathbb{E}^m$. If another set $\{\theta^i\}$ of r 1-forms is such that

$$\sum_{i=1}^{r} \alpha^i \wedge \theta^i = 0,$$

then the θ^i are linearly dependent on them:

$$\theta^i = \sum_{k=1}^{r} a^i{}_k \alpha^k,$$

with $a_{ij} = \delta_{ik}\, a^k{}_j = a_{ji}$. Using this lemma, it is possible to show that the set of forms $\omega^k{}_i$ satisfying both $\omega_{ij} = -\,\omega_{ji}$ and the above structure equations is unique.

22.2.3   *Adapted frames*

An imbedding

$$i : N \to \mathbb{E}^{n+d}$$

is a differentiable mapping whose differential

$$di_p : T_pN \to \mathbb{E}^{n+d}$$

is injective around any $p \in N$. The inverse function theorem says then that a neighbourhood U of p exists such that $i|_U$ (i restricted to U) is also injective. It is possible to show that there exists a neighbourhood

$$V \subset i(U) \subset \mathbb{E}^{n+d}$$

small enough for a basis $\{e_1, e_2, \ldots, e_n, e_{n+1}, \ldots, e_{n+d}\}$ to exist with the following property: the first n base members $(e_1, e_2, \ldots, e_n)$ are tangent to $i(U)$, and the remaining fields are normal to $i(U)$. Such a frame always exists and is called a *frame adapted to the imbedding*. Notice that this implies in particular that the commutators of the first n fields are written exclusively in terms of themselves.

Commentary 22.1   We are used, in the simple cases we commonly meet in current Physics, to take a space (preferably euclidean), there fix a frame once and for all, and refer everything to it. We even lose the notion that some frame is always involved. This simple-minded procedure fails, of course, whenever the space is somehow non-euclidean. The amazingly simple idea of Cartan was to consider instead a bunch of "moving" frames, valid actually in a neighbourhood of each point, and whose set will finally constitute the bundle of linear frames (Section 9.3). ◄

The imbedding i induces forms $i^*(\omega^j)$ and $i^*(\omega^j{}_i)$ on U. The pull-back i^* commutes with the operations of exterior product and exterior differentiation. The basic point of the method of moving frames comes thereof: the structure equations valid on the above mentioned "small enough" neighbourhood V will hold also on some open of N. It will be better to use the indices $i, j, k, \ldots$ from 1 to $n+d$, as above; say, let us take indices $\mu, \nu, \lambda, \ldots$ in the range 1 to n; and indices $a, b, c, \ldots$ from $n+1$ to $n+d$. Separating the structure equations in an obvious way, they become

$$d\omega^\mu = \omega^\mu{}_\nu \wedge \omega^\nu + \omega^\mu{}_a \wedge \omega^a, \tag{22.3}$$

$$d\omega^a = \omega^a{}_\nu \wedge \omega^\nu + \omega^a{}_b \wedge \omega^b, \tag{22.4}$$

$$d\omega^\nu{}_\mu = \omega^\lambda{}_\mu \wedge \omega^\nu{}_\lambda + \omega^c{}_\mu \wedge \omega^\nu{}_c, \tag{22.5}$$

$$d\omega^a{}_\mu = \omega^\lambda{}_\mu \wedge \omega^a{}_\lambda + \omega^c{}_\mu \wedge \omega^a{}_c, \tag{22.6}$$

$$d\omega^\nu{}_a = \omega^\lambda{}_a \wedge \omega^\nu{}_\lambda + \omega^c{}_a \wedge \omega^\nu{}_c, \tag{22.7}$$

$$d\omega^b{}_a = \omega^\lambda{}_a \wedge \omega^b{}_\lambda + \omega^c{}_a \wedge \omega^b{}_c. \tag{22.8}$$

These equations hold on $U \subset N$ but, if applied only on fields $u = u^\mu e_\mu$ on U, they lose some terms because $\omega^a(u) = 0$ for every a. Equation (22.4) reduces to $\omega^a{}_\nu \wedge \omega^\nu = 0$ which, by the Cartan lemma, means that

$$\omega^a{}_\nu = h^a{}_{\nu\lambda} \omega^\lambda \tag{22.9}$$

with $h^a{}_{\nu\lambda} = h^a{}_{\lambda\nu}$.

22.2.4 Second quadratic form

The second order symmetric form with the coefficients $h^a{}_{\mu\lambda}$ as components,

$$\Pi^a = h^a{}_{\mu\lambda}\omega^\mu\omega^\lambda, \tag{22.10}$$

is the second quadratic form of the imbedding along the direction "a".

22.2.5 First quadratic form

The first quadratic form is the metric on U induced by the imbedding: given u and $v \in T_pU$, this metric is defined by

$$< u, v >_p := < i_{p*}(u), i_{p*}(v) > . \tag{22.11}$$

The metric and the fields $\{e_\lambda\}$ determine the ω^μ and the $\omega^\nu{}_\lambda$. We say then that all these objects belong to the *intrinsic* geometry of U.

22.3 Riemannian structure

22.3.1 Curvature

Let us compare (22.5) with (22.2). The latter is valid for the euclidean space. The 2-forms

$$\Omega^\nu{}_\mu = \omega^c{}_\mu \wedge \omega^\nu{}_c$$

measure how much N departs from an euclidean space, they characterize its *curvature*. In terms of intrinsic animals, that is, in terms of objects on N itself, they are, from (22.5),

$$\Omega^\nu{}_\mu = d\omega^\nu{}_\mu - \omega^\lambda{}_\mu \wedge \omega^\nu{}_\lambda. \tag{22.12}$$

They are the curvature forms on N. With the forms acting on the space tangent to N, the structure equation (22.3) reduces to

$$d\omega^\mu = \omega^\mu{}_\nu \wedge \omega^\nu . \tag{22.13}$$

22.3.2 Connection

It is convenient to include the forms $\Omega^\nu{}_\mu$ in a matrix R, the $\omega^\mu{}_\nu$ in a matrix Γ and the ω^ν in a column ω. The equations above become

$$R = d\Gamma - \Gamma \wedge \Gamma, \tag{22.14}$$

and

$$d\omega = \Gamma \wedge \omega. \tag{22.15}$$

Consider an orthonormal basis transformation given by

$$e'_\mu = A^\nu{}_\mu e_\nu$$

and

$$\omega'^\nu = A_\mu{}^\nu \omega^\mu,$$

with $A_\mu{}^\nu$ the inverse of $A^\nu{}_\mu$, that is, $A_\mu{}^\nu = (A^{-1})^\nu{}_\mu$. In matrix language,

$$e' = A\,e \quad \text{and} \quad \omega = A\omega'.$$

Taking differentials in the last expression,

$$\begin{aligned} d\omega &= dA \wedge \omega' + A d\omega' \\ &= dA \wedge A^{-1}\omega + A\Gamma' \wedge \omega' \\ &= (dAA^{-1} + A\Gamma'A^{-1}) \wedge \omega. \end{aligned}$$

From the uniqueness of forms satisfying (22.15),

$$\Gamma = dAA^{-1} + A\Gamma'A^{-1}$$

or

$$\Gamma' = A^{-1}dA + A^{-1}\Gamma A = A^{-1}(d + \Gamma)A. \tag{22.16}$$

This is the very peculiar transformation behaviour of the connection form Γ. In the same way we find that the curvature form changes according to

$$\Omega' = A^{-1}\Omega A. \tag{22.17}$$

Matrix Ω behaves, under base changes, in the usual way matrices do under linear transformations.

22.3.3 *Gauss, Ricci and Codazzi equations*

Back to the structure equations, we notice the forms

$$\Omega^b{}_a = \omega^\lambda{}_a \wedge \omega^b{}_\lambda = d\omega^b{}_a - \omega^c{}_a \wedge \omega^b{}_c. \tag{22.18}$$

They are the normal curvature forms. This expression may be combined with (22.9) and (22.13) to give the Gauss equation

$$\Omega^\nu{}_\mu = \tfrac{1}{2}\sum_a (h^a{}_{\mu\lambda}h^{a\nu}{}_\rho - h^a{}_{\mu\rho}h^{a\nu}{}_\lambda)\,\omega^\rho \wedge \omega^\lambda \tag{22.19}$$

and the Ricci equation

$$\Omega^b{}_a = \tfrac{1}{2}(h_{a\mu\rho}h^{b\rho}{}_\nu - h^b{}_{\mu\rho}h_{a\nu}{}^\rho)\,\omega^\mu \wedge \omega^\nu. \tag{22.20}$$

The imbedding divides the geometry into two parts, which are related by the second quadratic forms and by Eq. (22.6), which is called the Codazzi equation:

$$d\omega^a{}_\mu = \omega^\lambda{}_\mu \wedge \omega^a{}_\lambda + \omega^c{}_\mu \wedge \omega^a{}_c. \qquad (22.21)$$

The above equations constitute the basis of classical differential geometry. The important point is that the geometrical objects on N (fields, forms, tensors, etc) may be given a treatment independent of the "exterior" objects. All the relations involving the indices μ, ν, ρ, etc may be written without making appeal to objects with indices a, b, c, etc. This fact, pointed out by Gauss, means that the manifold N has its own geometry, its intrinsic geometry, independently of the particular imbedding. This may be a matter of course from the point of view we have been following, but was far from evident in the middle of the nineteenth century, when every manifold was considered as a submanifold of an euclidean space. The modern approach has grown exactly from the discovery of such intrinsic character: the properties of a manifold ought to be described independently of references from without.

22.3.4   *Riemann tensor*

In components, the curvature 2-forms will be written

$$\Omega^\nu{}_\mu = \tfrac{1}{2} R^\nu{}_{\mu\rho\sigma} \omega^\rho \wedge \omega^\sigma. \qquad (22.22)$$

If the connection forms are written in some natural basis as

$$\omega^\nu{}_\mu = \Gamma^\nu{}_{\mu\rho} \, dx^\rho, \qquad (22.23)$$

the components in (22.22) are obtained from (22.12):

$$R^\nu{}_{\mu\rho\sigma} = \partial_\rho \Gamma^\nu{}_{\mu\sigma} - \partial_\sigma \Gamma^\nu{}_{\mu\rho} + \Gamma^\nu{}_{\lambda\rho}\Gamma^\lambda{}_{\mu\sigma} - \Gamma^\nu{}_{\lambda\sigma}\Gamma^\lambda{}_{\mu\rho}. \qquad (22.24)$$

These components constitute the Riemann curvature tensor. The components of the connection form (22.23) are the *Christoffel symbols*, which may be written in terms of derivatives of the components of the metric tensor. These metric components are, when restricted to the intrinsic sector,

$$g_{\mu\nu} = e_\mu \cdot e_\nu.$$

Of course, they are now point-dependent since the adapted basis vectors change from point to point.

The Ricci tensor $R_{\mu\nu} = R^\alpha{}_{\mu\alpha\nu}$ is symmetric on a riemannian manifold. A manifold whose Ricci tensor satisfies $R_{\mu\nu} = \lambda\, g_{\mu\nu}$, with λ a constant, is called an Einstein space. There are very interesting theorems concerning the immersion of Einstein spaces. One of them is the following:

> *If an Einstein space as above has dimension m and is immersed in $\mathbb{E}^{m+1}$, then necessarily $\lambda \geq 0$.*

Another curious result is the following:

> *Suppose that, on a connected manifold of dimension m, $R_{\mu\nu} = f\, g_{\mu\nu}$, with f a function; then, if $m \geq 3$, f is necessarily a constant.*

To get the Christoffel symbols, we start by differentiating the function $g_{\mu\nu}$,

$$
\begin{aligned}
dg_{\mu\nu} &= dx^\sigma \partial_\sigma g_{\mu\nu} = de_\mu e_\nu + e_\mu \cdot de_\nu \\
&= \omega^\lambda{}_\mu e_\lambda \cdot e_\nu + e_\mu \cdot \omega^\lambda{}_\nu e_\lambda \\
&= \omega^\lambda{}_\mu g_{\lambda\nu} + \omega^\lambda{}_\nu g_{\lambda\mu} \\
&= \left(g_{\lambda\nu} \Gamma^\lambda{}_{\mu\sigma} + g_{\mu\lambda} \Gamma^\lambda{}_{\nu\sigma} \right) dx^\sigma .
\end{aligned}
\tag{22.25}
$$

Defining $\Gamma_{\mu\nu\sigma} := g_{\mu\lambda} \Gamma^\lambda{}_{\nu\sigma}$, we see that

$$
\partial_\sigma g_{\mu\nu} = \Gamma_{\mu\nu\sigma} + \Gamma_{\nu\mu\sigma}.
$$

Calculating

$$
\partial_\mu g_{\nu\sigma} + \partial_\nu g_{\sigma\mu} - \partial_\sigma g_{\mu\nu},
$$

we arrive at

$$
\Gamma^\lambda{}_{\mu\nu} = g^{\lambda\sigma} \Gamma_{\sigma\mu\nu} = \tfrac{1}{2} g^{\lambda\sigma} \left(\partial_\mu g_{\sigma\nu} + \partial_\nu g_{\sigma\mu} - \partial_\sigma g_{\mu\nu} \right).
\tag{22.26}
$$

22.4 Geometry of surfaces

22.4.1 *Gauss theorem*

To get some more insight, as well as to make contact with the kernel of classical geometrical lore, let us examine surfaces imbedded in $\mathbb{E}^3$. Consider then some surface S, $\dim S = 2$, and an imbedding

$$
i : S \to \mathbb{E}^3.
$$

Two vectors $u, v \in T_p S$ will have an internal product given by (22.11), the metric on S induced through i by the euclidean metric of $\mathbb{E}^3$. To examine the local geometry around a point $p \in S$, take an open $U, S \supset U \ni p$, and an open V such that $\mathbb{E}^3 \supset V \supset i(U)$. Choose on V a moving frame (e_1, e_2, e_3) adapted to i in such a way that e_1 and e_2 are tangent to $i(U)$, and e_3 is normal. The orientation may be such that (e_1, e_2, e_3) is positive in $\mathbb{E}^3$. Here, to simplify the notation, we take the imbedding as a simple inclusion, all the geometrical objects on S being considered as restrictions to S of objects on $\mathbb{E}^3$. Given any vector

$$v = v^1 e_1 + v^2 e_2$$

on S, it follows that $\omega^3(v) = 0$. Equation (22.4) becomes

$$0 = d\omega^3 = \omega^3{}_1 \wedge \omega^1 + \omega^3{}_2 \wedge \omega^2.$$

It follows from the Cartan lemma that

$$\omega^3{}_1 = h_{11}\omega^1 + h_{12}\omega^2,$$
$$\omega^3{}_2 = h_{21}\omega^1 + h_{22}\omega^2, \qquad (22.27)$$

with

$$h_{12} = h_{21}, \qquad (22.28)$$

where we have used the simplified notation $h^3{}_{ij} = h_{ij}$. Notice that

$$h_{11} = \omega^3{}_1(e_1), \qquad h_{22} = \omega^3{}_2(e_2), \qquad h_{12} = \omega^3{}_1(e_2) = h_{21} = \omega^3{}_2(e_1).$$

As

$$\omega^j{}_i = -\omega_i{}^j \quad \text{and} \quad de_i = \omega^j{}_i e_j,$$

then

$$de_3(v) = -\omega^3{}_1(v)\, e_1 - \omega^3{}_2(v)\, e_2.$$

This may be put into the matrix form

$$de_3 \begin{pmatrix} v^1 \\ v^2 \end{pmatrix} = - \begin{pmatrix} h_{11} & h_{12} \\ h_{21} & h_{22} \end{pmatrix} \begin{pmatrix} v^1 \\ v^2 \end{pmatrix}. \qquad (22.29)$$

Thus, the matrix $(-h_{\mu\nu})$ represents on basis (e_1, e_2) the differential of the mapping

$$e_3 : U \to \mathbb{E}^3, \quad p \to e_{(p)3}.$$

As $|e_3| = 1$, this mapping takes values on a unit sphere of $\mathbb{E}^3$. It is called the *Gauss normal mapping*. The matrix $-(h_{\mu\nu})$ may be diagonalized with two real eigenvalues ρ_1 and ρ_2. These eigenvalues are the *principal curvature radii* of S at p. Its determinant is the *total curvature*, or *Gaussian curvature* of S at the point p:

$$K := \det(de_3) = \rho_1 \rho_2 = h_{11}h_{22} - (h_{12})^2. \qquad (22.30)$$

A quick calculation using (22.19) and (22.30) shows that

$$\Omega^2{}_1 = d\omega^2{}_1 = -K\,\omega^1 \wedge \omega^2. \qquad (22.31)$$

The form $\omega^1 \wedge \omega^2$ has a special meaning: applied to two vectors u and v, it gives the area of the parallelogram they define:

$$\omega^1 \wedge \omega^2(u, v) = u^1 v^2 - u^2 v^1.$$

It is the area element, which is in reality independent of the adapted frame and defined on the whole S. It will be denoted

$$\sigma = \omega^1 \wedge \omega^2 .$$

It corresponds, of course, to the volume form on S.

Unlike σ, the connection form $\omega^2{}_1$ depends on the adapted frame. Let us proceed to a change from the frame (e_1, e_2, e_3) to another frame (e_1', e_2', e_3'), related to it by

$$e_1' = \cos\theta\, e_1 + \sin\theta\, e_2, \tag{22.32}$$

$$e_2' = -\sin\theta\, e_1 + \cos\theta\, e_2. \tag{22.33}$$

The dual basis will change accordingly,

$$\omega'^1 = \cos\theta\, \omega^1 + \sin\theta\, \omega^2, \tag{22.34}$$

$$\omega'^2 = -\sin\theta\, \omega^1 + \cos\theta\, \omega^2. \tag{22.35}$$

Taking the differentials and using the structure equations, we get

$$d\omega'^1 = \omega'^2 \wedge (\omega^1{}_2 + d\theta), \tag{22.36}$$

$$d\omega'^2 = \omega'^1 \wedge (\omega^2{}_1 + d\theta). \tag{22.37}$$

As the forms satisfying such equations are unique, it follows that the connection form of the new basis is

$$\omega'^2{}_1 = \omega^2{}_1 + d\theta. \tag{22.38}$$

It follows that

$$d\omega'^2{}_1 = d\omega^2{}_1$$

and the curvature (22.31) is frame-independent. It depends only on the induced metric. This is the celebrated Gauss theorem of surface theory, which has lead its discoverer (we repeat ourselves) to the idea that the geometry of a space should be entirely described in terms of its own characteristics. This was shown to be possible in large generality and, although imbeddings were very helpful in finding fundamental properties and making them more easily understood, all of them can be arrived at in an intrinsic way (the "intrinsic geometry"), the only difficulty being the necessity of a more involved formalism. In higher dimensional spaces, the curvature is characterized by all the components $R^\nu{}_{\mu\rho\sigma}$ of the Riemann tensor which, for a n-dimensional space, amount to

$$n^2(n^2 - 1)/12$$

components. When $n = 2$, only one component is enough to characterize the curvature, as above. In this case, using equations (22.19), (22.21) and (22.30), we find that

$$K = \tfrac{1}{2}\, R_{1212}.$$

22.5   Relation to topology

22.5.1   *The Gauss–Bonnet theorem*

Suppose now that S is a compact surface. A field X on S will have a finite number of singular points p_i, given by $X_{p_i} = 0$. Consider around each singular point p_i an open set U_i small enough for p_i to be the only singular point inside it. To calculate the index at p_i, we should integrate the turning angle (Section 21.1) of X along the border ∂U_i. For that, we need to establish a starting direction, which we take to be e_1. A useful trick is the following: introduce on the complement $S' = S - \cup_i U_i$ another adapted frame $\{e_1', e_2', e_3'\}$, with

$$e_1' = \frac{X}{|X|}$$

and e_2', e_3' chosen so as to make the frame positively oriented. The angle θ to be integrated is then just that of equations (22.32), (22.34) and (22.38). The index at p_i will be

$$I_i = \frac{1}{2\pi} \oint_{\partial U_i} d\theta$$

or, by using (22.38),

$$2\pi I_i = \int_{\partial U_i} \omega'^2{}_1 - \int_{\partial U_i} \omega^2{}_1 = \int_{\partial U_i} \omega'^2{}_1 - \int_{U_i} d\omega^2{}_1 .$$

Keep in mind that $\omega^2{}_1$ is defined on the whole S, while $\omega'^2{}_1$ is only defined on S'. The integral $\int_{U_i} d\omega^2{}_1$ can be made to vanish by taking U_i smaller and smaller, so that $U_i \to \{p_i\}$. The form $\omega'^2{}_1$ keeps itself out from the p_i's. Recalling that the sum of all the indices is the Euler characteristic, we arrive, in the limit, at

$$2\pi \chi = \sum_i \int_{\partial U_i} \omega'^2{}_1 = - \sum_i \int_{S-U_i} d\omega'^2{}_1$$

because, in a compact surface, the union $\cup \partial U_i$ is also the boundary of S' $= S - \cup U_i$ with reversed orientation. From (22.31),

$$2\pi \chi = \int_{S-U_i} K\, \omega'^1 \wedge \omega'^2 .$$

However, the form $\omega'^1 \wedge \omega'^2$ is frame independent. In the limit $U_i \to \{p_i\}$, the integral leaves out only a set of zero measure — it is identical to the integral on the whole S. Consequently,

$$\chi = \frac{1}{2\pi} \int_S K\, \omega^1 \wedge \omega^2. \qquad (22.39)$$

As χ is independent of the induced metric and of the chosen field, this relation depends only on S. It holds clearly also for any manifold diffeomorphic to S. It is a relation between a differentiable characteristic of the manifold, the curvature, and the topological substratum. It is a famous result, the Gauss–Bonnet theorem. For a sphere S^2 of radius r, the Gaussian curvature is

$$K = 1/r^2, \quad \sigma = r^2 \sin\theta d\theta d\varphi$$

and we obtain $\chi = 2$ again. Notice that $r^2 \sin\theta = \sqrt{g}$, as on S^2 the metric is such that

$$g_{\mu\nu}dx^\mu dx^\nu = dl^2 = r^2(\sin^2\theta d\varphi^2 + d\theta^2).$$

In general, in a coordinate basis, (22.39) is written

$$\chi = \frac{1}{2\pi} \int_S K\sqrt{g}\,dx^1 dx^2. \tag{22.40}$$

22.5.2 The Chern theorem

The theorem above has been generalized to manifolds of dimension $2n$. First, imbeddings were used. Allendoerfer and Weil found its first intrinsic proof in 1943. Two years later, a simpler (for mathematicians) proof, using fiber bundles, was given by Chern. For these $2n$-manifolds, the Euler characteristic is

$$\chi = \frac{2}{a_{2n}} \int_S K_T\sqrt{g}\,dx^1 dx^2 \ldots dx^{2n}, \tag{22.41}$$

where a_{2n} is the area of the $2n$-dimensional unit sphere,

$$a_{2n} = \frac{\pi^n\, 2^{2n+1}n!}{(2n)!}, \tag{22.42}$$

and K_T is a generalization of the total curvature, given by

$$K_T = \left[(2n)!2^n g\right]^{-1} \varepsilon^{\mu_1\mu_2\ldots\mu_{2n}} \varepsilon^{\nu_1\nu_2\ldots\nu_{2n}}$$
$$\times R_{\mu_1\mu_2\nu_1\nu_2} R_{\mu_3\mu_4\nu_3\nu_4} \ldots R_{\mu_{2n-1}\mu_{2n}\nu_{2n-1}\nu_{2n}}. \tag{22.43}$$

General references

Additional references for this chapter are Lichnerowicz 1955, Kobayashi & Nomizu 1963, Spivak 1970, and Doubrovine, Novikov & Fomenko 1979.

Chapter 23

Non-Euclidean Geometries

After all we have said about euclidean spaces, we follow the historical trail and progressively introduce other spaces, deviating from them little-by-little. Even for small departures, the riches of new aspects turning up is overwhelming.

23.1 The age-old controversy

Euclid's postulate of the parallels (his "5-th postulate") stated that, given a straight line L and a point P not belonging to it, there was only one straight line going through P that never met L. The eon-long debate on this postulate was concerned with the question of its independence: is it an independent postulate, or can it be deduced from the other postulates? Euclid himself was aware of the problem and presented separately the propositions coming exclusively from the first four postulates. These propositions came to constitute "absolute geometry". Those dependent on the 5-th postulate, as for example the statement that the sum of the internal angles of a triangle is equal to π, he set apart. The debate was given a happy ending around the middle of the 19th century through the construction of spaces which kept in validity the first four postulates, but violated precisely the 5-th. On one side, the independence was thereby proved. On the other, the very way by which the solution had been found pointed to the existence of new, hitherto unsuspected kinds of space. Such new "non-euclidean" spaces are at present time called "riemannian spaces" and their character is deeply rooted in their metric properties. Though the word "space" has since then acquired a much more general, metric-independent meaning, we shall in this chapter follow the widespread usage of using it to denote a *metric* space. The main fact about non-euclidean spaces is spelled out by saying that "they are curved". On such spaces, the role of straight lines is played by geodesics.

23.2 The curvature of a metric space

As we have irritatingly repeated, curvature is not necessarily related to a metric. It is actually a property of a connection. It so happens that a metric does determine a very special connection, the Levi–Civita connection (represented by the Christoffel symbols in a convenient covector basis). It is that unique connection which has simultaneously two important properties: it parallel-transports the metric and it has vanishing torsion. By "curvature of a space" we understand the curvature of the Levi–Civita connection of its metric, and the space is said to be "curved" when the corresponding Riemann tensor is non-vanishing. A space is "flat" when $R^{\alpha\beta}{}_{\mu\nu} = 0$, from which it follows that also its scalar curvature $R = R^{\mu\nu}{}_{\mu\nu}$ is zero. Euclidean spaces are flat because the Levi–Civita connection of an euclidean metric has vanishing Riemann tensor.

A riemannian space is said to be of constant curvature when its scalar curvature R is a constant, and the simplest departures from the euclidean case are those spaces for which R is still constant but non-vanishing. It was only natural that the first "curved" spaces found were of this kind. When the constant $R > 0$, the manifold is said to have positive curvature, and when the constant $R < 0$ it is said to have negative curvature. A sphere S^2 provides an example of positive curvature, a sheet of a hyperboloid an example of negative curvature. We shall in what follows briefly sketch these 2-dimensional cases, though emphasizing the fact that the corresponding metrics can be attributed to the plane $\mathbb{R}^2$. In this way it becomes clear that distinct metrics can be defined on the point set $\mathbb{R}^2$, each one leading of course to different measures of distance. We shall privilege cartesian coordinates and also make some concessions to current language in this so much discussed subject.

The euclidean space $\mathbb{E}^3$, we recall, is the set $\mathbb{R}^3$, whose points are the ordered real triples like $p = (p_1, p_2, p_3)$ and $q = (q_1, q_2, q_3)$, with the metric topology given by the distance function

$$d(p,q) = +\sqrt{(p_1 - q_1)^2 + (p_2 - q_2)^2 + (p_3 - q_3)^2}\,. \qquad (23.1)$$

In cartesian coordinates (X, Y, Z), a sphere of radius L centered at the origin will be the set of points satisfying

$$X^2 + Y^2 + Z^2 = L^2.$$

It will have positive gaussian curvature. Hyperboloids, or "pseudospheres", will have negative gaussian curvature and are of two types. A single-sheeted hyperboloid will have its points specified by

$$X^2 + Y^2 - Z^2 = L^2.$$

A two-sheeted hyperboloid will be given by the equation

$$X^2 + Y^2 - Z^2 = -L^2.$$

The cone stands in between as a very special case of both, given by

$$X^2 + Y^2 - Z^2 = 0,$$

and being asymptotically tangent to them. These surfaces can be imbedded in $\mathbb{E}^3$ as differentiable manifolds and are illustrated in Figure 23.1.

Fig. 23.1   Two kinds of hyperboloids, with the cone in between.

The sphere S^2 is a riemannian positive curvature space in which each "straight line" (self-parallel curves, or geodesics, great circles in the case of S^2) meets each other sooner or later — so that there are actually no "straight" parallels to a previously given "straight" line. A hyperbolic sheet, on the contrary, may exhibit many parallels to a given geodesics. What we shall do will be to consider both S^2 and a hyperbolic space imbedded in $\mathbb{E}^3$, and project them into a plane, thereby obtaining curved spaces on $\mathbb{R}^2$. The projection to be used, the stereographic projection, has very nice properties, in special that of preserving circles. Geodesics are sent into geodesics, and angles are also preserved. It turns out that, if we want to preserve metric properties in the hyperbolic case, the imbedding space must be not $\mathbb{E}^3$, but the pseudo-euclidean space $\mathbb{E}^{2,1}$ instead.

The treatment is, up to the dimension, identical to that of the de Sitter spaces (Chapter 37). We shall, consequently, concentrate here on some basic aspects and refer to that chapter for others as, for example, the justification of the above statements on the values of the scalar curvature.

23.3 The spherical case

A point on the sphere S^2 with radius L will be fixed by the values X, Y, Z such that

$$Z = \pm\sqrt{L^2 - X^2 - Y^2}.$$

The relation to spherical coordinates are

$$X = L \sin\theta \cos\varphi; \qquad Y = L \sin\theta \sin\varphi; \qquad Z = L \cos\theta.$$

Consider now its stereographic projection into the plane. We choose the "north-pole" $N = (0,0,L)$ as projection center and project each point of S^2 on the plane tangent to the sphere at the "south-pole" $S = (0,0,-L)$, as indicated in Figure 23.2. Cartesian coordinates (x,y) are used on the plane.

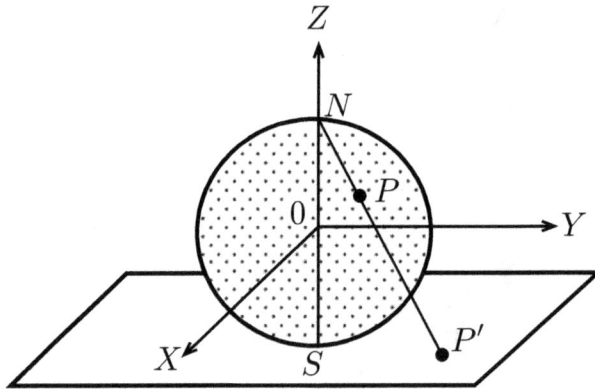

Fig. 23.2   Stereographic projection of the sphere S^2 on the plane.

Consider a point $P = (X,Y,Z)$ and its projection $P' = (x,y,-L)$. The points N, P and P' are on a straight line, so that the differences between their coordinates, $(N-P)$ and $(N-P')$, are proportional:

$$\frac{X}{x} = \frac{Y}{y} = \frac{L-Z}{2L} =: n.$$

The transformation is, thus,

$$X = nx, \qquad Y = ny, \qquad Z = L(1 - 2n).$$

We find then the solutions $n = 0$ (corresponding to the isolated point N), and

$$n = \frac{4L^2}{x^2 + y^2 + 4L^2} = \frac{L - Z}{2L}.$$

If we call

$$r^2 = \frac{x^2 + y^2}{4L^2},$$

the proportionality coefficient will be

$$n = \frac{1}{1 + r^2}. \tag{23.2}$$

The relations between the coordinates are thus

$$X = \frac{x}{1 + r^2}, \qquad Y = \frac{y}{1 + r^2}, \qquad Z = -L\frac{1 - r^2}{1 + r^2}. \tag{23.3}$$

The coordinates on the plane will be

$$x = \frac{2LX}{L - Z} \quad \text{and} \quad y = \frac{2LY}{L - Z}. \tag{23.4}$$

These (x, y) constitute a local system of coordinates with covering patch $S^2 \backslash \{N\}$. The metric, or the line element, will be given by

$$ds^2 = dX^2 + dY^2 + dZ^2 = n^2(dx^2 + dy^2) = \frac{dx^2 + dy^2}{(1 + r^2)^2}. \tag{23.5}$$

Commentary 23.1 Stereographic projections provide the most economical system of coordinates for the sphere S^2: only 2 charts are needed. One is the above one, the other is obtained by projecting from the south pole S onto the plane tangent to S^2 at the north pole N. Each projection is a homeomorphism of the plane with the chart $S^2 \backslash \{\text{projection center}\}$, which is thereby a locally euclidean set. Cartesian coordinates would need 8 charts to cover the sphere with locally euclidean sets. ◄

The important point is that this procedure may be seen alternatively as a means of defining a new, non-euclidean metric

$$g_{ij} = n^2(x, y)\, \delta_{ij},$$

on the plane, with

$$n(x, y) = 1/(1 + r^2)$$

and δ_{ij} the euclidean metric. With this metric, we agree to define as the distance between two points on the plane the length of the shortest arc connecting the corresponding points on the sphere. This is an example of riemannian structure on the plane $\mathbb{R}^2$. The curvature of the corresponding Levi–Civita connection will be constant and positive.

Given the interpretation of distance, we may expect that the geodesics on the plane be the projections of the spherical great circles. Indeed, there are two possible results of projecting circles: straight lines and circles. Notice to start with that

$$r^2 = \frac{L+Z}{L-Z} \, ,$$

so that lines at constant Z (horizontal circles) will be led into points satisfying $r^2 = $ constant. The equator ($Z = 0$), in particular, is taken into $r^2 = 1$, or

$$x^2 + y^2 = 4L^2.$$

Each great circle meeting the equator at (X, Y) will meet it again at $(-X, -Y)$, and this will correspond to (x, y) and $(-x, -y)$. In the general case, a great circle is the intersection of S^2 with a plane through the origin, with equation

$$uX + vY + wZ = 0.$$

This plane is orthogonal to the vector (u, v, w), whose modulus is irrelevant, so that actually the constants are not independent (if u, v, w are direction cosines, then $u^2 + v^2 + w^2 = 1$). Examine first the planes going through the axis OZ: $w = 0$ and then

$$Y = -(u/v)X =: \gamma X.$$

The intersection projection will satisfy the equation

$$y = \pm\gamma x,$$

representing straight lines through the plane origin. Now, $w \neq 0$ when the circle is in general position, which leads to

$$Z = \alpha X + \beta Y = n(\alpha x + \beta y)$$

with

$$\alpha = -u/w \quad \text{and} \quad \beta = -v/w.$$

We then obtain from

$$n = (L - Z)/2L$$

the equation

$$(x - 2L\alpha)^2 + (y - 2L\beta)^2 = 4L^2(1 + \alpha^2 + \beta^2),$$

representing circles centered at the point $(2L\alpha, 2L\beta)$.

Geodesics are great circles. Consequently, all the geodesics starting at a given point of S^2 will intersect again at its antipode. And we see in this way how Euclid's postulate of the parallels is violated: there are no parallels in such a space, as any two geodesics will meet at some point.

23.4 The Boliyai–Lobachevsky case

A point on a two-sheeted hyperboloid will be fixed by the values X, Y, Z of its coordinates satisfying the condition

$$X^2 + Y^2 - Z^2 = -L^2$$

or

$$Z = \pm\sqrt{X^2 + Y^2 + L^2} \,.$$

Again we choose the point $(0, 0, L)$ as projection center (it is now the lowest point of the upper branch) and project each point of the hyperboloid on the plane tangent at the point $(0, 0, -L)$ (which is now the highest point on the lower branch), as indicated in Figure 23.3. The same reasoning applied above to the spherical case will lead again to the relations

$$X/x = Y/y = (L - Z)/2L = n,$$

but the form

$$X^2 + Y^2 - Z^2 = -L^2$$

leads to another expression for the function n. Instead of Eq. (23.2), we have now

$$n(x, y) = \frac{1}{1 - r^2}, \tag{23.6}$$

so that the relations between the coordinates are

$$X = \frac{x}{1 - r^2}, \qquad Y = \frac{y}{1 - r^2}, \qquad Z = -L\frac{1 + r^2}{1 - r^2}. \tag{23.7}$$

Now we have a problem. We would like to have the equations defining the surfaces to represent relations between measured distances, and the interval to be

$$ds^2 = dX^2 + dY^2 - dZ^2.$$

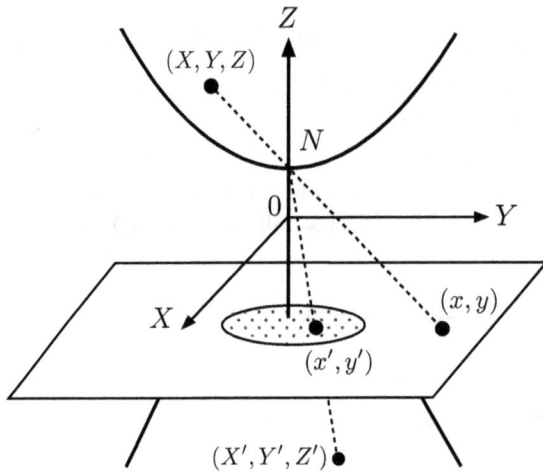

Fig. 23.3   Stereographic projection of the two-sheeted hyperboloid on the plane.

This is obviously impossible with the euclidean metric. In order to preserve that idea, we must change the ambient space and consider pseudo-euclidean spaces. The above non-compact surfaces will be (pseudo-) spheres in such spaces.

The pseudo-euclidean space $\mathbb{E}^{2,1}$ is the set point $\mathbb{R}^3$ with, instead of Eq. (23.1), the "pseudo-distance" function

$$d(p,q) = +\sqrt{(p_1 - q_1)^2 + (p_2 - q_2)^2 - (p_3 - q_3)^2}\,. \qquad (23.8)$$

Due to the negative sign before the last term, this is not a real distance function and does not define a topology. Taking the origin $q = (0,0,0)$ as the center, there are "spheres" of three types, according to the values of their radius L: precisely "spheres" of real radius, satisfying

$$X^2 + Y^2 - Z^2 = L^2,$$

those of vanishing radius,

$$X^2 + Y^2 - Z^2 = 0,$$

and those with purely imaginary radius, satisfying

$$X^2 + Y^2 - Z^2 = -L^2.$$

Thus, the above surfaces in $\mathbb{E}^3$ appear as the possible "spheres" in $\mathbb{E}^{2,1}$, with the extra bonus that now the equations have a metric sense.

Returning to the specific case of the two-sheeted hyperboloid, we have that the interval in the plane coordinates turns out to be

$$ds^2 = dX^2 + dY^2 - dZ^2 = n^2(dx^2 + dy^2) = \frac{dx^2 + dy^2}{(1 - r^2)^2}.$$

The metric is consequently

$$g_{ij} = \frac{1}{(1 - r^2)^2} \delta_{ij}. \tag{23.9}$$

Metrics like this one and that defined in Eq. (23.5), which are proportional to a flat metric, are called "conformally flat" metrics. Though they give measures of distance different from those of the euclidean metric, they give the same measure for angles.

The plane $\mathbb{R}^2$ with the above metric is called the "Lobachevsky plane". Notice that the metric actually "brings the hyperboloid down and up" to the plane. The lower sheet has $Z/L < 1$ and is mapped into the disc bounded by the circle $r^2 = 1$. The disc $r^2 < 1$ with the metric Eq. (23.9) is called the "Poincaré space". We shall come to it later. The upper sheet has $Z/L > 1$ and is mapped into the complement of the disc in the plane. As we go to infinity in the upper and lower sheets we approach the circle (respectively) from outside and from inside.

We may analyze the projections of intersections of the hyperboloids with planes (the now eventually open "great circles") in the same way used for the spherical case. Putting together the two metrics by writing

$$n = \frac{1}{1 \pm r^2},$$

we find on the plane the curves

$$(x \pm 2L\alpha)^2 + (y \pm 2L\beta)^2 = 4L^2(\alpha^2 + \beta^2 \pm 1). \tag{23.10}$$

23.5   On the geodesic curves

Given the metric $g_{ij} = n^2 \delta_{ij}$, the components of the Levi–Civita connection are

$$\Gamma^k{}_{ij} = \left(\delta^k_j \partial_i + \delta^k_i \partial_j - \delta_{ij}\delta^{kr}\partial_r\right)\ln n, \tag{23.11}$$

which is a general result for conformally flat metrics. The calculations are given in Chapter 37, where also everything concerning the curvature is to be found. Here, we shall rather comment on the geodesics. The geodesic equation is

$$\frac{d^2 x^k}{ds^2} + \Gamma^k{}_{ij}\frac{dx^i}{ds}\frac{dx^j}{ds} = 0 \tag{23.12}$$

which can be rewritten in the equivalent form

$$\frac{dv^k}{ds} + 2v^k \frac{d\ln n}{ds} + \tfrac{1}{2} \partial_k n^{-2} = 0. \tag{23.13}$$

It is a happy fact that, whenever we may define a "momentum" by

$$p_i = g_{ij} v^j, \tag{23.14}$$

the geodesics equation simplify because the first two contributions in the Christoffell (23.11) just cancel the term coming from the derivative of the metric. We remain with

$$\frac{dp_k}{ds} + \tfrac{1}{2} \left(\partial_k g^{ij} \right) p_i p_j = 0. \tag{23.15}$$

Notice that this is the same as $p(v) = p_k v^k = 1$, which happens always when the geodesics is parametrized by its length. Here we find a "force law"

$$\frac{dp_k}{ds} = -\tfrac{1}{2} \partial_k (\ln n). \tag{23.16}$$

This relates to Eq. (30.19) of the chapter devoted to Optics. There, n has the role of the refraction index. The situation is also analogous to the Poinsot construction of particle mechanics, described in Section 26.3.10. The geodesic motion in that case corresponds to that of a particle for which the quantity $\sqrt{\ln n}$ acts as a potential.

It seems simpler here to try to integrate just $p_k v^k = 1$, or

$$v^k v^k \equiv (\dot{x})^2 + (\dot{y})^2 = \frac{1}{n^2},$$

with $n = 1 \pm r^2$, which is equivalent to

$$\dot{r}^2 + r^2 \dot{\theta}^2 = \frac{(1 \pm r^2)^2}{4L^2}.$$

We shall prefer, however, to change coordinates before solving the equations.

The Jacobi equation, analogous to the case of de Sitter spaces, is

$$\frac{D^2 X^\alpha}{Ds^2} + \frac{1}{L^2} \left[X^\alpha - (X_\beta V^\beta) V^\alpha \right] = 0, \tag{23.17}$$

or simply

$$\frac{D^2 X}{Ds^2} + \frac{1}{L^2} \left[X - g(X, V) V \right] = 0. \tag{23.18}$$

Now,

$$X^\perp = \left[X - g(X, V) V \right]$$

is the component of X transversal to the curve. As the tangential part $X^{\|}$ has

$$\frac{D^2 X^{\|}}{Ds^2} = 0,$$

one arrives at

$$\frac{D^2 X^\perp}{Ds^2} + \frac{1}{L^2} X^\perp = 0. \tag{23.19}$$

23.6   The Poincaré space

Consider now the Poincaré space. We may consider the plane $\mathbb{E}^2$ as the complex plane. It is known that an open disc on the complex plane can be taken into the upper-half-plane by a homographic transformation, which is furthermore a conformal transformation. Actually, there is one transformation for each point of the half-plane. For each arbitrarily chosen point "a" on the half-plane, there will be one homographic half-plane $\rightarrow$ disc transformation taking "a" into the circle center.[1] Introducing the complex variables

$$z = x + iy \quad \text{and} \quad w = u + iv,$$

the transformation

$$z = K\,\frac{w - a}{w - a^*} \tag{23.20}$$

takes the open upper-half-plane onto a disc of radius K whose center is the transformed of a. We choose for convenience $a = iaL$. The relations between the coordinates are

$$x = 2L\,\frac{u^2 + v^2 - 4L^2}{u^2 + (v + 2L)^2} \tag{23.21}$$

and

$$y = -2L\frac{4Lu}{u^2 + (v + 2L)^2}\,, \tag{23.22}$$

with their inverses given by

$$u = \frac{-2y}{1 + r^2 - x/L} = \frac{-8L^2 y}{(x - 2L)^2 + y^2} \tag{23.23}$$

and

$$v = \frac{2L(1 - r^2)}{1 + r^2 - x/L} = \frac{8L^3(1 - r^2)}{(x - 2L)^2 + y^2}\,. \tag{23.24}$$

The last relation shows that

$$v = 0 \ \text{ corresponds to } \ r^2 = 1$$

and

$$v > 0 \ \text{ corresponds to } \ 1 > r^2.$$

The metric above becomes

$$ds^2 = (du^2 + dv^2)/v^2$$

---

[1] Lavrentiev & Chabat 1977.

on the upper-half-plane. Let us examine the geodesics in this case. With

$$g_{ij} = \frac{1}{v^2} \delta_{ij},$$

the Christoffell is (notation: $u^1 = u$; $u^2 = v$)

$$\Gamma^i{}_{jk} = -\frac{1}{v}\left(\delta_{ij}\delta_{k2} + \delta_{ik}\delta_{j2} - \delta_{jk}\delta_{i2}\right). \qquad (23.25)$$

The geodesic equations are

$$\ddot{u} - \frac{2}{v}\dot{u}\dot{v} = 0, \qquad (23.26)$$

and

$$\ddot{v} - \frac{1}{v}\left(\dot{v}^2 - \dot{u}^2\right) = 0. \qquad (23.27)$$

Now, we have two cases:

(i) If $\dot{u} = 0$, the solution $u = C$, with C a constant, will work if there is a solution for

$$\ddot{v} = \frac{\dot{v}^2}{v}.$$

This is solved by

$$v = \exp[At + B],$$

with A and B constants. Thus, all the vertical straight lines will be solutions.

(ii) If $\dot{u} \neq 0$, we find

$$\dot{u} = cv^2 \quad \text{and} \quad \frac{\ddot{v}}{v} - \frac{\dot{v}^2}{v} = -c^2 v^2 = -c\dot{u}.$$

Putting

$$z = \ln v, \qquad \dot{z} = \frac{\dot{v}}{v} = \frac{\ddot{v}}{v} - \frac{\dot{v}^2}{v}, \qquad \ddot{z} = -c\dot{u}$$

will lead to

$$\dot{z} = -cu + d.$$

The two remaining equations,

$$\frac{\dot{v}}{v} = d - cu \quad \text{and} \quad \dot{u} = cv^2$$

are put together in

$$\frac{dv}{du} = \frac{d - cu}{cv},$$

whose solutions are the circles

$$(u - c/2)^2 + v^2 = B^2$$

centered on the horizontal axis and orthogonal to it.

On the disc, these two families of solutions will have the following correspondence:

(1) The vertical straight lines $u = C$ will be taken into the circles

$$(x - 2L)^2 + (y + 4L^2/C)^2 = 16L^4/C^2.$$

These circles intersect at right angles the border $r^2 = 1$ when

$$y = (C/2L)x - C,$$

represented by lines a and b in Figure 23.4.

(2) The circles

$$(u - c/2)^2 + v^2 = B^2$$

will be transformed into

$$y = \pm(4L/c)x$$

represented by line c in Figure 23.4.

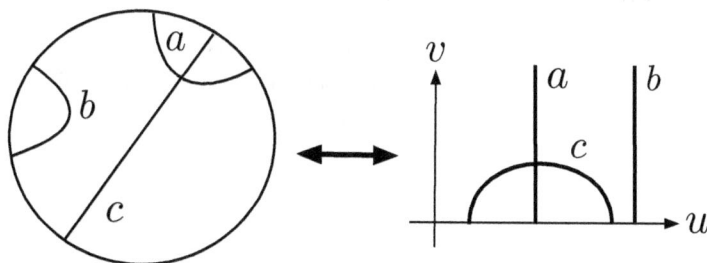

Fig. 23.4   Poincaré disk: circles.

The Poincaré disc has a very curious zoology of curves. Some are indicated in Figure 23.5. There are equidistant curves (as a and b), circumferences c, and circumferences which are tangent to the infinity circle $r^2 = 1$. The latter goes under the honest name of limiting circles, but also under those of oricycles, or still horocycles (curve h in Figure 23.5). They, and their higher-dimensional analogues (horispheres), are important in the study of representations of the groups of motions on hyperboloids.[2] Comparison with the corresponding curves in the half-plane is an amusing exercise. It serves at least to illustrate the vanity of a curve aspect, which depends heavily on the coordinate system.

---

[2] Vilenkin 1969.

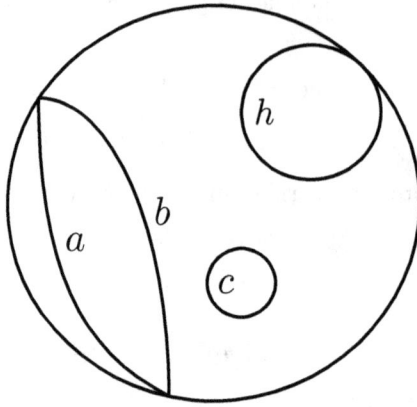

Fig. 23.5   Poincaré disk: parallels (a, b), circle (c), horicycle (h).

General references

Some relevant references for the contents of this chapter are Whittaker 1958, Rosenfeld & Sergeeva 1986, and Lavrentiev & Chabat 1977.

Chapter 24

Geodesics

A well-known aphorism says that there is only one experiment in Physics: Rutherford's. In order to acquire information on a "black-box" system (an atom, a molecule, a nucleus), we examine the resulting paths of a well-known probe (a particle, a light beam) sent against it. Comparison is made with the paths of the "free case", in which the "black-box" system is not there. That is to say that Physics dwells frequently with the study of trajectories. In this chapter we describe how to obtain information on a geometry through the study of its curves. It turns out that what is best probed by trajectories is always a connection.

24.1 Introductory remarks

The word "geodesic" is used in more than one sense. In the common lore, it means the shortest-length curve between two points. In a more sophisticated sense, it represents a Self-parallel curve, whose tangent vectors keep parallel to themselves along the curve. The first concept assumes a metric to measure the lengths, whereas the second supposes a well-defined meaning for the idea of parallelism. These are quite different concepts. The ambiguity comes from the fact that geodesics have been conceived in order to generalize to curved spaces the straight lines of euclidean spaces. They are "as straight as possible" on non-euclidean spaces. The trouble is that the starting notion of a straight line is itself ambiguous.

A straight line can be seen either as the shortest-length curve between two points or as a curve keeping a fixed direction all along. The two ideas coincide on euclidean spaces, but not on more general spaces. The "shortest-length" point of view is meaningful only on strictly riemannian manifolds (endowed with a positive-definite metric). The "self-parallel" idea has a meaning even for pseudo-riemannian spaces, and even in non-metric spaces.

As it happens, parallelism is an idea related to a connection, which may be or not related to a metric.

As the "self-parallel" concept is more general and reduces to the "shortest-length" point of view in the metric case, it is to be preferred as the general definition of geodesics. We shall here, however, start introducing the notion through its physical connotations. From the physical point of view, there are two main approaches to the idea of a geodesic. One relates to gravitation, the other to optics.

24.2 Self-parallel curves

24.2.1 *In General Relativity*

In the theory of gravitation, the notion of a geodesic curve comes from a particular use (and view) of the equivalence principle. Start with Newton's law for the force per unit mass, written of course in a cartesian coordinate system $\{y^k\}$ on the euclidean space $\mathbb{E}^3$. In that case, force is just acceleration,

$$F^k = A^k = \frac{d^2 y^k}{dt^2}.$$

Time is absolute, so that t has the same value for all events in space. Space-time is thus a direct product $\mathbb{E}^1 \otimes \mathbb{E}^3$. Now pass into another, arbitrary coordinate system $\{x^k\}$, in terms of which the distance between two infinitesimally close points in space-time is written

$$ds^2 = g_{ij}dx^i dx^j.$$

Then the expression for the acceleration becomes

$$A'^k = \frac{d^2 x^k}{dt^2} + \Gamma^k{}_{ij} \frac{dx^i}{dt} \frac{dx^j}{dt}, \tag{24.1}$$

where the $\Gamma^k{}_{ij}$'s are coefficients given in terms of the g_{ij}'s as

$$\Gamma^k{}_{ij} = \tfrac{1}{2} g^{kr} \left[\partial_i g_{rj} + \partial_j g_{ri} - \partial_r g_{ij} \right]. \tag{24.2}$$

There is nothing new in this change of appearance, only a change of coordinates. The equation stating the absence of forces,

$$\frac{d^2 x^k}{dt^2} + \Gamma^k{}_{ij} \frac{dx^i}{dt} \frac{dx^j}{dt} = 0, \tag{24.3}$$

gives simply a straight line written in the new coordinates. The set $\{g_{ij}\}$ contains the components of the euclidean metric of $\mathbb{E}^3$, which are $\{\delta_{ij}\}$ in cartesian coordinates $\{y^k\}$, written in the new coordinates $\{x^k\}$. It only happens that, in euclidean cartesian coordinates, the $\Gamma^k{}_{ij}$'s vanish, and so do their derivatives.

Suppose now that this is no more the case, that g_{ij} is another, non-euclidean metric. The $\Gamma^k{}_{ij}$'s are then the Christoffel symbols, representing a connection intimately related to the metric. It happens that, though it is still possible to make the $\Gamma^k{}_{ij}$'s to vanish at a point in a suitable ("normal") coordinate system, it is no more possible to make also their derivatives to vanish. The derivatives appear in the curvature of this connection which, unlike the euclidean case, does not vanish. But the equation (24.3) keeps its sense: it gives the path of a particle constrained to move on the new, "curved" space, but otherwise unsubmitted to any forces.

In the framework of General Relativity, not everything follow a geodesic. That a self-propelled rocket will not do it is a matter of course; but neither will a free but spinning particle follow a geodesic.[1]

24.2.2 *In optics*

In geometrical optics, geodesics turn up as light rays, which obey just the geodesic equation. There, actually, geodesics justify their primitive "geodesical" role. Not as the shortest length line between the two end-points, but as that line corresponding to the shortest *optical* length. The refraction index is essentially a 3-dimensional space metric, and the light ray follows that line which has the shortest length as measured in that metric. This is seen in some detail in Section 30.2. Light rays will follow geodesics also in the 4-dimensional space-time of General Relativity. We only notice here that, as the interval $ds^2 = 0$ for massless particles like the photon, four-velocities cannot be defined as they are [see eq.(24.6)] for massive particles following a time-like curve. Some other curve parameter must be used.

24.3 The absolute derivative

Take a differentiable manifold M and consider on it a linear connection Γ of curvature R and torsion T. The covariant derivative along a vector field X will be indicated by ∇_X. We recall the representation of curvature and

---

[1] Papapetrou 1951.

torsion as families of mappings (§ 9.4.15):

$$R(X,Y)Z = \left(\nabla_X\nabla_Y - \nabla_Y\nabla_X - \nabla_{[X,Y]}\right)Z$$
$$= \nabla_X(\nabla_Y Z) - \nabla_Y(\nabla_X Z) - \nabla_{[X,Y]}Z \qquad (24.4)$$

and

$$T(X,Y) = \nabla_X Y - \nabla_Y X - [X,Y]. \qquad (24.5)$$

Given a curve γ with parameter τ on M, and its tangent velocity field

$$V(\tau) = \dot{\gamma}(\tau) = \frac{d\gamma(\tau)}{d\tau}, \qquad (24.6)$$

the covariant derivative of a field or form W along the curve γ will be

$$\frac{DW}{D\tau} = \nabla_V W. \qquad (24.7)$$

The derivative

$$\frac{D}{D\tau} = \nabla_V \qquad (24.8)$$

along the curve is frequently called *absolute derivative*. The object W will be parallel-transported along γ if

$$\nabla_V W = 0.$$

The acceleration, for example, will have the invariant expression

$$A = \frac{DV}{D\tau} = \nabla_V V. \qquad (24.9)$$

This expression holds in any basis and differs from (24.1) only because here the connection involved is quite general. In the basis $\{e_a = \partial/\partial\gamma^a\}$, the velocity field V is $d/d\tau$ and the acceleration will be written

$$A = \left[\frac{d^2\gamma^a}{d\tau^2} + \Gamma^a{}_{bc}\frac{d\gamma^b}{d\tau}\frac{d\gamma^c}{d\tau}\right]e_a. \qquad (24.10)$$

The absolute derivative $DS/D\tau$ reduces to the simple derivative $dS/d\tau$ if S is an invariant, or a scalar. For instance,

$$\frac{D}{D\tau}(V_a V^a) = \frac{d}{d\tau}(V_a V^a).$$

In General (and Special) Relativity, the interval or (squared) proper time is given by

$$d\tau^2 = g_{ab}dx^a dx^b = dx_a dx^a.$$

If we use the proper time τ as the curve parameter, $V_a V^a = 1$. Then,

$$\frac{D}{D\tau}(V_a V^a) = 2V_a A^a = 0.$$

The acceleration, when non-vanishing, is orthogonal to the velocity. The property $||V||^2 = 1$ is valid everytime the curve is parametrized by the proper time. We have, in the simple calculation above, used

$$\frac{Dg_{ab}}{D\tau} = V^c g_{ab;c} = 0.$$

This supposes that the metric is parallel-transported by the connection. The property $g_{ab;c} = 0$ is sometimes called "metric compatibility" of the connection, and also "metricity condition". Notice that a different convention, with opposite sign for $d\tau^2$, is frequently used. This leads to $V_a V^a = -1$. The signs in some definitions given below (of transversal metric, of Fermi derivative) must be accordingly modified.

24.3.1 *Self-parallelism*

The curve γ will be a Self-parallel curve when the tangent velocity keeps parallel to itself along γ, that is, if

$$A = \frac{DV}{D\tau} = \nabla_V V = 0. \tag{24.11}$$

In the basis $\{e_a = \partial/\partial\gamma^a\}$,

$$\frac{d^2\gamma^a}{d\tau^2} + \Gamma^a{}_{bc} \frac{d\gamma^b}{d\tau} \frac{d\gamma^c}{d\tau} = 0$$

or

$$\frac{dV^a}{d\tau} + \Gamma^a{}_{bc} V^b V^c = 0. \tag{24.12}$$

A self-parallel curve is more suitably called a *geodesic* when the connection Γ is a metric connection (as said, there are fluctuations in this nomenclature: many people call geodesic any self-parallel curve). The equation above, spelling the vanishing of acceleration, is the *geodesic equation*.

Commentary 24.1 The parametrization

$$t \to \gamma(t) = (\gamma^0(t), \gamma^1(t), \gamma^2(t), \gamma^3(t))$$

is, when $\gamma(t)$ satisfies the geodesic equation, defined up to an affine transformation

$$t \to s = at + b,$$

where $a \neq 0$ and b are real constants. For this reason, such a t is frequently called an "affine parameter". ◄

In euclidean space and arbitrary coordinates, the left-hand side of the geodesic equation is simply the time (or curve parameter) second derivative of γ^a. The equation says that no force is exerted on a particle going along the curve, which is consequently the path followed by a "free" particle. A vector X is said to be parallel-transported along a curve if

$$V\nabla_V X = 0. \tag{24.13}$$

Commentary 24.2 If parallel-transported around a closed curve, X^k will come back to the initial point modified by an amount ΔX^k which is a measure of the curvature flux through the surface circumscribed by the loop:

$$\Delta X^k = \tfrac{1}{2} \int R^k{}_{ijl} X^i \, d\gamma^l \, d\gamma^j.$$

If in addition to curvature, the space has also torsion, X^k will not return to the original position, but will be displaced by a distance, which is a measure of the torsion of space. ◀

24.3.2 *Complete spaces*

A linear connection Γ is *complete* if every geodesic curve of Γ can be extended to a geodesic $\gamma(t)$ with $(-\infty < t < +\infty)$. Every geodesic on M is the projection on M of the integral curve of some standard horizontal field of $BLF(M)$, and vice-versa. Thus, a linear connection is complete *iff* every standard horizontal field of $BLF(M)$ is complete.

A riemannian manifold M is *geodesically complete* if every geodesic on M can be extended for arbitrarily large values of its parameter. This nomenclature is extended to the metric proper.

There are many results concerning such complete spaces. We shall quote three of them, only to give an idea of the interplay between geodesics and topological-differential properties.

(i) Let M be a riemannian complete manifold with non-positive curvature. Take a point $p \in M$ and consider the exponential mapping

$$\exp_p : T_p M \to M.$$

Then $\exp_p$ is a covering map. If M is simply-connected, $\exp_p$ is a diffeomorphism.

(ii) Every homogeneous riemannian manifold is complete.

(iii) Every compact riemannian manifold is complete.

24.3.3   *Fermi transport*

Derivatives different of the above absolute derivative are of interest on general curves. We only consider the Fermi derivative along $\gamma(t)$, which is

$$\frac{D_F X}{D\tau} = \frac{DX}{D\tau} - g\left(X, \frac{DV}{D\tau}\right) V + g(X, V) \frac{DV}{D\tau}. \tag{24.14}$$

A vector X such that $D_F X/D\tau = 0$ is said to be Fermi-propagated, or Fermi-transported, along the curve. This derivative has the following properties:

(i) When γ is a geodesic, then

$$\frac{D_F X}{D\tau} = \frac{DX}{D\tau}.$$

(ii) On any curve, we have

$$\frac{D_F V}{D\tau} = 0.$$

(iii) If both

$$\frac{D_F X}{D\tau} = 0 \quad \text{and} \quad \frac{D_F Y}{D\tau} = 0$$

hold, then $g(X, Y)$ is constant along the curve, meaning in particular that vectors which are orthogonal at some point remain so all along the curve.

(iv) Take an orthogonal basis $\{e_a\}$, such that $e_4 = V$ at some starting point. Then, if Fermi-transported, $\{e_a\}$ will be taken into another orthogonal basis at each point of the curve, with $e_4 = V$. The set $\{e_1, e_2, e_3\}$ will then constitute a non-rotating set of axes along $\gamma(t)$.

24.4   Congruences

Our objective in the following will be to give a general, qualitative description of the relative behaviour of neighbouring curves on a manifold. We shall concentrate on geodesics, and by "neighbouring" we shall mean those geodesics which are near to each other in some limited region of the manifold. A family of neighbouring curves is called a pencil or congruence (or still a ray bundle, or bunch).

24.4.1   *Jacobi equation*

If γ is a geodesic, its infinitesimal variation field X is called a Jacobi field (see Sec. 19.1 for general curves). Due to the particular conditions imposed on γ to make of it a geodesic, X will satisfy a second order ordinary differential equation, the Jacobi (or "deviation", or "second-variation", or in the case, "geodesic deviation") equation:

$$\nabla_V \nabla_V X + \nabla_V T(X,V) + R(X,V)V = 0. \qquad (24.15)$$

The field $X = 0$ is a trivial solution. Notice that $T(V,V) = 0$ and $R(V,V) = 0$. On a geodesic, as the acceleration $A = 0$, both A and V are Jacobi fields. On a non-geodesic curve, to impose that A is a Jacobi field is to say that the second acceleration vanishes:

$$\nabla_V \nabla_V V = \nabla_V A = \frac{D^2 V}{D\tau^2} = 0.$$

As

$$\frac{DW}{D\tau} = \nabla_V W = [V^a D_a W^c]e_c, \qquad (24.16)$$

equation (24.15) is the same as

$$\frac{D^2 X}{D\tau^2} + \frac{D}{D\tau}T(X,V) + R(X,V)V = 0. \qquad (24.17)$$

In a basis $\{e_a\}$, this equation reads

$$\frac{D^2 X}{D\tau^2} + V^b \frac{D}{D\tau}(X^a T^d{}_{ab})e_d + X^a V^b V^c R^d{}_{cab}\, e_d = 0. \qquad (24.18)$$

In components,

$$\left(\frac{D^2 X}{D\tau^2}\right)^d + V^b V^c \left[D_c(X^a T^d{}_{ab}) + X^a R^d{}_{cab}\right] = 0, \qquad (24.19)$$

or equivalently

$$\left(\frac{D^2 X}{D\tau^2}\right)^d + V^b \left[\frac{D}{D\tau}(X^a T^d{}_{ab}) + X^a R^d{}_{cab} V^c\right] = 0. \qquad (24.20)$$

On a geodesic, as $DV^b/D\tau = 0$,

$$\left(\frac{D^2 X}{D\tau^2}\right)^d + \left[V^c D_c(X^a T^d{}_{ab} V^b) + X^a R^d{}_{cab} V^b V^c\right] = 0, \qquad (24.21)$$

or equivalently

$$\left(\frac{D^2 X}{D\tau^2}\right)^d + \left[\frac{D}{D\tau}(X^a T^d{}_{ab} V^b) + X^a R^d{}_{cab} V^b V^c\right] = 0. \qquad (24.22)$$

In the study of the general behaviour of curves, one is most frequently interested in the "transversal" behaviour, on how things look on a plane (or space) orthogonal to the curve. More precisely, one considers the space tangent to the manifold at a point on the curve; one direction will be along the curve, colinear with its velocity vector. The remaining directions are transversal. The metric can be "projected" into a metric on that subspace (see Eq. (24.33) below):

$$h_{mn} = g_{mn} - V_m V_n.$$

Thus, to each point p on the curve will correspond a plane P_p orthogonal to the curve at p. The next step is to examine congruences of curves, together with their variations.

Consider a congruence of curves $\gamma(s)$ and a variation of it, $\gamma(s,t)$. At fixed s, these curves will cross $\gamma(s)$ and constitute another congruence, parametrized by t. Consider the fields

$$V = \frac{d}{ds} \quad \text{and} \quad U = \frac{d}{dt}$$

tangent to the respective congruences of curves and normalized to unity. Each one of these fields will be taken into itself by the other's congruence. That is, their Lie derivatives will vanish: $[V, U] = 0$. From this commutativity it follows that

$$V^a \partial_a U^b = U^a \partial_a V^b.$$

But then it follows also that

$$V^a D_a U^b = U^a D_a V^b,$$

or still

$$\frac{DU^b}{Ds} = \frac{DV^b}{Dt}.$$

This may be written as

$$\frac{DU^b}{Ds} = U^a D_a V^b$$

or

$$\nabla_V U = \nabla_U V. \qquad (24.23)$$

Notice the difference with respect to the usual invariant derivative

$$\frac{DU^b}{Ds} = V^a D_a U^b.$$

Equation (24.23) has important consequences. Notice that it is basically a matrix equation:

$$\frac{DU}{Ds} = VU.$$

Taking the absolute derivative D/Ds is the same as multiplying by the matrix $V = (V^b{}_{;a})$. It means that, while transported along $\gamma(s)$, the "transversal" vector field $U_{\gamma(s)}$ is taken from $U_{\gamma(s+ds)}$ by a simple matrix transformation. If we consider the orthogonal parts given by the projector, only

$$U^k = h^k{}_b U^b = g^{ka} h_{ab} U^b,$$

then

$$\frac{DU^k}{Ds} = V^k{}_{;j} U^j.$$

We shall come back to this "transversal" approach later.

Let us obtain the Jacobi equation in the strictly riemannian case. One sees that the torsion term appearing in the equation would anyhow vanish in this case: when $[U, V] = 0$ and $\nabla_U V = \nabla_V U$, then

$$T(U, V) = \nabla_U V - \nabla_V U - [U, V] = 0.$$

Going back to the definition of curvature, we find

$$R(U, V)V = \nabla_U(\nabla_V V) - \nabla_V(\nabla_V U) - \nabla_{[U,V]} V$$

$$= \nabla_U(\nabla_V V) - \nabla_V(\nabla_V U). \tag{24.24}$$

As for a geodesics $\nabla_V V = 0$, it follows that

$$\nabla_V(\nabla_V U) + R(U, V)V = 0, \tag{24.25}$$

which is the simplest case of the Jacobi equation. Other forms are

$$\frac{D^2 U}{D\tau^2} + R(U, V)V = 0 \tag{24.26}$$

and

$$\frac{D^2 U}{D\tau^2} + U^a V^b R(e_a, e_b) V \equiv \frac{D^2 U}{D\tau^2} + U^a V^b R^d{}_{cab} e_d = 0. \tag{24.27}$$

In a basis $\{e_a = \partial/\partial\gamma^a\}$, it is

$$\frac{D^2 U^d}{D\tau^2} + U^a R^d{}_{cab} \frac{d\gamma^b}{d\tau} \frac{d\gamma^c}{d\tau} \equiv \frac{D^2 U^d}{D\tau^2} + U^a R^d{}_{cab} V^b V^c = 0. \qquad (24.28)$$

The equation is sometimes given in terms of the sectional curvature. Given a plane in the tangent space $T_p M$ with $\{e_1, e_2\}$ a basis for it, then the sectional curvature on the plane is

$$K(\text{plane}) = g\big(R(e_1, e_2)e_2, e_1\big).$$

Thus, in the (X, V) plane, with $X \perp V$, it is just $g(X, R(X, V)V)$. We see that, if we project the equation along X, the curvature term is actually the sectional curvature.

There are two natural Jacobi fields along a geodesic $\gamma(s)$: they are $\dot\gamma(s)$ and $s\dot\gamma(s)$. It is a theorem that, on a riemannian manifold, any Jacobi field X can be uniquely decomposed as

$$X = a\dot\gamma + bs\dot\gamma + X^\perp, \qquad (24.29)$$

where a and b are real numbers, and $X^\perp$ is everywhere orthogonal to γ. If U is orthogonal to γ at two points, it will be orthogonal all along and will have the aspect of Figure 24.1.

Commentary 24.3 The set of all the Jacobi fields along a curve on an n-dimensional manifold constitutes a real vector space of dimension $2n$. ◂

Take a geodesic γ. Two points p and q on γ are *conjugate* to each other along γ if there exists some non-zero Jacobi field along γ which vanish at p and q. This means that infinitesimally neighbouring geodesics at p intersect at q. On a riemannian manifold with non-positive sectional curvature there are no conjugate points.

24.4.2 Vorticity, shear and expansion

It is clear that the study of Jacobi fields is the main source of information on the relative behaviour of neighbouring geodesics. They reflect all the qualitative behaviour: whether geodesics tend to crowd or to separate, to cross or to wind around one another. But there are some other tensor and scalar quantities which can be of great help. In particular, when dealing with the effect of curvature on families of curves in space-time, a hydrodynamic terminology is very convenient, as such curves (if timelike or null) represent possible flow lines (of a fluid constituted by test particles), or are histories of massless particles. We can introduce the notions of vorticity and

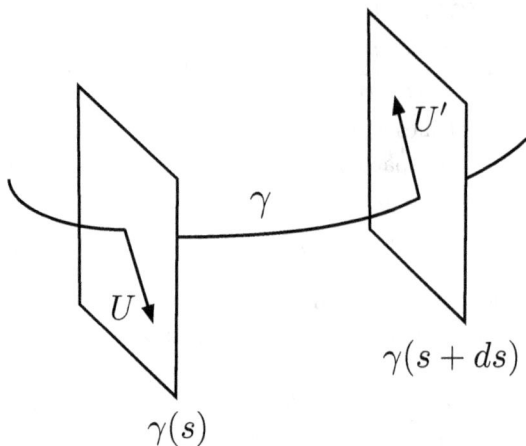

Fig. 24.1   A field U orthogonal to the curve γ.

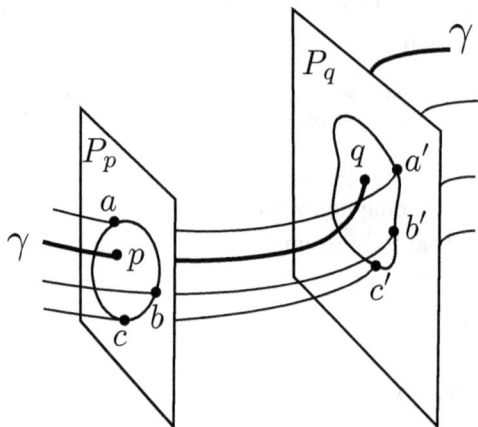

Fig. 24.2   Transversal view of the behaviour of a general congruence.

shear, expansion tensor and volume expansion, all of them duly projected into the transversal space.

The intuitive meaning of such tensor quantities is the following. Suppose given around the curve γ a congruence of curves, whose orthogonal cross section draw a circle on P_p. As we proceed along the curve from point p to another point q at a distance ds, the congruence will take the points on the circle at p into points on a line on P_q. *Volume expansion* will measure the enlargement of the circle, *tensor expansion* will measure its deformations, which may be larger or smaller in each direction (Figure 24.2). At vanishing tensor and volume expansion, the circle will be taken into an equal-sized circle, as shown in Figure 24.3. The same Figure illustrates the vorticity of the central curve γ, which measures how much the neighbouring geodesics turn around it. The procedure consists of looking at expansion, tensor and

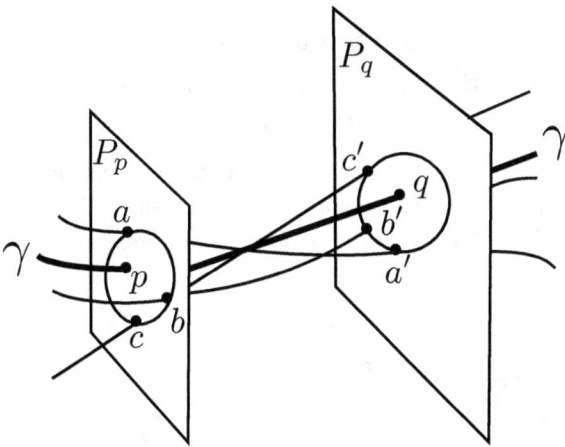

Fig. 24.3 In the absence of shear and volume expansion, a circle around the curve is taken into another circle of the same size. Vorticity will, however, cause a rotation.

volume, and see how such tensorial quantities behave along the curve. This means to calculate $D/D\tau$ acting on them.

Given a metric g, the general invariant definitions of vorticity and expansion are:

Vorticity:

$$\hat{\omega}(X,Y) = g(X, \nabla_Y V) - g(\nabla_X V, Y) \,.$$

Expansion:

$$\hat{\Theta}(X,Y) = \tfrac{1}{2} \left[g(X, \nabla_Y V) + g(\nabla_X V, Y) \right] \,.$$

They are related to a curve of tangent field V in the following way: vorticity is its covariant curl

$$\hat{\omega}(X, Y) = X^a Y^b V_{[a;b]} \tag{24.30}$$

or, in the basis $\{e_a\}$,

$$\hat{\omega}_{ab} = \hat{\omega}(e_a, e_b) = V_{a;b} - V_{b;a} . \tag{24.31}$$

The expansion tensor is

$$\hat{\Theta}(e_a, e_b) = \tfrac{1}{2}\left(V_{a;b} + V_{b;a}\right) . \tag{24.32}$$

Consequently,

$$V_{a;b} = \tfrac{1}{2}\hat{\omega}(e_a, e_b) + \hat{\Theta}(e_a, e_b).$$

If the tangent V to the curve is a Killing vector (§ 6.6.13), then

$$\hat{\omega}_{ab} = 2V_{a;b} \quad \text{and} \quad \hat{\Theta}_{ab} = 0.$$

This means that a Killing vector field admits no expansion.

In space-time, the four-velocity has been defined in such a way that $g(V, V) = 1$. This allows one to define a "transversal metric" by

$$h(X, Y) = g(X, Y) - g(X, V)\, g(Y, V) ,$$

which gives no components along V,

$$h(X, V) = g(X, V) - g(X, V)\, g(V, V) = 0 .$$

In components,

$$h_{ab} = g_{ab} - g_{ac} g_{bd} V^c V^d . \tag{24.33}$$

Then, the transversal vorticity and expansion are given, respectively, by

$$\omega(X, Y) = h(X, \nabla_Y V) - h(\nabla_X V, Y), \tag{24.34}$$

and

$$\Theta(X, Y) = \tfrac{1}{2}\left[h(X, \nabla_Y V) + h(\nabla_X V, Y)\right] . \tag{24.35}$$

At each point of a (timelike) curve, they give the vorticity and the expansion in space.

When $n = 4$, a vorticity vector is defined by

$$\omega^a = \tfrac{1}{2}\epsilon^{abcd} V_b\, \omega_{cd},$$

with

$$\omega_{ab} = h_a{}^c h_b{}^d V_{[c;d]} .$$

The volume expansion is $\Theta = V^a{}_{;a}$, and the shear tensor is the traceless part of the expansion tensor

$$\sigma_{ab} = \Theta_{ab} - \tfrac{1}{3}\, h_{ab}\Theta \tag{24.36}$$

24.4.3 *Landau–Raychaudhury equation*

We quote for completeness the Landau–Raychaudhury equation, which relates expansion to curvature, vorticity and shear:

$$\frac{d}{ds}\Theta = -R_{ab}V^aV^b + 2(\omega^2 - \sigma^2) - \tfrac{1}{3}\Theta^2 + \left(\frac{DV^a}{Ds}\right)_{;a}. \qquad (24.37)$$

As

$$2\omega^2 = \omega_{ab}\omega^{ab} \geq 0 \qquad \text{and} \qquad 2\sigma^2 = \sigma_{ab}\sigma^{ab} \geq 0,$$

we see that vorticity induces expansion, and shear induces contraction. Another important equation is (with everything transversal)

$$\frac{d\omega_{ab}}{ds} = 2\,\omega^\gamma{}_{[\alpha}\Theta_{\beta]\gamma} + \frac{d}{ds}V_{[\alpha;\beta]}. \qquad (24.38)$$

The use of the above concepts and techniques allowed Penrose and Hawking to show that Einstein's equations lead to a singularity in the primaeval universe. In General Relativity light rays follow null geodesics, and material objects follow time-like geodesics. In the Standard Cosmological Model,[2] the motion of the always receding galaxies are represented by an expanding time-like congruence of geodesics in space-time. If we start from the present-day (large scale) remarkably isotropic state of the universe and look backward in time, the above equations lead, under certain physically reasonable conditions, to a starting point of any geodesics. Cosmic space-time is consequently not a complete space.

General references

Some basic references for this chapter are Synge & Schild 1978, Kobayashi & Nomizu 1963, Synge 1960, and Hawking & Ellis 1973.

---

[2] See, for example, Weinberg 72.

Part 5: PHYSICAL TOPICS

Chapter 25

Hamiltonian Mechanics

This summary of hamiltonian mechanics has three main objectives: (i) to provide a good exercise on differential forms, (ii) to show how forms give a far more precise formulation of a well-known subject, and (iii) to present an introductory résumé of this most classical theory in what is nowadays the mathematicians colloquial language.

25.1 Introduction

Consider the classical phase space M of some conservative mechanical system with generalized coordinates $\boldsymbol{q} = (q^1, q^2, \ldots, q^n)$ and conjugate momenta $\boldsymbol{p} = (p_1, p_2, \ldots, p_n)$. States will be represented by points $(\boldsymbol{q}, \boldsymbol{p})$ on the $2n$-dimensional phase space M and any dynamical quantity will be a real function $F(\boldsymbol{q}, \boldsymbol{p})$. The time evolution of the state point $(\boldsymbol{q}, \boldsymbol{p})$ will take place along the integral curves of the velocity vector field on M,

$$X_H = \frac{dq^i}{dt} \frac{\partial}{\partial q^i} + \frac{dp_i}{dt} \frac{\partial}{\partial p_i}. \tag{25.1}$$

By using the Hamilton equations

$$\frac{dq^i}{dt} = \frac{\partial H}{\partial p_i} \quad \text{and} \quad \frac{dp_i}{dt} = -\frac{\partial H}{\partial q^i} \tag{25.2}$$

this evolution field takes up the form

$$X_H = \frac{\partial H}{\partial p_i} \frac{\partial}{\partial q^i} - \frac{\partial H}{\partial q^i} \frac{\partial}{\partial p_i}. \tag{25.3}$$

The *hamiltonian flow* is the one-parameter group (section 6.4) generated by X_H. The hamiltonian function $H(\boldsymbol{q}, \boldsymbol{p})$ will have as differential the 1-form

$$dH = \frac{\partial H}{\partial q^i} dq^i + \frac{\partial H}{\partial p_i} dp_i. \tag{25.4}$$

Applied to (25.3), this form gives

$$dH(X_H) = \frac{dH}{dt} = 0,$$

which says that the value of H is conserved along the integral curve of X_H. The hamiltonian formalism provides deep insights into the details of any mechanical system, even when the number of degrees of freedom is continuum-infinite, as they are in Field Theory. We are here confining ourselves to particle mechanics, but those willing to see an example of that kind can take a look at Section 32.2.3, where the formalism is applied to gauge fields.

25.2 Symplectic structure

Equations (25.3) and (25.4) suggest an intimate relationship between dH and X_H. In reality, a special structure is always present on phase spaces which is responsible for a general relation between vector fields and 1-forms on M. In effect, consider the 2-form

$$\Omega = dq^i \wedge dp_i. \tag{25.5}$$

It is clearly closed, and it can be shown to be also nondegenerate. The interior product $i_{X_H}\Omega$ is just

$$i_{X_H}\Omega = dH. \tag{25.6}$$

Through the interior product, the form Ω establishes a one-to-one relation between vectors and covectors on M.

Recall that a metric structure (see Section 6.6), defined on a manifold by a second-order symmetric, nondegenerate tensor, establishes a one-to-one relation between fields and 1-forms on the manifold, which is then called a "metric space". In an analogous way, the form Ω (which is an antisymmetric, nondegenerate second-order tensor) gives a one-to-one relation between fields and 1-forms on M:

$$X \leftrightarrow i_X\Omega. \tag{25.7}$$

This is a mere analogy — the structure defined by a closed 2-form differs deeply from a metric structure.

Formally, a *symplectic structure*, or *hamiltonian structure*,[1] is defined on a manifold M by any closed (not necessarily exact) nondegenerate 2-form Ω. The manifold M endowed with such a structure is a *symplectic*

---

[1] About this terminology, see Section 26.1.1.

manifold. The 2-cocycle Ω is the *symplectic form*. In the above case, it is also an exact form (a coboundary, or a trivial cocycle) as it is, up to a sign, the differential of the *canonical form*, or *Liouville form*[2]

$$\sigma = p_i \, dq^i. \tag{25.8}$$

With a clear introduction of the important notions in mind, we have actually been taking as a model for M the simplest example of phase space, the topologically trivial space $\mathbb{E}^{2n}$. Though a very particular example of symplectic manifold, the pair $(\mathbb{E}^{2n}, \Omega)$ is enough to model any case in which the configuration space is a vector space. It is good to have in mind, however, that many familiar systems have non-trivial phase spaces: for the mathematical pendulum, for example, it is a cylinder. In such cases, there may be no global coordinates such as the (q^i, p_i) supposed above. On generic, topologically non-trivial symplectic manifolds, not only global coordinates are absent, but the basic closed nondegenerate 2-form is not exact. Furthermore, in general, a vector field like X is not well defined on every point of M. In fact, it is generally supposed that H has at least one minimum, a critical point at which $dH = 0$. Comparing (25.3) and (25.4), we see that the field X_H vanishes at such a point: it has a singular point. Finally, integral curves of a given vector field are in general only locally extant and unique. Nevertheless, a theorem by Darboux ensures that

> *around any point on a $(2n)$-dimensional manifold M there exists a chart of "canonical", or "symplectic" coordinates (q, p) in which a closed nondegenerate 2-form Ω can be written as $\Omega = dq^i \wedge dp_i$.*

Consequently the above description, though sound for $M = \mathbb{E}^{2n}$, is in general valid only locally.

Commentary 25.1 Spaces of 1-forms defined on Lie groups have natural global symplectic structures (Section 27.1). ◄

25.3 Time evolution

To a contravariant field X, which is locally written as

$$X = X_{p_i} \frac{\partial}{\partial p_i} + X_{q^i} \frac{\partial}{\partial q^i}, \tag{25.9}$$

will correspond the covariant field

$$i_X \Omega = X_{q^i} dp_i - X_{p_i} dq^i. \tag{25.10}$$

---

[2] "Canonical" is a word as much abused in mathematics as in religion, and we shall profit to do homage to Liouville in the following.

Applying X_H to any given differentiable function $F(q, p)$ on M, we find that

$$X_H F = \frac{\partial F}{\partial q^i} \frac{\partial H}{\partial p_i} - \frac{\partial F}{\partial p_i} \frac{\partial H}{\partial q^i}$$

$$= \frac{\partial F}{\partial q^i} \frac{dq^i}{dt} + \frac{\partial F}{\partial p_i} \frac{dp_i}{dt} = \frac{dF}{dt}. \tag{25.11}$$

This expression says that

$$X_H F = \{F, H\} = \frac{\partial F}{\partial q^i} \frac{\partial H}{\partial p_i} - \frac{\partial F}{\partial p_i} \frac{\partial H}{\partial q^i}, \tag{25.12}$$

the *Poisson bracket* of F and H, and that the equation of motion is

$$\frac{dF}{dt} = X_H F = \{F, H\}, \tag{25.13}$$

that is to say, the*Liouville equation*. The field X_H "flows the function along time": this is precisely the role of a generator of infinitesimal transformations (section 6.4).

The operator X_H is known in Statistical Mechanics as the Liouville operator, or *liouvillian*. Functions like $F(q, p)$ are the classical *observables*, or *dynamical functions*. The hamiltonian function presides over the time evolution of the physical system under consideration: we shall say that $H(q, p)$ is the *generating function* of the field X_H. The time evolution of a dynamical quantity $F(q, p)$ is given by the solution of equation (25.13),

$$F(t) = F[q(t), p(t)] = e^{tX_H} F(0) = F(0) + t X_H F(0) + \frac{t^2}{2!} X_H X_H F(0) + \dots$$

$$= F(0) + t\{F(0), H\} + \frac{1}{2!} t^2 \{\{F(0), H\}, H\} + \dots$$

$$= F[e^{tX_H} q(0), e^{tX_H} p(0)]. \tag{25.14}$$

This is a purely formal expression, obtained by carelessly rearranging the series without checking its convergence. F is an *integral of motion* if its Lie derivative $L_{X_H} F = X_H F$ vanishes, or

$$\{F, H\} = 0.$$

The Lie derivative of Ω with respect to X_H vanishes,

$$L_{X_H} \Omega = 0, \tag{25.15}$$

because

$$L_X = d \circ i_X + i_X \circ d.$$

This means that the 2-form Ω is preserved by the hamiltonian flow, or by the time evolution. If M is two-dimensional, Ω is the volume form and its preservation is simply the Liouville's theorem for one degree of freedom. For $(2n)$-dimensional M, the property

$$L_X(\alpha \wedge \beta) = (L_X\alpha) \wedge \beta + \alpha \wedge (L_X\beta) \tag{25.16}$$

of the Lie derivative establishes with (25.15) the invariance of the whole series of Poincaré invariants $\Omega \wedge \Omega \wedge \Omega \wedge \ldots \wedge \Omega$, including that one with the number n of Ω's, which is proportional to the volume form of M:

$$\Omega^n = (-)^n dq^1 \wedge dq^2 \wedge \ldots \wedge dq^n \wedge dp_1 \wedge dp_2 \wedge \ldots \wedge dp_n. \tag{25.17}$$

The preservation of Ω^n by the hamiltonian flow is the general Liouville theorem. Dissipative systems will violate it.

Commentary 25.2 Another structure is usually introduced on the manifold M: an euclidean metric $(,)$ which, applied to two fields X and Y written in the manner of (25.9), is written

$$(X, Y) = X_{q^i} Y_{q^i} + X_{p_i} Y_{p_i}.$$

Amongst all the transformations on the $2n$-dimensional space M preserving this structure (that is, amongst all the isometries), those which are linear constitute a group, thef orthogonal group $O(2n)$. ◄

25.4 Canonical transformations

The properties of the hamiltonian function and its related evolution field are generalized as follows. A field X is a *hamiltonian field* if the form Ω is preserved by the local transformations generated by X. This is to say that

$$L_X\Omega = 0. \tag{25.18}$$

Such transformations leaving Ω invariant are the *canonical transformations*. When $n = 1$, as [25.5] is the phase space area form, canonical transformations appear as area-preserving diffeomorphisms.[3] The 1-form corresponding to such a field will be closed because, Ω being closed,

$$d(i_X\Omega) = L_X\Omega = 0. \tag{25.19}$$

If $i_X\Omega$ is also exact, so that

$$i_X\Omega = dF \tag{25.20}$$

for some $F(q, p)$, then X is said to be a *strictly hamiltonian field*[4] (or *globally hamiltonian field*) and F is its *generating function*. In a more

---

[3] Arnold 1966. Area-preserving diffeomorphisms are more general, as they may act on non-symplectic manifolds.

[4] We follow here the terminology of Kirillov 1974.

usual language, F is the generating function of the corresponding canonical transformation. In reality, most hamiltonian fields do not correspond to a generating function at all, as $i_X \Omega$ is not exact in most cases — a generating function exists only locally. On the other hand, as a closed form is always locally exact, around any point of M there is a neighbourhood where some $F(q, p)$ does satisfy $i_X \Omega = dF$.

Notice that any field X_F related to some dynamical function F by (25.20) automatically fulfills

$$L_{X_F} \Omega = 0. \tag{25.21}$$

This happens because

$$L_{X_F} \Omega = d \circ i_{X_F} \Omega + i_{X_F} \circ d\Omega = d^2 F = 0.$$

The one-to-one relationship established by Ω allows one to adopt for fields the same language used for forms. Any dynamical function $F(q, p)$ will correspond to a strictly hamiltonian field X_F by

$$i_{X_F} \Omega = dF.$$

We may say that X_F is exact. In this adapted language, every field is closed.

Suppose now we have a field X satisfying the hamiltonian condition $L_X \Omega = 0$ at some point of M, but such that $i_X \Omega$ is not exact. If we force the existence of a generating function F beyond its local natural validity, it will be a multivalued function and the corresponding canonical transformation will not be unique. We could talk in this case of *non-integrable canonical transformations*. Of course, when the first real homology group $H_1(M, \mathbb{R})$ is trivial, every hamiltonian field will be exact.

Let us consider strictly hamiltonian fields. The dynamical function F generates the strictly hamiltonian field

$$X_F = \frac{\partial F}{\partial p_i} \frac{\partial}{\partial q^i} - \frac{\partial F}{\partial q^i} \frac{\partial}{\partial p_i}. \tag{25.22}$$

Given two functions F and G, their Poisson bracket will have several different though equivalent expressions:

$$\begin{aligned}
\{F, G\} &= \frac{\partial F}{\partial q^i} \frac{\partial G}{\partial p_i} - \frac{\partial F}{\partial p_i} \frac{\partial G}{\partial q^i} \\
&= \Omega(X_F, X_G) \\
&= -X_F(G) = X_G(F) = -i_{X_F} i_{X_G} \Omega \\
&= dF(X_G) = -dG(X_F).
\end{aligned}$$

The simplest examples of generating functions are given by

$$F(q, p) = q^i,$$

corresponding to the field

$$X_F = -\frac{\partial}{\partial p_i},$$

and $G(q, p) = p_i$, whose field is

$$X_G = \frac{\partial}{\partial q^i}.$$

They lead to

$$\{q^i, p_k\} = \delta^i_k.$$

Next in simplicity are the dynamical functions of the type

$$f_{ab} = aq + bp, \tag{25.23}$$

with a, b real constants. The corresponding fields are

$$J_{ab} = -a\frac{\partial}{\partial p} + b\frac{\partial}{\partial q}.$$

The commutator of two such fields is

$$[J_{ab}, J_{cd}] = 0,$$

and consequently the corresponding generating function

$$F_{[J_{ab}, J_{cd}]} = F_0$$

is a constant. The Poisson brackets are the determinants

$$\{f_{ab}, f_{cd}\} = \Omega(J_{ab}, J_{cd}) = ad - bc. \tag{25.24}$$

Each dynamical function G will generate canonical transformations in a way analogous to the time evolution [25.13]. The field X_G will be the infinitesimal generator of the corresponding local one-parameter group, the transformations taking place "along" its local integral curve. Under a transformation generated by G, another observable F will change according to

$$\frac{dF}{dr} = X_G F = \{F, G\}, \tag{25.25}$$

r being the parameter along the integral curve of X_G. The formal solution of this equation is alike to [25.14],

$$F[q(r), p(r)] = e^{rX_G}F(0) = F(0) + rX_GF(0) + \frac{1}{2!}r^2X_GX_GF(0) + \dots$$
$$= F(0) + r\{F(0), G\} + \frac{1}{2!}r^2\{\{F(0), G\}, G\} + \dots$$
$$= F[e^{rX_G}q(0), e^{rX_G}p(0)]. \tag{25.26}$$

We should furthermore insist on its local character: it has a meaning only as long as X_G has a unique integral curve. As long as this holds, the fields X_G extend the notion of liouvillian to generators of general canonical transformations.

Let us sum it up: to a contravariant field X like [25.9],

$$X = X_{p_i} \frac{\partial}{\partial p_i} + X_{q^i} \frac{\partial}{\partial q^i},$$

Ω will make to correspond the covariant field

$$i_X \Omega = X_{q^i} dp_i - X_{p_i} dq^i. \tag{25.27}$$

Suppose another field is given,

$$Y = Y_{p_i} \frac{\partial}{\partial p_i} + Y_{q^i} \frac{\partial}{\partial q^i}.$$

The action of the 2-form Ω on X and Y will give

$$\Omega(X,Y) = X_{q^i} Y_{p_i} - X_{p_i} Y_{q^i}. \tag{25.28}$$

This is twice the area of the triangle defined on M by X and Y, as it is easy to see in the example given by eqs.[25.23, 25.24].

Commentary 25.3   A free particle in $\mathbb{E}^3$ will be described by a phase space $\mathbb{E}^6$, with q^i and $p_i^{(0)} = mv^i$ as Darboux coordinates. The symplectic form is simply

$$\Omega^{(0)} = dq^i \wedge dp_i^{(0)}.$$

A charged particle will have as conjugate momentum components

$$p_i = p_i^{(0)} - \frac{e}{c} A_i(q)$$

and the consequent symplectic form is

$$\Omega = \Omega^{(0)} + \frac{e}{2c} F_{ij} \, dq^i \wedge dq^j.$$

Notice that the condition $dF = 0$ is essential for Ω to be closed. ◀

25.5   Phase spaces as bundles

As said above, phase spaces are very particular cases of symplectic manifolds.[5] Not all symplectic manifolds are phase spaces. Think for instance of the sphere S^2. The area form endows S^2 with a symplectic structure which does not correspond to a phase space. Phase spaces are the cotangent bundles (§ 6.4.7) of the configuration spaces. As such, they have, for each point in configuration space, a non-compact subspace, the cotangent space,

---

[5] See for instance Godbillon 1969.

dual and isomorphic to the tangent space. Consequently, phase spaces are always non-compact spaces, which excludes, for instance, all the spheres. Actually, the cotangent bundle T^*N of any differentiable manifold N has a natural symplectic structure. Forms defined on T^*N are members of the space $T^*(T^*N)$. Now, it so happens that $T^*(T^*N)$ has a "canonical element" σ which is such that, given any section $s : N \to T^*N$, the relation $s^*\sigma = s$ holds.

Recall now how we have introduced topology and charts (see for example § 6.4.3) on a tensor bundle. Take a chart (U, q) on N with coordinates $q^1, q^2, \ldots, q^n$. Let π be the natural projection

$$\pi : T^*N \to N, \quad \pi : T_q^*N \to q \in N.$$

Then, on $\pi^{-1}(U)$ in T^*N a chart is given by $(q^1, q^2, \ldots, q^n, p_1, p_2, \ldots, p_n)$. These coordinates represent on T^*N the Pfaffian form, which in the natural basis on U is written

$$\sigma_U = p_i dq^i. \tag{25.29}$$

We recognize here the Liouville form [25.8]. The differentiable structure allows the extension of σ_U to the whole T^*N, giving a 1-form σ (the *Liouville form* of the cotangent bundle) which reduces to σ_U in $\pi^{-1}(U)$, to some σ_V in $\pi^{-1}(V)$, etc. Define then the 2-form

$$\Omega = - d\sigma.$$

It will be closed and nondegenerate and, in the chart (U, q), it will be given by expression [25.5]. Such a 2-form establishes a bijection between $T_q N$ and T_q^*N at each point $q \in N$. To each $X \in TN$ corresponds $i_X\Omega \in T^*N$.

A particularly simple case appears when the infinitesimal transformations preserve the Liouville form σ itself, that is, when

$$L_X\sigma = 0. \tag{25.30}$$

The transformations are automatically canonical, as

$$L_X\Omega = - L_X d\,\sigma = - d\,L_X\sigma = 0.$$

It follows from

$$L_X\sigma = i_X[d\sigma] + d[i_X\sigma] = 0$$

that

$$i_X\Omega = d[\sigma(X)].$$

A good example is the angular momentum on the euclidean plane $\mathbb{E}^2$: the rotation generator

$$X = q^1 \frac{\partial}{\partial q^2} - q^2 \frac{\partial}{\partial q^1}$$

is related to an integral of motion if $X(H) = 0$. The integral of motion will then be

$$\sigma(X) = q^1 p_2 - q^2 p_1.$$

The same holds for the linear momenta p_1 and p_2, generating functions for the translation generators $\partial/\partial q^1$ and $\partial/\partial q^2$. If we calculate directly the Poisson brackets, we find

$$\{q^1 p_2 - q^2 p_1, p_1\} = p_2, \qquad \{q^1 p_2 - q^2 p_1, p_2\} = -p_1,$$

and

$$\{q^1 p_2 - q^2 p_1, q^1\} = q^2, \qquad \{q^1 p_2 - q^2 p_1, q^2\} = -q^1.$$

We see that they reproduce the algebra of the plane euclidean group. The field algebra, nevertheless, is

$$\left[X, \frac{\partial}{\partial q^1}\right] = -\frac{\partial}{\partial q^2}; \qquad \left[X, \frac{\partial}{\partial q^2}\right] = \frac{\partial}{\partial q^1}; \qquad \left[X, \frac{\partial}{\partial p_j}\right] = 0.$$

In reality,

$$\Omega(X, -\partial/\partial p_1) = q^2 \quad \text{and} \quad \Omega(X, -\partial/\partial p_2) = -q^1$$

as it should be, but

$$\Omega(X, \frac{\partial}{\partial q^1}) = \Omega(X, \frac{\partial}{\partial q^2}) = 0,$$

so that

$$\Omega(X, \frac{\partial}{\partial q^1}) \neq \{q^1 p_2 - q^2 p_1, p_1\}.$$

Commentary 25.4 This is related to lagrangian manifolds. An n-dimensional subspace Γ of the $2n$-dimensional phase space M is a Lagrange manifold if $\Omega(X, Y) = 0$ for any two vectors X, Y tangent to it. That is, the restriction Ω_Γ of Ω to Γ is zero. Examples are the configuration space itself, or the momentum space. The angular momentum X above is a field on configuration space. One must be careful when Lagrange manifolds are present, since Ω may be degenerate. Of course, canonical transformations preserve Lagrange manifolds, that is, they take a Lagrange manifold into another one.

◄

25.6   The algebraic structure

We have been using above the holonomic base $\{\partial/\partial q^i, \partial/\partial p_j\}$ for the vector fields on phase space. In principle, any set of $2n$ linearly independent fields may be taken as a basis. Such a general basis $\{e_i\}$ will have its dual, the base $\{\omega^j\}$ with $\omega^j(e_i) = \delta_i^j$, and its members will have commutators

$$[e_i, e_j] = c^k{}_{ij} e_k,$$

where the structure coefficients $c^k{}_{ij}$ give a measure of the basis anholonomicity. A general field will be written

$$X = X^i e_i = \omega^i(X) e_i,$$

a general 1-form will be

$$\sigma = \sigma_i \omega^i = \sigma(e_i) \omega^i,$$

the differential of a function F will be

$$dF = e_i(F) \omega^i,$$

and so on. The symplectic 2-form will be

$$\Omega = \tfrac{1}{2}\Omega_{ij}\, \omega^i \wedge \omega^j = \tfrac{1}{2}\Omega(e_i, e_j)\, \omega^i \wedge \omega^i. \tag{25.31}$$

Consider the (antisymmetric) matrix $\Omega = (\Omega_{ij})$ and its inverse $\Omega^{-1} = (\Omega^{ij})$, whose existence is identical to the nondegeneracy condition:

$$\Omega_{ij}\, \Omega^{jk} = \Omega^{kj}\, \Omega_{ji} = \delta_i^k. \tag{25.32}$$

As the interior product is that 1-form satisfying

$$i_X\Omega(Y) = \Omega(X, Y)$$

for any field Y, its general expression is

$$i_X\Omega = X^i\Omega_{ij}\, \omega^j. \tag{25.33}$$

The component of a strictly hamiltonian field can then be extracted:

$$X_F^j = e_k(F)\, \Omega^{kj}. \tag{25.34}$$

The Poisson bracket is

$$\{F, G\} = \Omega(X_F, X_G) = X_F^i \Omega_{ij} X_G^j = e_k(G)\, \Omega^{kj} e_j(F). \tag{25.35}$$

This gives the Poisson bracket in terms of the inverse to the symplectic matrix. An interesting case occurs when the Liouville form is preserved by all the basis elements, which are consequently all strictly hamiltonian: from $L_{e_k}\sigma = 0$ it follows that

$$i_{e_k}\Omega = df_k,$$

with $f_k = \sigma(e_k)$. As $\Omega = -d\sigma$, then

$$\Omega_{ij} = -d\sigma(e_i, e_j) = \tfrac{1}{2}\left[e_j(f_i) - e_i(f_j) + c^k{}_{ij}f_k\right].$$

As also $e_i(f_j) = \{f_i, f_j\}$, it follows that

$$\{f_i, f_j\} = c^k{}_{ij}f_k.$$

The Poisson brackets of the generating functions mimic the algebra of the corresponding fields.

If we come back to the Darboux holonomic basis related to the coordinates $\{x^k\} = \{q^i, p_j\}$, the vector base $\{\partial/\partial x^k\}$ will be

$$e_k = \{\partial/\partial q^k\} \quad \text{for} \quad k = 1, 2, \ldots, n$$

and

$$e_k = \{\partial/\partial p_k\} \quad \text{for} \quad k = (n+1), (n+2), \ldots, (2n).$$

The matrices Ω and Ω^{-1} will have the forms

$$\Omega = \begin{pmatrix} \mathbf{0} & I_n \\ -I_n & \mathbf{0} \end{pmatrix} \quad \text{and} \quad \Omega^{-1} = \begin{pmatrix} \mathbf{0} & -I_n \\ I_n & \mathbf{0} \end{pmatrix}, \tag{25.36}$$

where I_n is the n-dimensional unit matrix. In terms of the collected coordinates $\{x^k\}$, Hamilton's equations are then

$$\frac{dx_k}{dt} = \Omega\, dH(e_k). \tag{25.37}$$

Modern approaches frequently *define* the hamiltonian formalism via the Poisson bracket, introduced as

$$\{F, G\} = h^{ij}\, e_i(F)\, e_j(G) \quad \text{with} \quad h^{ij} = -\Omega^{ij}.$$

This is advantageous (see 26.1.1) in the study of the relationship of the Poisson algebra to another Lie algebra of specific interest in a particular problem. Such a relationship may be better seen in some special basis. For example, the configuration space for a rigid body turning around one of its points is the group $SO(3)$ and it is convenient to choose a hamiltonian structure in which the group Lie algebra coincides with the Poisson bracket algebra. This happens if

$$\{M_i, M_j\} = \epsilon_{ijk}M_k,$$

in which case we must choose

$$h^{ij}(x) = \epsilon_{ijk} M_k.$$

This means that h depends only on the coordinates, not on the conjugate momenta. In the holonomic basis related to Darboux coordinates, $h^{ij}(x)$ is a constant. The case above is a special example of a wide class of problems in which $h^{ij}(x)$ is linearly dependent on the x's.[6]

Notice that $\Omega^2 = -\boldsymbol{I}_{2n}$. In terms of the euclidean scalar product

$$(X, Y) = X_{q^i} Y_{q^i} + X_{p_i} Y_{p_i},$$

we see that

$$\Omega(X, Y) = X_{q^i} Y_{p_i} - X_{p_i} Y_{q^i} = X^T \Omega\, Y = (X, \Omega\, Y).$$

A *complex structure* may be introduced by using the complex representation

$$X = (X_q, X_p) \Rightarrow X = X_q + iX_p, \tag{25.38}$$

by which the manifold M becomes locally the complex n-dimensional space $\mathbb{C}^n$. In this representation we see immediately that $\Omega X = iX$, coherently with the above remark that $\Omega^2 = -\boldsymbol{I}$. The linear transformations preserving this complex structure constitute the complex linear group $GL(n, \mathbb{C})$.

Matrix Ω may be seen as a twisted metric, defining a skew-symmetric inner product

$$\langle X, Y \rangle := \Omega(X, Y) = \Omega_{ij} X^i X^j = (X, \Omega\, Y) = -\langle Y, X \rangle,$$

which is quite equivalent to the symplectic structure. Of course, canonical transformations preserve all that. In particular, we may consider *linear canonical transformations*, given by those matrices S preserving the "metric" Ω: $S^{-1}\Omega S = \Omega$. Such linear transformations on the $2n$-dimensional manifold M are the *symplectic transformations* and also constitute a group, the *symplectic group* $Sp(2n)$.

Let us recapitulate: the linear transformations which preserve the euclidean scalar product form the group $O(2n)$; those preserving the symplectic structure constitute the symplectic group $Sp(2n)$; and those preserving the complex structure, the group $GL(n, \mathbb{C})$. Transformations preserving simultaneously these three structures will be in the intersection of the three groups. As it happens,

$$O(2n) \cap Sp(2n) = Sp(2n) \cap GL(n, \mathbb{C}) = GL(n, \mathbb{C}) \cap O(2n) = U(n),$$

the unitary group of the $n \times n$ unitary complex matrices. Summarizing, the unitary transformations preserve the hermitian scalar product

$$H(X, Y) = (X, Y) + i(X, \Omega\, Y). \tag{25.39}$$

---

[6] Novikov 1982.

Commentary 25.5 A short additional comment on the lagrangian manifolds of Commentary 25.4. Consider a simple phase space $M = \mathbb{E}^{2n}$. Given an n-plane $\mathbb{E}^n$ in M, it is called a *lagrangian plane* if, for any two vectors X and Y in $\mathbb{E}^n$,

$$\Omega(X, Y) = 0.$$

An alternative definition of lagrangian manifold is just as follows: any n-dimensional submanifold of M whose tangent spaces are all lagrangian planes. An interesting point is that the set of all such planes is itself a manifold, called the *lagrangian grassmannian* $\Lambda(n)$ of M. This manifold has important topological characteristics:

$$H_1(\Lambda(n), \mathbb{Z}) \approx H^1(\Lambda(n), \mathbb{Z}) \approx \pi_1(\Lambda(n), \mathbb{Z}) \approx \mathbb{Z}.$$

Actually, it happens that $\Lambda(n) = U(n)/O(n)$. The group $U(n)$ acts transitively on $\Lambda(n)$ and makes here a rare intrusion in Classical Mechanics. ◄

25.7 Relations between Lie algebras

The Poisson bracket is antisymmetric and satisfies the Jacobi identity. It is an operation defined on the space $C^\infty(M, \mathbb{R})$ of real differentiable functions on M. Consequently, $C^\infty(M, \mathbb{R})$ is an infinite-dimensional Lie algebra with the operation defined by the Poisson bracket. Actually, $F \to X_F$ is a Lie algebra homomorphism (that is, a representation, see Section 18.2) of $C^\infty(M, \mathbb{R})$ into the algebra of strictly hamiltonian fields on M. Let us now say a few words on the relations between these two Lie algebras.

Unless it is an isomorphism (a "faithful" representation), a homomorphism such as the above $F \to X_F$ loses information. The connection between commutators of fields and the Poisson brackets of the corresponding generating functions is not immediate. This may be guessed from the most trivial one-dimensional case:

$$F = q, \quad G = p, \quad X_F = -\partial/\partial p, \quad X_G = \partial/\partial q.$$

The commutator $[X_F, X_G]$ vanishes while the corresponding Poisson bracket $\{q, p\} = 1$. Notice, however, that the commutator of two hamiltonian fields is always strictly hamiltonian, as, for any function K,

$$\begin{aligned}
[X_F, X_G]K &= X_F\{K, G\} - X_G\{K, F\} \\
&= \{\{K, G\}, F\} + \{\{F, K\}, G\} \\
&= -\{\{G, F\}, K\} \\
&= -X_{\{F,G\}} K.
\end{aligned} \tag{25.40}$$

Therefore,

$$[X_F, X_G] = -X_{\{F,G\}} \tag{25.41}$$

and

$$[X_F, X_G]K = d\{F, G\}(X_K).\tag{25.42}$$

This means that

$$dF_{[X,Y]} = d\{F_X, F_Y\},\tag{25.43}$$

which can be alternatively obtained by using the identity

$$i_{[X,Y]} = L_X i_Y - i_X L_Y.$$

It then follows that

$$F_{[X,Y]} = \{F_X, F_Y\} + \omega(X, Y),\tag{25.44}$$

$\omega(X, Y)$ being a constant (as $d\omega(X, Y) = 0$) depending antisymmetrically on the two argument fields. Such a constant is thus typically the effect of applying a 2-form ω on X and Y. Unless $\omega(X, Y)$ vanishes, the generating function corresponding to the commutator is not the Poisson bracket of the corresponding generating functions. Application of the Jacobi identity to both sides of (25.44), in a way analogous to the above reasoning concerning Ω, shows that ω must be a closed form. The appearance of a cocycle like ω is rather typical of relations between distinct "representations". It would be better to use the word "action", as things may become very different from the relationship usually denoted by the word "representation". Because its presence frustrates an anticipation of simple formal algebraic likeness, we might venture to call ω an *anomaly*, a word which became popular for analogous failures in quantization procedures.

The presence of this 2-cocycle is related to the cohomology of the field Lie algebra. Generating functions are defined only up to a constant and the cohomology classes are connected with this freedom of choice. In effect, choose new functions

$$F'_X = F_X + \alpha(X), \ F'_Y = F_Y + \alpha(Y), \ F'_{[X,Y]} = F_{[X,Y]} + \alpha([X, Y]),$$

with $\alpha(X)$, $\alpha(Y)$ and $\alpha([X, Y])$ constants corresponding to the argument fields. Then, (25.44) becomes

$$F'_{[X,Y]} = \{F'_X, F'_Y\} + \omega'(X, Y).$$

where

$$\omega'(X, Y) = \omega(X, Y) + \alpha([X, Y]).$$

Now, α may be seen as a 1-form on the Lie algebra, which gives the constants when applied to the fields. In this particular case, from the general expression for the derivative of a 1-form,

$$2d\alpha(X, Y) = X[\alpha(Y)] - Y[\alpha(X)] - \alpha([X, Y]),$$

we see that

$$\omega'(X,Y) = \omega(X,Y) - 2\,\alpha([X,Y]).$$

Consequently, ω' may be put equal to zero if an α can be found such that $\omega = 2\,d\alpha$. The 2-cocycle ω is then exact, that is, a coboundary.

Summing up: the general relationship between generating functions related to global hamiltonian fields and the generating functions related to their commutators is given by (25.44), with ω a cocycle on the Lie algebra of fields. When ω is also a coboundary, the relationship becomes a direct translation of commutators into Poisson brackets if convenient constants are added to the generating functions. The cocycle ω defines a cohomology class on the field Lie algebra. Field commutators and Poisson brackets are interchangeable only if this class is trivial.[7]

Let us finally comment on the closedness of the symplectic form. Why have we insisted so much that Ω be a cocycle? Using the above relations, we find that

$$\Omega(X,[Y,Z]) = [Y,Z]F_X = -\,\{F_X,\{F_Y,F_Z\}\}. \qquad (25.45)$$

Combined with the general expression for the differential of a 2-form,

$$3!\,d\Omega(X,Y,Z) = X(\Omega(Y,Z)) + Z(\Omega(X,Y)) + Y(\Omega(Z,X))$$
$$+ \Omega(X,[Y,Z]) + \Omega(Z,[X,Y]) + \Omega(Y,[Z,X]),$$

that equation gives

$$3!\,d\Omega(X,Y,Z) = -\,\{F_X,\{F_Y,F_Z\}\}$$
$$-\,\{F_Z,\{F_X,F_Y\}\} - \{F_Y,\{F_Z,F_X\}\} = 0.$$

We see in this way the meaning of the closedness of Ω: it is equivalent to the Jacobi identity for the Poisson bracket.

25.8   Liouville integrability

A hamiltonian system with n degrees of freedom, whose flow is given by the hamiltonian function H, is integrable if there exists a set of n independent integrals of motion $\{F_i(q,p),\ i = 1,2,\ldots,n\}$ in involution, that is, such that

$$\{H,F_i\} = 0 \qquad \text{and} \qquad \{F_i,F_j\} = 0. \qquad (25.46)$$

---

[7] Arnold 1976, Appendix 5.

This is the most widely used formulation of integrability, due to Liouville. Of course, the first equation above declares simply that the F_i's are integrals of motion. The second is the involution condition. Notice that H is not independent of the F_i's. From (25.42) it follows that the corresponding fields commute:

$$[X_i, X_j] = 0.$$

Equation (25.12) will say that $X_i(F_j) = 0$, which is the same as

$$dF_i(X_j) = 0.$$

All this means that there will be n-dimensional integral manifolds tangent to the X_j's, which are furthermore level manifolds $F_k(q, p) = $ constant. As the involution condition is the same as

$$\Omega(X_i, X_j) = 0,$$

such level manifolds are actually lagrangian manifolds.

General references

Highly recommended general texts covering the topics of this chapter are Arnold 1976, Goldstein 1980, the mathematical "Bible" Abraham & Marsden 1978, Arnold, Kozlov & Neishtadt 1988 and McCauley 1997.

Chapter 26

More Mechanics

Basic notions of the Hamilton–Jacobi approach to classical mechanics are described. The Lagrange derivative is introduced to provide an example of "covariant derivative" in standard classical Physics. And the rigid body exhibits, in that same context, deep relationships which can turn up between groups, algebras and dynamics. Some abstract mathematical notions acquire mechanical versions.

26.1   Hamilton–Jacobi

26.1.1   *Hamiltonian structure*

Modern authors prefer to define a hamiltonian structure as follows. Take a space M and the set $R(M)$ of real functions defined on M (we do not fix the degree of differentiability for the moment: it may be $C^\infty(M)$, to fix the ideas). Suppose that $R(M)$ is equipped with a Poisson bracket

$$\{\ ,\ \}_M : R(M) \times R(M) \to R(M),$$

given for any two functions F and G as

$$\{F, G\}_M = h^{ij}(x)\ \partial_i F\ \partial_j G\,, \tag{26.1}$$

where the functions $h^{ij}(x)$ are such that the bracket is antisymmetric and satisfies the Jacobi identity. The space M is then called a *Poisson manifold*, and the bracket is said to endow M with a hamiltonian structure. Of course, $h^{ij}(x)$ is just the inverse symplectic matrix $\Omega^{ij}(x)$ *when that one is invertible* (compare Eq. (26.1) with Eq. (25.35)). The preference given to this definition of hamiltonian structure, based on the bracket, comes from the study of systems for which the matrix $[h^{ij}(x)]$ is itself well-defined but not invertible at every point x.

Consider two Poisson manifolds, M and N. A mapping $f: M \to N$ is a *Poisson mapping* if

$$\{f^*F, f^*G\}_M = \{F, G\}_N \tag{26.2}$$

where f^* is the pullback, $f^*F = F \circ f$. Take two symplectic manifolds (M_1, Ω_1) and (M_2, Ω_2). Call π_1 and π_2 respectively the projections of $M_1 \times M_2$ into M_1 and M_2,

$$\pi_1 : M_1 \times M_2 \to M_1 \quad \text{and} \quad \pi_2 : M_1 \times M_2 \to M_2.$$

Consider also a mapping $f: M_1 \to M_2$, with graph Γ_f. Let further

$$i_f : \Gamma_f \to M_1 \times M_2$$

be the inclusion. Then,[1]

$$\Omega = \pi_1^* \Omega_1 - \pi_2^* \Omega_2$$

is a symplectic form on the product $M_1 \times M_2$. If the inclusion is such that $i_f^* \Omega = 0$, the mapping f is said to be a *symplectic mapping*.

Commentary 26.1 We shall not in the following make any distinction between Poisson and symplectic manifolds and/or mappings. ◄

The graph Γ_f is a *lagrangian submanifold*. Write locally $\Omega = -d\sigma$ (it might be such that $\sigma = \pi_1^*\sigma_1 - \pi_2^*\sigma_2$, but not necessarily). Then,

$$i_f^* d\sigma = d i_f^* \sigma,$$

so that f is symplectic *iff* $i_f^*\sigma$ is closed. In that case, locally,

$$i_f^* \sigma = -dS$$

for some $S: \Gamma_f \to R$. The function S is the generating function for the mapping f. Thus, given f, S is a σ-dependent real function defined on the lagrangian submanifold. If

$$(q^1, q^2, \ldots, q^n, p_1, p_2, \ldots, p_n)$$

are coordinates on M_2 and

$$(Q^1, Q^2, \ldots, Q^n, P_1, P_2, \ldots, P_n)$$

are coordinates on M_1, then there are many ways to chart the graph Γ_f on which the function S is defined. Examples are

$$S(q^1, q^2, \ldots, q^n, Q^1, Q^2, \ldots, Q^n)$$

and

$$S(q^1, q^2, \ldots, q^n, P_1, P_2, \ldots, P_n).$$

Notice that S is only locally defined.

Commentary 26.2 Lagrangian submanifolds are of great import to the semi-classical approximation to Quantum Mechanics. One of their global properties, the Maslov index,[2] appears as the ground state contribution.[3] ◄

---

[1] Abraham & Marsden 1978.

[2] Arnold (Appendix), in Maslov 1972.

[3] Arnold 1976, Appendix 11.

26.1.2 *Hamilton–Jacobi equation*

Take again the same coordinates as above on M_1 and on M_2:[4]

$$f(Q^1, Q^2, \ldots, Q^n, P_1, P_2, \ldots, P_n) = (q^1, q^2, \ldots, q^n, p_1, p_2, \ldots, p_n).$$

If we now consider

$$S(q^1, q^2, \ldots, q^n, Q^1, Q^2, \ldots, Q^n),$$

the expression

$$i_f^* \sigma = -dS \tag{26.3}$$

enforces

$$p_k = \frac{\partial S}{\partial q^k} \quad \text{and} \quad P_k = -\frac{\partial S}{\partial Q^k}.$$

Suppose we find a symplectic mapping f such that S is independent of the Q^j's. Then, S becomes simply a function on the configuration space. In this case all the $P_j = 0$ and the hamiltonian is a constant, $H = E$ or, in the remaining variables,

$$H\left(q^k, \frac{\partial S}{\partial q^k}\right) = E. \tag{26.4}$$

This is the time-independent *Hamilton–Jacobi equation*. A curve $c(t)$ in configuration space such that

$$\frac{dc(t)}{dt} = dS(c(t))$$

is an integral curve of the field X_H. The surfaces $S = $ constant are characteristic surfaces and $c(t)$ is a gradient line of S, orthogonal to them.

The time evolution of a mechanical system of hamiltonian H is given by the Liouville field

$$X_H = \frac{\partial H}{\partial p_i} \frac{\partial}{\partial q^i} - \frac{\partial H}{\partial q^i} \frac{\partial}{\partial p_i}.$$

The hamiltonian flow is precisely the one-parameter group generated by X_H. Any dynamical function $F(q, p, t)$ will evolve according to the equation of motion

$$\frac{dF}{dt} = \{F, H\} = X_H F,$$

the Liouville equation Eq. (25.13).

---

[4] See Babelon & Viallet 1989.

Let us consider this equation in some more detail: in terms of the coordinates

$$\{x^k\} = \{q^1, q^2, \ldots, q^n, p_1, p_2, \ldots, p_n\},$$

it is a partial differential equation

$$\frac{d}{dt} F(q, p, t) = \sum_{i=1}^{2n} X_H^i(x) \frac{\partial}{\partial x^i} F(x, t) , \tag{26.5}$$

to be solved with some given initial condition

$$F(q, p, 0) = f_0(q, p) . \tag{26.6}$$

If $F_t(x)$ is a flow, the solution will be $F(x, t) = f_o(F_t(x))$. The orbits of the vector field X_H are the characteristics of the above differential equation. Notice that they will fix the evolution of *any* function. The curve solving Hamilton equations (which are ordinary differential equations) in configuration space is the characteristic curve of the solutions of the Hamilton–Jacobi equations (which are partial differential equations). More will be said on characteristics in Section 29.1.

A famous application comes out in Quantum Mechanics, where S appears as the phase of the wavefunction:

$$\Psi = \exp[iS/\hbar].$$

Then, with

$$H(q, p) = \frac{p^2}{2m} + V(q),$$

the Schrödinger equation

$$-\frac{\hbar^2}{2m} \Delta\Psi + V\Psi = E\,\Psi$$

leads to the "quantum-corrected" Hamilton–Jacobi equation

$$\frac{1}{2m} (\nabla S)^2 + V = E + \frac{i\hbar}{2m} \nabla^2 S , \tag{26.7}$$

where the last term is purely quantal.

Mathematicians tend to call "Hamilton–Jacobi equation" any equation of the type $H(x, \psi_x) = 0$, that is, any equation in which the dependent quantity f does not appear explicitly.[5] The eikonal equation

$$\left(\frac{\partial\psi}{\partial x^1}\right)^2 + \left(\frac{\partial\psi}{\partial x^2}\right)^2 + \ldots + \left(\frac{\partial\psi}{\partial x^n}\right)^2 = 1 \tag{26.8}$$

---

[5] Arnold 1980.

is, of course, an example (see Section 29.4). Function ψ is the optical length, and the level surfaces of ψ are the wave fronts. The characteristics of $H(x, \psi_x) = 0$ obey differential equations which are just the Hamilton equations

$$\dot{x} = \frac{\partial H}{\partial p} \quad \text{and} \quad \dot{p} = -\frac{\partial H}{\partial x} \, .$$

The projections of the trajectories on the x-space are the rays. Given a hypersurface Σ in $\mathbb{E}^n$, define $\psi(x)$ as the distance of the point x to Σ. Then ψ satisfies the eikonal equation. Actually, any solution of this equation is, locally and up to an additive constant, the distance of x to some hypersurface.

26.2 The Lagrange derivative

We shall not really examine the lagrangian formalism, which is summarized in other sections, such as 20.2 and 31.4, and furthermore pervades many other chapters. The intention here is only to present the Lagrange derivative of Classical Mechanics as an example of "covariant derivative".

26.2.1 *The Lagrange derivative as a covariant derivative*

The configuration space M is the space spanned by the values of the degrees of freedom. Its points are described by a coordinate set $x = \{x^k\}$, one x^k for each degree. The velocity space is described accordingly by $\dot{x} = \{\dot{x}^k\}$, where

$$\dot{x}^k = \frac{dx^k}{dt} \, .$$

The lagrangian function $L(x, \dot{x}, t)$ is defined on the combined configuration-velocity space, which is actually the tangent bundle $T(M)$ of M. The extremals of $L(x, \dot{x}, t)$ are curves $\gamma(t)$ satisfying the Euler–Lagrange equations, or equations of motion.

Commentary 26.3 That the combined space is $T(M)$ is a simplicity assumption. No system has been exhibited where this hypothesis has been found wanting. There is no problem in identifying the point set, but the supposition means also that there is a projection, with the charts of the configuration space being related to those of the combined space, etc. ◀

Commentary 26.4 When the configuration space is non-trivial, that is, when the degrees of freedom take values in a non-euclidean space, many local systems of coordinates may be necessary to cover it, but the lagrangian function should be independent of the number and choice of the charts. ◄

Consider a time-independent change of coordinates in configuration space,

$$x^j \rightarrow y^k = y^k(x^j),$$

which has the following properties:

(i) Is invertible, that is, we can find $x^j = x^j(y^k)$.
(ii) Takes time as absolute, so that the velocities simply follow the configuration transformation, $\dot{y}^k = dy^k/dt$.
(iii) Leaves invariant the lagrangian function: $L(x, \dot{x}, t) = L(y, \dot{y}, t)$.

In the lagrangian formalism, the coordinates of the configuration and velocity spaces are independent in each chart but, once a coordinate transformation is performed, the new velocities may depend on (say) the old coordinates. A first important thing is that, despite the highly arbitrary character of the transformation, the velocities (tangent vectors) are linearly transformed:

$$\dot{y}^k = \frac{dy^k}{dt} = \frac{\partial y^k}{\partial x^j} \dot{x}^j.$$

This simply says that they are indeed vectors under coordinate transformations on the configuration space. Notice that

$$\frac{\partial \dot{x}^n}{\partial \dot{y}^j} = \frac{\partial x^n}{\partial y^j}. \tag{26.9}$$

From

$$\dot{x}^k = \frac{\partial x^k}{\partial y^n} \dot{y}^n$$

it follows that

$$\frac{\partial \dot{x}^k}{\partial y^j} = \frac{\partial^2 x^k}{\partial y^j \partial y^n} \dot{y}^n. \tag{26.10}$$

The Euler–Lagrange equations are conditions fixing the extrema of L, so that it is necessary to examine $\partial L/\partial x^j$. If L were a simple function of the coordinates, the derivative would be a vector and all should be well, but the velocity dependence embroils the things. Notice first that

$$\frac{\partial L}{\partial \dot{y}^j} = \frac{\partial L}{\partial \dot{x}^n} \frac{\partial \dot{x}^n}{\partial \dot{y}^j} = \frac{\partial L}{\partial \dot{x}^n} \frac{\partial x^n}{\partial y^j}. \tag{26.11}$$

This shows that the conjugate momentum

$$p_n = \frac{\partial L}{\partial \dot{x}^n}$$

is also a good "vector", though in a converse way: it transforms by the inverse of the matrix transforming the velocity. The velocity is a contravariant vector, or simply vector. The "converse" behaviour shows that the momentum is a covariant vector, or covector. Anyhow, it behaves covariantly under coordinate transformations on the configuration space. Its time-derivative, however, does not:

$$\frac{d}{dt}\frac{\partial L}{\partial \dot{y}^j} = \left[\frac{d}{dt}\frac{\partial L}{\partial \dot{x}^n}\right]\frac{\partial \dot{x}^n}{\partial \dot{y}^j} + \frac{\partial L}{\partial \dot{x}^k}\frac{\partial^2 x^k}{\partial y^n \partial y^j}\, \dot{y}^n. \tag{26.12}$$

The first term on the right-hand side would be all right, but the second represents a serious deviation from tensorial behaviour. As to $\partial L/\partial x^j$ itself, it is ill-behaved from the start:

$$\frac{\partial L}{\partial y^j} = \frac{\partial L}{\partial x^n}\frac{\partial x^n}{\partial y^j} + \frac{\partial L}{\partial \dot{x}^k}\frac{\partial \dot{x}^k}{\partial y^j} = \frac{\partial L}{\partial x^n}\frac{\partial x^n}{\partial y^j} + \frac{\partial L}{\partial \dot{x}^k}\frac{\partial^2 x^k}{\partial y^j \partial y^n}\, \dot{y}^n. \tag{26.13}$$

The same offending term of (26.12) turns up, and we conclude that

$$\frac{\partial L}{\partial y^j} - \frac{d}{dt}\frac{\partial L}{\partial \dot{y}^j} = \frac{\partial x^k}{\partial y^j}\left[\frac{\partial L}{\partial x^k} - \frac{d}{dt}\frac{\partial L}{\partial \dot{x}^k}\right]. \tag{26.14}$$

The operator acting on L has the formal properties of a derivative: it is linear and obeys the Leibniz rule. This modified derivative is the Lagrange derivative. Unlike the usual derivative, it gives a well-behaved, tensorial object under coordinate changes in configuration space. The terms

$$\frac{d}{dt}\frac{\partial L}{\partial \dot{y}^j} \quad \text{and} \quad \frac{d}{dt}\frac{\partial L}{\partial \dot{x}^k}$$

are compensating terms, alike to the compensating contributions of the gauge fields or connections in their covariant derivatives.

Actually, we have not used any characteristic of L as a lagrangian. The above results say simply that derivatives on configuration space of any function $F(x, \dot{x})$ on the combined configuration-velocity space (which is the tangent bundle of the configuration space) must be supplemented with an extra term in order to have a coordinate-independent meaning. The "good", coordinate-independent derivative of any function is thus the Lagrange derivative

$$\frac{\delta}{\delta x^k} := \frac{\partial}{\partial x^k} - \frac{d}{dt}\frac{\partial}{\partial \dot{x}^k}. \tag{26.15}$$

The reasoning remains true for higher-order lagrangians, dependent on the acceleration, second acceleration, etc. The Lagrange derivative must be accordingly adapted, the compensating terms being then an alternate sum of higher-order contributions:

$$\frac{\delta}{\delta x^k} := \frac{\partial}{\partial x^k} - \frac{d}{dt}\frac{\partial}{\partial \dot{x}^k} + \frac{d^2}{dt^2}\frac{\partial}{\partial \ddot{x}^k} - \frac{d^3}{dt^3}\frac{\partial}{\partial \dddot{x}^k} + \cdots. \tag{26.16}$$

Commentary 26.5 The Lagrange derivative is not invariant under a general transformation in velocity space. Notice that, in the simple $L = L(x, \dot{x})$ case, the non-covariant compensating term

$$\frac{d}{dt}\frac{\partial L}{\partial \dot{x}^k}$$

is just dp^k/dt, the newtonian expression for the force. That force has, consequently, no coordinate-independent meaning. ◄

The complete Lagrange derivative is essential to obtain consistent expressions[6] for the force Q_i coming from velocity-dependent potentials $V(q, \dot{q})$, such as the Lorentz force of electrodynamics. The "generalized force", appearing in the expression

$$W = \sum_i Q_i dq^i$$

for the work, is

$$Q_k := - \frac{\delta V}{\delta q^k} = - \frac{\partial V}{\partial q^k} + \frac{d}{dt}\frac{\partial V}{\partial \dot{q}^k}. \tag{26.17}$$

In this case, with $L = T - V$, the Lagrange equations of motion retain the form

$$\frac{\delta L}{\delta q^k} = 0. \tag{26.18}$$

An example is Weber's law[7] of attraction, which comes from

$$V = \frac{1}{r}\left[1 + \frac{\dot{r}^2}{c^2}\right]$$

and leads to the complete Ampère's law. The same holds for the conjugate momentum, if L depends on $\ddot{q}$.

The degrees of freedom x^k are indexed here by the discrete labels "k". When there is a continuum of degrees of freedom, each degree becomes a function ϕ of the labels, which are then indicated (say) by "x". Each degree is then a "field" $\phi(x)$. Systems with a continuous infinity of degrees of freedom are discussed, for example, in Sections 20.2 and 31.4.

---

[6] Whittaker 1944, p. 44 on.
[7] Whittaker 1953, p. 226 on.

Commentary 26.6 We have been cheating a little. Recall that "t" is not necessarily "time": it is actually a curve parameter, and all the above derivatives are concerned with points on a curve. What really happens is that the Lagrange derivative is the good derivative on a functional space, the space of functionals on the space of trajectories. A well-known example of such a functional is the action functional (see Section 20.2.3). ◄

In a system with several degrees of freedom, the individual Euler–Lagrange equations are not necessarily invariant under coordinate transformations. It is their set which is invariant. Each equation is said to be "covariant", not "invariant". This reminds us of the components of vector fields and forms: components are covariant, though vector and covector fields are invariant. In effect, the Euler–Lagrange equations can be assembled in a certain functional differential form, which is invariant and can be defined also for sets of equations which do not come as extremal conditions on a Lagrange function (Section 20.2 and on).

26.3   The rigid body

The study of rigid body motion has many points of great interest:

(i) It gives an example of non-trivial configuration space, which is furthermore a Lie group whose Maurer–Cartan form has a clear physical interpretation.

(ii) It gives a simple example of a metric-induced canonical isomorphism between a vector space and its dual.

(iii) It illustrates the use of moving frames.

(iv) It shows the difference between left- and right-action of a group.

26.3.1   *Frames*

A rigid body is defined as a set of material points in the metric space $\mathbb{E}^3$ such that the distance between any pair of points is fixed. Any set of three non-colinear points on the so defined rigid body defines a frame, in general non-orthogonal, in $\mathbb{E}^3$. This frame attached to the body will be indicated by F'. It is called the "body frame", in oposition to a fixed frame F given *a priori* in the vector space $\mathbb{E}^3$, called the "space frame", or "laboratory frame".

We are taking advantage of the double character of $\mathbb{E}^3$, which is both a manifold and a vector space. Given the position of the three points, the position of any other point of the rigid body will be completely determined.

We say then that the "configuration" is given. Thus, in order to specify the position of any body point, we need only to give the positions of the three points. These would require 9 coordinates, which are however constrained by the 3 conditions freezing their relative distances. Consequently, the position of any point will be given by 6 coordinates.

We may take one of the three points as the origin O of the frame F'. From an arbitrary starting configuration, any other may be obtained by performing two types of transformations: (i) a translation taking O into any other point of $\mathbb{E}^3$, and (ii) a rotation around O. There is consequently a one-to-one correspondence between the set of configurations and the set of these transformations. This set of transformations is actually a 6-dimensional group, denoted by $SO(3) \otimes \mathbb{R}^3$, the direct product of the rotation group $SO(3)$ by the translation group $\mathbb{R}^3$. Thus, the configuration space of a rigid body is (the manifold of) $SO(3) \otimes \mathbb{R}^3$. This manifold may be identified to the tangent bundle $TSO(3)$, which is a direct product because $SO(3)$, as any Lie group, is parallelizable.

This is not to be mistaken by the euclidean group, the group of transformations (motions, or isometries) in our ambient $\mathbb{E}^3$, which is not a direct product: it is the *semi-direct* product $SO(3) \oslash \mathbb{R}^3$, a non-trivial bundle. The difference comes from the fact that, in the latter, rotations and translations do not commute in general.

26.3.2   *The configuration space*

We shall consider the case in which the body has a fixed point, and take that point as the origin O for both frames F and F'. There are no translations anymore: the configuration space of a rigid body moving around a fixed point reduces to $SO(3)$. We are supposing a fixed orientation for the body, otherwise the configuration space would be $O(3)$. As a manifold, $SO(3)$ is a 3-dimensional half-sphere:

$$SO(3) \approx S^3/Z_2.$$

When we use angles as coordinates on $SO(3)$ (say, the Euler angles), the members of the tangent space will be angular velocities. We may go from such starting coordinates to other generalized coordinates, of course. It is of practical interest to identify the frame origin O also to the group identity element. The space tangent to the configuration space at O will then be identifiable to the Lie algebra $so(3)$ of the group, whose generators $\{L_i\}$

obey

$$[L_i, L_j] = \epsilon_{ijk} L_k.$$

As the structure constants are just given by the Kronecker symbol ϵ_{ijk}, the Lie operation coincides with the usual vector product in $\mathbb{E}^3$

$$L_i \times L_j = \epsilon_{ijk} L_k.$$

26.3.3   The phase space

The phase space will be the cotangent bundle $T^*SO(3)$. The members of the cotangent space will be the duals to the angular velocities, that is, the angular momenta. In problems involving rotational symmetry, one frequently starts with phase space coordinates (q^i, p_k) and find the angular momenta

$$M_1 = q^2 p_3 - q^3 p_2, \quad \text{etc,}$$

as conserved quantities. They satisfy the Poisson-bracket algebra

$$\{M_i, M_j\} = \epsilon_{ijk} M_k.$$

This means that the M_k's provide a representation of $so(3)$ (the Lie algebra of the group $SO(3)$) in the Poisson-bracket Lie algebra of all functions on the phase space. In some cases, it is convenient to use the M_k's themselves as generalized coordinates. The Poisson bracket

$$\{F, G\} = h^{ij}(x)\, \partial_i F\, \partial_j G$$

in this case has

$$h^{ij}(x) = \epsilon_{ijk} M^k \quad \text{for} \quad i, j = 1, 2, 3,$$
$$h^{ij}(x) = 1 \quad \text{when} \quad j = i - 3 \quad \text{for } i, j = 4, 5, 6,$$
$$h^{ij}(x) = 0 \quad \text{for all the remaining cases.}$$

Furthermore, the invariant Cartan–Killing metric of $SO(3)$ is constant and euclidean, so that in this metric we are tempted to write $M_i = M^i$. Nevertheless, another metric is present, the moment of inertia I_{ij}. A comparison with the case of optics is helpful here. The Cartan–Killing metric plays the same role played by the euclidean $\mathbb{E}^3$ metric in optics, while the moment of inertia has some analogy to the refractive metric (see Section 30.3 and on).

26.3.4   *Dynamics*

Dynamics is presided over by the hamiltonian

$$H = \tfrac{1}{2} \sum_{ij} I^{ij} M_i M_j, \tag{26.19}$$

where $I^{ij} = (I^{-1})_{ij}$ are the entries of the inverse to the moment of inertia matrix. The hamiltonian H can be diagonalized as

$$H = \tfrac{1}{2} \sum_i a_i M_i^2,$$

and the angular velocity is defined as

$$\omega^k = \partial H / \partial M_k.$$

The Liouville equation becomes

$$\dot{\boldsymbol{M}} = \{\boldsymbol{M}, H\} = \boldsymbol{M} \times \boldsymbol{\omega},$$

which is Euler's equation for the rigid body motion. Actually, all these quantities are referred to the body frame, and will be indexed with "b". Thus, when a rigid body moves freely around a fixed point O, its angular momentum $\boldsymbol{M}_b$ and angular velocity $\boldsymbol{\omega}_b$ with respect to O are related by Euler's equation

$$\frac{d\boldsymbol{M}_b}{dt} + \boldsymbol{\omega}_b \times \boldsymbol{M}_b = 0. \tag{26.20}$$

The same equation follows from the lagrangian

$$L = \sum_i M_i \omega^i - H.$$

Commentary 26.7   With respect to space, the angular momentum $\boldsymbol{M}_s$ satisfies $d\boldsymbol{M}_s/dt = 0$, which expresses the conservation of overall angular momentum. ◀

26.3.5   *The "space" and the "body" derivatives*

The "space" and the "body" derivatives of the components of a vector quantity $\boldsymbol{G}$ are related by

$$\left(\frac{d\boldsymbol{G}}{dt}\right)_s = \left(\frac{d\boldsymbol{G}}{dt}\right)_b + \boldsymbol{\omega}_b \times \boldsymbol{G}. \tag{26.21}$$

An example is the relation between the velocities

$$\boldsymbol{v}_s = \boldsymbol{v}_b + \boldsymbol{\omega}_b \times \boldsymbol{r}.$$

Another is given by the above relationship between the rates of variation of the angular momenta. The linear velocity of a point at position r is given by

$$v_b = \omega_b \times r.$$

The points on the axis instantaneously colinear with the angular velocity ω, given by $r = a\,\omega_b$ for any a, have vanishing velocities. They constitute the *instantaneous axis of rotation*.

26.3.6   The reduced phase space

There are, therefore, 4 integrals of motion: the three components of M and the energy E. The reduced phase space, in which the motion forcibly takes place, will be a 2-dimensional subspace of $T^*SO(3)$, determined by the constraints $M = $ constant and $E = $ constant. This 2-dimensional subspace is the torus T^2. That this is so comes from a series of qualitative considerations:[8]

(i) The subspace admits "global motions", i.e., given the initial conditions, the system will evolve indefinitely along the flow given by a vector field, which is consequently complete, without singularities.

(ii) It is connected, compact and orientable.

(iii) The only 2-dimensional connected, compact and orientable manifolds are the sphere and the multiple toruses with genus $n = 1, 2, \ldots$.

The torus T^2 is the case $n = 1$. In order to have a complete vector field, the manifold must have vanishing Euler number. Here,

$$\chi = b_0 - b_1 + b_2.$$

Connectedness implies $b_0 = 1$, Poincaré duality implies $b_2 = b_0$, so that we must have $b_1 = 2$. The genus is just $b_1/2$, so that we are forced to have $n = 1$. On the torus, we may choose two angular coordinates, α_1 and α_2, and find the equations of motion as

$$\frac{d\alpha_1}{dt} = \omega_1 \quad \text{and} \quad \frac{d\alpha_2}{dt} = \omega_2.$$

This means that the motion of a rigid body with a fixed point can be reduced to two periodic motions with independent, possibly incommensurate frequencies. In the last case, the body never comes back to a given state and we have an example of deterministic chaotic motion (see Comment 1.3.3 and Section 15.2.2).

---

[8] See Arnold 1976.

An Introduction to Geometrical Physics

26.3.7 Moving frames

Let us consider again the two frames F and F' with the same origin O. Take a cartesian system of coordinates in each one. A point will have coordinates $x = (x^1, x^2, \ldots, x^n)$ in F and $x' = (x'^1, x'^2, \ldots, x'^n)$ in F'. Let us first simply consider the motion of an arbitrary particle with respect to both frames. The coordinates will be related by transformations

$$x'^i = A^{ij} x^j \quad \text{and} \quad x^i = (A^{-1})^{ij} x'^j.$$

Compare now the velocities in the two frames. To begin with, the point will have velocity $\boldsymbol{v}_F = \dot{\boldsymbol{x}}$ of components

$$\dot{x}^k = \frac{dx^k}{dt}$$

in the space frame F, and $\boldsymbol{v}'_{F'} = \dot{\boldsymbol{x}}'$ of components

$$\dot{x}'^k = \frac{dx'^k}{dt}$$

in the space frame F'. Here, of course, the absolute character of time (t is the same in both frames) is of fundamental importance. But there is more: we may want to consider the velocity with respect to F as seen from F', and vice-versa. Let us call $\boldsymbol{v}'_F$ the first and $\boldsymbol{v}_{F'}$ the velocity with respect to F' as seen from F. Using the convenient notation

$$v^{(\text{seen from})}_{(\text{with respect to})}$$

let us list the possible velocities:

$\boldsymbol{v}_F = \dot{\boldsymbol{x}} \rightarrow$ velocity with respect to, and seen from, the space frame F.
$\boldsymbol{v}'_F \rightarrow$ velocity with respect to F as seen from F'.
$\boldsymbol{v}_{F'} \rightarrow$ velocity with respect to F' as seen from F.
$\boldsymbol{v}'_{F'} = \dot{\boldsymbol{x}}' \rightarrow$ velocity with respect to, and seen from the rotating frame F'.

As velocities are vectors with respect to coordinates transformations,

$$\boldsymbol{v}'_F = A\,\boldsymbol{v}_F.$$

Also,

$$\boldsymbol{v}'_{F'} = A\,\boldsymbol{v}_{F'} \quad \text{and} \quad \boldsymbol{v}_{F'} = A^{-1}\boldsymbol{v}'_{F'}.$$

But

$$\dot{\boldsymbol{x}}' = \boldsymbol{v}'_{F'} = A\dot{\boldsymbol{x}} + \dot{A}\boldsymbol{x},$$

so that

$$\boldsymbol{v}_{F'} = A^{-1}\boldsymbol{v}'_{F'} = \boldsymbol{v}_F + A^{-1}\dot{A}\,\boldsymbol{x}.$$

It will be useful to write this in components,

$$v_{F'}^k = v_F^k + (A^{-1})^{kj}\frac{dA^{ji}}{dt}x^i \tag{26.22}$$

which means that

$$v_{F'}^k dt = dx^k + (A^{-1})^{kj}dA^{ji}x^i.$$

For a particle belonging to the rigid body, consequently fixed in F', we have that $\boldsymbol{v}'_{F'} = 0$ and $\boldsymbol{v}_{F'} = 0$. Thus,

$$v_F^k = -(A^{-1})^{kj}\dot{A}^{ji}x^i. \tag{26.23}$$

Let us call

$$\omega^{ki} = (A^{-1})^{kj}\dot{A}^{ji} \tag{26.24}$$

the angular velocity tensor. Then we have

$$v_F^k = -\omega^{ki}x^i, \tag{26.25}$$

which is the equation $\boldsymbol{v} = \boldsymbol{\omega} \times \boldsymbol{r}$ when $n = 3$. In this case, the usual relationship of antisymmetric tensors to vectors allows one to define the vector angular velocity ω^j from the tensor angular velocity by $\omega^{ki} = \epsilon^{kij}\omega^j$, or

$$\omega^j = \tfrac{1}{2}\epsilon^{jki}(A^{-1})^{kn}\dot{A}^{ni}. \tag{26.26}$$

The well-known consequence is that matrix action on column vectors turns into vector product. We actually find

$$\boldsymbol{v}_F = \boldsymbol{\omega} \times \boldsymbol{r}.$$

We might invert all the discussion, taking F' as fixed and F as turning. The whole kinematics is equivalent, with only an obvious change of sign in the angular velocity. The same treatment holds, of course, for other vectors under coordinate transformations.

26.3.8 *The rotation group*

Each matrix A taking a vector given in F into the same vector in F' represents a rotation. It is a member of the rotation group $SO(3)$, and the form

$$\Omega = A^{-1}dA$$

is the $SO(3)$ canonical (or Maurer–Cartan) form. The angular velocity tensor is the result of applying this form to the field d/dt, tangent to the particle trajectory,

$$A^{-1}dA\left(\frac{d}{dt}\right) = A^{-1}\frac{dA}{dt}.$$

Thus, the angular velocity is the canonical form "along" the trajectory. But the role of the canonical form is to take any vector field on the group into a vector field at the identity. That is, into the Lie algebra of the group. If

$$A = \exp[\alpha^i J_i],$$

then

$$\Omega = \exp[-\alpha^i J_i]\, d\alpha^k J_k \exp[\alpha^j J_j]$$
$$= \exp[-\alpha^i J_i]\, J_k \exp[\alpha^j J_j]\, d\alpha^k,$$

so that

$$\Omega\,(d/dt) = \exp[-\alpha^i J_i]\, J_k\, \exp[\alpha^j J_j] d\alpha^k\,(d/dt)$$
$$= [(Ad_{A^{-1}})_k{}^j J_j]\,(d\alpha^k/dt)$$
$$= J'_k\, \frac{d\alpha^k}{dt}$$

belongs to the Lie algebra. We see in this way how the angular velocities turn up in the Lie algebra $so(3)$ of $SO(3)$. By the way, the above considerations show also that

$$\Omega = (Ad_{A^{-1}}J)_k\, d\alpha^k = Ad^*_{A^{-1}}(d\alpha).$$

26.3.9 Left- and right-invariant fields

We can transport a tangent vector into the group identity by two other means: left-translation and right-translation. To each position of the body corresponds an element of the group. Take an initial position of the body and identify it (arbitrarily) to the identity element. One obtains every other position by applying group elements. Take some J in the algebra and consider the one-parameter group of elements

$$g(\tau) = \exp[\tau J].$$

As we have seen that angular velocities belong to the Lie algebra, it will be the group of rotations with angular velocity J. Now,

$$\dot{g} = \left[\frac{d}{d\tau}\exp[\tau J]\right]_{\tau=0} g = J\,g$$

is a tangent vector, and we see that $J = \dot{g}\,g^{-1}$, that is, the angular velocity J is obtained by right-translation. Another angular velocity is obtained by left-translation. The first is identified to the "space" angular velocity, and the latter to the "body" angular velocity. Let us see how it happens.

A point in configuration space is a point of the group manifold. Let us try to use both the differential and the group structure simultaneously. If $g \in SO(3)$, then

$$\Omega_L = g^{-1}dg$$

is left-invariant (it is just the Maurer–Cartan canonical form Ω) and

$$\Omega_R = dgg^{-1}$$

is right invariant. Direct calculations show that:

$$\Omega_R = g\,\Omega_L\,g^{-1} = Ad_g^*(\Omega_L), \tag{26.27}$$

$$\Omega_L = g^{-1}\Omega_R\,g = Ad_{g^{-1}}^*(\Omega_R), \tag{26.28}$$

$$d\Omega_L + \Omega_L \wedge \Omega_L = 0, \tag{26.29}$$

and

$$d\Omega_R - \Omega_R \wedge \Omega_R = 0. \tag{26.30}$$

We may use the holonomic "group parameter basis", writing g in terms of the group generators $\{J_i\}$, that is $g = \exp[\alpha^i J_i]$, to obtain

$$\Omega_R = J_i d\alpha^i$$

and

$$\Omega_L = g^{-1}J_i\,g\,d\alpha^i = (Ad_{g^{-1}}J_i)\,d\alpha^i.$$

The left-invariant form is, as repeatedly said, the Maurer–Cartan canonical form, which we write simply

$$\Omega_L = \Omega = J_i\,\Omega^i.$$

If we write

$$Ad_{g^{-1}}J_i = h_i{}^j J_j$$

for the adjoint representation, the Maurer–Cartan basis $\{\Omega^i\}$ will be related to the parameter basis $\{d\alpha^i\}$ by $\Omega^j = h_i{}^j d\alpha^i$. The parameters α^i are angles, and the usual angular rate of change is

$$\dot{\alpha}^i = \frac{d\alpha^i}{dt} = d\alpha^i\left(\frac{d}{dt}\right).$$

Notice that

$$\frac{d}{dt} = \dot{\alpha}^i \frac{\partial}{\partial \alpha^i}$$

and that the time variation of the anholonomic form is given by

$$\Omega^j \left(d/dt \right) = h_i{}^j \dot{\alpha}^i = h_i{}^j w_R^i.$$

In matrix notation, $g = (g_{ij})$ and $g^{-1} = (g^{ij})$ implies

$$\Omega_R = dg \, g^{-1},$$

the relation with vector notation being

$$\Omega_R^{ij} = \epsilon^{ijk} \Omega_R^k.$$

We check that $\dot{\boldsymbol{\alpha}} = \Omega(\dot{\boldsymbol{\alpha}})$ and then that

$$\dot{\alpha}^k = \epsilon^{ijk} \dot{g}_{ir} (g^{-1})_{rj},$$

as it should. The formula

$$\dot{\boldsymbol{\alpha}} = d\boldsymbol{\alpha} \, (d/dt)$$

hints that we might use also the notation (a suggestive convention, though basically incorrect) "$\dot{\Omega}$" for $\Omega \, (d/dt)$. Notice finally that, with that convention,

$$\frac{d}{dt} = \dot{\alpha}^i \frac{\partial}{\partial \alpha^i} = \dot{\Omega}^k J_k = \Omega \left(\frac{d}{dt} \right) = \dot{\Omega}.$$

Right- and left-invariance are related to the presence of two distinct derivatives on the group. Given a function $F(g)$, a field X can derive it from the left,

$$X^L F(g) = \left[\frac{d}{ds} F(e^{sX} \, g) \right]_{s=0} = d^L F(g)(X),$$

and from the right,

$$X^R F(g) = \left[\frac{d}{ds} F(g \, e^{sX}) \right]_{s=0} = d^R F(g)(X).$$

One sees that

$$d^R F(g)(X) = d^L F(g)(Ad_g X).$$

In particular, for the field $X = d/dt$ tangent to a curve,

$$(d/dt)^R F(g) = \left[g \, (d/dt)^L \, g^{-1} \right] F(g)$$

$$= - dg \, (d/dt) \, g^{-1} F(g) + (d/dt)^L \, F(g),$$

or

$$\frac{d^R F}{dt} = \frac{d^L F}{dt} - \Omega_R F.$$

For a vector component,

$$\frac{d^R F^i}{dt} = \frac{d^L F^i}{dt} - \Omega_R^{ij} F^j = \frac{d^L F^i}{dt} - \epsilon^{ijk} \Omega_R^k F^j,$$

or

$$\frac{d^R \boldsymbol{V}}{dt} = \frac{d^L \boldsymbol{V}}{dt} + \boldsymbol{\Omega}_R \times \boldsymbol{V}.$$

Comparison with (26.21) shows that:

(i) Ω_R (d/dt) is the usual "space" velocity;

(ii) Ω_L (d/dt) is the usual "body" velocity;

(iii) (d^R/dt) is the usual "space" derivative;

(iv) (d^L/dt) is the usual "body" derivative.

26.3.10 *The Poinsot construction*

The tangent space is constituted by the angular velocities. The cotangent space is the space of the angular momenta. Thus, the angular momentum belongs to the coadjoint representation. The well known property by which the angular momentum is related to the angular velocity through the inertia operator reveals the metric. Indeed, the inertia operator shows up as a left-invariant metric, relating as usual the tangent and the cotangent spaces. And just as above, given a covector at a point of the group, the angular momentum in "space" is obtained by the right-action, and the angular momentum in "body" by left-action. The metric appears in the kinetic energy, which is given by

$$T = \tfrac{1}{2}(\boldsymbol{M}_b, \boldsymbol{\omega}_b).$$

In the absence of external forces, the rigid body motion is a geodesic on $SO(3)$ with this metric.

Given the hamiltonian

$$H = \tfrac{1}{2} \sum_{ij} I_{ij}\, \omega^i \omega^j = f(\omega),$$

the angular momentum is $\boldsymbol{M} = \mathrm{grad}\, f$. The inertia ellipsoid is given by

$$f = \text{constant} = E.$$

Using the euclidean metric (,) and

$$M_i = I_{ij}\omega^j,$$

we may define the inertia ellipsoid as the point set

$$\{\omega \text{ such that } (\boldsymbol{M}, \boldsymbol{\omega}) = 1\}.$$

This means that we stay at an energy level-surface $H = 1/2$. To each point on the ellipsoid will correspond an angular velocity. Draw ω as the position vector, and get the tangent at the point. Then $\boldsymbol{M}$ will be the vector perpendicular to the tangent from the origin taken at the ellipsoid center (Figure 26.1, left-side).

Suppose now that a metric g_{ij} is present, which relates fields and cofields. The case

$$g_{ij} = \text{diag}(1/a^2, 1/b^2)$$

is of evident interest, as $f(\boldsymbol{v}) = g(\boldsymbol{v}, \boldsymbol{v})$, with $\boldsymbol{v}$ the position vector (x, y). To the vector of components (x^j) will correspond the covector of components $(p_k = g_{kj}x^j)$ and

$$p(v) = p_k v^k = g_{ij}x^i x^j.$$

As we are also in an euclidean space, the euclidean metric $m_{ij} = \delta_{ij}$ may be

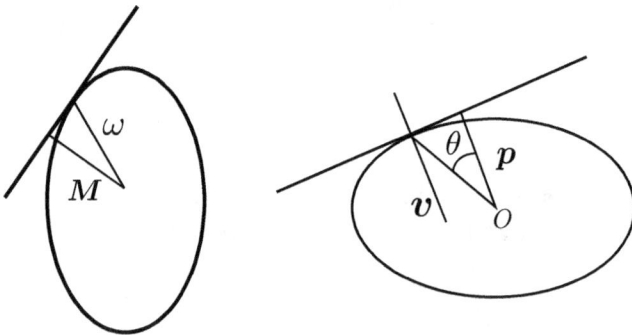

Fig. 26.1   The Poinsot construction.

used to help intuition. We may consider $\boldsymbol{p}$ and $\boldsymbol{r}$ as two euclidean vectors of components (p_k) and (x^k). Comparison of the two metrics is made by using

$$g(\boldsymbol{v}, \boldsymbol{v}) = m(\boldsymbol{p}, \boldsymbol{v}).$$

Our eyes are used to the metric m, and we shall use it to measure angles and define (now, really metric) orthogonality. In the right-side of Figure 26.1 we show the vector v giving a point on the ellipse and the covector p, now assimilated to an euclidean vector. The vector p is orthogonal to the curve at each point, or to its tangent at the point. It has the direction given by the thin line and we draw it from the origin O at the ellipse center. The curve equation is

$$p(v) = g(v, v) = m(p, v) = C.$$

As

$$|p| = m(p, p)^{1/2} = |df|,$$

we can take

$$p = v \cos \theta.$$

This construction to relate a form to a field in the presence of a non-trivial metric, mainly in its 3-dimensional version, is very much used in Physics. For rigid bodies, the metric is the inertia tensor, the vector is the angular velocity and its covector is the angular momentum. The ellipsoid is the inertia ellipsoid and the whole construction goes under the name of Poinsot. An analogous case, the Fresnel ellipsoid, turns up in crystal optics (see Section 30.7).

General references

Commendable general texts covering the subject of this chapter are Arnold 1976, Lovelock & Rund 1975, Lanczos 1986, Goldstein 1980 and Westenholz 1978. More advanced, with topics of recent interest, are: Fetter &Walecka 2003 and 2006, as well as Charap 1995.

Chapter 27

Symmetries on Phase Space

The study of the action of symmetry groups on phase space is an opportunity to introduce some topics of contemporary research: non-linear representations, cohomology of Lie algebras, anomalies, etc. It is, however, a theme of fundamental importance by itself. It leads to a partial but significant classification of phase spaces and opens a road to the general problem of quantization and its relationship to representation theory through the "orbit method".

27.1 Symmetries and anomalies

Suppose that a group G acts transitively on a phase space M (so that M is homogeneous under G), in such a way that its transformations are canonical. This means that G will act through a representation in a subgroup of the huge group of canonical transformations. In this case, M is said to be a *symplectic homogeneous manifold*. Suppose the generators J_a of the Lie algebra G' of G will have commutation relations

$$[J_a, J_b] = f^c{}_{ab} J_c. \tag{27.1}$$

Each J_a will be represented on M by a fundamental field, a hamiltonian field X_a. There will be a representation ρ of G' by vector fields $X_a = \rho(J_a)$. We would expect that the representative fields X_a satisfy the same relations,

$$[X_a, X_b] = f^c{}_{ab} X_c. \tag{27.2}$$

In this case, the algebra representation ρ is said to be *linear*. But this is not what happens in general. Usually, the actions of groups on manifolds are typically non-linear (non-linear representations are even sometimes *defined* as these actions). The very use of the word "representation" is an abuse (to which we shall nevertheless stick for the sake of simplicity), as ρ is no true homomorphism: the word "action", less stringent, would be more correct. A special case occurring with some frequency comes out when the action is given by

$$\rho(J_a) = X_a - \xi_a, \tag{27.3}$$

where the ξ_a's are functions. Recall that a field like X_a acts on functions defined on M, producing other functions. Equation (27.3) says that the action of each generator of G has an extra contribution which can be accounted for through multiplication by a function. A dynamical quantity F will change according to

$$\rho(J_a)F = X_a F - \xi_a F.$$

As to the function ξ_a, it can be interpreted as the result of the action of some form ξ on X_a : $\xi_a = \xi(X_a)$. This situation corresponds to the minimum departure from the simplest expected case (27.2). It comes immediately that

$$[\rho(J_a), \rho(J_b)] = [X_a, X_b] - X_a[\xi_b] + X_b[\xi_a]. \tag{27.4}$$

This kind of action, in which each generator is represented by a field-action plus multiplication by a function, has a nice property: the whole group action can work in that way, because the representative Lie algebra is "closed" in this kind of action: as seen in (27.4), the commutator of two such actions is also a field plus a multiplicative function. The action is a *projective representation* of the algebra G' when (27.4) can be rewritten as

$$[\rho(J_a), \rho(J_b)] = f^c{}_{ab}\, \rho(J_c) - K_{ab}, \tag{27.5}$$

where the K_{ab}'s constitute an antisymmetric set of functions, which can be taken as the components of a 2-form K:

$$K_{ab} = K(X_a, X_b).$$

We put it in this way because (27.5) is just the textbook definition of a projective representation. That it represents the slightest departure from linear representations may be seen by a simple cohomological reasoning. To begin with, we can impose the Jacobi identity on the commutator, to ensure the Lie algebra character. We find easily that the condition is equivalent to $dK = 0$ in the subspace generated by the X_a's. General representations satisfying (27.5) will require that K be a cocycle. We might ask to which one of the cohomology classes a projective representation given by (27.3) would belong. As

$$[\rho(J_a), \rho(J_b)] = f^c{}_{ab}\rho(J_c) - \{X_a[\xi_b] + X_b[\xi_a] - f^c{}_{ab}\xi_c\},$$

K will be

$$K_{ab} = X_a[\xi_b] + X_b[\xi_a] - f^c{}_{ab}\,\xi_c. \qquad (27.6)$$

Always in the subspace of the X_a's, this expression says that the 2-form K of components K_{ab} is the (exterior) differential of the 1-form ξ of components ξ_c. The condition reduces to $K = d\xi$, a coboundary. The cohomology class of K is trivial for representations like (27.3). From (27.2), (27.5) and (27.6), the linear case is seen to require

$$X_a[\xi_b] + X_b[\xi_a] - f^c{}_{ab}\,\xi_c = 0. \qquad (27.7)$$

This is just $d\xi = 0$ written in components. Projective representations reduce to linear representations when also ξ is a cocycle. It is in this sense that they represent a minimal departure from linearity.

By the way, we may give this discussion a contemporary flavor by calling *anomalies* both K_{ab} in (27.5) and ξ_a in (27.3), as they represent an expectation failure analogous to that coming out in the quantization processes,[1] for which this terminology has been introduced. The expression (27.7) is quite analogous to the Wess–Zumino condition and typical of "anomaly removal" (Section 32.3.4). The above kind of procedure is typical of the applications of cohomology to representations. In particular, it is an example of the so called "cohomology of Lie algebra representations".

Unfortunately, even when the representation $\rho(G')$ is linear, further anomalies insist in showing up. Suppose the representative fields to be all strictly hamiltonian, so that to each X_a corresponds a function F_a (called in the present case the "hamiltonian related to J_a") such that

$$i_{X_a}\Omega = dF_a, \qquad (27.8)$$

where Ω is, of course, the symplectic form. In this expression we recognize F_a as the generating function of the canonical transformation whose infinitesimal generator is X_a. But this means that there is still another representation at work here, that of the algebra of hamiltonian fields in the algebra of differentiable functions on M, with the Poisson bracket as algebra operation. With a simplified notation, it is the homomorphism

$$\varphi : \rho(G') \to \mathbb{C}^\infty(M, \mathbb{R}), \quad \varphi : X_a \to F_a,$$

---

[1] There is more than a mere analogy here. See, for instance, Faddeev & Shatashvilli 1984.

with the corresponding operations [ ] → { }. From what is seen in Section 25.7,

$$dF_{[X_a,X_b]}(Y) = i_{[X_a,X_b]}\Omega(Y)$$

for every field Y. The last expression is the same as

$$\Omega([X_a, X_b], Y) = f^c{}_{ab}\Omega(X_c, Y),$$

so that

$$dF_{[X_a,X_b]} = f^c{}_{ab}\, dF_c.$$

As also

$$d\{F_a, F_b\} = dF_{[X_a,X_b]},$$

we have

$$d\{F_a, F_b\} = f^c{}_{ab}dF_c,$$

from which

$$\{F_a, F_b\} = f^c{}_{ab}F_c + \beta(X_a, X_b). \tag{27.9}$$

The presence of the term $\beta(X_a, X_b)$, which comes out from applying a 2-form β to the two fields X_a and X_b, says that, in principle, also φ is a projective representation. This is related to the fact that generating functions are defined up to constants. We may proceed in a way quite analogous to the previous case. The Jacobi identity applied to (27.9) will say that the 2-form β is a cocycle. Let us add a constant to each of the above functions, and consider the modified functions:

$$F'_a = F_a + \alpha_a, \quad F'_b = F_b + \alpha_b, \text{ etc.}$$

The relation becomes

$$\{F'_a, F'_b\} = f^c{}_{ab}\, dF'_c + \beta'(X_a, X_b)$$

with

$$\beta'(X_a, X_b) = \beta(X_a, X_b) - \alpha_c f^c{}_{ab}.$$

The constants α_a may be seen as the result of applying an invariant 1-form α to the respective fields, $\alpha_a = \alpha(X_a)$, etc. As

$$(d\alpha)_{ab} = -\alpha_c f^c{}_{ab},$$

we see that

$$\beta' = \beta + d\alpha.$$

If some α exists whose choice leads to $\beta' = 0$, showing β as the exact form $\beta = - d\alpha$, then the functions may be displaced by arbitrary constants so that the algebra reduces to

$$\{F_a, F_b\} = f^c{}_{ab} F_c.$$

The projective representation reduces to a linear representation when the cohomology class of β is trivial. In this case, we have

$$\{F_a, F_b\} = F_{[Xa, Xb]},$$

and the symplectic manifold M is said to be *strictly homogeneous* (or *Poisson manifold*) under the action of G.

27.2 The Souriau momentum

When the above F_a's exist, we can also consider directly the composite mapping

$$F : G' \to C^\infty(M, \mathbb{R}), \quad F = \varphi \circ \rho,$$

such that $F(J_a) = F_a(x)$. We are supposing that G is a symmetry group of the system. The transformations generated by its representative generators will preserve the hamiltonian function H. As

$$\{F_a(x), H\} = - X_a H = - L_{X_a} H,$$

the invariance of H under the transformations whose generating function is F_a, the expression $L_{X_a} H = 0$ gives just the usual

$$\{F_a(x), H\} = 0.$$

Each F_a is a constant of motion. This is the hamiltonian version of *Noether's theorem* (Section 31.4.2). Each symmetry yields a conserved quantity.

Let us place ourselves in the particular case in which the Liouville form is also preserved: $L_{X_a} \sigma = 0$. Then,

$$dF_a = i_{X_a} \Omega = - i_{X_a} d\sigma = - (L_{X_a} - d i_{X_a}) \sigma = d [i_{X_a} \sigma],$$

and with the generating functions defined up to constants,

$$F_a(x) = [i_{X_a} \sigma](x) = [\sigma(X_a)](x). \tag{27.10}$$

The composite mapping F, such that $F(J_a) = F_a(x)$, can be realized as a cofield on G. Take the Maurer–Cartan basis $\{\omega^a\}$ for G'^* and define

$$P : M \to G'^*$$

such that

$$P : x \to P(x) = P_x = F_a(x)\,\omega^a. \tag{27.11}$$

The hamiltonians are, of course, $F_a(x) = P_x(J_a)$. The mapping P is the *Souriau momentum* and is defined up to an arbitrary constant in G'^*. Notice that its existence presupposes the global hamiltonian character of the X_a's. Given the action of G on M, it provides the constants of motion related to its generators. There is more: one can show that

(i) The mapping P commutes with the group action, so that $P(x)$ is a G-orbit in G'^*.

(ii) P is a local homeomorphism of the phase space M into one of the orbits of G in G'^*; this will have a beautiful consequence.

27.3   The Kirillov form

Consider an n-dimensional Lie group G acting on itself. This action is a diffeomorphism and consequently fields and cofields on G will be preserved, that is, taken into themselves. A set of n left-invariant fields J_a may be taken as a basis for the Lie algebra G'. Such a basis will be preserved and will keep the same commutation relations

$$[J_a, J_b] = f^c{}_{ab} J_c$$

at any point of G, so that the $f^c{}_{ab}$'s will be constant. G acts on the J_a's according to the adjoint representation

$$g^{-1} J_a g = K_a{}^b J_b.$$

The dual basis to $\{J_a\}$ is formed by the Maurer–Cartan 1-forms ω^c such that $\omega^c(J_a) = \delta^c_a$, which satisfy

$$d\omega^c = -\tfrac{1}{2} f^c{}_{ab}\, \omega^a \wedge \omega^b.$$

The group G acts on the ω^c's according to the coadjoint representation

$$g^{-1} \omega^b g = K_a{}^b \omega^a.$$

Now, each 1-form $\zeta = \zeta_a \omega^a$ on G defines a 2-form Ω_ζ, the *Kirillov form*, by the relation

$$\Omega_\zeta(J_a, J_b) = \zeta[J_a, J_b]) = \zeta_c f^c{}_{ab}.$$

That is,

$$\Omega_\zeta = \tfrac{1}{2} \zeta_c f^c{}_{ab} \omega^a \wedge \omega^b = -\zeta_c \, d\omega^c. \qquad (27.12)$$

This form is closed, nondegenerate and G-invariant. It defines a symplectic structure. As ζ is preserved by the group action, the same Ω is defined along all its orbit,

$$\text{Orb}\,(\zeta) = \{\text{Ad}^*{}_g \zeta,\ \text{all}\ g \in G\},$$

by G in the coadjoint representation. So, on each such orbit (usually called coorbit) there is a symplectic structure, which is furthermore strictly homogeneous.

The important point is the following: orbits in the coadjoint representation of Lie groups are, in reality, the only symplectic strictly homogeneous manifolds. Any other strictly homogeneous manifold is locally homeomorphic to one of these orbits and, consequently, is a covering of it. The local homeomorphism is precisely the Souriau mapping P. In this way, such orbits classify all symplectic strictly homogeneous manifolds.

27.4 Integrability revisited

Consider[2] on a phase space two functions $L(q, p)$ and $M(q, p)$ with values in some Lie algebra G' of a Lie group G:

$$L = J_a L^a(q, p) \quad \text{and} \quad M = J_a M^a(q, p). \qquad (27.13)$$

They are said to constitute a *Lax pair* if the evolution equation for L is

$$\frac{dL}{dt} = [L, M] = J_a f^a{}_{bc} L^b M^c. \qquad (27.14)$$

If now we take for the "hamiltonian" $M = g^{-1}\frac{d}{dt}g$, the solution of this equation is

$$L(t) = g^{-1}(t) L(0) g(t). \qquad (27.15)$$

Recognizing the action of the adjoint representation, we can write this also as

$$L(t) = Ad_{g^{-1}(t)} L(0). \qquad (27.16)$$

---

[2] Babelon & Viallet 1989.

The evolution is governed by the adjoint action. Consider now any polynomial $I(L)$ of L which is invariant under the adjoint representation. It will not change its form under the group action, and therefore

$$\frac{d}{dt}I(L) = 0. \tag{27.17}$$

Lax pairs provide consequently a very convenient means to find integrable systems. To obtain integrals of motion, one chooses a representation in which L and M are well known matrices and check candidate invariants of the type $I_j = \mathrm{tr}\ (L^j)$, which are adjoint-invariant, verifying whether or not they are in involution. The secular equation fixing the eigenvalue spectrum of L,

$$\det(L - \lambda I) = 0,$$

is a polynomial in λ with coefficients which are themselves polynomials in the traces of powers of L. The eigenvalues are thus also adjoint-invariant. The evolution equation

$$\frac{dL}{dt} = [L, M]$$

is for this reason called an "isospectral evolution".

27.5   Classical Yang–Baxter equation

The classical Yang–Baxter equation is the Jacobi identity for the Poisson bracket for phase spaces defined on Lie groups, written in terms of the inverse to the symplectic matrix. Let us see how that comes out.[3] The Poisson bracket of two functions F and G is related to the symplectic cocycle Ω by

$$\{F, G\} = \Omega(X_F, X_G) = e_k(G)\Omega^{kj}e_j(F), \tag{27.18}$$

where X_F and X_G are the hamiltonian fields corresponding to the functions, $\{e_k\}$ is a vector basis and the matrix (Ω^{ij}) is inverse to the matrix (Ω_{ij}) formed with the components $\Omega_{ij} = \Omega(e_i, e_j)$ of the symplectic form. The requirement that Ω be a closed form, or a cocycle, is equivalent to the Jacobi identity for the Poisson bracket.

We look for the Jacobi identity written in terms of the inverse matrix (Ω^{ij}). Any differentiable manifold may in principle become a symplectic

[3] Drinfel'd 1983.

manifold, provided there exists defined on it a closed nondegenerate two-form, leading to a Poisson bracket. In the case of interest, the manifold is a Lie group endowed with a hamiltonian structure consistent with the group structure.

Consider then as symplectic manifold a Lie group G with Lie algebra G'. Choose a basis $\{J_a\}$ of generators, with

$$[J_a, J_b] = f^c{}_{ab} J_c.$$

Such generators correspond to smooth left-invariant (or right-invariant) complete fields on the manifold G acting on the functions $F \in C^\infty(G, \mathbb{R})$ as derivations. A curve on G will be given by a one-parameter set of elements

$$g(t) = \exp[tX],$$

where $X = X^a J_a$ is the generator corresponding to

$$g \equiv g(1) = \exp[X].$$

Each generator J_a is represented on the group manifold by a left-invariant field e_a. The set $\{e_a\}$ provides a basis for the vector fields on G, with

$$[e_a, e_b] = f^c{}_{ab} e_c.$$

The consistency between the Lie group structure and the symplectic structure of G is obtained by imposing that the fields e_a be hamiltonian fields, that is, that the infinitesimal transformations they generate are canonical. This means that they preserve the symplectic cocycle Ω, which is expressed by the vanishing of the Lie derivative

$$L_{e_a} \Omega = (d\, i_{e_a} + i_{e_a}\, d)\, \Omega = 0, \tag{27.19}$$

where i_{e_a} is the interior product. With the cocycle condition $d\Omega = 0$, this implies the closedness of the form $i_{e_a}\Omega$. It follows that there exists locally a function F_a such that $i_{e_a}\Omega = dF_a$. The function F_a will be the generating function of the canonical transformation generated by e_a. Consequently,

$$\{F_a, F_b\} = -e_a(F_b) = \Omega_{ab}.$$

The Jacobi identity is then

$$\{\{F_a, F_b\}, F_c\} + \{\{F_c, F_a\}, F_b\} + \{\{F_b, F_c\}, F_a\} =$$
$$= \{\Omega_{ab}, F_c\} + \{\Omega_{ca}, F_b\} + \{\Omega_{bc}, F_a\}$$
$$= e_c(\Omega_{ab}) + e_b(\Omega_{ca}) + e_a(\Omega_{bc}) = 0. \tag{27.20}$$

Using

$$e_c(\Omega_{ab}) = \tfrac{1}{2}\left[e_c e_b(F_a) - e_c e_a(F_b)\right]$$

for each term of (27.20), we find

$$f^d{}_{cb}\Omega_{ad} + f^d{}_{ac}\Omega_{bd} + f^d{}_{ba}\Omega_{cd} = 0.$$

Contracting with the product $\Omega^{ka}\Omega^{jb}\Omega^{ic}$, we arrive finally at

$$f^i{}_{ab}\Omega^{ka}\Omega^{jb} + f^j{}_{ab}\Omega^{ia}\Omega^{kb} + f^k{}_{ab}\Omega^{ja}\Omega^{ib} = 0. \tag{27.21}$$

To change to the standard notation in the literature, put $r^{ab} = \Omega^{ab}$ for a symplectic structure defined on a Lie group. Thus,

$$f^i{}_{ab}\, r^{ka} r^{jb} + f^j{}_{ab}\, r^{ia} r^{kb} + f^k{}_{ab}\, r^{ja} r^{ib} = 0. \tag{27.22}$$

This is just the classical Yang–Baxter equation, though not in its most usual form, which is given in direct product notation. The contravariant tensors (r^{kj}) may be seen as a map

$$G \to TG \otimes TG, \quad g \to r(g) = r^{kj}\, e_k \otimes e_j.$$

This represents on the group manifold a general member of the direct product $G' \otimes G'$ which will have the form

$$r = r^{ab}\, J_a \otimes J_b.$$

In this notation, the algebra is included in higher product spaces by adjoining the identity algebra. For an element of G', we write, for example,

$$X_1 = X \otimes 1 = X^a(J_a \otimes 1), \quad X_2 = 1 \otimes X = X^a(1 \otimes J_a),$$

or

$$X_1 = X \otimes 1 \otimes 1, \quad X_2 = 1 \otimes X \otimes 1, \quad X_3 = 1 \otimes 1 \otimes X,$$

and so on. Elements of $G' \otimes G'$ may then be written

$$r_{12} = r^{ab}\, J_a \otimes J_b \otimes 1, \quad r_{13} = r^{ab}\, J_a \otimes 1 \otimes J_b, \quad r_{23} = r^{ab}\, 1 \otimes J_a \otimes J_b.$$

We can make use of the multiple index notation:

$$< ij|A \otimes B|mn > \ = \ < i|A|m >< j|B|n >,$$
$$< ijk|A \otimes B \otimes C|mnr > \ = \ < i|A|m >< j|B|n >< k|C|r >,$$

and so on. If r belongs to $G' \otimes G'$, the matrix elements are

$$< ij|r|mn > = r^{ij}{}_{mn}$$

and

$$< ijr|r \otimes E|mns >= \delta^r_s \, r^{ij}{}_{mn}.$$

We can then calculate to find

$$[r_{12}, r_{13}] = r^{ab} r^{cd} [J_a, J_c] \otimes J_b \otimes J_d = r^{db} r^{ec} f^a{}_{de} J_a \otimes J_b \otimes J_c,$$

$$[r_{12}, r_{23}] = r^{ab} r^{cd} J_a \otimes [J_b, J_c] \otimes J_d = r^{ad} r^{ec} f^b{}_{de} J_a \otimes J_b \otimes J_c,$$

$$[r_{13}, r_{23}] = r^{ab} r^{cd} J_a \otimes J_c \otimes [J_b, J_d] = r^{ad} r^{be} f^c{}_{de} J_a \otimes J_b \otimes J_c.$$

Equation (27.22) takes, therefore, the standard form

$$[r_{12}, r_{13}] + [r_{12}, r_{23}] + [r_{13}, r_{23}] = 0. \qquad (27.23)$$

The name "classical" comes from the fact that, when conveniently parametrized, this equation is the limit $h \to 0$ of the Yang–Baxter equation (14.22).

General references

Basic texts on the above matter are Arnold 1976, Kirillov 1974 and Babelon & Viallet 1989. More recent, talking about points of contemporary research, is Fadell & Husseini 2001.

Chapter 28

Statistics and Elasticity

The internal constitutions of gases, crystals and rubber provide surprising examples of the use of geometry. Historically these objects have, like the systems described by elementary mechanics, dropped many a hint of basic necessary concepts.

28.1 Statistical mechanics

28.1.1 *Introduction*

The objective of Statistical Mechanics is to describe the behaviour of macroscopic systems, composed by a large number of elements, assuming the knowledge of the underlying dynamics of these individual constituents. In the effort to describe real systems, usually very involved objects, it is forced to resort to simplified models. Some models are actually reference models, supposed to give a first approximation to a whole class of systems and playing the role of guiding standards. They are fundamental to test calculation methods and as starting points for more realistic improvements. For low-density gases, for instance, the main reference models are the ideal gases, classical and quantal, and the hard-sphere gas. For solids, lattices with oscillators in the vertices are standard when the involved atoms or molecules have no structure. The next step involves attributing some simple "internal" structure to the atoms or molecules. It is of course very tempting to look at the lattice as a "space" of which the cells are building blocks. And then consider the case of negligible spacing between the atoms as a model for the continuous media.

As discussed in Section 2.3, the structure of a space is at least in part revealed by its building blocks. However, it is in general difficult to find out which ones are the necessary blocks. We have there used irregular tetrahedra to cover $\mathbb{E}^3$. It would have been impossible to do it with regular tetrahedra. Because historically this problem was at first studied in the 2-dimensional case, we refer to it as the "problem of the tilings (or pavings) of a space". Thus, we say that we cannot pave $\mathbb{E}^3$ with regular tetrahedra, though we can do it with irregular ones; and we cannot pave the plane $\mathbb{E}^2$ with regular pentagons, though we can pave a sphere with them, and then project into $\mathbb{E}^2$, as we do in order to endow the plane with a spherical metric (see Section 23.3). This procedure is helpful in modeling some distortions in crystals, caused by defects which, in the continuum limit, induce a curvature in the medium. Indeed, once the lattices become very tightly packed, some continuity and differentiability can be assumed. Vectors become vector fields, and tensors alike. Whether or not the system "is a continuum" is a question of the physical scales involved. When we can only see the macroscopic features, we may look at the limiting procedure as either an approximation (the "continuum approximation") or as a real description of the medium. Elasticity theory treats the continuum case, but the lattice picture is too suggestive to be discarded even in the continuum approximation.

We are thus led to examine continuum media, elastic or not. Introducing defects into regular lattices can account for many properties of amorphous media. The addition of defects to regular model crystals, for example, provide good insights into the qualitative structure of glasses.

Modern theory of glasses[1] sets up a bridge between lattice models and Elasticity Theory. Adding defects changes the basic euclidean character of regular lattices. It turns out that some at least of the 'amorphous' aspects of glasses can be seen as purely geometrical and that adding defects amounts to attributing torsion and/or curvature to the medium.

28.1.2   *General overview*

Statistical Mechanics starts by supposing that each constituent follows the known particle mechanics of Chapter 25. Basically, this is the "mechanics" involved in its name, though the most interesting systems require Quantum Mechanics instead. To fix the ideas, we shall most of time consider the classical case. But we are none the wiser after assuming microscopic hamiltonian dynamics, which supposes the knowledge of the boundary values. It is clearly impossible to have detailed information on the boundary conditions for all the particles in any realistic situation, such as a commonplace

---

[1] A beautiful modern text is Kerner 2007.

sample of gas with around 10^{23} particles. It is essential to take averages on these boundary conditions, and that is where the "statistical" comes forth. Different assumptions, each one related to a different physical situation, lead to different ways of taking the average. This is the subject of the "ensemble theory" of Statistical Mechanics. The "microcanonical" ensemble is used when all the energy values are equally probable; the "canonical" ensemble describes systems plunged in a thermal bath, for which only the average energy of the system is conserved; the "grand-canonical" ensemble describes systems for which only the average number of particles is preserved; and so on. From a more mathematical point of view, Statistical Mechanics is a privileged province of Measure Theory. Each ensemble defines a measure on phase space, making of it a probability space (Chapter 15).

The volume of a $2n$-dimensional phase space M is given by the Lebesgue measure

$$dq\,dp \equiv dq^1 dq^2 \ldots dq^n dp_1 dp_2 \ldots dp_n.$$

If we want it to be non-dimensional, the measure (the volume) of a domain D can be taken as

$$m(D) = h^{-n} \int_D dqdp\,,$$

where we have divided the usual volume by a convenient power of Planck's constant h.[2] There are, however, two converging aspects leading to the choice of another measure. First, M can be a non-compact space and in that case $m(M)$ is infinite. Second, in physical situations there are preferred regions on phase space. An n-particle gas with total hamiltonian $h(q,p)$, for example, will have a distribution proportional to the Boltzmann factor

$$\exp[-h(q,p)/kT].$$

In general, the adopted measure on phase space includes a certain non-negative distribution function $F(q,p) \geq 0$ which, among other qualities, cuts down contributions from high q's and p's. Such a measure will attribute to D the value

$$m(D) = \int_D F(q,p)dqdp.$$

If we want to have a probability space, we have to normalize $F(q,p)$ so that $m(M) = 1$. The canonical ensemble, for example, adopts the measure

$$F(q,p) = \frac{\exp[-h(q,p)/kT]}{\int_M \exp[-h(q,p)/kT]dqdp}\,. \tag{28.1}$$

---

[2] Unavoidable notation, not to be confused with the equally unavoidable hamiltonian $h(q,p)$ just below.

In Classical Statistical Mechanics, macroscopic quantities are described by piecewise continuous functions of the time variable and of the position in the physical space. Thus, the energy density $\mathcal{H}(x,t)$ of an n-particle gas around the point $x = (x^1, x^2, x^3)$ at instant t will be a functional of the microscopical hamiltonian $h(q, p; x, t)$:

$$\mathcal{H}(x,t) = \int_M F(q,p) h(q,p; x,t) dqdp.$$

This is quite general: given any microscopic quantity $r(q, p; x, t)$, its macroscopic correspondent $R(x, t)$ will be given by the average

$$R(x,t) = \int_M F(q,p)\, r(q,p; x,t)\, dqdp. \tag{28.2}$$

The normalizing denominator in F is the partition function, from which thermodynamical quantities can be calculated. All this means that the state of the system, as far as Statistical Mechanics is concerned, is fixed by the probability measure

$$dm = m(dqdp) = F(q,p)\, dqdp,$$

which is interpreted as the probability for finding the system in a region of volume $dqdp$ around the point (q, p) of the phase space. The measure "translates" microscopic into macroscopic quantities. Actually, $F(q, p)$ is a dynamical quantity $F(q, p, t)$ satisfying the Liouville equation

$$\partial_t F = \{F, \mathcal{H}\},$$

and the time evolution of the whole system is fixed by the behaviour of $F(q, p, t)$. Systems in equilibrium are described by time-independent solutions, for which F is an integral of motion. Different conditions lead to various solutions for F, each one an *ensemble*. In the general case, the measure then defines the evolution of each physical quantity through the weighted averages similar to Eq. (28.2),

$$R(x,t) = \int_M F(q,p,t)\, r(q,p; x)\, dqdp, \tag{28.3}$$

now with the time dependence centralized in $F(q, p, t)$. Thus, the state of the system is given by $F(q, p, t)$. In equilibrium, F is constant in time and so is each $R(x, t)$ — the system has fixed values for all observable quantities. In practice, time-averages instead are observed. That the expectancy (average on phase space) is equivalent to the time average is the content of the famous Boltzmann's ergodic theorem (Section 15.2).

Of course, when quantum effects are important,

$$h^{-n} \int dq \, dp$$

is only a first approximation. We should actually sum over discrete values, as in the lattice models of Section 28.2 below. Everything lies on the density matrix,

$$\rho = \frac{\exp[-\mathcal{H}/kT]}{\mathrm{tr} \, \exp[-\mathcal{H}/kT]}. \tag{28.4}$$

The expectation of an observable represented by the operator A will be the trace

$$< A > = \mathrm{tr} \, \rho \, A. \tag{28.5}$$

The state of the system is now centralized in the density matrix, and the partition function is the denominator in (28.4).

The canonical ensemble is particularly convenient, as the fixed number of particles makes it easier to define the hamiltonian $\mathcal{H}$. Consider for a moment a gas with N particles of the same species. All average values can be obtained from the partition function. The semi-classical case is obtained as the limit of Planck's constant going to zero, but it is wise to preserve one quantum characteristic incorporated in the Gibbs' rule: *the particles are indistinguishable.* For the state of the system, fixed by the expectation values of the observables, it is irrelevant *which* individual particle is in this or that position in phase space. This means that the partition function, in terms of which the average values can be obtained, must be invariant under the action of the symmetric group S_N, which presides over the exchange of particles. The canonical partition function $Q_N(\beta, V)$ of a real non-relativistic gas of N particles contained in a d-dimensional volume V at temperature

$$T = 1/k\beta$$

is an invariant polynomial of S_N, just the S_N cycle indicator polynomial C_N (Section 14.1.4). If λ is the mean thermal wavelength and b_j is the j-th cluster integral which takes into account the interactions of j particles at a time, then

$$Q_N(\beta, V) = \frac{1}{N!} C_N \left(b_1 \frac{V}{\lambda^d}, 2b_2 \frac{V}{\lambda^d}, 3b_3 \frac{V}{\lambda^d}, \dots \right). \tag{28.6}$$

On 2-dimensional manifolds the exchange of particles is not governed by the symmetric group, but by the braid group, and the statistics changes accordingly. Instead of the usual statistics, which leads to particles behaving either as bosons or as fermions, the so-called braid statistics[3] is at work, giving to the particles a continuum of possible intermediate behaviours (Section 14.2.5).

In the quantum case, we have actually to consider a space (an algebra) of operators, of which the density matrices are the most important. Basically, as long as N is finite, we have a finite von Neumann algebra. Once the partition function, as well as its logarithm and its derivatives, are obtained, thermodynamical quantities are arrived at by taking the thermodynamical limit of large N and V, the volume of the system. The background algebra will then be a more general, infinite-dimensional von Neumann algebra (see Section 17.5).

Commentary 28.1 Basic texts on Statistical Mechanics are many; maybe the best for our purposes are Pathria 1972 and Balescu 1975. ◄

28.2 Lattice models

28.2.1 *The Ising model*

Most of the lattice models suppose a d-dimensional lattice with N vertices ("sites") and some spacing (the "lattice parameter") between them. In each site is placed a molecule, endowed with some "internal" discrete degree of freedom, generically called "spin". The lattice can be cubic, hexagonal, cubic centered in the faces, etc. The spin at the site "k" is described by a q-dimensional vector σ_k (Figure 28.1, right). The interaction takes place along the edges (the "bonds") and is given by a general ("Stanley") hamiltonian of the form

$$\boldsymbol{H} = -\sum_{i<j} J_{ij}\,\sigma_i \cdot \sigma_j - \boldsymbol{H} \cdot \sum_k \sigma_k. \qquad (28.7)$$

The factor J_{ij} represents the coupling between the molecules situated in the i-th and the j-th sites. "To solve" such a model means to obtain the explicit form of the partition function. A reasonably realistic case is the Heisenberg model, for which $d = q = 3$. Though no analytic solution has been obtained, many results can be arrived at by numerical methods. Some other cases have been solved, none of them realistic enough. The problem is less difficult in the nearest-neighbour approximation, which supposes that $J_{ij} \neq 0$ only when i and j are immediate neighbours. For 1-dimensional

---

[3] A rather detailed treatment of two-dimensional quantum gases can be found in Aldrovandi 1992.

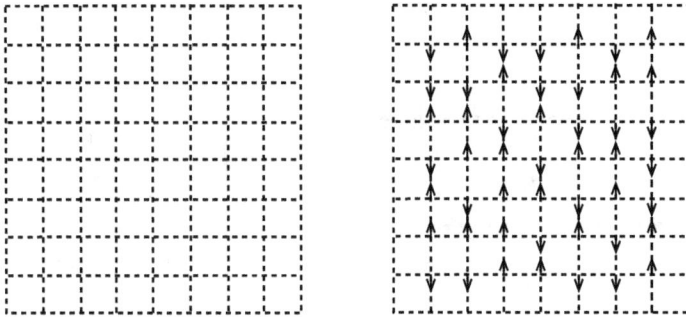

Fig. 28.1   A cubic lattice of spins.

systems with no external magnetic field ($H = 0$), exact solutions are known for all values of the "spin dimension" q. For higher dimensions, the best known model is the celebrated Lenz–Ising model, for which $q = 1$ ($s_k = \sigma_k = \pm 1$) and the same interaction is assumed for each pair of neighbours:

$$\mathcal{H} = - J \sum_{<ij>} s_i s_j - H \cdot \sum_k s_k. \qquad (28.8)$$

The symbol $< ij >$ recalls that the summation takes place only on nearest neighbours. The partition function is

$$Q_N(\beta, H) = \sum_{s_k = \pm 1} \exp\left[K \sum_{<ij>} s_i s_j + h \sum_k s_k \right], \qquad (28.9)$$

where $K = \beta J$ and $h = \beta H$. The 1-dimensional case was solved by Ising in 1925. For $H = 0$, the partition function is

$$Q_N(\beta) = (2 \cosh K)^N = (2 \cosh(\beta J))^N. \qquad (28.10)$$

To illustrate the general method of solution and to introduce the important concept of transfer matrix, let us see how to arrive at this result. The model consists of a simple line of spins $1/2$, disposed in N sites. The left segment of Figure 28.2 shows it twice, one to exhibit the site numeration and the other to give an example of possible spin configuration. Identification of the sites "1" and "$N + 1$" ($\sigma_{N+1} = \sigma_1$) corresponds to a periodic boundary condition. We have in this case an "Ising chain" (Figure 28.2, right part), which becomes a torus in the 2-dimensional case. The total energy will be

$$E(\sigma) = - J \sum_{i=1}^{N} \sigma_i \sigma_{i+1} - H \sum_{i=1}^{N} \sigma_i, \qquad (28.11)$$

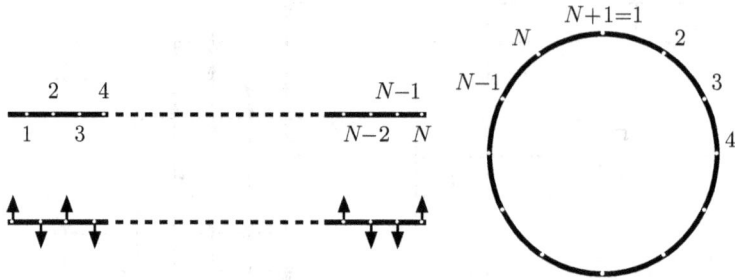

Fig. 28.2   Ising chain with, at the left, a chosen spin configuration.

with the partition function given by

$$Q_M(H,T) = \sum_\sigma \exp[-\beta E(\sigma)]$$

$$= \sum_\sigma \exp\left[K \sum_i \sigma_i \sigma_{i+1} + h \sum_i \sigma_i\right]. \qquad (28.12)$$

The symbol σ represents a configuration of spins, $\sigma = (\sigma_1, \sigma_2, \sigma_3, \ldots, \sigma_N)$ and the summation is over the possible values of σ.

There are many methods to solve the problem. We shall here introduce, as announced, the "transfer matrix" V, a symmetric matrix with elements

$$V_{\sigma\sigma'} = \exp\left[K\sigma\sigma' + (h/2)(\sigma + \sigma')\right]. \qquad (28.13)$$

Instead of the usual labeling, we use as labels the spin values $\{+, -\}$. The matrix elements will be

$$V_{++} = e^{K+h}, \quad V_{+-} = e^{-K}, \quad V_{-+} = e^{-K}, \quad V_{--} = e^{K-h}$$

so that

$$V = \begin{pmatrix} e^{K+h} & e^{-K} \\ e^{-K} & e^{K-h} \end{pmatrix}. \qquad (28.14)$$

One finds then from (28.12) that the partition function can be written as

$$Q_N(H,T) = \sum_\sigma V_{\sigma_1\sigma_2} V_{\sigma_2\sigma_3} V_{\sigma_3\sigma_4} \cdots V_{\sigma_N\sigma_1} = \operatorname{tr} V^N. \qquad (28.15)$$

The partition function is the trace of the N-power of the transfer matrix. In other cases, more than one transfer matrix can be necessary, as in the Potts model (Sec. 28.2.3), in which each transfer matrix is related to a generator of the braid group. Here, V alone suffices. It can be diagonalized with two eingenvalues λ_1 and λ_2. Then, (28.15) will say that

$$Q_N(H,T) = \lambda_1^N + \lambda_2^N. \tag{28.16}$$

The eingenvalues are solutions of the secular equation

$$\begin{vmatrix} e^{K+h} - \lambda & e^{-K} \\ e^{-K} & e^{K-h} - \lambda \end{vmatrix} = \lambda^2 + \lambda\,(2e^K \cosh h) + 2\sinh(2K) = 0. \tag{28.17}$$

One finds

$$\lambda = e^K \cosh h \pm \sqrt{e^{2K}\cosh^2 h - 2\,\sinh(2\,K)}\,,$$

so that

$$Q_N(H,T) = \left(e^K \cosh h + \sqrt{e^{2K}\cosh^2 h - 2\,\sinh(2\,K)} \right)^N$$
$$+ \left(e^K \cosh h - \sqrt{e^{2K}\cosh^2 h - 2\,\sinh(2\,K)} \right)^N. \tag{28.18}$$

For $N = 1$,

$$Q_1(H,T) = 2\,e^K \cosh h\,,$$

and for $N = 2$,

$$Q_2(H,T) = 2\,e^{2K}\cosh 2h + 2e^{-2K}.$$

When $h = 0$,

$$Q_N(0,T) = \left(e^K + e^{-K} \right)^N + \left(e^K - e^{-K} \right)^N$$
$$= \left(2\cosh K \right)^N + \left(2\sinh K \right)^N. \tag{28.19}$$

As $\cosh x > \sinh x$, the first term will dominate for large values of N, and the result (28.10) comes forth.

For the 2-dimensional case, the solution[4] for a square lattice and external field $H = 0$ has been found by Onsager in 1944. It is given by

$$Q_N(\beta) = \left\{ 2\cosh(2\beta J) \exp\left[\frac{1}{2\pi} \int_0^\pi d\alpha \ln\left[1 + (1 - b^2 \sin^2 \alpha)^{1/2} \right]^{1/2} \right] \right\}^N$$

where

$$b = \frac{2\sinh(2\beta J)}{\cosh^2(2\beta J)}.$$

The procedure to find it was extremely difficult. A simpler derivation, due to Vdovichenko,[5] has been known since the nineteen-sixties. No analytic solution has been found as yet for the d = 3 case.

Commentary 28.2 If we go to the continuum limit, with the lattice parameter going to zero, the "spins" constitute a spin field. We can also go to the classical limit, so that "spin" is a variable taking continuum values at each point: it becomes a field whose character depends on the range of values it is allowed to assume. ◀

---

[4] See Huang 1987.
[5] Landau & Lifshitz 1969.

28.2.2 *Spontaneous breakdown of symmetry*

The main interest in these models lies in the study of phase transitions. The 1-dimensional solution (28.10) shows no transition at all, which is an example of a general result, known as the van Hove theorem: 1-dimensional systems with short-range interactions between constituents exhibit no phase transition. The 2-dimensional Onsager solution, however, shows a beautiful transition, signaled by the behaviour of the specific heat C, whose derivative exhibits a singularity near the critical ("Curie") temperature

$$kT_c \approx 2,269 \text{ J.}$$

The specific heat itself behaves, near this temperature, as

$$C \approx \text{constant} \times \ln |T - T_c|^{-1}.$$

A logarithmic singularity is considered to be a weak singularity. The magnetization M, however, has a more abrupt behaviour, of the form

$$\frac{M}{N\mu} \approx 1.2224 \left(\frac{T_c - T}{T_c} \right)^{\beta}, \quad \text{with} \quad \beta = 1/8. \tag{28.20}$$

Numerical studies show that in the 3-dimensional case the phase transition is more accentuated. For a cubic lattice the specific heat near the critical temperature behaves as

$$C \approx |T - T_c|^{-\alpha}, \text{ with } \alpha \approx 0,125. \tag{28.21}$$

All this is to say that the general behaviour depends strongly on the dimension. This transition may be thought of as a ferromagnetic transition, with an abrupt change from a state in which the magnets are randomly oriented, at high temperatures, to a microscopically anisotropic state in lower temperatures, in which the magnets are aligned. It is also an order-disorder transition. The magnetization of some real metals is qualitatively described by the Ising model. The fractional values of the exponents in the behaviours (28.20) and (28.21) tell us that these critical points are not singular points of the simple Morse type, which would have a polynomial aspect (Section 21.6). They point to degenerate points, and are actually obtained via the far more sophisticated procedures of the renormalization group. With an energy of the form

$$E = -J \sum_{<ij>} s_i s_j,$$

there are two configurations with the (same) minimum energy: all the spins up (+1) and all spins down (−1). Thus, at temperature zero, there are two possible states, and the minimal entropy is not zero, but

$$S_0 = k \ln 2.$$

When the temperature is high, the system is in complete microscopic disorder, with its spins pointing toward all directions. Even small domains (large enough if compared to molecular dimensions) of the medium exhibit this isotropy, or rotational symmetry. As the temperature goes down, there is a critical value at which the system chooses one of the two possible orientations and becomes "spontaneously" magnetized while proceeding towards the chosen fundamental state.

The original macroscopic rotational symmetry of the system breaks down. Notice that the hamiltonian is, and remains, rotationally symmetric. The word "spontaneous" acquired for this reason a more general meaning. We call nowadays "spontaneous breakdown of symmetry" every symmetry breaking which is due to the existence of more than one ground state. The fundamental state is called "vacuum" in field theory. When it is multiple, we say that the vacuum is *degenerate*. There is thus spontaneous breakdown of symmetry whenever the vacuum is degenerate. A quantity like the above magnetization, which vanishes above the critical temperature and is different from zero below it, is an "order parameter". The presence of an order parameter is typical of phase transitions of the second kind, more commonly called *critical phenomena*.

28.2.3 The Potts model

The Potts model[6] may be defined on any graph (see Section 2.2), that is, any set of vertices (sites) with only (at most) one edge between each pair. This set of sites and edges constitutes the basic lattice, which in principle models some crystalline structure. A variable s_i, taking on N values, is defined on each site labelled i. For simplicity, we call this variable *spin*. Dynamics is introduced by supposing that only adjacent spins interact, and that with an interaction energy

$$e_{ij} = -J\,\delta_{s_i s_j},$$

where δ is a Kronecker delta. The total energy will be

$$E = -J\sum_{(ij)} \delta_{s_i s_j},$$

the summation being on all edges (ij). Then, with $K = J/kT$, the partition function for a M-site lattice will be

$$Q_M = \sum_s \exp\left[K\,\Sigma_{(ij)}\delta_{s_i s_j}\right],$$

---

[6] We follow here the Bible of lattice models: Baxter 1982.

the summation being over all the possible configurations

$$s = (s_1, s_2, \ldots, s_M).$$

The Ising model, with cyclic boundary conditions, is the particular case with $N = 2$ and K replaced by $2K$. Despite the great generality of this definition on generic graphs, we shall only talk of lattices formed with squares. The main point for what follows is that Q_M may be obtained as the sum of all the entries of a certain transfer matrix T analogous to (28.14). This matrix T turns out to be factorized into the product of simpler matrices, the "local transfer matrices", which are intimately related to the projectors E_i of the Temperley–Lieb algebra (Section 17.6).

A surprising outcome is that the partition function for the Potts model can be obtained as a Jones polynomial for a knot related to the lattice in a simple way. Given a square lattice as that of Figure 28.1, with the interactions just defined, consider the $N^n \times N^n$ matrices $E_1, E_2, \ldots, E_{2n-1}$, with

$$(E_{2i-1})_{s,s'} = \frac{1}{\sqrt{N}} \prod_{j \neq i=1}^{n} \delta_{s_j s'_j} \tag{28.22}$$

and

$$(E_{2i})_{s,s'} = \sqrt{N}\, \delta_{s_i s_{i+1}} \prod_{j=1}^{n} \delta_{s_j s'_j}. \tag{28.23}$$

Let us give some examples, with the notation $|s> = |s_1, s_2, s_3, \ldots, s_n>$:

$$< s|E_2|s' > = \delta_{s_1 s_2}(\delta_{s_1 s'_1} \delta_{s_2 s'_2} \cdots \delta_{s_n s'_n}),$$

$$< s|E_4|s' > = \delta_{s_2 s_3}(\delta_{s_1 s'_1} \delta_{s_2 s'_2} \cdots \delta_{s_n s'_n}),$$

$$\cdots$$

$$< s|E_{2n-2}|s' > = \delta_{s_{n-1} s_n}(\delta_{s_1 s'_1} \delta_{s_2 s'_2} \cdots \delta_{s_n s'_n}),$$

$$< s|E_1|s' > = \tfrac{1}{N} \delta_{s_2 s'_2} \cdots \delta_{s_n s'_n},$$

$$< s|E_3|s' > = \delta_{s_1 s'_1} \tfrac{1}{N} \delta_{s_3 s'_3} \delta_{s_4 s'_4} \cdots \delta_{s_n s'_n},$$

$$< s|E_5|s' > = \delta_{s_1 s'_1} \delta_{s_2 s'_2} \tfrac{1}{N} \delta_{s_4 s'_4} \cdots \delta_{s_n s'_n},$$

$$\cdots$$

$$< s|E_{2n-1}|s' > = \delta_{s_1 s'_1} \delta_{s_2 s'_2} \cdots \delta_{s_{n-1} s'_{n-1}} \tfrac{1}{N}.$$

Thus, in the direct product notation, if E is the identity $N \times N$ matrix, the even-indexed matrices are

$$E_{2i} = \sqrt{N}\, \delta_{s_i s_{i+1}} E \otimes E \otimes E \otimes E \otimes E \ldots \otimes E$$
$$= \sqrt{N}\, \delta_{s_i s_{i+1}} (E^{\otimes n}). \tag{28.24}$$

Matrix E_{2i} is, thus, a diagonal matrix, with entries $\sqrt{N}\, \delta_{s_i s_{i+1}}$. The odd-indexed matrices are

$$E_{2i-1} = E \otimes E \otimes E \otimes E \otimes E \otimes \ldots \otimes \left[1/\sqrt{N}\right] \otimes \ldots \otimes E \otimes E$$
$$= E^{\otimes(i-1)} \otimes \left[1/\sqrt{N}\right] \otimes E^{\otimes(n-i)}, \tag{28.25}$$

where $[1/\sqrt{N}]$, which is in the i-th position, is a $N \times N$ matrix (also a projector) with all the entries equal to $1/\sqrt{N}$. The notation is purposeful: such E_k's satisfy just the defining relations of the above mentioned Temperley–Lieb algebra with $M = 2n - 1$, and Jones index $= N$. By the way, we see that the Jones index is in this case just the dimension of the "spin" space.

We introduce the local transfer matrices

$$V_j = I + \frac{v}{\sqrt{N}} E_{2j} \quad \text{and} \quad W_j = \frac{v}{\sqrt{N}} I + E_{2j-1} \tag{28.26}$$

with I the identity matrix and $v = e^K - 1$. We can also introduce the Kauffman decomposition (see Section 14.3.5)

$$\asymp \; = \;)(\; + \; \frac{v}{\sqrt{N}} \; \overset{\cup}{\cap} \, , \tag{28.27}$$

which means that the inverse is

$$\asymp \; = \; \frac{v}{\sqrt{N}} \;)(\; + \; \overset{\cup}{\cap} \, . \tag{28.28}$$

The bubble normalization is

$$\bigcirc = \sqrt{N} \, . \tag{28.29}$$

There will be two global transfer matrices, given by

$$V = \exp\{K(E_2 + E_4 + \ldots + E_{2n-2})\} = \exp\left[K \sum_{j=1}^{n-1} \delta_{s_j s_{j+1}}\right] (E^{\otimes n})$$

and

$$W = \prod_{j=1}^{n} \left[vI + \sqrt{N} E_{2j-1}\right] .$$

They can be rewritten as

$$V = \prod_{j=1}^{n-1} \left[I + \frac{v}{\sqrt{N}} E_{2j} \right]$$

$$= \prod_{j=1}^{n-1} \left[\overset{2j}{)}\overset{2j+1}{(} + \frac{v}{\sqrt{N}} \overset{2j}{\cup} \overset{2j+1}{\cap} \right]$$

$$= \prod_{j=1}^{n-1} \left[\overset{2j}{\times} \right] \tag{28.30}$$

and

$$W = N^{n/2} \prod_{j=1}^{n} \left[\frac{v}{\sqrt{N}} \overset{2j-1}{)}\overset{2j}{(} + \overset{2j-1}{\cup}\overset{2j}{\cap} \right]$$

$$= N^{n/2} \prod_{j=1}^{n} \left[\overset{2j-1}{\times} \right] . \tag{28.31}$$

We now look at these transfer matrices in terms of the $(2n-1)$ generators of the braid group B_{2n}. They are

$$V = \sigma_2 \sigma_4 \ldots \sigma_{2n-2} \quad \text{and} \quad W = \sigma_1^{-1} \sigma_3^{-1} \ldots \sigma_{2n-1}^{-1}.$$

In the case of a $n \times m$ Potts lattice, the partition function is

$$Q_{nm} = \xi^T V W V W \ldots V \xi = \xi^T T \xi,$$

where ξ is a column vector whose all entries are equal to 1. There are m V's and $(m-1)$ W's in the product. To sandwich the matrix T between ξ^T and ξ is a simple trick: it means that we sum all the entries of T.

Let us now try to translate all this into the diagrammatic language. The sum over all configurations is already accounted for in the matrix product, as the index values span all the possible spin values. The question which remains is: how to put into the matrix-diagrammatic language the summation over the entries of the overall transfer matrix? The solution comes from the use of the projectors. In order to see it, let us take for instance the case $n = m = 2$. In this case the diagrams have 4 strands,

$$V = \left(I + \frac{v}{\sqrt{N}} E_2 \right) \quad \text{and} \quad W = \left(vI + \sqrt{N} E_1 \right)\left(vI + \sqrt{N} E_3 \right).$$

The matrix involved will be $T = VWV$, which is an element of B_4:

$$T = VWV = \sigma_2 \sigma_1^{-1} \sigma_3^{-1} \sigma_2.$$

We have to sum over all the values of the indices a, b, c, d, e, and f in Figure 28.3 (left). The result wished for is obtained by adding projectors before and after the diagram as in the center diagram of Figure 28.3, and then "taking the trace", that is, closing the final diagram. This closure is represented by taking identical labels for the corresponding extreme points — which is just closure in the sense of knot theory. Of course, there will be

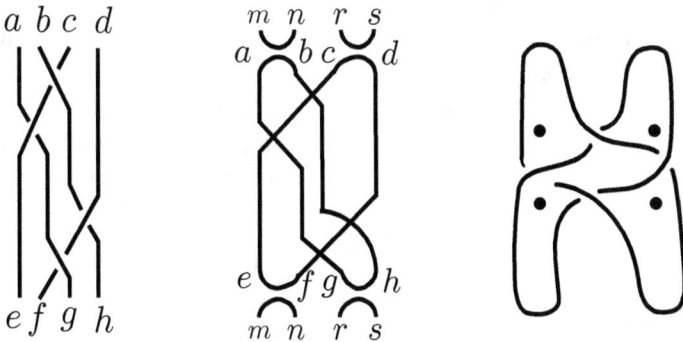

Fig. 28.3   Matrix T with indices (left), with added projectors (center), and its closure (right).

two extra factors from the bubbles, which must be extracted. This solution is general: for $M = 2n$ vertices, we add $2n$ projectors and then close the result, obtaining n extra bubbles which must then be extracted. Thus, the partition function is

$$Q = N^{n/2} < K > . \qquad (28.32)$$

The general relationship is thus the following: given a lattice, draw its "medium alternate link" K, which weaves itself around the vertices going alternatively up and down the edges. Figure 28.3 (right) shows the case $m = n = 2$, which corresponds simply to T closed by pairs of cups. To each edge of the lattice will correspond a crossing, a generator of B_n (or its inverse). Vertices will correspond to regions circumvented by loops. With the convenient choice of variables given above, the partition function is the Jones polynomial of the link.

In these lattice models, the lattice itself has been taken as fixed and regular, dynamics being concentrated in the interactions between the spin variables in the vertices. In the study of elastic media and glasses, this regularity is weakened and the variables at the vertices acquire different values and meanings.

28.2.4 *Cayley trees and Bethe lattices*

Suppose we build up a graph in the following way: take a point p_0 as an original vertex and draw q edges starting from it. To each new extremity add again $(q-1)$ edges. Thus q is the coordination number, or degree of each vertex in the terminology of Section 2.2. The first q vertices constitute the "first shell", the added $q(q-1)$ ones form the "second shell". Proceed iteratively in this way, adding $(q-1)$ edges to each point of the r-th shell to obtain the "$(r+1)$-th shell". There are $q(q-1)^{r-1}$ vertices in the r-th shell. Suppose we stop in the n-th shell. The result is a tree with

$$V = \frac{q\big[(q-1)^n - 1\big]}{q-2}.$$

This graph is called a Cayley tree. It is used in Statistical Mechanics, each vertex being taken as a particle endowed with spin. The partition function will be the sum over all possible spin configurations. There is a problem, though. The number of vertices in the n-th shell is not negligible with respect to V, so that one of the usual assumptions of the thermodynamical limit — that border effects are negligible — is jeopardized. One solution is to take $n \to \infty$, consider averages over large regions not including last-shell vertices, and take them as representative of the whole system. The tree so obtained is called the Bethe lattice and the model is *the Ising model* on the Bethe lattice. Its interest is twofold: it is exactly solvable and it is a first approximation to models with more realistic lattices (square, cubic, etc).

28.2.5 *The four-color problem*

The intuitive notion of a map on the plane may be given a precise definition in the following way. A *map* M is a connected planar graph G and an embedding (a drawing) of G in the plane $\mathbb{E}^2$. The map divides the plane into components, the *regions* or *countries*. G is the *underlying graph* of M, each edge corresponding to a piece of the boundary between two countries. Actually, one same graph corresponds to different maps, it can be drawn in different ways.

However, there is another graph related to a given map M: place a vertex in each country of M and join vertices in such a way that to each common border correspond an edge (as we have done in drawing the graph for the Königsberg bridges in § 2.2.6). This graph is the *dual graph* of M, denoted $D(M)$. When we talk of coloring a map we always suppose that no two regions with a common border have the same color. The celebrated four-color conjecture says that 4 colors are *sufficient*. That 4 colors are the minimum necessary number is easily seen from some counter-examples. That they are also sufficient is believed to have been demonstrated in the 1970's by Appel and Haken. The "proof", involving very lengthy computer checkings and some heuristic considerations, originated a warm debate.[7] What matters here is that the question is a problem in graph theory and related to lattice models. In effect, the problem is equivalent to that of coloring the vertices of the dual graph with different colors whenever the vertices are joined by a common edge. Or, if we like, to consider the dual graph as a lattice, and colors as values of a spin variable, with the proviso that neighbouring spins be different. Given a graph G, the number $P(G, t)$ of colorings of G using t or fewer colors may be extended to any value of the variable t. It is then called a *chromatic polynomial*. It comes not as a great surprise that such polynomials are related to partition functions of some lattice models in Statistical Mechanics.

28.3   Elasticity

28.3.1   *Regularity and defects*

Despite its position as a historical source of geometrical terminology, the language of Elasticity Theory takes nowadays some liberties with respect to current geometrical jargon. There are differences concerning basic words, as happens already with "torsion", taken in a more prosaic sense, and also some shifting in the nomenclature, even inside the Elasticity community. Texts on elasticity keep much of lattice language in the continuum limit and make use of rather special names for geometrical notions. Thus, "local system of lattice vectors" is used for Cartan moving frames and "lattice correspondence functions" for "moving frame components". "Distant parallelism" is the eloquent expression for "asymptotic flatness". And differentials are frequently supposed to be integrable.

We try here to present a simple though general formulation, the simplest we have found seemingly able to accommodate coherently the main

---

[7] For a thorough account, see Saaty & Kainen 1986.

concepts. The formalism is in principle applicable to crystals, elastic bodies and glasses in the continuum limit. When talking about "crystals", we think naturally of some order or periodicity at the microscopic level. Amorphous media like glasses, however, can be considered in the same approach, provided some defects are added to the previous crystalline regularity. We start thus from the usual supposition about microscopic regularity in crystals and deform the medium to obtain a description of amorphous solids. The continuum approximation to an elastic body is taken as the limit of infinitesimal lattice parameter. We shall consequently use the word "elasticity" in a very broad sense, so as to include general continuum limits of regular and irregular crystals, with preference for the latter. Some at least of the "amorphous" aspects of glasses can be seen as purely geometrical, and adding defects amounts to attributing torsion and/or curvature to the medium, which makes of such systems physical gateways into these geometrical concepts.

Commentary 28.3   Regularity means symmetry, usually under translations and/or rotations. Take the simple example of a 2-dimensional square lattice (Figure 28.1). Translations and rotations are discrete: the regular "crystal" is invariant under discrete translations of a multiple of the lattice parameter, and rotations of angles which are integer multiples of $\pi/2$. These rotations constitute the so-called rotation group of order 4: the only generator is $\exp[i\pi/2]$, so that it is a cyclic group (Section 14.1.3). ◄

The sources of deformations may be external or internal. The first case is the main subject of the classical texts on elasticity. Forces are applied on the system through their surface. The main objective then is to find the relation between the applied stress and the internal strain, which for small, reversible deformations, is given by a Hooke's law. The interaction between the atoms (or molecules) at the vertices should be represented by realistic potential wells, but a simple view is given by their first approximation. The first approximation to any reasonable potential well is a harmonic oscillator, so that a rough qualitative model is obtained by replacing the bonds by springs.[8]

Internal deformations are the principal concern of the theory of glasses and amorphous media. They arise from defects, and defects are of many kinds, but there are two main types of internal deformations: dislocations and disclinations.

There are some fluctuations in the very definitions of these concepts. Some authors define a dislocation simply as any linear defect, and a discli-

---

[8] For an illuminating discussion, see Askhar 1985.

nation as a defect leading to non-integrability of vector fields. For other, dislocations are failures of microscopic translational invariance and, in the same token, disclinations are related to failures of microscopic rotational invariance. In this line of thought, (geometric) torsion is then related to dislocations, and curvature to disclinations. In our inevitable geometrical bias, we shall rather adopt this point of view.

Commentary 28.4 Such notions are not always equivalent. You can, for example, distort a space to become S^3, which is curved and has rotation invariance. ◀

Commentary 28.5 Beauty is sometimes related to slightly broken symmetry. And some masterpieces are what they are because they are slightly uncomfortable to the eye. Some of Escher's woodprints, such as the celebrated "Waterfall", are good illustrations of torsion as engendered by a line of dislocations. Euclidean perspective goes wrong in such a space and our euclidean eyes are at a loss. ◀

In order to help ideas sinking in, let us see in a simple 2-dimensional example how such deformations can bring about curvature. We shall talk of an imaginary 2-dimensional semi-conductor. It is possible to pave the plane with regular hexagons because the internal angle at each hexagon vertex is $2\pi/3$. Regular pentagons would not do it because the angles at the vertices are not of the form $2\pi/N$, with N an integer. If an edge collapses so that one of the original hexagons becomes a pentagon, an angle defect would come out (Figure 28.4). Nevertheless, it is possible to tile a *sphere* with pentagons: a possible polyhedron (§ 2.3.4) of S^2 is the pentagon-faced dodecahedron (Figure 28.5). On the other hand, we can pave a *hyperboloid*

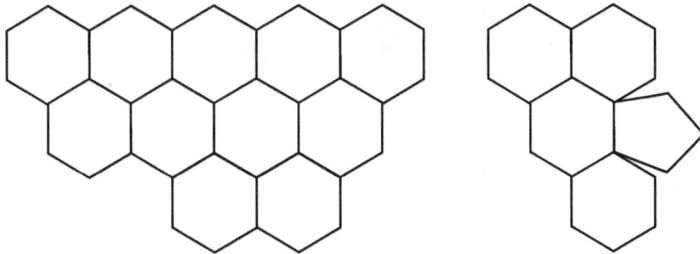

Fig. 28.4   A perfect tiling with hexagons, and one with a defect.

with heptagons, which signals to positive curvature when we remove an edge, and negative curvature if we add one. The crystal would be globally transformed if all the edges were changed, but the presence of a localized defect would only change the curvature locally. The local curvature will thus be either positive or negative around a defect of this kind. It is a general rule

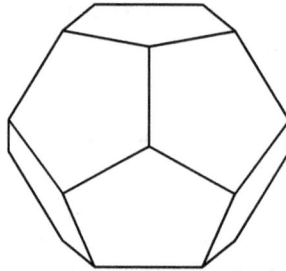

Fig. 28.5   The pentagon-faced dodecahedron.

that, when a localized defect is inserted in a lattice, it deforms the region around but its effect dies out progressively with distance. Figure 28.6 shows a typical case of (two-dimensional) dislocation through the insertion of a limited extra line. Experiments with a paper sheet can be of help here.

Well, in real media we have to do with atoms placed at the vertices. It happens in some amorphous semi-conductors[9] that, due to the presence of impurities (sometimes simply hydrogen in the realistic 3-dimensional cases), some of the edges do collapse, so that two adjoining hexagons become pentagons with only a common vertex. In the rough model with springs, some of then acquire a large spring constant and the oscillators become very steep. We might think that this would lead to a situation in which some of the cells would be irregular, with sides of different lengths. Nevertheless, at the microscopic level, the edges — distances between the atoms — are fixed by the inter-atomic potentials. In principle, they correspond to minimal values of the energy in these potentials. It happens in some cases that the distances are kept the same. In the more realistic 3-dimensional case, the suggestion to drop or to add a wedge comes from the experimental evidence of the presence of rings with one-less or one-more atoms in an otherwise regular lattice in amorphous semi-conductors.[10] Thus, the inter-atomic potentials require regular polyhedra to tile the system, which is consequently deformed into some spherical geometry. Some other physical systems require instead that an extra edge is added so as to form heptagons

---

[9] Harris 1975; Kléman & Sadoc 1979.
[10] Sadoc & Mosseri 1982.

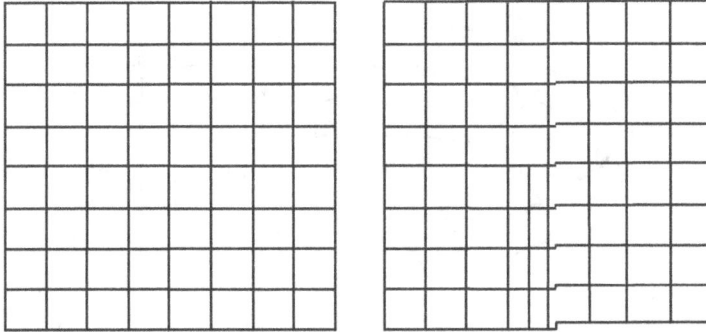

Fig. 28.6   A dislocation through the insertion of a limited extra line.

and leading to a hyperbolic geometry.

The original flat space is thus curved by the presence of "impurities". Remember that, at 2 dimensions, the sign of the curvature $1/rr'$ at a given point is very easy to see. Trace two tangent circles perpendicular to each other at the point. Their radii r and r' are the so called curvature radii. If both have the same sign, the curvature is positive. If they have opposite signs, negative. If you prefer, in the first case there is an osculating sphere at the point, and in the second, a hyperboloid. Figure 28.7 shows the case of square-to-triangle collapse. At the right, the vector field represented by the crosses comes back to itself after a trip around a loop circumventing no defect. At the left, the vector field is taken along a loop around the defect, and is rotated of an angle θ at the end of the trip. This vindicates the view of non-integrable vector fields, which anyhow lies behind the very notion of curvature. Recall that, taken along an infinitesimal geodesic loop, a vector field V is changed by (§ 9.4.13)

$$\delta V^k = - R^k{}_{rij} V^r dx^i \wedge x^j.$$

28.3.2   *Classical elasticity*

There are two main properties characterizing the "euclidean crystal" we started from. First, we can measure the distance between two neighbouring points using the euclidean metric,

$$dl = (\delta_{ab} dx^a dx^b)^{1/2}. \tag{28.33}$$

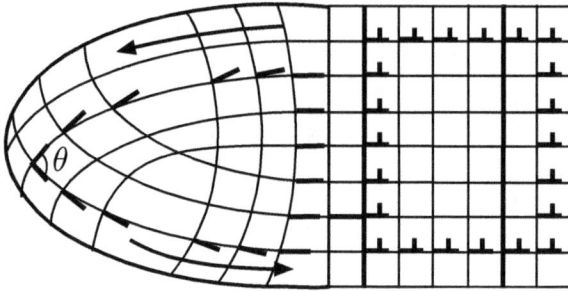

Fig. 28.7   A square-to-triangle collapse.

Second, we can say whether or not two vectors at distinct points are parallel. We do it by transporting one of them along the lines defining the lattice into the other's position, in such a way that the angles it forms with the lines are kept the same. If their direction and sense coincide when they are superposed, they are parallel. This corresponds to parallel-transporting according to the trivial euclidean connection (the Levi–Civita connection for the metric δ_{ab}), whose covariant derivatives coincide with usual derivatives in a Cartesian natural basis, and for which the lattice lines are the geodesics. The lattice itself play the role of a geodesic grid. We can place at each vertex a set of "lattice vectors" $\{e_a\}$, oriented along the lines and parallel-transported all over the lattice. Given the set at one vertex, we know it at any other vertex. The euclidean metric will fix $\delta_{ab} = (e_a, e_b)$ for this initial dreibein.

The first clear visible signal of a deformation is that the measure of distance between the points changes. In the general case, the change is different in different regions. Two neighbouring points initially separated by an euclidean distance (28.33) will, after the deformation, be at a different distance, though we keep measuring it with an euclidean rule. We represent this by

$$dl' = \left[g_{ij}(x)dx^i dx^j\right]^{1/2}, \qquad (28.34)$$

as if the distance were given by some other, point-dependent metric g_{ij}. The $\{dx^i\}$ are the same as the previous $\{dx^a\}$: we are simply concentrating the deformation in the metric. The euclidean and the new metric tensors are related by some point-dependent transformation $h^a{}_i(x)$,

$$g_{ij} = \delta_{ab} h^a{}_i h^b{}_j = h^a{}_i h^b{}_j (e_a, e_b) = (e_i, e_j). \qquad (28.35)$$

Each $e_i = h^a{}_i e_a$ is a member of the new dreibein. The new metric is given by the relative components of these e_i's, measured in the old euclidean metric. This is to say that the initial covector basis $\{dx^a\}$ is related to another covector basis $\{\omega^i\}$, dual to $\{e_i\}$, by

$$dx^a = h^a{}_i \omega^i.$$

Of course,

$$g_{ij}\,\omega^i \omega^j = dl^2.$$

We have, so, just a 3-dimensional example of "repére mobile" (§ 7.3.12), with all its proper relationships and a metric defined by it (see also Sections 9.3.5 and 22.2.1). As the deformations are supposed to be contiguous to the identity, the h^a_i's are always of the form

$$h^a{}_i = \delta^a_i + b^a{}_i, \tag{28.36}$$

for some fields $b^a{}_i$, which represent the departure from the trivial dreibein related to the unstrained state. The new metric has the form

$$g_{ij} = \delta_{ij} + 2u_{ij},$$

where

$$u_{ij} = \tfrac{1}{2}\left(b_{ij} + b_{ji} + \delta_{ab}b^a{}_i b^b{}_j\right) \tag{28.37}$$

is the *strain tensor*. If the new basis is holonomic, then

$$b^a{}_i = \partial_i u^a$$

for some field $u^a(x)$, called the *deformation field*. In this case the field of deformations is the variation in the coordinates, given by

$$x'^k = x^k + u^k(x).$$

This gives

$$dx'^k = dx^k + du^k(x) = dx^k + \partial_j u^k(x)dx^j,$$

so that the length element changes by

$$dl^2 \to dl'^2 = dx^k dx^k + \partial_j u^k(x)dx^k dx^j$$
$$+ \partial_j u^k(x)dx^j dx^k + \partial_i u^k(x)\partial_j u^k(x)dx^i dx^j.$$

The derivative $w_{ij} = \partial_i u_j$ is the *distortion tensor*. To first order in u and its derivatives,

$$dl'^2 = (\delta_{jk} + 2u_{jk})dx^j dx^k,$$

where

$$u_{jk} = w_{(jk)} = \tfrac{1}{2}\left(\partial_j u_k + \partial_k u_j\right) \tag{28.38}$$

is the strain tensor for this holonomic case. Notice from

$$g_{ij} = \delta_{ij} + 2u_{ij} = \delta_{ab}h^a{}_i h^b{}_j$$

that the Cartan frames (dreibeine) are

$$h^a{}_i = \delta^a_i + w^a{}_i. \tag{28.39}$$

The fields b^a_i of (28.36), consequently, generalize the distortion tensor to the anholonomic case. The holonomic case comes up when they are the derivatives of some deformation field. Some authors define defects as the loci of singular points, and dislocations as lines of singularity of the deformation field. This only has a meaning, of course, when this field exists.

Summing up, a deformation creates a new metric and new dreibeine. It changes consequently also the connection. We might think at first the new connection to be the Levi–Civita connection of the new metric, but here comes a novelty. The connection is, in principle, a metric-independent object and can acquire proper characteristics. For example, it can develop a non-vanishing torsion. Impurities, besides changing the distances between the atoms, can also disrupt some of the bonds, in such a way that the original (say) hexagon is no more closed. They may become open rings in

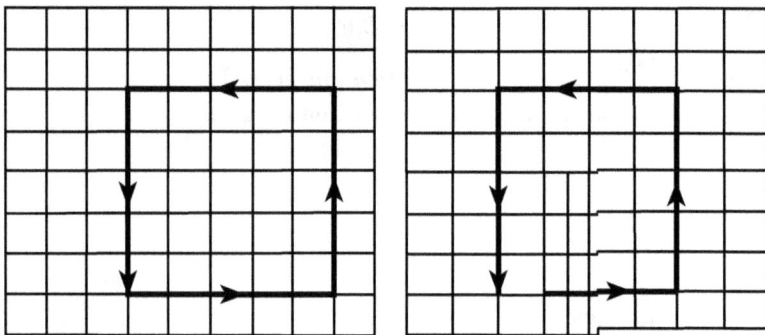

Fig. 28.8   Effect of a dislocation on the grid of Figure 28.6.

the plane, but they may also acquire a helicoidal aspect in 3 dimensional media. The euclidean geodesic grid collapses. These deformations are called "dislocations" and are of different kinds. Figure 28.8 shows what happens to a loop in the case of the dislocation of Figure 28.6. If we keep using the original geodesic grid, we find a breach: the grid is destroyed and there are no more "infinitesimal geodesic parallelograms" (§ 9.4.14). The new dreibeine fail to be parallel-transported. The torsion T mesures precisely this failure, because there is a theorem (the Ricci theorem) which says that, given g as above and a torsion T, the connection Γ is unique.

The presence of (geometrical) torsion in amorphous media is confirmed by experimental measurements, and is, in physical terminology, related to two physical quantities: the Nye index and the Burgers vector.[11] These quantities measure the cleavage and are related to torsion as follows. Take a loop in the undeformed crystal as in the left diagram of Figure 28.8 (such a loop is called a Burgers circuit in elasticity jargon). Once the crystal is deformed, it fails to close into a loop. The vector from the starting point to the final point of a curve, which would be a loop in the undeformed crystal, is called the *Burgers vector*. The situation is simpler when a deformation field does exist. Consider in this case a closed line γ before the deformation, and a point p on it, which we shall take as its coincident initial and final endpoints. After the deformation, when the deformed γ' is no more closed, p goes to two distinct points, p' and p''. The Burgers vector is then defined as

$$B^k := - \int_{\gamma'} du^k = - \int_{\gamma'} \frac{\partial u^k}{\partial x^i}\, dx^i = - \int_{\gamma'} w_i{}^k dx^i = u^k(p'') - u^k(p').$$

It is clearly a measure of the disruption. The distortion tensor appears as a linear density for the Burgers vector. The Nye index $\alpha^k{}_i$ is introduced by the expression

$$B^k = 2 \int_{\gamma} \alpha^k{}_i dx^i \qquad (28.40)$$

in general, eventually also in the non-integrable case. It is a line integral of the form $b^k{}_i dx^i$. In the continuum case, the classical notion of torsion is related precisely to this disruption of infinitesimal geodesic parallelograms. The failure δx^k, not necessarily integrable, of such closure is precisely measured by

$$\delta x^k = T^k{}_{ij}\, dx^i \wedge dx^j. \qquad (28.41)$$

Recall that on a 3-dimensional euclidean space,

$$dx^i \wedge dx^j = \varepsilon^{ij}{}_k dx^k.$$

[11] Burgers 1940.

Consequently,

$$\delta x^k = T^k{}_{ij}\varepsilon^{ij}{}_s dx^s = 2\alpha^k{}_s dx^s$$

and

$$\alpha^{rk} = \tfrac{1}{2}\,\varepsilon^{rij}T^k{}_{ij}. \tag{28.42}$$

Thus, *the Nye index is precisely the dual to the torsion field.* The torsion is

$$T^a{}_{ij} = \partial_i h^a{}_j - \partial_j h^a{}_i + \Gamma^a{}_{bi}h^b{}_j - \Gamma^a{}_{bj}h^b{}_i,$$

so that

$$T^k{}_{ij} = h_a{}^k T^a{}_{ij} = \Gamma^k{}_{ji} - \Gamma^k{}_{ij}.$$

Thus,

$$\alpha^{rk} = -\,\varepsilon^{rij}\Gamma^k{}_{[ij]}, \tag{28.43}$$

where $\Gamma^k{}_{[ij]}$ is the antisymmetric part of $\Gamma^k{}_{ij}$.

What happens to the connection in the deformed case? As we also want to keep vector moduli and angles with a coherent meaning, we should ask that the connection preserves the metric:

$$dg_{ij} = (\Gamma_{ijr} + \Gamma_{jir})\,dx^r. \tag{28.44}$$

The metric can only fix a piece of the symmetric part of Γ_{ijr} in the two first indices. To determine Γ_{ijr} completely we need to know the torsion through experimental measurements of the Nye index.

28.3.3 Nematic systems

In the above discussion of torsion and curvature, only positional degrees of freedom were supposed for the constituent molecules. They have been treated as punctual, only the positions of their centers of gravity have been considered. Deformations were supposed to introduce new metrics and connections on the lattice-manifold. The situation is consequently related only to the linear frame bundle (Section 9.4).

We can imagine that adding internal degrees, like in the spin lattice models of Section 28.2 above, gives a situation more akin to that of gauge fields, with internal spaces as fibers,[12] and in which general principal bundles (Sections 9.5 and 9.6) are at work. Molecules have in general internal degrees, the simplest of which is, for non-spherical molecules, orientation.

---

[12] On such "spin glasses" and gauge field theories, see Toulouse & Vannimenus 1980.

In a solid crystal there is perfect positional order: it represents the case in which the centers of gravity are perfectly established at the fixed sites. Classical elasticity theory studies precisely small departures from this regular case. Melting takes the crystal into a state of positional disorder, a liquid.

There are systems in which the solid-liquid transition is not so simple, but takes place in a series of steps in which order is progressively lost. And in some situations a system can be stable in some intermediate state. It is in this case a "liquid crystal". For instance, the system can lose the ordering in 2 dimensions while retaining a periodic order in the third. Such a phase is called *smectic*. There are phase transitions related to change of orientation in liquids, solids and smectic media.

The quantum Ising model considers in each site a two-valued spin. One might imagine cases more "classical", in which "spin" takes on values in a continuous range. In the nematic crystals we consider, instead of spin, an "internal" variable describing the orientation of the molecules. This case is more involved, as different orientations between neighbouring molecules imply distinct couplings between them. And also, a "direction" is less than a vector, because two opposite vectors correspond to the same direction.

Some organic systems, as well as solid hydrogen, show a high-temperature phase which is positionally ordered but orientationally disordered (*plastic crystals*).

Certain organic liquids have a low-temperature phase which is positionally disordered but orientationally ordered, with the molecules oriented along a preferential direction. Such systems usually have a parity symmetry: sufficiently large subdomains do not change if all the three axes are reversed. This is a consequence of the requirement that the molecules orientation be the only origin of anisotropy. Parity invariant systems (crystals, liquids, or liquid crystals) with an orientation degree of freedom are generically called nematic systems. In most cases the molecules are of ellipsoidal shape ("batons"), so that only the direction, not the sense of the molecule orientation, is of import (Figure 28.9). The state of a molecule can be characterized by (say) a versor $\mathbf{n}$ along its major axis (called the *director*). Thus, the director can be seen, in the continuum limit, as a field. As states $\mathbf{n}$ and $-\mathbf{n}$ coincide (Figure 28.9, right), the director field has values in the half-sphere

$$S^2/Z_2 = PR^2,$$

a projective space. It is not a vector field, it is a direction field. The average value of the direction field has the role of an order parameter, which vanishes above the critical temperature and becomes significant below it.

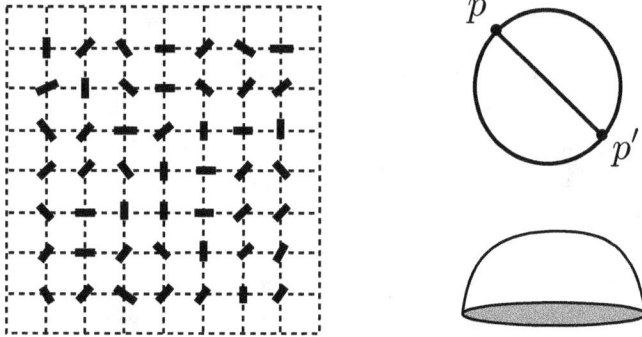

Fig. 28.9 A baton distribution is characterized by a director field, which has values in the projective space PR^2.

Commentary 28.6 Systems with the same general characteristics, but without parity invariance, are thermodynamically unstable — they "decay" into stable systems of another kind, "cholesterics". ◄

Consider a finite nematic system. By that we mean that the system covers a compact domain V in $\mathbb{E}^3$ with boundary ∂V. The distribution of the direction field will be fixed up when the values of n are given on ∂V.

Two standard examples[13] come to the scene when we consider an infinite cylindrical system with axis (say) along the axis Oz. In cylindrical coordinates (ρ, z, φ), the field $n(r)$ will not depend on z because it is infinite in that direction, and will not depend on ρ because there is not in the system any parameter with the dimension of length, in terms of which we could write a non-dimensional variable like $n(r)$. Thus, we have only to consider a plane transverse section of the cylinder and the only significant variable is the angle φ: $n(r) = n(\varphi)$. The two cases correspond to two distinct kinds of boundary conditions: the values of $n(r)$ at the boundary, $n(r)|_{\partial V}$, are either orthogonal to ∂V or parallel to ∂V. Continuity will then fix the field all over the section. It is clear then, by symmetry, that $n(r)$ is in both cases ill-defined on the Oz axis, which is supposed perpen-

---

[13] Landau & Lifchitz 1990; the last editions have chapters concerning dislocations and nematic systems, written respectively in collaboration with A.M. Kossevitch and L.P. Pitayevski.

dicular to the plane at the origin in Figure 28.10. This line of singularity Oz is a disclination line.

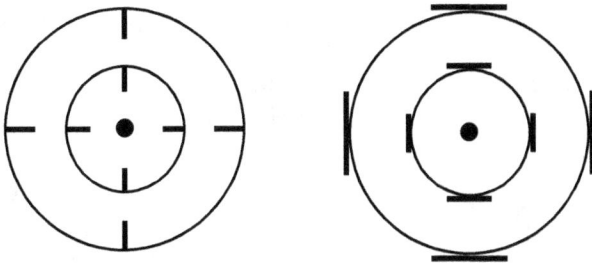

Fig. 28.10   Relations between topology and symmetry plus boundary conditions.

28.3.4   *The Franck index*

An indubitable beautiful physical example of topological number is the Franck index.[14] It would be a pure case of winding number were not for the fact that not a true *vector* field is involved, but a *direction* field. As a consequence, the resulting number ν can take half-integer values. Let us

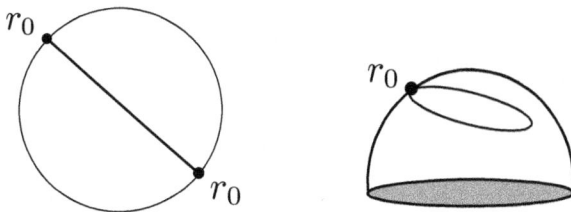

Fig. 28.11   A curve on RP^2 is a closed loop provided, in the correspond sphere, it crosses two antipodes.

proceed now to the general "winding number" definition. We shall traverse a closed path in the system and look at each point at the value of $n(r)$ in RP^2. Consider the value of $n(r_0)$ at the starting and final point r_0 in the system. As $n(r)$ is a physical, necessarily single-valued field, we start

---

[14] Franck 1951.

at a certain value $n(r_0)$ of RP^2, go around for a trip on RP^2 and come necessarily back to the same value $n(r_0)$. Thus, a closed curve in V is led into a closed curve in RP^2. But the curve on RP^2 can make a certain number of turns before coming back to the original point. The Franck index ν is precisely this number, the number of loops of the curve on RP^2 corresponding to one loop on the system.

It is easier to visualize things on the sphere S^2, provided we are attentive to the antipodes (see Figure 28.11). A closed curve on RP^2 has this difference with respect to a closed curve in S^2, that when we say "the same point" we can mean not only the same point on the sphere, but also its antipode, so that on RP^2 a curve is closed also if it connects two antipodes on S^2. A curve looping to the same point on the sphere will have integer index. But a curve connecting two antipodes will have a half-integer value of ν.

Looking at Figure 28.10, we see the particularly clear relationship between the topological characteristic, on one hand, and symmetry plus boundary conditions, on the other.

General references

Some additional basic references for this chapter are Pathria 1972, Toda, Kubo & Saitô 1998 and Timoshenko & Goodier 1970. Are also of interest Love 1944, Landau & Lifchitz 1990, Nabarro 1987 and de Gennes & Prost 1993.

Chapter 29

Propagation of Discontinuities

The relationship between a light-ray trajectory and the propagation of
the corresponding wavefront, locally normal to it, illustrates a beautiful
relationship between the ordinary and the partial differential equations
which — respectively — describe them. Looking at the wavefronts as
discontinuities, the trajectories appear as just the "characteristic" lines
along which they propagate.

29.1 Characteristics

Given a system of first-order *partial* differential equations, its solution can
always be obtained from the solution of a certain system of *ordinary* differential equations. The latter are called the 'characteristic equations' of the
original system. This is also true for some important higher-order equations. Such a conversion from partial-differential into ordinary equations
can be seen as a mere method for finding solutions. For some physical
problems, however, it is much more than that. The physical trajectory,
solution of Hamilton equations (which are ordinary differential equations)
in configuration space, is the characteristic curve of the solutions of the
Hamilton–Jacobi equations (which are partial differential equations). This
points to their main interest for us: the solutions of the characteristic equations (frequently called *the* characteristics) have frequently a clear physical meaning: particle trajectories, light rays, etc. The classical example is
Hamilton's approach to geometrical optics (Section 30.2).

There are two main views on characteristics:

(i) They are lines (or surfaces, or still hypersurfaces, depending on the
dimensionality of the problem) along which disturbances or discontinuities
propagate, in the limit of short wavelength (geometric acoustics and/or
geometric optics); thus, they appear as lines of propagation of the "quickest

perturbations". The surface (or line, or still hypersurface) bordering the region attained by the disturbance originated at some point, called the characteristic surface, is conditioned by causality, which reigns sovereign in this point of view.

(ii) They are lines "perpendicular" to the wavefronts; this approach relates to the Cauchy problem of partial differential equations.

The first view has been the traditional one, but the second won the front scene in the forties, with Luneburg's approach to Geometrical Optics. He emphasized the identity of the eikonal equation and the equation of characteristics of Maxwell's equations which, as we shall see, governs the propagation of discontinuous solutions.

Let us mention two of the noblest physical examples. One appears in the above mentioned relationship between Hamilton equations and Hamilton–Jacobi equations, described in Section 26.1. The hamiltonian flow is generated by a field X_H, which gives the time evolution of a dynamical function $F(q, p, t)$ according to the Liouville equation (Section 25.3), a partial differential equation

$$\frac{d}{dt} F(q, p, t) = \sum_{i=1}^{2n} X_H^i(x) \frac{\partial}{\partial x^i} F(x, t) \tag{29.1}$$

with some initial condition

$$F(q, p, 0) = f_0(q, p). \tag{29.2}$$

If $F_t(x)$ is a flow, the solution will be $F(x, t) = f_0(F_t(x))$. The orbits of the vector field X_H are the characteristics of equation (29.1).

The other example is given by the eikonal equation

$$\left(\frac{\partial \Psi}{\partial x^1}\right)^2 + \left(\frac{\partial \Psi}{\partial x^2}\right)^2 + \cdots + \left(\frac{\partial \Psi}{\partial x^n}\right)^2 = 1, \tag{29.3}$$

in which Ψ is the optical length and the level surfaces of Ψ are the wavefronts (see Chapter 30).

29.2 Partial differential equations

The classical lore[1] on the characteristics of partial differential equations runs as follows (to make things more visible, we shall talk most of time about the two-dimensional case, so that we shall meet characteristic curves

---

[1] See Sommerfeld 1964b.

instead of characteristic surfaces or hypersurfaces). The most general second order linear partial differential equation will be of the form

$$A \frac{\partial^2 f}{\partial x^2} + 2B \frac{\partial^2 f}{\partial x \partial y} + C \frac{\partial^2 f}{\partial y^2} - F\left(x, y, f, \frac{\partial f}{\partial x}, \frac{\partial f}{\partial y}\right) = 0, \tag{29.4}$$

where the "source" term F is a general expression, not necessarily linear in f, $\partial f/\partial x$ and $\partial f/\partial y$. It is convenient to introduce the notations

$$p = \frac{\partial f}{\partial x}, \quad q = \frac{\partial f}{\partial y}, \quad r = \frac{\partial^2 f}{\partial x^2}, \quad s = \frac{\partial^2 f}{\partial x \partial y}, \quad t = \frac{\partial^2 f}{\partial y^2},$$

in terms of which the equation is written

$$Ar + 2Bs + Ct = F. \tag{29.5}$$

It follows also that

$$dp = r dx + s dy \tag{29.6}$$

and

$$dq = s dx + t dy. \tag{29.7}$$

Now we ask the following question: given a curve $\gamma = \gamma(\tau)$ in the (xy) plane, on which f and its derivative $\partial f/\partial n$ in the normal direction are given, does a solution exist? If f is given along γ, then $\partial f/\partial \tau$ is known. As from $\partial f/\partial n$ and $\partial f/\partial \tau$ we can obtain $\partial f/\partial x$ and $\partial f/\partial y$, f and its first derivatives, p and q, are known on γ. In order to find the solution in a neighborhood of γ, we should start by (in principle at least) determining the second derivatives r, s and t on γ. In order to obtain them we must solve Eqs. (29.6) and (29.7). The condition for that is that the determinant

$$\Delta = A dy^2 - 2B dx dy + C dx^2$$

be different from zero. There are then two directions dy/dx at each point (x, y) for which there are no solutions. The two families of curves on which $\Delta = 0$ are the characteristic curves. Along them one cannot find r, s and t from the knowledge of f, p and q. Thus, in this line of attack, one must require that γ be nowhere tangent to the characteristics. We shall see later the opposite case, in which γ just coincides with a characteristic. Once the non-tangency condition is fulfilled, there must be a solution in a neighbourhood of γ. The miracle of the story is that, when looking for the higher order derivatives in terms of the preceding ones, one finds, step by step, always the same condition, with the same determinant. Thus, if Δ is different from zero, f can be obtained as a Taylor series.

The characteristic equation

$$Ady^2 - 2Bdxdy + Cdx^2 = 0, \tag{29.8}$$

whose solutions correspond to

$$\frac{\partial y}{\partial x} = \frac{B \pm \sqrt{B^2 - AC}}{C}, \tag{29.9}$$

determines, in principle, two families of curves on the plane (xy), which are the characteristic curves.

There are three quite distinct cases, according to the values of the discriminant $B^2 - AC$, and this leads to the classification of the equations and of the corresponding differential operators appearing in (29.4):

$B^2 - AC < 0$: the equation is of elliptic type.
$B^2 - AC > 0$: the equation is of hyperbolic type.
$B^2 - AC = 0$: the equation is of parabolic type.

Notice that A, B and C depend at least on x and y, so that the character may be different in different points of the plane. Thus, the above conditions are to be thought of in the following way: if the discriminant is negative in all the points of a region D, then the equation is elliptic on D. And in an analogous way for the other two cases.

Only for the hyperbolic type, for which there are two real roots λ_1 and λ_2, is the above process actually applied. If the coefficients A, B, C are functions of x and y only, then these curves are independent of the specific solution of the differential equation (see the Klein–Gordon example below). The families of curves are then given by $\xi(x, y) = c_1$, which is the integral of

$$y' + \lambda_1(x, y) = 0,$$

and $\chi(x, y) = c_2$, which is the integral of

$$y' + \lambda_2(x, y) = 0.$$

Suppose the differential equation has a fixed solution $f = f_0(x, y)$. In order to pass into geometric acoustics and/or optics,

(i) we add to it a perturbation f_1. Usually certain conditions are imposed on such perturbations, conditions related to geometric acoustics or optics: f_1 is small, their first derivatives are small, but their second derivatives are relatively large. This means that f_1 varies strongly at small distances. It obeys the "linearized equation",

$$A\frac{\partial^2 f_1}{\partial x^2} + 2B\frac{\partial^2 f_1}{\partial x \partial y} + C\frac{\partial^2 f_1}{\partial y^2} = 0,$$

where, in the coefficients A, B and C, the function f is replaced by the solution f_0.

(ii) Then, we write f_1 in the form $f_1 = ae^{i\psi}$, with a large function (which is the eikonal), and "a" a very slowly varying function (small derivatives), to find the eikonal equation

$$A \left(\frac{\partial \psi}{\partial x} \right)^2 + 2B \left(\frac{\partial \psi}{\partial x} \right) \left(\frac{\partial \psi}{\partial y} \right) + C \left(\frac{\partial \psi}{\partial y} \right)^2 = 0. \qquad (29.10)$$

(iii) Finally, we find the ray propagation by putting

$$k = \frac{\partial \psi}{\partial x}, \quad \omega = -\frac{\partial \psi}{\partial y}, \quad \frac{dx}{dy} = \frac{d\omega}{dk},$$

the latter being the group velocity. The eikonal equation turns into the "dispersion relation"

$$Ak^2 - 2Bk\omega + C\omega^2 = 0,$$

and then

$$\frac{dx}{dy} = \frac{B\omega - Ak}{C\omega - Bk},$$

which is an alternative form of (29.9).

Commentary 29.1 Actually, we have been considering "scalar optics": f is a scalar. In the real case of optics, the procedure above must be followed for each component of the electric and magnetic fields, as well as the four-potential. ◄

As a very simple though illustrative case, consider the Klein–Gordon equation,

$$\frac{\partial^2 f}{\partial x^2} - \frac{1}{c^2} \frac{\partial^2 f}{\partial t^2} + m^2 c^2 f = 0.$$

As the equation is already linear, the perturbations will obey the same equation. The characteristics are

$$\frac{dx}{dt} = \pm c,$$

that is, the light cone. The dispersion relation for the eikonal equation will be

$$\omega = + kc.$$

The "source" term F of (29.4) is here the term $- m^2 c^2 f$, but it does not influence the characteristics. This is general: whenever the coefficients A, B

and C depend only on the independent variables x and t, the characteristics are independent of the special solution of the starting equation.

Only a few words on geometric acoustics.[2] In a medium with sound velocity c, the differential equations of the two families of characteristic curves C_+ and C_- are

$$\frac{dx}{dt} = v + c \quad \text{and} \quad \frac{dx}{dt} = v - c.$$

The disturbances propagate with the sound velocity with respect to a local frame moving with the fluid. The velocities $v + c$ and $v - c$ are the velocities with respect to the fixed reference frame. Along each characteristic, the fluid velocity v remains constant. Some perturbations are simply transported with the fluid, that is, propagate along a third characteristic C_0, given by

$$\frac{dx}{dt} = v.$$

In general, a disturbance propagates along the three characteristics passing through a certain point on the plane (x, t). It can nevertheless be decomposed into components, each one going along one of them.

Commentary 29.2 What we gave here is a telegraphic sketch of a large, wonderful theory. Systems of first order partial differential equations are equivalent to systems of Pfaffian forms, from which a systematic theory of characteristics can be more directly formulated.[3] ◄

29.3 Maxwell's equations in a medium

Geometrical optics was traditionally regarded as an asymptotic approximation, for large wave numbers, of the wave solutions of Maxwell's equations. In two series of lectures delivered in the nineteen-forties, Luneburg changed the tune. He noticed the identity of the eikonal equation and the equation of characteristics of Maxwell's equations. Think of a light signal emitted at $t = 0$, which attains at an instant t the points of a surface defined by some function $\psi(x, y, z) = ct$. The surface is a border, separating the region already attained by the waves from the region not yet reached by any field. In the "inner" side of the surface, the field has some nonvanishing value; at the other side, the field is zero. The wavefront $\psi(x, y, z)$ represents a discontinuity of the field, propagating at speed c, which may be point-dependent.

---

[2] Landau & Lifshitz 1989.
[3] Westenholz 1978; Choquet–Bruhat, DeWitt–Morette & Dillard–Bleick 1977.

Though a little more involved at the start, this point of view has the great advantage of treating light propagation no more as an approximation, but as a particular class of *exact* solutions of Maxwell's equations, light rays appearing in this view as lines along which discontinuities propagate. The equations, of course, coincide with those obtained in the short wavelength treatment.

Consider an isotropic but non-homogeneous medium which is otherwise electromagnetically inert (neither macroscopic magnetization nor electric polarization). This means that the electric and the magnetic permeabilities depend on the positions but not on the directions. In anisotropic media ϵ and μ become symmetric tensors, but we shall not consider this case here. Then, with

$$\boldsymbol{D} = \epsilon\,\boldsymbol{E} \quad \text{and} \quad \boldsymbol{B} = \mu\,\boldsymbol{H},$$

the sourceless Maxwell equations are

$$c\,\operatorname{rot}\boldsymbol{H} - \frac{\partial\boldsymbol{D}}{\partial t} = 0; \tag{29.11a}$$

$$c\,\operatorname{rot}\boldsymbol{E} + \frac{\partial\boldsymbol{B}}{\partial t} = 0; \tag{29.11b}$$

$$\operatorname{div}\boldsymbol{D} = 0; \tag{29.11c}$$

$$\operatorname{div}\boldsymbol{B} = 0. \tag{29.11d}$$

The second and the fourth may be obtained respectively from the first and the third by the duality symmetry:

$$\epsilon \leftrightarrow \mu, \quad \boldsymbol{E} \to -\boldsymbol{H}, \quad \boldsymbol{H} \to \boldsymbol{E}.$$

The energy is

$$W = \frac{1}{8\pi}\,(\boldsymbol{E}\cdot\boldsymbol{D} + \boldsymbol{H}\cdot\boldsymbol{B}),$$

and the Poynting vector (energy flux vector) is

$$\boldsymbol{S} = \frac{c}{4\pi}\,(\boldsymbol{E}\times\boldsymbol{H}).$$

Energy conservation is written

$$\frac{\partial W}{\partial t} + \operatorname{div}\boldsymbol{S} = 0.$$

A hypersurface[4] in $\mathbb{E}^n$, we recall, is an $(n-1)$-dimensional space immersed in $\mathbb{E}^n$. Consider the wavefront defined by ψ as a closed surface $\Gamma = \partial D \subset \mathbb{E}^3$, circumscribing the domain D whose characteristic function

---

[4] Gelfand & Shilov 1964. More details are given in § 7.5.15.

is a "step-function" $\theta(\psi)$. For later purposes, let us compute some integrals. The first one is:

$$\int_D \partial_k f = \int_{\mathbb{E}^3} \theta(\psi) \partial_k f = \partial_k \int_{\mathbb{E}^3} \theta(\psi) f - \int_{\mathbb{E}^3} \partial_k \theta(\psi) f$$

$$= \int_{\mathbb{E}^3} (\partial_k \psi) \delta(\psi) f = \int_{\Gamma} f(\partial_k \psi). \tag{29.12}$$

The second one is:

$$\int_D \partial_i V_j = \int_{\mathbb{E}^3} \theta(\psi) \partial_i V_j = -\int_{\mathbb{E}^3} \partial_i \theta(\psi) V_j$$

$$= \int_{\mathbb{E}^3} V_j(\partial_i \psi) \delta(\psi) = \int_{\Gamma} V_j(\partial_i \psi). \tag{29.13}$$

We compute also

$$\int_D (\text{rot } V)_k = \varepsilon_{kij} \int_D \partial_i V_j = \varepsilon_{kij} \int_{\mathbb{E}^3} \theta(\psi) \partial_i V_j = -\varepsilon_{kij} \int_{\mathbb{E}^3} \partial_i \theta(\psi) V_j$$

$$= \varepsilon_{kij} \int_{\mathbb{E}^3} V_j(\partial_i \psi) \delta(\psi) f = \varepsilon_{kij} \int_{\Gamma} (\partial_i \psi) V_j, \tag{29.14}$$

which in vector notation reads

$$\int_D \text{rot} \mathbf{V} = \int_{\Gamma} (\text{grad } \psi) \times \mathbf{V}.$$

The components $\partial_i \psi$ are proportional to the direction cosines of the normal to the surface. The unit normal will have, along the direction "k", the component

$$\frac{\partial_k \psi}{|\text{grad } \psi|} = \frac{\partial_k \psi}{\sqrt{\sum_i (\partial_i \psi)^2}}.$$

These results are in general of local validity, the surfaces being supposed to be piecewise differentiable.

Take the Maxwell equation $\text{div} \mathbf{D} = 0$. Its integral form will be obtained by integrating it on a domain D and using Eq. (29.13) above:

$$\int_D \text{div} \mathbf{D} = \int_{\Gamma} D_i(\partial_i \psi) = \int_{\Gamma} \mathbf{D} \cdot \text{grad } \psi = \int (\mathbf{D} \cdot \text{grad } \psi) \, \omega_{\Gamma}.$$

The same holds for the Maxwell equation $\text{div} \mathbf{B} = 0$, so that these equations say that both

$$\int (\mathbf{D} \cdot \text{grad } \psi) \quad \text{and} \quad \int (\mathbf{B} \cdot \text{grad } \psi)$$

vanish for arbitrary closed surfaces Γ. The other equations,

$$c \, \text{rot} \mathbf{H} - \frac{\partial \mathbf{D}}{\partial t} = 0 \quad \text{and} \quad c \, \text{rot} \mathbf{E} + \frac{\partial \mathbf{B}}{\partial t} = 0,$$

because of the time dependence, are better approached by considering D as a domain in $\mathbb{E}^4$ instead of $\mathbb{E}^3$. One could have done it for the above Gauss theorems, with the simplifying fact that all timelike components vanish. The closed surface Γ is now 3-dimensional and the same expressions above lead to the result that both

$$\int_D (c\,\mathrm{rot}\,\boldsymbol{H} - \partial_t \boldsymbol{D}) = \int_\Gamma (c\,\mathrm{grad}\,\psi \times \boldsymbol{H} - \partial_t \psi \boldsymbol{D})\,\omega_\Gamma$$

and

$$\int_D (c\,\mathrm{rot}\,\boldsymbol{E} + \partial_t \boldsymbol{B}) = \int_\Gamma (c\,\mathrm{grad}\,\psi \times \boldsymbol{E} + \partial_t \psi\,\boldsymbol{B})\,\omega_\Gamma$$

vanish for any closed surface Γ in $\mathbb{E}^4$.

Finally, all this may be used to study the case in which the fields are discontinuous on a surface. We consider the domain D in $\mathbb{E}^4$ to be divided into two subdomains D_1 and D_2 (Figure 29.1) by the spacelike surface Γ_0, defined by $\psi(x^1, x^2, \ldots, x^4) = 0$. Integrating the equations on D_1,

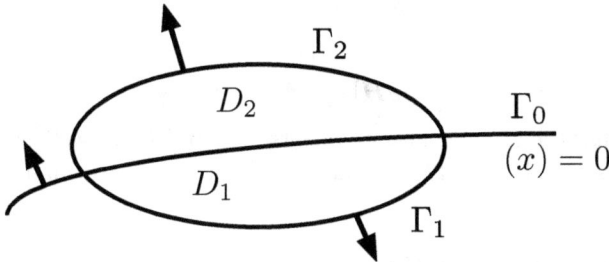

Fig. 29.1 Domain D in $\mathbb{E}^4$, divided by Γ_0 into two pieces D_1 and D_2.

D_2 and $D = D_1 + D_2$, using the above results separately for each region, and comparing the results, one arrives at the following conditions for the discontinuities of the fields, indicated by the respective brackets:

$$[\boldsymbol{H}] \times c\,\mathrm{grad}\,\psi + [\boldsymbol{D}]\,\partial_t \psi = 0; \tag{29.15a}$$

$$[\boldsymbol{E}] \times c\,\mathrm{grad}\,\psi - [\boldsymbol{B}]\,\partial_t \psi = 0; \tag{29.15b}$$

$$[\boldsymbol{D}] \cdot \mathrm{grad}\,\psi = 0; \tag{29.15c}$$

$$[\boldsymbol{B}] \cdot \mathrm{grad}\,\psi = 0. \tag{29.15d}$$

It might seem that these are conditions on the field discontinuities. But usually the discontinuities are given, or supposed, so that actually these are conditions on the surface, on the function ψ.

The formulae contain two main possibilities. Either the fields are discontinuous, or the permeabilities are. Using

$$\boldsymbol{D} = \epsilon\, \boldsymbol{E} \quad \text{and} \quad \boldsymbol{B} = \mu\, \boldsymbol{H},$$

the equations become:

$$[\boldsymbol{H}] \times c \operatorname{grad} \psi + [\epsilon\boldsymbol{E}]\, \partial_t \psi = 0; \qquad (29.16\text{a})$$

$$[\boldsymbol{E}] \times c \operatorname{grad} \psi - [\mu\boldsymbol{H}]\, \partial_t \psi = 0; \qquad (29.16\text{b})$$

$$[\epsilon\boldsymbol{E}] \cdot \operatorname{grad} \psi = 0; \qquad (29.16\text{c})$$

$$[\mu\boldsymbol{H}] \cdot \operatorname{grad} \psi = 0. \qquad (29.16\text{d})$$

The case in which the permeabilities are not continuous is the usual one in optical instruments. In such instruments, the hypersurface is fixed in time,

$$\psi(x^1, x^2, x^3, x^4) = \varphi(x^1, x^2, x^3) = 0,$$

so that $\partial_t \psi = 0$ and the conditions become:

$$[\boldsymbol{H}] \times c \operatorname{grad} \psi = 0; \qquad (29.17\text{a})$$

$$[\boldsymbol{E}] \times c \operatorname{grad} \psi = 0; \qquad (29.17\text{b})$$

$$[\epsilon\boldsymbol{E}] \cdot \operatorname{grad} \psi = 0; \qquad (29.17\text{c})$$

$$[\mu\boldsymbol{H}] \cdot \operatorname{grad} \psi = 0. \qquad (29.17\text{d})$$

This says that the tangencial components of $\boldsymbol{E}$ and $\boldsymbol{H}$, as well as the normal components of $\epsilon\boldsymbol{E}$ and $\mu\boldsymbol{H}$, are continuous on the discontinuity surface.

29.4   The eikonal equation

Suppose the permeabilities are continuous. In this case, Eqs. (29.16) become

$$[\boldsymbol{H}] \times c \operatorname{grad} \psi + \epsilon[\boldsymbol{E}]\, \partial_t \psi = 0; \qquad (29.18\text{a})$$

$$[\boldsymbol{E}] \times c \operatorname{grad} \psi - \mu[\boldsymbol{H}]\, \partial_t \psi = 0; \qquad (29.18\text{b})$$

$$[\boldsymbol{E}] \cdot \operatorname{grad} \psi = 0; \qquad (29.18\text{c})$$

$$[\boldsymbol{H}] \cdot \operatorname{grad} \psi = 0. \qquad (29.18\text{d})$$

Vector-multiplying by $(c \operatorname{grad} \psi)$ the first equation and using the second, one obtains

$$[\boldsymbol{H}] \times c^2 \operatorname{grad} \psi \times \operatorname{grad} \psi + \epsilon\mu[\boldsymbol{H}]\, (\partial_t \psi)^2 = 0,$$

from which it follows that

$$\left((\operatorname{grad} \psi)^2 - \frac{\epsilon\mu}{c^2}(\partial_t \psi)^2 \right)[\boldsymbol{H}] = 0.$$

Using the equations in the inverse order we find instead

$$\left((\operatorname{grad}\psi)^2 - \frac{\epsilon\mu}{c^2}(\partial_t\psi)^2\right)[\boldsymbol{E}] = 0.$$

As $[\boldsymbol{E}]$ and $[\boldsymbol{H}]$ are nonvanishing, the eikonal equation is forcible

$$(\operatorname{grad}\psi)^2 - \frac{\epsilon\mu}{c^2}(\partial_t\psi)^2 = 0. \tag{29.19}$$

With the refraction index $n = \sqrt{\epsilon\mu}$, it becomes

$$(\operatorname{grad}\psi)^2 - \frac{n^2}{c^2}(\partial_t\psi)^2 = 0. \tag{29.20}$$

Because one starts by integrating all over D, the terms without derivatives just compensate and disappear. This will always be the case with discontinuities: only the derivative terms contribute to the conditions on the surface. Comparing with the extra terms and looking back to those terms really giving some contribution to the ultimate result, one sees that only those constituting a wave equation contribute (which, by the way, answers for its ubiquity in Physics). What happens to the large-frequency approach is then clear: the hypotheses made in the asymptotic approach are such that only the higher derivative terms are left, so that the results coincide.

Of course, once the results are obtained, we may consider the surface to be, at the beginning, the source of a disturbance, taking the field to be zero at one side. The disturbance which is propagated is then the field itself. Notice that the equation

$$[\boldsymbol{H}] \times c\operatorname{grad}\psi + \epsilon[\boldsymbol{E}]\partial_t\psi = 0$$

comes from

$$c\operatorname{rot}\boldsymbol{H} - \epsilon\frac{\partial\boldsymbol{E}}{\partial t} = 0,$$

and that

$$[\boldsymbol{E}] \times c\operatorname{grad}\psi - \mu[\boldsymbol{H}]\partial_t\psi = 0$$

comes from

$$c\operatorname{rot}\boldsymbol{E} + \mu\frac{\partial\boldsymbol{H}}{\partial t} = 0.$$

Remember how we get the wave equation: we take the curl of

$$c\operatorname{rot}\boldsymbol{H} - \epsilon\frac{\partial\boldsymbol{E}}{\partial t} = 0$$

to get

$$c^2 \operatorname{rot} \operatorname{rot} \boldsymbol{H} - c\epsilon \frac{\partial}{\partial t}(\operatorname{rot}\boldsymbol{E}) = 0,$$

and use the second,

$$c \operatorname{rot}\boldsymbol{E} = -\mu \frac{\partial \boldsymbol{H}}{\partial t},$$

to arrive at

$$\operatorname{rot} \operatorname{rot} \boldsymbol{H} + \frac{\epsilon\mu}{c^2} \frac{\partial^2 \boldsymbol{H}}{\partial t^2} = -\Delta \boldsymbol{H} + \frac{\epsilon\mu}{c^2} \frac{\partial^2 \boldsymbol{H}}{\partial t^2} = 0.$$

We see thus the parallelism of the two procedures, the operation $(\times \operatorname{grad} \psi)$ playing a role dual to taking the curl.

General references

Basic contributions and texts on the above matter are Luneburg 1966, Abraham & Marsden 1978, Guillemin & Sternberg 1977 and Choquet–Bruhat, DeWitt–Morette & Dillard–Bleick 1977.

Chapter 30

Geometrical Optics

The central equation of Geometrical Optics is the light ray equation. The usual approach to it is to start by looking for asymptotic solutions of Maxwell's equations, fall upon the eikonal equation and then examine the ray curvature. The result is an equation in euclidean 3-dimensional space, which may be interpreted as the geodesic equation in a metric defined by the refractive index. We shall here proceed from the eikonal equation as obtained in the previous chapter, therefore parting with this traditional, asymptotic approach. Consequently, we look at rays as characteristics, curves along which certain electromagnetic discontinuities, the wave fronts, propagate.

30.1 Introduction

The characteristic equation for Maxwell's equations in an isotropic (but not necessarily homogeneous) medium of dielectric function $\epsilon(r)$, magnetic permeability $\mu(r)$ and refractive index

$$n = \sqrt{\epsilon(r)\,\mu(r)} \tag{30.1}$$

is given by

$$\left(\frac{\partial \varphi}{\partial x}\right)^2 + \left(\frac{\partial \varphi}{\partial y}\right)^2 + \left(\frac{\partial \varphi}{\partial z}\right)^2 - \frac{\epsilon\mu}{c^2}\left(\frac{\partial \varphi}{\partial t}\right)^2 = 0,$$

or equivalently

$$(\operatorname{grad} \varphi)^2 - \frac{n^2}{c^2}\left(\frac{\partial \varphi}{\partial t}\right)^2 = 0. \tag{30.2}$$

Looking for a solution in the form

$$\varphi(x,y,z,t) = \psi(x,y,z) - ct,$$

we fall upon the eikonal equation, or equation for the wave fronts, under the form

$$(\operatorname{grad} \psi)^2 = n^2(r). \tag{30.3}$$

30.2 The light-ray equation

The wave fronts are surfaces given by $\psi(x, y, z) = \text{constant}$, consequently integrals of the equation

$$\text{grad } \psi = 0.$$

A light ray is defined as a path "conjugate" to the wave front in the following sense: if $\boldsymbol{r}$ is the position vector of a point on the path and

$$ds = \sqrt{dx^2 + dy^2 + dz^2}$$

is the element of arc length, then $\boldsymbol{u} = (d\boldsymbol{r}/ds)$ is the tangent velocity normalized to unity (with the path length s as the curve parameter). The light ray is then fixed by

$$n\frac{d\boldsymbol{r}}{ds} = d\psi = \text{grad } \psi. \qquad (30.4)$$

This means that applying the 1-form $d\psi$ to the tangent vector gives the refractive index, $d\psi(u) = n$. We shall later give another characterization of this "conjugacy". We may eliminate ψ by taking the derivative, while noticing that $d/ds = u^j \partial_j$:

$$\begin{aligned}
\frac{d}{ds}\left(n\frac{d\boldsymbol{r}}{ds}\right) &= \frac{d}{ds}\text{grad } \psi = \frac{d\boldsymbol{r}}{ds} \cdot \text{grad}\,(\text{grad } \psi) \\
&= \frac{1}{n}(\text{grad } \psi) \cdot \text{grad}\,(\text{grad } \psi) \\
&= \frac{1}{2n}\text{grad}\,[(\text{grad } \psi)^2] \\
&= \frac{1}{2n}\text{grad}\,(n^2). \qquad (30.5)
\end{aligned}$$

Thus,

$$\frac{d}{ds}\left(n\frac{d\boldsymbol{r}}{ds}\right) = \text{grad } n. \qquad (30.6)$$

This is the same as

$$\frac{du^k}{ds} - \frac{1}{n}(\partial^k n - u^k u^j \partial_j n) = 0, \qquad (30.7)$$

which is the differential equation for the light rays. The Poynting vector, and thus the energy flux, is oriented along the direction of $\boldsymbol{u}$. The equations for the light rays are the characteristic equations for the eikonal equations, which are themselves the characteristic equations of Maxwell's equations. For this reason they are called the *bicharacteristic equations* of Maxwell's equations.

The curvature of a curve at one of its points is $1/R$, with R the radius of the osculating circle. As a vector, it is $\hat{n}/R$, where $\hat{n}$ is the unit vector along the radius. When $\boldsymbol{u}$ is the unit tangent vector and s is the length parameter,

$$\hat{n}/R = (d\boldsymbol{u}/ds).$$

The ray curvature (or curvature vector of a ray) is thus the vector field $\boldsymbol{K} = (d\boldsymbol{u}/ds)$. Equation (30.7) is equivalent to

$$\boldsymbol{K} = \frac{1}{n}\,(\boldsymbol{u} \times \operatorname{grad} n) \times \boldsymbol{u}, \tag{30.8}$$

which is the form most commonly found in textbooks.

30.3 Hamilton's point of view

We are looking at things in $\mathbb{E}^3$, of course. More insight comes up from the following analogue in particle mechanics. Let us consider the cotangent bundle $T^*\mathbb{E}^3$ and there take coordinates (x, y, z, p_x, p_y, p_z), with p_i the conjugate coordinate to x^i. This means that the natural symplectic form of $T^*\mathbb{E}^3$ may be written locally in the form (see Section 25.2)

$$\Omega = dx^k \wedge dp_k.$$

When necessary, we may use the euclidean metric to relate vectors and covectors, which in the cartesian coordinates we are using is the same as identifying them. We shall see that another metric will be simultaneously present, and playing a fundamental role.

Consider the hamiltonian

$$H = \frac{1}{2n^2}\sum_j p_j^2. \tag{30.9}$$

Write $\boldsymbol{p} = \operatorname{grad}\psi$ and rephrase the eikonal equation as $H(p) = 1/2$. We may think of the "particle" as endowed with a position-dependent mass n^2. The corresponding hamiltonian field ("bicharacteristic field") is

$$X = \frac{1}{n^2}\sum_k \left[p_k \frac{\partial}{\partial x^k} + \frac{1}{n}\left(\sum_j p_j^2\right)\frac{\partial n}{\partial x^k}\frac{\partial}{\partial p_k} \right]. \tag{30.10}$$

This field is symplectically dual to the form

$$dH = i_X\Omega = i_X(dx^k \wedge dp_k).$$

The bicharacteristic curve is an integral curve of X lying on the "characteristic manifold" $H(p) = 1/2$. Notice that we are here talking about bicharacteristic or characteristic objects (fields, equations and curves) on the phase space, of which the characteristic objects on the configuration space are projections. As

$$\sum_k p_k \dot{x}^k = 2H,$$

the lagrangian corresponding to H above is

$$L[\gamma] = \tfrac{1}{2} n^2 \left(\dot{x}^2 + \dot{y}^2 + \dot{z}^2 \right). \tag{30.11}$$

This gives the kinetic energy which is related to the Riemann metric

$$n\sqrt{dx^2 + dy^2 + dz^2} = n \, ds. \tag{30.12}$$

We have thus a metric

$$g_{ij} = n^2(x, y, z)\delta_{ij}, \tag{30.13}$$

which we call the *refractive metric*.

Now an important point: with this new metric, a new relationship between vector and covectors comes up. Let us examine the velocity vector $\boldsymbol{v} = (\dot{x}, \dot{y}, \dot{z})$. It is

$$\boldsymbol{v} = \dot{\boldsymbol{r}} = (d\boldsymbol{r}/d\tau),$$

where τ is the "proper time", given by

$$d\tau^2 = n^2(dx^2 + dy^2 + dz^2).$$

Its relation to the unit tangent above is $\boldsymbol{u} = n\boldsymbol{v}$, and its components satisfy

$$\sum_i v^i v^i = 1/n^2.$$

Just as $(d/ds) = u^j \partial_j$, the derivative with respect to the new parameter is $(d/d\tau) = v^j \partial_j$. The momentum $\boldsymbol{p}$ above is just its covariant image by the refractive metric,

$$p_i = g_{ij} v^j = n^2 v^i.$$

We may even write, as usual in geometry, the contravariant version as $p^i = g^{ij} p_j = v^i$. We arrive in this way at a better characterization of the above mentioned "conjugacy" between the wavefront and the trajectory: it is summed up in

$$d\psi(\boldsymbol{v}) = \sum_i p_i v^i = \sum_i g_{ij} v^i v^j = 1.$$

The gradient defining the family of surfaces $\psi = $ constant, which is the differential form $d\psi$, applied to the tangent velocity to the path, gives 1. As shown in the Fresnel construction, which will be given below, this means that, seen as an euclidean vector, the covector $\boldsymbol{p}$ at each point of the path is orthogonal (in the euclidean metric) to the plane tangent to the wavefront.

Let us repeat that there are two different velocities at work here:

$$v^j = (dx^j/d\tau) \quad \text{and} \quad u^j = (dx^j/ds),$$

with $d\tau = nds$ and $\boldsymbol{u} = n\boldsymbol{v}$. As it happens whenever the curve is parametrized by the length ("s" here), the corresponding velocity is unitary: $|\boldsymbol{u}| = 1$. Thus, the velocity $\boldsymbol{v}$ along a ray and the one normal to the wavefront, the gradient $\boldsymbol{p}$ of the surface $\psi = $ constant, are respectively a vector and a covector related by the metric given by the refraction index:

$$\boldsymbol{p} = n\,\boldsymbol{u} = n^2\boldsymbol{v}.$$

30.4 Relation to geodesics

Let us show that light rays are simply the geodesics of the refractive metric (30.13). The corresponding Christoffel symbols are

$$
\begin{aligned}
\Gamma^k{}_{ij} &= \frac{1}{n}\left[\delta^k_i \partial_j n + \delta^k_j \partial_i n - \delta_{ij}\delta^{kr}\partial_r n\right]\\
&= \left[\delta^k_{(i}\partial_{j)} - \delta_{ij}\delta^{kr}\partial_r\right](\ln n),
\end{aligned}
\tag{30.14}
$$

where the notation $(i \ldots j)$ indicates index symmetrization. The geodesic equation in this case is

$$\frac{Dv^k}{D\tau} \equiv \frac{dv^k}{d\tau} + \Gamma^k{}_{ij}v^i v^j = 0,$$

or

$$\frac{dv^k}{d\tau} + \frac{1}{n}\left[2v^k v^j \partial_j n - \sum_i (v^i v^i)\partial^k n\right] = 0. \tag{30.15}$$

Changing variables from τ and $\boldsymbol{v}$ to s and $\boldsymbol{u}$, we find just the light ray equation. In the inverse way, the equation for the light rays is

$$
\begin{aligned}
\operatorname{grad} n &= \frac{d}{ds}\left(n\frac{d\mathbf{r}}{ds}\right) = \frac{d}{ds}\left(n^2\frac{d\mathbf{r}}{d\tau}\right)\\
&= \frac{d}{ds}\left(n^2\,\boldsymbol{v}\right) = \boldsymbol{v}\frac{dn^2}{ds} + n^2\frac{d\boldsymbol{v}}{ds}\\
&= 2vn\frac{dn}{ds} + n^2\frac{d\boldsymbol{v}}{ds}
\end{aligned}
\tag{30.16}
$$

or

$$\frac{dv}{ds} + \frac{2}{n} v \frac{dn}{ds} - \frac{1}{n^2} \operatorname{grad} n = 0. \tag{30.17}$$

Using

$$\frac{dv^k}{d\tau} = \frac{1}{n} \frac{dv^k}{ds},$$

we arrive at

$$\frac{dv^k}{d\tau} + \frac{2}{n} v^k \frac{dn}{d\tau} - \frac{1}{n^3} \partial_k n = 0, \tag{30.18}$$

which is the same as the geodesic equation above. By the way, this equation becomes particularly simple and significant when written in terms of p_k:

$$\frac{dp_k}{d\tau} = \partial_k (\ln n). \tag{30.19}$$

The logarithm of the refractive index acts as (minus) the potential in the mechanical picture.

Thus, the equation for the ray curvature just states that the light ray is a geodesic curve in the refractive metric

$$g_{ij} = n^2 \delta_{ij}.$$

The procedure above is general. If we write $p_i = g_{ij} v^j = \partial_i \psi$, then the calculation of $dp_k/d\tau$ leads automatically to

$$\frac{dv^k}{d\tau} + \Gamma^k{}_{ij} v^i v^j = 0, \tag{30.20}$$

with Γ the Levi–Civitta connection of the refractive metric. The inverse procedure works if p is an exact 1-form, that is, a gradient of some ψ, because in a certain moment we are forced to use $\partial_i p_j = \partial_j p_i$. Anyhow, one always finds

$$\frac{dp_k}{d\tau} = \tfrac{1}{2} g^{ij} \partial_k (p_i p_j) = -\tfrac{1}{2} \partial_k (g^{ij}) \, p_i p_j. \tag{30.21}$$

The condition

$$\operatorname{rot} (n\boldsymbol{u}) = \operatorname{rot} (n^2 \boldsymbol{v}) = 0,$$

known in Optics as the "condition for the existence of the eikonal", is an obvious consequence of the Poincaré lemma, as $p = d\psi$.

As seen, the relationship between optical media and metrics is deep indeed. Mathematicians go as far as identifying the expressions "optical instrument" and "riemannian manifold".[1]

---

[1] See, for example, Guillemin & Sternberg 1977.

Metric (30.13) is the euclidean metric multiplied by a function. This kind of metric, related to a flat metric (that is, to a metric whose Levi–Civitta connection has vanishing Riemann tensor) by the simple product of a function, is said to be conformally flat (see Section 23.4). This means that, though measurements of lengths differ from those made with the flat metric, the measurements of angles coincide. The refractive metric, as a consequence, has a strong analogy with the conformally flat metrics (the de Sitter spaces, see Section 37.2 and on) appearing in General Relativity. There, the corresponding flat space is Minkowski. A consequence of this common character is found in similar behaviour of geodesics, as in the fact that anti-de Sitter universes have focusing properties quite analogous to that of an optical ideal apparatus, Maxwell's fish-eye (Section 30.6).

We may look for the Euler–Lagrange equation for the lagrangian (30.11). However, we find that

$$\frac{\delta L}{\delta x^k} = -n^2 \frac{Dv^k}{D\tau}, \tag{30.22}$$

so that $\delta L/\delta x^k = 0$ is equivalent to the geodesic equation. In this way the equivalence is established between the "mechanical" and the "optical" points of view, at least in what concerns the equations.

30.5 The Fermat principle

We have seen that the differential equations given by the hamiltonian field (30.10), in the form of Hamilton equations, are equivalent to the geodesic equation for the refractive metric, or still to the Lagrange equations written in hamiltonian form. The geodesics extremize the arc length $\int nds$ for this metric. Now, the integral

$$\int_\gamma nds = \int_\gamma d\tau$$

along a path γ is the optical length of γ, or its "time of flight". Thus, the light rays are those paths between two given points which extremize the optical length. This is Fermat's principle.

The surfaces $\psi = $ constant may be seen as surfaces of discontinuity (Chapter 29), or as wavefronts. The higher the value of $|\text{grad } \psi|$, the closer are these surfaces packed together. The eikonal equation would say that the refractive index is a measure of the density of such surfaces. Thus, the discontinuities propagate more slowly in regions of higher index. If we interpret grad ψ as a vector, it will be tangent to some curve, it will be a velocity which is larger in higher-index regions.

30.6   Maxwell's fish-eye

Consider the unit sphere S^2 and its stereographic projection into the plane. A point on S^2 will be fixed by the values X, Y, Z of its coordinates, with

$$X^2 + Y^2 + Z^2 = 1.$$

The relation to spherical coordinates are

$$X = \sin\theta\cos\varphi, \quad Y = \sin\theta\sin\varphi, \quad Z = \cos\theta.$$

We choose the "north pole" $(0, 0, 1)$ as projection center and project each point of the sphere on the plane $Z = 0$. The corresponding plane coordinates (x, y) will be given by

$$x = \frac{X}{1 - Z} \quad \text{and} \quad y = \frac{Y}{1 - Z}. \tag{30.23}$$

Call

$$r^2 = x^2 + y^2 = \frac{X^2 + Y^2}{(1 - Z)^2}. \tag{30.24}$$

The line element will then be

$$ds^2 = dX^2 + dY^2 + dZ^2 = 4\frac{dx^2 + dy^2}{(1 + r^2)^2}. \tag{30.25}$$

This corresponds to a 2-dimensional medium with refractive index

$$n = \frac{2}{1 + r^2}. \tag{30.26}$$

It is found that the geodesics are all given by

$$(x^2 - \sqrt{R^2 - 1})^2 + y^2 = C^2,$$

with some constants R and C. Thus, they are all the circumferences through the points $(0, \pm 1)$. All the light rays starting at a given point will intersect again at another point, corresponding to its antipode on S^2. This is an example of perfect focusing.

Commentary 30.1   A manifold such that all points have this property is called a *wiedersehen manifold*. The sphere S^2 is a proven example, but it is speculated that others exist, and also conjectured that only (higher dimensional) spheres may be 'wiedersehen manifolds'. ◄

But things become far more exciting in anisotropic media. Let us say a few words on crystal optics.

30.7 Fresnel's ellipsoid

In an electrically anisotropic medium, the electric displacement $\boldsymbol{D}$ is related to the electric field $\boldsymbol{E}$ by

$$D_i = \sum_j \epsilon_{ij} \, E_j,$$

where ϵ_{ij} is the electric permitivity (or dielectric) tensor. The electric energy density is

$$W_e = \tfrac{1}{2} \, \boldsymbol{E} \cdot \boldsymbol{D} = \tfrac{1}{2} \sum_{ij} \epsilon_{ij} \, E_i E_j. \tag{30.27}$$

The Fresnel ellipsoid is given by

$$\sum_{ij} \epsilon_{ij} \, x^i x^j = \text{constant} = 2W_e.$$

The 2-tensor $\epsilon = (\epsilon_{ij})$ is non-degenerate and symmetric, the latter property being a consequence of the requirement that the work done in building up the field,

$$dW_e = \tfrac{1}{2} \, \boldsymbol{E} \cdot d\boldsymbol{D} \,,$$

be an exact differential form. Consequently, ϵ is a metric, $\boldsymbol{E}$ is a field and $\boldsymbol{D}$ its covariant version according to this metric. Now, the metric can be diagonalized, in which case the field and cofield are colinear in the 3-space $\mathbb{E}^3$, or parallel: $D_i = \epsilon_i E_i$. The ellipsoid then becomes

$$\sum_i \epsilon_i (x^i)^2 = 2\,W_e. \tag{30.28}$$

The metric eigenvalues ϵ_i are the *principal dielectric constants*. The construction to get $\boldsymbol{D}$ from $\boldsymbol{E}$ using the ellipsoid is analogous to the Poinsot construction for obtaining the angular momentum of a rigid body from its angular velocity (Section 26.3.10). It is also the same given above to obtain $\boldsymbol{p}$ from $\boldsymbol{v}$. Diagonalizing the metric corresponds to taking the three cartesian axes along the three main axes of the ellipsoid.

Usual crystals are magnetically isotropic, or insensitive, so that the magnetic permeability may be taken as constant: $\mu = \mu_0$. The magnetic induction $\boldsymbol{B}$ and the magnetic field $\boldsymbol{H}$ are simply related by

$$\boldsymbol{B} = \mu \boldsymbol{H} = \mu_0 \boldsymbol{H}.$$

Thus, the (squared) refraction index is given by the tensor

$$(n^2)_{ij} = \mu_0 \, \epsilon_{ij}.$$

By diagonalization as above, Fresnel's ellipsoid becomes

$$\sum_i n_i^2 E_i^2 = constant = 2\,W_e.$$

The n_i's are the *principal refraction indices*. Along the principal axes of the ellipsoid, of size $1/n_i$, light will travel with the so called principal light velocity,

$$u_i = \frac{c}{n_i} = \frac{1}{\sqrt{\epsilon_i\,\mu_0}}. \qquad (30.29)$$

In the above procedure, $\boldsymbol{D}$ is seen as a form, a covector, while $\boldsymbol{E}$ is a vector.[2]

Of course there is an arbitrariness in the above choice: we might instead have chosen to use the inverse metric ϵ^{-1}, with $\boldsymbol{D}$ as the basic field. In this case another ellipsoid comes up, given by

$$\sum_i \left(\frac{x^i}{n_i}\right)^2 = 2\,W_e, \qquad (30.30)$$

which has the advantage that the principal axes are just the principal refraction indices. This is called the *index ellipsoid, reciprocal, ellipsoid of wave normal, Fletcher's ellipsoid*, or still *optical indicatrix*. .

General references

Commendable texts covering the subject of this chapter are: Synge 1937, Sommerfeld 1954, Born & Wolf 1975, Gel'fand & Shilov 1964 and Luneburg 1966.

---

[2] That is why Sommerfeld 1954 (p. 139, footnote) talks of the $\boldsymbol{E}$ components as "point coordinates" and of those of $\boldsymbol{D}$ as "plane coordinates".

Chapter 31

Classical Relativistic Fields

In Quantum Field Theory, elementary particles come up as field quanta. Their quantum numbers, such as mass and helicity, are fixed by the corresponding free fields. This means that, to be related to a particle, a field must exist in free state, or to be well defined far away from any region of interaction. It is not clear that every field has such "asymptotic" behaviour. It is not clear, in particular, that fields such as those corresponding to quarks and gluons do describe real particles. There is, consequently, a modern tendency to give priority to fields with respect to particles. We shall here talk about fields.

31.1 The fundamental fields

Elementary particles must have a well-defined behaviour under changes of inertial frames in Minkowski spacetime. Such changes constitute the Poincaré group (or inhomogeneous Lorentz group). In order to have a well-defined behaviour under the transformations of a group, an object must belong to a representation, to a multiplet. If the representation is reducible, the object is composite, in the sense that it can be decomposed into more elementary objects belonging to irreducible representations. Thus, truly elementary particles must belong to irreducible representations and be classified in multiplets of the Poincaré group. Each multiplet will have well-defined mass and helicity.

The word "field", as used here, is of course not to be mistaken by the algebraic structure of § 13.3.3. Neither is it to be taken as the geometrical fields which are natural denizens of the tangent structure of any smooth manifold and are, at each point, vectors of the linear group of basis transformations. Those are *linear vector* fields. In Relativistic Field Theory, fields (scalar, vector, spinor, tensor, etc) belong to representations of the

Lorentz group, as specified below (Section 31.2.2). They are *Lorentz* fields. In particular, the "vector" fields of Field Theory (like the electromagnetic 4-vector potential) are mostly represented as 1-forms, covectors of the Lorentz group seen as a subgroup of the linear group in Minkowski space.

31.2   Spacetime transformations

31.2.1   *The Poincaré group*

The Poincaré group is the group of motions (isometric automorphisms) of Minkowski spacetime. This means that what was said above holds for free systems, in the absence of interaction. Only the total quantum numbers are preserved in interactions. Individual particles are identified only "far from the interaction region", where their momenta (consequently, masses) and helicities are measured. The status of confined particles like quarks, as said, is not clear.

Acting on spacetime, the Poincaré group P is the semi-direct product of the (homogeneous) Lorentz group $L = SO(3, 1)$ by the translation group T, $P = L \oslash T$, its transformations being given in cartesian coordinates as

$$x'^{\mu} = \Lambda^{\mu}{}_{\nu} x^{\nu} + a^{\mu}. \tag{31.1}$$

Let us first consider the Lorentz group. If η is the Lorentz metric matrix, the matrices $\Lambda = (\Lambda^{\mu}{}_{\nu})$ satisfy $\Lambda^{T} \eta \Lambda = \eta$, with Λ^{T} standing for the transpose matrix. This is the defining condition for the Lorentz group. The matrices for a general Lorentz transformation will have the form

$$\Lambda = \exp\left[\tfrac{1}{2} \omega^{\alpha\beta} J_{\alpha\beta} \right],$$

where $\omega^{\alpha\beta} = -\omega^{\beta\alpha}$ are the transformation parameters, and $J_{\alpha\beta}$ are the generators obeying the commutation relations

$$[J_{\alpha\beta}, J_{\gamma\delta}] = \eta_{\beta\gamma} J_{\alpha\delta} + \eta_{\alpha\delta} J_{\beta\gamma} - \eta_{\beta\delta} J_{\alpha\gamma} - \eta_{\alpha\gamma} J_{\beta\delta}. \tag{31.2}$$

Each representation will have as generators some matrices $J_{\alpha\beta}$. Expression (31.2) holds for anti-adjoint generators. There is no special reason to use self-adjoint operators, as anyhow the group $SO(3, 1)$, being non-compact, will have no unitary finite-dimensional representations. Furthermore, $SO(3, 1)$ cannot accommodate half-integer spin particles. The true Lorentz group of Nature is $SL(2, C)$, which has also spinor representations and is the covering group of $SO(3, 1)$. We put, consequently, $L = SL(2, C)$. Thus, the classifying group of elementary particles is the covering group of the Poincaré group. It is more practical to rechristen the "Poincaré group" and write $P = SL(2,C) \oslash T$. The translation group has generators T_α, and the remaining commutation relations are the following:

$$[J_{\alpha\beta}, T_\varepsilon] = \eta_{\beta\varepsilon} T_\alpha - \eta_{\alpha\varepsilon} T_\beta, \tag{31.3}$$

$$[T_\alpha, T_\beta] = 0. \tag{31.4}$$

Besides mass and spin (or helicity), particles (and their fields) carry other quantum numbers (charges in general: electric charge, flavor, isotopic spin, color, etc), related to other symmetries, not concerned with spacetime. For simplicity, we shall call them "internal" symmetries. If G is their group, the total symmetry is the direct product $P \otimes G$. A particle (a field) is thus characterized by being put into multiplets of P and G. A particle with a zero charge is invariant under the respective transformation and is accommodated in a singlet (zero-dimensional) representation of the group.

There are two kinds of internal symmetries. They may be global (as that related to the conservation of baryon number), independent of the point in spacetime; or they may be local (those involved in gauge invariance). In the last case, the above direct product is purely local, the fields being either in a principal (if a gauge potential) or in an associated fiber bundle (if a source field).

31.2.2 The basic cases

Relativistic fields are defined according to their behaviour under Lorentz transformations, that is, according to the representation they belong to. Fortunately, Nature seems to use only the lowest representations: the scalar, the vector and the spinor representations.

Scalar fields

Are those that remain unchanged (belonging to a singlet representation):

$$\varphi'(x') = \varphi(x). \tag{31.5}$$

They obey the Klein–Gordon equation

$$\Box\varphi(x) + m^2\varphi(x) = 0. \tag{31.6}$$

Vector fields

Are those that transform according to

$$\varphi'^{\mu}(x') = \Lambda^{\mu}{}_{\nu}\varphi^{\nu}(x) = \left(\exp\left[\tfrac{i}{2}\omega^{\alpha\beta}M_{\alpha\beta}\right]\right)^{\mu}{}_{\nu}\varphi^{\nu}(x), \tag{31.7}$$

where each $M_{\alpha\beta}$ is a 4×4 matrix with elements

$$[M_{\alpha\beta}]^{\mu}{}_{\nu} = i(\eta_{\alpha\nu}\delta^{\mu}_{\beta} - \delta^{\mu}_{\alpha}\eta_{\beta\nu}). \tag{31.8}$$

This matrix basis is such that

$$\Lambda^{\mu}{}_{\nu} = \left(\exp\left[\tfrac{i}{2}\omega^{\alpha\beta}M_{\alpha\beta}\right]\right)^{\mu}{}_{\nu} = \exp[\omega^{\mu}{}_{\nu}],$$

that is to say, the components $\omega^{\mu}{}_{\nu}$ coincide with the matrix elements.

An example of vector field is the electromagnetic field, a zero mass case involving furthermore a local gauge invariance. The basic equations are Maxwell's equations, examined in Sections 29.3 and 32.2.1:

$$\partial_{\lambda}F_{\mu\nu} + \partial_{\nu}F_{\lambda\mu} + \partial_{\mu}F_{\nu\lambda} = 0, \tag{31.9}$$

$$\partial^{\lambda}F_{\lambda\nu} = J_{\nu}. \tag{31.10}$$

As $F_{\mu\nu} = \partial_{\mu}A_{\nu} - \partial_{\nu}A_{\mu}$, the potential A_{μ} submits to the wave equation

$$\partial^{\mu}\partial_{\mu}A_{\nu} + \partial_{\nu}\partial^{\mu}A_{\mu} = J_{\nu}. \tag{31.11}$$

Spinor fields

Are those that transform according to

$$\psi'(x') = \exp\left[-\tfrac{i}{4}\omega^{\alpha\beta}\sigma_{\alpha\beta}\right]\psi(x), \tag{31.12}$$

where $\tfrac{1}{2}\sigma_{\alpha\beta}$ are 4×4 matrices generating the bispinor representation. Under infinitesimal Lorentz transformations with parameters $\delta\omega^{\alpha\beta}$, bispinor wavefunctions and their conjugates will change according to

$$\delta\psi = -\tfrac{i}{4}\sigma_{\alpha\beta}\,\psi\,\delta\omega^{\alpha\beta} \quad \text{and} \quad \delta\overline{\psi} = \tfrac{i}{4}\overline{\psi}\,\sigma_{\alpha\beta}\,\delta\omega^{\alpha\beta}. \tag{31.13}$$

Spinor fields in the absence of interactions obey the Dirac equation

$$i\gamma^{\mu}\partial_{\mu}\psi - m\psi = 0. \tag{31.14}$$

31.3   Internal transformations

Fields will further belong to representations $U(G)$ of "internal" transformation groups G. Under a transformation given by the element g of such a group G, their behaviour is generically represented by

$$\varphi'_i(x) = [U(g)]^j_i \varphi_j(x) = [\exp\{\omega^a T_a\}]^j_i \varphi_j(x). \qquad (31.15)$$

The T_a's are generators in the U-representation.

If the transformation parameters ω^a are independent of spacetime points, the above expression represents *global* gauge transformations. If the transformation parameters ω^a are point-dependent, it represents *local* gauge transformations.

The fields $\varphi_j(x)$ are "source fields". Gauge potentials (which will have the roles of interaction mediators) may be written as

$$A_\mu = J_a A^a{}_\mu,$$

with J_a the generators in the adjoint representation of G. Under a local transformation generated by the group element

$$g = \exp\{-\omega^a J_a\},$$

they behave according to

$$A'_\mu = g^{-1}A_\mu g + g^{-1}\partial_\mu g. \qquad (31.16)$$

The corresponding infinitesimal transformation is

$$\bar{\delta}A^b{}_\nu = -f^b{}_{cd}A^c{}_\nu \delta\omega^d - \partial_\nu \delta\omega^b. \qquad (31.17)$$

with $f^b{}_{cd}$ the structure constants of the group. Field strengths

$$F_{\mu\nu} = J_a F^a{}_{\mu\nu}$$

transform according to

$$F'_{\mu\nu} = g^{-1}F_{\mu\nu}g, \qquad (31.18)$$

with the corresponding infinitesimal version given by

$$\delta F^b{}_{\mu\nu} = -f^b{}_{cd}F^c{}_{\mu\nu}\delta\omega^d. \qquad (31.19)$$

Gauge fields are the special subject of the next Chapter.

31.4 Lagrangian formalism

31.4.1 *The Euler–Lagrange equation*

A physical system will be characterized as a whole by the symmetry-invariant action functional

$$S[\varphi] = \int d^4x \, \mathcal{L}[\varphi(x)], \tag{31.20}$$

where "φ" represents collectively all the involved fields. The variation of a field φ_i may be decomposed as

$$\delta\varphi_i(x) = \bar{\delta}\varphi_i(x) + \delta x^\mu \partial_\mu \varphi_i(x). \tag{31.21}$$

The second term in the right-hand side is the variation due to changes in the coordinate argument

$$\delta x^\mu = x'^\mu - x^\mu.$$

As to the first,

$$\bar{\delta}\varphi_i(x) = \varphi'_i(x) - \varphi_i(x)$$

is the change in the functional form of φ_i at a fixed value of the argument. Notice that

$$[\partial_\mu, \bar{\delta}] = 0. \tag{31.22}$$

The general variation of $S[\varphi]$ is given by:

$$\delta S[\varphi] = \int \delta(d^4x) \, \mathcal{L}[\varphi(x)] + \int d^4x \, \delta\mathcal{L}[\varphi(x)] = \int \delta[d^4x]\mathcal{L}[\varphi(x)]$$

$$+ \int d^4x \left\{ \frac{\partial \mathcal{L}[\varphi(x)]}{\partial \varphi_i(x)} \bar{\delta}\varphi_i(x) + \frac{\partial \mathcal{L}[\varphi(x)]}{\partial \partial_\mu \varphi_i(x)} \bar{\delta}\partial_\mu \varphi_i(x) + \partial_\mu \mathcal{L}[\varphi(x)]\delta x^\mu \right\},$$

with $\delta(d^4x)$ the variation of the integration measure, which is

$$\delta(d^4x) = \partial_\mu(\delta x^\mu) \, d^4x$$

in cartesian coordinates. Collecting terms,

$$\delta S[\varphi] = \int d^4x \left\{ \frac{\delta \mathcal{L}[\varphi(x)]}{\delta \varphi_i(x)} \bar{\delta}\varphi_i(x) + \partial_\mu \left[\frac{\partial \mathcal{L}[\varphi(x)]}{\partial \partial_\mu \varphi_i(x)} \bar{\delta}\varphi_i(x) + \mathcal{L}[\varphi(x)]\delta x^\mu \right] \right\},$$
$$\tag{31.23}$$

where the Lagrange derivative (Section 26.2) is simply

$$\frac{\delta \mathcal{L}[\varphi(x)]}{\delta \varphi_i(x)} = \frac{\partial \mathcal{L}[\varphi(x)]}{\partial \varphi_i(x)} - \partial_\mu \frac{\partial \mathcal{L}[\varphi(x)]}{\partial \partial_\mu \varphi_i(x)}, \tag{31.24}$$

because we are supposing the lagrangian density to depend only on φ and on the first derivative $\partial_\mu \varphi$. The Euler–Lagrange equations are then

$$\frac{\delta \mathcal{L}[\varphi(x)]}{\delta \varphi_i(x)} = 0.$$

31.4.2 *First Noether's theorem*

The first Noether's theorem is concerned with the action invariance under a global transformation: it imposes the vanishing of the derivative of S with respect to the corresponding constant (but otherwise arbitrary) parameter ω^a. The condition for that, from Eq. (31.23), is

$$\frac{\delta \mathcal{L}[\varphi(x)]}{\delta \varphi_i(x)} \frac{\bar{\delta} \varphi_i(x)}{\delta \omega^a} = -\partial_\mu \left[\frac{\partial \mathcal{L}[\varphi(x)]}{\partial \partial_\mu \varphi_i(x)} \frac{\bar{\delta} \varphi_i(x)}{\delta \omega^a} + \mathcal{L}[\varphi(x)] \frac{\delta x^\mu}{\delta \omega^a} \right]. \qquad (31.25)$$

For φ_i satisfying the Euler–Lagrange equation, the current

$$J_a{}^\mu = - \left[\frac{\partial \mathcal{L}[\varphi(x)]}{\partial \partial_\mu \varphi_i(x)} \frac{\bar{\delta} \varphi_i(x)}{\delta \omega^a} + \mathcal{L}[\varphi(x)] \frac{\delta x^\mu}{\delta \omega^a} \right] \qquad (31.26)$$

is conserved. Internal symmetries are concerned with the first term, while spacetime symmetries are concerned with the last one. Let us then examine some examples.

Translations

Translations are given by:

$$x'^\mu = x^\mu + \delta x^\mu = x^\mu + \frac{\delta x^\mu}{\delta a^\alpha} \delta a^\alpha. \qquad (31.27)$$

If we take the x^μ themselves as parameters, then

$$\frac{\delta x^\mu}{\delta a^\alpha} = \delta^\mu_\alpha.$$

Fields are Lorentz tensors and spinors, and as that they are unaffected by translations: $\delta \varphi_i / \delta a^\alpha = 0$. Consequently,

$$\bar{\delta} \varphi_i = - (\partial_\alpha \varphi_i) \delta x^\alpha.$$

The Noether current related to spacetime translations in the energy-momentum tensor density

$$\Theta^\alpha{}_\mu = \frac{\partial \mathcal{L}[\varphi]}{\partial \partial^\mu \varphi_i} \partial^\alpha \varphi_i - \delta^\alpha_\mu \mathcal{L}. \qquad (31.28)$$

As an example, let us consider a fermion field, for which

$$\mathcal{L} = \tfrac{i}{2} \left\{ \overline{\psi} \gamma_\mu \partial^\mu \psi - [\partial^\mu \overline{\psi}] \gamma_\mu \psi \right\} - m \overline{\psi} \psi, \qquad (31.29)$$

and consequently

$$\Theta^\alpha{}_\mu = \tfrac{i}{2} \left\{ \overline{\psi} \gamma_\mu \partial^\alpha \psi - [\partial^\alpha \overline{\psi}] \gamma_\mu \psi \right\}. \qquad (31.30)$$

Lorentz transformations

Lorentz transformations are, in their infinitesimal form, given by

$$\delta x^\mu = \tfrac{i}{2}\left[\delta\omega^{\alpha\beta}M_{\alpha\beta}\right]^\mu{}_\nu x^\nu = -\tfrac{1}{2}\left[\delta\omega^{\nu\mu} - \delta\omega^{\mu\nu}\right]x_\nu = \delta\omega^{\mu\nu}x_\nu, \qquad (31.31)$$

where use was made of (31.8). Consequently,

$$\frac{\delta x^\lambda}{\delta\omega^{\alpha\beta}} = (\delta^\lambda_\alpha x_\beta - \delta^\lambda_\beta x_\alpha),$$

and the Noether current will be the total angular momentum current density:

$$M^\mu{}_{\alpha\beta} = -\frac{\partial\mathcal{L}[\varphi]}{\partial\partial_\mu\varphi_i}\frac{\bar{\delta}\varphi_i}{\delta\omega^{\alpha\beta}} - \mathcal{L}\frac{\delta x^\mu}{\delta\omega^{\alpha\beta}}$$

$$= \Theta_\alpha{}^\mu x_\beta - \Theta_\beta{}^\mu x_\alpha - \frac{\partial\mathcal{L}}{\partial\partial_\mu\varphi_i}\frac{\bar{\delta}\varphi_i}{\delta\omega^{\alpha\beta}}. \qquad (31.32)$$

The last term is the spin current density

$$S^\mu{}_{\alpha\beta} = -\frac{\partial\mathcal{L}}{\partial\partial_\mu\varphi_i}\frac{\bar{\delta}\varphi_i}{\delta\omega^{\alpha\beta}}, \qquad (31.33)$$

which appears only when the field is not a Lorentz singlet. From the conservation laws

$$\partial_\mu M^\mu{}_{\alpha\beta} = 0 \quad \text{and} \quad \partial_\mu\Theta_\alpha{}^\mu = 0,$$

it follows that

$$\partial_\mu S^\mu{}_{\alpha\beta} = \Theta_{\beta\alpha} - \Theta_{\alpha\beta}. \qquad (31.34)$$

If a vector field has Lagrangean density

$$\mathcal{L} = \tfrac{1}{2}\left\{\partial_\mu\varphi^\nu\partial^\mu\varphi_\nu - m^2\varphi^\nu\varphi_\nu\right\}, \qquad (31.35)$$

its spin current density reads

$$S^\mu{}_{\alpha\beta} = \varphi_\beta\partial^\mu\varphi_\alpha - \varphi_\alpha\partial^\mu\varphi_\beta. \qquad (31.36)$$

For a fermion field

$$\mathcal{L} = i\,\overline{\psi}\gamma_\mu\partial^\mu\psi - m\overline{\psi}\psi \qquad (31.37)$$

and the spin current density is

$$S^\mu{}_{\alpha\beta} = -\tfrac{1}{4}\,\overline{\psi}\left\{\gamma_\mu\sigma_{\alpha\beta} + \sigma_{\alpha\beta}\gamma^\mu\right\}\psi. \qquad (31.38)$$

Phase transformations

Phase transformations related to abelian groups are of the form

$$\Psi' = \exp[iq\alpha]\Psi \quad \text{and} \quad \overline{\psi}' = \overline{\psi}\exp[-iq\alpha]. \tag{31.39}$$

In this case, the Noether current will be the "electric" current

$$J^\mu = q\,\overline{\psi}\gamma^\mu\Psi.$$

For non-abelian transformations like (31.15), the conserved current is

$$J_a^\mu = -\frac{\partial\mathcal{L}[\varphi(x)]}{\partial\partial_\mu\varphi_i(x)}\frac{\overline{\delta}\varphi_i(x)}{\delta\omega^a} = -\frac{\partial\mathcal{L}[\varphi(x)]}{\partial\partial_\mu\varphi_i(x)}\,[T_a]_i^j\varphi_j = -\frac{\partial\mathcal{L}[\varphi(x)]}{\partial\partial_\mu\varphi(x)}\,T_a\varphi. \tag{31.40}$$

Under global transformations, obtained from (31.17) with $\partial_\nu\delta\omega^b = 0$, the gauge potentials obey

$$\overline{\delta}A^b{}_\nu = -f^b{}_{cd}A^c{}_\nu\delta\omega^d. \tag{31.41}$$

The corresponding Noether current will in this case be the self-current

$$j^{a\nu} = -f^a{}_{bc}A^b{}_\mu F^{c\mu\nu}. \tag{31.42}$$

The conservation law will be

$$\partial_\mu(J_a{}^\mu + j_a{}^\mu) = 0, \tag{31.43}$$

which means that the total current is conserved.

31.4.3 Minimal coupling prescription

These currents may also be obtained from the free lagrangians $\mathcal{L}_\varphi$ for φ and $\mathcal{L}_G$ for $A^b{}_\mu$ through the minimal coupling rule $\mathcal{L}_\varphi \to \mathcal{L}'_\varphi$, where $\mathcal{L}'_\varphi$ has the form of $\mathcal{L}_\varphi$ but with usual derivatives ∂_μ replaced by covariant derivatives:

$$\partial_\mu \to D_\mu = \partial_\mu + A^a{}_\mu\frac{\delta}{\delta\omega^a}\,, \tag{31.44}$$

with the derivative

$$\frac{\delta}{\delta\omega^a} = \frac{\partial}{\partial\omega^a} - \partial_\mu\frac{\partial}{\partial\partial_\mu\omega^a} \tag{31.45}$$

taking into account the total (Lagrange) functional, and not simply $\partial/\partial\omega^a$. As a consequence,

$$D_\mu = \partial_\mu + A^a{}_\mu\frac{\partial}{\partial\omega^a} - \partial_\lambda A^a{}_\mu\frac{\partial}{\partial\partial_\lambda\omega^a}\,. \tag{31.46}$$

The currents can then be written as

$$J_a{}^\mu = -\frac{\delta \mathcal{L}'_\varphi}{\delta A^a{}_\mu} \quad \text{and} \quad j_a{}^\mu = -\frac{\partial \mathcal{L}_G}{\partial A^a{}_\mu}. \tag{31.47}$$

In general, source fields transformations do not depend on the parameter derivatives and for them the last term of (31.46) does not contribute, but that term is essential when we take the covariant derivative of the gauge potentials to obtain the field strength: under local transformations, the gauge potential transforms according to (31.17) and, consequently,

$$D_\mu A^b{}_\nu = \partial_\mu A^b{}_\nu - \partial_\nu A^b{}_\mu + f^b{}_{ac} A^a{}_\mu A^c{}_\nu = F^b_{\mu\nu}. \tag{31.48}$$

31.4.4   *Local phase transformations*

The self-current, as given by (31.42), is covariant under global transformations, but not under local transformations. In fact, it is just

$$j_a{}^\mu = -\frac{\partial \mathcal{L}_G}{\partial A^a{}_\mu}, \tag{31.49}$$

as we can see by comparing the Yang–Mills equation

$$E^{a\nu} = \frac{\delta \mathcal{L}_G}{\delta A^a{}_\nu} = \partial_\mu F^{a\mu\nu} + f^a{}_{bc} A^b{}_\mu F^{c\mu\nu} = J^{a\nu} \tag{31.50}$$

with equation (31.42). This is an example of the well known result of Mechanics (Section 26.2), which says that simple functional derivatives are not covariant, whereas Lagrange derivatives are. As the time component of a current is the charge density, the total charge is its space-integral. The continuity equation

$$\partial_\mu(J^{a\mu} + j^{a\nu}) = 0$$

then implies the time-conservation of the total charge. In the above case, as

$$J^{a\nu} + j^{a\nu} = \partial_\mu F^{a\mu\nu},$$

the continuity equation is automatically satisfied and the corresponding charge

$$Q = \int_V d^3x \, J_a \partial_\mu F^{a\mu0} = \int_V d^3x \, \partial_\mu F^{\mu0} = \int_V d^3x \, \partial_k F^{k0} = \int_{\partial V} d^2\sigma_i F^{i0}$$

is conserved: $dQ/dt = 0$. To ensure the covariance of the charge, we should have

$$Q' = \int_{\partial V} d^2\sigma F' = \int_{\partial V} d^2\sigma g^{-1} F g = g^{-1} \left(\int_{\partial V} d^2\sigma F \right) g = g^{-1} Q g.$$

As

$$g = g(x) = \exp[\omega^a(x)J_a],$$

in order to extract g from inside the integral, it is necessary that $\omega^a(x)$ be constant at the surface ∂V.

Finally, we should mention that, if we had insisted in using the complete derivative

$$j_a{}^\mu = -\frac{\delta\mathcal{L}_G}{\delta A^a{}_\mu}$$

instead of the "ill-defined" j_a^μ given by (31.49), the very expression $E^{a\nu}$ in (31.50) would result (with opposite sign) and the total current would be covariant but vanishing for solutions of the field equation. We are consequently forced to keep on working with the non-covariant current and consequently restricting the gauge transformations to become global beyond a certain distance. In that case, though the currents are not covariant, the charges are.

31.4.5   *Second Noether's theorem*

The second Noether's theorem is concerned with local transformations, those with point-dependent parameters. We shall consider here only the case of a gauge field $A^a{}_\mu$ and a generic source field φ belonging to a representation with generators $\{T_a\}$. The total lagrangian, supposed invariant, will be

$$\mathcal{L} = \mathcal{L}_G + \mathcal{L}_\varphi. \tag{31.51}$$

The important point is that the pure gauge lagrangian

$$\mathcal{L}_G = -\tfrac{1}{4}F_a{}^{\mu\nu}F^a{}_{\mu\nu}$$

is invariant under gauge transformations. This means that also the source lagrangian $\mathcal{L}_\varphi$, which is the free lagrangian with the derivatives replaced by covariant derivatives, is invariant by itself. More than that: if there are many source fields, each one will contribute with a lagrangian chosen so as to be isolatedly gauge invariant.

Consider again Eq. (31.23). A *local* invariance means that the action remains unmoved under transformations in a small region around any point in the system. Take a point "y" inside the system and calculate from (31.15) and (31.17) the following functional derivatives

$$\frac{\delta A^a{}_\mu(x)}{\delta\omega^b(y)} = \delta^4(x-y)f^a{}_{bc}A^c{}_\mu(x) - \delta^a_b\partial_\mu\delta^4(x-y) \tag{31.52}$$

and

$$\frac{\bar{\delta}\varphi(x)}{\delta\omega^a(y)} = T_a\varphi(x)\delta^4(x-y).\qquad(31.53)$$

We see that the deltas concentrate the variations at the interior point y. The last, derivative term of (31.23) is an integration on the surface of the system, which for the generic field φ is

$$\int d^4x\,\partial_\mu\left[\frac{\partial\mathcal{L}[\varphi(x)]}{\partial\partial_\mu\varphi_i(x)}\,\bar{\delta}\varphi_i(x)\right] = \int_S d^3\sigma_\mu\left[\frac{\partial\mathcal{L}[\varphi(x)]}{\partial\partial_\mu\varphi_i(x)}\,\bar{\delta}\varphi_i(x)\right].$$

The integration variable "x" will be at the surface of the system, whereas "y" represents a point inside the system. Due to the deltas, this term will vanish. The remaining terms will give, after integration,

$$\frac{\delta S_\varphi}{\delta\omega^a(y)} = \frac{\partial\mathcal{L}_\varphi(y)}{\partial\varphi_i(y)}(T_a)_{ij}\varphi_j(y) + \frac{\delta\mathcal{L}_G(y)}{\delta A^b{}_\mu(y)}\,f^b{}_{ac}A^c{}_\mu(y) + \partial_\mu\frac{\delta\mathcal{L}_G(y)}{\delta A^b{}_\mu(y)}\ .$$
$$(31.54)$$

When there is a local invariance,

$$\frac{\delta S_\varphi}{\delta\omega^a(y)} = 0.$$

We have said that each piece of $\mathcal{L}$ in (31.51) is independently gauge invariant. This means that actually the variation vanishes for each field, so that

$$\frac{\delta S_\varphi}{\delta\omega^a(y)} = \frac{\partial\mathcal{L}_\varphi(y)}{\partial\varphi_i(y)}(T_a)_{ij}\varphi_j(y) = 0\qquad(31.55)$$

for each source field φ, and

$$\frac{\delta S_G}{\delta\omega^a(y)} = \frac{\delta\mathcal{L}_G(y)}{\delta A^b{}_\mu(y)}\,f^b{}_{ac}A^c{}_\mu(y) + \partial_\mu\frac{\delta\mathcal{L}_G(y)}{\delta A^b{}_\mu(y)} = 0\qquad(31.56)$$

for the gauge field. Relations like (31.55) and (31.56), coming solely from the invariance requirement and independent of the field equations, are said to be "strong relations". Consider first the latter. The last expression is

$$\left[\delta^b_a\partial_\mu + f^b{}_{ac}A^c{}_\mu\right]\frac{\delta\mathcal{L}_G(y)}{\delta A^b{}_\mu(y)} = 0.\qquad(31.57)$$

We can introduce the notation

$$D^b{}_{a\mu} = \delta^b_a\partial_\mu + f^b{}_{ac}A^c{}_\mu\qquad(31.58)$$

for the covariant derivative, and write the strong relation for the gauge field as

$$D_\mu D_\nu F_a{}^{\mu\nu} = 0.\qquad(31.59)$$

For the source field, (31.55) gives

$$\frac{\delta \mathcal{L}_\varphi(y)}{\delta \varphi_i(y)}(T_a)_{ij}\varphi_j(y) = \left[\frac{\partial \mathcal{L}_\varphi(y)}{\partial \varphi_i(y)} - \nabla_\mu \frac{\partial \mathcal{L}_\varphi(y)}{\partial \nabla_\mu \varphi_i(y)}\right](T_a)_{ij}\varphi_j(y) = 0. \quad (31.60)$$

Take, for example, the case of a fermion field, whose lagrangian is given by

$$\mathcal{L}_\psi = \tfrac{i}{2}\left\{\overline{\psi}\gamma^\mu \nabla_\mu \psi - [\nabla_\mu \overline{\psi}]\gamma^\mu \psi\right\} - m\overline{\psi}\psi. \quad (31.61)$$

The source current is

$$J_a{}^\mu = i\left\{\overline{\psi}\gamma^\mu T_a \psi - T_a \overline{\psi}\gamma^\mu \psi\right\}, \quad (31.62)$$

which is also

$$J_a{}^\mu = \frac{\delta \mathcal{L}_\psi}{\delta A^a{}_\mu}, \quad (31.63)$$

and the strong relation (31.60) takes the form

$$D_\mu J_a{}^\mu = 0. \quad (31.64)$$

The strong relations are not real conservation laws, but mere manifestations of the local invariance. As said in Section 31.4.3, only the sum of the source current and the gauge field self-current has zero divergence:

$$\partial_\mu(J_a{}^\nu + j_a{}^\nu) = 0.$$

This is a matter of consistency: *since the self-current is not gauge covariant, the derivative cannot be gauge covariant either, but in such a way that the conservation law itself is gauge covariant — and consequently physically meaningful.* However, a gauge invariant conserved charge can be obtained only under the additional proviso that the local transformations become constant transformations at the border of the system.

General references

Basic contribution texts on the above matters are: Bjorken & Drell 1964 and 1965, Itzykson & Zuber 1980, Bogoliubov & Shirkov 1980 and Konopleva & Popov 1981.

Chapter 32

Gauge Fields: Fundamentals

Universality gives to gravitation an immediate geometric nature, and this is very clear in its description by General Relativity: the underlying geometry is felt equally by every particle. The basic field for gravitation is the spacetime metric itself — whence its universality. The strong, weak and electromagnetic interactions are described by gauge theories, which also have a geometric background, but are not universal. The basic fields are connections, different connections for different kinds of particle. Notice that even these differences are of geometrical character. We let gravitation to later chapters, and give below a sketchy presentation of the gauge scheme of things.

32.1 Introductory remarks

The badge of the geometrical character of General Relativity is given by the fact that, under the action of a gravitational field, a test particle moves freely but in a curved spacetime, the curvature being that of the Levi–Civita connection of the gravitational field, which is a metric. The gravitational interaction is thereby "geometrized".

Gauge theories are also of geometrical character, but in a different sense: the basic fields, the gauge potentials, are connections on general principal bundles. In both cases, a significant geometric stage set is supposed as a kind of background to which dynamics is added. Gauge theories are, in a sense, still more geometric than General Relativity, because there is a duality symmetry between their dynamics and the geometric background. Trautman and Yang have greatly emphasized the "geometric" approach to gauge theories. This has been roughly presented in the main text. We shall give in what follows a résumé of the rather intuitive "physicist's approach". For simplicity of notation, we use in this chapter units in which $\hbar = c = 1$.

Commentary 32.1 In addition to the geometric description, gravitation has also a gauge description, in which the interaction is not so much "geometrized". This theory is knowns as Teleparallel Gravity, and will presented in Chapter 35. ◄

32.2 The gauge tenets

32.2.1 *Electromagnetism*

The spinor lagrangian in the presence of an electromagnetic field

$$\mathcal{L} = \tfrac{1}{2}\left[i\overline{\psi}\gamma^\mu\partial_\mu\psi - i(\partial_\mu\overline{\psi})\gamma^\mu\psi\right] - m\overline{\psi}\psi + eA_\mu\overline{\psi}\gamma^\mu\psi - \tfrac{1}{4}F^{\mu\nu}F_{\mu\nu} \quad (32.1)$$

is invariant under transformations of the type

$$(x) \to \psi'(x) = e^{i\alpha}\psi(x), \quad (32.2)$$

where α is a constant phase. Such transformations have been called "gauge transformations of the first kind", and the invariance leads to charge conservation by Noether's first theorem. The phase factor $e^{i\alpha}$ is an element of the unitary group $U(1)$ of 1×1 matrices. Yang and Mills[1] noticed that $\mathcal{L}$ is also invariant when $\alpha = \alpha(x)$ is point-dependent, provided that simultaneously the potential changes according to the well-known gauge transformations "of the second kind",

$$A_\mu(x) \to A'_\mu(x) = A_\mu(x) + \frac{1}{e}\partial_\mu\alpha(x). \quad (32.3)$$

The field A_μ must have this behaviour in order to compensate the derivative terms $\partial_\mu\alpha(x)$ turning up in the point-dependent case. The electromagnetic potential appears in this way as a "compensating field", with some analogy to the extra term needed in the Lagrange derivative in mechanics (Section 26.2). The phase factors $e^{i\alpha(x)}$ are now point-dependent elements of the group $U(1)$, which is the *gauge group*. The phase symmetry then requires the second Noether's theorem to be related to the charge, besides some extra assumptions concerning asymptotic behaviour (see Sections 31.4.3 and 31.4.4).

Lagrangian (32.1) can be rearranged as

$$\mathcal{L} = \tfrac{1}{2}\left[i\overline{\psi}\gamma^\mu(\partial_\mu - ieA_\mu)\psi - i(\partial_\mu + ieA_\mu)\overline{\psi}\,\gamma^\mu\psi\right]$$
$$- m\overline{\psi}\psi - \tfrac{1}{4}F^{\mu\nu}F_{\mu\nu}. \quad (32.4)$$

---

[1] Yang & Mills 1954.

It all works as if the effect of the electromagnetic field is solely to change the derivative acting on the source field:

$$\partial_\mu \rightarrow \partial_\mu - ieA_\mu. \qquad (32.5)$$

The change of derivative gives a rule to introduce the electromagnetic interaction: given a source field, we write down its free field equations and then change the derivatives. This is the "minimal coupling prescription", which will be generalized next for the case of non-abelian theories.

32.2.2 *Non-abelian theories*

Gauge theories[2] in their modern sense were really inaugurated when Yang and Mills proceeded to consider the isospin group $SU(2)$, whose non-abelian character made a lot of difference.

Commentary 32.2 Weyl's pioneering version, in which *gauge invariance* appears as an indifferent scaling in ambient space, is summarized in Weyl 1932 and in Synge & Schild 1978. The original literature includes Weyl 1919 and Weyl 1929. Additional important references are London 1927, Fock 1926, and Klein 1938; excerpts can be found as addenda in Okun 1984. ◄

A fermionic source will now be a multiplet, a field transforming in a well-defined way under the action of the group. Utiyama[3] generalized their procedure to other Lie groups and we shall prefer to recall once and for all the case of a general group G. The phase factors, elements of the gauge group G in that representation, are now operators of the form

$$U(x) = \exp[i\alpha^a(x)T_a],$$

and the source multiplets will transform according to

$$(x) \rightarrow \psi'(x) = U(x)\psi(x) = \exp[i\alpha^a(x)T_a]\,\psi(x). \qquad (32.6)$$

The indices $a, b, c, \ldots = 1, 2, 3, \ldots \dim G$, are Lie algebra indices, which are lowered by the Killing–Cartan metric γ_{ab} of the gauge group, supposed to be semi-simple. The T_a's are the generators of the group Lie algebra, written in the representation to which ψ belongs.

The vector potentials, now one for each group generator, will have a behaviour more involved than (32.6). We write

$$A_\mu = T_a A^a{}_\mu,$$

---

[2] The subject is treated in every modern text on Field Theory, from the classical treatise Bogoliubov & Shirkov 1980 to the more recent Itzykson & Zuber 1980. There are many other very good books, such as Faddeev & Slavnov 1978. For a short introduction covering practically all the main points, see Jackiw 1980.

[3] Utiyama 1955.

which makes of A_μ a matrix in the representation of ψ. In order to keep its role of compensating field, this matrix gauge potential will have to transform according to

$$A_\mu \to A'_\mu = U(x)A_\mu U^{-1}(x) + i\,U(x)\partial_\mu U^{-1}(x). \qquad (32.7)$$

A certain confusion comes up here. Acting on a field ψ belonging to the representation T_a, the covariant derivative will be

$$D_\mu\psi = \partial_\mu\psi - iA^a{}_\mu T_a\psi.$$

Suppose that another field φ, belonging to some other representation of the gauge group G, appears as the source field. If the group generators are T'_a in the representation of φ, then $A_\mu(x)$ must be written

$$A_\mu = T'_a A^a{}_\mu,$$

and the covariant derivative will assume the form

$$D_\mu\varphi = \partial_\mu\varphi - iA^a{}_\mu T'_a\varphi.$$

The particular expression of the covariant derivative changes from one representation to the other.

The gauge potential A_μ belongs to the adjoint representation of G. By this representation, the group acts on its own Lie algebra. Denoting these generators by J_a, they will satisfy the general commutation relations

$$[J_a, J_b] = f^c{}_{ab}\,J_c,$$

with $f^c{}_{ab}$ the structure constants, and will transform by

$$J_a \to U^{-1}(x)J_a U(x).$$

The vector potential will then be

$$A_\mu = J_a A^a{}_\mu. \qquad (32.8)$$

The field strength $F_{\mu\nu}$, given by the covariant derivative of the the gauge potential $A^a{}_\mu$, is then written in the form

$$F_{\mu\nu} = \partial_\mu A_\nu - \partial_\nu A_\mu + [A_\mu, A_\nu], \qquad (32.9)$$

which is a matrix

$$F_{\mu\nu} = J_a F^a{}_{\mu\nu}$$

in the adjoint representation. On account of (32.7), it transforms according to

$$F_{\mu\nu} \to F'_{\mu\nu} = U^{-1}(x)\,F_{\mu\nu}\,U(x) = U^{-1}J_a U\,F^a_{\mu\nu}. \qquad (32.10)$$

The lagrangian for a fermionic source field will now be

$$\mathcal{L} = \tfrac{1}{2} \left[i\overline{\psi}\gamma^\mu (\partial_\mu - iA^a{}_\mu T_a)\,\psi - i(\partial_\mu + iA^a{}_\mu T_a)^- \gamma^\mu \psi \right]$$
$$- m\,\overline{\psi}\psi - \tfrac{1}{4} F^{a\mu\nu} F_{a\mu\nu}. \qquad (32.11)$$

The gauge field strength is the 2-form

$$F = \tfrac{1}{2}\, J_a F^a{}_{\lambda\mu} dx^\lambda \wedge dx^\mu, \qquad (32.12)$$

given in terms of the 1-form gauge potential

$$A = J_a A^a{}_\mu dx^\mu \qquad (32.13)$$

as

$$F = dA + \tfrac{1}{2}\,[A, A] = dA + A \wedge A. \qquad (32.14)$$

In components, this is

$$F = \tfrac{1}{2} J_a \big(\partial_\mu A^a{}_\nu - \partial_\nu A^a{}_\mu + f^a{}_{bc} A^b{}_\mu A^c{}_\nu\big) dx^\mu \wedge dx^\nu, \qquad (32.15)$$

from which one gets back expression (32.9),

$$F^a{}_{\mu\nu} = \partial_\mu A^a{}_\nu - \partial_\nu A^a{}_\mu + f^a{}_{bc} A^b{}_\mu A^c{}_\nu. \qquad (32.16)$$

The *Bianchi identity*

$$\partial_{[\lambda} F^a{}_{\mu\nu]} + f^a{}_{bc} A^b{}_{[\lambda} F^c{}_{\mu\nu]} = 0 \qquad (32.17)$$

is an automatic consequence of definition (32.16). If we define the *dual* tensor

$$\tilde{F}^{a\rho\lambda} = \tfrac{1}{2}\varepsilon^{\rho\lambda\mu\nu} F^a{}_{\mu\nu},$$

it may be written as

$$\partial^\mu \tilde{F}^a{}_{\mu\nu} + f^a{}_{bc} A^{b\mu} \tilde{F}^c{}_{\mu\nu} = 0. \qquad (32.18)$$

Notice that the dual presupposes a metric, so that the former version is in principle to be preferred.

The field equations of gauge theories are the *Yang–Mills equations*

$$\partial^\mu F^a{}_{\mu\nu} + f^a{}_{bc} A^{b\mu} F^c{}_{\mu\nu} = J^a{}_\nu, \qquad (32.19)$$

where $J^a{}_\nu$ is the source current. It is equivalent to

$$\partial_{[\lambda} \tilde{F}^a{}_{\mu\nu]} + f^a{}_{bc} A^b{}_{[\lambda} \tilde{F}^c{}_{\mu\nu]} = \tilde{J}^a{}_{\lambda\mu\nu}. \qquad (32.20)$$

Observe that, in the sourceless case, the field equations are just the Bianchi identities written for the dual of F. This is the duality symmetry. If we know the geometrical background, we know the dynamics. For this reason we have said that gauge theories are fundamentally geometric theories.

Commentary 32.3 It is opportune to mention a crucial difference in relation to General Relativity. Since this theory does not have duality symmetry, its dynamics must be introduced independently of the Bianchi identities. In the linear approximation, however, General Relativity does have a duality symmetry, whose dynamics can then be related to the linearized Bianchi identity. ◄

32.2.3   *The gauge prescription*

We arrive in this way at the general gauge prescription. In order to introduce an interaction invariant under the local symmetry given by a group G, we change all ordinary derivatives in the free equations into covariant derivatives acting on each source field φ,

$$\partial_\mu \to D_\mu = \partial_\mu - iA^a{}_\mu T_a, \qquad (32.21)$$

where the T_a's are the group generators in the representation of G to which φ belongs.

Commentary 32.4   The coupling constant g, which would here take the place of the electric charge e of electromagnetism, can at this level of the theory be absorbed in A. We have ignored it up to this point, and shall continue to ignore it in the formulae which follow. ◀

The covariant derivative D_μ is a particular case of the general covariant derivative given in Section 31.4.3. The field equation for each source field will involve the Lagrange derivative with the usual derivatives replaced by the covariant ones:

$$\frac{\delta \mathcal{L}_\phi(y)}{\delta \phi(y)} = \frac{\partial \mathcal{L}_\phi(y)}{\partial \phi(y)} - D_\mu \frac{\partial \mathcal{L}_\phi(y)}{\partial D_\mu \phi(y)} = 0. \qquad (32.22)$$

Because the covariant derivative is different when acting on fields in different representations, it is important to pay careful attention to its form on each object.

32.2.4   *Hamiltonian approach*

There are many reasons to prefer the hamiltonian approach[4] to Yang–Mills equations.[5] Space and time have very distinct roles in this formalism. The electric and magnetic fields, given respectively by

$$E_a{}^i = F_a{}^{i0} \quad \text{and} \quad B_a{}^i = \tfrac{1}{2}\varepsilon^{ijk}F_{ajk},$$

also play different roles. In the hamiltonian formalism, the canonical coordinates are the $A^a{}_k$'s and, given the sourceless lagrangian

$$\mathcal{L}_G = -\tfrac{1}{4} F^{a\mu\nu}F_{a\mu\nu}, \qquad (32.23)$$

the conjugate momenta are the electric fields

$$\Pi^{ai} = \frac{\partial \mathcal{L}_G}{\partial_0 A_{ai}} = E^{ai} = F^{ai0}. \qquad (32.24)$$

---

[4] Itzykson & Zuber 1980.

[5] About its advantages, as well as for a beautiful general discussion, see Jackiw 1980.

The hamiltonian can be written in the form

$$H = \int d^4x \, \mathrm{tr}\Big[\big(\partial^0 \boldsymbol{A} + \boldsymbol{\nabla} A^0 - [A^0, \boldsymbol{A}]\big) \cdot \boldsymbol{E} + \tfrac{1}{2}(\boldsymbol{E}^2 + \boldsymbol{B}^2)\Big], \qquad (32.25)$$

and the action can be written as

$$S = \int d^4x \, \mathrm{tr}\Big[\partial^0 \boldsymbol{A} \cdot \boldsymbol{E} + \tfrac{1}{2}(\boldsymbol{E}^2 + \boldsymbol{B}^2) - A_a{}^0 G^a(x)\Big], \qquad (32.26)$$

where $G^a(x)$ is the expression representing the non-abelian Gauss law, which reads

$$G^a(x,t) \equiv D_k E^{ak} = \partial_k E^{ak} + f^a{}_{bc} A^b{}_k E^{ck} = 0. \qquad (32.27)$$

We see in (32.26) that A^0 acts as a Lagrange multiplier enforcing Gauss law, which appears as a constraint. Another constraint is the magnetic field expression

$$B^{ai} = \tfrac{1}{2}\varepsilon^{ijk} F^a{}_{jk} = \varepsilon^{ijk}\Big(\partial_j A^a{}_k + \tfrac{1}{2}\varepsilon^{abc} A_{bj} A_{ck}\Big). \qquad (32.28)$$

The ensuing dynamical equations are Hamilton's equations: Ampère's law

$$\frac{1}{c}\frac{\partial \boldsymbol{E}^i}{\partial t} = (\boldsymbol{\nabla} \times \boldsymbol{B})^i + \varepsilon^{ijk}[A_j, B_k] + [A^0, E^i], \qquad (32.29)$$

and the time variation of the vector potential,

$$\frac{1}{c}\frac{\partial \boldsymbol{A}}{\partial t} = -\boldsymbol{E} - \boldsymbol{\nabla} A^0 + [A^0, \boldsymbol{A}]. \qquad (32.30)$$

The nonvanishing canonical commutation relations are:

$$\{A^a{}_i(x), E_b{}^j(y)\} = i\,\delta^a_b\,\delta^j_i\,\delta^3(x-y). \qquad (32.31)$$

The hamiltonian formalism is of special interest to quantization. In the Schrödinger picture of field theory, the state is given by a functional $\Psi[A]$, a kind of wave function on the coordinates $A^a{}_i$. Applied to a general state functional $\Psi[A]$, the A^a_i are to be seen as multiplication-by-function operators, while their conjugate momenta are operators,

$$E_a{}^k \Psi[A] = -i\,\frac{\delta \Psi[A]}{\delta A^a{}_k}, \qquad (32.32)$$

in complete analogy with elementary Quantum Mechanics.

32.2.5 *Exterior differential formulation*

We have given the basic formulae in the main text. Let us only repeat a few of them. The field strength (32.14) is the curvature of the gauge potential A, which is a connection. Taking the differential of F leads directly to the Bianchi identity

$$dF + [A, F] = 0. \tag{32.33}$$

The Yang–Mills equations (32.19) are, in invariant notation,

$$\tilde{d}F + *^{-1}[A, *F] = J. \tag{32.34}$$

Thus, in the sourceless case, we see clearly the invariant version of the duality symmetry. A self-dual (or antiself-dual) 2-form in a 4-dimensional space, solution of

$$F = \pm *F, \tag{32.35}$$

will respect

$$F = \pm *F = \pm *[\pm *F] = **F = (-)^{(4-s)/2}F = (-)^{s/2}F. \tag{32.36}$$

In Minkowski space, the signature is $s = 2$ and the self-duality implies the vanishing of F. However, in an euclidean 4-dimensional space there may exist non-trivial self-dual solutions. Such self-dual euclidean fields are called *instantons*. Any self-dual F of the form

$$F = dA + A \wedge A$$

will solve automatically the sourceless field equations.

It comes out clearly from the differential approach that the gauge prescription must be improved. There are covariant derivatives, but there are also covariant coderivatives. The latter are to replace the usual coderivatives of the free case. This is the case, for example, in the continuity equation. Functional forms (see Section 20.4.1, to which we refer for the calculations) enlarge the geometrical meaning of gauge fields.

32.3 Functional differential approach

32.3.1 *Functional forms*

The Euler Form for a sourceless gauge field is

$$E = \left(\partial^\mu F^a{}_{\mu\nu} + f^a{}_{bc}A^{b\mu}F^c{}_{\mu\nu}\right)\delta A_a{}^\nu = (D^\mu F^a_{\mu\nu})\delta A_a{}^\nu. \tag{32.37}$$

The coefficient, whose vanishing gives the Yang–Mills equations, is the covariant coderivative of the curvature F of the connection A according to that same connection. Each component $A^a{}_\mu$ is a variable labelled by the double index (a, μ), and $f^a{}_{bc}$ are the gauge group structure constants. Let us examine the condition for the existence of a lagrangian. Taking the (functional) differential, we find

$$\delta E = \left(\partial^\mu \delta F^a{}_{\mu\nu} + f^a{}_{bc} A^{b\mu} \delta F^c{}_{\mu\nu} + f^a{}_{bc} \delta A^{b\mu} F^c{}_{\mu\nu}\right) \wedge \delta A_a{}^\nu. \qquad (32.38)$$

The last term vanishes if we use the complete antisymmetry (or cyclic symmetry) of $f^a{}_{bc}$: the coefficients become symmetric under the change $(a, \nu) \leftrightarrow (b, \mu)$. Integrating by parts the first term, using again the cyclic symmetry and conveniently antisymmetrizing in (μ, ν), we arrive at the necessary condition for the existence of a lagrangian:

$$\delta E = -\tfrac{1}{2} dF^a{}_{\mu\nu} \wedge dF_a{}^{\mu\nu} = 0. \qquad (32.39)$$

The cyclic symmetry used above holds for semisimple groups, for which the Cartan–Killing form is an invariant metric well-defined on the group. Actually, we have been using this metric to raise and lower indices all along. No lagrangian exists in the nonsemisimple case.[6] In the semisimple case, we obtain

$$\mathcal{L}_G = \tfrac{1}{2} A_a{}^\nu D^\mu F^a_{\mu\nu} = -\tfrac{1}{4} F^{a\mu\nu} F_{a\mu\nu}. \qquad (32.40)$$

The action is just that of Eq. (7.160), which is

$$S = \int F \wedge *F.$$

Conversely, if we assume an action of the form

$$C_G = \int F \wedge F, \qquad (32.41)$$

it is not difficult to find that there exists a 3-form K such that

$$F \wedge F = dK.$$

This means that C_G is a surface term, that would not lead to local equations by variation — though a naïve variation would lead to the Bianchi identities. In the euclidean case,

$$\int_{\mathbb{E}^4} F \wedge F = \int_{S^3} K.$$

We now consider the field F concentrated in a limited region, that is, on a sphere S^3 of large enough radius — so that only the vacuum exists far

---

[6] Aldrovandi & Pereira 1986, 1988.

enough. The potential will there be given by the last term of equation (32.7). Examination of the detailed form of K shows that, with a convenient normalization, C_G can be put in the form

$$n = \frac{1}{24\pi^2} \int_{S^3} d^3x \, \mathrm{tr}\left[\varepsilon_{ijk}\left(g^{-1}(x)\partial^i g(x)\right)\left(g^{-1}(x)\partial^j g(x)\right)\left(g^{-1}(x)\partial^k g(x)\right)\right].$$

When the gauge group is $SU(2)$, whose manifold is also S^3, the function $g(x)$ takes S^3 into S^3. Once more, we can show[7] that the integrand is actually a volume form on S^3, so that in the process of integration we are counting the number of times the values $g(x)$ cover $SU(2) = S^3$ while the variable "x" covers S^3 one time. The normalization above is chosen so that just such integer number "n" comes out. This is the winding number (see § 3.4.12 and § 6.2.14) of the function g. The values of the number n can be used to classify the vacua, which are of the form $g^{-1}dg$. This topological number is a generalization of the Chern number (Section 22.5.2), introduced in the bundle of frames, to general bundles on $\mathbb{E}^4$ with structure group $SU(2)$.

32.3.2   *The space of gauge potentials*

The functional approach gives an important role to the space of the gauge potentials, on which the state functional $\Psi[A]$ is defined. This space is usually called the A-space and will be denoted by Σ. More about it will be said in Chapter 33. We shall here concentrate on some aspects which constitute a natural suite for the above sections.

As a function, the potential A depends on some fixed starting value a, and on the group element g by whose transformation A is obtained from a. Thus, we rewrite (32.7) as

$$A(a,g) = g^{-1}ag + g^{-1}dg. \tag{32.42}$$

This decomposition corresponds to some specially convenient coordinates on A-space. The vacuum term, $v = g^{-1}dg$, corresponds to the Maurer–Cartan form of the group: one checks easily that

$$dv + v \wedge v = 0.$$

In the functional case, A itself becomes a functional of the functions $g(x)$ and $a(x)$. Notice that, in this case, also $g(x)$ is to be seen, not as an element of the gauge group G, but as a member g of the space of G-valued functions ($g(x)$ is actually a chiral field, see below). This space, an infinite

----
[7] See Coleman 1979.

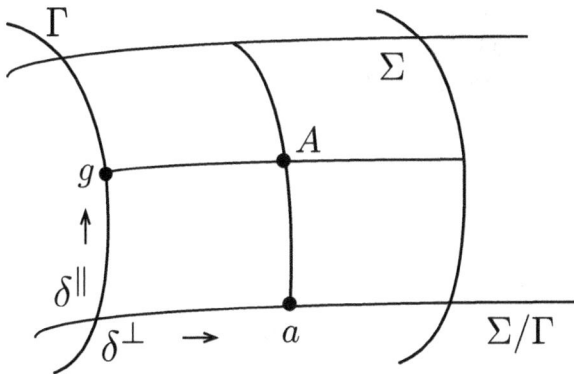

Fig. 32.1 Local decomposition of Σ into components "along" and "orthogonal" to the large group.

group formed by all the gauge transformations on spacetime, is called the *large group*, and will be denoted by Γ.

The space Σ is starshaped,[8] so that the use of the homotopy formula to get the Lagrangian $\mathcal{L}_G$ of (32.40) is straightforward, and $\mathcal{L}_G$ will be valid on the whole Σ as far as no subsidiary gauge condition is imposed. Of course this is not the real physical space, which requires a choice of gauge and is far more complicated.[9] Given the large group Γ, the physical space is formed by the gauge-inequivalent points of Σ, the quotient space Σ/Γ constituted by the gauge orbits, or the space of "points" $a(x)$. This leads to a (local!) decomposition of Σ into components along and perpendicular to the functional large group (Figure 32.1). Variations on Σ may be locally decomposed into a part "along" Γ and a part "orthogonal" to Γ,

$$\delta A^a{}_\mu = \delta^\| A^a{}_\mu + \delta^\perp A^a{}_\mu. \tag{32.43}$$

The part $\delta^\|$ parallel to Γ is a gauge transformation. Defined on Σ there are entities which act as representatives of the geometrical entities defined on the gauge group G. Such representatives are, however, dependent on the point in spacetime.

We have here an opportunity to apply functional exterior calculus. As the exterior differential is that given by the variational differential, we may think of the form

$$\omega = \omega^a J_a = g^{-1}\delta g$$

[8] Singer 1981.
[9] Wu & Zee 1985.

as a functional version of the Maurer–Cartan form $v = g^{-1}dg$. We shall find below that a small correction is necessary to this interpretation. The functional version of the Maurer–Cartan form will be

$$\Omega = g^{-1}\delta^{\|}g,$$

which stands "along" the large group. We obtain from (32.42)

$$\delta A(a, g) = d\omega + [A, \omega] + g^{-1}\delta ag = D\omega + g^{-1}\delta ag, \qquad (32.44)$$

from which we can interpret

$$\delta^{\|}A = D\omega \qquad (32.45)$$

as the usual gauge transformation of A:

$$\delta^{\|}A_\mu(a, g) = \partial_\mu\omega + [A_\mu, \omega]. \qquad (32.46)$$

But we see also that

$$\delta^\perp A = g^{-1}\delta ag, \qquad (32.47)$$

which says that the perpendicular variation is the transformation of the "physical" variation. It is the real variation, to the exclusion of any gauge transformation. Gauge transformations will be given by

$$\delta^{\|} = \omega^a X_a, \qquad (32.48)$$

where the X_a's are the generators in the functional representation, to be found in the following. Another interesting result is

$$\delta v = d\omega + [v, \omega]. \qquad (32.49)$$

It is sometimes more convenient to work with components. Take

$$g = \exp[\alpha] = \exp[\alpha^a J_a].$$

Then,

$$\begin{aligned} \omega &= g^{-1}\delta g = g^{-1}(\delta\alpha)g \\ &= g^{-1}(\delta\alpha^a J_a)g = (\delta\alpha^a)g^{-1}J_a g \\ &= \delta\alpha^a K_a{}^b J_b, \end{aligned} \qquad (32.50)$$

or equivalently, since $\omega = \omega^b J_b$,

$$\omega^b = \delta\alpha^a K_a{}^b. \qquad (32.51)$$

Here, the $K_a{}^b$'s are the coefficients of the adjoint representation.

Take the functional forms $\{\omega^b\}$ as basis and consider its dual basis $\{X_a\}$. On any functional Ψ,

$$
\begin{aligned}
\delta\Psi &\equiv \delta^{\|}\Psi + \delta^{\perp}\Psi \\
&= \frac{\delta\Psi}{\delta A^a{}_\mu}\,\delta^{\|}A^a{}_\mu + \frac{\delta\Psi}{\delta A^a{}_\mu}\,\delta^{\perp}A^a{}_\mu \\
&= \frac{\delta\Psi}{\delta A^a{}_\mu}\,D_\mu\omega^a + \frac{\delta\Psi}{\delta A^a{}_\mu}\,\delta^{\perp}A^a{}_\mu \\
&= -D_\mu\!\left(\frac{\delta\Psi}{\delta A^a{}_\mu}\right)\omega^a + \frac{\delta\Psi}{\delta A^a{}_\mu}\,\delta^{\perp}A^a{}_\mu.
\end{aligned}
\tag{32.52}
$$

On the other hand, this must also be

$$
\delta\Psi = X_a(\Psi)\,\omega^a + \frac{\delta\Psi}{\delta A^a{}_\mu}\,\delta^{\perp}A^a{}_\mu.
$$

We find in this way that the group generators acting on the functionals are

$$
X_a = -D_\mu\,\frac{\delta}{\delta A^a{}_\mu}.
\tag{32.53}
$$

Also the Euler Form (32.37) can be decomposed:

$$
\begin{aligned}
E &\equiv \operatorname{tr}(E_\mu\delta A^\mu) = \operatorname{tr}(E_\mu\delta^{\|}A^\mu + E_\mu\delta^{\perp}A^\mu) \\
&= \operatorname{tr}(E^\mu D_\mu\omega + E_\mu\delta^{\perp}A^\mu) \\
&= -\operatorname{tr}\!\left[(D_\mu E^\mu)\,\omega - E_\mu\,g^{-1}\delta a^\mu\,g\right] \\
&= -\operatorname{tr}\!\left[(D_\mu E^\mu)\,\omega - gE_\mu g^{-1}\,\delta a^\mu\right],
\end{aligned}
$$

or equivalently,

$$
E = -\operatorname{tr}\!\left[(D_\mu E^\mu)\,\omega\right] + \operatorname{tr}\!\left[e_\mu\,\delta a^\mu\right],
\tag{32.54}
$$

where we have introduced the notation $e_\mu = gE_\mu g^{-1}$, which stands for the expression appearing in the equation when a stands for the potential. As

$$
D_\mu E^\mu = 0
$$

identically, $e_\mu = 0$ is the true equation. It is noteworthy that

$$
\begin{aligned}
\delta\mathcal{L}_G &\equiv \delta^{\|}\mathcal{L}_G + \delta^{\perp}\mathcal{L}_G \\
&= E^a{}_\mu\delta^{\perp}A_a{}^\mu + \delta^{\|}\mathcal{L}_G E_a{}^\mu \\
&= E^a{}_\mu\delta^{\perp}A_a{}^\mu - \omega^a D_\mu E_a{}^\mu \\
&= E^a{}_\mu\delta^{\perp}A_a{}^\mu - \omega^a X_a\mathcal{L}_G.
\end{aligned}
\tag{32.55}
$$

We can also recognize, by integrating by parts, that

$$
0 = \delta^{\|}\mathcal{L}_G = (D_\mu\omega^a)E_a{}^\mu = \delta^{\|}A^a{}_\mu E_a{}^\mu,
\tag{32.56}
$$

which again says that the equation is "orthogonal" to Γ.

The group parameters η^a, in terms of which an element of G is written as

$$g = \exp[\eta^a T_a]$$

in some representation generated by $\{T_a\}$, become fields $\eta^a(x)$. The canonical Maurer–Cartan 1-forms $v = g^{-1}dg$ on G are represented by cofields

$$\Omega(x) = g^{-1}(x)\delta^{||}g(x),$$

1-Forms on Γ, whose expression is enough to ensure that Ω satisfies a functional version of the Maurer–Cartan equations,

$$\delta^{||}\Omega = -\,\Omega \wedge \Omega \tag{32.57}$$

or

$$\delta^{||}\Omega^a = -\tfrac{1}{2}f^a{}_{bc}\,\Omega^b \wedge \Omega^c. \tag{32.58}$$

The components Ω^a, or the matrix

$$\Omega^i{}_j = \Omega^a(J_a)^i{}_j = \Omega^a f^i{}_{aj},$$

are used when convenient, the same holding for A_μ, $F_{\mu\nu}$, δA_μ, etc.

32.3.3 *Gauge conditions*

Gauge subsidiary conditions correspond to 1-Forms along Γ. Take, for example, the one-dimensional abelian case of electromagnetism, for which the Maxwell Euler Form is

$$E = (\partial^\mu F_{\mu\nu})\delta A^\nu . \tag{32.59}$$

As

$$\delta A^\nu = \delta^{||}A^\nu + \delta^{\perp}A^\nu$$

with

$$\delta^{||}A^\mu = \partial^\mu \eta$$

for some parameter field η, an integration by parts shows that the contribution along Γ vanishes. The Lorenz gauge condition is specified by the 1-Form

$$H = \lambda(\partial^\mu A_\mu)\delta\eta = -\lambda A_\mu \partial^\mu \delta\eta - \lambda A_\mu \delta^{||}A^\mu. \tag{32.60}$$

The complete Euler Form governing electromagnetism in the Lorenz gauge is consequently

$$E^* = (\partial^\mu F_{\mu\nu})\delta^\perp A^\nu - \frac{\lambda}{2}\,\delta^{\parallel}(A_\mu A^\mu). \tag{32.61}$$

The Form H is exact only along Γ, so that we cannot say that the transgression

$$TH = -\frac{\lambda}{2}\,(A_\mu A^\mu)$$

is a lagrangian in the usual sense. But (32.61) is an eloquent expression: the equation goes along the physical space, whereas the gauge condition lies along the large group.

The above considerations can be transposed without much ado to the nonabelian case. Putting

$$\delta A_a{}^\nu = D^\nu \delta\eta_a + \delta^\perp A_a{}^\nu \tag{32.62}$$

in (32.37), the contribution along the group vanishes again. The Lorenz Form is now

$$\begin{aligned} H &= \lambda(\partial^\mu A^a{}_\mu)\delta\eta_a = -\lambda A^a{}_\mu(D^\mu\delta\eta_a - [A^\mu, \delta\eta]_a) \\ &= -\lambda A^a{}_\mu\delta^{\parallel}A_a{}^\mu = -\tfrac{1}{2}\lambda\,\delta^{\parallel}(A^a{}_\mu A_a{}^\mu). \end{aligned} \tag{32.63}$$

Supposing a convenient normalization for the Cartan–Killing metric that we have been using implicitly, the total Euler Form can be written

$$E^* = \mathrm{tr}\left[D^\mu F_{\mu\nu}\delta^\perp A^\nu - \tfrac{1}{2}\lambda\,\delta^{\parallel}(A_\mu A^\mu)\right]. \tag{32.64}$$

We have used, for the sector along Γ, the holonomic (or "coordinate") basis $\{\delta\eta_a\}$, composed of exact Forms. We could likewise have used a nonholonomic basis. With differential forms, the choice of basis is in general dictated by the symmetry of the problem.

32.3.4   Gauge anomalies

The expressions for the gauge anomaly are components of 1-Forms along Γ in the anholonomic basis $\{\Omega^a\}$:

$$U = U_a\,\Omega^a. \tag{32.65}$$

Using (32.58), we find

$$\delta^{\parallel}U = \tfrac{1}{2}\left[T_a U_b - T_b U_a - U_c f^c{}_{ab}\right]\Omega^a \wedge \Omega^b. \tag{32.66}$$

The vanishing of the expression inside the brackets is the usual Wess–Zumino consistency condition, which in this language becomes simply

$$\delta^{\|}U = 0. \tag{32.67}$$

Again, U must be locally an exact Form, but only along Γ, so that TU is not a lagrangian.

Notice that, unlike the case of the action, the last equation does not express the invariance of U under gauge transformations. Only when acting on 0-Forms does $\delta^{\|}$ represent gauge transformations. The situation is again analogous to differential geometry, where transformations are represented by Lie derivatives. Let us consider vector fields on Σ, say entities such as

$$\eta = \eta^a T_a \qquad \text{or} \qquad X = X^a \delta/\delta\eta^a.$$

Transformations on Forms will be given by the Lie derivatives

$$L_X = \delta \circ i_X + i_X \circ \delta.$$

For 0-Forms, only the last term remains, but for U the first will also contribute. The invariance of a Form W under a transformation whose generator is represented by a "Killing Field" X is expressed by $L_X W = 0$. In the case of an Euler Form coming from a lagrangian, $E = \delta S$. The commutativity between the Lie derivative and the differential operator leads to

$$L_X E = \delta L_X S,$$

a well known result: the invariance of S ($L_X S = 0$) implies the invariance of E ($L_X E = 0$), but not vice-versa (see Section 20.4.3). The invariance of E only implies the closeness of $L_X S$, and the equations may have symmetries which are not in the lagrangian.[10]

32.3.5 *BRST symmetry*

A final remark concerning gauge fields: we have already used

$$\delta^{\|} A_a{}^\mu = D^\mu \delta\eta_a.$$

As $A_a{}^\mu$ is a 0-Form, this measures to first order its change under a group transformation given by

$$g(x) = \exp[-\delta\eta(x)] \sim 1 - \delta\eta(x).$$

---

[10] Okubo 1980.

By using that $\Omega = g^{-1}(\delta\eta)g$, we can write

$$\delta^{||}A^\mu = D^\mu\Omega. \tag{32.68}$$

A fermionic field ψ will transform according to

$$\delta^{||}\psi' = \delta\eta\,\psi' = g^{-1}\Omega\,g\,\psi',$$

or

$$\delta^{||}\psi = \Omega\,\psi. \tag{32.69}$$

Let us repeat equation (32.57),

$$\delta^{||}\Omega = -\,\Omega\wedge\Omega. \tag{32.70}$$

Equations (32.68), (32.69) and (32.70) express the BRST transformations[11] provided the Maurer–Cartan Form Ω is interpreted as the ghost field,[12] and Slavnov's operator is recognized as $\delta^{||}$. The use of $\delta^{||}$ to obtain topological results[13] is a fine suggestion of the convenience of variational Forms to treat global properties in functional spaces, although it remains, to our knowledge, the only such application to the present.

32.4   Chiral fields

We make now a few comments on pure chiral fields, here understood simply as the group-valued fields $g(x)$ met above.

(1) The functional space reduces to Γ, and δ will coincide with the previous $\delta^{||}$. Neither G nor Γ are starshaped spaces, so that we must work with tensor fields on the Lie algebra (which, being a vector space, is starshaped) and their functional counterparts. The variation of the Maurer–Cartan form

$$\omega_\mu = g^{-1}\partial_\mu g$$

is the covariant derivative of its corresponding Form:

$$\begin{aligned}\delta\omega_\mu &= -\,g^{-1}(\delta g)g^{-1}\partial_\mu g + g^{-1}\partial_\mu(gg^{-1}\delta g)\\ &= \partial_\mu\Omega + [\omega_\mu,\Omega] = D_\mu\Omega.\end{aligned} \tag{32.71}$$

---

[11] Stora 1984; Baulieu 1984.
[12] Stora 1984; Leinaas & Olaussen 1982.
[13] Mañes 1985.

(2) To obtain the Euler Form corresponding to the 2-derivative contribution to the chiral field dynamics, we start from the usual action

$$S = -\tfrac{1}{2}\,\mathrm{tr}(\omega_\mu \omega^\mu),\tag{32.72}$$

from which

$$
\begin{aligned}
E = \delta S &= -\,\mathrm{tr}\left(\omega_\mu \delta \omega^\mu\right)\\
&= -\,\mathrm{tr}\left[\omega_\mu\left(\partial^\mu \Omega + [\omega^\mu, \Omega]\right)\right]\\
&= -\,\mathrm{tr}\left[\omega_\mu \partial^\mu \Omega\right] = \mathrm{tr}\left[\left(\partial^\mu \omega_\mu\right)\Omega\right]\\
&= \mathrm{tr}\left[\partial^\mu(g^{-1}\partial_\mu g)g^{-1}\delta g\right].
\end{aligned}\tag{32.73}
$$

(3) The existence of a lagrangian here is a consequence of the functional Maurer–Cartan equation. In effect,

$$
\begin{aligned}
\delta E &= \delta\left[(\partial_\mu \omega_a{}^\mu)\Omega^a\right]\\
&= \delta\omega_a^\mu \wedge \partial_\mu \Omega^a + (\partial_\mu \omega_a{}^\mu)\delta\Omega^a\\
&= -\left(\partial_\mu \Omega^a + f^a{}_{bc}\,\omega_\mu^b\,\Omega^c\right)\wedge \partial^\mu \Omega_a + (\partial_\mu \omega_a{}^\mu)\delta\Omega^a\\
&= (\partial_\mu \omega_a{}^\mu)\left[\delta\Omega^a + \tfrac{1}{2}f^a{}_{bc}\,\Omega^b \wedge \Omega^c\right] = 0.
\end{aligned}\tag{32.74}
$$

The presence of Ω in the trace argument in (32.72) would not be evident from the field equation

$$\partial^\mu\left(g^{-1}\partial_\mu g\right) = 0.\tag{32.75}$$

The variation was entirely made in terms of ω_μ and Ω, which belong to starshaped spaces, and not in terms of $g(x)$. We can consequently follow the inverse way: put (32.73) in the form

$$E = -\,\mathrm{tr}\left(\omega_\mu \delta\omega^\mu\right)$$

and get (32.72) back.

The above considerations are examples of the power of exterior variational calculus, on which more is said in Chapter 20.

General references

Basic references suggested for this chapter are Trautman 1970, Trautman 1979, Yang 1974, Popov 1975, and Faddeev & Shatashvilli 1984.

Chapter 33

More Gauge Fields: Wu–Yang Copies

Two distinct gauge potentials can have the same field strength, in which case they are said to be "copies" of each other. When we examine this ambiguity in the general affine space $\mathcal{A}$ of gauge potentials we find an interesting result: any two potentials are connected by a straight line in $\mathcal{A}$, but a straight line going through two copies either contains no other copy or is entirely formed by copies.

33.1 The ambiguity

Consider two connection forms A and $A^\sharp$, with curvatures

$$F = dA + A \wedge A \quad \text{and} \quad F^\sharp = dA^\sharp + A^\sharp \wedge A^\sharp. \tag{33.1}$$

Under a gauge transformation produced by a member g of the gauge group, the connections transform non-covariantly, according to

$$A \Rightarrow A' = gAg^{-1} + gdg^{-1} \quad \text{and} \quad A^\sharp \Rightarrow A^{\sharp\,\prime} = gA^\sharp g^{-1} + gdg^{-1}. \tag{33.2}$$

Their difference

$$K = A^\sharp - A, \tag{33.3}$$

is a 1-form transforming covariantly:

$$K' = A^{\sharp\,\prime} - A' = g(A^\sharp - A)g^{-1} = gKg^{-1}.$$

The curvatures will be covariant 2-forms:

$$F \to F' = gFg^{-1} \quad \text{and} \quad F^\sharp \to F^{\sharp\prime} = gF^\sharp g^{-1}. \tag{33.4}$$

The Wu–Yang ambiguity[1] turns up when

$$F^\sharp = F \quad \text{but} \quad K \neq 0. \tag{33.5}$$

---

[1] Wu & Yang 1975

693

Connections like A and $A^\sharp$, which are different but correspond to one same field strength, are called "copies". Notice that this can happen in any non-abelian gauge theory: two or more distinct connections — or gauge potentials in the physicist's language — can correspond to the same field strength. In what follow we will be using the invariant notation

$$A = J_a A^a{}_\mu dx^\mu \tag{33.6}$$

for a connection related to a gauge group with generators $\{J_a\}$,

$$F = \frac{1}{2} J_a F^a{}_{\mu\nu} \, dx^\mu \wedge dx^\nu \tag{33.7}$$

for the field strength, and

$$K = J_a K^a{}_\mu dx^\mu \tag{33.8}$$

for a covector in the adjoint representation of the gauge algebra. As only field strengths are measurable, it is impossible to know exactly which gauge potential is actually at work in a given physical system.

The importance of this ambiguity comes from the fact that, in Quantum Field Theory, the fundamental fields are the gauge potentials and not the field strengths. In particular, the corresponding particles are the quanta of the gauge potentials. Discussions on this subject are not, in general, concerned with solutions and mostly ignore the dynamic classical equations. Nevertheless, non-solutions are crucial precisely because they appear as off-shell contributions in the quantum case.

A complete understanding of this ambiguity, in all its aspects and consequences, does not seem to have been achieved. Before starting on our real concern here, let us say a few words on General Relativity: there are no copies in the relationship between the torsionless Levi–Civita spin connection $\overset{\circ}{A}$ and its curvature $\overset{\circ}{R}$. There exists always, around each point p, a system of coordinates in which $\overset{\circ}{A} = 0$ at p, so that the usual expression

$$\overset{\circ}{R} = d\,\overset{\circ}{A} + \overset{\circ}{A} \wedge \overset{\circ}{A} \tag{33.9}$$

reduces to $\overset{\circ}{R} = d\overset{\circ}{A}$, which can be integrated to give locally $\overset{\circ}{A}$ in terms of $\overset{\circ}{R}$. Actually, the equivalence principle "protects" General Relativity from ambiguity, but — we insist — holds only for torsionless connections. On the other hand, general linear connections with curvature and torsion exhibit copies in a quite natural way, as they can, in principle, have the same curvature and different torsions.

33.2 The connection space $\mathcal{A}$

Let us be unashamedly pedagogical and repeat that the geometric background for any gauge theory is a principal bundle with spacetime as base space and the gauge group as structure group. This bundle includes connections and curvatures, and its tangent structure contains the adjoint representation. Other fields, belonging to other representations, behave as sources and inhabit associated bundles. A section on the principal bundle is a "gauge" and picks out, for each point x of the base space, a point $g(x)$ on the bundle proper. A section is taken into another by a gauge transformation, induced precisely by such a group element $g(x)$. As these group elements are point-dependent, it is more convenient to take a functional point of view[2] and consider as principals the mappings g from the base space into the group.

In the same token, each connection A can be seen as the mapping taking point x of the base space into its value $A(x)$ in space $\mathcal{A}$. The set

$$\mathcal{G} = \{g : x \to g(x)\}$$

of all mappings from the base space into the group is frequently called the "large group". Gauge covariance divides $\mathcal{A}$ into equivalence classes,[3] each class representing a potential up to gauge transformations as those given in Eq. (33.2). The space of gauge inequivalent connections is the quotient $\mathcal{A}/\mathcal{G}$. An element of $\mathcal{A}$ can be locally represented by

$$A = (a, g), \quad \text{with} \quad a \in \mathcal{A}/\mathcal{G} \quad \text{and} \quad g \in \mathcal{G}.$$

Only variations along $\mathcal{A}/\mathcal{G}$ are of interest for copies, as variations along $\mathcal{G}$ are mere gauge transformations.

The ambiguity turns up because F, given as in (33.1), does not determine A. At each point of spacetime a gauge can be chosen in which $A = 0$ and consequently $F = dA$. This is true also along a line.[4] One might think of integrating by the homotopy formula (§ 7.2.11 and Section 20.4.1) to obtain A from F. This is impossible because the involved homotopy requires the validity of $F = dA$ on a domain of the same dimension of spacetime and the alluded gauge cannot exist (unless $F = 0$) on a domain of dimension 2 or higher.[5]

---

[2] Faddeev & Shatashvilli 1984
[3] See for example Wu & Zee 1985.
[4] Iliev 1996, 1997, 1998.
[5] Aldrovandi, Barros & Pereira 2003.

33.3 Copies in terms of covariant derivative

Covariant differentials have different expressions for different representations and different differential forms. For example, the gauge group element g can be seen as a matrix acting on column-vectors V belonging to an associated vector representation. The covariant differentials in the connections A and $A^\sharp$ will have, in that representation, the forms

$$DV = dV + A \wedge V$$

and

$$D^\sharp V = dV + A^\sharp \wedge V = dV + (A^\sharp - A) \wedge V + A \wedge V,$$

from which it follows

$$D^\sharp V = DV + K \wedge V. \tag{33.10}$$

For a 1-form in the adjoint representation, like the difference 1-form K of (33.3), we have

$$DK = dK + A \wedge K + K \wedge A = dK + \{A, K\}$$

and

$$D^\sharp K = dK + A^\sharp \wedge K + K \wedge A^\sharp = dK + \{A^\sharp, K\}.$$

It is immediately found that

$$D^\sharp K = DK + 2\,K \wedge K \tag{33.11}$$

and the relation between the two curvatures is

$$F^\sharp = F + DK + K \wedge K. \tag{33.12}$$

A direct calculation gives

$$DDK + [K, F] = 0, \tag{33.13}$$

which actually holds for any covariant 1-form in the adjoint representation. Equation (33.12) leads to a general result: given a connection A defining a covariant derivative D_A, each solution K of

$$D_A K + K \wedge K = 0 \tag{33.14}$$

will provide a copy.

33.4 The affine character of $\mathcal{A}$

We have been making implicit use of one main property of the space $\mathcal{A}$ of connections, namely: $\mathcal{A}$ is a convex affine space, homotopically trivial.[6] Being an affine space means that any connection $A^\sharp$ is related to a given connection A by $A^\sharp = A + K$, for some covariant covector K. Being convex means that through any two connections A and $A^\sharp$ there exists a straight line of connections A_t, given by

$$A_t = tA^\sharp + (1-t)A. \tag{33.15}$$

Of course, $A_0 = A$ and $A_1 = A^\sharp$. In terms of the difference form K,

$$A_t = A + tK = A^\sharp - (1-t)K. \tag{33.16}$$

Notice that $dA_t/dt = K$. Denoting by D_t the covariant derivative in the connection A_t, we find

$$D_t K = DK + 2tK \wedge K. \tag{33.17}$$

The curvature of A_t is

$$F_t = dA_t + A_t \wedge A_t = tF^\sharp + (1-t)F + t(t-1)K \wedge K, \tag{33.18}$$

or

$$F_t = F + tDK + t^2 K \wedge K = F + tD_t K - t^2 K \wedge K. \tag{33.19}$$

Observe that $F_0 = F$ and $F_1 = F^\sharp$. It thus follows that

$$\frac{dF_t}{dt} = DK + 2tK \wedge K = D_t K. \tag{33.20}$$

33.5 The copy-structure of space $\mathcal{A}$

Now on the question of copies: from (33.12), the necessary and sufficient condition to have $F^\sharp = F$ is

$$DK + K \wedge K = 0. \tag{33.21}$$

From the Bianchi identity $D^\sharp F^\sharp = 0$ applied with $A^\sharp = A + K$ it follows that

$$[K, F] = 0. \tag{33.22}$$

These conditions[7] lead to some determinantal conditions[8] for the non-existence of copies. Notice that (33.13) and (33.22) imply $DDK = 0$.

---

[6] Singer 1978.

[7] Deser & Wilczek 1976.

[8] Roskies 1977; Calvo 1977.

Commentary 33.1   We remark again that copies are of interest only for non-abelian theories. In fact, in the abelian case $K \wedge K \equiv 0$, which implies that $DK \equiv dK$. In this case, condition (33.21) reduces to $dK = 0$, which in turn means that locally $K = d\phi$ for some ϕ. Then $A^\sharp = A + d\phi$ is a mere gauge transformation. ◄

A first consequence of the conditions above is

$$\frac{dF_t}{dt} = D_t K = (1 - 2t)DK = (2t - 1)K \wedge K.$$

A second consequence is that now the line through F and $F^\sharp$ takes the form

$$F_t = F + t(t - 1)K \wedge K = F + t(1 - t)DK. \qquad (33.23)$$

We have thus the curvatures of all the connections linking two copies along a line in connection space. Are there other copies on this line? In other words, is there any $s \neq 0, 1$ for which $F_s = F$? The existence of one such copy would imply, by the two expressions in Eq. (33.23),

$$DK = 0 \qquad \text{and} \qquad K \wedge K = 0.$$

But then, by the first equality of Eq. (33.19), all points on the line

$$A_t = A + tK$$

are copies. Three colinear copies imply that A_t is a line entirely formed of copies.

As $DK = 0$ implies $K \wedge K = 0$ by (33.21), it also implies $D^\sharp K = 0$ by (33.11), and vice-versa. Consequently, *every point of the line*

$$A_t = tA^\sharp + (1 - t)A$$

through two copies A and $A^\sharp$ represents a copy when the difference tensor

$$K = A^\sharp - A$$

is parallel–transported by either A or $A^\sharp$. In this case $dF_t/dt = 0$, that is, $F_t = F$ for all values of t. Also, $D_t K = 0$ for all t, so that K is parallel-transported by each connection on the line. Notice that an arbitrary finite K such that $DK = 0$ does not necessarily engender a line of copies. It is necessary that K be *a priori* the difference between two copies.

The above condition is necessary and sufficient: if $F_t \neq F$ for some $t \neq 0, 1$, Eqs.(33.23) imply both

$$DK \neq 0 \qquad \text{and} \qquad K \wedge K \neq 0.$$

If the line joining two copies includes one point which is not a copy, then all other points for $t \neq 0, 1$ correspond to non-copies. The basic result can thus be stated in the following form:

Given two copies and the straight line joining them, either there is no other copy on the line or every point of the line represents a copy.

As a consequence, if there are copies for a certain F, and one of them (say, A) is isolated, then there are no copies on the lines joining A to the other copies. Notice, however, that the existence of families of copies dependent on continuous parameters is known.[9] Thus, certainly not every copy is isolated.

The question of isolated copies is better understood by considering, instead of the above finite K, infinitesimal translations on $\mathcal{A}$. In effect, consider the variation of F,

$$\delta F = d\delta A + \delta A \wedge A + A \wedge \delta A = D_A \delta A.$$

In order to have $\delta F = 0$ it is enough that $D_A \delta A = 0$. Consequently, no copy is completely isolated. There can be copies close to any A: each variation satisfying $D_A \delta A = 0$ leads to a copy. Taken together with what has been said above on the finite case, this means that there will be lines of copies along the "directions" of the parallel–transported δA's.

Notice that a line through copies of the vacuum is necessarily a line of copies. In effect, given A and $A^\sharp$ with $F = F^\sharp = 0$, there is a gauge in which $A = 0$ and another gauge in which $A^\sharp = 0$. Using the first of these gauges, we see that $A_t = tK$ along the line. On the other hand,

$$F_t = t(t-1)K \wedge K = 0$$

by (33.23). Since

$$DK + K \wedge K = 0,$$

we can write

$$\begin{aligned}
K \wedge K &= DK + K \wedge K + K \wedge K \\
&= dK + A \wedge K + K \wedge A + K \wedge K + K \wedge K \\
&= dK + A^\sharp \wedge K + K \wedge A^\sharp \\
&= D^\sharp K = D^\sharp A^\sharp = F^\sharp = 0.
\end{aligned}$$

It then follows that

$$F_t = 0.$$

The overall picture can be summed up in the following form. From any A will emerge lines of three kinds:[10]

---

[9] Freedman & Khuri 1994.

[10] Aldrovandi & Barbosa 2005.

- Lines of copies, given by those δA which are parallel-transported by A.
- Lines of non-copies, given by those δA which are not parallel-transported by A.
- Lines along covariant matrix 1-forms K satisfying $D_A K + K \wedge K = 0$, which will meet one copy at $A + K$, and only that one.

General refrences

There has been a great activity on the subject in the nineteen-seventies. Among many others, see Coleman 1977, Halpern 1977 and 1978, Solomon 1979, Brown & Weisberger 1979, Deser & Drechsler 1979, Mostow 1979, Bollini, Giambiagi & Tiomno 1979, Doria 1981. The interest in this topic declined afterwards, although with some interesting results from time to time. See, for example, Majumdar & Sharatchandra 2001.

Chapter 34

General Relativity

Einstein's theory for gravitation is the only example of a theory in which the dynamics of an interaction is (almost) entirely geometrized. What follows is little more than a topical formulary, with emphasis on some formal points.

34.1 Introductory remarks

As known since a famous experiment at the leaning tower of Pisa all particles, independently of their masses, feel equally the gravitational field. More precisely, particles with different masses experience a gravitational field in such a way that all of them acquire the same acceleration and, given the same initial conditions, will follow the same path. This *universality* of response is the most fundamental characteristic of the gravitational interaction. It is a unique property, peculiar to gravitation: no other basic interaction of nature has it.

Due precisely to such universality of response, the gravitational interaction admits a description which makes no use of the concept of *gravitational force*. According to this description, the presence of a gravitational field is supposed to produce a *curvature* in spacetime itself, but no other kind of deformation. This deformation preserves the pseudo-riemannian character of the *flat* Minkowski spacetime, the non-deformed spacetime that represents absence of gravitation.

A free particle in flat space follows a straight line, that is, a curve keeping a constant direction. A geodesic is a curve keeping a constant direction on a curved space. As the only effect of the gravitational interaction is to bend spacetime so as to endow it with a curvature, a particle submitted exclusively to gravity will follow a geodesic of the deformed spacetime.

This is the approach of Einstein's General Relativity, according to which the responsibility of describing the gravitational interaction is transferred to the spacetime geometry.

34.2 The equivalence principle

Equivalence is a guiding principle, which inspired Einstein in his construction of General Relativity. It is firmly rooted on experience.[1] In its most usual form, the Principle includes three sub-principles: the weak, the strong, and the so-called "Einstein's equivalence principle". Let us shortly list them with a few comments.

Weak equivalence principle

It states the universality of free fall, or the equality between *inertial* and *gravitational* masses: $m_i = m_g$. It can be stated in the form:

> *In a gravitational field, all pointlike structureless particles follow one same path. That path is fixed once given (i) an initial position $x(t_0)$ and (ii) the correspondent velocity $\dot{x}(t_0)$.*

This leads to an equation of motion which is a second-order ordinary differential equation. No characteristic of any special particle, no particular property appears in the equation. Gravitation is consequently universal. Being universal, it can be seen as a property of space itself. It determines geometrical properties which are common to all particles.

The weak equivalence principle goes back — as above mentioned — to Galilei. It raises to the status of fundamental principle a deep experimental fact: the equality of inertial and gravitational masses of all bodies. If these masses were not equal, Newton's second law would be written as

$$\boldsymbol{F} = m_i\,\boldsymbol{a},$$

whereas the law of gravitation would be

$$\boldsymbol{F} = m_g\,\boldsymbol{g},$$

with $\boldsymbol{g}$ the acceleration produced by a gravitational field. The acceleration at a given point would then be

$$a = \frac{m_g}{m_i}\,\boldsymbol{g},$$

---

[1] Those interested in the experimental status will find an appraisal in Will 2005.

and would be different for different bodies. Along the history, many different experiments have been performed to test for this difference, all of them yielding a negative result.

Strong equivalence principle

Despite well known qualms,[2] we shall choose a naïve course, whose main advantage is that of being short. The strong equivalence, also known as Einstein's lift, says basically that

> *gravitation can be made to vanish locally through*
> *an appropriate choice of frame.*

It requires that, for any and every particle, and at each point x_0, there exists a frame in which $\ddot{x}^\mu = 0$.[3]

Einstein's equivalence principle

It requires, besides the weak principle, the local validity of Poincaré invariance — that is, of Special Relativity. It can be stated in the form:

> *Every law of physics reduces locally to that of Special Relativity*
> *through an appropriate choice of frame.*

Special-relativistic invariance is, in Minkowski space, summed up in the Lorentz metric. The requirement suggests that the above deformation caused by gravitation could be a change in that metric.

34.3 Pseudo-riemannian metric and tetrad

Forces equally felt by all bodies were known since long. They are the inertial forces, whose name comes from their turning up in non-inertial frames. Examples on Earth (not an inertial system) are the centrifugal force and the Coriolis force. Universality of inertial forces has been the first hint towards General Relativity. A second ingredient is the notion of field. The concept allows the best approach to interactions coherent with Special Relativity. All known forces are mediated by fields on spacetime. Now, if gravitation is to be represented by a field, it should, by the considerations above, be

---

[2] See the Preface of Synge 1960.

[3] A precise, mathematically sound formulation of the strong principle can be found in Aldrovandi, Barros & Pereira 2003.

a universal field, equally felt by every particle. A natural possibility would then be to change spacetime itself. And, of all the fields present in a space, the metric — the first fundamental form, as it is also called — seemed to be the basic one. The simplest way to change spacetime would be to change its metric. The gravitational field, therefore, could be represented by the spacetime metric.

Each spacetime is a 4-dimensional pseudo–riemannian manifold. Its main character is the fundamental form, or metric. For example, the space-time of special relativity is the flat Minkowski spacetime. Minkowski space is the simplest, standard spacetime, and its metric, called the Lorentz metric, is denoted (we are going to use $a, b, c, \ldots = 0, \ldots, 3$ for the Minkowski spacetime indices)

$$\eta(x) = \eta_{ab}\, dx^a dx^b. \tag{34.1}$$

Up to the signature, the Minkowski space is an Euclidean space, and as such can be covered by a single, global coordinate system. This system — the cartesian system — is the mother of all coordinate systems, and just puts η in the diagonal form

$$\eta = \begin{pmatrix} +1 & 0 & 0 & 0 \\ 0 & -1 & 0 & 0 \\ 0 & 0 & -1 & 0 \\ 0 & 0 & 0 & -1 \end{pmatrix}. \tag{34.2}$$

The Minkowski line element, therefore, is

$$ds^2 = \eta_{ab} dx^a dx^b,$$

or equivalently

$$ds^2 = c^2 dt^2 - dx^2 - dy^2 - dz^2. \tag{34.3}$$

On the other hand, the metric of a general 4-dimensional pseudo-riemannian spacetime will be denoted by (we are going to use indices $\mu, \nu, \rho, \ldots = 0, \ldots, 3$ for a pseudo-riemannian spacetime)

$$g(x) = g_{\mu\nu} dx^\mu dx^\nu. \tag{34.4}$$

Like the Minkowski metric, it has signature 2. Being symmetric, the matrix $g(x)$ can be diagonalized. In the convention we shall adopt this means that, at any selected point P, it is possible to choose coordinates $\{x^\mu\}$, in terms of which $g_{\mu\nu}$ acquires the form

$$g(P) = \begin{pmatrix} +|g_{00}| & 0 & 0 & 0 \\ 0 & -|g_{11}| & 0 & 0 \\ 0 & 0 & -|g_{22}| & 0 \\ 0 & 0 & 0 & -|g_{33}| \end{pmatrix}. \tag{34.5}$$

Nontrivial tetrads (or frames) are defined by

$$h_a = h_a{}^\mu \, \partial_\mu \quad \text{and} \quad h^a = h^a{}_\mu dx^\mu \ . \tag{34.6}$$

These tetrad fields relate the metric tensor $g_{\mu\nu}$ of a general pseudo-riemannian spacetime to the Minkowski metric η_{ab} of the tangent space through

$$g_{\mu\nu} = \eta_{ab} \, h^a{}_\mu h^b{}_\nu \ . \tag{34.7}$$

A tetrad field is a linear frame whose members h_a are pseudo-orthogonal by the pseudo-riemannian metric $g_{\mu\nu}$. The components of the dual basis members $h^a = h^a{}_\nu dx^\nu$ satisfy

$$h^a{}_\mu h_a{}^\nu = \delta^\nu_\mu \quad \text{and} \quad h^a{}_\mu h_b{}^\mu = \delta^a_b \ , \tag{34.8}$$

so that Eq. (34.7) has the inverse

$$\eta_{ab} = g_{\mu\nu} \, h_a{}^\mu h_b{}^\nu. \tag{34.9}$$

Notice that the determinant

$$g = \det(g_{\mu\nu}) \tag{34.10}$$

is negative because of the signature of metric η_{ab}. This means that

$$h = \det(h^a{}_\mu) = \sqrt{-g}. \tag{34.11}$$

Commentary 34.1 A tetrad $h^a{}_\mu$ is called nontrivial if it relates a general pseudo-riemennian metric $g_{\mu\nu}$ to the Minkowski spacetime metric η_{ab}. A trivial tetrad e^a, on the other hand, is a tetrad that relates two forms of the same spacetime metric. In a coordinate basis, for example, a trivial tetrad is written in the form $e^a = dx^a$. This means that

$$\eta_{\mu\nu} = \eta_{ab} \, e^a{}_\mu e^b{}_\nu$$

is still the Minkowski metric, though written in the same coordinate system in which $e^a{}_\mu$ is written. ◄

34.4 Lorentz connections

Connections related to the linear group $GL(4, \mathbb{R})$ and its subgroups — such as the Lorentz group $SO(3, 1)$ — are called *linear* connections.[4] They have a larger degree of intimacy with spacetime because they are defined on the bundle of linear frames, which is a constitutive part of its manifold structure. That bundle has some properties not found in the bundles related to *internal* gauge theories. Mainly, it exhibits soldering, which leads to the existence of *torsion* for every connection. Linear connections — in

---

[4] Greub, Halperin & Vanstone 1973.

particular, Lorentz connections — always have curvature *and* torsion, while internal gauge potentials have only curvature.

A *Lorentz connection* A_μ, frequently referred to also as a *spin connection*, is a 1-form assuming values in the Lie algebra of the Lorentz group,

$$A_\mu = \tfrac{1}{2} A^{ab}{}_\mu S_{ab}, \tag{34.12}$$

with S_{ab} a given representation of the Lorentz generators. As these generators are antisymmetric in the algebraic indices, $A^{ab}{}_\mu$ must be equally antisymmetric in order to be lorentzian. This connection defines the Fock–Ivanenko covariant derivative[5]

$$\mathcal{D}_\mu = \partial_\mu - \tfrac{i}{2} A^{ab}{}_\mu S_{ab}, \tag{34.13}$$

whose second part acts only on the algebraic, or tangent space indices. For a scalar field ϕ, for example, the generators are

$$S_{ab} = 0. \tag{34.14}$$

For a Dirac spinor ψ, they are spinorial matrices of the form

$$S_{ab} = \tfrac{i}{4} [\gamma_a, \gamma_b], \tag{34.15}$$

with γ_a the Dirac matrices. A Lorentz vector field ϕ^c, on the other hand, is acted upon by the vector representation of the Lorentz generators, matrices S_{ab} with entries

$$(S_{ab})^c{}_d = i \left(\eta_{bd} \delta^c_a - \eta_{ad} \delta^c_b \right). \tag{34.16}$$

The Fock–Ivanenko derivative is, in this case,

$$\mathcal{D}_\mu \phi^c = \partial_\mu \phi^c + A^c{}_{d\mu} \phi^d, \tag{34.17}$$

and so on for any other fundamental field.

On account of the soldered character of the tangent bundle, a tetrad field relates tangent space (or "internal") tensors to spacetime (or "external") tensors. For example, if ϕ^a is an internal, or Lorentz vector, then

$$\phi^\rho = h_a{}^\rho \phi^a \tag{34.18}$$

will be a spacetime vector. Conversely, we can write

$$\phi^a = h^a{}_\rho \phi^\rho. \tag{34.19}$$

On the other hand, due to its non-tensorial character, a connection will acquire a vacuum, or non-homogeneous term, under the same operation.

---

[5] Fock & Ivanenko 1929.

For example, to each spin connection $A^a{}_{b\mu}$, there is a corresponding general linear connection $\Gamma^\rho{}_{\nu\mu}$, given by

$$\Gamma^\rho{}_{\nu\mu} = h_a{}^\rho \partial_\mu h^a{}_\nu + h_a{}^\rho A^a{}_{b\mu} h^b{}_\nu \equiv h_a{}^\rho \mathcal{D}_\mu h^a{}_\nu, \qquad (34.20)$$

where $\mathcal{D}_\mu$ is the covariant derivative (34.17), in which the generators act on internal (or tangent space) indices only. The inverse relation is, consequently,

$$A^a{}_{b\mu} = h^a{}_\nu \partial_\mu h_b{}^\nu + h^a{}_\nu \Gamma^\nu{}_{\rho\mu} h_b{}^\rho \equiv h^a{}_\nu \nabla_\mu h_b{}^\nu, \qquad (34.21)$$

where ∇_μ is the standard covariant derivative in the connection $\Gamma^\nu{}_{\rho\mu}$, which acts on external indices only. For a spacetime vector ϕ^ν, for example, it is given by

$$\nabla_\mu \phi^\nu = \partial_\mu \phi^\nu + \Gamma^\nu{}_{\rho\mu} \phi^\rho. \qquad (34.22)$$

Using relations (34.18) and (34.19), it is easy to verify that

$$\mathcal{D}_\mu \phi^d = h^d{}_\rho \nabla_\mu \phi^\rho. \qquad (34.23)$$

Commentary 34.2 Observe that, whereas the Fock–Ivanenko derivative $\mathcal{D}_\mu$ can be defined for all fields — tensorial and spinorial — the covariant derivative ∇_μ can be defined for tensorial fields only. In order to describe the interaction of spinor fields with gravitation, therefore, the use of Fock–Ivanenko derivatives is mandatory.[6] ◄

Equations (34.20) and (34.21) are simply different ways of expressing the property that the total covariant derivative of the tetrad — that is, a covariant derivative with connection terms for both internal and external indices — vanishes identically:

$$\partial_\mu h^a{}_\nu - \Gamma^\rho{}_{\nu\mu} h^a{}_\rho + A^a{}_{b\mu} h^b{}_\nu = 0. \qquad (34.24)$$

On the other hand, a connection $\Gamma^\rho{}_{\lambda\mu}$ is said to be "metric compatible" if the so-called *metricity condition*

$$\nabla_\lambda g_{\mu\nu} \equiv \partial_\lambda g_{\mu\nu} - \Gamma^\rho{}_{\mu\lambda} g_{\rho\nu} - \Gamma^\rho{}_{\nu\lambda} g_{\mu\rho} = 0 \qquad (34.25)$$

is fulfilled. From the tetrad point of view, and using Eqs. (34.20) and (34.21), this equation can be rewritten in the form

$$\partial_\mu \eta_{ab} - A^d{}_{a\mu} \eta_{db} - A^d{}_{b\mu} \eta_{ad} = 0, \qquad (34.26)$$

or equivalently

$$A_{ba\mu} = - A_{ab\mu}. \qquad (34.27)$$

The underlying content of the metric-preserving property, therefore, is that the spin connection is lorentzian — that is, antisymmetric in the algebraic indices. Conversely, when $\nabla_\lambda g_{\mu\nu} \neq 0$, the corresponding connection $A^a{}_{b\mu}$ does not assume values in the Lie algebra of the Lorentz group — it is not a Lorentz connection.

---

[6] Dirac 1958.

34.5   Curvature and torsion

Curvature and torsion are properties of Lorentz connections.[7] This becomes evident if we note that many different connections can be defined on the very same metric spacetime. Of course, when restricted to the specific case of General Relativity, where only the zero-torsion spin connection is present, universality of gravitation allows its curvature to be interpreted — together with the metric — as part of the spacetime definition, and one can then talk about "spacetime curvature". However, in the presence of connections with different curvatures and torsions, it seems far more convenient to follow the mathematicians and take spacetime simply as a manifold, and connections (with their curvatures and torsions) as additional structures.

The curvature of a Lorentz connection $A^a{}_{b\mu}$ is a 2-form assuming values in the Lie algebra of the Lorentz group,

$$\boldsymbol{R} = \tfrac{1}{4} R^a{}_{b\nu\mu} \, S_a{}^b \, dx^\nu \wedge dx^\mu. \tag{34.28}$$

Torsion is also a 2-form, but assuming values in the Lie algebra of the translation group,

$$\boldsymbol{T} = \tfrac{1}{2} T^a{}_{\nu\mu} \, P_a \, dx^\nu \wedge dx^\mu, \tag{34.29}$$

with $P_a = \partial_a$ the translation generators. The curvature and torsion components are defined, respectively, by

$$R^a{}_{b\nu\mu} = \partial_\nu A^a{}_{b\mu} - \partial_\mu A^a{}_{b\nu} + A^a{}_{e\nu} A^e{}_{b\mu} - A^a{}_{e\mu} A^e{}_{b\nu} \tag{34.30}$$

and

$$T^a{}_{\nu\mu} = \partial_\nu h^a{}_\mu - \partial_\mu h^a{}_\nu + A^a{}_{e\nu} h^e{}_\mu - A^a{}_{e\mu} h^e{}_\nu. \tag{34.31}$$

Through contraction with tetrads, these tensors can be written in spacetime-indexed forms:

$$R^\rho{}_{\lambda\nu\mu} = h_a{}^\rho \, h^b{}_\lambda \, R^a{}_{b\nu\mu}, \tag{34.32}$$

and

$$T^\rho{}_{\nu\mu} = h_a{}^\rho \, T^a{}_{\nu\mu}. \tag{34.33}$$

Using relation (34.21), their components are found to be

$$R^\rho{}_{\lambda\nu\mu} = \partial_\nu \Gamma^\rho{}_{\lambda\mu} - \partial_\mu \Gamma^\rho{}_{\lambda\nu} + \Gamma^\rho{}_{\eta\nu} \Gamma^\eta{}_{\lambda\mu} - \Gamma^\rho{}_{\eta\mu} \Gamma^\eta{}_{\lambda\nu} \tag{34.34}$$

and

$$T^\rho{}_{\nu\mu} = \Gamma^\rho{}_{\mu\nu} - \Gamma^\rho{}_{\nu\mu}. \tag{34.35}$$

---

[7] Kobayashi & Nomizu 1996.

Considering that the spin connection $A^a{}_{b\nu}$ is a vector in the last index, we can write

$$A^a{}_{bc} = A^a{}_{b\nu} h_c{}^\nu. \tag{34.36}$$

It can thus be verified that, in the anholonomic basis $\{h_a\}$, the curvature and torsion components are given respectively by

$$R^a{}_{bcd} = h_c \left(A^a{}_{bd} \right) - h_d \left(A^a{}_{bc} \right) + A^a{}_{ec} A^e{}_{bd} - A^a{}_{ed} A^e{}_{bc} - f^e{}_{cd} A^a{}_{be} \tag{34.37}$$

and

$$T^a{}_{bc} = A^a{}_{cb} - A^a{}_{bc} - f^a{}_{bc}, \tag{34.38}$$

where h_c stands for $h_c{}^\mu \partial_\mu$ and the $f^c{}_{ab}$'s are the coefficients of anholonomy of the frame $\{h_a\}$.

34.6 General Relativity connection

Given the commutation relation

$$[h_a, h_b] = f^c{}_{ab} h_c, \tag{34.39}$$

the $f^c{}_{ab}$'s are given by the curls of the basis members:

$$f^c{}_{ab} = h_a{}^\mu h_b{}^\nu (\partial_\nu h^c{}_\mu - \partial_\mu h^c{}_\nu). \tag{34.40}$$

The spin connection $\overset{\circ}{A}{}^a{}_{bc} = \overset{\circ}{A}{}^a{}_{b\mu} h_c{}^\mu$ of General Relativity can be written solely in terms of these coefficients, as

$$\overset{\circ}{A}{}^a{}_{bc} = \tfrac{1}{2} \left(f_b{}^a{}_c + f_c{}^a{}_b - f^a{}_{bc} \right). \tag{34.41}$$

The corresponding spacetime-indexed linear connection, defined by

$$\overset{\circ}{\Gamma}{}^\rho{}_{\nu\mu} = h_a{}^\rho \partial_\mu h^a{}_\nu + h_a{}^\rho \overset{\circ}{A}{}^a{}_{b\mu} h^b{}_\nu, \tag{34.42}$$

is the Christoffel, or Levi–Civita connection

$$\overset{\circ}{\Gamma}{}^\rho{}_{\nu\mu} = \tfrac{1}{2} g^{\rho\lambda} \left(\partial_\nu g_{\lambda\mu} + \partial_\mu g_{\lambda\nu} - \partial_\lambda g_{\mu\nu} \right). \tag{34.43}$$

The inverse relation is, of course,

$$\overset{\circ}{A}{}^a{}_{b\mu} = h^a{}_\nu \partial_\mu h_b{}^\nu + h^a{}_\nu \overset{\circ}{\Gamma}{}^\nu{}_{\rho\mu} h_b{}^\rho. \tag{34.44}$$

In General Relativity the Christoffel is assumed to be the fundamental connection.

Commentary 34.3 In gauge theories of the Yang–Mills type, connections are also the basic field. However, there is a crucial difference in relation to General Relativity. The Levi–Civita connection is a tributary field: it is completely fixed by the metric, which is thus the true fundamental field. ◄

The Levi–Civita connection (34.43) is the single connection which satisfies two conditions: (i) the metricity condition

$$\overset{\circ}{\nabla}_\mu g_{\rho\sigma} \equiv \partial_\mu g_{\rho\sigma} - \overset{\circ}{\Gamma}{}^\lambda{}_{\rho\mu} g_{\lambda\sigma} - \overset{\circ}{\Gamma}{}^\lambda{}_{\sigma\mu} g_{\rho\lambda} = 0, \tag{34.45}$$

and (ii) the symmetry in the last two indices

$$\overset{\circ}{\Gamma}{}^\lambda{}_{\mu\nu} = \overset{\circ}{\Gamma}{}^\lambda{}_{\nu\mu}. \tag{34.46}$$

This symmetry has a deep meaning: torsion is vanishing.

Let us insist, and repeat ourselves: the Levi–Civita connection has a special relationship to the metric and is the only metric-preserving connection with zero torsion. Its curvature,

$$\overset{\circ}{R}{}^\kappa{}_{\lambda\rho\sigma} = \partial_\rho \overset{\circ}{\Gamma}{}^\kappa{}_{\lambda\sigma} - \partial_\sigma \overset{\circ}{\Gamma}{}^\kappa{}_{\lambda\rho} + \overset{\circ}{\Gamma}{}^\kappa{}_{\nu\rho} \overset{\circ}{\Gamma}{}^\nu{}_{\lambda\sigma} - \overset{\circ}{\Gamma}{}^\kappa{}_{\nu\sigma} \overset{\circ}{\Gamma}{}^\nu{}_{\lambda\rho} , \tag{34.47}$$

has some special symmetries in the indices, which can be obtained from its detailed expression in terms of the metric:

$$\overset{\circ}{R}_{\kappa\lambda\rho\sigma} = - \overset{\circ}{R}_{\kappa\lambda\sigma\rho} = \overset{\circ}{R}_{\lambda\kappa\sigma\rho} , \tag{34.48}$$

as well as

$$\overset{\circ}{R}_{\kappa\lambda\rho\sigma} = \overset{\circ}{R}_{\rho\sigma\kappa\lambda}. \tag{34.49}$$

As a consequence, the Ricci tensor, which is defined by

$$\overset{\circ}{R}_{\mu\nu} = \overset{\circ}{R}{}^\rho{}_{\mu\rho\nu}, \tag{34.50}$$

is also symmetric: $\overset{\circ}{R}_{\mu\nu} = \overset{\circ}{R}_{\nu\mu}$. The scalar curvature is

$$\overset{\circ}{R} = g^{\mu\nu} \overset{\circ}{R}_{\mu\nu}. \tag{34.51}$$

Commentary 34.4 Spaces whose Ricci tensors are proportional to the metric are usually called "Einstein spaces". They have been classified by Petrov[8] in terms of their groups of motion, that is, the groups formed by their Killing vectors. ◄

---

[8] Petrov 1951; Petrov 1954; Petrov 1969.

34.7   Geodesics

The action describing the motion of a free particle of mass m in Minkowski spacetime is

$$S = - mc \int_a^b ds, \tag{34.52}$$

where

$$ds = (\eta_{ab}\, dx^a dx^b)^{1/2}. \tag{34.53}$$

In the presence of gravitation, the action describing a particle of mass m is still that given by Eq. (34.52), but now with

$$ds = (g_{\mu\nu} dx^\mu dx^\nu)^{1/2}. \tag{34.54}$$

Taking the variation of S, the condition $\delta S = 0$ yields as equation of motion the *geodesic equation*

$$\frac{du^\rho}{ds} + \overset{\circ}{\Gamma}{}^\rho{}_{\mu\nu}\, u^\mu\, u^\nu = 0, \tag{34.55}$$

where $u^\rho = dx^\rho/ds$ is the particle four-velocity. The solution of this equation gives the trajectory of the particle in the presence of gravitation.

An important property of the geodesic equation is that it does not involve the mass of the particle, a natural consequence of universality. Another important property is that it represents the vanishing of the covariant derivative of the four-velocity u^ρ along the trajectory of the particle, that is to say, the vanishing of the four-acceleration a^ρ:

$$a^\rho \equiv u^\lambda \overset{\circ}{\nabla}_\lambda u^\rho = 0. \tag{34.56}$$

In General Relativity, therefore, the *gravitational force* vanishes. The question then arises: if the gravitational force vanishes, how is the gravitational interaction described? The answer is quite simple. The presence of gravitation produces a curvature in spacetime, and all particles moving in such spacetime will follow its curvature. In other words, they will move along a geodesic. The gravitational interaction is thereby geometrized: the responsibility for describing the interaction is transferred to the geometry of the space. Note that in this way the universality of gravitation is automatically achieved as all particles, independently of their masses and constitution, will follow the same trajectory.

Commentary 34.5   There are currently some experimental projects aiming to probe the universality of gravitation at an unprecedented level. If some violation is found, the geometric description of General Relativity breaks down: in this theory, which is fully based on the equivalence principle, there is no room for non-universality. We recall that this is different from Newtonian gravity which, although able to comply to universality, remains a consistent theory in its absence. One has just to re-insert back the inertial and gravitational masses in its equations. ◄

34.8   Bianchi identities

Consider a vector field U with components U^α, and take twice the covariant derivative: $\overset{\circ}{\nabla}_\gamma \overset{\circ}{\nabla}_\beta U^\alpha$. Reverse then the order to obtain $\overset{\circ}{\nabla}_\beta \overset{\circ}{\nabla}_\gamma U^\alpha$. A simple algebra shows that

$$\overset{\circ}{\nabla}_\gamma \overset{\circ}{\nabla}_\beta U^\alpha - \overset{\circ}{\nabla}_\beta \overset{\circ}{\nabla}_\gamma U^\alpha = - \overset{\circ}{R}{}^\alpha{}_{\epsilon\beta\gamma} U^\epsilon. \tag{34.57}$$

Curvature turns up in the commutator of two covariant derivatives:

$$[\overset{\circ}{\nabla}_\gamma, \overset{\circ}{\nabla}_\beta] U^\alpha = - \overset{\circ}{R}{}^\alpha{}_{\epsilon\beta\gamma} U^\epsilon. \tag{34.58}$$

A detailed calculation leads also to some identities. One of them is

$$\overset{\circ}{R}{}^\kappa{}_{\lambda\rho\sigma} + \overset{\circ}{R}{}^\kappa{}_{\sigma\lambda\rho} + \overset{\circ}{R}{}^\kappa{}_{\rho\sigma\lambda} = 0 . \tag{34.59}$$

Another one is

$$\overset{\circ}{\nabla}_\mu \overset{\circ}{R}_{\kappa\lambda\rho\sigma} + \overset{\circ}{\nabla}_\sigma \overset{\circ}{R}_{\kappa\lambda\mu\rho} + \overset{\circ}{\nabla}_\rho \overset{\circ}{R}_{\kappa\lambda\sigma\mu} = 0 . \tag{34.60}$$

Notice, in both cases, the cyclic permutation of three of the indices. These expressions are called, respectively, the first and the second Bianchi identities.

Now, as the metric tensor has zero covariant derivative, it can be inserted in the second identity to contract indices in a convenient way. Contracting the last equation above with $g^{\kappa\rho}$, it comes out

$$\overset{\circ}{\nabla}_\mu \overset{\circ}{R}_{\lambda\sigma} - \overset{\circ}{\nabla}_\sigma \overset{\circ}{R}_{\lambda\mu} + \overset{\circ}{\nabla}_\rho \overset{\circ}{R}{}^\rho{}_{\lambda\sigma\mu} = 0.$$

A further contraction with $g^{\lambda\sigma}$ yields

$$\overset{\circ}{\nabla}_\mu \overset{\circ}{R} - \overset{\circ}{\nabla}_\sigma \overset{\circ}{R}{}^\sigma{}_\mu - \overset{\circ}{\nabla}_\rho \overset{\circ}{R}{}^\rho{}_\mu = 0,$$

which is the same as

$$\overset{\circ}{\nabla}_\mu \left(\overset{\circ}{R}{}^\mu{}_\nu - \tfrac{1}{2} \delta^\mu_\nu \overset{\circ}{R} \right) = 0. \tag{34.61}$$

This expression is the *contracted form* of the second Bianchi identity. The covariantly conserved tensor

$$G_{\mu\nu} = \overset{\circ}{R}_{\mu\nu} - \tfrac{1}{2} g_{\mu\nu} \overset{\circ}{R}, \tag{34.62}$$

has an important role in gravitation, and is called *Einstein tensor*. Its contraction with the metric gives (up to a sign) the scalar curvature

$$g^{\mu\nu} G_{\mu\nu} = - \overset{\circ}{R}. \tag{34.63}$$

34.9   Einstein's field equations

Historic approach

The Einstein tensor (34.62) is a purely geometrical second-rank tensor which has vanishing covariant derivative. It is actually possible to prove that it is the only one. The source energy-momentum tensor $\Theta_{\mu\nu}$, on the other hand, is also a covariantly conserved second-rank tensor. Einstein was convinced that some physical characteristic of the sources of a gravitational field should engender the deformation in spacetime, that is, in its geometry. He looked for a dynamical equation which, in the non-relativistic limit, should for consistency reasons reduce to newtonian gravity, represented by the Poisson equation

$$\Delta V = 4\pi G\rho \qquad (34.64)$$

where Δ is the Laplacian, V is Newton's gravitational potential, and ρ is the energy density.

Considering that the source energy-momentum tensor $\Theta_{\mu\nu}$ contains, as one of its components, the energy density ρ, Einstein took then the bold step of equating (up to a constant) $G_{\mu\nu}$ to $\Theta_{\mu\nu}$, obtaining what is known today as the simplest possible generalization of the Poisson equation in a riemannian context:

$$\overset{\circ}{R}_{\mu\nu} - \tfrac{1}{2} g_{\mu\nu} \overset{\circ}{R} = \frac{8\pi G}{c^4} \Theta_{\mu\nu}. \qquad (34.65)$$

This is the *Einstein equation*, which fixes the dynamics of the gravitational field. The constant in the right-hand side was in principle unknown, but he fixed it by requiring that, in the due limit, the Poisson equation of newtonian theory was obtained. The tensor in the right-hand side is the *symmetric*, or Belinfante–Rosenfeld[9] energy-momentum tensor of a matter field, which is the source of the gravitational field.

Einstein tensor (34.62) is not actually the most general geometrical second-rank tensor with vanishing covariant derivative. Since the metric has also vanishing covariant derivative, it is in principle possible to add a term $\Lambda g_{\mu\nu}$ to $G_{\mu\nu}$, with Λ a constant. Equation (34.65) then becomes

$$\overset{\circ}{R}_{\mu\nu} - \tfrac{1}{2} \overset{\circ}{R} g_{\mu\nu} + \Lambda\, g_{\mu\nu} = \frac{8\pi G}{c^4} \Theta_{\mu\nu}. \qquad (34.66)$$

From the point of view of covariantly preserved objects, this equation is as valid as (34.65). In his first trial to apply his theory to cosmology, Einstein

---

[9] Belinfante 1939; Rosenfeld 1940.

looked for a static solution to comply with the cosmological ideas of that time. He found it, but it was unstable. He then added the term $\Lambda g_{\mu\nu}$ to make it stable, and gave to Λ the name *cosmological constant*. Later, when evidence for an expanding universe became clear, he abandoned the idea and dropped the term. Recent cosmological observations, however, points to an accelerated universe expansion, which not only asks again for a non-vanishing Λ but actually makes of it the dominant contribution to the whole cosmic dynamics (see Chapter 11).

Variational approach

Einstein's equations (34.65) can be obtained also from a variational principle using the action $S = S_{EH} + S_m$, where

$$S_{EH} = -\frac{c^3}{16\pi G} \int \sqrt{-g}\, \overset{\circ}{R}\, d^4x \qquad (34.67)$$

is the so-called Einstein–Hilbert action, and

$$S_m = \frac{1}{c} \int \sqrt{-g}\, \mathcal{L}_m\, d^4x \qquad (34.68)$$

a matter field action, with $\mathcal{L}_m$ the corresponding lagrangian. By considering arbitrary variations in the metric tensor $\delta g_{\mu\nu}$, we get for the gravitational action

$$\delta S_{EH} = -\frac{c^3}{16\pi G} \int d^4x \left[\delta\left(\sqrt{-g}\, g^{\mu\nu}\right) \overset{\circ}{R}_{\mu\nu} + \sqrt{-g}\, g^{\mu\nu} \delta \overset{\circ}{R}_{\mu\nu} \right].$$

Performing these variations, we obtain

$$\delta S_{EH} = -\frac{c^3}{16\pi G} \int d^4x \left[\sqrt{-g} \left(\overset{\circ}{R}_{\mu\nu} - \tfrac{1}{2} g_{\mu\nu} \overset{\circ}{R} \right) \delta g^{\mu\nu} + \partial_\mu(\sqrt{-g}\, \omega^\mu) \right].$$

Using the Gauss theorem, the last term on the right-hand side can be transformed into a surface term, and consequently does not contribute to the field equations. We are then left with

$$\delta S_{EH} = -\frac{c^3}{16\pi G} \int d^4x \sqrt{-g} \left(\overset{\circ}{R}_{\mu\nu} - \tfrac{1}{2} g_{\mu\nu} \overset{\circ}{R} \right) \delta g^{\mu\nu}. \qquad (34.69)$$

On the other hand, the variation of the matter action is

$$\delta S_m = \frac{1}{2c} \int d^4x \sqrt{-g}\, \Theta_{\mu\nu} \delta g^{\mu\nu}, \qquad (34.70)$$

where

$$\frac{\sqrt{-g}}{2} \Theta_{\mu\nu} = \frac{\partial L_m}{\partial g^{\mu\nu}} - \partial_\rho \frac{\partial L_m}{\partial(\partial_\rho g^{\mu\nu})} \qquad (34.71)$$

is the energy-momentum tensor of the matter fields, with $L_m = \sqrt{-g}\,\mathcal{L}_m$. Thus, from the invariance of the total action, we find

$$-\frac{c^3}{16\pi G}\int d^4x\,\sqrt{-g}\left(\overset{\circ}{R}_{\mu\nu} - \tfrac{1}{2}g_{\mu\nu}\overset{\circ}{R} - \frac{8\pi G}{c^4}\,\Theta_{\mu\nu}\right)\delta g^{\mu\nu} = 0, \qquad (34.72)$$

from where, due to the arbitrariness of $\delta g^{\mu\nu}$, we get

$$\overset{\circ}{R}_{\mu\nu} - \tfrac{1}{2}g_{\mu\nu}\overset{\circ}{R} = \frac{8\pi G}{c^4}\,\Theta_{\mu\nu}, \qquad (34.73)$$

just the Einstein equation.

Potential form of Einstein equation

There exists no *invariant* lagrangian for General Relativity that depends on the metric and its first derivatives only — or equivalently, on the tetrad and its first derivatives only. What exists is the second-order Einstein–Hilbert invariant lagrangian L_{EH}, in which the second-derivative terms reduce to a total divergence. This lagrangian can be split in the form ($\kappa = 8\pi G/c^4$)

$$L_{EH} \equiv -\frac{h}{2\kappa}\overset{\circ}{R} = L_g + \partial_\mu(h\,w^\mu), \qquad (34.74)$$

where L_g is a (non-invariant) lagrangian that depends on the tetrad and its first derivatives only, w^μ is a four-vector, and $h = \det(h^a{}_\mu)$ with $h^a{}_\mu$ the tetrad field. Of course, since the divergence term does not contribute to the field equations, they can be obtained either from L_{EH} or L_g. There is a difference though: in the second case the field equations can be obtained directly from the Euler–Lagrange equations

$$\frac{\delta L_g}{\delta h^a} \equiv \frac{\partial L_g}{\partial h^a} - d\,\frac{\partial L_g}{\partial dh^a} = 0 \qquad (34.75)$$

where, we recall, $h^a = h^a{}_\mu dx^\mu$. The functional derivative $\delta L_g/\delta h^a$ is just the Lagrange derivative (see Section 26.2). A prominent example of first-order lagrangians is Møller's lagrangian[10]

$$L_M = -\frac{c^3}{16\pi G}\,g^{\mu\nu}\left(\overset{\circ}{\Gamma}{}^\rho{}_{\mu\lambda}\,\overset{\circ}{\Gamma}{}^\lambda{}_{\nu\rho} - \overset{\circ}{\Gamma}{}^\lambda{}_{\mu\nu}\,\overset{\circ}{\Gamma}{}^\rho{}_{\lambda\rho}\right). \qquad (34.76)$$

Denoting by L_m the source Lagrangian, the Euler–Lagrange equation obtained from $L = L_g + L_m$ is the potential (or lagrangian) form of Einstein equation[11]

$$d\,(h\overset{\circ}{S}_a) - \kappa\,h\,\overset{\circ}{t}_a = \kappa\,h\,\Theta_a, \qquad (34.77)$$

---

[10] Møller 1961.
[11] Møller 1958.

where

$$\overset{\circ}{S}_a = -\frac{\kappa}{h}\frac{\partial L_g}{\partial dh^a} \tag{34.78}$$

is the gravitational field excitation 2-form, also known as superpotential. In addition,

$$\overset{\circ}{t}_a = -\frac{1}{h}\frac{\partial L_g}{\partial h^a} \tag{34.79}$$

stands for the gravitational self-current, which in this case represents the gravitational energy-momentum pseudo-tensor, and

$$\Theta_a = -\frac{1}{h}\frac{\delta L_m}{\delta h^a} \equiv -\frac{1}{h}\left(\frac{\partial L_m}{\partial h^a} - d\frac{\partial L_m}{\partial dh^a}\right) \tag{34.80}$$

is the source energy-momentum current. One should note the crucial difference between the matter and the gravitational energy-momentum currents: whereas the matter current Θ_a is defined with a Lagrange functional derivative, the gravitational pseudo-current $\overset{\circ}{t}_a$ is defined with a partial functional derivative. In this form, Einstein equation (34.77) is similar, in structure, to the Yang–Mills equation. Its main property is to explicitly exhibit the complex defining the energy-momentum pseudo-current of the gravitational field. From the Poincaré lemma $dd = 0$, it follows from (34.77) that the total energy-momentum density is conserved:

$$d\left[h(\overset{\circ}{t}_a + \Theta_a)\right] = 0. \tag{34.81}$$

Commentary 34.6   There is an important question concerning the energy-momentum as a current: it has not the usual current-density dimension and, consequently, the constant κ must have a compensating dimension. As a coupling constant, κ appears also in the self-interactions of the gravitational field, included in the term $\overset{\circ}{t}_a$ of Eq. (34.77) above. The problems of renormalizability with a non-dimensionless coupling constant are well known and will appear in any theory with energy-momentum as a source. Quantization of the gravitational field is thereby jeopardized. ◄

Let us consider again the Einstein–Hilbert lagrangian

$$L_{EH} = L_g + \partial_\mu(h\,w^\mu). \tag{34.82}$$

Although both sides are invariant under general coordinate transformations, each piece of the right-hand side is not invariant. This means that, by performing successive transformations, we arrive at

$$L_{EH} = L'_g + \partial'_\mu(h\,w'^\mu) = L''_g + \partial''_\mu(h\,w''^\mu) + \cdots . \tag{34.83}$$

From (34.79) we see that it is thus possible to define infinitely many energy-momentum pseudo-tensors for the gravitational field

$$\overset{\circ}{t}'_a,\ \overset{\circ}{t}''_a,\ \overset{\circ}{t}'''_a,\ \ldots$$

each one connected to a particular first-order lagrangian L_g.

It is important to remark that the matter energy-momentum tensor Θ_a is not uniquely defined either. In fact, it is defined up to the divergence of an anti-symmetric tensor.[12] However, only the symmetric energy-momentum tensor, which is that appearing in the right-hand side of Einstein equation (34.77), is physically relevant. Notice that this selection rule cannot be applied to the gravitational energy-momentum current t_a, because in this case only the left-hand side as a whole must be symmetric: each one of its pieces does not need to be symmetric. As a matter of fact, they do not need to be covariant (or tensorial) either: only the left-hand side of the equation must be covariant. This means that in the general case, the gravitational energy-momentum current $\overset{\circ}{t}_a$ is neither symmetric nor tensorial.

Commentary 34.7 The reason for the gravitational energy-momentum current in General Relativity not to be a tensor is that it includes, in addition to the energy-momentum density of gravitation itself, also the energy-momentum density of inertial effects, which are non-tensorial by their very nature. In the context of Teleparallel Gravity, due to the possibility of separating inertial effects from gravitation, it turns out possible to define a tensorial quantity for the gravitational energy-momentum density. See Chapter 35 for additional details. ◄

34.10 Spin and (the break of) universality

All particles (or fields) of nature can be classified according to the representations of the Poincaré group, the semi-direct product of the Lorentz and the translation groups:

$$\mathcal{P} = \mathcal{L} \otimes \mathcal{T}.$$

This is possible because the eingenvalues of the corresponding Casimir invariants are, respectively, the *spin* and the *mass* of the representation. These physical quantities establish thus a direct relation between any existing particle and a representation of the Poincaré group (Chapter 31).

The translation generators $P_\alpha = \partial_\alpha$ act upon fields $\Psi(x)$ by changing their very argument. As a result, every field of nature, independently of its mass, will feel gravitation the same through this representation, a property that — as repeatedly said — goes under the name of *universality*. Such universality, however, does not extend to the Lorentz sector of the Poincaré group, whose eigenvalues are the spin of the particles. In fact, particles with different spin feel gravity differently, a property that follows from the

---

[12] Belinfante 1939.

Papapetrou equation[13]

$$\frac{\overset{\circ}{\mathcal{D}}\mathcal{P}_\mu}{ds} = -\tfrac{1}{2}\,\overset{\circ}{R}{}^{\alpha\beta}{}_{\mu\nu}\,s_{\alpha\beta}\,u^\nu, \tag{34.84}$$

where $\overset{\circ}{\mathcal{D}}$ is the Fock–Ivanenko covariant derivative in the Levi–Civita spin connection $\overset{\circ}{A}{}^a{}_{b\mu}$,

$$\mathcal{P}_\mu = m\,c\,u_\mu + u^\rho\,\frac{\overset{\circ}{\mathcal{D}}s_{\mu\rho}}{ds} \tag{34.85}$$

is a generalized four-momentum, with $s_{\mu\rho}$ the spin angular momentum tensor, which satisfies the constraints

$$s_{\mu\rho}\,s^{\mu\rho} = 2\,\mathsf{s}^2 \qquad \text{and} \qquad s_{\mu\rho}\,u^\mu = 0, \tag{34.86}$$

with s the particle spin vector. Such tensor, as can be seen from (34.84), couples to the Riemann curvature of spacetime. Universality of gravitation is, therefore, a macroscopic property. In the microscopic realm, where spins become relevant, in addition to the universal coupling related to translations, there is also the non-universal coupling of the particle's spin with gravitation. Particles with different spins, therefore, couple differently to gravitation, breaking in this way universality. Such breaking of universality, of course, is not harmful to General Relativity because the spinless part of the coupling — which is the part related to the equivalence principle — remains universal.

General references

There are many books on General Relativity, and we mention here just a few of them. Although somewhat old, a highly recommended textbook is Synge 1960. More recent textbooks, equally commendable, are Misner, Thorne & Wheeler 1973, Weinberg 1972, Wald 1984 and Carrol 2004.

---

[13] Papapetrou 1951.

Chapter 35

Teleparallel Gravity

A review of the foundations and achievements of Teleparallel Gravity, a theory that is fully equivalent to General Relativity in what concerns observational results, but conceptually quite inequivalent to it.

35.1 Introductory remarks

General Relativity is a geometric theory for the gravitational interaction. According to it, the presence of a gravitational field manifests itself by creating a curvature in spacetime. When a particle moves, it naturally follows freely that spacetime curvature: there is no gravitational force. It is important to notice that, by transferring to spacetime the responsibility of describing the gravitational interaction, it automatically incorporates the universality property of gravitation. As a geometric theory, however, General Relativity does not have the geometrical structure of a gauge theory.

The question then arises: is there a gauge theory for gravitation? The answer can be obtained from the gauge paradigm. First, remember that the source of gravitation is energy and momentum. From Noether's theorem, an instrumental piece in gauge theories,[1] we know that the energy-momentum tensor is conserved provided the source lagrangian is invariant under spacetime translations. If gravity is to be described by a gauge theory with energy-momentum as source, therefore, it must be a gauge theory for the translation group. This theory, as is well-known, is just Teleparallel Gravity. Its most prominent property is to be equivalent to General Relativity. For this reason it is sometimes called the Teleparallel Equivalent of General Relativity.

Although equivalent to General Relativity, however, Teleparallel Grav-

---

[1] Konopleva & Popov 1980.

ity is, conceptually speaking, a completely different theory. For example, the gravitational field in this theory is represented by torsion, not by curvature. Furthermore, in General Relativity curvature is used to *geometrize* the gravitational interaction: geometry replaces the concept of gravitational force, and the trajectories are determined by geodesics — trajectories that follow the curvature of spacetime. Teleparallel Gravity, on the other hand, attributes gravitation to torsion, which acts as a *force*, not geometry. In Teleparallel Gravity, therefore, the trajectories are not described by geodesics, but by force equations quite similar to the Lorentz force of Electrodynamics.[2]

The reason for gravitation to present two equivalent descriptions is related to its most peculiar property: *universality*. Like all other fundamental interactions of nature, gravitation can be described in terms of a gauge theory. This is, as said, just Teleparallel Gravity, a gauge theory for the translation group. Universality of free fall, on the other hand, allows a second, geometric description based on the equivalence principle, just General Relativity. As the unique universal interaction, it is the only one to allow a geometric interpretation, and hence two equivalent descriptions. From this point of view, curvature and torsion are alternative ways of representing the very same gravitational field, accounting for the same degrees of freedom of gravity.

The notion of teleparallel structure, also known as absolute or distant parallelism, is characterized by a particular Lorentz connection that parallel-transports everywhere the tetrad field. This structure was used by Einstein in his unsuccessful attempt to construct a unified field theory for electromagnetism and gravitation.[3] The birth of Teleparallel Gravity as a gravitational theory, however, took place in the late nineteen-fifties and early sixties with the pioneering works by Møller.[4] Since then many contributions from different authors have been incorporated into the theory, giving rise to what is known today as Teleparallel Gravity.[5] In this chapter the fundamentals of this theory will be presented, and some of the new insights it provides into gravitation discussed.

---

[2] Andrade & Pereira 1997.
[3] Einstein 1930; Sauer 2006.
[4] Møller 1961.
[5] Aldrovandi & Pereira 2012.

35.2 Fundamentals of Teleparallel Gravity

35.2.1 *Geometrical setting*

The geometrical setting of Teleparallel Gravity is the tangent bundle: at each point p of a general riemannian spacetime $\mathbb{R}^{3,1}$ — the base space — there is "attached" a Minkowski tangent-space $M = T_p\mathbb{R}^{3,1}$ — the fiber — on which the gauge transformations take place. In Fig. 35.1, the tangent space at x^μ is indicated perpendicularly. A gauge transformation will be a point-dependent translation of the $T_p\mathbb{R}^{3,1}$ coordinates x^a,

$$x^a \to x'^a = x^a + \varepsilon^a(x^\mu) \tag{35.1}$$

with $\varepsilon^a(x^\mu)$ the transformation parameters. The generators of infinitesimal translations are the differential operators

$$P_a = \frac{\partial}{\partial x^a} \equiv \partial_a \tag{35.2}$$

which satisfy the commutation relations

$$[P_a, P_b] = 0. \tag{35.3}$$

The corresponding infinitesimal transformation can then be written in the form

$$\delta x^a = \varepsilon^b(x^\mu) P_b\, x^a. \tag{35.4}$$

It is important to remark that, due to the peculiar character of translations, any gauge theory including them will differ from the usual internal — Yang–Mills type — gauge models in many ways, the most significant being the presence of a tetrad field. The gauge bundle will then present the soldering property, and the "internal" and "external" sectors of the theory will be closely linked to each other. For example, if the spacetime metric is denoted by $g_{\mu\nu}$ and the Minkowski tangent space metric is denoted by η_{ab}, they are soldered by the relation

$$g_{\mu\nu} = \eta_{ab}\, h^a{}_\mu h^b{}_\nu\,, \tag{35.5}$$

with $h^a{}_\mu$ the tetrad field.

35.2.2 *Frames and inertial effects*

In terms of a coordinate basis, an inertial frame in special relativity is represented by

$$e^a{}_\mu = \partial_\mu x^a. \tag{35.6}$$

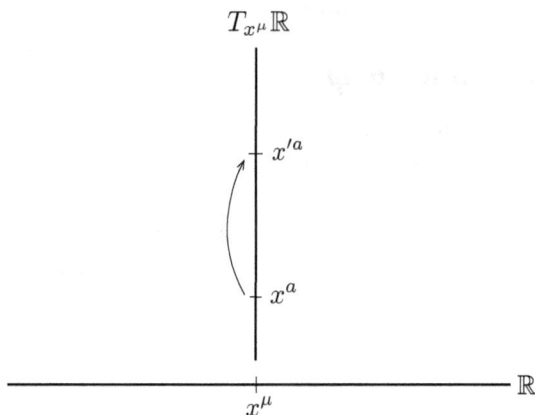

Fig. 35.1   Spacetime with the Minkowski tangent space at x^μ.

Within this class, different inertial frames are related by *global* Lorentz transformations $\Lambda^a{}_c$. Different classes of non-inertial frames are, however, obtained by performing *local* Lorentz transformations

$$x^a \to \Lambda^a{}_c(x)\, x^c. \tag{35.7}$$

As a simple computation shows, these non-inertial frames are given by

$$e^a{}_\mu \equiv \overset{\bullet}{\mathcal{D}}_\mu x^a = \partial_\mu x^a + \overset{\bullet}{A}{}^a{}_{b\mu}\, x^b, \tag{35.8}$$

where

$$\overset{\bullet}{A}{}^a{}_{b\mu} = \Lambda^a{}_c(x)\, \partial_\mu \Lambda_b{}^c(x) \tag{35.9}$$

is a Lorentz connection that represents inertial effects present in the frames $e^a{}_\mu = \overset{\bullet}{\mathcal{D}}_\mu x^a$. Conversely, there is a special class of frames in which no inertial effects are present, and the inertial spin connection vanishes:

$$\overset{\bullet}{A}{}^a{}_{b\mu} = 0. \tag{35.10}$$

Commentary 35.1   In special relativity, these frames in which $\overset{\bullet}{A}{}^a{}_{b\mu} = 0$ are called *inertial frames*. In the presence of gravitation, where no inertial frames can be defined, they are called *proper frames*.[6] ◄

---

[6] Gribl Lucas, Obukhov & Pereira 2009.

35.2.3 Translational gauge potential

As a gauge theory for the translation group, the gravitational field in Teleparallel Gravity is represented by a translational gauge potential B_μ, a 1-form assuming values in the Lie algebra of the translation group,

$$B_\mu = B^a{}_\mu P_a. \tag{35.11}$$

The frame $e_\mu = e^a{}_\mu P_a$ is actually a translational covariant derivative in absence of gravitation. The corresponding covariant derivative in the presence of gravitation is obtained by using the translational coupling prescription

$$e_\mu \to h_\mu = e_\mu + B_\mu \tag{35.12}$$

where $h_\mu = h^a{}_\mu P_a$, with

$$h^a{}_\mu = \overset{\bullet}{\mathcal{D}}_\mu x^a + B^a{}_\mu. \tag{35.13}$$

We see in this way that the translational gauge potential $B^a{}_\mu$ appears as the non-trivial part of the tetrad.

Under an infinitesimal gauge translation

$$\delta x^a = \varepsilon^b(x^\mu) P_b \, x^a \equiv \varepsilon^a(x^\mu), \tag{35.14}$$

the gravitational potential $B^a{}_\mu$ transforms according to

$$\delta B^a{}_\mu = - \overset{\bullet}{\mathcal{D}}_\mu \varepsilon^a(x^\nu). \tag{35.15}$$

The tetrad (35.13) is, of course, gauge invariant:

$$\delta h^a{}_\mu = 0. \tag{35.16}$$

35.2.4 Teleparallel spin connection

The fundamental Lorentz connection of Teleparallel Gravity is the purely inertial connection (35.9). As a vacuum connection, it has vanishing curvature:

$$\overset{\bullet}{R}{}^a{}_{b\mu\nu} = \partial_\mu \overset{\bullet}{A}{}^a{}_{b\nu} - \partial_\nu \overset{\bullet}{A}{}^a{}_{b\mu} + \overset{\bullet}{A}{}^a{}_{e\mu} \overset{\bullet}{A}{}^e{}_{b\nu} - \overset{\bullet}{A}{}^a{}_{e\nu} \overset{\bullet}{A}{}^e{}_{b\mu} = 0. \tag{35.17}$$

Nevertheless, for a tetrad involving a non-trivial translational gauge potential, that is, for a gauge potential satisfying

$$B^a{}_\mu \neq \overset{\bullet}{\mathcal{D}}_\mu \varepsilon^a, \tag{35.18}$$

its torsion will be non-vanishing:

$$\overset{\bullet}{T}{}^a{}_{\mu\nu} = \partial_\mu h^a{}_\nu - \partial_\nu h^a{}_\mu + \overset{\bullet}{A}{}^a{}_{e\mu} h^e{}_\nu - \overset{\bullet}{A}{}^a{}_{e\nu} h^e{}_\mu \neq 0. \tag{35.19}$$

Since the spin connection (35.9) has vanishing curvature, the corresponding Fock–Ivanenko derivative is commutative:

$$[\overset{\bullet}{\mathcal{D}}_\rho, \overset{\bullet}{\mathcal{D}}_\sigma] = 0. \tag{35.20}$$

Using this property, torsion can be rewritten in the form

$$\overset{\bullet}{T}{}^a{}_{\mu\nu} = \partial_\mu B^a{}_\nu - \partial_\nu B^a{}_\mu + \overset{\bullet}{A}{}^a{}_{b\mu} B^b{}_\nu - \overset{\bullet}{A}{}^a{}_{b\nu} B^b{}_\mu, \tag{35.21}$$

which is the field strength of Teleparallel Gravity.

In this theory, therefore, gravitation is represented by torsion, not by curvature. On account of the gauge invariance of the tetrad, the field strength is also invariant under gauge transformations:

$$\overset{\bullet}{T}{}'^a{}_{\mu\nu} = \overset{\bullet}{T}{}^a{}_{\mu\nu}. \tag{35.22}$$

This is an expected result. In fact, recalling that the generators of the adjoint representation are the structure coefficients of the group taken as matrices, and considering that these coefficients vanish for abelian groups, fields belonging to the adjoint representation of abelian gauge theories will always be gauge invariant — a well-known property of electromagnetism.

The spacetime-indexed linear connection corresponding to the inertial spin connection (35.9) is

$$\overset{\bullet}{\Gamma}{}^\rho{}_{\nu\mu} = h_a{}^\rho \partial_\mu h^a{}_\nu + h_a{}^\rho \overset{\bullet}{A}{}^a{}_{b\mu} h^b{}_\nu \equiv h_a{}^\rho \overset{\bullet}{D}_\mu h^a{}_\nu. \tag{35.23}$$

This is the so-called Weitzenböck connection. Its definition is equivalent to the identity

$$\partial_\mu h^a{}_\nu + \overset{\bullet}{A}{}^a{}_{b\mu} h^b{}_\nu - \overset{\bullet}{\Gamma}{}^\rho{}_{\nu\mu} h^a{}_\rho = 0. \tag{35.24}$$

In the class of frames in which the spin connection $\overset{\bullet}{A}{}^a{}_{b\mu}$ vanishes (see Section 35.2.2), it reduces to

$$\partial_\mu h^a{}_\nu - \overset{\bullet}{\Gamma}{}^\rho{}_{\nu\mu} h^a{}_\rho = 0, \tag{35.25}$$

which is the so-called absolute, or distant parallelism condition, from where Teleparallel Gravity got its name.

35.2.5 *Teleparallel lagrangian*

As for any gauge theory,[7] the action functional of Teleparallel Gravity can be written in the form

$$\overset{\bullet}{S} = \frac{1}{2c\kappa} \int \eta_{ab} \overset{\bullet}{T}{}^a \wedge \star \overset{\bullet}{T}{}^b, \tag{35.26}$$

---

[7] Faddeev & Slavnov 1978.

where

$$\overset{\bullet}{T}{}^a = \tfrac{1}{2}\, \overset{\bullet}{T}{}^a{}_{\mu\nu}\, dx^\mu \wedge dx^\nu \qquad (35.27)$$

is the torsion 2-form, $\star \overset{\bullet}{T}{}^a$ is its dual form, and $\kappa = 8\pi G/c^4$. More explicitly,

$$\overset{\bullet}{S} = \frac{1}{8c\kappa} \int \eta_{ab}\, \overset{\bullet}{T}{}^a{}_{\mu\nu}\, \star \overset{\bullet}{T}{}^b{}_{\rho\sigma}\, dx^\mu \wedge dx^\nu \wedge dx^\rho \wedge dx^\sigma . \qquad (35.28)$$

Taking into account the identity

$$dx^\mu \wedge dx^\nu \wedge dx^\rho \wedge dx^\sigma = -\, \epsilon^{\mu\nu\rho\sigma}\, h\, d^4 x, \qquad (35.29)$$

with $h = \det(h^a{}_\mu)$, the action functional assumes the form

$$\overset{\bullet}{S} = -\frac{1}{8c\kappa} \int \overset{\bullet}{T}{}_{a\mu\nu}\, \star \overset{\bullet}{T}{}^a{}_{\rho\sigma}\, \epsilon^{\mu\nu\rho\sigma}\, h\, d^4 x. \qquad (35.30)$$

Using the generalized dual definition for soldered bundles[8]

$$\star \overset{\bullet}{T}{}^a{}_{\mu\nu} = \frac{h}{2}\, \epsilon_{\mu\nu\alpha\beta}\, S^{a\alpha\beta}, \qquad (35.31)$$

as well as the identity

$$\epsilon_{\mu\nu\alpha\beta}\, \epsilon^{\mu\nu\rho\sigma} = -2(\delta^\rho_\alpha \delta^\sigma_\beta - \delta^\sigma_\alpha \delta^\rho_\beta), \qquad (35.32)$$

it reduces to

$$\overset{\bullet}{S} = \frac{1}{4c\kappa} \int \overset{\bullet}{T}{}^a{}_{\rho\sigma}\, \overset{\bullet}{S}{}_a{}^{\rho\sigma}\, h\, d^4 x, \qquad (35.33)$$

where

$$\overset{\bullet}{S}{}_a{}^{\rho\sigma} \equiv -\overset{\bullet}{S}{}_a{}^{\sigma\rho} = h_a{}^\nu \left(\overset{\bullet}{K}{}^{\rho\sigma}{}_\nu - \delta_\nu{}^\sigma\, \overset{\bullet}{T}{}^{\theta\rho}{}_\theta + \delta_\nu{}^\rho\, \overset{\bullet}{T}{}^{\theta\sigma}{}_\theta \right) \qquad (35.34)$$

is the superpotential, and

$$\overset{\bullet}{K}{}^{\rho\sigma}{}_\nu = \tfrac{1}{2} \left(\overset{\bullet}{T}{}^{\sigma\rho}{}_\nu + \overset{\bullet}{T}{}_\nu{}^{\rho\sigma} - \overset{\bullet}{T}{}^{\rho\sigma}{}_\nu \right) \qquad (35.35)$$

is the contortion of the teleparallel torsion. The lagrangian density corresponding to the above action is

$$\overset{\bullet}{\mathcal{L}} = \frac{h}{4\kappa}\, \overset{\bullet}{T}{}_{a\mu\nu}\, \overset{\bullet}{S}{}^{a\mu\nu}. \qquad (35.36)$$

Using Ricci's theorem for the specific case of the teleparallel spin connection

$$\overset{\bullet}{A}{}^a{}_{b\mu} = \overset{\circ}{A}{}^a{}_{b\mu} + \overset{\bullet}{K}{}^a{}_{b\mu}, \qquad (35.37)$$

it is possible to show that

$$\overset{\bullet}{\mathcal{L}} = \overset{\circ}{\mathcal{L}} - \partial_\mu \left(2h\kappa^{-1}\, \overset{\bullet}{T}{}^{\nu\mu}{}_\nu \right), \qquad (35.38)$$

---

[8] Gribl Lucas & Pereira 2009.

where

$$\overset{\circ}{\mathcal{L}} = -\frac{\sqrt{-g}}{2\kappa}\,\overset{\circ}{R} \tag{35.39}$$

is the Einstein–Hilbert lagrangian of General Relativity. Up to a divergence, therefore, the teleparallel lagrangian is equivalent to the lagrangian of General Relativity.

One may wonder why the lagrangians are equivalent up to a divergence term. To understand that, let us recall that the Einstein–Hilbert lagrangian (35.39) depends on the tetrad, as well as on its first and second derivatives. The terms containing second derivatives, however, reduce to a divergence term.[9] It is then possible to rewrite the Einstein–Hilbert lagrangian in a form explicitly showing this property,

$$\overset{\circ}{\mathcal{L}} = \overset{\circ}{\mathcal{L}}_1 + \partial_\mu(\sqrt{-g}\,w^\mu), \tag{35.40}$$

where $\overset{\circ}{\mathcal{L}}_1$ is a lagrangian that depends solely on the tetrad and on its first derivatives, and w^μ is a four-vector. On the other hand, the teleparallel lagrangian (35.36) depends only on the tetrad and on its first derivative. The divergence term in the equivalence relation (35.38) is then necessary to account for the different orders of the teleparallel and the Einstein–Hilbert lagrangians. We mention in passing that in classical field theory — as well as in classical mechanics — the lagrangians depend only on the field and its first derivative (see Chapter 31). We can then say that the gauge structure of Teleparallel Gravity is more akin to a field theory than the geometric approach of General Relativity.

35.2.6 *Field equations*

Consider the lagrangian

$$\mathcal{L} = \overset{\bullet}{\mathcal{L}} + \mathcal{L}_s, \tag{35.41}$$

with $\mathcal{L}_s$ the lagrangian of a general source field. Variation with respect to the gauge potential $B^a{}_\rho$ yields the teleparallel version of the gravitational field equation

$$\partial_\sigma(h\overset{\bullet}{S}_a{}^{\rho\sigma}) - \kappa\,h\overset{\bullet}{J}_a{}^\rho = \kappa\,h\,\Theta_a{}^\rho. \tag{35.42}$$

In this equation,

$$h\overset{\bullet}{J}_a{}^\rho \equiv -\frac{\partial\overset{\bullet}{\mathcal{L}}}{\partial h^a{}_\rho} = \frac{1}{\kappa}\,h_a{}^\mu\,\overset{\bullet}{S}_c{}^{\nu\rho}\,\overset{\bullet}{T}^c{}_{\nu\mu} - \frac{h_a{}^\rho}{h}\,\overset{\bullet}{\mathcal{L}} + \frac{1}{\kappa}\,\overset{\bullet}{A}^c{}_{a\sigma}\,\overset{\bullet}{S}_c{}^{\rho\sigma} \tag{35.43}$$

---

[9] Landau & Lifshitz 1975.

stands for the Noether gauge pseudo-current,[10] which in this case represents the energy-momentum density of both gravitation and inertial effects, and

$$h\Theta_a{}^\rho = -\frac{\delta \mathcal{L}_s}{\delta h^a{}_\rho} \equiv -\left(\frac{\partial \mathcal{L}_s}{\partial h^a{}_\rho} - \partial_\mu \frac{\partial \mathcal{L}_s}{\partial_\mu \partial h^a{}_\rho} \right) \qquad (35.44)$$

is the source energy-momentum tensor. Due to the anti-symmetry of the superpotential in the last two indices, the total (gravitational plus inertial plus source) energy-momentum density is conserved in the ordinary sense:

$$\partial_\rho \left(h\overset{\bullet}{J}_a{}^\rho + h\,\Theta_a{}^\rho \right) = 0. \qquad (35.45)$$

Using the identity (35.37) in the left-hand side of the gravitational field equation (35.42), it is easy to verify that

$$\partial_\sigma \left(h\overset{\bullet}{S}_a{}^{\rho\sigma} \right) - \kappa\, h\overset{\bullet}{J}_a{}^\rho = h \left(\overset{\circ}{R}_a{}^\rho - \tfrac{1}{2}\, h_a{}^\rho\, \overset{\circ}{R} \right). \qquad (35.46)$$

As expected, due to the equivalence between the lagrangians, the teleparallel field equation (35.42) is equivalent to Einstein's field equation

$$\overset{\circ}{R}_a{}^\rho - \tfrac{1}{2}\, h_a{}^\rho \overset{\circ}{R} = \kappa\,\Theta_a{}^\rho. \qquad (35.47)$$

Observe that the energy-momentum tensor appears as the source in both theories: as the source of curvature in General Relativity, and as the source of torsion in Teleparallel Gravity. This means that, from the teleparallel point of view, curvature and torsion are related to the same degrees of freedom of gravity.

35.3   Some achievements of Teleparallel Gravity

Despite being equivalent to General Relativity, Teleparallel Gravity shows many conceptually distinctive features. In what follows we discuss some of these features, as well as explore their possible consequences.

35.3.1   *Separating inertia and gravitation*

In Teleparallel Gravity, the tetrad field has the form

$$h^a{}_\mu = \partial_\mu x^a + \overset{\bullet}{A}{}^a{}_{b\mu}\, x^b + B^a{}_\mu. \qquad (35.48)$$

The second term on the right-hand side represents the inertial effects present in the frame $h^a{}_\mu$. The third term, given by the translational gauge

---
[10] Andrade, Guillen & Pereira 2000.

potential, represents pure gravitation. This means that, in Teleparallel Gravity, inertial effects and gravitation are represented by different variables — and can consequently be separated. On the other hand, since both inertia and gravitation are included in the tetrad $h^a{}_\mu$, its coefficient of anholonomy

$$f^c{}_{ab} = h_a{}^\mu h_b{}^\nu (\partial_\nu h^c{}_\mu - \partial_\mu h^c{}_\nu) \tag{35.49}$$

will also represent both inertia and gravitation. Of course, the same is true of the spin connection of General Relativity, which is

$$\mathring{A}^a{}_{b\mu} = \tfrac{1}{2}\, h^c{}_\mu \left(f_b{}^a{}_c + f_c{}^a{}_b - f^a{}_{bc} \right). \tag{35.50}$$

Differently from Teleparallel Gravity, therefore, in General Relativity inertia and gravitation are both represented by one and the same variable, and cannot be separated. They are blended in an inextricable way.

According to the identity (35.37), the spin connection $\mathring{A}^a{}_{b\mu}$ can be decomposed in the form

$$\mathring{A}^a{}_{b\mu} = \overset{\bullet}{A}{}^a{}_{b\mu} - \overset{\bullet}{K}{}^a{}_{b\mu}. \tag{35.51}$$

Since $\overset{\bullet}{A}{}^a{}_{b\mu}$ represents inertial effects only, whereas $\overset{\bullet}{K}{}^a{}_{b\mu}$ represents the gravitational field, the above identity amounts actually to a decomposition of the General Relativity spin connection (35.50) into inertial and gravitational parts. To see that this is in fact the case, let us consider a locally inertial frame in which the spin connection of General Relativity vanishes:

$$\mathring{A}^a{}_{b\mu} \doteq 0. \tag{35.52}$$

Making use of identity (35.51), the local vanishing of $\mathring{A}^a{}_{b\mu}$ is equivalent to

$$\overset{\bullet}{A}{}^a{}_{b\mu} \doteq \overset{\bullet}{K}{}^a{}_{b\mu}. \tag{35.53}$$

This expression shows explicitly that, in such a local frame, inertial effects (left-hand side) exactly compensate for gravitation (right-hand side), and gravitation becomes locally undetectable.

The possibility of separating inertial effects from gravitation is one of the most outstanding properties of Teleparallel Gravity. It opens up many interesting new roads for the study of gravitation, which are not possible in the context of General Relativity.

35.3.2 Geometry versus force

In General Relativity, the trajectories of spinless particles are described by the geodesic equation

$$\frac{du^a}{ds} + \overset{\circ}{A}{}^a{}_{b\mu}\, u^b u^\mu = 0, \qquad (35.54)$$

where $ds = (g_{\mu\nu}\, dx^\mu dx^\nu)^{1/2}$ is the Lorentz invariant spacetime interval. This equation says that the four-acceleration of the particle vanishes:

$$\overset{\circ}{a}{}^a = 0. \qquad (35.55)$$

This means that in General Relativity *there is no concept of gravitational force*. Using identity (35.51), the geodesic equation can be rewritten in the form

$$\frac{du^a}{ds} + \overset{\bullet}{A}{}^a{}_{b\mu}\, u^b\, u^\mu = \overset{\bullet}{K}{}^a{}_{b\mu}\, u^b\, u^\mu. \qquad (35.56)$$

This is the teleparallel equation of motion of a spinless particle as seen from a general Lorentz frame $h^a{}_\mu$. Of course, it is equivalent to the geodesic equation (35.54). There are conceptual differences, though. In General Relativity, a theory fundamentally based on the equivalence principle, curvature is used to *geometrize* the gravitational interaction. The gravitational interaction in this case is described by letting (spinless) particles to follow the curvature of spacetime. Geometry replaces the concept of force, and the trajectories are determined, not by force equations, but by geodesics. Teleparallel Gravity, on the other hand, attributes gravitation to torsion, which accounts for gravitation not by geometrizing the interaction, but by acting as a force.[11] In consequence, there are no geodesics in Teleparallel Gravity, only force equations similar to the Lorentz force equation of electrodynamics. Notice that the inertial forces coming from the frame non-inertiality are represented by the connection on the left-hand side of (35.56), which is non-covariant by its very nature. In Teleparallel Gravity, therefore, whereas the gravitational effects are described by a covariant force, the non-inertial effects of the frame remain *geometrized* in the sense of General Relativity. In the geodesic equation (35.54), both inertial and gravitational effects are described by the connection term on the left-hand side.

---

[11] Andrade & Pereira 1997.

Commentary 35.2 Although the force equation (35.56) was obtained substituting identity (35.51) in the geodesic equation (35.54), it can also be deduced directly from a variational principle, by using the action

$$S = -mc \int_p^q ds \equiv -mc \int_p^q u_a \, h^a{}_\mu \, dx^\mu \qquad (35.57)$$

with $h^a{}_\mu$ given by (35.48). ◀

35.3.3 *Gravitational energy-momentum density*

All fundamental fields of nature have a well-defined local energy-momentum density. It would be natural to expect that the same should happen to the gravitational field. However, no tensorial expression for the gravitational energy-momentum density can be defined in the context of General Relativity. The basic reason for this impossibility is that the gravitational and the inertial effects are mixed in the spin connection of the theory, and cannot be separated. Even though some quantities, like curvature, are not affected by inertial effects, some others turn out to depend on it. For example, the energy-momentum density of gravitation will necessarily include both the energy-momentum density of gravity and the energy-momentum density of the inertial effects present in the frame. Since the inertial effects are non-tensorial by their very nature — they depend, of course, on the frame — the quantity defining the energy-momentum density of the gravitational field in this theory always shows up as a non-tensorial object.

On the other hand, owing to the possibility of separating gravitation from inertial effects, in Teleparallel Gravity it turns out that it is possible to define an energy-momentum density for pure gravitation, excluding the contribution from inertia. As it would be expected, that quantity is a tensorial object. To see how this is possible, let us consider the sourceless version of the teleparallel field equation (35.42)

$$\partial_\sigma(h\overset{\bullet}{S}_a{}^{\rho\sigma}) - \kappa \, h\overset{\bullet}{J}_a{}^\rho = 0, \qquad (35.58)$$

where

$$h\overset{\bullet}{J}_a{}^\rho = \frac{1}{\kappa} h_a{}^\mu \overset{\bullet}{S}_c{}^{\nu\rho} \overset{\bullet}{T}{}^c{}_{\nu\mu} - \frac{h_a{}^\rho}{h}\overset{\bullet}{\mathcal{L}} + \frac{1}{\kappa} \overset{\bullet}{A}{}^c{}_{a\sigma}\overset{\bullet}{S}_c{}^{\rho\sigma} \qquad (35.59)$$

is the usual gravitational energy-momentum pseudo-current, which is conserved in the ordinary sense:

$$\partial_\rho(h\overset{\bullet}{J}_a{}^\rho) = 0. \qquad (35.60)$$

This is actually a matter of necessity: since the derivative is not covariant, the conserved current cannot be covariant either, so that the conservation law itself is covariant — and consequently physically meaningful.

Using now the fact that the last term of the pseudo-current (35.59) together with the potential term of the field equation (35.58) make up a Fock–Ivanenko covariant derivative,

$$\partial_\sigma(h\overset{\bullet}{S}_a{}^{\rho\sigma}) - \overset{\bullet}{A}{}^c{}_{a\sigma}(h\,\overset{\bullet}{S}_c{}^{\rho\sigma}) \equiv \overset{\bullet}{\mathcal{D}}_\sigma(h\overset{\bullet}{S}_a{}^{\rho\sigma}), \qquad (35.61)$$

that field equation can be rewritten in the form

$$\overset{\bullet}{\mathcal{D}}_\sigma(h\overset{\bullet}{S}_a{}^{\rho\sigma}) - \kappa\,h\,\overset{\bullet}{t}_a{}^\rho = 0 \qquad (35.62)$$

where

$$\overset{\bullet}{t}_a{}^\rho = \frac{1}{\kappa}\, h_a{}^\lambda\,\overset{\bullet}{S}_c{}^{\nu\rho}\,\overset{\bullet}{T}{}^c{}_{\nu\lambda} - \frac{h_a{}^\rho}{h}\,\overset{\bullet}{\mathcal{L}} \qquad (35.63)$$

is a tensorial current that represents the energy-momentum of gravity alone, to the exclusion of the inertial effects.[12] Remembering that the covariant derivative $\overset{\bullet}{\mathcal{D}}_\sigma$ is commutative, and taking into account the anti-symmetry of the superpotential in the last two indices, it follows from the field equation (35.62) that the tensorial current (35.63) is conserved in the covariant sense:

$$\overset{\bullet}{\mathcal{D}}_\rho(h\overset{\bullet}{t}_a{}^\rho) = 0. \qquad (35.64)$$

This is again a matter of necessity: a covariant current can only be conserved with a covariant derivative. In the class of frames in which the teleparallel spin connection vanishes, it is conserved in the ordinary sense:

$$\partial_\rho(h\overset{\bullet}{t}_a{}^\rho) = 0. \qquad (35.65)$$

It should be added that the use of pseudotensors to compute the energy of a gravitational system requires some amount of handwork to get the physically relevant result. The reason is that, since any pseudotensor includes the contribution from the inertial effects, which is in general divergent for large distances (remember of the centrifugal force, for example), the space integration of the energy density usually yields divergent results. It is then necessary to make use of a regularization process to eliminate the spurious contributions coming from the inertial effects. In Teleparallel Gravity, on the other hand, owing to the possibility of separating gravitation from inertial effects, the elimination of the spurious contribution from the inertial effects is straightforward. In fact, notice that to each tetrad there is naturally associated a spin connection. Provided the same spin connection is used in all covariant derivatives, the inertial effects are automatically removed from the theory. The use of the appropriate spin

---

[12] Andrade, Guillen & Pereira 2000.

connection can thus be viewed as a regularization process in the sense that the computation of energy and momentum naturally yields the physically relevant values, no matter the coordinate or frame used to perform the computation.[13]

35.3.4   *A genuine gravitational variable*

Owing to the fact that the spin connection of General Relativity represents both gravitation and inertial effects, it is always possible to find a local frame — called *locally inertial frame* — in which inertial effects exactly compensate for gravitation, and the connection vanishes at that point:

$$\overset{\circ}{A}{}^{a}{}_{b\mu} \doteq 0. \tag{35.66}$$

Since there is a gravitational field at that point, such connection cannot be considered a genuine gravitational variable in the usual sense of classical field theory. Strictly speaking, therefore, General Relativity is not a true classical field theory. Considering furthermore that the non-covariant behavior of $\overset{\circ}{A}{}^{a}{}_{b\mu}$ under local Lorentz transformations is due uniquely to its inertial content, not to gravitation itself, it is not a gravitational, but just an inertial connection. Accordingly, General Relativity cannot be interpreted as a gauge theory for the Lorentz group because Lorentz is a kinematic, not a dynamic symmetry.

In Teleparallel Gravity, on the other hand, the gravitational field is represented by a translational-valued gauge potential

$$B_{\mu} = B^{a}{}_{\mu} P_{a}, \tag{35.67}$$

which shows up as the non-trivial part of the tetrad. Considering that the translational gauge potential represents gravitation only, not inertial effects, it can be considered a true gravitational variable in the sense of classical field theory. Notice, for example, that it is not possible to find a local frame in which it vanishes at a point. Furthermore, it is also a genuine gravitational connection: its connection behavior under gauge translations is related uniquely to its gravitational content. Teleparallel Gravity — with its gauge structure — can thus be considered to be much more akin to a classical field theory than General Relativity. It should be, for this reason, the theory to be used in any approach to quantum gravity.

---

[13] Krššák & Pereira 2015.

35.3.5 *Gravitation without the equivalence principle*

Universality of gravitation means that everything feels gravity the same. Provided the initial conditions are the same, all particles — independently of their masses and constitutions — will follow the same trajectory. In the motion of a massive particle, universality of free fall is directly connected with the equality between inertial and gravitational masses: $m_i = m_g$. In fact, in order to be eliminated from the classical equation of motion — so that the motion become universal — they must necessarily coincide. It is important to remark that, even though newtonian gravity can comply with universality, it remains a consistent theory for $m_i \neq m_g$. General Relativity, on the other hand, is a theory fundamentally based on the universality of free fall — that is, on the weak equivalence principle. There is no room at all for any violation of universality. Any violation of the weak equivalence principle would lead to its conceptual breakdown.

Gravitation is the only universal interaction of nature. It is, consequently, the only one to allow a geometric description, as is done by General Relativity. On account of the non-universal character of electromagnetism, it is not possible to construct a geometric description for the electromagnetic interaction. In fact, as is well-known, that interaction requires a gauge formulation. On the other hand, as a gauge theory for the translation group, Teleparallel Gravity does not depend on the validity of the weak equivalence principle to describe the gravitational interaction.[14] Like newtonian gravity, it remains a consistent theory in the absence of universality.

To see that this is in fact the case, let us consider the motion of a spinless particle of mass m in a gravitational field. Its action integral is given by

$$S = - m c \int_{\mathsf{p}}^{\mathsf{q}} ds \equiv - m c \int_{\mathsf{p}}^{\mathsf{q}} u_a h^a{}_\mu \, dx^\mu. \tag{35.68}$$

Substituting the teleparallel tetrad (35.13), it assumes the form

$$S = - m c \int_{\mathsf{p}}^{\mathsf{q}} u_a \left[\partial_\mu x^a + \overset{\bullet}{A}{}^a{}_{b\mu} + B^a{}_\mu \right] dx^\mu. \tag{35.69}$$

Since inertial and gravitational effects are represented by different variables, this action can be easily generalised for the case in which inertial and gravitational masses are different: it is given by[15]

$$S = - m_i c \int_{\mathsf{p}}^{\mathsf{q}} u_a \left[\partial_\mu x^a + \overset{\bullet}{A}{}^a{}_{b\mu} + (m_g/m_i) B^a{}_\mu \right] dx^\mu. \tag{35.70}$$

---

[14] Aldrovandi, Pereira & Vu 2004.
[15] Aldrovandi & Pereira 2012, page 121.

The invariance of the action under a general spacetime variation δx^μ yields the equation of motion

$$\frac{du_a}{ds} - \overset{\bullet}{A}{}^b{}_{a\rho}\, u_b\, u^\rho = -\, \overset{\bullet}{K}{}^b{}_{a\rho}\, u_b\, u^\rho + F_a, \qquad (35.71)$$

where

$$F_a = -\left[(m_g/m_i) - 1\right] h_a{}^\mu \left[P^\rho{}_\mu B^b{}_\rho\, \frac{du_b}{ds} - \left(\partial_\mu B^b{}_\rho - \partial_\rho B^b{}_\mu\right) u_b\, u^\rho\right]$$

is a new gravitational force, with

$$P^\rho{}_\mu = \delta^\rho_\mu - u^\rho u_\mu \qquad (35.72)$$

a velocity-projection tensor. The first term on the right-hand side of (35.71) represents the universal gravitational force. The second term, on the other hand, represents the effects coming from the lack of universality. This means that the breaking of the weak equivalence principle would correspond to the discovery of a new force, not predicted by General Relativity. When $m_g = m_i$, this new force vanishes and the equation of motion reduces to

$$\frac{du_a}{ds} - \overset{\bullet}{A}{}^b{}_{a\rho}\, u_b\, u^\rho = -\, \overset{\bullet}{K}{}^b{}_{a\rho}\, u_b\, u^\rho, \qquad (35.73)$$

which is just the universal teleparallel force equation (35.56).

Commentary 35.3 It is important to note that both forces appearing on the right-hand side of (35.71) are orthogonal to the four-velocity u^a, as it should be for a relativistic force. ◀

35.4 Final remarks

Although equivalent to General Relativity, Teleparallel Gravity provides a completely different approach to gravitation. As a consequence, many gravitational phenomena acquire a new perspective when analysed from the teleparallel point of view. For example, on account of the geometric description of General Relativity, which makes use of the torsionless Levi–Civita connection, there is a general understanding that gravity produces a curvature in spacetime. The universe as a whole, therefore, should also be curved. However, the advent of Teleparallel Gravity broke this paradigm: it became a matter of convention to describe the gravitational interaction in terms of curvature or in terms of torsion. This means that the attribution of curvature to spacetime is not an absolute, but a model-dependent statement. Notice furthermore that, according to Teleparallel Gravity, torsion has already been detected: it is responsible for all gravitational phenomena, including the physics of the solar system, which can be re-interpreted in terms of a force equation with torsion (or contortion) playing the role of force.

Other achievements of Teleparallel Gravity include the possibility of defining a tensorial expression for the energy-momentum density of gravity alone, to the exclusion of inertial effects. Furthermore, similarly to the teleparallel gauge potential, a fundamental spin-2 field should be interpreted, not as symmetric second-rank tensor, but as a translational-valued vector field.[16] In other words, a fundamental spin-2 field should be interpreted as a perturbation of the tetrad, instead of a perturbation of the metric. We can then say that Teleparallel Gravity is not just a theory equivalent to General Relativity, but a whole new way to understand and describe gravitation.

General references

The basic references on Teleparallel Gravity can be traced back from Aldrovandi & Pereira 2012. For a short review, see Pereira 2014.

---

[16] Arcos, Gribl Lucas & Pereira 2010.

Chapter 36

Einstein–Cartan Theory

In Teleparallel Gravity torsion and curvature are related to the same gravitational degrees of freedom. In Einstein–Cartan theory, on the other hand, curvature and torsion represent different gravitational degrees of freedom. For the sake of comparison, a brief review of this theory is presented, and some of its properties discussed.

36.1 Introduction

The basic motivation for the Einstein–Cartan construction[1] is the fact that, at the microscopic level, matter is represented by elementary particles, which in turn are characterized by mass *and* spin. If one adopts the same *geometrical spirit of General Relativity*, not only mass but also spin should be source of gravitation at that level.[2] In this line of thought, energy-momentum should keep its General Relativity role of source of curvature, whereas spin should appear as source of torsion.

The Lorentz connection of the Einstein–Cartan theory is a general Cartan connection $A^a{}_{b\mu}$ with non-vanishing curvature and torsion:

$$R^a{}_{b\nu\mu} \equiv \partial_\nu A^a{}_{b\mu} - \partial_\mu A^a{}_{b\nu} + A^a{}_{e\nu} A^e{}_{b\mu} - A^a{}_{e\mu} A^e{}_{b\nu} \neq 0 \qquad (36.1)$$

and

$$T^a{}_{\nu\mu} \equiv \partial_\nu h^a{}_\mu - \partial_\mu h^a{}_\nu + A^a{}_{e\nu} h^e{}_\mu - A^a{}_{e\mu} h^e{}_\nu \neq 0. \qquad (36.2)$$

The corresponding spacetime-indexed linear connection is

$$\Gamma^\rho{}_{\nu\mu} = h_a{}^\rho \partial_\mu h^a{}_\nu + h_a{}^\rho A^a{}_{b\mu} h^b{}_\nu. \qquad (36.3)$$

---

[1] Cartan 1922; Cartan 1923; Cartan 1924.
[2] For a textbook reference, see de Sabbata & Gasperini 1985.

According to Ricci's theorem, this connection can be decomposed in the form[3]

$$\Gamma^{\rho}{}_{\nu\mu} = \overset{\circ}{\Gamma}{}^{\rho}{}_{\nu\mu} + K^{\rho}{}_{\nu\mu}, \qquad (36.4)$$

where

$$\overset{\circ}{\Gamma}{}^{\rho}{}_{\nu\mu} = \tfrac{1}{2} g^{\rho\sigma} \left(\partial_{\nu} g_{\mu\sigma} + \partial_{\mu} g_{\nu\sigma} - \partial_{\sigma} g_{\mu\nu} \right) \qquad (36.5)$$

is the Levi–Civita connection of General Relativity, and

$$K^{\rho}{}_{\nu\mu} = \tfrac{1}{2} \left(T_{\nu}{}^{\rho}{}_{\mu} + T_{\mu}{}^{\rho}{}_{\nu} - T^{\rho}{}_{\nu\mu} \right) \qquad (36.6)$$

is the contortion tensor.

36.2 Field equations

The lagrangian of the Einstein–Cartan theory is

$$\mathcal{L}_{EC} = - \frac{c^4}{16\pi G} \sqrt{-g}\, R. \qquad (36.7)$$

Although it formally coincides with the Einstein–Hilbert lagrangian of General Relativity, the scalar curvature

$$R = g^{\mu\nu} R^{\rho}{}_{\mu\rho\nu} \qquad (36.8)$$

now refers to the curvature of a general Cartan connection:

$$R^{\rho}{}_{\mu\sigma\nu} = \partial_{\sigma} \Gamma^{\rho}{}_{\mu\nu} - \partial_{\nu} \Gamma^{\rho}{}_{\mu\sigma} + \Gamma^{\rho}{}_{\eta\sigma} \Gamma^{\eta}{}_{\mu\nu} - \Gamma^{\rho}{}_{\eta\nu} \Gamma^{\eta}{}_{\mu\sigma}. \qquad (36.9)$$

Since the connection $\Gamma^{\rho}{}_{\mu\nu}$ has a non-vanishing torsion,

$$T^{\rho}{}_{\nu\mu} = \Gamma^{\rho}{}_{\mu\nu} - \Gamma^{\rho}{}_{\nu\mu}, \qquad (36.10)$$

in contrast to the Levi–Civita connection $\overset{\circ}{\Gamma}{}^{\rho}{}_{\nu\mu}$ of General Relativity, it is not symmetric in the last two indices. As a consequence, the Ricci curvature tensor $R_{\mu\nu} = R^{\rho}{}_{\mu\rho\nu}$ is not symmetric either:

$$R_{\mu\nu} \neq R_{\nu\mu}.$$

Considering the total lagrangian

$$\mathcal{L} = \mathcal{L}_{EC} + \mathcal{L}_m, \qquad (36.11)$$

with $\mathcal{L}_m$ the lagrangian of a source field Ψ, the gravitational field equations are obtained by taking variations with respect to both the metric $g^{\mu\nu}$ and the contortion tensor $K_{\rho}{}^{\mu\nu}$. The resulting field equations are

$$R_{\mu\nu} - \tfrac{1}{2} g_{\mu\nu} R = \frac{8\pi G}{c^4} \theta_{\mu\nu} \qquad (36.12)$$

---

[3] Kobayashi & Nomizu 1996

and

$$T^\rho{}_{\mu\nu} + \delta^\rho_\mu T^\alpha{}_{\nu\alpha} - \delta^\rho_\nu T^\alpha{}_{\mu\alpha} = \frac{8\pi G}{c^4} s^\rho{}_{\mu\nu}. \tag{36.13}$$

In these equations,

$$\sqrt{-g}\, \theta_\mu{}^\nu = \frac{\partial \mathcal{L}_m}{\partial(\mathcal{D}_\nu \Psi)} \, h^a{}_\mu \partial_a \Psi - \delta^\nu_\mu \mathcal{L}_m \tag{36.14}$$

is the *canonical* energy-momentum tensor and

$$\sqrt{-g}\, s^\rho{}_{\mu\nu} = \frac{1}{2} \frac{\partial \mathcal{L}_m}{\partial(\mathcal{D}_\rho \Psi)} \, h^a{}_\mu h^b{}_\nu S_{ab} \Psi \tag{36.15}$$

is the *canonical* spin tensor of the source, with S_{ab} the Lorentz generators taken in the representation to which Ψ belongs. Equation (36.13) can be recast in the form

$$T^\rho{}_{\mu\nu} = \frac{8\pi G}{c^4} \left(s^\rho{}_{\mu\nu} + \tfrac{1}{2} \delta^\rho_\mu s^\alpha{}_{\nu\alpha} - \tfrac{1}{2} \delta^\rho_\nu s^\alpha{}_{\mu\alpha} \right). \tag{36.16}$$

For spinless ($s^\rho{}_{\mu\nu} = 0$) matter, torsion vanishes. Concomitantly, the canonical energy-momentum tensor $\theta_\mu{}^\nu$ reduces to the symmetric energy-momentum tensor $\Theta_\mu{}^\nu$, and the field equation (36.12) coincides with the ordinary Einstein equation. In the presence of spinning matter, however, there will be a non-vanishing torsion, given by Eq. (36.16). As this equation is purely algebraic, torsion is a non-propagating field.

36.3 Gravitational coupling prescription

The coupling prescription in Einstein–Cartan theory is defined by

$$\partial_\mu \rightarrow \mathcal{D}_\mu = \partial_\mu - \tfrac{i}{2} A^{ab}{}_\mu S_{ab}, \tag{36.17}$$

with S_{ab} an appropriate representation of the Lorentz generators. Acting on a spinor field ψ, for example, it assumes the form

$$\partial_\mu \psi \rightarrow \mathcal{D}_\mu \psi = \partial_\mu \psi - \tfrac{i}{2} A^{ab}{}_\mu S_{ab} \psi, \tag{36.18}$$

with S_{ab} the spinor representation

$$S_{ab} = \tfrac{i}{4} \left[\gamma_a, \gamma_b \right]. \tag{36.19}$$

Using the Ricci decomposition

$$A^a{}_{b\mu} = \overset{\circ}{A}{}^a{}_{b\mu} + K^a{}_{b\mu}, \tag{36.20}$$

with $\overset{\circ}{A}{}^a{}_{b\mu}$ the spin connection of General Relativity, it becomes

$$\partial_\mu \psi \rightarrow \mathcal{D}_\mu \psi = \overset{\circ}{\mathcal{D}}_\mu \psi - \tfrac{i}{2} K^{ab}{}_\mu S_{ab} \psi, \tag{36.21}$$

where

$$\overset{\circ}{\mathcal{D}}_\mu \psi = \partial_\mu \psi - \tfrac{i}{2} \overset{\circ}{A}{}^{ab}{}_\mu S_{ab} \psi \qquad (36.22)$$

is the covariant derivative defining the coupling prescription of General Relativity.

We see from Eq. (36.21) that physical phenomena which would be new with respect to General Relativity are expected in the presence of spin, which is in consonance with the fact that curvature and torsion represent, in this theory, independent gravitational degrees of freedom.

For a Lorentz scalar field ϕ the generators are null, $S_{ab}\phi = 0$. Furthermore, since torsion vanishes for the (spinless) scalar field, the coupling prescription coincides with that of General Relativity:

$$\partial_\mu \phi \;\rightarrow\; \mathcal{D}_\mu \phi = \partial_\mu \phi. \qquad (36.23)$$

In the case of a Lorentz vector ϕ^a, on the other hand, for which the generators S_{ab} are given by

$$(S_{ab})^c{}_d = i \left(\eta_{bd}\, \delta^c_a - \eta_{ad}\, \delta^c_b \right), \qquad (36.24)$$

the coupling prescription (36.18) reads

$$\partial_\mu \phi^a \;\rightarrow\; \mathcal{D}_\mu \phi^a = \partial_\mu \phi^a + A^a{}_{b\mu}\, \phi^b. \qquad (36.25)$$

For the corresponding spacetime vector $\phi^\rho = h_a{}^\rho \phi^a$, the coupling prescription has the form

$$\partial_\mu \phi^\rho \;\rightarrow\; \nabla_\mu \phi^\rho = \partial_\mu \phi^\rho + \Gamma^\rho{}_{\nu\mu}\, \phi^\nu. \qquad (36.26)$$

36.4 Particle equations of motion

We consider now the question of the equation of motion of a spinning particle in the context of Einstein–Cartan theory. A possible procedure is to begin by observing that the geodesic equation of General Relativity,

$$\frac{du^\rho}{ds} + \overset{\circ}{\Gamma}{}^\rho{}_{\mu\nu}\, u^\mu\, u^\nu = 0, \qquad (36.27)$$

can be obtained from the lagrangian

$$S = -\int_a^b h^a{}_\mu\, p_a\, dx^\mu, \qquad (36.28)$$

where $p_a = mcu_a$ is the particle four-momentum. It is then natural to assume that the action integral describing a spinning particle minimally coupled to a general Lorentz connection $A^a{}_{b\mu}$ be written in the form

$$S = \int_a^b \left(- h^a{}_\mu\, p_a + \tfrac{1}{2} A^{ab}{}_\mu\, s_{ab} \right) dx^\mu, \qquad (36.29)$$

where s_{ab} is the spin angular momentum of the particle. The corresponding routhian can then be written as[4]

$$\mathcal{R} = -h^a{}_\mu\, p_a\, u^\mu + \tfrac{1}{2}\, A^{ab}{}_\mu\, s_{ab}\, u^\mu - \frac{\mathcal{D}u^a}{ds}\,\frac{s_{ab}u^b}{u^2}\,, \qquad (36.30)$$

where

$$\frac{\mathcal{D}u^a}{ds} = u^\mu\, \mathcal{D}_\mu u^a, \qquad (36.31)$$

with $\mathcal{D}_\mu$ the covariant derivative (36.25). This term is a constraint introduced to ensure that the four-velocity and the spin angular momentum density satisfy

$$s_{ab}\, s^{ab} = 2\,\mathbf{s}^2 \qquad \text{and} \qquad s_{ab}\, u^a = 0, \qquad (36.32)$$

with $\mathbf{s}$ the particle spin vector. We recall that the spin angular momentum $s_{\mu\nu}$ is related to spin tensor through

$$s_{\mu\nu} = \int s^0{}_{\mu\nu}\, d^3x. \qquad (36.33)$$

Considering the hamiltonian formalism, the equation of motion for the spin is found to be

$$\frac{\mathcal{D}s_{ab}}{ds} = (u_a\, s_{bc} - u_b\, s_{ac})\, \frac{\mathcal{D}u^c}{ds}. \qquad (36.34)$$

On the other hand, making use of the lagrangian formalism, the equation of motion for the trajectory of the particle is found to be

$$\frac{\mathcal{D}\mathcal{P}_\mu}{ds} = T^a{}_{\mu\nu}\, \mathcal{P}_a\, u^\nu - \tfrac{1}{2}\, R^{ab}{}_{\mu\nu}\, s_{ab}\, u^\nu, \qquad (36.35)$$

where $\mathcal{P}_\mu = h_\mu{}^c\, \mathcal{P}_c$ is a generalized momentum, with

$$\mathcal{P}_c = m\,c\,u_c + u^a\, \frac{\mathcal{D}s_{ca}}{ds}. \qquad (36.36)$$

Equation (36.35) is the Einstein–Cartan version of the Papapetrou equation.[5] In addition to the usual Papapetrou coupling between the particle spin and the Riemann tensor, there is also a coupling between torsion and the generalized momentum $\mathcal{P}_\rho$, which gives rise to new phenomena not predicted by General Relativity. For a spinless particle, torsion vanishes and the equation of motion (36.35) reduces to the geodesic equation (36.27) of General Relativity.

Commentary 36.1 The Einstein–Cartan theory shows a strange feature: the source currents and the field strengths are inversely related. More precisely, whereas energy-momentum (a current associated with translations) appears as source of the Lorentz-valued curvature, spin (a current associated with Lorentz transformations) appears as source of the translational-valued torsion. However, in the equation of motion of a spinning particle this inverse relation disappears: the four-momentum of the particle couples to torsion, whereas the particle's spin couples to curvature. ◀

---

[4] We follow here the approach described in Yee & Bander 1993.
[5] Trautman 2006.

36.5   Conceptual issues

The Einstein–Cartan theory, briefly described in this chapter, shows a series of conceptual difficulties. The first one refers to its coupling prescription, defined by Eq. (36.17): it violates the strong equivalence principle, according to which the gravitational coupling prescription is *minimal* only in the spin connection of General Relativity.[6] Another problem of the same coupling prescription is that, when used to describe the interaction of the electromagnetic field with gravitation, it violates the gauge invariance of the coupled Maxwell equations. In fact, under a gauge transformation of the electromagnetic potential

$$A'_\mu = A_\mu - \partial_\mu \epsilon(x^\alpha),		(36.37)$$

the gravitationally-coupled electromagnetic field strength

$$F_{\mu\nu} = \nabla_\mu A_\nu - \nabla_\nu A_\mu,		(36.38)$$

with ∇_μ the covariant derivative (36.26), is easily seen not to be gauge invariant:

$$F'_{\mu\nu} = F_{\mu\nu} + T^\rho{}_{\mu\nu} \partial_\rho \epsilon(x^\alpha).		(36.39)$$

Furthermore, since the canonical energy-momentum tensor of the electromagnetic field, given by

$$\theta_{\mu\nu} = -\partial_\mu A^\rho F_{\nu\rho} + \tfrac{1}{4} g_{\mu\nu} F_{\rho\sigma} F^{\rho\sigma},		(36.40)$$

is not gauge invariant, the gravitational field equation (36.12) will not be gauge invariant either when the electromagnetic field is considered as a source. Of course, the symmetric energy-momentum tensor

$$\Theta_{\mu\nu} = -F_\mu{}^\rho F_{\nu\rho} + \tfrac{1}{4} g_{\mu\nu} F_{\rho\sigma} F^{\rho\sigma}		(36.41)$$

is gauge invariant, but since the left-hand side of the field equation (36.12) is not symmetric in the presence of torsion, such tensor cannot appear as source in the Einstein–Cartan theory. These problems are usually circumvented by *postulating* that the electromagnetic field does not couple nor produce torsion.[7] This solution, however, is highly questionable.

For example, from a quantum point of view, one should always expect an interaction between photons and torsion.[8] The reason is that a photon,

---

[6] Arcos & Pereira 2004.

[7] See, for example, Benn, Dereli & Tucker 1980; Hehl, McCrea, Mielke & Ne'eman 1995; Shapiro 2002.

[8] de Sabbata & Sivaram 1994.

perturbatively speaking, can virtually disintegrate into an electron-positron pair. Considering that these particles are massive fermions that do couple to torsion, at the quantum level a photon will necessarily feel the presence of torsion. Considering furthermore that all macroscopic phenomena have an interpretation based on an average of microscopic phenomena, and taking into account the strictly attractive character of gravitation, which eliminates the possibility of a vanishing average, one should expect the electromagnetic field to interact with torsion also at the classical (non-quantum) level.

With independent works by Sciama[9] and Kibble[10] in the early nineteensixties, the original Einstein–Cartan theory was further developed in the form of a gauge theory for the Poincaré group, in which torsion becomes a propagating field. This generalization is known as Einstein–Cartan–Sciama–Kibble theory, in reference to those who have most contributed to Cartan's generalization of Einstein's theory. Since then these models have attracted considerable attention in the literature.

Commentary 36.2 It is important to remark that torsion in Teleparallel Gravity plays a completely different role from that played in Einstein–Cartan theory. However, from a conceptual point of view, torsion cannot play two different roles in nature. This means that, if the role played by torsion in Einstein–Cartan is assumed to be correct, the role played in Teleparallel Gravity is necessarily wrong, and vice versa. The answer to this puzzle can only be given by experiment. ◀

General references

The relevant literature about Einstein–Cartan theories can be found in the reprint volume (with commentaries) Blagojevic & Hehl 2013. Additional commendable references are Trautman 2006 and de Sabbata & Sivaram 1994.

---

[9] Sciama 1964.
[10] Kibble 1961.

Chapter 37

de Sitter Invariant Special Relativity

The foundations of a de Sitter invariant Special Relativity are introduced. Some of its immediate consequences for the notions of space, time, motion and conserved quantities are established. Possible implications for gravitation and cosmology are briefly discussed. They exhibit to a high degree the deep relationship that Physics can have with Geometry.

37.1 Towards a de Sitter special relativity

37.1.1 *de Sitter as a Quotient Space*

Spacetimes with constant sectional curvature are maximally symmetric in the sense that they can lodge the highest possible number of Killing vectors.[1] Their curvature tensor is completely specified by the scalar curvature R, which is constant all the way through. Minkowski space M, with vanishing curvature, is the simplest case. Its kinematic group is the Poincaré group $\mathcal{P} = \mathcal{L} \oslash \mathcal{T}$, the semi-direct product of the Lorentz group $\mathcal{L}$ and the translation group $\mathcal{T}$. It is actually a homogeneous space under the Lorentz group:

$$M = \mathcal{P}/\mathcal{L}.$$

The invariance of M under the transformations of $\mathcal{P}$ reflects its uniformity. The Lorentz subgroup provides an isotropy around a given point of M, and the translation symmetry enforces this isotropy around any other point. This is the meaning of homogeneity: all the points of spacetime are ultimately equivalent under some kind of transformation.

In addition to Minkowski, there are two other maximally symmetric

---

[1] See, for example, Weinberg 1972, page 371.

four-dimensional spacetimes with constant sectional curvature. One is the de Sitter space, with topology $R^1 \times S^3$ and (let us say) positive scalar curvature. The other is the anti-de Sitter space, with topology $S^1 \times R^3$ and negative scalar curvature. Of course, as hyperbolic spaces, both have negative Gaussian curvature. Our interest here will be restricted to the de Sitter spacetime, denoted dS, whose kinematic group is the de Sitter group $SO(4,1)$. Like Minkowski, it is a homogeneous space under the Lorentz group:

$$dS = SO(4,1)/\mathcal{L}.$$

Its homogeneity property, however, is completely different from Minkowski. Of course, in order to be physically relevant, the de Sitter spacetime must be a solution to the Einstein equation. As a quotient space, however, and similarly to Minkowski, it is more fundamental than the Einstein equation in the sense that it is known *a priori*, independently of any field equation.

In spite of these properties, the de Sitter spacetime is usually interpreted as the simplest *dynamical* solution of the sourceless Einstein equation in the presence of a cosmological constant, standing on an equal footing with all other gravitational solutions — like for example Schwarzschild's and Kerr's. As a non-gravitational spacetime, the de Sitter solution should instead be interpreted as a fundamental background for the construction of physical theories, standing on an equal footing with the Minkowski solution. General relativity, for instance, can be built up *on any one of them*. Of course, in either case gravitation will have the same dynamics, only their local kinematics will be different. If the underlying spacetime is Minkowski, the local kinematics will be ruled by the Poincaré group of ordinary special relativity. If the underlying spacetime is de Sitter space, the local kinematics will be ruled by the de Sitter group, which amounts then to replace ordinary special relativity by a de Sitter-ruled special relativity.[2]

37.1.2 *Rationale for a de Sitter special relativity*

The existence of an invariant length parameter at the Planck scale constitutes a clear evidence that the spacetime kinematics at that scale cannot be described by ordinary special relativity, as the Poincaré group does not allow the existence of such an invariant parameter. One then has to look for a *modified* special relativity. An interesting attempt in this direction

---

[2] Aldrovandi, Beltrán Almeida & Pereira 2007.

is the so-called "doubly special relativity",[3] obtained by introducing into the dispersion relation of special relativity scale-suppressed terms of higher order in the momentum, in a way such as to allow the existence of an invariant length at the Planck scale. The importance of these terms is controlled by a parameter κ, which changes the kinematic group of special relativity from the Poincaré to a κ-deformed Poincaré group, in which Lorentz symmetry is explicitly violated. Far away from the Planck scale these terms are suppressed, Lorentz symmetry is recovered and one obtains back ordinary special relativity.

A different solution to the same problem shows up by noting that *Lorentz transformations do not change the curvature of the homogeneous spacetime in which they are performed.* Considering that the scalar curvature R of any homogeneous spacetime is of the form

$$R \sim l^{-2},$$

with l the pseudo-radius (we are restricting ourselves to the case of the de Sitter spacetime, which has a positive scalar curvature), *Lorentz transformations are then found to leave the length parameter l invariant.* Although somewhat hidden in Minkowski space, because what is left invariant in this case is an infinite length — corresponding to a vanishing scalar curvature — in de Sitter and anti-de Sitter spacetimes, whose pseudo-radii are finite, this property becomes manifest. One then sees that, contrary to the usual belief, *Lorentz transformations do leave invariant a very particular length invariant: that defining the scalar curvature of the homogeneous spacetime.*

Now, if the Planck length l_P is to be invariant under Lorentz transformations, it is natural to assume that it represents the pseudo-radius of spacetime at the Planck scale. In this case, spacetime at the Planck scale will then be a de Sitter space with the scalar curvature given by

$$R \sim l_P^{-2} \simeq 10^{66}\,\mathrm{cm}^{-2}. \tag{37.1}$$

As one moves away from the Planck scale, the pseudo-radius l will be larger and larger, and for a sufficient large value the de Sitter special relativity will approach ordinary special relativity, which is ruled by the Poincaré group. One can then say that, in the same way ordinary (Poincaré-ruled) special relativity can be seen as a generalisation of Galileo relativity for velocities comparable to the speed of light, the de Sitter-ruled special relativity can be seen as a generalisation of ordinary special relativity for energies comparable to the Planck energy.

---

[3] Amelino-Camelia 2002.

Before going through the details of the de Sitter special relativity, we are going to review the de Sitter spacetime and group. In particular, we are going to see how the stereographic coordinates can be introduced. A study of the possible contraction limits[4] of the de Sitter spacetime and group will also be performed.

Commentary 37.1 One should note that, even though there is an invariant length-parameter related to the cosmological term, according to the de Sitter special relativity the Lorentz group remains part of the spacetime kinematics, which means that this symmetry is not broken at any energy scale. Taking into account the connection between Lorentz symmetry and causality,[5] this theory implies that causality is preserved at all energy scales, even at the Planck scale. ◄

37.2   de Sitter spacetime and stereographic coordinates

The maximally symmetric de Sitter spacetime, denoted dS, can be seen as a hypersurface in a host pseudo-Euclidean 5−space with metric $\eta_{AB} = (+1, -1, -1, -1, -1)$ $(A, B, \ldots = 0, \ldots, 4)$, whose points in Cartesian coordinates χ^A satisfy the relation[6]

$$\eta_{AB}\, \chi^A \chi^B = -l^2, \tag{37.2}$$

or equivalently, in four-dimensional coordinates,

$$\eta_{\mu\nu}\, \chi^\mu \chi^\nu - \left(\chi^4\right)^2 = -l^2. \tag{37.3}$$

It has the de Sitter group $SO(4,1)$ as group of motions, and is homogeneous under the Lorentz group $\mathcal{L} = SO(3,1)$, that is,

$$dS = SO(4,1)/\mathcal{L}. \tag{37.4}$$

In Cartesian coordinates χ^A, the generators of the infinitesimal de Sitter transformations are written in the form

$$L_{AB} = \eta_{AC}\, \chi^C \frac{\partial}{\partial \chi^B} - \eta_{BC}\, \chi^C \frac{\partial}{\partial \chi^A}. \tag{37.5}$$

They satisfy the commutation relations

$$[L_{AB}, L_{CD}] = \eta_{BC} L_{AD} + \eta_{AD} L_{BC} - \eta_{BD} L_{AC} - \eta_{AC} L_{BD}. \tag{37.6}$$

Among all coordinate systems used to describe the de Sitter metric, the four-dimensional stereographic coordinates $\{x^\mu\}$ emerge as a special one, in the sense that, for a vanishing cosmological term, they reduce to the

---

[4] Inönü & Wigner 1953.
[5] Zeeman 1964.
[6] Hawking & Ellis 1973.

Cartesian coordinates of Minkowski spacetime. They are obtained through a stereographic projection from the de Sitter hypersurface into a target Minkowski spacetime. However, in order to obtain the stereographic coordinates, it is necessary to use two different parameterizations: one appropriate for large values of the de Sitter parameter l, and another appropriate for small values of l. Considering that the cosmological term Λ depends on l according to $\Lambda \sim l^{-2}$, a large l means small Λ whereas a small l means a large Λ. In what follows we are going to consider separately each one of these parameterizations.

Commentary 37.2 The reference value for defining small and large l is the Planck length l_P. Accordingly, a large l is represented by the condition

$$l \gg l_P ,$$

which is equivalent to

$$\Lambda \, l_P^2 \ll 1 .$$

On the other hand, assuming that the Planck length is the smallest possible length in nature, a small l is represented by

$$l \gtrsim l_P ,$$

which is equivalent to

$$\Lambda \, l_P^2 \lesssim 1 .$$

This last expression can be rephrased as

$$\Lambda \lesssim \Lambda_P$$

with $\Lambda_P \sim l_P^{-2}$ the Planck cosmological constant. ◀

37.2.1 *Large pseudo-radius parameterization*

In the parameterization appropriate to deal with large values of l, the stereographic coordinates are defined by[7]

$$\chi^\mu = \Omega \, x^\mu \qquad (37.7)$$

and

$$\chi^4 = -\, l\, \Omega \left(1 + \sigma^2/4l^2\right) , \qquad (37.8)$$

where $\Omega \equiv \Omega(x)$ is given by

$$\Omega = (1 - \sigma^2/4l^2)^{-1} \qquad (37.9)$$

with σ^2 the Lorentz invariant quadratic form $\sigma^2 = \eta_{\mu\nu}\, x^\mu x^\nu$. In these coordinates, the infinitesimal de Sitter quadratic interval

$$ds^2 = g_{\alpha\beta}\, dx^\alpha dx^\beta \qquad (37.10)$$

---

[7] Gürsey 1962.

can be seen as the conformally flat metric

$$g_{\alpha\beta} = \Omega^2 \, \eta_{\alpha\beta}. \qquad (37.11)$$

The corresponding Christoffel connection is

$$\Gamma^\lambda{}_{\mu\nu} = \frac{\Omega}{2l^2} \left(\delta^\lambda_\mu \, \eta_{\nu\alpha} \, x^\alpha + \delta^\lambda_\nu \, \eta_{\mu\alpha} \, x^\alpha - \eta_{\mu\nu} \, x^\lambda \right), \qquad (37.12)$$

with the Riemann tensor given by

$$R^\mu{}_{\nu\rho\sigma} = \frac{\Omega^2}{l^2} \left(\delta^\mu_\rho \, \eta_{\nu\sigma} - \delta^\mu_\sigma \, \eta_{\nu\rho} \right). \qquad (37.13)$$

The Ricci tensor and the scalar curvature are, consequently,

$$R_{\nu\sigma} = \frac{3\Omega^2}{l^2} \, \eta_{\nu\sigma} \qquad \text{and} \qquad R = \frac{12}{l^2}. \qquad (37.14)$$

In terms of the stereographic coordinates $\{x^\mu\}$, the de Sitter generators (37.5) are written in the form

$$L_{\mu\nu} = \eta_{\mu\rho} \, x^\rho \, P_\nu - \eta_{\nu\rho} \, x^\rho \, P_\mu \qquad (37.15)$$

and

$$L_{4\mu} = l \, P_\mu - \frac{1}{4l} \, K_\mu, \qquad (37.16)$$

where

$$P_\mu = \partial_\mu \quad \text{and} \quad K_\mu = \left(2\eta_{\mu\nu} x^\nu x^\rho - \sigma^2 \delta^\rho_\mu \right) \partial_\rho \qquad (37.17)$$

are, respectively, the generators of translations and proper conformal transformations.[8] Generators $L_{\mu\nu}$ refer to the Lorentz subgroup, whereas the elements $L_{4\mu}$ define the transitivity on the homogeneous space. From Eq. (37.16) it follows that the de Sitter spacetime is transitive under a combination of translations and proper conformal transformations — usually called de Sitter "translations". The relative importance of these two transformations is clearly determined by the value of the pseudo-radius l.

In order to study the limit of large values of l, it is necessary to parameterise the generators (37.16) according to

$$\Pi_\mu \equiv \frac{L_{4\mu}}{l} = P_\mu - \frac{1}{4l^2} \, K_\mu. \qquad (37.18)$$

In terms of these generators, the de Sitter algebra (37.6) assumes the form

$$[L_{\mu\nu}, L_{\rho\sigma}] = \eta_{\nu\rho} \, L_{\mu\sigma} + \eta_{\mu\sigma} \, L_{\nu\rho} - \eta_{\nu\sigma} \, L_{\mu\rho} - \eta_{\mu\rho} \, L_{\nu\sigma}, \qquad (37.19)$$

$$[\Pi_\mu, L_{\rho\sigma}] = \eta_{\mu\rho} \Pi_\sigma - \eta_{\mu\sigma} \Pi_\rho, \qquad (37.20)$$

$$[\Pi_\mu, \Pi_\rho] = l^{-2} L_{\mu\rho}. \qquad (37.21)$$

The last commutator shows that the de Sitter "translation" generators are not really translations, but rotations — hence the quotation marks.

---

[8] Callan, Coleman & Jackiw 1970.

The contraction limit $l \to \infty$

In the limit $l \to \infty$, we see from Eq. (37.18) that the de Sitter generators Π_μ reduce to generators of ordinary translations

$$\Pi_\mu \to P_\mu. \tag{37.22}$$

Concomitantly, the de Sitter algebra (37.19-37.21) contracts to

$$[L_{\mu\nu}, L_{\rho\sigma}] = \eta_{\nu\rho} L_{\mu\sigma} + \eta_{\mu\sigma} L_{\nu\rho} - \eta_{\nu\sigma} L_{\mu\rho} - \eta_{\mu\rho} L_{\nu\sigma} \tag{37.23}$$

$$[P_\mu, L_{\rho\sigma}] = \eta_{\mu\rho} P_\sigma - \eta_{\mu\sigma} P_\rho \tag{37.24}$$

$$[P_\mu, P_\rho] = 0 \tag{37.25}$$

which is the Lie algebra of the Poincaré group $\mathcal{P} = \mathcal{L} \oslash \mathcal{T}$, the semi-direct product of the Lorentz ($\mathcal{L}$) and the translation ($\mathcal{T}$) groups. As a result of this algebra and group deformations, the de Sitter spacetime dS contracts to the flat Minkowski space M:

$$dS \to M = \mathcal{P}/\mathcal{L}. \tag{37.26}$$

In fact, as a simple inspection shows, the de Sitter metric (37.11) reduces to the Minkowski metric

$$g_{\mu\nu} \to \eta_{\mu\nu}, \tag{37.27}$$

and the Riemann, Ricci and scalar curvatures vanish identically:

$$R^\mu{}_{\nu\rho\sigma} \to 0, \quad R_{\nu\sigma} \to 0, \quad R \to 0. \tag{37.28}$$

Furthermore, we see from these relations that Minkowski spacetime is transitive under ordinary translations.

37.2.2 Small pseudo-radius parameterisation

To deal with small values of l, it is convenient to define the 'inverse' host space coordinates

$$\bar{\chi}^A = \chi^A/4l^2, \tag{37.29}$$

in terms of which relation (37.3) assumes the form

$$\eta_{\mu\nu} \bar{\chi}^\mu \bar{\chi}^\nu - \left(\bar{\chi}^4\right)^2 = -\frac{1}{16l^2}. \tag{37.30}$$

The stereographic projection is now defined by

$$\bar{\chi}^\mu = \bar{\Omega}\, x^\mu \tag{37.31}$$

and

$$\bar{\chi}^4 = -l\,\bar{\Omega}\left(1 + \sigma^2/4l^2\right),\tag{37.32}$$

where

$$\bar{\Omega} \equiv \frac{\Omega}{4l^2} = \frac{1}{4l^2 - \sigma^2}.\tag{37.33}$$

In these coordinates, the infinitesimal de Sitter quadratic interval

$$d\bar{s}^2 = \bar{g}_{\alpha\beta}\,dx^\alpha dx^\beta \tag{37.34}$$

is written with metric components

$$\bar{g}_{\alpha\beta} = \bar{\Omega}^2\,\eta_{\alpha\beta}.\tag{37.35}$$

Considering that $\bar{g}_{\alpha\beta}$ and $g_{\alpha\beta}$ differ by a constant, the corresponding Christoffel connections will coincide

$$\bar{\Gamma}^\lambda{}_{\mu\nu} \equiv \Gamma^\lambda{}_{\mu\nu} = 2\bar{\Omega}\left(\delta^\lambda_\mu\,\eta_{\nu\alpha}\,x^\alpha + \delta^\lambda_\nu\,\eta_{\mu\alpha}\,x^\alpha - \eta_{\mu\nu}\,x^\lambda\right).\tag{37.36}$$

Of course, the same happens to the Riemann tensor

$$\bar{R}^\mu{}_{\nu\rho\sigma} \equiv R^\mu{}_{\nu\rho\sigma} = 16\,l^2\,\bar{\Omega}^2\left(\delta^\mu_\rho\,\eta_{\nu\sigma} - \delta^\mu_\sigma\,\eta_{\nu\rho}\right),\tag{37.37}$$

as well as to the Ricci tensor

$$\bar{R}_{\nu\sigma} \equiv R_{\nu\sigma} = 16\,l^2\,\bar{\Omega}^2\,\eta_{\nu\sigma}.\tag{37.38}$$

The scalar curvature, however, due to a further contraction with the metric tensor, assumes a different form

$$\bar{R} \equiv 16\,l^4 R = 192\,l^2.\tag{37.39}$$

The cosmological constant depends on l according to

$$\Lambda \sim l^{-2}.\tag{37.40}$$

Use of (37.39) shows that it relates to the scalar curvature according to

$$\Lambda \sim \frac{\bar{R}}{l^4}.\tag{37.41}$$

In terms of the stereographic coordinates (37.31-37.32), the de Sitter generators (37.5) assume the same form as in the previous parameterization

$$L_{\mu\nu} = \eta_{\mu\rho}\,x^\rho\,P_\nu - \eta_{\nu\rho}\,x^\rho\,P_\mu \tag{37.42}$$

and

$$L_{4\mu} = l\,P_\mu - \frac{1}{4l}\,K_\mu.\tag{37.43}$$

However, in order to study the limit of large values of l, it is necessary to rewrite generators (37.43) in the form

$$\bar{\Pi}_\mu \equiv 4l\,L_{4\mu} = 4l^2 P_\mu - K_\mu.\tag{37.44}$$

In this case, the de Sitter algebra (37.6) becomes

$$[L_{\mu\nu}, L_{\rho\sigma}] = \eta_{\nu\rho}\,L_{\mu\sigma} + \eta_{\mu\sigma}\,L_{\nu\rho} - \eta_{\nu\sigma}\,L_{\mu\rho} - \eta_{\mu\rho}\,L_{\nu\sigma};\tag{37.45}$$

$$[\bar{\Pi}_\mu, L_{\rho\sigma}] = \eta_{\mu\rho}\bar{\Pi}_\sigma - \eta_{\mu\sigma}\bar{\Pi}_\rho;\tag{37.46}$$

$$[\bar{\Pi}_\mu, \bar{\Pi}_\rho] = 16\,l^2 L_{\mu\rho}.\tag{37.47}$$

The contraction limit $l \to 0$

In the contraction limit $l \to 0$, the generators $\bar{\Pi}_\mu$ reduce to (minus) the proper conformal generators:

$$\bar{\Pi}_\mu \to -K_\mu. \tag{37.48}$$

Accordingly, the de Sitter group $SO(4,1)$ contracts to the *conformal* Poincaré group[9]

$$\bar{\mathcal{P}} = \mathcal{L} \oslash \bar{\mathcal{T}},$$

the semi-direct product between the Lorentz $\mathcal{L}$ and the proper conformal group $\bar{\mathcal{T}}$, whose Lie algebra is

$$[L_{\mu\nu}, L_{\lambda\rho}] = \eta_{\nu\lambda} L_{\mu\rho} + \eta_{\mu\rho} L_{\nu\lambda} - \eta_{\nu\rho} L_{\mu\lambda} - \eta_{\mu\lambda} L_{\nu\rho}; \tag{37.49}$$

$$[K_\mu, L_{\lambda\rho}] = \eta_{\mu\lambda} K_\rho - \eta_{\mu\rho} K_\lambda; \tag{37.50}$$

$$[K_\mu, K_\lambda] = 0. \tag{37.51}$$

Concomitant with the group contraction, on account of their quotient character, the de Sitter spacetime dS reduces to the homogeneous space $\bar{M}$

$$dS \to \bar{M} = \bar{\mathcal{P}}/\mathcal{L}. \tag{37.52}$$

The kinematic group $\bar{\mathcal{P}}$, like the Poincaré group, has the Lorentz group $\mathcal{L}$ as the subgroup accounting for the spacetime isotropy. The homogeneity, however, is completely different: instead of ordinary translations, all points of $\bar{M}$ are equivalent under special conformal transformations. In other words, the point-set of $\bar{M}$ is that determined by special conformal transformations.

In the limit $l \to 0$, the de Sitter metric (37.35) assumes the conformal invariant form

$$\bar{g}_{\mu\nu} \to \bar{\eta}_{\mu\nu} = \sigma^{-4} \eta_{\mu\nu}, \tag{37.53}$$

which is the metric on $\bar{M}$. The Christoffel connection, on the other hand, reduces to

$$\bar{\Gamma}^\lambda{}_{\mu\nu} \to {}^0\bar{\Gamma}^\lambda{}_{\mu\nu} = -2\sigma^{-2} \left(\delta^\lambda_\mu \eta_{\nu\alpha} x^\alpha + \delta^\lambda_\nu \eta_{\mu\alpha} x^\alpha - \eta_{\mu\nu} x^\lambda \right). \tag{37.54}$$

The corresponding Riemann, Ricci, and scalar curvatures vanish identically:

$$\bar{R}^\mu{}_{\nu\rho\sigma} \to {}^0\bar{R}^\mu{}_{\nu\rho\sigma} = 0, \quad \bar{R}_{\nu\sigma} \to {}^0\bar{R}_{\nu\sigma} = 0, \quad \bar{R} \to {}^0\bar{R} = 0. \tag{37.55}$$

The cosmological term, however, goes to infinity:

$$\Lambda \to \infty. \tag{37.56}$$

---

[9] Aldrovandi & Pereira 1998.

From these properties one can infer that $\bar{M}$ is a singular, four-dimensional cone spacetime (see Figure 37.1), transitive under proper conformal transformations.[10]

It is important to remark that the contraction limit $l \to 0$ is not continuous — actually a general property of any group contraction. This can be seen by observing that, whereas the de Sitter group is semi-simple, the conformal Poincaré group is not, which means that it is not possible to continuously deform the former into the latter. It is this singular character that produces a decoupling between the scalar curvature and the cosmological term: whereas the former goes to zero, the latter goes to infinity.

37.3 Some properties of the cone spacetime $\bar{M}$

As is well-known, in $(3+1)$ dimensions there are three homogeneous spaces: Minkowski, de Sitter and anti-de Sitter. The cone spacetime $\bar{M}$ is an additional four-dimensional maximally symmetric spacetime, with the conformal Poincaré group $\bar{P}$ as kinematic group, which is however singular. In this section we explore some of its properties.

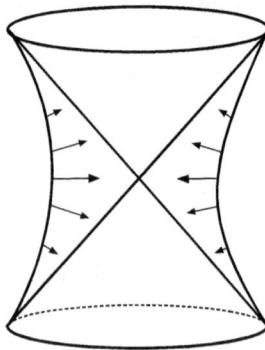

Fig. 37.1   *Pictorial view of the de Sitter spacetime in the contraction limit $l \to 0$, showing its deformation into a four-dimensional cone spacetime.*

---

[10] Aldrovandi, Beltrán Almeida & Pereira 2006.

37.3.1 Geometric relation between M and $\bar{M}$

Under the spacetime inversion

$$x^\mu \to -\frac{x^\mu}{\sigma^2} \qquad (37.57)$$

the translation generators are led into the proper conformal transformations, and vice versa[11]:

$$P_\mu \to K_\mu \qquad \text{and} \qquad K_\mu \to P_\mu. \qquad (37.58)$$

The Lorentz generators, on the other hand, remain unchanged:

$$L_{\mu\nu} \to L_{\mu\nu}. \qquad (37.59)$$

This means that, under such inversion, the Poincaré group $\mathcal{P}$ is led to the conformal Poincaré group $\bar{\mathcal{P}}$, and vice versa. Concomitantly, Minkowski M is transformed into the four-dimensional cone-spacetime $\bar{M}$, and vice versa. The corresponding spacetime metrics are also transformed into each other:

$$\eta_{\mu\nu} \to \bar{\eta}_{\mu\nu} = \sigma^{-4}\eta_{\mu\nu} \qquad \text{and} \qquad \bar{\eta}_{\mu\nu} = \sigma^{-4}\eta_{\mu\nu} \to \eta_{\mu\nu}. \qquad (37.60)$$

Minkowski and the cone spacetimes can be considered a kind of *dual* to each other, in the sense that their geometries are determined, respectively, by a vanishing and an infinite cosmological term. Observe that, in the limit $l \to 0$, the stereographic projection (37.31)–(37.32) reduces to the spacetime inversion (37.57).

37.3.2 The nature of the cone singularity

As shown in Section 37.2.2, the cone spacetime $\bar{M}$ has vanishing Riemann, Ricci and scalar curvature tensors. The cosmological term, on the other hand, is infinite, pointing to a singular spacetime. In fact, its metric tensor $\bar{\eta}_{\mu\nu}$, given by Eq. (37.53), is singular at the origin, located at $x^\mu = 0$, in which case $\sigma^2 = 0$. However, if we perform a conformal re-scaling of the metric

$$\bar{\eta}_{\mu\nu} \to \bar{\eta}'_{\mu\nu} = \omega^2(x)\,\bar{\eta}_{\mu\nu}, \qquad (37.61)$$

with the conformal factor given by

$$\omega^2(x) = \sigma^4\,\alpha^2(x), \qquad (37.62)$$

the resulting metric tensor

$$\bar{\eta}'_{\mu\nu} = \alpha^2(x)\,\bar{\eta}_{\mu\nu} \qquad (37.63)$$

is no longer singular. This kind of singularity, in which the metric is singular but the conformal equivalence class of the metric is not, is called a *conformal gauge singularity*.[12]

---

[11] Coleman 1985, p. 75.
[12] Tod & Luebbe 2006.

37.3.3 Transitivity and the notions of distance and time

Minkowski and de Sitter are both isotropic and homogeneous spaces, but their homogeneity properties differ substantially: whereas Minkowski is transitive under translations, de Sitter is transitive under a combination of translations and proper conformal transformations. Transitivity is intimately related to the notions of space distance and time interval. Since any two points of Minkowski spacetime are connected by a spacetime translation, the corresponding notions of space distance and time interval will be essentially translational.

On the other hand, the cone spacetime $\bar{M}$ is transitive under proper conformal transformations, which means that our conventional (translational) notions of space and time do not exist on $\bar{M}$, only the conformal notions exist.[13] As a consequence, local clocks cannot be defined, a result that is somehow in agreement with the general idea that (our conventional notion of) time should not exist at (or beyond) the Planck scale.[14] In fact, the cone spacetime metric $\bar{\eta}_{\mu\nu}$ does not have the conventional meaning of a physical metric. It is not dimensionless, and the interval it defines,

$$d\bar{s}^2 = \sigma^{-4}\,\eta_{\mu\nu}\,dx^\mu dx^\nu, \tag{37.64}$$

has to do with proper conformal distances, not with our usual notion of translational distances.

37.4 de Sitter transformations and Killing vectors

In terms of the five-dimensional ambient space coordinates, an infinitesimal de Sitter transformation is written as

$$\delta\chi^A = \tfrac{1}{2}\,\epsilon^{CD} L_{CD}\,\chi^A\,, \tag{37.65}$$

where $\epsilon^{CD} = -\epsilon^{DC}$ are the transformation parameters, and

$$L_{CD} = \eta_{CE}\,\chi^E\,\partial/\partial\chi^D - \eta_{DE}\,\chi^E\,\partial/\partial\chi^C \tag{37.66}$$

are the de Sitter generators. In terms of stereographic coordinates, it assumes the form

$$\delta x^\mu \equiv \delta_L x^\mu + \delta_\Pi x^\mu = \tfrac{1}{2}\epsilon^{\rho\sigma} L_{\rho\sigma} x^\mu + \epsilon^{4\rho} L_{4\rho} x^\mu, \tag{37.67}$$

where

$$L_{\rho\sigma} = \eta_{\rho\lambda}\,x^\lambda\,P_\sigma - \eta_{\sigma\lambda}\,x^\lambda\,P_\rho \tag{37.68}$$

are the Lorentz generators, and $L_{4\rho}$ are the generators that define the transitivity on de Sitter spacetime.

---

[13] Isham 1992; Kuchař 2011.
[14] Rovelli 2003.

Large values of l

For large values of l, it is convenient to introduce the so-called de Sitter "translation" generators

$$\Pi_\rho \equiv \frac{L_{4\rho}}{l} = P_\rho - \frac{1}{4l^2}\,K_\rho. \tag{37.69}$$

In this case a de Sitter transformation assumes the form

$$\delta x^\mu \equiv \delta_L x^\mu + \delta_\Pi x^\mu = \tfrac{1}{2}\epsilon^{\rho\sigma}L_{\rho\sigma}x^\mu + \epsilon^\rho\Pi_\rho x^\mu, \tag{37.70}$$

where $\epsilon^{\rho\sigma}$ and $\epsilon^\rho \equiv l\,\epsilon^{4\rho}$ are the transformation parameters. Substituting the generators (37.68) and (37.69) in the transformation (37.70), the Lorentz transformation acquires the form

$$\delta_L x^\mu = \tfrac{1}{2}\,\xi^\mu_{\rho\sigma}\,\epsilon^{\rho\sigma}, \tag{37.71}$$

where

$$\xi^\mu_{\rho\sigma} = \left(\eta_{\rho\lambda}\,\delta^\mu_\sigma - \eta_{\sigma\lambda}\,\delta^\mu_\rho\right)x^\lambda \tag{37.72}$$

are the Killing vectors of the Lorentz group. The de Sitter "translation", on the other hand, becomes

$$\delta_\Pi x^\mu = \xi^\mu_\rho\,\epsilon^\rho, \tag{37.73}$$

where

$$\xi^\mu_\rho = \delta^\mu_\rho - \frac{1}{4l^2}\,\bar\delta^\mu_\rho \tag{37.74}$$

are the corresponding Killing vectors, with δ^μ_ρ the Killing vector of translations, and

$$\bar\delta^\mu_\rho = 2\eta_{\rho\nu}\,x^\nu x^\mu - \sigma^2\delta^\mu_\rho \tag{37.75}$$

the Killing vectors of proper conformal transformations. In the contraction limit $l \to \infty$, they reduce to the transformations of the Poincaré group:

$$\delta_L x^\mu = \tfrac{1}{2}\,\xi^\mu_{\rho\sigma}\,\epsilon^{\rho\sigma} \quad \text{and} \quad \delta_P x^\mu = \delta^\mu_\rho\,\epsilon^\rho. \tag{37.76}$$

Of course, the ten Killing vectors $\xi^{(\rho\sigma)}_\mu$ and $\xi^{(\rho)}_\mu$ of the de Sitter spacetime satisfy the Killing equations

$$-\nabla_\nu\xi^{(\rho\sigma)}_\mu - \nabla_\mu\xi^{(\rho\sigma)}_\nu = 0 \tag{37.77}$$

and

$$-\nabla_\nu\xi^{(\rho)}_\mu - \nabla_\mu\xi^{(\rho)}_\nu = 0 \tag{37.78}$$

with ∇_ν a covariant derivative in the connection (37.12) of the de Sitter metric.

Small values of l

For small values of l, it is convenient to redefine the de Sitter "translation" generators in the form

$$\bar{\Pi}_\rho \equiv 4l\, L_{4\rho} = 4l^2 P_\rho - K_\rho. \tag{37.79}$$

In this case a de Sitter transformation assumes the form

$$\delta x^\mu \equiv \delta_L x^\mu + \delta_{\bar{\Pi}} x^\mu = \tfrac{1}{2} \epsilon^{\rho\sigma} L_{\rho\sigma} x^\mu + \epsilon^\rho \bar{\Pi}_\rho x^\mu \tag{37.80}$$

with $\epsilon^{\rho\sigma}$ and $\epsilon^\rho \equiv \epsilon^{4\rho}/4l$ the transformation parameters. Substituting the generators (37.68) and (37.79) in the transformation (37.80), the Lorentz transformation remains given by

$$\delta_L x^\mu = \tfrac{1}{2}\, \xi^\mu_{\rho\sigma}\, \epsilon^{\rho\sigma} \tag{37.81}$$

with $\xi^\mu_{\rho\sigma}$ the Killing vectors (37.72) of the Lorentz group. The de Sitter "translation", on the other hand, becomes

$$\delta_{\bar{\Pi}} x^\mu = \bar{\xi}^\mu_\rho\, \epsilon^\rho \tag{37.82}$$

where

$$\bar{\xi}^\mu_\rho = 4l^2\, \delta^\mu_\rho - \bar{\delta}^\mu_\rho \tag{37.83}$$

are the corresponding Killing vectors. In the contraction limit $l \to 0$, they reduce to the transformations of the conformal Poincaré group:

$$\delta_L x^\mu = \tfrac{1}{2}\, \xi^\mu_{\rho\sigma}\, \epsilon^{\rho\sigma} \quad \text{and} \quad \delta_{\bar{P}} x^\mu = \bar{\delta}^\mu_\rho\, \epsilon^\rho. \tag{37.84}$$

Of course, the ten Killing vectors $\xi^{(\rho\sigma)}_\mu$ and $\xi^{(\rho)}_\mu$ of the de Sitter spacetime satisfy the Killing equations

$$-\bar{\nabla}_\nu \xi^{(\rho\sigma)}_\mu - \bar{\nabla}_\mu \xi^{(\rho\sigma)}_\nu = 0 \tag{37.85}$$

and

$$-\bar{\nabla}_\nu \bar{\xi}^{(\rho)}_\mu - \bar{\nabla}_\mu \bar{\xi}^{(\rho)}_\nu = 0 \tag{37.86}$$

with $\bar{\nabla}_\nu$ a covariant derivative in the connection (37.36) of the de Sitter metric.

37.5 Transitivity and the notion of motion

Transitivity — the property that specifies how one moves from one point
to another in a given spacetime — is intimately related to the notion of
motion. For example, any two points of Minkowski spacetime are connected
by a spacetime translation. As a consequence, motion in this spacetime
is described by trajectories whose points are connected to each other by
ordinary translations. On the other hand, any two points of a de Sitter
spacetime are connected to each other by a combination of translation and
proper conformal transformations — the so-called de Sitter "translations".
As a consequence, motion in this spacetime will be described by trajectories
whose points are connected to each other by a combination of translation
and proper conformal transformations. However, in the usual procedure to
obtain the geodesics of the de Sitter spacetime,

$$\frac{du_\rho}{ds} - \Gamma^\nu{}_{\rho\gamma}\, u_\nu\, u^\gamma = 0, \tag{37.87}$$

with $\Gamma^\nu{}_{\rho\gamma}$ the de Sitter Christoffel connection (37.12), the homogeneity
properties of the de Sitter spacetime are not taken into account in the
variational principle. As a consequence, there are points in this spacetime
which are not connected by any one of these geodesics.[15]

It should be remarked that the trajectories described by the solutions
to the geodesic equation (37.87) are able to connect all points of the de Sitter spacetime separated by a timelike distance.[16] They do not represent,
however, the most general trajectories of the de Sitter spacetime. By taking into account the transitivity properties of the de Sitter spacetime, we
obtain now these general trajectories. As in the previous sections, we consider separately the cases of large and small values of the de Sitter length
parameter l.

Large values of l

In the parameterization appropriate to large values of l, the action functional for a particle of mass m is given by

$$S = -mc \int_a^b ds, \tag{37.88}$$

---

[15] Hawking & Ellis 1973.
[16] Pérez–Nadal, Roura & Verdaguer 2010.

with $ds = (g_{\alpha\beta} \, dx^\alpha dx^\beta)^{1/2}$. In the case of de Sitter spacetime, a variation of the spacetime coordinates has the form

$$\delta_\Pi x^\beta = \xi^\beta_\rho \, \delta x^\rho \tag{37.89}$$

where ξ^β_ρ represent the de Sitter Killing vector (37.74), and $\delta x^\rho = \epsilon^\rho(x)$ is an ordinary (translational) parameter. Accordingly, the action variation assumes the form

$$\delta S = -mc \int_b^a \left[\tfrac{1}{2} \partial_\beta(g_{\alpha\gamma}) \, u^\alpha dx^\gamma \, \delta_\Pi x^\beta + g_{\alpha\beta} \, u^\alpha \, \delta_\Pi(dx^\beta) \right] \tag{37.90}$$

with $u^\alpha = dx^\alpha/ds$ the ordinary particle four-velocity. Using the identity

$$\delta_\Pi(dx^\beta) = d(\delta_\Pi x^\beta) \tag{37.91}$$

in the last term, we get

$$\delta S = -mc \int_b^a \left[\tfrac{1}{2} \partial_\beta(g_{\alpha\gamma}) \, u^\alpha dx^\gamma \, \delta_\Pi x^\beta + u_\beta \, d(\delta_\Pi x^\beta) \right] . \tag{37.92}$$

Integrating the second term by parts, and considering that $\delta_\Pi x^\beta$ vanishes at the extrema of integration, the variation becomes

$$\delta S = -mc \int_b^a \left[\frac{1}{2} \xi^\beta_\rho \frac{\partial g_{\alpha\gamma}}{\partial x^\beta} \, u^\alpha u^\gamma - \xi^\beta_\rho \frac{du_\beta}{ds} \right] \epsilon^\rho(x) \, ds. \tag{37.93}$$

After some algebraic manipulation, this can be rewritten in the form

$$\delta S = mc \int_b^a \left[u^\gamma \nabla_\gamma u^\beta \, \xi^\rho_\beta \right] \epsilon_\rho(x) \, ds. \tag{37.94}$$

Using the Killing equations (37.78), we arrive at

$$\delta S = mc \int_b^a u^\gamma \nabla_\gamma (\xi^\rho_\beta \, u^\beta) \, \epsilon_\rho(x) \, ds. \tag{37.95}$$

Taking into account the arbitrariness of $\epsilon_\rho(x)$, the invariance of the action yields the equation of motion[17]

$$\frac{d\vartheta^\rho}{ds} - \Gamma^\rho{}_{\mu\gamma} \, \vartheta^\mu \, u^\gamma = 0, \tag{37.96}$$

where

$$\vartheta^\rho \equiv \xi^\rho_\beta \, u^\beta = \left[\delta^\rho_\beta - \frac{1}{4l^2} \left(2g_{\beta\nu} \, x^\nu x^\rho - x^2 \delta^\rho_\beta \right) \right] u^\beta \tag{37.97}$$

is an anholonomic four-velocity, which takes into account the translational and the proper conformal 'degrees of freedom' of the de Sitter spacetime. As a consequence, the corresponding trajectories include both notions of motion: translational and proper conformal. They are, consequently, able to connect any two points of de Sitter spacetime. Defining the de Sitter four-momentum

$$\pi^\mu = mc \, \vartheta^\mu, \tag{37.98}$$

the equation of motion (37.96) assumes the form

$$\frac{d\pi^\mu}{ds} + \Gamma^\mu{}_{\nu\lambda} \, \pi^\nu \, u^\lambda = 0, \tag{37.99}$$

which is the conservation law of the de Sitter four-momentum.

---

[17] Pereira & Sampson 2012.

Small values of *l*

In the parameterization appropriate for small values of l, the action functional for a particle of mass m in a de Sitter spacetime is

$$\bar{S} = -mc \int_a^b d\bar{s}, \tag{37.100}$$

with $d\bar{s} = (\bar{g}_{\alpha\beta} \, dx^\alpha dx^\beta)^{1/2}$. In this spacetime, a variation of the spacetime coordinates has the form

$$\delta_{\bar{\Pi}} x^\beta = \bar{\xi}_\rho^\beta \, \delta x^\rho, \tag{37.101}$$

where $\bar{\xi}_\rho^\beta$ is the de Sitter Killing vector (37.83), and $\delta x^\rho = \epsilon^\rho(x)$ is an ordinary (translational) parameter. The corresponding action variation is

$$\delta\bar{S} = -mc \int_b^a \left[\tfrac{1}{2} \partial_\beta(\bar{g}_{\alpha\gamma}) \, \bar{u}^\alpha dx^\gamma \, \delta_{\bar{\Pi}} x^\beta + \bar{g}_{\alpha\beta} \, u^\alpha \, \delta_{\bar{\Pi}}(dx^\beta) \right], \tag{37.102}$$

with $\bar{u}^\alpha = dx^\alpha/d\bar{s}$ the particle four-velocity. Using the identity

$$\delta_{\bar{\Pi}}(dx^\beta) = d(\delta_{\bar{\Pi}} x^\beta) \tag{37.103}$$

in the last term, we get

$$\delta\bar{S} = -mc \int_b^a \left[\tfrac{1}{2} \partial_\beta(\bar{g}_{\alpha\gamma}) \, \bar{u}^\alpha dx^\gamma \, \delta_{\bar{\Pi}} x^\beta + \bar{u}_\beta \, d(\delta_{\bar{\Pi}} x^\beta) \right]. \tag{37.104}$$

Integrating the second term by parts, and considering that $\delta_{\bar{\Pi}} x^\beta$ vanishes at the extrema of integration, the variation becomes

$$\delta\bar{S} = -mc \int_b^a \left[\frac{1}{2} \bar{\xi}_\rho^\beta \frac{\partial \bar{g}_{\alpha\gamma}}{\partial x^\beta} \bar{u}^\alpha \bar{u}^\gamma - \bar{\xi}_\rho^\beta \frac{d\bar{u}_\beta}{d\bar{s}} \right] \epsilon^\rho(x) \, d\bar{s}. \tag{37.105}$$

After some algebraic manipulation, it can be rewritten in the form

$$\delta\bar{S} = mc \int_b^a \left[\bar{u}^\gamma \bar{\nabla}_\gamma \bar{u}^\beta \, \bar{\xi}_\beta^\rho \right] \epsilon_\rho(x) \, d\bar{s} \tag{37.106}$$

with $\bar{\nabla}_\gamma$ a covariant derivative in the connection (37.36). Using the Killing equations (37.86), we arrive at

$$\delta\bar{S} = mc \int_b^a \bar{u}^\gamma \bar{\nabla}_\gamma(\bar{\xi}_\beta^\rho \, \bar{u}^\beta) \, \epsilon_\rho(x) \, d\bar{s}. \tag{37.107}$$

Taking into account the arbitrariness of $\epsilon_\rho(x)$, the invariance of the action yields the equation of motion

$$\frac{d\bar{\vartheta}^\rho}{d\bar{s}} - \bar{\Gamma}^\rho{}_{\mu\gamma} \, \bar{\vartheta}^\mu \, \bar{u}^\gamma = 0, \tag{37.108}$$

where

$$\bar{\vartheta}^\rho \equiv \bar{\xi}^\rho_\beta \, \bar{u}^\beta = \left[4l^2 \delta^\rho_\beta - (2\bar{g}_{\beta\nu} \, x^\nu x^\rho - \bar{x}^2 \delta^\rho_\beta) \right] u^\beta \qquad (37.109)$$

is an anholonomic four-velocity, which takes into account the translational and the proper conformal 'degrees of freedom' of the de Sitter spacetime. As a consequence, the corresponding trajectories include both notions of motion: translational and proper conformal. They are, consequently, able to connect any two points of de Sitter spacetime. Defining the de Sitter four-momentum

$$\bar{\pi}^\mu = mc \, \bar{\vartheta}^\mu, \qquad (37.110)$$

the equation of motion (37.108) assumes the form

$$\frac{d\bar{\pi}^\mu}{d\bar{s}} + \bar{\Gamma}^\mu{}_{\nu\lambda} \, \bar{\pi}^\nu \, \bar{u}^\lambda = 0, \qquad (37.111)$$

which is the conservation law of the de Sitter four-momentum.

37.6 Dispersion relation of de Sitter relativity

We proceed now to obtain the dispersion relation of a de Sitter invariant special relativity. For the sake of completeness, we obtain first the usual dispersion relation of ordinary special relativity.

37.6.1 *Ordinary special relativity revisited*

The dispersion relation of ordinary special relativity can be obtained from the first Casimir operator of the Poincaré group, which is the kinematic group of Minkowski spacetime. As is well known, it is given by

$$\hat{C}_P = \eta^{\mu\nu} P_\mu P_\nu, \qquad (37.112)$$

with $P_\mu = \partial_\mu$ the translation generators. Its eigenvalue c_P, on the other hand, is

$$c_P = -\frac{m^2 c^2}{\hbar^2}. \qquad (37.113)$$

From the identity $\hat{C}_P = c_P$, we obtain

$$\eta^{\mu\nu} P_\mu P_\nu = -\frac{m^2 c^2}{\hbar^2}. \qquad (37.114)$$

Applied to a scalar field ϕ, it yields the Klein–Gordon equation

$$\Box \phi + \frac{m^2 c^2}{\hbar^2} \phi = 0, \qquad (37.115)$$

with □ the Minkowski spacetime d'Alembertian operator.

The dispersion relation of ordinary special relativity can be obtained from (37.114) by replacing

$$P_\mu \to -\frac{i}{\hbar} p_\mu \tag{37.116}$$

with $p_\mu = mcu_\mu$ the four-momentum of a particle of mass m and four-velocity u_μ. As can be easily verified, the result is

$$\eta_{\mu\nu} p^\mu p^\nu = m^2 c^2, \tag{37.117}$$

which is the ordinary special relativity dispersion relation.

37.6.2   de Sitter special relativity

In order to obtain the dispersion relation of a de Sitter-ruled kinematics, it is necessary to consider again different parameterisations for large and small values of the de Sitter length parameter l.

Large values of l

For large values of the de Sitter length parameter l, the first Casimir invariant operator of the de Sitter group is written in the form[18]

$$\hat{\mathcal{C}}_{dS} = -\frac{1}{2l^2} \eta^{AC} \eta^{BD} J_{AB} J_{CD}, \tag{37.118}$$

where

$$J_{AB} = L_{AB} + S_{AB}, \tag{37.119}$$

with L_{AB} the orbital generators (37.5), and S_{AB} the matrix spin generators. For the sake of simplicity, we assume the case of spinless particles, in which case the Casimir operator reduces to

$$\hat{\mathcal{C}}_{dS} = -\frac{1}{2l^2} \eta^{AC} \eta^{BD} L_{AB} L_{CD}. \tag{37.120}$$

In terms of stereographic coordinates, it assumes the form

$$\hat{\mathcal{C}}_{dS} = \eta^{\alpha\beta} \Pi_\alpha \Pi_\beta - \frac{1}{2l^2} \eta^{\alpha\beta} \eta^{\gamma\delta} L_{\alpha\gamma} L_{\beta\delta}. \tag{37.121}$$

with Π_α the generators (37.69). Through a lengthy, but otherwise straightforward computation, one can verify that

$$\hat{\mathcal{C}}_{dS} = g^{\mu\nu} \nabla_\mu \nabla_\nu \equiv \square \tag{37.122}$$

---

[18] Gürsey 1962.

with $\Box$ the Laplace–Beltrami operator in the de Sitter metric (37.11).

On the other hand, for representations of the *principal series*, the eigenvalues of the Casimir operator are given by[19]

$$c_{dS} = -\frac{m^2 c^2}{\hbar^2} - \frac{1}{l^2}\left[\mathsf{s}(\mathsf{s}+1) - 2\right], \tag{37.123}$$

with s the spin of the particle under consideration. For the case of spinless particles ($\mathsf{s} = 0$), they assume the form

$$c_{dS} = -\frac{m^2 c^2}{\hbar^2} + \frac{2}{l^2}. \tag{37.124}$$

From the identity $\hat{\mathcal{C}}_{dS} = c_{dS}$, we obtain

$$g^{\mu\nu}\nabla_\mu\nabla_\nu = -\frac{m^2 c^2}{\hbar^2} + \frac{2}{l^2}. \tag{37.125}$$

Commentary 37.3 Applied to a scalar field ϕ, it yields the Klein–Gordon equation in a de Sitter spacetime

$$\Box\phi + \frac{m^2 c^2}{\hbar^2}\phi - \frac{1}{6}R\phi = 0 \tag{37.126}$$

where $R = 12/l^2$ is the scalar curvature of the de Sitter spacetime [see Eq. (37.14)]. For $m = 0$, it is the conformal invariant equation for a scalar field. ◂

Similarly to ordinary special relativity, the dispersion relation of the de Sitter special relativity can be obtained from (37.125) by replacing

$$\nabla_\mu \rightarrow -\frac{i}{\hbar}\pi_\mu \tag{37.127}$$

with $\pi_\mu = \xi_\mu^\alpha p_\alpha$ the de Sitter four-momentum of a particle of mass m. As an easy computation shows, the dispersion relation of de Sitter special relativity is found to be

$$g_{\mu\nu}\pi^\mu\pi^\nu = m^2 c^2 - \frac{2\hbar^2}{l^2}. \tag{37.128}$$

In the limit $l \to \infty$, the de Sitter spacetime contracts to Minkowski, and it reduces to the dispersion relation (37.117) of ordinary special relativity.

Small values of l

For small values of the de Sitter length parameter l, and considering the case of spinless particles, the first Casimir invariant operator of the de Sitter group is written in the form

$$\hat{\mathcal{C}}_{dS} = -8\,\mathsf{s}l^2\,\eta^{AC}\,\eta^{BD}\,L_{AB}L_{CD}, \tag{37.129}$$

---

[19] Dixmier 1961.

with L_{AB} the orbital generators (37.5). In terms of stereographic coordinates, it assumes the form

$$\hat{\mathcal{C}}_{dS} = \eta^{\alpha\beta}\,\bar{\Pi}_\alpha\,\bar{\Pi}_\beta - 8l^2\,\eta^{\alpha\beta}\,\eta^{\gamma\delta}\,L_{\alpha\gamma}\,L_{\beta\delta}\,, \tag{37.130}$$

with $\bar{\Pi}_\alpha$ the generators (37.79). Through a lengthy, but straightforward computation, one can verify that

$$\hat{\mathcal{C}}_{dS} = \bar{g}^{\mu\nu}\bar{\nabla}_\mu\bar{\nabla}_\nu \equiv \square \tag{37.131}$$

with $\square$ the Laplace–Beltrami operator in the de Sitter metric (37.35). We remark that, since the Levi–Civita connection is the same for the metrics $g^{\mu\nu}$ and $\bar{g}^{\mu\nu}$ [see Eq. (37.36)], then we have $\nabla_\mu = \bar{\nabla}_\mu$.

On the other hand, for representations of the *principal series*, the eigenvalues of the Casimir operator are given by

$$c_{dS} = -\frac{16l^4 m^2 c^2}{\hbar^2} - 16l^2\,[s(s+1)-2]\,, \tag{37.132}$$

with s the spin of the particle under consideration. For the case of spinless particles ($s = 0$), they assume the form

$$c_{dS} = -\frac{16l^4 m^2 c^2}{\hbar^2} + 32l^2. \tag{37.133}$$

From the identity $\hat{\mathcal{C}}_{dS} = c_{dS}$, we obtain

$$\bar{g}^{\mu\nu}\bar{\nabla}_\mu\bar{\nabla}_\nu = -\frac{16l^4 m^2 c^2}{\hbar^2} + 32l^2. \tag{37.134}$$

Commentary 37.4 Applied to a scalar field ϕ, it yields the Laplace equation in de Sitter spacetime

$$\square\phi + \frac{16l^4 m^2 c^2}{\hbar^2}\phi - \frac{1}{6}\bar{R}\phi = 0 \tag{37.135}$$

where $\bar{R} = 192l^2$ is the scalar curvature of the de Sitter spacetime [see Eq. (37.39)]. For the massless case ($m = 0$), it is the conformal invariant equation for a scalar field. ◀

The dispersion relation of the de Sitter special relativity for small values of l can be obtained from (37.134) by replacing

$$\bar{\nabla}_\mu \to -\frac{i}{\hbar}\,\bar{\pi}_\mu \tag{37.136}$$

with $\bar{\pi}_\mu = \bar{\xi}_\mu^\alpha p_\alpha$ the de Sitter four-momentum of a particle of mass m. As an easy computation shows, the dispersion relation of de Sitter special relativity is then found to be

$$\bar{g}_{\mu\nu}\,\bar{\pi}^\mu\,\bar{\pi}^\nu = 16l^4 m^2 c^2 - 32l^2\hbar^2. \tag{37.137}$$

In the contraction limit $l \to 0$, the de Sitter spacetime contracts to the four-dimensional conic spacetime, and it reduces to

$$\bar{\eta}_{\mu\nu} \, k^\mu \, k^\nu = 0 \qquad (37.138)$$

with $\bar{\eta}_{\mu\nu}$ the cone spacetime metric (37.53), and k^μ the proper conformal four-current. This is the dispersion relation valid in the cone spacetime $\bar{M}$.

It should be remarked that $l \to 0$ is just a formal limit, in the sense that quantum effects preclude it to be fully performed. It is actually a contraction limit, on an equal footing with the classical contraction limit in which the speed of light goes to infinity $c \to \infty$. The cone spacetime $\bar{M}$, which emerges as the output of this limit, should then be thought of as a kind of *frozen geometric structure* behind the spacetime quantum fluctuations taking place at the Planck scale. The dispersion relation (37.138) holds in such idealized frozen spacetime.

37.7   Gravitation in Cartan geometry

The de Sitter spacetime is usually interpreted as the simplest *dynamical* solution of the sourceless Einstein equation in the presence of a cosmological constant, standing on an equal footing with all other gravitational solutions — like, for example, the Schwarzschild or Kerr solutions. However, as a non-gravitational spacetime, the de Sitter solution should instead be interpreted as a fundamental background for the construction of physical theories, standing on an equal footing with the Minkowski solution. When General Relativity is constructed on a de Sitter space, spacetime will no longer present a riemannian structure, but will be described by a more general structure called Cartan geometry.[20] As a matter of fact, it will be described by a Cartan geometry that reduces locally to de Sitter — and for this reason called de Sitter–Cartan geometry.[21] Accordingly, in a locally inertial frame, where inertial effects exactly compensate for gravitation, the spacetime metric reduces to the de Sitter metric. One should note that such construction does not change the dynamics of the gravitational field. The only change will be in the strong equivalence principle, which passes to state that *in a locally inertial frame, where gravitation goes unnoticed, the laws of physics reduce to those of de Sitter invariant special relativity.*

---

[20] Sharpe 1997.
[21] Wise 2010.

37.7.1 General Relativity in locally-de Sitter spacetimes

In a de Sitter–Cartan geometry, de Sitter represents the underlying space-time, which defines the local kinematics. In a more geometric language, the usual tangent bundle, with the tangent Minkowski space at each spacetime point playing the role of fiber, is replaced by a generalised "tangent" bundle in which the fiber is an osculating de Sitter spacetime. We could then talk of an *osculating bundle*.

As repeatedly said, in order to study such geometry it is necessary to consider separately the cases of large and small values of the de Sitter length parameter l. We begin by considering the case of large values.

Large values of l

In a spacetime that reduces locally to de Sitter, a local diffeomorphism for large values of l is defined by

$$\delta_\Pi x^\mu = \xi^\mu_\rho \, \epsilon^\rho(x),\qquad(37.139)$$

with ξ^μ_ρ the de Sitter "translation" Killing vectors (37.74). When the space-time itself is de Sitter, it reduces to the transformation (37.73). Let us then consider the action integral of a general matter field

$$S_m = \frac{1}{c} \int \mathcal{L}_m \, \sqrt{-g} \, d^4x,\qquad(37.140)$$

with $\mathcal{L}_m$ the corresponding lagrangian density. Invariance of this action under the transformation (37.139) yields, through Noether's theorem, the conservation law[22]

$$\nabla_\mu \Pi^{\rho\mu} = 0,\qquad(37.141)$$

where the conserved current has the form

$$\Pi^{\rho\mu} \equiv \xi^\rho_\alpha \, T^{\alpha\mu} = T^{\rho\mu} - (2l)^{-2} \, K^{\rho\mu},\qquad(37.142)$$

with $T^{\rho\mu}$ the symmetric energy-momentum current and

$$K^{\rho\mu} = \left(2\eta_{\alpha\gamma} \, x^\gamma x^\rho - \sigma^2 \delta^\rho_\alpha\right) T^{\alpha\mu}\qquad(37.143)$$

the proper conformal current.[23] The current $\Pi^{\rho\mu}$ represents a generalised notion of energy-momentum, which is consistent with the local transitivity of spacetime.

---

[22] Pereira & Sampson 2012.
[23] Coleman 1985.

It is important to observe that, in locally-de Sitter spacetimes, ordinary energy-momentum $T^{\alpha\mu}$ is allowed to transform into proper conformal current $K^{\alpha\mu}$ and vice versa, while keeping the total current co conserved. This is an important additional freedom, which is not present in locally-Minkowski spacetimes. In fact, in the formal limit $l \to \infty$, the underlying de Sitter spacetime contracts to Minkowski, and the de Sitter Killing vectors ξ_ρ^μ reduce to the translation Killing vectors δ_ρ^μ. Concomitantly, the diffeomorphism assumes the usual form of riemannian spacetimes

$$\delta_P x^\mu = \delta_\rho^\mu \, \epsilon^\rho(x), \tag{37.144}$$

and we recover the ordinary conservation law of locally-Minkowski spacetimes

$$\nabla_\mu T^{\rho\mu} = 0, \tag{37.145}$$

where no proper conformal current appears.

The question then arises: what is the form of Einstein equation in a locally-de Sitter spacetime? To answer this question we are going to use the same method used by Einstein to obtain the gravitational field equations (see Section 34.9). Combining the contracted form of the the second Bianchi identity

$$\nabla_\mu\left(R^\mu{}_\nu - \tfrac{1}{2}\,\delta_\nu^\mu R\right) = 0 \tag{37.146}$$

with the covariant conservation law (37.141), the gravitational field equation turns out to be written in the form

$$R_{\mu\nu} - \tfrac{1}{2}\,g_{\mu\nu} R = \frac{8\pi G}{c^4}\,\Pi_{\mu\nu}. \tag{37.147}$$

It should be noted that, since in this theory the cosmological term Λ is encoded in the underlying local kinematics, it does not appear explicitly in Einstein equation. As a consequence, the Bianchi identity (37.146) does not restricted Λ to be constant, as it happens in ordinary General Relativity. In the contraction limit $l \to \infty$, the gravitational field equation (37.147) reduces to the ordinary Einstein equation for a vanishing Λ.

The curvature tensor in this theory represents both the General Relativity *dynamic curvature*, whose source is the energy-momentum current, and the *kinematic curvature* implied by the underlying de Sitter spacetime. This means that the cosmological term Λ does not appear explicitly in Einstein equation, and consequently the second Bianchi identity (37.146) *does not require Λ be constant*.[24] Notice furthermore that the *local* notion of de

---

[24] Beltrán Almeida, Mayor & Pereira 2012.

Sitter spacetime is different from the usual *global*, or homogeneous notion. In fact, considering that the "source" of the kinematic de Sitter curvature is the proper conformal current of matter, it must vanish outside the region where the system is located. The local kinematic curvature can thus be thought of as a kind of *asymptotically flat* de Sitter spacetime — in the sense that Λ vanishes outside the source.

Commentary 37.5 This notion of local de Sitter space has already been discussed by F. Mansouri some years ago.[25] According to his proposal, a high energy experiment could modify the local structure of spacetime for a short period of time, so that the immediate neighborhood of a collision would depart from the Minkowski and become a de Sitter or anti-de Sitter spacetime. ◄

Any solution to the above kinematic-modified Einstein equation will be a de Sitter–Cartan spacetime. For example, the Schwarzschild solution turns out to be given by the so-called Schwarzschild–de Sitter solution

$$ds^2 = f(r)\, dt^2 - \frac{dr^2}{f(r)} - r^2 \left(d\theta^2 + \sin^2\theta\, d\phi^2\right), \tag{37.148}$$

where

$$f(r) = 1 - \frac{r^2}{l^2} - \frac{2M}{r}. \tag{37.149}$$

Conceptually speaking, however, it is completely different from the usual Schwarzschild–de Sitter solution. In fact, whereas the Schwarzschild part of the solution represents the dynamical gravitational field, the de Sitter part represents now just the underlying kinematics. In a locally inertial frame, where inertial effects exactly compensate for gravitation, making gravitation out of sight, it reduces to the background de Sitter metric. Conversely, in the contraction limit $l \to \infty$, the underlying de Sitter spacetime reduces to Minkowski, and the above metric reduces to the Schwarzschild solution of ordinary (locally Minkowski) General Relativity.

Small values of l

In the parameterization appropriate for small values of l, a diffeomorphism is defined by

$$\delta_{\bar{\Pi}} x^\mu = \bar{\xi}^\mu_\rho\, \epsilon^\rho(x), \tag{37.150}$$

where

$$\bar{\xi}^\mu_\rho = 4l^2\, \delta^\mu_\rho - \bar{\delta}^\mu_\rho \tag{37.151}$$

---

[25] Mansouri 2002.

are the corresponding Killing vectors. The invariance of the source lagrangian (37.140) under such diffeomorphism yields the conservation law

$$\bar{\nabla}_\mu \bar{\Pi}^{\mu\nu} = 0, \qquad (37.152)$$

where

$$\bar{\Pi}^{\mu\nu} = 4l^2 T^{\mu\nu} - K^{\mu\nu} \qquad (37.153)$$

is the de Sitter-modified conserved energy-momentum current. On account of this covariant conservation law, Einstein equation assumes the form

$$\bar{R}_{\mu\nu} - \tfrac{1}{2}\, \bar{g}_{\mu\nu} \bar{R} = \frac{8\pi G}{c^4} \bar{\Pi}_{\mu\nu} \qquad (37.154)$$

with $\bar{\Pi}_{\mu\nu}$ the covariantly conserved current (37.153). This is the form of Einstein equation valid for small values of the background de Sitter length parameter l. As one can see from (37.153), it describes gravitation when the energy of the universe is preponderantly in the form of proper conformal current.

37.7.2   *Some cosmological consequences*

Any solution to the above kinematically-modified Einstein equations will be a de Sitter–Cartan spacetime. In the contraction limit $l \to \infty$, corresponding to a vanishing cosmological term Λ, the underlying de Sitter spacetime contracts to Minkowski, and the de Sitter–Cartan geometry reduces to the usual riemannian geometry of General Relativity, which is consistent with the ordinary Poincaré-ruled Special Relativity. We consider now the contraction limit $l \to 0$, which corresponds to an infinite cosmological term Λ. To begin with, we see from the de Sitter "translation" generators (37.44) that in this limit the translational (that is, gravitational) degrees of freedom are switched off: only the proper conformal degrees of freedom remain active. Since the gravitational degrees of freedom are switched off, the contraction limit $l \to 0$ of such spacetime will be the same that follows from a pure de Sitter spacetime. More precisely, it will be the four-dimensional, flat, singular, conic spacetime $\bar{M}$, which is transitive under proper conformal transformations.

Due to this peculiar transitivity property, the metric tensor $\bar{\eta}_{\mu\nu}$ of the cone spacetime $\bar{M}$ does not have the conventional meaning of a physical metric: it is not dimensionless, and the interval it defines,

$$d\bar{s}^2 = \sigma^{-4}\, \eta_{\mu\nu}\, dx^\mu dx^\nu, \qquad (37.155)$$

has to do with the proper conformal notions of time interval and space distance. In such spacetime, therefore, the usual notions of space and time cannot be defined. In particular, our conventional (translational) notion of time does not exist on $\bar{M}$.[26] This means that local clocks cannot be defined, a result that is somehow in agreement with the general idea that (our conventional notion of) time should not exist at the Planck scale.[27] One should note, however, that the proper conformal notion of time does exist.

We remark that $l \to 0$ is just a formal limit in the sense that quantum effects preclude it to be fully performed. It is actually a contraction limit, on an equal footing with the classical contraction limit in which the speed of light goes to infinity $c \to \infty$. The cone spacetime $\bar{M}$, which emerges as the output of this limit, should then be thought of as a kind of *frozen geometric structure* behind the spacetime quantum fluctuations taking place at the Planck scale. It is an idealized universe in which all energy content is in the form of proper conformal current. It can accordingly be assumed to represent an initial condition for the universe, a kind of "nothing" from where our universe came from.[28]

In what follows, we present a speculative discussion on how such scenario points to a cyclic view of the universe.[29] The quantum fluctuations from the cone spacetime with $l = 0$ to a de Sitter spacetime with $l = l_P$ could give rise to a non-singular de Sitter universe with a huge cosmological term $\Lambda_P \simeq 10^{66}$ cm^{-2}, which would drive inflation. Once this transition occurs, the translational degrees of freedom are turned on, and our usual notions of time and space emerge, as can be seen from the generators (37.18). At this point, however, owing to the tiny value of l, space and time will still be preponderantly determined by proper conformal transformations. Concomitantly, the proper conformal current begins transforming into energy-momentum current, giving rise to the baryonic matter present in the universe.[30]

As the proper conformal current transforms into energy-momentum current, the cosmological term experiences a decaying process, which means that the universe expansion is decelerating. This process continues until the cosmological term assumes a tiny value, in a way such as to allow

---

[26] For a general discussion on the problem of time in Quantum Gravity, see Isham 1992, Kuchař 2011.

[27] Rovelli 2009.

[28] Aldrovandi, Beltrán Almeida & Pereira 2006.

[29] Penrose 2011.

[30] How this transformation can be accomplished in nature is as yet an open question.

the formation of the cosmological structures we see today. In this period, most of the proper conformal current has already been transformed into energy-momentum current, and the underlying de Sitter spacetime differs slightly from Minkowski. Accordingly, space and time in this period are preponderantly determined by ordinary translations.

Now, cosmological observations indicate that the universe has entered (a few billion years ago) an accelerated expansion era.[31] From the point of view of a de Sitter-ruled Special Relativity, this means that the energy-momentum current began transforming back into proper conformal current, which produces an increase in the value of the cosmological term. This view is corroborated by more recent observations, which has measured the energy generated within a large portion of space, discovering that it is only half what it was two billion years ago and fading.[32] According to the de Sitter-invariant special relativity, however, the energy is not actually fading away, but simply transforming back into proper conformal current. Such process means that the universe began moving back towards its initial state, characterised by a huge cosmological term — and represented by the cone spacetime $\bar{M}$. This scenario points to a new kind of cyclic cosmology: although seemingly paradoxical, it is an ever expanding cyclic universe.[33]

We shall not cease from exploration
And the end of all our exploring
Will be to arrive where we started
And know the place for the first time.[34]

---

[31] Riess *et al.* 1998, Perlmutter *et al.* 1999, de Bernardis *et al.* 2000
[32] Driver 2015.
[33] Araujo, Jennen, Pereira, Sampson & Savi 2015.
[34] T. S. Eliot, *Little Gidding.*

References

- Abraham R & Marsden J 1978: *Foundations of Mechanics*, 2nd edition, Benjamin–Cummings, Reading.
- Adams C C 1994: *The Knot Book*, W. H. Freeman, New York.
- Agarwal G S & Wolf E 1970: Phys. Rev. D **2** 2161.
- Aharonov Y & Anandan J 1987: Phys. Rev. Lett. **58** 1593.
- Aharonov Y & Anandan J 1988: Phys. Rev. D **38** 1863.
- Aharonov Y & Bohm D 1959: Phys. Rev. **115** 485.
- Aharonov Y & Bohm D 1961: Phys. Rev. **123** 1511.
- Ahluwalia D V 2002: Mod. Phys. Lett. A **17**, 1135.
- Albert J *et al* (for the MAGIC Collaboration), Ellis J, Mavromatos N E, Nanopoulos D V, Sakharov A S & Sarkisyan E K G 2008: Phys. Lett. B **668** 253.
- Aldrovandi R 1992: Fort. Physik **40** 631.
- Aldrovandi R 1995: *Gauge Theories, and Beyond*, in Barret T W & Grimes D M, *Advanced Electromagnetism: Foundations, Theory and Applications*, World Scientific, Singapore.
- Aldrovandi R 2001: *Special Matrices of Mathematical Physics*, World Scientific, Singapore.
- Aldrovandi R & Barbosa A L 2005: Int. J. Math. Mathematical Sci. **15** 2365.
- Aldrovandi R, Barros P B & Pereira J G 2003: Found. Phys. **33** 545.
- Aldrovandi R, Beltrán Almeida J P & Pereira J G 2006: J. Geom. Phys. **56** 1042.
- Aldrovandi R, Beltrán Almeida J P & Pereira J G 2007: Class. Quantum Grav. **24** 1385.
- Aldrovandi R & Galetti D 1990: J. Math. Phys. **31** 2987.
- Aldrovandi R & Kraenkel R A 1988: J. Phys. A **21** 1329.

- Aldrovandi R & Pereira J G 1986: Phys. Rev. D **33** 2788.
- Aldrovandi R & Pereira J G 1988: Rep. Math. Phys. **26** 237.
- Aldrovandi R & Pereira J G 1991: *Global and Local Transformations: an Overview*, in MacDowell S, Nussenzveig H M & Salmeron R A (eds.), *Frontier Physics: Essays in Honour of J. Tiomno*, World Scientific, Singapore.
- Aldrovandi R & Pereira J G 1998: *A Second Poincaré Group*, in *Topics in Theoretical Physics: Festschrift for A. H. Zimerman*, ed. by Aratyn H *et al*, Fundação IFT, São Paulo.
- Aldrovandi R, Pereira J G & Vu K H 2004: Gen. Rel. Grav. **36** 101.
- Aldrovandi R & Pereira J G 2009: Grav. Cosmol. **15**, 287.
- Aldrovandi R & Pereira J G 2012: *Teleparallel Gravity: An Introduction* (Springer, Dordrecht).
- Aldrovandi R 2014: *Tetration: an iterative approach* (arXiv:1410.3896).
- Alexandrov P 1977: *Introduction à la Théorie Homologique de la Dimension et la Topologie Combinatoire*, MIR, Moscow.
- Amelino-Camelia G 2000: Lect. Not. Phys. **541**, 1.
- Amelino-Camelia G 2002a: Nature **418**, 34.
- Amelino-Camelia G 2002: Int. J. Mod. Phys. D **11**, 35.
- Anderson I M and Duchamp T 1980: Am. J. Math. **102** 781.
- Andrade V C & J. G. Pereira J G 1997: Phys. Rev. D **56**, 4689.
- Andrade V C, Guillen L C T & Pereira J G 2000: Phys. Rev. Lett. **84** 4533.
- Appel K & Haken W 1977: Sci. Amer. **237** 108.
- Arafune J , Freund P G O & Göbel C J 1975: J. Math. Phys. **16** 433.
- Araujo A, Jennen H, Pereira J G, Sampson A C & Savi L L 2015: Gen. Rel. Grav. **47** 151.
- Arcos H I, Gribl Lucas T & Pereira J G 2010: Class. Quantum Grav. **27** 145007.
- Arcos H I & Pereira J G 2004: Class. Quantum Grav. **21**, 5193.
- Aristotle 1952: *Physics*, Great Books **8**, vol. **I**, p. 259. Encyclopaedia Britannica, Chicago.
- Arkhangel'skii A V & Fedorchuk V V 1990: in Arkhangel'skii A V & Pontryagin L S (eds.), *General Topology I, Encyclopaedia of Mathematical Sciences*, Springer, Berlin.
- Arnold V I 1966: Ann. Inst. Fourier **16** 319.
- Arnold V I 1973: *Ordinary Differential Equations*, MIT Press, Cambridge.
- Arnold V I 1976: *Les Méthodes Mathématiques de la Méchanique Clas-*

sique, MIR, Moscow.

- Arnold V I 1980: *Chapitres Supplémentaires de la Théorie des Équations Différentielles Ordinaires*, MIR, Moscow.

- Arnold V I, Kozlov V V & Neishtadt A I 1988: in Arnold V I (ed.), *Dynamical Systems III, Encyclopaedia of Mathematical Sciences*, Springer, Berlin.

- Askar A 1985: *Lattice Dynamical Foundations of Continuum Theories*, World Scientific, Singapore.

- Atherton R W and Homsy G M 1975: Stud. Appl. Math. **54** 31.

- Atiyah M F, Drinfeld V G, Hitchin N J & Manin Yu. I 1978: Phys. Lett. A **65** 185.

- Atiyah M F 1979: *Geometry of Yang-Mills Fields*, Acad. Naz. Lincei, Pisa.

- Atiyah M F 1991: *The Geometry and Physics of Knots*, Cambridge University Press, Cambridge.

- Avis S J & Isham C J 1978: Proc. Roy. Soc. **362** 581.

- Avis S J & Isham C J 1979: in Lévy M & Deser S, *Recent Developments in Gravitation*, Proceedings of the 1978 Cargèse School, Gordon and Breach, New York.

- Babelon O & Viallet C-M 1989: *Integrable Models, Yang-Baxter Equation, and Quantum Groups*, SISSA-ISAS preprint 54 EP.

- Bacry H & Lévy-Leblond J-M 1968: J. Math. Phys. **9**, 1605.

- Baker G A 1958: Phys. Rev. **109** 2198.

- Balescu R 1975: *Equilibrium and Nonequilibrium Statistical Mechanics*, J. Wiley, New York.

- Baulieu L 1984: Nucl. Phys. **B241** 557.

- Baxter R J 1982: *Exactly Solved Models in Statistical Mechanics*, Academic Press, London.

- Bayen F, Flato M, Fronsdal C, Lichnerowicz A & Sternheimer D 1978: Ann. Phys. (NY) **111** 61.

- Becchi C, Rouet A & Stora R 1974: Phys. Lett. **B52** 344.

- Becchi C, Rouet A & Stora R 1975: Commun. Math. Phys. **42** 127.

- Becchi C, Rouet A & Stora R 1976: Ann. Phys. **98** 287.

- Becher P & Joos H 1982: Z. Phys. C **15** 343.

- Belinfante F J 1939: Physica **6**, 687.

- Beltrán Almeida J P, Mayor C S O & Pereira J G 2012: Grav. Cosm. **18** 181.

- Benn I M, Dereli T & Tucker R W 1980: Phys. Lett. B **96**, 100.

- Berge C 2001: *The Theory of Graphs*, Dover, New York.

- Berry M V 1976: Advances in Physics **25** 1.
- Berry M V 1984: Proc. Roy. Soc. A **392** 45.
- Biggs N L, Lloyd E K & Wilson R J 1977: *Graph Theory 1736-1936*, Oxford University Press, Oxford.
- Birman J S 1975: *Braids, Links, and Mapping Class Groups*, Princeton University Press, Princeton.
- Birman J S 1991: Math. Intell. **13** 52.
- Bjorken J D & Drell S D 1964: *Relativistic Quantum Mechanics*, McGraw-Hill, New York.
- Bjorken J D & Drell S D 1965: *Relativistic Quantum Fields*, McGraw-Hill, New York.
- Blagojevic M & Hehl F (eds.) 2013: *Gauge Theories of Gravitation: A Reader with Commentaries* (Imperial College Press, London.)
- Bogoliubov N N & Shirkov D V 1980: *Introduction to the Theory of Quantized Fields*, 3rd edition, J. Wiley, New York.
- Bollini C G, Giambiagi J J & Tiomno J 1979, Phys. Lett. B **83** 185.
- Bonora L & Cotta-Ramusino P 1983: Comm. Math. Phys. **87** 589.
- Boothby W M 1975: *An Introduction to Differentiable Manifolds and Riemannian Geometry*, Academic Press, New York.
- Born M & Wolf E 1975: *Principles of Optics*, Pergamon, Oxford.
- Born M 1964: *Natural Philosophy of Cause and Chance*, Dover, New York.
- Born M & Infeld L 1934: Proc. Roy. Soc. A **144** 425.
- Bott R & Tu L W 1982: *Differential Forms in Algebraic Topology*, Springer, New York.
- Boya L J, Cariñena J F & Mateos J 1978: Fort. Physik **26** 175.
- Bratelli O & Robinson D W 1979: *Operator Algebras and Quantum Statistical Mechanics I*, Springer, New York.
- Brandenberger R H & Marti J 2002: Int. J. Mod. Phys. A **17**, 3663.
- Broadbent S R & Hammersley J M 1957: Proc. Camb. Phil. Soc. **53** 629.
- Brown L S & Weisberger W I 1979: Nucl. Phys. B **157** 285.
- Budak B M & Fomin S V 1973: *Multiple Integrals, Field Theory and Series*, MIR, Moscow.
- Burgers J M 1940: Proc. Phys. Soc. **52** 23.
- Burke W L 1985: *Applied Differential Geometry*, Cambridge University Press, Cambridge.
- Carrol S 2004: *Spacetime and Geometry: An Introduction to General Relativity*, Addison Wesley, San Francisco.

- Cartan É 1922: C. R. Acad. Sci. (Paris) **174** 593.
- Cartan É 1923: Ann. Ec. Norm. Sup. **40**, 325.
- Cartan É 1924: Ann. Ec. Norm. Sup. **41**, 1.
- Callan C G, Coleman S & Jackiw R 1970: Ann. Phys. (NY) **59** 42.
- Calvo M 1977: Phys. Rev. **D15** 1733.
- Chandrasekhar S 1939: *An Introduction to the Study of Stellar Structure*, U. Chicago Press, Chicago.
- Chandrasekhar S 1972: Am. J. Phys. **40** 224.
- Charap J M (ed.) 1995: *Geometry of Constrained Dynamical Systems*, Cambridge University Press.
- Chillingworth D 1974: in *Global Analysis and its Applications*, Vol. 1, IAEA, Vienna.
- Chinn W G & Steenrod N E 1966: *First Concepts of Topology*, Random House, New York.
- Cho Y M 1975: J. Math. Phys. **16** 2029.
- Choquet-Bruhat Y, DeWitt-Morette C & Dillard-Bleick M 1977: *Analysis, Manifolds and Physics*, North-Holland, Amsterdam.
- Christie D E 1976: *Basic Topology*, MacMillan, New York.
- Coleman S 1977: in Zichichi A (ed.), *New Phenomena in Subnuclear Physics*, Proceedings of the 1975 International School of Subnuclear Physics (Erice, Sicily), Plenum Press, New York.
- Coleman S 1979 : *The Uses of Instantons*, in Zichichi A (ed.), *The Whys of Subnuclear Physics* , Proceedings of the 1977 International School of Subnuclear Physics (Erice, Sicily), Plenum Press, New York.
- Coleman S 1985: *Aspects of Symmetry* (Cambridge University Press, Cambridge).
- Connes A 1980: C. R. Acad. Sci. Paris A **290** 599.
- Connes A 1986: *Noncommutative Differential Geometry*, Pub. IHES (Paris) **62** 257.
- Connes A 1990: *Géométrie Non-Commutative*, InterEditions, Paris.
- Connes A 1994: *Non-Commutative Geometry*, Academic Press, New York.
- Connes A, Douglas M R & Schwarz A 1998: J. High Energy Phys. **02** 003.
- Comtet L 1974: *Advanced Combinatorics*, Springer, Dordrecht.
- Coquereaux R 1989: J. Geom. Phys. **6** 425.
- Croom F H 1978: *Basic Concepts of Algebraic Topology*, Springer, Berlin.

- Crowell R H & Fox R H 1963: *Introduction to Knot Theory*, Springer, Berlin.
- Daniel M & Viallet C M l980: Rev. Mod. Phys. **52** 175.
- Das A & Kong O W C 2006: Phys. Rev. D **73**, 124029.
- Davis W R & Katzins G H 1962: Am. J. Phys. **30** 750.
- de Bernardis P *et al.* 2000: Nature **404** 955.
- De Gennes P G & Prost J 1993: *The Physics of Liquid Crystals*, 2nd edition, Clarendon Press, Oxford.
- De Rham G 1960: *Varietés Differentiables*, Hermann, Paris.
- de Sabbata V & Gasperini M 1985: *Introduction to Gravitation*, World Scientific, Singapore.
- de Sabbata V & C. Sivaram C 1994: *Spin and Torsion in Gravitation*, World Scientific, Singapore.
- Deser S & Wilczek F 1976: Phys. Lett. B **65** 391.
- Deser S & Drechsler W 1979: Phys. Lett. B **86** 189.
- DeWitt-Morette C 1969: Ann. Inst. Henri Poincaré **XI** 153.
- DeWitt-Morette C 1972: Comm. Math. Phys. **28** 47.
- DeWitt-Morette C, Masheshwari A & Nelson B 1979: Phys. Rep. **50** 255.
- Dirac P A M 1926: Proc. Roy. Soc. A **109** 642.
- Dirac P A M 1958: in Kockel B, Macke W & Papapetrou A, *Planck Festschrift*, VEB Deutscher Verlag der Wissenschaften, Berlin.
- Dirac P A M 1958b: Proc. Roy. Soc. A **246** 333.
- Dirac P A M 1959: Phys. Rev. **114** 924.
- Dirac P A M 1984: *The Principles of Quantum Mechanics*, Clarendon Press, Oxford.
- Dittrich J 1979: Czech. J. Phys. B **29** 1342.
- Dixmier J 1961: Bul. Soc. Math. Franc. **89** 9.
- Dixmier J 1981: *von Neumann Algebras*, North-Holland, Amsterdam.
- Dixmier J 1982: *C*-Algebras*, North-Holland, Amsterdam.
- Donaldson S K & Kronheimer P B 1991: *The Geometry of Four-Manifolds*, Clarendon Press, Oxford.
- Donaldson S K 1996: Bull. Amer. Math. Soc. **33** 45.
- Doria F A 1981: Commun. Math. Phys. **79** 435.
- Doubrovine B, Novikov S & Fomenko A 1979: *Géométrie Contemporaine*, MIR, Moscow.
- Dowker J S 1979: *Selected Topics in Topology and Quantum Field Theory*, lectures delivered at the Center for Relativity, Austin, Texas.
- Drinfeld V G 1983: Sov. Math. Dokl. **27** 222.

- Driver S P *et al* 2015: MNRAS **455** 3911.
- Dubois-Violette M 1988: C. R. Acad. Sci. Paris **307I** 403.
- Dubois-Violette M 1991: in Bartocci C, Bruzzo U & Cianci R (eds.), *Differential Geometric Methods in Theoretical Physics*, Lect. Notes Phys. **375**, Springer, Berlin.
- Dubois-Violette M, Kerner R & Madore J 1990: J. Math. Phys. **31** 316.
- Dubois-Violette M & Launer G 1990: Phys. Lett. B **245** 175.
- Efros A L 1986: *Physics and Geometry of Disorder*, MIR, Moscow.
- Einstein A 1930: *Auf die Riemann-Metrik und den Fern-Parallelismus gegründete einheitliche Feldtheorie*, Math. Annal. **102**, 685. For an english translation, together with three previous essays, see A. Unzicker and T. Case, *Unified Field Theory based on Riemannian Metrics and distant Parallelism*.
- Eisenhart L P 1949: *Riemannian Geometry*, Princeton University Press, Princeton.
- Elworthy K D, Le Jan Y & Li X M 2010: *The Geometry of Filtering*, Birkhäuser, Basel.
- Endo M, Iijima S and Dresselhaus M S (Eds.) 1996: *Carbon Nanotubes*, Pergamon, Oxford.
- Enoch M & Schwartz J-M 1992: *Kac Algebras and Duality of Locally Compact Groups*, Springer, Berlin.
- Essam J W 1972: in Domb C & Green M S (eds.), *Phase Transitions and Critical Phenomena*, vol.2, Academic Press, London.
- Ezawa Z F 1978: Phys. Rev. D **18** 2091.
- Ezawa Z F 1979: Phys. Lett. B **81** 325.
- Ezawa M 2009: Eur. Phys. J. B **67** 543.
- Faddeev L D 1982: Sov. Phys. Usp. **25** 130.
- Faddeev L D & Shatashvilli S L 1984: Theor. Math. Phys. **60** 770.
- Faddeev L D & Slavnov A A 1978: *Gauge Fields. Introduction to Quantum Theory*, Benjamin/Cummings, Reading.
- Fadell E R & Husseini S V 2001: *Geometry and Topology of COnfigurationnSpaces*, Springer, Berlin.
- Fairlie D B, Fletcher P & Zachos C Z 1989: Phys. Lett. B **218** 203.
- Farmer J D, Ott E & Yorke J E 1983: Physica D **7** 153.
- Fetter A L &Walecka J D 2003: *Theoretical Mechanics of Particles and Continua*, Dover, New York.
- Fetter A L &Walecka J D 2006: *Nonlinear Mechanics*, Dover, New York.

- Feynman R P, Leighton R B & Sands M 1965: *The Feynman Lectures in Physics*, Addison-Wesley, Reading.
- Finkelstein D 1966: J. Math. Phys. **7** 1218.
- Finlayson B A 1972: Phys. Fluids **13** 963.
- Flanders H 1963: *Differential Forms*, Academic Press, New York.
- Fock V 1926: Z. Physik **39** 226.
- Fock V A 1964: *The Theory of Space, Time and Gravitation*, 2nd edition, Pergamon, New York.
- Fock V A & Ivanenko D 1929 Z. Phys. **54**, 798.
- Forsyth A R 1965: *The Theory of Functions*, Dover, New York.
- Fraleigh J B 1974: *A First Course in Abstract Algebra*, Addison-Wesley, Reading.
- Franck F C 1951: Phil. Mag. **42** 809.
- Freed D S & Uhlenbeck K K 1984: *Instantons and Four-Manifolds*, Springer, Berlin.
- Freedman D Z & Khuri R R 1994: Phys. Lett. B **329** 263.
- Friedmann A 1922: Z. Phys. **10** 377.
- Friedmann A 1924: Z. Phys. **21** 326.
- Fullwood D T 1992: J. Math. Phys. **33** 2232.
- Furry W H 1963: in Brittin W E, Downs B W & Downs J (eds.), *Lectures in Theoretical Physics*, 1962 Summer Institute for Theoretical Physics (Boulder), Interscience, New York.
- Galetti D & Toledo Piza A F 1988: Physica A **149** 267.
- Gardner M 1997: *Penrose Tiles to Trapdoor Ciphers*, The Mathematical Association of America, Washington.
- Gelfand I M & Shilov G E 1964: *Generalized Functions*, Vol.1, Academic Press, New York.
- Gelfand I M, Graev I M & Vilenkin N Ya 1966: *Generalized Functions*, Vol. 5, Academic Press, New York.
- Geroch R P 1968: J. Math. Phys. **9** 1739.
- Gerstenhaber M & Stasheff J (eds.) 1992: *Deformation Theory and Quantum Groups with Applications to Mathematical Physics*, AMS, Providence.
- Gibling P J 1977: *Graphs, Surfaces and Homology*, Chapman & Hall, London.
- Gilmore R 1974: *Lie Groups, Lie Algebras, and Some of Their Applications*, J. Wiley, New York.
- Godbillon G 1971: *Eléments de Topologie Algébrique*, Hermann, Paris.
- Göbel R 1976a: J. Math. Phys. **17** 845.

- Göbel R 1976b: Comm. Math. Phys. **46** 289.
- Goldberg S I l962: *Curvature and Homology*, Dover, New York.
- Goldstein H 1980: *Classical Mechanics*, 2nd edition, Addison-Wesley, Reading.
- Graver J E & Watkins M E 1977: *Combinatorics with Emphasis on the Theory of Graphs*, Springer, Berlin.
- Greenberg M J 1967: *Lectures on Algebraic Topology*, Benjamin, Reading.
- Grebogi C, Ott E & Yorke J A 1987: Science **238** 585.
- Greub W, Halperin S & Vanstone R 1973: *Connections, Curvature, and Cohomology: Lie Groups, Principal Bundles, and Characteristic Classes*, Academic Press, New York.
- Gribl Lucas T & Pereira J G 2009: J. Phys. A **42** 035402.
- Gribl Lucas T, Obukhov Yu N & Pereira J G 2009: Phys. Rev. D **80** 064043.
- Guckenheimer J & Holmes P 1986: *Nonlinear Oscillations, Dynamical Systems, and Bifurcations of Vector Fields*, Springer, Berlin.
- Guillemin V & Sternberg S 1977: *Geometric Asymptotics*, AMS, Providence.
- Gurarie D 1992: *Symmetries and Laplacians*, North Holland, Amsterdam.
- Gürsey F 1962: in Gürsey F (ed.), *Group Theoretical Concepts and Methods in Elementary Particle Physics*, Istanbul Summer School of Theoretical Physics, Gordon and Breach, New York.
- Haag R 1993: *Local Quantum Physics*, Springer, Berlin.
- Hajicek P 1971: Comm. Math. Phys. **21** 75.
- Halmos P R 1957: *Introduction to Hilbert Spaces*, Chelsea Pub. Co., New York.
- Halpern M B 1977: Phys. Rev. D **16** 1798.
- Halpern M B 1978: Nucl. Phys. B **139** 477.
- Hamermesh M 1962: *Group Theory and its Applications to Physical Problems*, Addison–Wesley, Reading.
- Harris W F 1975: Phil. Mag. B **32** 37.
- Harrison E 2001: *Cosmology*, 2nd ed., reprinted, Cambridge University Press, Cambridge.
- Hassanabadi H, Derakhsani Z & Zarrinkamar S 2016: Acta Physica Polonica A **129** 3.
- Hawking S W & Ellis G F R 1973: *The Large Scale Structure of Space–Time*, Cambridge University Press, Cambridge.

- Hawking S W & Israel W (eds.) 1979: *General Relativity: An Einstein Centenary Survey*, Cambridge University Press, Cambridge.
- Hawking S W, King A R & McCarthy P J 1976: J. Math. Phys. **17** 174.
- Hehl F W, von der Heyde P, Kerlick G D & Nester J M 1976: Rev. Mod. Phys. **48** 393.
- Hehl F W, McCrea J D, Mielke E W & Ne'eman Y 1995: Phys. Rep. **258**, 1.
- Hilton P J 1953: *An Introduction to Homotopy Theory*, Cambridge University Press, Cambridge.
- Hilton P J & Wylie S 1967: *Homology Theory*, Cambridge University Press, Cambridge.
- Hocking J C & Young G S 1961: *Topology*, Addison–Wesley, Reading.
- Hu S T 1959: *Homotopy Theory*, Academic Press, New York.
- Huang K 1987: *Statistical Mechanics*, 2nd edition, J. Wiley, New York.
- Hurewicz W & Wallman H 1941: *Dimension Theory*, Princeton University Press, Princeton.
- Hwa R W & Teplitz V L 1966: *Homology and Feynman Integrals*, Benjamin, Reading.
- Inönü E & Wigner E P 1953: *On the Contraction of Groups and Their Representations*, Proc. Natl. Acad. Sci. **39**, 510.
- Iliev B Z 1996: *J. Phys. A* **29** 6895.
- Iliev B Z 1997: *J. Phys. A* **30** 4327.
- Iliev B Z 1998: *J. Phys. A* **31** 1287.
- Isham C J, Penrose R & Sciama D W (eds.) 1975: *Quantum Gravity*, Oxford University Press, Oxford.
- Isham C J 1978: Proc. Roy. Soc. **362** 383.
- Isham C J 1984: in DeWitt B S & Stora R (eds.), *Relativity, Groups and Topology II*, Les Houches Summer School, North-Holland, Amsterdam.
- Isham C J 1992: *Canonical quantum gravity and the problem of time*, arXiv:gr-qc/9210011.
- Itzykson C & Zuber J-B 1980: *Quantum Field Theory*, McGraw-Hill, New York.
- Jackiw R 1980: Rev. Mod. Phys. **52** 661.
- Jackson J D 1975: *Classical Electrodynamics*, 2nd edition, John Wiley, New York.
- Jacobson N 1962: *Lie Algebras*, Dover, New York.
- Jacobson T, Liberati S & Mattingly D 2002: Phys. Rev. D **66**, 081302.
- Jancel R 1969: *Foundations of Classical and Quantum Statistical Me-*

chanics, Pergamon, Oxford.
- Jones V F R 1983: Invent. Math. **72** 1.
- Jones V F R 1985: Bull. Amer. Math. Soc. **12** 103.
- Jones V F R 1987: Ann. Math. **126** 335.
- Jones V F R 1991: *Subfactors and Knots*, AMS, Providence.
- Katznelson Y 1976: *An Introduction to Harmonic Analysis*, Dover, New York.
- Kauffman L H 2012: *Knots and Physics*, 4th edition, World Scientific, Singapore.
- Kerner R 2007: *Models of Agglomeration and Glass Transition*, Imperial College Press, London.
- Khavin V P 1991: in Khavin V P & Nikol'skij N K (eds.), *Commutative Harmonic Analysis I, Encyclopaedia of Mathematical Sciences*, vol.15, Springer, Berlin.
- Kibble T W B 1961: J. Math. Phys. **2**, 212.
- Kirillov A 1974: *Élements de la Théorie des Représentations*, MIR, Moscow.
- Klein O 1938: in *New Theories in Physics, Conference Organized in Collaboration with the International Union of Physics and the Polish Intellectual Co-Operation Committee, Warsaw, May 30-June 3, 1938*, (Scientific Collections), Warsaw.
- Kléman M & Sadoc J F 1979: Journal de Physique Lettres **40** 569.
- Kohno T 1990: *New Developments in the Theory of Knots*, World Scientific, Singapore.
- Kobayashi S & Nomizu K l996: *Foundations of Differential Geometry*, 2nd edition, Wiley–Intersciense, New York.
- Kolb E W & Turner M S 1994: *The Early Universe*, Perseus Books, New York.
- Kolmogorov A N & Fomin S V 1970: *Introductory Real Analysis*, Prentice-Hall, Englewood Cliffs.
- Kolmogorov A N & Fomin S V 1977: *Éléments de la Théorie des Fonctions et de l'Analyse Fonctionelle*, MIR, Moscow.
- Konopleva N P & Popov V N 1980: *Gauge Fields*, Harwood, New York.
- Kowalski-Glikman J 2005: Lect. Notes Phys. **669**, 131.
- Kowalski-Glikman J 2006: in Oriti (ed.), *Approaches to quantum gravity — toward a new understanding of space, time, and matter*, Cambridge University Press, Cambridge.
- Kršśák M & Pereira J G 2015: *Spin Connection and Renormalization of Teleparallel Action*, Eur. Phys. J. C **75** 519.

- Kuchař K V 2011, *Time and interpretations of quantum gravity*, in Kunstatter G, Vincent D & Williams J (eds.), Proceedings of the 4th Canadian Conference on General Relativity and Relativistic Astrophysics, World Scientific, Singapore. Reprinted in Int. J. Mod. Phys. Proc. Suppl. D **20**, 3 (2011).
- Kugo T & Ojima I 1978: Progr. Theor. Phys. **60** 1869.
- Laidlaw M G G & DeWitt-Morette C 1971: Phys. Rev. D **3** 1375.
- Lanczos C 1986: *The Variational Principles of Mechanics*, 4th edition, Dover, New York.
- Landau L D & Lifshitz E M 1969: *Statistical Physics*, Pergamon, Oxford.
- Landau L D & Lifshitz E M 1975: *The Classical Theory of Fields*, Pergamon, Oxford.
- Landau L D & Lifshitz E M 1989: *Méchanique des Fluides*, MIR, Moscow.
- Landau L D & Lifshitz E M 1990: *Theory of Elasticity*, MIR, Moscow.
- Lavrentiev M & Chabat B 1977: *Méthodes de la Théorie des Fonctions d'Une Variable Complexe*, MIR, Moscow.
- Leinaas J M & Myrheim J 1977: Nuovo Cimento B **37** 1.
- Lemaître A G 1931: MNRAS **91**, 483.
- Levi B G 1993: Physics Today, **46** 17.
- Lévy M & Deser S (eds.) 1979: *Recent Developments in Gravitation*, Cargèse Lectures 1978, Plenum Press, New York.
- Lichnerowicz A 1955: *Théorie Globale des Connexions et des Groupes d'Holonomie*, Dunod, Paris.
- London F 1927: Z. Physik **42** 375.
- Loos H G 1967: J. Math. Phys. **8** 2114.
- Love A E H 1944: *A Treatise on the Mathematical Theory of Elasticity*, Dover, New York.
- Lovelock P & Rund J 1975: *Tensors, Differential Forms and Variational Principles*, J. Wiley, New York.
- Ludvigsen M 1999: *General Relativity*, Cambridge University Press, Cambridge.
- Lu Qi-keng 1975: Chin. J. Phys. **23** 153.
- Luneburg R K 1966: *Mathematical Theory of Optics*, University of California Press, Berkeley.
- Mackey G W 1955: *The Theory of Group Representations*, Lecture Notes, Department of Mathematics, University of Chicago.
- Mackey G W 1968: *Induced Representations of Groups and Quantum*

Mechanics, Benjamin, New York.

- Mackey G W 1978: *Unitary Group Representations in Physics, Probability and Number Theory*, Benjamin/Cummings, Reading.
- Magueijo J & Smolin L 2002: Phys. Rev. Lett. **88**, 190403.
- Majid S 1990: Int. J. Mod. Phys. A **5** 1.
- Majumdar P & Sharatchandra H S 2001, Phys. Rev. D **63** 067701.
- Malament D B 1977: J. Math. Phys. **18** 1399.
- Mandelbrot B B 1977: *The Fractal Geometry of Nature*, W. H. Freeman, New York.
- Mañes J, Stora R & Zumino B 1985: Comm. Math. Phys. **108** 157.
- Manin Yu I 1989: Comm. Math. Phys. **123** 163.
- Mansouri F 2002: Phys. Lett. B **538** 239.
- Marsden J 1974: *Applications of Global Analysis in Mathematical Physics*, Publish or Perish, Boston.
- Maslov V P 1972: *Théorie des Perturbations et Méthodes Asymptotiques*, Dunod–Gauthier–Villars, Paris.
- Maurer J 1981: *Mathemecum*, Vieweg & Sohn, Wiesbaden.
- McCauley J L 1997: *Classical Mechanics*, Cambridge University Press.
- McGuire J B 1964: J. Math. Phys. **5** 622.
- Michel L 1964: in Gürsey F (ed.), in *Group Theoretical Concepts and Methods in Elementary Particle Physics*, Istanbul Summer School of Theoretical Physics (1963), Gordon and Breach, New York.
- Milnor J 1973: *Morse Theory*, Princeton University Press, Princeton.
- Milnor J 1997: *Topology from the Differentiable Point of View*, Princeton University Press, Princeton.
- Misner C W & Wheeler J A 1957: Ann. Phys. **2** 525.
- Misner C W, Thorne K & Wheeler J A 1973: *Gravitation*, W. H. Freeman, San Francisco.
- Møller C 1958: Ann. Phys. (NY) **4** 347.
- Møller C 1961: K. Dan. Vidensk. Selsk. Mat. Fys. Skr. 1, bind 10.
- Monerat G A, Corrêa Silva E V, Neves C, Oliveira–Neto G, Rezende Rodrigues L G & Silva de Oliveira M 2015, *Can noncommutativity affect the whole history of the Universe?*, arXiv:1109.3514v2 [gr-qc].
- Morgan J W 2004: Bull. Am. Math. Soc. **42** 57
- Mostow M 1979: Commun. Math. Phys. **78** 137.
- Mrugala R 1978: Rep. Math. Phys. **14** 419.
- Munkres J R 1975: *Topology: a First Course*, Prentice Hall, New York.
- Myers R C & Pospelov M 2003: Phys. Rev. Lett. **90**, 211601.
- Nabarro F R N 1987: *Theory of Crystal Dislocations*, Dover, New York.

- Nakahara M 1992: *Geometry, Topology and Physics*, Institute of Physics Publishing, Bristol. Second edition 2003, Taylor & Francis, Boca Raton, FL, USA.
- Narlikar J V 2002: *An Introduction to Cosmology*, third edition, Cambridge University Press, Cambridge.
- Nash C 1991: *Differential Topology and Quantum Field Theory*, Academic Press, London.
- Nash C & Sen S 1983: *Topology and Geometry for Physicists*, Academic Press, London.
- Neuwirth L P 1965: *Knot Groups*, Princeton University Press, Princeton.
- Neuwirth L P 1979: Scient. Am. **240** 84.
- Nomizu K 1956: *Lie Groups and Differential Geometry*, Math. Soc. Japan, Tokyo.
- Novikov S P 1982: Sov. Math. Rev. **3** 3.
- Okubo S 1980: Phys. Rev. D **22** 919.
- Okun L B 1984: *Introduction to Gauge Theories*, ITEP Publications, Moscow.
- Olver P J 1986: *Applications of Lie Groups to Differential Equations*, Springer, New York.
- Ore O 1963: *Graphs and their Uses*, The Mathematical Association of America, New York.
- Ott E 1994: *Chaos in Dynamical Systems*, Cambridge University Press, Cambridge.
- Papapetrou A 1951: Proc. Roy. Soc. A **209** 248.
- Pasquier V 2007: *Quantum Hall Effect and Noncommutative Geometry*, Séminaire Poincaré **X** 1-14.
- Pathria R K 1972: *Statistical Mechanics*, Pergamon, Oxford.
- Penrose R & MacCallum M A H 1972: Phys. Rep. **6** 241.
- Penrose R 1977: Rep. Math. Phys. **12** 65.
- Penrose R 2011: *Cycles of Time*, Alfred A. Knopf, New York.
- Pereira J G 2014: *Teleparallelism: a new insight into gravity*, in Ashtekar A & Petkov V (eds.), Springer Handbook of Spacetime, Springer, Dordrecht.
- Pereira J G & Sampson A C 2012: Gen. Rel. Grav. **44** 1299.
- Perelman G 2002: *The entropy formula for the Ricci flow and its geometric applications*, arXiv:math/0211159.
- Perelman G 2003: *Ricci flow with surgery on three-manifolds*, arXiv:math/0303109.

- Perlmutter S *et al.* 1999: Ap. J. **517** 565.
- Pérez-Nadal G, Roura A & Verdaguer E 2010: JCAP 1005 036, see Section 2.1.
- Petrov A Z 1951: Dokl. Akad. Nauk **31** 149.
- Petrov A Z 1954: Scientific Proceedings of Kazan State University **114** 55; reprinted in Gen. Rel. Grav. **32**, 1665 (2000).
- Petrov A Z 1969: *Einstein Spaces*, Pergamon, Oxford.
- Pontryagin L S 1939: *Topological Groups*, Princeton University Press, Princeton.
- Popov D A 1975: Theor. Math. Phys. **24** 347.
- Porteous I R 1969: *Topological Geometry*, van Nostrand, London.
- Protheroe R J & Meyer H 2000: Phys. Lett. B **493**, 1.
- Quinn F 1982: J. Diff. Geom. **17** 503.
- Rasband S N 1990: *Chaotic Dynamics of Nonlinear Systems*, J. Wiley, New York.
- Regge T 1961: Nuovo Cimento **19** 558.
- Riess A G *et al.* 1998: Astron. J. **116** 1009.
- Rindler W 2007: *Relativity*, 2nd. ed., Oxford University Press, Oxford.
- Robertson H P 1935: Ap. J. **82** 248.
- Rolfsen D 1976: *Knots and Links*, Publish or Perish, Berkeley.
- Rosenfeld L 1940: Mém. Acad. Roy. Belg. Sci. **18**, 1.
- Rosenfeld B A & Sergeeva N D 1986: *Stereoghaphic Projection*, MIR, Moscow.
- Roskies R 1977: Phys. Rev. D **15** 1731.
- Rourke C & Stewart I 1986: New Scientist, 4th September, pg. 41
- Rovelli C 2009, *Forget time* [arXiv:0903.3832].
- Saaty T L & Kainen P C 1986: *The Four-Color Problem*, Dover, New York.
- Sadoc J F & Mosseri R 1982: Phil. Mag. B **45** 467.
- Sauer T: *Field equations in teleparallel spacetime: Einsteins Fernparallelismus approach towards unified field theory*, Einsteins Papers Project [arXiv:physics/0405142].
- Schmidt H-J 1993: Fort. Physik **41** 179.
- Schreider Ju A 1975: *Equality, Resemblance and Order*, MIR, Moscow.
- Schulman L 1968: Phys. Rev. **176** 1558.
- Schutz B 1985: *Geometrical Methods of Mathematical Physics*, Cambridge University Press, Cambridge.
- Shapiro I L 2002: Phys. Rep. **357**, 113.
- Sharpe R 1997: *Differential Geometry: Cartan's Generalization of*

Klein's Erlangen Program, Springer, Berlin.

- Sciama D W 1964: Rev. Mod. Phys. **36**, 463.
- Sierpiński W 1956: *General Topology*, University of Toronto Press, Toronto (Dover edition, New York, 2000).
- Simmons G F 1963: *Introduction to Topology and Modern Analysis*, McGraw–Hill/Kogakusha, New York/Tokyo.
- Simon B 1983: Phys. Rev. Lett. **51** 2167.
- Singer I M & Thorpe J A 1967: *Lecture Notes on Elementary Topology and Geometry*, Scott Foresman, Glenview.
- Singer I M 1978: Commun. Math. Phys. **60** 7.
- Singer I M 1981: Physica Scripta **24** 817.
- Skyrme T H R 1962: Nucl. Phys. **31** 556.
- Slebodzinski W. 1970: *Exterior Forms and their Applications*, PWN, Warszawa.
- Solomon S 1979: Nucl. Phys. B **147** 174.
- Sommerfeld A 1954: *Optics*, Academic Press, New York.
- Sommerfeld A 1964a: *Mechanics of Deformable Bodies*, Academic Press, New York.
- Sommerfeld A 1964b: *Partial Differential Equations in Physics*, Academic Press, New York.
- Spivak M 1970: *Comprehensive Introduction to Differential Geometry*, Publish or Perish, Brandeis.
- Steen L A & Seebach, J A 1970: *Counterexamples in Topology*, Holt, Rinehart and Winston, New York.
- Steenrod N 1970: *The Topology of Fibre Bundles*, Princeton University Press, Princeton.
- Stora R 1984: in t'Hooft G et al. (eds.), *Progress in Gauge Field Theory*, Plenum, New York.
- Streater R F & Wightman A S 1964: *PCT, Spin and Statistics, and all That*, Benjamin, New York.
- Synge J L 1937: *Geometrical Optics*, Cambridge University Press, Cambridge.
- Synge J L 1960: *Relativity: The General Theory*, J. Wiley, New York.
- Synge J L & Schild A 1978: *Tensor Calculus*, Dover, New York.
- Takesaki M 1978: in Dell'Antonio G et al (eds.), Lect. Notes Phys. **80**, Springer, Berlin.
- Temperley H N V & Lieb E H 1971: Proc. Roy. Soc. **322** 251.
- Thom R 1972: *Stabilité Structurelle et Morphegenèse* , Benjamin, New York.

- Thom R 1974: *Modèles Mathématiques de la Morphegenèse*, Union Générale d' Editions, Paris.
- Timoshenko S & Goodier J N 1970: *Theory of Elasticity*, 3rd edition, McGraw-Hill, New York.
- Tod K P & Luebbe C 2006: *Conformal gauge singularities*, Oberwalfach Reports **3** 91.
- Toda M, Kubo R & Saitô N 1998: *Statistical Physics I and II*, Springer, Berlin. **3** 91.
- Toulouse G & Vannimenus J 1980: Phys. Rep. **67** 47.
- Trautman A 1970: Rep. Math. Phys. **1** 29.
- Trautman A 1979: Czech. J. Phys. B **29** 107.
- Trautman A 2006: *Einstein-Cartan Theory*, in Françoise J P, Naber G L & Tsun T S, *Encyclopedia of Mathematical Physics*, Vol. 2, page 189, Elsevier/Academic Press, Amsterdam.
- Trudeau R J 2001: *Introduction to Graph Thory*, Dover, New York.
- Tyutin I V 1975:*Gauge Invariance in Field Theory and Statistical Physics in Operator Formalism*, Lebedev Physics Institute preprint 39 - arXiv:0812.0580.
- Utiyama R 1955: Phys. Rev. **101** 1597.
- Vainberg M M 1964: *Variational Methods for the Study of Nonlinear Operators*, Holden-Day, San Francisco.
- Vilenkin N Ya 1969: *Fonctions Spéciales et Théorie de la Représentation des Groupes*, Dunod, Paris.
- Wald R M 1984: *General Relativity*, The University of Chicago Press, Chicago.
- Walker A G 1936: *Proc. Lond. Math. Soc.* **42** 90.
- Warner F W 1983: *Foundations of Differential Manifolds and Lie Groups*, Scott-Foreman, Glenview.
- Warner G 1972: *Harmonic Analysis on Semi-Simple Lie Groups I*, Springer, Berlin.
- Weinberg S 1972: *Gravitation and Cosmology*, J. Wiley, New York.
- Weinberg S 2008: *Cosmology*, Oxford University Press.
- Wenzl H 1988: Invent. Math. **92** 349.
- Westenholz C 1978: *Differential Forms in Mathematical Physics*, North-Holland, Amsterdam.
- Weyl H 1919: Ann. Phys. (Berlin) **59** 101.
- Weyl H 1929: Z. Physik **56** 330.
- Weyl H 1932: *Theory of Groups and Quantum Mechanics*, E. P. Dutton, New York.

- Wheeler J A & Zurek W H (eds.) 1983: *Quantum Theory and Measurement*, Princeton University Press, Princeton.
- Whittaker E T 1944: *Analytical Dynamics*, Dover, New York.
- Whittaker E T 1953: A *History of the Theories of Aether and Electricity*, Thomas Nelson, London.
- Wickramasekara S 2001: Class. Quantum Grav. **18** 5353.
- Will C M 2005: *The Confrontation between General Relativity and Experiment*, Living Rev. Rel. **9**, 3.
- Wise D K 2010: Class. Quantum Grav. **27** 155010.
- Woodcock A & Davis M 1980: *Catastrophe Theory*, Penguin Books, Harmondsworth.
- Wootters W & Zurek W H 1979: Phys. Rev. D **19** 473.
- Woronowicz S L 1987: Commun. Math. Phys. **111** 613.
- Wu T T & Yang C N 1975: Phys. Rev. D **12** 3843.
- Wu Y S & Zee A 1985: Nucl. Phys. B **258** 157.
- Yang C N & Mills R L 1954: Phys. Rev. **96** 191.
- Yang C N 1974: Phys. Rev. Lett. **33** 445.
- Yang C N & Ge M L 1989: *Braid Group, Knot Theory and Statistical Mechanics*, World Scientific, Singapore.
- Yee K & Bander M 1993: Phys. Rev. D **48** 2797.
- Zeeman E C 1964: J. Math. Phys. **5** 490.
- Zeeman E C 1967: Topology **6** 161.
- Zeeman E C 1977: *Catastrophe Theory*, Addison-Wesley, Reading.
- Zumino B, Wu Y S & Zee A 1984: Nucl. Phys. B **241** 557.
- Zwanziger J W, Koenig M & Pines A 1990: Ann. Rev. Phys. Chem. **41** 601.

Index